Gabriel Nguetseng
Faculty of Science
University of Yaounde
Yaounde, Cameroon

ISSN 1439-7382 ISSN 2196-9922 (electronic)
Springer Monographs in Mathematics
ISBN 978-3-031-84704-2 ISBN 978-3-031-84705-9 (eBook)
https://doi.org/10.1007/978-3-031-84705-9

This Springer imprint is published by the registered company Springer Nature Switzerland AG
The registered company address is: Gewerbestrasse 11, 6330 Cham, Switzerland

Gabriel Nguetseng

Homogenization Algebras and Applications

A Deterministic Homogenization Theory

Springer

This series publishes advanced monographs giving well-written presentations of the "state-of-the-art" in fields of mathematical research that have acquired the maturity needed for such a treatment. They are sufficiently self-contained to be accessible to more than just the intimate specialists of the subject, and sufficiently comprehensive to remain valuable references for many years. Besides the current state of knowledge in its field, an SMM volume should ideally describe its relevance to and interaction with neighbouring fields of mathematics, and give pointers to future directions of research.

Springer Monographs in Mathematics

Un peu de science éloigne de Dieu,
mais beaucoup y ramène.
Louis PASTEUR

To the memory of my parents
To my brother Claude Joseph M'bafou and
my beloved wife Françoise Nguetseng

Preface

A careful survey of the history of mathematics reveals that every young domain grows day after day in two main directions which are complementary to one another: on the one hand, say a pragmatic direction in which available results are turned to the best account in favour of applications within or outside the discipline, e.g., in science and engineering; on the other hand, a direction devoted to searching after new solutions to be generally found in abstract settings beyond the classical theory, in the same manner as the weak solutions of Partial differential equations are framed in abstract settings including among other things the theory of Distributions (with that of Topological vector spaces, of course) and the theory of Sobolev spaces. This now routine approach was suspect or even subjected to much recrimination some 60 years ago before it becomes a rather classical matter.

Each mathematical domain not only has its own history but also involves the germs of an abstract development. The examples of Functional analysis and Harmonic analysis are quite illustrative of such a fact.

The enormous success of periodic homogenization theory is due to the properties of the spaces of periodic continuous real (or complex)-valued functions on numerical spaces. Each of such spaces is a Banach algebra whose numerous main properties have been put to account by different homogenization procedures, and especially the well-known two-scale convergence method (see the introduction). In fact, the latter is tailored on the specificities of algebras of periodic scalar functions.

The homogenization procedure proposed by the book, namely the Σ-convergence method, is in every respect similar to the two-scale convergence method of which it is actually the natural extension. This is needed for tackling those numerous real-life multiscale problems whose settings are reasonably beyond the scope of periodic homogenization. The method is straightforward, easy to handle, and global in that it preserves the concomitance of interactions between microscopic and macroscopic phenomena, all such things inherited from two-scale convergence theory. Many concrete examples of nonperiodic deterministic homogenization problems are worked out, the ansatz consisting in determining which homogenization algebra is appropriate to the problem under consideration.

A quick survey of the literature on homogenization reveals that the still more enormous success of homogenization theory as a whole is due to the multiple facets of the domain. Indeed, homogenization is open to mathematicians of various specialities (Analysis, Functional analysis, Probability theory, Harmonic analysis, Applied mathematics, Numerical analysis, Geometry), to scientists from various origins (Physics, Mechanics, Chemistry), to engineers. In short, homogenization is open to whomsoever is interested in sailing from the microscopic to the macroscopic. This is confirmed by the diversity of the theories of homogenization available in the literature (see the introduction), which theories contribute, each in its own direction, to the capacity building of the domain; which theories are very far from conflicting with each other. Surely there is room for whatever initiative may enable homogenization theory to go ahead.

Yaounde, Cameroon
December 2024

Gabriel Nguetseng

Acknowledgments

I am indebted to a number of friends and colleagues for the interest they took in my research activities during the long period of cogitation upon the ins and outs of homogenization theory. In particular my hearty thanks go to Professor *Etienne Tsamo* who constantly suggested that I write the book and who had every confidence in my capability of carrying out the project. I am greatly obliged to Professor *Maurice Tchuenté* for his valuable support, and to Professor *Robert Nzengwa* for learning me to identify precisely the eventual physical phenomena underlying a homogenization problem. My special thanks are due to Professor *Noël Lohoue* who opened my eyes to the beauty of Harmonic analysis. I also wish to thank the anonymous reviewers for their remarks and suggestions that help to improve the manuscript, and the editors of Springer Nature and Straive for cheerful and expert cooperation during the revision of the manuscript as well as the production of the book. Last but not least I would like to express my deep gratitude to Dr *Rodrigue Kenne* for both his kindness and the tedious work of typewriting the manuscript.

Introduction

The homogenization algebras enable classical periodic homogenization theory to extend to a general deterministic setting framed to develop a rigorous mathematical analysis of multiscale problems in partial differential equations with periodic as well as nonperiodic coefficients.

The aim of the book is not only to interest the reader in the abstract theory of homogenization algebras. Actually, one of the author's main concerns is to fill the gap between periodic homogenization and stochastic homogenization. Specifically, first of all, the deterministic homogenization theory provided by the book is twofold: there is, on one hand a self-contained periodic theory, on the other hand an abstract theory which is intermediate between periodic homogenization and stochastic homogenization. Next, the corner-stone of the said deterministic homogenization theory is precisely the concept of so-called sigma-convergence (see Chaps. 3 and 4) justifiably considered as the natural generalization of two-scale convergence. Furthermore, it turns out that the concept of sigma-convergence is likely to underlie a stochastic homogenization theory generalizing the well-known two-scale convergence in the mean, as is readily seen by a careful survey of, e.g., Bourgeat et al. [26]. However, such an undertaking towards an eventual theory of sigma-convergence in the mean is outside the scope of the book. Concerning stochastic homogenization we refer the reader to, e.g., Blanc et al. [19, 20], Pardoux and Piatnitski [115], Wright [145, 146], and more recently, Sango [123], Mohamed and Sango [86, 87], Deugoue and Woukeng [43], Fouetio and Woukeng [51], and Fouetio et al. [52].

It is worth perhaps recalling beforehand what is known as homogenization theory.

In its «physical» interpretation, homogenization theory is concerned with the study of the macroscopic behaviours of nonhomogeneous phenomena induced by given microstructures in various fields of Science and Engineering. It should however be stated that such phenomena are generally assumed to be governed by partial differential equations or by integral functionals, and their common feature is that they exhibit multiple scales in space or (and) time.

From the mathematical viewpoint, homogenization in a broad sense may be defined as the study of the behaviours of differential operators (and functional integrals) under actions of the multiplicative group $\mathbb{R}_+^* = \{\varepsilon|\varepsilon \in \mathbb{R}, \varepsilon > 0\}$ on numerical spaces $\mathbb{R}^N$. It is precisely in this direction that the book is framed.

Homogenization began to gain its patent of nobility as from the early 1970s with De Giorgi and Spagnolo [42] who were the first to give a proof of convergence of the homogenization process (see also Spagnolo [128] and Sbordone [124]). The major advance was the historically important book by Bensoussan et al. [14] published in 1978 and followed shortly after by the well-known lecture notes by Sanchez-Palencia [121] in 1980. Since then the literature on homogenization has been enormous (see in particular Jikov et al. [63], Tartar [132], and the following together with their bibliographies: Pavliotis and Stuart [116], Cioranescu and Donato [34], Blanc and Le Bris [17], Geers et al. [56], Lukkassen et al. [77], Damlamian [41]), and several methods of homogenization have risen here and there to solve problems in various fields of science and engineering, but also to give interesting results in other branches of mathematics. Thus, in addition to pioneering methods such as that of G-convergence (see De Giorgi and Spagnolo [42], Spagnolo [128], Pankov [114], Marcellini and Sbordone [79, 80]) or that of multiscale asymptotic expansions in combination with the method of oscillation test functions usually called Tartar's energy method (see Tartar [131, 132], Bensoussan et al. [14]), there are the methods of Γ-convergence (see, e.g., Dal Maso [40]), H-convergence (see Murat [88], Murat and Tartar [89]), two-scale convergence (see Nguetseng [92], Allaire [6], Zhikov [150], Lukkassen et al. [77], Frénod [54]), and more recently the method of periodic unfolding (see Cioranescu et al. [36]) inspired by the seminal work of Arbogast et al. [11].

The book falls into three parts. Part I (Chaps. 1 and 2) is concerned with the discussion of two basic tools we shall be dealing with throughout the study: on the one hand, some specific actions of the multiplicative group $\mathbb{R}_+^*$ on numerical spaces $\mathbb{R}^N$, on the other hand, the so-called homogenization algebras and their Gelfand representation. In Part II (Chaps. 3 and 4) the concept of Σ-convergence is discussed in the periodic setting as well as in a general deterministic setting relevant to homogenization algebras. Finally, Part III (Chaps. 5 up to 8) is devoted to the homogenization of differential operators in a general deterministic setting including the classical periodic setting. Let us go further into details.

Chapter 1 comprises two main sections. In Sect. 1.1, on the one hand are summarized some of the classical background notions and results from Banach algebras theory that are needed to develop the theory of homogenization algebras, on the other hand is proved one theorem of major interest about the closed product of a finite family of C^*-algebras of bounded continuous complex functions. This seems to be a nonclassical result whose originality is left to the reader's appreciation. Section 1.2 deals with so-called absorptive continuous actions of $\mathbb{R}_+^*$ on numerical spaces. Each of such actions generates a universal mean value that unifies most of the concepts of a mean value in use in mathematical analysis. The section includes some trace results needed to provide the text with all the necessary rigour. Section 1.2 ends with a series of three lemmas culminating in a result of considerable interest with regard to the applications.

In Chap. 2 is developed the theory of homogenization algebras (H-*algebra* for brevity). The keystone of the said theory is without doubt the Gelfand representation theory. Thanks to the Gelfand transformation on an H-algebra A, the following facts among other things prove true: (**1**) The partial derivatives in $\mathbb{R}^N$ carry over to the spectrum $\Delta(A)$, so that one can define various spaces of smooth complex functions on $\Delta(A)$, in particular the space $\mathcal{D}(\Delta(A))$ (equipped with a suitable locally convex topology) of $\mathcal{C}^\infty$ complex-valued functions on $\Delta(A)$, hence the distributions on $\Delta(A)$. (**2**) The compact space $\Delta(A)$ can be equipped with a (unique) positive Radon measure (of total mass 1) enabling the universal mean value on $\mathbb{R}^N$ to admit an integral representation. Consequently, suitable L^p spaces on $\Delta(A)$ are defined. Joining together these two facts yields various Sobolev spaces on $\Delta(A)$, thereby framing an appropriate setting for a local analysis of homogenization problems. Several examples of H-algebras are exhibited and thoroughly discussed.

Concerning Chaps. 3 and 4, attention is focused on the notion of sigma-convergence from which is derived the eponymous homogenization approach we will be following throughout the next chapters. The pertinence of sigma-convergence originates from the fact that it is tailored for the purpose of capturing the rapid oscillations pertaining to the microstructures, which oscillations are beyond the reach of usual weak convergence in L^p spaces. For example, suppose we have to compute a limit such as

$$\lim_{0<\varepsilon\to 0}\int_\Omega u_\varepsilon(x)\psi\left(x,\frac{x}{\varepsilon}\right)dx \quad (\Omega \text{ an open set in } \mathbb{R}^N_x)$$

of frequent use in the mathematical analysis of multiscale problems, where $u_\varepsilon \to u$ in $L^2(\Omega)$-weak as $\varepsilon \to 0$, and $\psi \in L^2(\Omega; \mathcal{C}_{per}(Y))$ with $Y = (-\frac{1}{2}, \frac{1}{2})^N, \mathcal{C}_{per}(Y)$ (equipped with the supremum norm on $\mathbb{R}^N_y$) being the space of those continuous complex-valued functions f on $\mathbb{R}^N_y$ that are Y-periodic, i.e., that verify $f(y+k) = f(y)$ for $y \in \mathbb{R}^N$ and $k \in \mathbb{Z}^N$ ($\mathbb{Z}$ denotes the integers). It is of interest to note that $\psi^\varepsilon \to \widetilde{\psi}$ in $L^2(\Omega)$-weak as $\varepsilon \to 0$, where ψ^ε and $\widetilde{\psi}$ denote the functions in $L^2(\Omega)$ given by $\psi^\varepsilon(x) = \psi(x, \frac{x}{\varepsilon})$ and $\widetilde{\psi}(x) = \int_Y \psi(x, y)dy,\ x \in \Omega$, respectively. Unless ψ is reduced to a constant function in y, which would lead to a triviality, it is hopeless trying to have strong convergence in place of latter weak convergence. Thus, it is beyond the classical resources of Analysis to decide whether the limit under consideration thereinbefore exists in one way or another. However, this is within the scope of sigma-convergence, as we will see in Chap. 3.

Chapter 5 is entirely devoted to the homogenization of elliptic differential operators in a general deterministic setting including the usual periodic setting. Linear as well as nonlinear operators, second order as well as higher order operators, are considered. Also, homogenization problems in open sets with holes and homogenization of Navier-Stokes equations are thoroughly studied.

Chapters 6 and 7 are concerned with the homogenization of quasi-linear parabolic monotone operators. The case where the operators have periodic coefficients is exclusively investigated in Chap. 6 whereas Chap. 7 considers the general case, with several examples by way of illustration.

Chapter 8 is devoted to a thorough study of reiterated homogenization in a general deterministic setting. An abstract framework appropriate to reiterated homogenization is constructed. In particular the notion of a reiteration H-algebra and that of reiterated sigma-convergence are introduced and discussed in the fullest detail. The differential operator chosen to illustrate the theoretical studies is a quasi-linear elliptic monotone operator. Several examples of concrete homogenization problems associated to this operator are studied. Particular attention is paid to the periodic setting which is investigated in a self-contained fashion.

It should be noted that the sigma-convergence method is not the first procedure able to address deterministic homogenization problems beyond the periodic setting. For example, G-convergence, Γ-convergence and H-convergence were basically framed on purpose. The literature on homogenization reveals that the corresponding homogenization methods, in particular G-convergence and H-convergence methods, have enabled to solve a number of complex problems governed by differential operators whose coefficients may have very general behaviours.

Anyhow, nonperiodic homogenization is booming, as can be seen by the great variety of problems addressed on the topic following alternative approaches. In this connection, attention is attracted to some recent advances among which will be found, e.g., Blanc and Wolf [18] for perforated domains, Aiyappan et al. [5] for nonlinear problems in a locally periodic domain via unfolding method, Zhuge [152] for almost-periodic homogenization, Mawassy et al. [82] for quasi-periodic media, and finally, Su [129], Bahlali et al. [12] for stochastic homogenization.

Also, some recent advances having a close connection with the book are worth pointing out: for corrector problems we refer to Woukeng [144], Cardone and Woukeng [28]; for linear Boltzmann equations, see Fouegap et al. [50]. We refer to Kenne et al. [66] for Vlasov equations and to Pokam Kakeu and Woukeng [117] for nonlinear parabolic equations with hysteresis. Finally, a flow in thin heterogeneous layers is studied by Cardone et al. [29] whereas homogenized coefficients are approximated by Jäger et al. [62].

There is a long list of differential operators and other topics relevant to homogenization theory about which nothing has been said in this book. To mention only a few: integral functionals (see Nnang [110], Visintin [137], Nguetseng et al. [106]), Hyperbolic operators (see Nguetseng et al. [105, 108]), Shape optimization (see Allaire [7]), Numerical homogenization (see Abdulle [1, 2]), Fissured media (see Douanla et al. [48], Nguetseng et al. [109]), Eigenvalue problems (see Douanla and Svanstedt [47]).

Before embarking upon mathematical developments as such it seems opportune to call the reader's attention to a basic standard of wording. Except where otherwise stated, we are here concerned with vector spaces over $\mathbb{C}$ (the complex numbers) and with scalar functions assuming values in $\mathbb{C}$. If X is, say a locally compact space and F is, for example, a Banach space (this is indeed of frequent occurrence throughout), one writes $\mathcal{C}(X; F)$ for the space of continuous mappings of X into F, $\mathcal{B}(X; F)$ for that of bounded continuous mappings of X into F, and $\mathcal{K}(X; F)$ for the space of those functions in $\mathcal{C}(X; F)$ having compact supports. One often says «function» instead of «mapping» provided this entails no risk of confusion. $\mathcal{C}(X; F)$ is provided with the compact convergence topology (i.e., the topology of uniform convergence on the compact subsets of X), $\mathcal{B}(X; F)$ is equipped with the supremum norm $\|u\|_\infty = \sup_{x\in X} \|u(x)\|$ ($\|\cdot\|$ the norm in F), and $\mathcal{K}(X; F)$ is provided with the usual inductive limit topology. So topologized, each of these spaces is a Hausdorff locally convex space, $\mathcal{B}(X; F)$ being in particular a Banach space. One has $\mathcal{K}(X; F) \subset \mathcal{B}(X; F) \subset \mathcal{C}(X; F)$ with continuous embeddings. In the sequel the locally compact spaces we will be dealing with will have most of the time the particularity of being either compact or σ-compact, i.e., expressible as union of a countable family of compact subspaces. Assuming X to be σ-compact, it follows that $\mathcal{C}(X; F)$ is a Fréchet space and $\mathcal{K}(X; F)$ is complete. For further details see Bourbaki [21, 25] (see also Bourbaki [22, pp. 40–46]), Vo-Khac [138]. On the other hand, assuming X to be equipped with a positive Radon measure, we denote by $L^p_{loc}(X; F)$ and $L^p(X; F)$ $(1 \leq p \leq \infty)$ the usual Lebesgue spaces. We mostly refer to Bourbaki [22, 23] for integration theory.

For short we will write $\mathcal{C}(X), \mathcal{B}(X)$ and $\mathcal{K}(X)$ for $\mathcal{C}(X; \mathbb{C}), \mathcal{B}(X; \mathbb{C})$ and $\mathcal{K}(X; \mathbb{C})$, respectively. Likewise we put $L^p_{loc}(X)$ and $L^p(X)$ for $L^p_{loc}(X; \mathbb{C})$ and $L^p(X; \mathbb{C})$, respectively.

It is implicitly assumed that the numerical space $\mathbb{R}^N$ and its open sets are each equipped with Lebesgue measure $dx = dx_1 \cdots dx_N$ again denoted in certain occurrences by λ.

Classical topics such as Sobolev Spaces and Linear Functional Analysis with which the reader is supposed to be familiar (in some measure) are deliberately passed over in silence. In this connection, see, e.g., Brézis [27], Lions and Magenes [74], Larsen [70], Edwards [49], Adams [3].

The author hopes that this book will encourage the study, in the framework of homogenization theory, of those multiscale problems whose natural setting is reasonably outside the scope of periodic homogenization.

Contents

Chapter 1
Preliminaries

This preliminary chapter is divided into two main sections. The first section presents some basic facts about Banach algebras, especially the Gelfand representation of commutative C^*-algebras with identity. After basic notions and notation that must be familiar to the reader, we exhibit what is justifiably considered as the corner-stone of the theory of homogenization algebras, namely the commutative Gelfand-Naimark representation theorem. Attention is focused on a few examples of those Banach algebras that underly the concept of a homogenization algebra, that is, on commutative C^*-algebras with identity. Finally, we conclude the section by discussing the Gelfand representation of the closed product of a finite family of C^*-algebras of continuous complex functions. We show that the Cartesian product of the spectra of such C^*-algebras is homeomorphic to the spectrum of the said closed product. The result thus established will play a crucial role in applications, especially in the study of evolution equations and in reiterated homogenization, all that in the deterministic setting.

In the next section we discuss a family of continuous actions of the multiplicative group $\mathbb{R}_+^* = \{\varepsilon | \varepsilon \in \mathbb{R}, \varepsilon > 0\}$ on the numerical space $\mathbb{R}^N$ characterized by the so-called absorptiveness property. Such actions are closely connected to the micro-scaling in use in homogenization. Each of them generates an appropriate kind of mean value and on the other hand they enable to frame a class of function spaces that prove essential in the sequel. Finally, a fundamental lemma is proved for further needs.

1.1 About C^*-algebras

1.1.1 Basic Notions and Notation

We recall below some rudiments of Banach algebras we shall be dealing with throughout the rest of the chapter and beyond.

G. Nguetseng, *Homogenization Algebras and Applications*, Springer Monographs in Mathematics, https://doi.org/10.1007/978-3-031-84705-9_1

One calls an *algebra* (over $\mathbb{C}$) any vector space A equipped with a binary operation $(u, v) \mapsto uv$ from $A \times A$ to A, referred to as *multiplication*, such that

(1) $u(vw) = (uv)w$
(2) $u(v + w) = uv + uw$, $(v + w)u = vu + wu$
(3) $a(uv) = (au)v = u(av)$
for $u, v, w \in A$ and $a \in \mathbb{C}$.

An algebra A is said to be:

(4) a *commutative algebra* if $uv = vu \quad (u, v \in A)$
(5) an *algebra with identity (or unit)* if there is some $e \in A$ such that $eu = ue = u$ for all $u \in A$.

Such an e is unique and referred to as the *identity* (or the *unit*).

An algebra A is called a *Banach algebra* if A is equipped with a norm $\| \cdot \|$ satisfying $\|uv\| \leq \|u\| \|v\|$ for all $u, v \in A$, and further if A is a Banach space under the said norm.

An algebra A is said to be an *algebra with involution* if there exists a mapping $u \mapsto u^*$ of A into itself called *involution* and denoted by $*$, such that for $u, v \in A$ and $a \in \mathbb{C}$, one has

(6) $(u + v)^* = u^* + v^*$
(7) $(au)^* = \overline{a}u^*$ ($\overline{a}$ denotes the complex conjugate of a)
(8) $(u^*)^* = u^{**} = u$.

By a *C^*-algebra* is meant a Banach algebra with involution $*$ such that $\|uu^*\| = \|u\|^2$ for all $u \in A$.

If A and B are C^*-algebras with involutions $*$ and $+$, respectively, an algebra homomorphism $f : A \to B$ is called a *C^*-homomorphism* if $f(u^*) = f(u)^+$ for all $u \in A$.

1.1.2 Gelfand Representation of C*-algebras. Basic Examples

Having fixed the preceding standard of wording, we now come to the heart of the matter. A statement without proof will be either a simple fact whose verification is left to the reader, or a classical fundamental result proven in a published work such as Dieudonné [44], Dixmier [46] or Larsen [69], to which the reader is referred for a detailed presentation.

In all that follows, A denotes a commutative Banach algebra with identity, e.

By a *character* of A is meant any algebra homomorphism of A into $\mathbb{C}$; that is, a linear form s on A such that $s(uv) = s(u)s(v)$ for all $u, v \in A$. It is an easy exercise to check that a character is actually a mapping of A **onto** $\mathbb{C}$.

Any character s of A is continuous with norm $\|s\| \leq 1$. If $\|e\| = 1$ then $\|s\| = 1$.

The set of all nonzero characters of A is denoted by $\Delta(A)$, and is called the *spectrum* of A. We have $\Delta(A) \neq \emptyset$ (empty set). On the other hand, as pointed out above, $\Delta(A)$ is contained in the strong dual A' of A. More precisely, $\Delta(A)$ is a subset of the closed unit ball of A', which unit ball is a compact space under the relative weak $*$ topology on A' (Banach-Alaoglu theorem). Since $\Delta(A)$ with the relative weak $*$ topology turns out to be closed, it follows that $\Delta(A)$ with the said topology is a compact space. It is only with this topology, sometimes referred to as the *Gelfand topology*, that we shall consider the spectrum $\Delta(A)$ of A.

In the sequel we will write $\langle s, u\rangle$ for $s(u)$, where $s \in A'$ and $u \in A$, and where $\langle , \rangle$ denotes the duality pairing between A' and A.

By the *Gelfand transformation* on A is meant the mapping $\mathcal{G} : A \to \mathcal{C}(\Delta(A))$ defined by $\mathcal{G}(u)(s) = \langle s, u\rangle$ for $u \in A, s \in \Delta(A)$. The function $\mathcal{G}(u)$, sometimes denoted by $\widehat{u}$, is called the *Gelfand transform* of $u \in A$. On the other hand, it is clear that $\mathcal{G}$ is an algebra homomorphism.

We turn now to the statement of the so-called commutative Gelfand-Naimark theorem. The reader is referred to Larsen [69, p. 277], Dieudonné [44, p. 322], for the proof.

Theorem 1.1 (Gelfand-Naimark) *Let A be a commutative C^*-algebra with identity. The Gelfand transformation on A is an isometric C^*-isomorphism of A onto $\mathcal{C}(\Delta(A))$.*

Remark 1.1 If A is a commutative C^*-algebra without identity, the above theorem is still valid provided the algebra $\mathcal{C}(\Delta(A))$ is substituted by the algebra $\mathcal{B}_0(\Delta(A))$ of those continuous complex-valued functions on $\Delta(A)$ (a locally compact space) vanishing at infinity.

We will conclude the present section with a few examples of commutative C^*-algebras with identity together with their Gelfand representation. It should be added that the interest which we take in the forthcoming examples is not simply to provide a matter for illustration of the preceding development. We anticipate that they have connection with the applications, as will later become apparent.

1.1.2.1 The C^*-algebra $\mathcal{B}(X)$

Example 1.1 Let X be a locally compact space. We denote by $\mathcal{B}(X)$ the algebra of bounded continuous complex-valued functions on X, the algebra operations being the obvious ones of pointwise addition, multiplication and scalar multiplication. With the supremum norm

$$\|u\|_\infty = \sup_{x \in X} |u(x)| \quad (u \in \mathcal{B}(X))$$

$\mathcal{B}(X)$ is a commutative Banach algebra with identity the constant function 1. That $\mathcal{B}(X)$ is a Banach space (and therefore obviously a Banach algebra) follows by a classical result that can be found in, e.g., Schwartz [127, p. 196]. Furthermore, $\mathcal{B}(X)$ is a Banach algebra with involution, the latter being the obvious one of complex conjugation. In fine, $\mathcal{B}(X)$ is a commutative C^*-algebra with identity. Hence, according to the commutative Gelfand-Naimark theorem, the Gelfand transformation $\mathcal{G}$ on $\mathcal{B}(X)$ is an isometric C^*-isomorphism of $\mathcal{B}(X)$ onto $\mathcal{C}(\Delta(\mathcal{B}(X)))$. Moreover, the spectrum $\Delta(\mathcal{B}(X))$ turns out to be in a sense the Stone-Čech compactification of the locally compact space X (see Bourbaki [25, IX, p. 9] and Larsen [69, p. 90]).

The following classical result will be needed later on.

Lemma 1.1 *Let A be a closed subalgebra of $\mathcal{B}(X)$ that contains the constants and preserves complex conjugation. Let j be the mapping of X into $\Delta(A)$ defined by $j(x) = \delta_x$ $(x \in X)$, where $\langle \delta_x, u \rangle = u(x)$ for $u \in A$. Then*

(i) j is continuous.
(ii) $j(X)$ is dense in $\Delta(A)$.
(iii) If A separates the points of X, then j is injective.

Proof First of all, with the algebra structure of $\mathcal{B}(X)$ and the supremum norm on X, A is clearly a commutative C^*-algebra with identity. Consequently, the Gelfand transformation $\mathcal{G}$ on A is an isometric C^*-isomorphism of A onto $\mathcal{C}(\Delta(A))$.

This being so, let us verify that j is continuous at some arbitrary given $a \in X$. Clearly it is enough to show that to any arbitrary set $V = \{s | s \in \Delta(A), |\langle s, \varphi \rangle - \varphi(a)| \leq \varepsilon \text{ for all } \varphi \in \mathcal{F}\}$ $(\varepsilon > 0, \mathcal{F} \subset A, \mathcal{F} \text{ finite})$ there is attached a neighbourhood U of a in X such that one has $j(U) \subset V$. This is straightforward. Indeed, to each $\varphi \in \mathcal{F}$ there is assigned by continuity a neighbourhood U_φ of a in X such that $|\varphi(x) - \varphi(a)| \leq \varepsilon$ for $x \in U_\varphi$. Therefore the set $U = \cap_{\varphi \in \mathcal{F}} U_\varphi$ answers the question, thereby proving (i). Let us turn now to the proof of (ii). Assuming $j(X)$ not to be dense in $\Delta(A)$ yields some $r \in \Delta(A)$, r lying off $\overline{j(X)}$ (closure of $j(X)$ in $\Delta(A)$) and, subsequently, some function $f \in \mathcal{C}(\Delta(A))$ such that $f(r) = 1$ and $f(j(x)) = 0$ for all $x \in X$ (see, e.g., Bourbaki [25, IX, p. 41]). It follows readily that $\psi(x) = 0$ for $x \in X$, where $\psi \in A$ and $f = \mathcal{G}(\psi)$. Hence $f = 0$, a contradiction, and so (ii) is established. Part (iii) being trivial, the lemma is proven. ∎

1.1.2.2 The C^*-algebra $\mathcal{B}_\infty(X)$

Example 1.2 Let X be a metrizable σ-compact locally compact space. It is perhaps worth recalling that the locally compact space X is said to be σ-compact if it is expressible as union of a countable family of compact subspaces. Finally, we assume that X is noncompact.

This being so, we denote by $\mathcal{B}_\infty(X)$ the algebra of continuous complex-valued functions on X admitting a limit at *infinity*, where the latter term refers to the point at infinity, denoted here by ∞, of the Alexandroff compactification of X. It is an easy exercise (left to the reader) to check that $\mathcal{B}_\infty(X)$ is a closed subalgebra of $\mathcal{B}(X)$ that contains the constants and preserves complex conjugation. Hence, it follows that $\mathcal{B}_\infty(X)$ with the same algebra structure and the same norm as on $\mathcal{B}(X)$ is a commutative C^*-algebra with identity. Consequently, Theorem 1.1 applies and therefore the Gelfand transformation $\mathcal{G}$ is an isometric C^*-isomorphism of $\mathcal{B}_\infty(X)$

onto $\mathcal{C}(\Delta(\mathcal{B}_\infty(X)))$. Moreover it is of interest to note that $\mathcal{B}_\infty(X)$ is separable (see Bourbaki [25, IX, p. 18, prop. 12, and X, p. 25, corol.]). Consequently, the compact space $\Delta(\mathcal{B}_\infty(X))$ is metrizable, thanks to a classical result that can be found in Bourbaki [25, X, p. 24, and IX, p. 18, prop. 12].

To complete this discussion, let us show what follows.

Proposition 1.1 *The compact space $\Delta(\mathcal{B}_\infty(X))$ is homeomorphic to the Alexandroff compactification of X.*

Proof Let X' be the Alexandroff compactification of X, ∞ its point at infinity, and θ a homeomorphism of $X' \setminus \{\infty\}$ (the complement of ∞ relative to X') onto X (see Bourbaki [24, I, p. 67] for details). Introduce the mapping Φ of $\mathcal{B}_\infty(X)$ into $\mathcal{C}(X')$ defined on each $f \in \mathcal{B}_\infty(X)$ by $\Phi f(x') = f(\theta(x'))$ if $x' \in X' \setminus \{\infty\}$, $\Phi f(\infty) = \lim_{x'\to\infty} f(\theta(x'))$. It is not hard to check that Φ is a bijective C^*-homomorphism of $\mathcal{B}_\infty(X)$ onto $\mathcal{C}(X')$. In fact, more can be said about this mapping, namely that Φ is an isometric C^*-isomorphism of $\mathcal{B}_\infty(X)$ onto $\mathcal{C}(X')$. It is clear that one only needs to verify that Φ is an isometry. For that purpose, fix freely some $f \in \mathcal{B}_\infty(X)$. We have $|\Phi f(x')| \leq \|f\|_\infty$ for all $x' \in X' \setminus \{\infty\}$. On the other hand, given any arbitrary $\varepsilon > 0$, let K be some compact set in X such that $|f(x) - \Phi f(\infty)| \leq \varepsilon$ for $x \in X \setminus K$. Then $|\Phi f(\infty)| \leq \varepsilon + \|f\|_\infty$, hence $|\Phi f(\infty)| \leq \|f\|_\infty$ by reason of the arbitrariness of ε. Consequently $\|\Phi f\|_\infty \leq \|f\|_\infty$. But we have also $\|f\|_\infty \leq \|\Phi f\|_\infty$, as is readily seen by the equality $f(x) = \Phi f(\theta^{-1}(x))$ for $x \in X$. Hence, it follows that $\|\Phi f\|_\infty = \|f\|_\infty$ for all $f \in \mathcal{B}_\infty(X)$, as claimed.

The immediate corollary of all this is that the mapping $\mathcal{G} \circ \Phi^{-1}$ is an algebra isomorphism (or even an isometric C^*-isomorphism) of $\mathcal{C}(X')$ onto $\mathcal{C}(\Delta(\mathcal{B}_\infty(X)))$. We deduce by a classical result (see Larsen [69, p. 90]) that $\Delta(\mathcal{B}_\infty(X))$ is homeomorphic to X'. ■

Thanks to the preceding proposition we may identify $\Delta(\mathcal{B}_\infty(X))$ with X' and then justifiably state that $\Delta(\mathcal{B}_\infty(X))$ **is the Alexandroff compactification of the locally compact space** X.

1.1.2.3 The C^*-algebra $\mathcal{C}_{per}(Y)$

Example 1.3 By $\mathcal{C}_{per}(Y)$ with $Y = (-\frac{1}{2}, \frac{1}{2})^N$ we denote the algebra of continuous complex-valued functions on $\mathbb{R}^N$ (integer $N \geq 1$) that are Y-periodic. Let us clarify this terminology. We say that a continuous complex-valued function u on $\mathbb{R}^N$ is Y-periodic if $u(y+k) = u(y)$ for $y \in \mathbb{R}^N$, $k \in \mathbb{Z}^N$, where $\mathbb{Z}$ denotes the set of integers. Instead of Y-periodic one usually says *periodic with period* 1 (in each coordinate). The notation $\mathcal{C}_{per}(Y)$ is justified by the fact that a Y-periodic continuous function is entirely determined by its restriction to $\overline{Y}$ (closure of Y).

Now, it is evident that $\mathcal{C}_{per}(Y)$ is a closed subalgebra of $\mathcal{B}(\mathbb{R}^N)$ that contains the constants and preserves complex conjugation. Therefore, with the usual pointwise algebra operations and the supremum norm, $\mathcal{C}_{per}(Y)$ is a commutative C^*-algebra with identity. Consequently, the Gelfand transformation on $\mathcal{C}_{per}(Y)$ is an isometric C^*-isomorphism of $\mathcal{C}_{per}(Y)$ onto $\mathcal{C}(\Delta(\mathcal{C}_{per}(Y)))$. As a corollary of this, we have that the compact space $\Delta(\mathcal{C}_{per}(Y))$ is homeomorphic to the N-torus $\mathbb{T}^N = (\mathbb{R}/\mathbb{Z})^N$.

Indeed, let φ denote the canonical mapping of $\mathbb{R}^N$ (as an additive topological group) onto the quotient group $\mathbb{R}^N/\mathbb{Z}^N$ (a compact additive group). Let L denote the mapping $f \mapsto Lf = f \circ \varphi$ of $\mathcal{C}(\mathbb{R}^N/\mathbb{Z}^N)$ into $\mathcal{C}_{per}(Y)$ (this mapping is well defined by virtue of the classical fact that $f \circ \varphi$ is continuous if and only if f is). It is an easy matter to verify that L is an isometric C^*-isomorphism of $\mathcal{C}(\mathbb{R}^N/\mathbb{Z}^N)$ onto $\mathcal{C}_{per}(Y)$. This implies that $\Delta(\mathcal{C}_{per}(Y))$ is homeomorphic to $\mathbb{R}^N/\mathbb{Z}^N$ (see Larsen [69, p. 90]). Hence, since $\mathbb{R}^N/\mathbb{Z}^N$ is homeomorphic to $\mathbb{T}^N$, the desired conclusion follows. Consequently we are justified in stating

$$\Delta(\mathcal{C}_{per}(Y)) = \mathbb{T}^N. \tag{1.1}$$

For further needs the preceding development is worth generalizing. Let S be a *network* in $\mathbb{R}^N$. This means precisely that S is a discrete subgroup of rank N of the additive group $\mathbb{R}^N$, or equivalently that S is a discrete subgroup of $\mathbb{R}^N$ isomorphic to the discrete group $\mathbb{Z}^N$ (see, e.g., Bourbaki [25, VII, pp. 1–4]). Therefore, according to Bourbaki [25, VII, p. 4, Thm. 1], there exists a basis $\{b_1, \cdots, b_N\}$ of $\mathbb{R}^N$ (viewed as an N-dimensional vector space over $\mathbb{R}$) such that S coincides with the set of all $k \in \mathbb{R}^N$ of the form

$$k = \sum_{i=1}^{N} t_i b_i \quad (t_i \in \mathbb{Z}). \tag{1.2}$$

To this there is attached

$$Y = \left\{ y | y \in \mathbb{R}^N, y = \sum_{i=1}^{N} t_i b_i, -\frac{1}{2} < t_i < \frac{1}{2} \right\}, \tag{1.3}$$

which is an open set in $\mathbb{R}^N$. Concretely, Y is the open parallelepiped with centre $\omega = (0, \cdots, 0)$ constructed on $\{\frac{b_1}{2}, \cdots, \frac{b_N}{2}\}$. Reciprocally, to any given basis $\{b_1, \cdots, b_N\}$ of the N-dimensional vector space $\mathbb{R}^N$ there is attached a unique couple made up of the network $S = \{k | k$ given by (1.2)$\}$ and the parallelepiped Y defined by (1.3).

Remark 1.2 Note that the basis $\{b_1, \cdots, b_N\}$ is not necessarily orthogonal. However, if $S = \mathbb{Z}^N$ then it is clear that we may take $\{b_1, \cdots, b_N\}$ to be the canonical basis of $\mathbb{R}^N$, in which case Y is nothing else than the unit cube $(-\frac{1}{2}, \frac{1}{2})^N$; and this carries us back to the preceding framework.

Having made these preliminaries, we say that a continuous complex-valued function u on $\mathbb{R}^N$ is Y-periodic (Y given by (1.3)) if

$$u(y + k) = u(y) \text{ for all } y \in \mathbb{R}^N \text{ and all } k \in S,$$

where S is the network in $\mathbb{R}^N$ made up of all $k \in \mathbb{R}^N$ of the form (1.2). Such a function is entirely determined by its restriction to $\overline{Y}$, and is therefore bounded.

We will still write $\mathcal{C}_{per}(Y)$ for the set of such functions. With the usual pointwise algebra operations and the supremum norm, $\mathcal{C}_{per}(Y)$ is a commutative C^*-algebra with identity. We will later establish in the general case the analogue of (1.1).

1.1.2.4 The C^*-algebra $AP(G)$

Example 1.4 In the next chapter and in a number of concrete homogenization problems beyond we will have to deal with almost periodic functions on the additive group $\mathbb{R}^N$ as well as on some of its specific subgroups. Thus, such notions as a *continuous character* or a *dual group* will be unavoidable. So, there is much to be gained by discussing those notions in an abstract setting instead of reasoning from particular cases at the risk of sacrificing the exactness. We mostly refer to Larsen [69], Guichardet [57], Hewitt and Ross [58].

So, in what follows, G denotes a locally compact Abelian group additively written, with neutral element ω. For the benefit of the reader we recall that by a (continuous) character of G is meant any continuous (group) homomorphism of G into the multiplicative group $\mathbb{U} = \{z|z \in \mathbb{C}, |z| = 1\}$. The product of two characters γ and γ', denoted by $\gamma\gamma'$, is defined by $\gamma\gamma'(y) = \gamma(y)\gamma'(y)$, $y \in G$. This yields a binary operation referred to as *multiplication* under which the characters of G constitute a multiplicative Abelian group. Moreover, with the topology of compact convergence, the characters of G form an Abelian topological group denoted by $\widehat{G}$, and referred to as the *dual group* of G. Furthermore, $\widehat{G}$ is locally compact, and compact if and only if G is discrete. It is customary to write $\langle \gamma, y \rangle$ or $\langle y, \gamma \rangle$ in place of $\gamma(y)$, where $\gamma \in \widehat{G}$, $y \in G$.

Let us turn now to the notion of an almost periodic continuous function. A continuous complex-valued function u on G is said to be *almost periodic* if it is bounded (in other words, if $u \in \mathcal{B}(G)$), and further if the set of translates $\{\tau_a u | a \in G\}$ has compact closure in $\mathcal{B}(G)$. We recall that $\tau_a u(y) = u(y - a)$ $(a \in G)$. The set of almost periodic continuous complex-valued functions on G is denoted by $AP(G)$. This is a closed subalgebra of $\mathcal{B}(G)$ that contains the constants and that preserves complex conjugation. Therefore, with further the supremum norm, $AP(G)$ is a commutative C^*-algebra with identity. Consequently, the Gelfand transformation on $AP(G)$ is an isometric C^*-isomorphism of $AP(G)$ onto $\mathcal{C}(\Delta(AP(G)))$. Furthermore, the group structure of G can be carried over to $\Delta(AP(G))$ so as to make the latter into a compact Abelian group. Thus structured, $\Delta(AP(G))$ is often referred to as the *Bohr compactification* of G, denoted by bG.
We conclude this section with the important notion of a trigonometric polynomial. By a trigonometric polynomial on G is meant any function $u : G \to \mathbb{C}$ of the form $u = \sum_{\gamma \in F} a_\gamma \gamma$ $(a_\gamma \in \mathbb{C})$, i.e.,

$$u(y) = \sum_{\gamma \in F} a_\gamma \langle \gamma, y \rangle \quad (y \in G),$$

where F is an arbitrary finite subset of $\widehat{G}$. The set of all trigonometric polynomials on G will be denoted by $T(G)$. Any trigonometric polynomial lies in $AP(G)$ and further $T(G)$ is a dense subalgebra of $AP(G)$. We shall mostly be dealing with $G = \mathbb{R}^N$.

1.1.3 A Structure Theorem About the Closed Product of C^*-algebras of Continuous Functions

Let $(X_j)_{1\leq j\leq m}$ be a finite family of locally compact spaces. Let $X = X_1 \times \cdots \times X_m$ be their Cartesian product provided with the product topology, which makes it into a locally compact space. Points in X will be denoted by $x = (x_1, \cdots, x_m)$ with $x_j \in X_j$ $(j = 1, \cdots, m)$, of course.

Now, let $(A_j)_{1\leq j\leq m}$ be a finite family of Banach algebras constituted as follows. For each index $j = 1, \cdots, m$, A_j is a separable closed subalgebra of $\mathcal{B}(X_j)$ that contains the constants and preserves complex conjugation. Thus, with the algebra structure induced by the usual one of $\mathcal{B}(X_j)$ and the supremum norm on X_j, A_j is a separable commutative C^*-algebra with identity.

This being so, we define A to be the closure in $\mathcal{B}(X)$ of the tensor product $\bigotimes_{j=1}^m A_j$. It is well to recall that $\bigotimes_{j=1}^m A_j$ denotes the set of all finite sums of the form

$$\sum_{i\in I}\bigotimes_{j=1}^{m} f_{ij},\ I \text{ an arbitrary finite set,}$$

where $f_{ij} \in A_j$ $(j = 1, \cdots, m)$. It is clear that $\bigotimes_{j=1}^m A_j$ is a subalgebra of $\mathcal{B}(X)$ that contains the constants and preserves complex conjugation. It follows that A is a closed subalgebra of $\mathcal{B}(X)$ with the same properties. The result is that A is a commutative C^*-algebra with identity. Our main purpose in the present section is to establish the equality (in a sense to be specified)

$$\Delta(A) = \Delta(A_1) \times \cdots \times \Delta(A_m). \tag{1.4}$$

Let us prove beforehand some simple but useful lemmas.

Lemma 1.2 *Let $f_j \in \mathcal{B}(X_j)$ $(j = 1, \cdots, m)$. Then*

$$\left|\prod_{j=1}^{m} f_j(x_j) - \prod_{j=1}^{m} f_j(y_j)\right| \leq \prod_{j=1}^{m}(1 + \|f_j\|_\infty)\sum_{j=1}^{m}|f_j(x_j) - f_j(y_j)| \tag{1.5}$$

for any couple of points $x = (x_j)$, $y = (y_j)$ in $\prod_{j=1}^m X_j$.

Proof This can be shown by induction on $m \geq 2$. The case $m = 2$ being evident, let us assume $m > 2$, and suppose that for any given $(m-1)$-tuple $f_j \in \mathcal{B}(X_j)$ $(j = 1, \cdots, m-1)$, we have

$$|\prod_{j=1}^{m-1} f_j(x_j) - \prod_{j=1}^{m-1} f_j(y_j)| \leq \prod_{j=1}^{m-1}(1 + \|f_j\|_\infty) \sum_{j=1}^{m-1} |f_j(x_j) - f_j(y_j)|$$

for all $x = (x_j)$, $y = (y_j)$ in $\prod_{j=1}^{m-1} X_j$.

Let then $f_j \in \mathcal{B}(X_j)$ $(j = 1, \cdots, m)$. If $x = (x_j)$ and $y = (y_j)$ are given in $\prod_{j=1}^{m} X_j$, then by the equality

$$\prod_{j=1}^{m} f_j(x_j) - \prod_{j=1}^{m} f_j(y_j) = \left[\prod_{j=1}^{m-1} f_j(x_j) - \prod_{j=1}^{m-1} f_j(y_j)\right] f_m(x_m) + [f_m(x_m) - f_m(y_m)] \prod_{j=1}^{m-1} f_j(y_j)$$

and use of

$$|f_m(x_m)| \leq (1 + \|f_m\|_\infty),$$

and

$$|\prod_{j=1}^{m-1} f_j(y_j)| \leq \prod_{j=1}^{m}(1 + \|f_j\|_\infty),$$

one quickly arrives at (1.5). ∎

Lemma 1.3 *Let $f_j, g_j \in \mathcal{B}(X_j)$ $(j = 1, \cdots, m)$ be such that*

$$\|g_j\|_\infty \leq 1 + \|f_j\|_\infty \quad (j = 1, \cdots, m). \tag{1.6}$$

Then

$$|\prod_{j=1}^{m} f_j(x_j) - \prod_{j=1}^{m} g_j(x_j)| \leq \prod_{j=1}^{m}(1 + \|f_j\|_\infty) \sum_{j=1}^{m} |f_j(x_j) - g_j(x_j)| \tag{1.7}$$

for all $x = (x_j) \in \prod_{j=1}^{m} X_j$.

Proof We proceed by induction on $m \geq 2$. The case $m = 2$ being obvious, we let $m > 2$, and suppose that for any $f_j, g_j \in \mathcal{B}(X_j)$ $(j = 1, \cdots, m-1)$ such that $\|g_j\|_\infty \leq 1 + \|f_j\|_\infty$ $(j = 1, \cdots, m-1)$, we have

$$|\prod_{j=1}^{m-1} f_j(x_j) - \prod_{j=1}^{m-1} g_j(x_j)| \leq \prod_{j=1}^{m-1}(1 + \|f_j\|_\infty) \sum_{j=1}^{m-1} |f_j(x_j) - g_j(x_j)| \tag{1.8}$$

for all $x = (x_j) \in \prod_{j=1}^{m-1} X_j$.

This being so, fix freely $f_j, g_j \in \mathcal{B}(X_j)$ $(j = 1, \cdots, m)$ such that (1.6) holds. Let $x = (x_j) \in \prod_{j=1}^{m} X_j$. From

$$\begin{aligned}\prod_{j=1}^{m} f_j(x_j) - \prod_{j=1}^{m} g_j(x_j) &= [f_m(x_m) - g_m(x_m)] \prod_{j=1}^{m-1} f_j(x_j) \\ &\quad + g_m(x_m) \left[\prod_{j=1}^{m-1} f_j(x_j) - \prod_{j=1}^{m-1} g_j(x_j) \right]\end{aligned}$$

and use of the inequalities

$$|\prod_{j=1}^{m-1} f_j(x_j)| \leq \prod_{j=1}^{m}(1 + \|f_j\|_\infty)$$

and

$$|g_m(x_m)| \leq (1 + \|f_m\|_\infty),$$

we see that (1.7) follows readily by (1.8). ■

Lemma 1.4 *The Banach algebra A is separable.*

Proof First of all, fix freely some $f = \bigotimes_{j=1}^{m} f_j$, $f_j \in A_j$. Recalling that A_j is separable, let D_j be a dense countable subset of A_j $(j = 1, \cdots, m)$. Let $0 < \varepsilon \leq 1$ be given in an arbitrary fashion. Put $c = m \prod_{j=1}^{m}(1 + \|f_j\|_\infty)$. Then, for each j, there is some $g_j \in D_j$ such that $\|f_j - g_j\|_\infty \leq \frac{\varepsilon}{c}$. Put $g = \bigotimes_{j=1}^{m} g_j$. Thanks to Lemma 1.3, we have $\|f - g\|_\infty \leq \varepsilon$.

With this in mind, let

$$\bigodot_{j=1}^{m} D_j = \{g | g = \bigotimes_{j=1}^{m} g_j, g_j \in D_j\}$$

and denote by $\mathcal{D}$ the set of all finite sums of elements of $\bigodot_{j=1}^{m} D_j$. Then, based on the preceding result, it is a routine exercise to check that $\mathcal{D}$ is dense in A.

Furthermore, $\mathcal{D}$ is countable, since $\bigodot_{j=1}^{m} D_j$ is a countable subset of A. This completes the proof. ∎

Remark 1.3 The above lemma implies via the Gelfand transformation on A, that $\mathcal{C}(\Delta(A))$ is separable. We deduce by a classical result that the compact space $\Delta(A)$ is metrizable.

Having done this, define

$$\mathring{x} = (\delta_{x_1}, \cdots, \delta_{x_m}) \text{ for } x = (x_1, \cdots, x_m) \in X,$$

where δ_{x_j} is the Dirac measure on X_j at x_j, and then set

$$D = \{\mathring{x} | x = (x_1, \cdots, x_m) \in X\}.$$

Clearly $D \subset \Delta(A_1)\times\cdots\times\Delta(A_m)$. In the sequel the compact space $\Delta(A)$ is provided with its natural uniform structure (Bourbaki [24, II, p. 27]) and the same is true of each compact space $\Delta(A_j)$ $(j = 1, \cdots, m)$. The product space $\Delta(A_1)\times\cdots\times\Delta(A_m)$ is provided with the corresponding product uniform structure, and D is viewed as a uniform subspace of $\Delta(A_1) \times \cdots \times \Delta(A_m)$ (see Bourbaki [24, II]).

We are now in a position to prove (1.4).

Theorem 1.2 *The mapping $h : D \to \Delta(A)$ defined by $h(\mathring{x}) = \delta_x$ $(x \in X)$ is uniformly continuous, and it extends by continuity to a homeomorphism H of $\Delta(A_1) \times \cdots \times \Delta(A_m)$ onto $\Delta(A)$.*

Proof The first point is to show the uniform continuity of h. So, let V be a vicinity of the uniform structure of $\Delta(A)$. Of course, we may assume without loss of generality that

$$V = \{(r, s) | r, s \in \Delta(A), |\langle r, \varphi\rangle - \langle s, \varphi\rangle| \leq \alpha \text{ for all } \varphi \in \Phi\},$$

where Φ is a nonempty finite subset of A, $\alpha > 0$, and $\langle , \rangle$ denotes the duality pairing between A' and A. This being so, fix freely some $\varphi \in \Phi$, and consider some $f = \sum_{i=1}^{n}(\bigotimes_{j=1}^{m} u_{ij})$, with $u_{ij} \in A_j$, such that $\|\varphi - f\|_\infty \leq \frac{\alpha}{4}$. Then

$$|\varphi(x) - \varphi(y)| \leq |f(x) - f(y)| + \frac{\alpha}{2}$$

for all $x, y \in X$. But

$$\begin{aligned} |f(x) - f(y)| &\leq \sum_{i=1}^{n} |\prod_{j=1}^{m} u_{ij}(x_j) - \prod_{j=1}^{m} u_{ij}(y_j)| \\ &\leq \sum_{i=1}^{n} c_i \sum_{j=1}^{m} |u_{ij}(x_j) - u_{ij}(y_j)| \text{ (Lemma 1.2)}, \end{aligned}$$

where $c_i = \prod_{j=1}^m (1 + \|u_{ij}\|_\infty)$. Letting $c = \max_{1 \le i \le n} c_i$, we deduce that

$$|\varphi(x) - \varphi(y)| \le c \sum_{i=1}^{n} \sum_{j=1}^{m} |u_{ij}(x_j) - u_{ij}(y_j)| + \frac{\alpha}{2}$$

for all $x = (x_1, \cdots, x_m)$ and $y = (y_1, \cdots, y_m)$ in X. Hence

$$|\varphi(x) - \varphi(y)| \le \alpha \tag{1.9}$$

for all $x, y \in X$ such that $(\delta_{x_j}, \delta_{y_j}) \in W_\varphi^j \ (1 \le j \le m)$, where

$$W_\varphi^j = \{(r, s) | r, s \in \Delta(A_j), |\langle r, u_{ij}\rangle - \langle s, u_{ij}\rangle| \le \frac{\alpha}{2mnc} \ (1 \le i \le n)\}.$$

Now, let pr_j denote the natural projection of $\Delta(A_1) \times \cdots \times \Delta(A_m)$ onto $\Delta(A_j)$. Put $q_j = pr_j \times pr_j$, that is, q_j is the mapping

$$q_j : [\Delta(A_1) \times \cdots \times \Delta(A_m)] \times [\Delta(A_1) \times \cdots \times \Delta(A_m)] \to \Delta(A_j) \times \Delta(A_j)$$

$$q_j(r, s) = (pr_j(r), pr_j(s)) \text{ for } r, s \in \Delta(A_1) \times \cdots \times \Delta(A_m).$$

Finally, let

$$W = \bigcap_{j=1}^{m} q_j^{-1}(W^j) \text{ with } W^j = \bigcap_{\varphi \in \Phi} W_\varphi^j.$$

The set W is a vicinity of the uniform structure of $\Delta(A_1) \times \cdots \times \Delta(A_m)$ and we have, thanks to (1.9), $(h(\mathring{x}), h(\mathring{y})) \in V$ whenever $(\mathring{x}, \mathring{y}) \in W$. This shows that h is uniformly continuous.

Consequently, since D is dense in $\Delta(A_1) \times \cdots \times \Delta(A_m)$ (use Lemma 1.1), the mapping h extends by continuity to a continuous mapping H of $\Delta(A_1) \times \cdots \times \Delta(A_m)$ into $\Delta(A)$ (see Bourbaki [24, II, p. 29, Corol.]). Thus, by appealing to a well-known argument, the theorem is proved if we can verify that H is bijective.

Let us first show the surjectivity. Fix freely some $s \in \Delta(A)$. By noting that $h(D)$ is dense in $\Delta(A)$ and recalling that $\Delta(A)$ is metrizable (see Lemma 1.4 and Remark 1.3), we see that there is a sequence $(\xi_n)_{n \in \mathbb{N}}$ in X such that $h(\mathring{\xi}_n) \to s$ in $\Delta(A)$ as $n \to \infty$. Furthermore, since $\Delta(A_1) \times \cdots \times \Delta(A_m)$ is compact and metrizable, a subsequence, still denoted by (ξ_n) for simplicity, can be extracted such that $\mathring{\xi}_n \to r$ in $\Delta(A_1) \times \cdots \times \Delta(A_m)$ as $n \to \infty$. Hence $H(r) = s$. This shows that H is surjective.

It remains to show that H is injective. First, let $u_j \in A_j \ (1 \leq j \leq m)$. For any $x = (x_j) \in X$, we have

$$\langle H(\mathring{x}), \bigotimes_{j=1}^{m} u_j \rangle = \prod_{j=1}^{m} \langle \delta_{x_j}, u_j \rangle.$$

By combining the continuity of H with the density of D in $\Delta(A_1) \times \cdots \times \Delta(A_m)$, we deduce that

$$\langle H(s), \bigotimes_{j=1}^{m} u_j \rangle = \prod_{j=1}^{m} \langle s_j, u_j \rangle \text{ for } s = (s_j) \in \prod_{j=1}^{m} \Delta(A_j). \tag{1.10}$$

Having made this point, we now consider $s = (s_j)$ and $r = (r_j)$ in $\prod_{j=1}^{m} \Delta(A_j)$ such that $H(r) = H(s)$. Then

$$\prod_{j=1}^{m} \langle s_j, u_j \rangle = \prod_{j=1}^{m} \langle r_j, u_j \rangle \text{ for any } u_j \in A_j \ (j = 1, \cdots, m).$$

Fix freely an integer $i \ (1 \leq i \leq m)$ and let $\psi \in A_i$. In (1.10) choose $u_j = \psi$ if $j = i, u_j = 1$ if $j \neq i$. By recalling that $\langle s_j, 1 \rangle = \langle r_j, 1 \rangle = 1$, it follows $\langle s_i, \psi \rangle = \langle r_i, \psi \rangle$. Hence $s = r$. Therefore H is one to one and so the proof is complete. ■

Thanks to Theorem 1.2, $\Delta(A)$ may be identified with the product $\Delta(A_1) \times \cdots \times \Delta(A_m)$ by means of H. Thus, we are justified in setting equality (1.4), and it should be borne in mind that the use of the said equality will be systematic throughout.

Theorem 1.2 has the following corollary of considerable interest.

Corollary 1.1 *We have $\mathcal{G}(\bigotimes_{j=1}^{m} u_j) = \bigotimes_{j=1}^{m} \mathcal{G}(u_j)$ for $u_j \in A_j$, where $\mathcal{G}$ denotes the Gelfand transformation on A and A_j as well.*

Proof This follows directly by (1.10) and use of (1.4). ■

1.2 Absorptive Continuous Actions of $\mathbb{R}_+^*$ on $\mathbb{R}^N$

1.2.1 Definitions and Basic Results

Let $\mathbb{R}_+^*$ denote the positive real numbers. In what follows, $\mathbb{R}_+^*$ is viewed as an Abelian topological group, the group operation being usual multiplication (but the dot is omitted) and the topology on $\mathbb{R}_+^*$ being that induced by the usual one on the numerical line $\mathbb{R}$.

For the benefit of the reader we recall that by an action of $\mathbb{R}_+^*$ on the N-dimensional numerical space $\mathbb{R}^N$ is understood any family $\mathcal{H} = (H_\varepsilon)_{\varepsilon>0}$ of permutations of $\mathbb{R}^N$ (i.e., of bijections of $\mathbb{R}^N$ onto itself) such that

$$H_\varepsilon \circ H_{\varepsilon'} = H_{\varepsilon\varepsilon'} \text{ and } H_{\varepsilon=1} = Id_N \quad (\varepsilon, \varepsilon' \in \mathbb{R}_+^*),$$

where $\circ$ and Id_N denote the usual composition and the identity of $\mathbb{R}^N$, respectively. The action $\mathcal{H} = (H_\varepsilon)_{\varepsilon>0}$ is said to be *continuous* if the mapping $(\varepsilon, x) \mapsto H_\varepsilon(x)$ sends continuously $\mathbb{R}_+^* \times \mathbb{R}^N$ (with the product topology) into $\mathbb{R}^N$.

In the sequel, $\omega = (0, \cdots, 0)$ denotes the origin in $\mathbb{R}^N$. On the other hand, for the sake of convenience, we will often write ε^{-1} for $\frac{1}{\varepsilon}(\varepsilon > 0)$.

The notion of an absorptive action is not classical.

Definition 1.1 The action $\mathcal{H} = (H_\varepsilon)_{\varepsilon>0}$ is said to be *absorptive* if for any neighbourhood V of ω and any point $x \in \mathbb{R}^N$, there are some neighbourhood U of x and some real $\alpha > 0$ such that $H_{\varepsilon^{-1}}(U) \subset V$ for $0 < \varepsilon \leq \alpha$.

Remark 1.4 Suppose the action $\mathcal{H} = (H_\varepsilon)_{\varepsilon>0}$ is absorptive. Then $\lim_{\varepsilon\to 0} H_{\varepsilon^{-1}}(x) = \omega$ for any $x \in \mathbb{R}^N$.

The following proposition provides a concrete illustration of the absorptiveness.

Proposition 1.2 *Let $\mathcal{H} = (H_\varepsilon)_{\varepsilon>0}$ be an action of $\mathbb{R}_+^*$ on $\mathbb{R}^N$. The following two assertions are equivalent.*

(i) $\mathcal{H}$ is absorptive.
(ii) If V is any arbitrary neighbourhood of ω, then for any compact set K there is some $\alpha > 0$ such that

$$H_{\varepsilon^{-1}}(K) \subset V \text{ for } 0 < \varepsilon \leq \alpha. \tag{1.11}$$

Proof Let us first check that (ii) implies (i). Suppose a neighbourhood V of ω and an arbitrary point $x \in \mathbb{R}^N$ are given. If we consider any compact neighbourhood K of x, then (1.11) follows at once according to the assumption (ii). This shows the absorptiveness. Conversely, assume (i), and fix freely a compact set K in $\mathbb{R}^N$. On the other hand, let V be any given neighbourhood of ω. By hypothesis, to each $x \in K$ there is associated some neighbourhood $U(x)$ of x and some $\alpha_x > 0$ such that $H_{\varepsilon^{-1}}(U(x)) \subset V$ for $0 < \varepsilon \leq \alpha_x$. It is quite clear that we may assume $U(x)$ to be open. Therefore, we have at our disposal an open covering $\{U(x)\}_{x\in K}$ of K, from which we can extract a finite covering $\{U(x_i)\}_{1\leq i\leq n}$ to which corresponds a finite set of positive numbers α_{x_i} $(i = 1, \cdots, n)$ such that

$$H_{\varepsilon^{-1}}(U(x_i)) \subset V \text{ for } 0 < \varepsilon \leq \alpha_{x_i} \quad (1 \leq i \leq n).$$

Finally, denoting by U the union of the sets $U(x_i)$ $(1 \leq i \leq n)$ and putting $\alpha = \min_{1\leq i\leq n} \alpha_{x_i}$, we have $H_{\varepsilon^{-1}}(U) \subset V$ for $0 < \varepsilon \leq \alpha$. Hence (1.11) follows and so (ii) is established. ■

Remark 1.5 We may express assertion (ii) of Proposition 1.2 by saying that any compact set is absorbed by any neighbourhood of the origin. This terminology is evidently borrowed from the theory of Topological Vector Spaces.

As a consequence of the preceding proposition, we have the following

Corollary 1.2 *For $x \in \mathbb{R}^N$ with $x \neq \omega$, we have $\lim_{\varepsilon \to 0} H_\varepsilon(x) = \infty$, where ∞ denotes the point at infinity of the Alexandroff compactification of $\mathbb{R}^N$.*

Proof Let K be any arbitrary compact set in $\mathbb{R}^N$. Let V be a neighbourhood of ω not containing x. According to Proposition 1.2, some real $\alpha > 0$ exists such that (1.11) holds. But it is clear that $H_{\varepsilon^{-1}}(K)$ does not contain x, and that for $0 < \varepsilon \leq \alpha$. Whence we conclude that $H_\varepsilon(x) \in \mathbb{R}^N \setminus K$ for $0 < \varepsilon \leq \alpha$. This proves the corollary. ∎

Remark 1.6 The two assertions $\lim_{\varepsilon \to 0} H_\varepsilon(x) = \infty$ and $\lim_{\varepsilon \to 0} |H_\varepsilon(x)| = +\infty$ are equivalent ($|\cdot|$ denotes the Euclidean norm on $\mathbb{R}^N$). Indeed, this is immediate by the fact that any closed ball in $\mathbb{R}^N$ is compact.

We will pursue farther our study by turning our attention to proving that the product of a finite family of absorptive continuous actions of $\mathbb{R}_+^*$ is still continuous and absorptive. More precisely, let

$$\mathbb{R}^N = \mathbb{R}^{N_1} \times \cdots \times \mathbb{R}^{N_m} \tag{1.12}$$

with $N_i \in \mathbb{N}^*$ $(i = 1, \cdots, m)$ and $N = N_1 + \cdots + N_m$. For each integer i $(1 \leq i \leq m)$, let $\mathcal{H}_i = (H_\varepsilon^i)_{\varepsilon > 0}$ be an action of $\mathbb{R}_+^*$ on $\mathbb{R}^{N_i}$. Put $H_\varepsilon = H_\varepsilon^1 \times \cdots \times H_\varepsilon^m$ (direct product). It is worth recalling that H_ε is the mapping of $\mathbb{R}^N$ into itself defined by

$$H_\varepsilon(x) = (H_\varepsilon^1(pr_1(x)), \cdots, H_\varepsilon^m(pr_m(x))) \quad (x \in \mathbb{R}^N), \tag{1.13}$$

where pr_i denotes the natural projection of $\mathbb{R}^N$ onto $\mathbb{R}^{N_i}$. The family $\mathcal{H} = (H_\varepsilon)_{\varepsilon > 0}$ thus defined is an action of $\mathbb{R}_+^*$ on $\mathbb{R}^N$, referred to as the product of the actions $\mathcal{H}_i$ $(1 \leq i \leq m)$, sometimes denoted by $\mathcal{H} = \mathcal{H}_1 \times \cdots \times \mathcal{H}_m$.

Proposition 1.3 *Let the notation be as above. Assume that each action $\mathcal{H}_i$ is continuous and absorptive. Then the product action $\mathcal{H}$ is continuous and absorptive.*

Proof We begin by showing the continuity. To do this, we introduce the mappings $H : \mathbb{R}_+^* \times \mathbb{R}^N \to \mathbb{R}^N$, $H^i : \mathbb{R}_+^* \times \mathbb{R}^{N_i} \to \mathbb{R}^{N_i}$ $(1 \leq i \leq m)$ and $f_i : \mathbb{R}_+^* \times \mathbb{R}^N \to \mathbb{R}_+^* \times \mathbb{R}^{N_i}$ $(1 \leq i \leq m)$ defined respectively by

$$H(\varepsilon, x) = H_\varepsilon(x),\ H^i(\varepsilon, t) = H_\varepsilon^i(t)$$

and $f_i(\varepsilon, x) = (\varepsilon, pr_i(x))$ for $\varepsilon > 0$, $x \in \mathbb{R}^N$ and $t \in \mathbb{R}^{N_i}$. Clearly

$$H(\varepsilon, x) = (H^1(f_1(\varepsilon, x)), \cdots, H^m(f_m(\varepsilon, x)))$$

for $(\varepsilon, x) \in \mathbb{R}_+^* \times \mathbb{R}^N$, and the claimed continuity follows at once.

We next check the absorptiveness. Let V be a neighbourhood of the origin ω in $\mathbb{R}^N$. Let $V_1 \times \cdots \times V_m \subset V$, where V_i is some neighbourhood of the origin in $\mathbb{R}^{N_i}$ $(1 \le i \le m)$. Finally, let $x \in \mathbb{R}^N$. For fixed i, the absorptiveness of $\mathcal{H}_i$ yields some $\alpha_i > 0$ and some neighbourhood U_i of $pr_i(x)$ in $\mathbb{R}^{N_i}$ such that $U_i \subset H_\varepsilon^i(V_i)$ for $0 < \varepsilon \le \alpha_i$. Hence $U \subset H_\varepsilon(V)$ for $0 < \varepsilon \le \alpha$, where $\alpha = \min_{1 \le i \le m} \alpha_i$ and $U = U_1 \times \cdots \times U_m$ (a neighbourhood of x in $\mathbb{R}^N$). This shows the absorptiveness of $\mathcal{H}$. ■

We will conclude this section with two examples of great interest in homogenization theory.

Example 1.5 Let $\mathcal{H} = (H_\varepsilon)_{\varepsilon>0}$ be the action of $\mathbb{R}_+^*$ on $\mathbb{R}^N$ defined by

$$H_\varepsilon(x) = \frac{x}{\varepsilon^r}, x \in \mathbb{R}^N, \quad (\varepsilon > 0),$$

where the real $r > 0$ is given. It is an easy matter to verify that $\mathcal{H}$ is continuous and absorptive.

Example 1.6 Let the action $\mathcal{H} = (H_\varepsilon)_{\varepsilon>0}$ of $\mathbb{R}_+^*$ on $\mathbb{R}^N$ be given by

$$H_\varepsilon(x) = \left(\frac{x_1}{\varepsilon^{r_1}}, \cdots, \frac{x_N}{\varepsilon^{r_N}}\right), x = (x_1, \cdots, x_N) \in \mathbb{R}^N \quad (\varepsilon > 0)$$

where $r_i > 0$ $(1 \le i \le N)$. According to (1.13) we have $\mathcal{H} = \mathcal{H}_1 \times \cdots \times \mathcal{H}_N$, where for each integer i $(1 \le i \le N)$, $\mathcal{H}_i = (H_\varepsilon^i)_{\varepsilon>0}$ is the action of $\mathbb{R}_+^*$ on $\mathbb{R}$ defined by

$$H_\varepsilon^i(t) = \frac{t}{\varepsilon^{r_i}} \quad (\varepsilon > 0) \text{ for } t \in \mathbb{R}.$$

The action $\mathcal{H}$ is continuous and absorptive, as is easily seen by Example 1.5 and use of Proposition 1.3.

1.2.2 *Mean Values: Definition and Examples*

One of the facts in favour of the continuous absorptive actions is that each of them generates a typical mean value which turns out to be a right one needed in deterministic homogenization.

However, before embarking upon discussing the mean value attached to a continuous absorptive action of $\mathbb{R}_+^*$ on $\mathbb{R}^N$, let us first draw attention to what is generically meant by a mean value on a topological space.

Let X be a Hausdorff topological space, and let $\mathcal{B}(X)$ denote the usual Banach space of bounded continuous complex-valued functions on X.

Definition 1.2 A *mean value* on X is defined to be an unbounded linear operator m from $\mathcal{B}(X)$ to $\mathbb{C}$ with the following properties:

(MV1) The domain $D(m)$ of m is a closed vector subspace of $\mathcal{B}(X)$ that contains the constants and preserves complex conjugation.

(MV2) We have

(1) $\overline{m(f)} = m(\overline{f})$ for $f \in D(m)$,
(2) $m(f) \geq 0$ for $f \in D(m)$ with $f \geq 0$,
(3) $m(1) = 1$.

(MV3) Furthermore, if X is an Abelian Hausdorff topological group, then $D(m)$ is required to preserve translation (i.e., $\tau_a f \in D(m)$ for $f \in D(m)$ and $a \in X$) with further

(4) $m(\tau_a f) = m(f)$ for $f \in D(m)$ and $a \in X$.

Remark 1.7 A mean value m on X is continuous in the sense that there is some constant $c > 0$ such that

$$|m(f)| \leq c\|f\|_\infty \quad (f \in D(m)). \tag{1.14}$$

Indeed, if f is real-valued, then (1.14) (with $c = 1$) follows immediately by the inequalities $-\|f\|_\infty \leq f \leq \|f\|_\infty$. If f assumes complex values, then (1.14) (with $c = 2$) is readily deduced from the preceding case by noting that the real and imaginary parts of each function in $D(m)$ still lie in $D(m)$.

This is worth illustrating.

Example 1.7 (The Mean Value of the Convergence at Infinity) Let X be a noncompact locally compact space, and let m be the unbounded linear operator from $\mathcal{B}(X)$ to $\mathbb{C}$ defined by

$$D(m) = \mathcal{B}_\infty(X) \text{ (Example 1.2)}, m(f) = \lim_{x\to\infty} f(x),\ f \in D(m).$$

It is worth noting by the way that, as is customary, X is here (and in the sequel) identified with $X' \setminus \{\infty\}$. It is an easy exercise to check that m is a mean value on X.

Example 1.8 (The Mean Value for Periodic Functions) Let $\mathcal{C}_{per}(Y)$ be the Banach algebra in Example 1.3 with $Y = (-\frac{1}{2}, \frac{1}{2})^N$. We define $m : D(m) \subset \mathcal{B}(\mathbb{R}^N) \to \mathbb{C}$ by

$$D(m) = \mathcal{C}_{per}(Y), \quad m(f) = \int_Y f(y)dy, \quad f \in D(m).$$

The unbounded operator thus defined is clearly a mean value on $\mathbb{R}^N$. More generally, given a basis $\{b_1, \cdots, b_N\}$ of $\mathbb{R}^N$, if we consider $\mathcal{C}_{per}(Y)$ with Y given by (1.3), the corresponding mean value is defined as above save that we have in this case

$$m(f) = \frac{1}{|Y|}\int_Y f(y)dy \quad (f \in D(m),$$

where $|Y|$ denotes the measure of Y for Lebesgue measure on $\mathbb{R}^N$. Furthermore, if $\mathbb{R}^N$ is viewed as a locally compact Abelian group, it is obvious that condition (MV3) is also fulfilled.

Example 1.9 (The Mean Value for Almost Periodic Functions) Let G be a locally compact Abelian group additively written. We consider the unbounded linear form $m : D(m) \subset \mathcal{B}(G) \to \mathbb{C}$ defined as follows. The domain of m is $D(m) = AP(G)$ (Example 1.4), and for $f \in D(m)$, $m(f)$ is the unique constant (function) lying in the closed convex hull (in $\mathcal{B}(G)$) of the translates $\tau_a f$ (a ranging over G). As is classical, this yields a linear mapping m that satisfies (MV1),(MV2) and (MV3) with $X = G$; see, e.g., Guichardet [57], Jacobs [60], Hewitt and Ross [58].

1.2.3 *Mean Values Generated by Some Specific Actions*

We shall now concentrate on the non-classical notion of a mean value generated by a suitable action of $\mathbb{R}_+^*$ on $\mathbb{R}^N$. In what follows, λ denotes Lebesgue measure on $\mathbb{R}^N$, and $\mathcal{H} = (H_\varepsilon)_{\varepsilon>0}$ denotes an absorptive continuous action of $\mathbb{R}_+^*$ on $\mathbb{R}^N$ satisfying the condition

$$(Q) \qquad \begin{array}{l}\text{To each real } \varepsilon > 0 \text{ there is attached some constant}\\ q(\varepsilon) \text{ such that } H_\varepsilon(\lambda) = q(\varepsilon)\lambda.\end{array}$$

Remark 1.8 It is worth, perhaps, recalling that $H_\varepsilon(\lambda)$ (the image of λ under H_ε) denotes the Radon measure on $\mathbb{R}^N$ defined by

$$H_\varepsilon(\lambda)(\varphi) = \int \varphi(H_\varepsilon(x))d\lambda(x)$$

for $\varphi \in \mathcal{K}(\mathbb{R}^N)$ (space of compactly supported continuous complex functions on $\mathbb{R}^N$). In order that a function f of $\mathbb{R}^N$ into a Banach space B be integrable for $H_\varepsilon(\lambda)$, it is necessary and sufficient that the function $f \circ H_\varepsilon$ be integrable for λ, in which case

$$\int f(H_\varepsilon(x))d\lambda(x) = \int f dH_\varepsilon(\lambda).$$

For more details the reader is referred to Bourbaki [23, pp. 68–80].

In view of property (Q), Lebesgue measure is said to be *quasi-invariant* under the action $\mathcal{H}$ (however, in a restricted sense); see Nguetseng [99]; see also Dieudonné [45, p. 23] for a detailed analysis of quasi-invariant measures, and Walters [141, p. 237] for an ergodic setting. Attention must be drawn to the fact that there is no hope of Lebesgue measure being invariant under the action $\mathcal{H}$, that is, of λ satisfying the equality $H_\varepsilon(\lambda) = \lambda$ (see Nguetseng [99]).

It is possible to compute the constant $q(\varepsilon)$ in (Q). We will need the following lemma which is based on the sole absorptiveness and continuity hypotheses.

Lemma 1.5 *There exists a set $B \subset \mathbb{R}^N$ with the following properties:*

(i) B is a bounded open neighbourhood of $\omega = (0, \cdots, 0)$
(ii) $H_{\varepsilon^{-1}}(B) \subset B$ for $0 < \varepsilon \leq 1$.

Proof Let us begin by noting that the origin ω in $\mathbb{R}^N$ is $\mathcal{H}$-invariant, that is, $H_\varepsilon(\omega) = \omega$ for all $\varepsilon > 0$. To see this, fix freely $x \in \mathbb{R}^N$. For arbitrary given $r > 0$, one has $\lim_{\varepsilon \to 0} H_r(H_{\varepsilon^{-1}}(x)) = H_r(\omega)$, as is straightforward by Remark 1.4 and use of the continuity of H_r on $\mathbb{R}^N$. But $\lim_{\varepsilon \to 0} H_r(H_{\varepsilon^{-1}}(x)) = \lim_{\varepsilon \to 0} H_{(r^{-1}\varepsilon)^{-1}}(x) = \omega$ (Remark 1.4). Hence the claimed invariance property follows.

With this in mind, let V be a compact neighbourhood of ω. Let U_0 be a bounded open neighbourhood of ω (e.g., U_0 is the open unit ball). Since U_0 is relatively compact, according to Proposition 1.2 we can find some $\alpha > 0$ such that $H_{\varepsilon^{-1}}(U_0) \subset V$ for $0 < \varepsilon \leq \alpha$. Clearly the set $U_1 = H_{\alpha^{-1}}(U_0)$ is an open neighbourhood of ω and moreover $H_{\varepsilon^{-1}}(U_1) \subset V$ for $0 < \varepsilon \leq 1$. Finally, let $B = \bigcup_{0<\varepsilon\leq 1} H_{\varepsilon^{-1}}(U_1)$. This set has the required properties (the boundedness of B is ensured by the relation $B \subset V$). ■

A set such as B will be referred to as an open *elementary set* (for the action $\mathcal{H}$).

Remark 1.9

(1) If B is an open elementary set, then so also is $H_r(B)$ for any arbitrary real $r > 0$.
(2) There is a neighbourhood base at ω consisting of open elementary sets.
(3) Each compact set has a neighbourhood which is an open elementary set (this follows by combining point (1) with Proposition 1.2).

We can now restate condition (Q) in a more explicit form.

Proposition 1.4 *Let the absorptive continuous action $\mathcal{H} = (H_\varepsilon)_{\varepsilon>0}$ satisfy condition (Q). Then, there exists a (unique) real number $s > 0$ such that $H_\varepsilon(\lambda) = \varepsilon^s\lambda$ for all $\varepsilon > 0$.*

Proof To begin, let us note that the number $q(\varepsilon)$ attached to each $\varepsilon > 0$ is unique and positive. We may then consider the mapping $\varepsilon \mapsto q(\varepsilon)$ of $\mathbb{R}_+^*$ into itself, where it must be borne in mind that $\mathbb{R}_+^*$ is an Abelian topological group. This mapping, q, is a homomorphism, that is, $q(\varepsilon\varepsilon') = q(\varepsilon)q(\varepsilon')$ for $\varepsilon, \varepsilon' > 0$, as is immediate by the equality $H_{\varepsilon\varepsilon'}(\lambda) = H_{\varepsilon'}(H_\varepsilon(\lambda))$ (see Bourbaki [23, p. 72]). Let us show that this homomorphism is continuous. It is enough to check the continuity at $\varepsilon = 1$, of course. For this purpose, fix $\varphi \in \mathcal{K}(\mathbb{R}^N)$ with $\lambda(\varphi) = 1$. Then

$$q(\varepsilon) = \int \varphi(H_\varepsilon(x))dx \quad (\varepsilon > 0).$$

Now, let B be an open elementary set containing the support of φ (see Remark 1.9), and let $\alpha > 1$. The set $I = (0, \alpha]$ is a neighbourhood of 1 in $\mathbb{R}_+^*$ and we have $H_{\varepsilon^{-1}}(B) \subset H_{\alpha^{-1}}(B)$ for $\varepsilon \in I$. Hence

$$|\varphi(H_\varepsilon(x))| \leq \|\varphi\|_\infty f(x) \quad (x \in \mathbb{R}^N, \varepsilon \in I),$$

where f is the characteristic function of $H_{\alpha^{-1}}(B)$ in $\mathbb{R}^N$. Therefore the claimed result follows by a classical argument (see, e.g., Bourbaki [22, p. 144]). The result is that the mapping q is a continuous homomorphism of $\mathbb{R}_+^*$ into $\mathbb{R}_+^*$. Consequently, there is some real s such that $q(\varepsilon) = \varepsilon^s$ for all $\varepsilon > 0$ (see Bourbaki [25, V, p. 12] for details). To finish, we must verify that $s > 0$. It is clear that we will be through if we have shown that $\lim_{\varepsilon \to 0} q(\varepsilon) = 0$. To this end, let $x \in \mathbb{R}^N$, $x \neq \omega$. Then $\lim_{\varepsilon \to 0} \varphi(H_\varepsilon(x)) = 0$, as is readily seen by Corollary 1.2. Hence the desired result follows by the preceding inequality and use of the dominated convergence theorem. ∎

In the sequel and throughout the next chapters we shall be dealing exclusively, without further ado, with absorptive continuous actions $\mathcal{H} = (H_\varepsilon)_{\varepsilon>0}$ of $\mathbb{R}_+^*$ on $\mathbb{R}^N$ under which Lebesgue measure λ on $\mathbb{R}^N$ is quasi-invariant in the preceding restricted sense, that is, satisfies $H_\varepsilon(\lambda) = \varepsilon^s \lambda$ $(\varepsilon > 0)$ for some suitable $s > 0$. This allows us to state that the transformation $u \mapsto u \circ H_\varepsilon$ maps the space $L^1_{loc}(\mathbb{R}^N)$ (resp. $L^1(\mathbb{R}^N)$) into itself. Finally, it is worth noting that the action in Example 1.5 (resp. Example 1.6) satisfies $H_\varepsilon(\lambda) = \varepsilon^s \lambda$ with $s = Nr$ (resp. $s = r_1 + \cdots + r_N$).

We turn now to the statement of a fundamental theorem.

Theorem 1.3 *Let $\Pi^\infty_{\mathcal{H}}(\mathbb{R}^N)$ be the set of all functions $u \in \mathcal{B}(\mathbb{R}^N)$ for which a constant (function) $\widetilde{u} \in \mathbb{C}$ exists such that, as $\varepsilon \to 0$, $u \circ H_\varepsilon \to \widetilde{u}$ in $L^\infty(\mathbb{R}^N)$-weak∗. Then*

(i) $\Pi^\infty_{\mathcal{H}}(\mathbb{R}^N)$ is a closed vector subspace of $\mathcal{B}(\mathbb{R}^N)$ that contains the constants and preserves complex conjugation.
(ii) If the N-dimensional numerical space $\mathbb{R}^N$ is merely considered as a topological space, then the unbounded operator $M : D(M) \subset \mathcal{B}(\mathbb{R}^N) \to \mathbb{C}$ defined by

$$\begin{cases} D(M) = \Pi^\infty_{\mathcal{H}}(\mathbb{R}^N) \ \textit{(domain of definition of } M\textit{)} \\ M(u) = \widetilde{u} \ \textit{for } u \in D(M) \end{cases}$$

is a mean value on $\mathbb{R}^N$.

(iii) *In case $\mathbb{R}^N$ is viewed as an Abelian topological group, if the action $\mathcal{H}$ is compatible with addition in $\mathbb{R}^N$, that is, if $\mathcal{H}$ satisfies*

$$H_\varepsilon(x+y) = H_\varepsilon(x) + H_\varepsilon(y) \ \ (\varepsilon > 0) \textit{ for } x, y \in \mathbb{R}^N,$$

then $\Pi^\infty_{\mathcal{H}}(\mathbb{R}^N)$ preserves translation and further

$$M(\tau_a u) = M(u) \ \ (a \in \mathbb{R}^N) \textit{ for } u \in \Pi^\infty_{\mathcal{H}}(\mathbb{R}^N),$$

so that M is a mean value on the Abelian topological group $\mathbb{R}^N$.

Remark 1.10 Before embarking upon the proof of this theorem it seems useful to clarify the compatibility condition required of $\mathcal{H}$ in part (iii). First of all, let us note that any continuous homomorphism of the additive topological group $\mathbb{R}^N$ into itself is an endomorphism of the vector space $\mathbb{R}^N$ (see, e.g., Bourbaki [25, VII, p. 11]). Thus, the (absorptive) continuous action $\mathcal{H} = (H_\varepsilon)_{\varepsilon>0}$ of $\mathbb{R}_+^*$ on $\mathbb{R}^N$ will be compatible with addition in $\mathbb{R}^N$ if and only if for each $\varepsilon > 0$, H_ε is an endomorphism of the N-dimensional vector space $\mathbb{R}^N$, i.e., if and only if it satisfies $H_\varepsilon(x+y) = H_\varepsilon(x) + H_\varepsilon(y)$ and $H_\varepsilon(tx) = tH_\varepsilon(x)$ for all $x, y \in \mathbb{R}^N$ and all $t \in \mathbb{R}$. Whenever the latter property is not necessary we will nevertheless keep on talking of the compatibility of $\mathcal{H}$ with addition in $\mathbb{R}^N$.

Remark 1.11 Suppose the absorptive continuous action $\mathcal{H} = (H_\varepsilon)_{\varepsilon>0}$ is compatible with addition in $\mathbb{R}^N$. Then, according to Remark 1.10, to each real number $\varepsilon > 0$ there is attached an $N \times N$ matrix $\mathcal{M}(\varepsilon)$ with entries $a_{ij}(\varepsilon)$ $(1 \leq i, j \leq N)$ such that

$$H^i_\varepsilon(x) = \sum_{j=1}^N a_{ij}(\varepsilon) x_j \ \ (1 \leq i \leq N)$$

for $x = (x_1, \cdots, x_N) \in \mathbb{R}^N$, where $H_\varepsilon = (H^1_\varepsilon, \cdots, H^N_\varepsilon)$. It is clear that $\det \mathcal{M}(\varepsilon) \neq 0$ for all $\varepsilon > 0$. Therefore, from classical integration theory it follows that λ is quasi-invariant under $\mathcal{H}$. More precisely, $H_\varepsilon(\lambda) = |\det \mathcal{M}(\varepsilon^{-1})| \lambda$ $(\varepsilon > 0)$. One deduces by Proposition 1.4 that there is some real $s > 0$ such that $H_\varepsilon(\lambda) = \varepsilon^s \lambda$ for all $\varepsilon > 0$.

Proof of Theorem 1.3 It is clear that $\Pi^\infty_{\mathcal{H}}(\mathbb{R}^N)$ is a vector subspace of $\mathcal{B}(\mathbb{R}^N)$. Let us show that it is closed in $\mathcal{B}(\mathbb{R}^N)$. Let $(u_n)_{n\in\mathbb{N}}$ be an ordinary sequence of elements of $\Pi^\infty_{\mathcal{H}}(\mathbb{R}^N)$ that converges uniformly (i.e., in the supremum norm) to some $u \in \mathcal{B}(\mathbb{R}^N)$. The aim is to check that u lies in $\Pi^\infty_{\mathcal{H}}(\mathbb{R}^N)$. First, it is an easy matter to verify that the complex numbers $\widetilde{u}_n (n \in \mathbb{N})$ form a Cauchy sequence. Thus $\widetilde{u}_n \to \xi$ as $n \to \infty$. Next, let us fix freely some $f \in L^1(\mathbb{R}^N)$ and write

$$\int u^\varepsilon f\,dx - \xi \int f\,dx = \int u^\varepsilon f\,dx - \int u_n^\varepsilon f\,dx + \int u_n^\varepsilon f\,dx - \widetilde{u}_n \int f\,dx$$
$$+ \widetilde{u}_n \int f\,dx - \xi \int f\,dx,$$

where, for convenience, we put $\psi^\varepsilon = \psi \circ H_\varepsilon$ for any $\psi \in \mathcal{B}(\mathbb{R}^N)$. Next, fix any arbitrary $\eta > 0$. By noting on one hand that

$$\left| \int u^\varepsilon f\,dx - \int u_n^\varepsilon f\,dx \right| \le \|u_n - u\|_\infty \int |f|\,dx,$$

and recalling on the other hand that $\widetilde{u}_n \to \xi$ as $n \to \infty$, we see that we can fix an integer $n > 0$ (n large enough) such that

$$\left| \int u^\varepsilon f\,dx - \int u_n^\varepsilon f\,dx \right| \le \frac{\eta}{3} \text{ for all } \varepsilon > 0$$

and

$$\left| \widetilde{u}_n \int f\,dx - \xi \int f\,dx \right| \le \frac{\eta}{3}.$$

Finally, it is clear that there is some $\alpha > 0$ such that

$$\left| \int u_n^\varepsilon f\,dx - \widetilde{u}_n \int f\,dx \right| \le \frac{\eta}{3} \text{ for } 0 < \varepsilon \le \alpha.$$

From all that we deduce

$$\left| \int u^\varepsilon f\,dx - \xi \int f\,dx \right| \le \eta \text{ for } 0 < \varepsilon \le \alpha,$$

which shows that $u^\varepsilon = u \circ H_\varepsilon \to \xi$ in $L^\infty(\mathbb{R}^N)$-weak$*$; thus $u \in \Pi_{\mathcal{H}}^\infty(\mathbb{R}^N)$ with $\widetilde{u} = \xi$. The rest of part (i) of the theorem being trivial, let us next turn to part (ii) to observe that condition (MV2) in Definition 1.2 is evidently fulfilled by the operator M, so that the latter is a mean value on $\mathbb{R}^N$ as stated.

It remains to prove (iii). But all that amounts to showing that for any given $u \in \Pi_{\mathcal{H}}^\infty(\mathbb{R}^N)$ and $a \in \mathbb{R}^N$, one has

$$\int \varphi(x) u(H_\varepsilon(x) - a) dx \to M(u) \int \varphi(x) dx \text{ as } \varepsilon \to 0 \tag{1.15}$$

for all $\varphi \in L^1(\mathbb{R}^N)$. However, for an obvious reason it is quite enough to prove (1.15) for $\varphi \in \mathcal{K}(\mathbb{R}^N)$. So, fix freely one such φ, and let $K = Supp(\varphi)$ (the support of φ). On the other hand, let B_N stand for the open unit ball in $\mathbb{R}^N$ and finally, fix a constant $c > 0$ such that $\|u\|_\infty \lambda(K + B_N) \leq c$, where $K + B_N$ denotes the set of all $x + y$ with $x \in K$ and $y \in B_N$. This being so, let $\eta > 0$. By uniform continuity, consider some $0 < r \leq 1$ such that

$$|\varphi(x) - \varphi(z)| \leq \frac{\eta}{c} \text{ for } x, z \in \mathbb{R}^N, x - z \in rB_N.$$

Finally, on account of the absorptiveness of $\mathcal{H}$, one can find some real $\alpha > 0$ such that $H_{\varepsilon^{-1}}(a) \in rB_N$ for $0 < \varepsilon \leq \alpha$. Hence

$$\left| \int \varphi(x)u(H_\varepsilon(x) - a)dx - \int \varphi(x)u(H_\varepsilon(x))dx \right| =$$

$$\left| \int u(H_\varepsilon(x))[\varphi(x + H_{\varepsilon^{-1}}(a)) - \varphi(x)]dx \right| \leq \eta$$

for $0 < \varepsilon \leq \alpha$, where we have made use of the compatibility of $\mathcal{H}$ with addition in $\mathbb{R}^N$. From which (1.15) follows, and so the proof is complete. ■

Definition 1.3 The operator constructed in Theorem 1.3 is termed the *mean value on $\mathbb{R}^N$ generated by the action $\mathcal{H}$.*

In the sequel, instead of $\Pi^\infty_{\mathcal{H}}(\mathbb{R}^N)$ we will sometimes write $\Pi^\infty(\mathbb{R}^N)$ or even Π^∞ when there is no risk of ambiguity.

Let us show that the mean value M is consistent with convolution.

Proposition 1.5 *Suppose the action $\mathcal{H}$ is compatible with addition in $\mathbb{R}^N$. Let $f \in L^1(\mathbb{R}^N)$. For $u \in \Pi^\infty(\mathbb{R}^N)$, we have $f * u \in \Pi^\infty(\mathbb{R}^N)$ with*

$$M(f * u) = M(u) \int f(y)dy.$$

Proof Let $u \in \Pi^\infty(\mathbb{R}^N)$. We have $f * u \in \mathcal{B}(\mathbb{R}^N)$ (this is a classical result). Next, fix freely $\varphi \in L^1(\mathbb{R}^N)$. It is clear that

$$\int (f * u)^\varepsilon(x)\varphi(x)dx = \int \left[\int (\tau_y u)^\varepsilon(x)\varphi(x)dx \right] f(y)dy,$$

where, for the sake of convenience, we put $\psi^\varepsilon = \psi \circ H_\varepsilon$ for $\psi \in \mathcal{B}(\mathbb{R}^N)$. But for almost all $y \in \mathbb{R}^N$, one has on one hand

$$f(y) \int (\tau_y u)^\varepsilon(x)\varphi(x)dx \to M(u)f(y) \int \varphi(x)dx \text{ as } \varepsilon \to 0$$

according to (1.15), on the other hand,

$$\left| f(y) \int (\tau_y u)^{\varepsilon}(x)\varphi(x)dx \right| \leq \|u\|_{\infty} \|\varphi\|_{L^1} |f(y)|.$$

Hence, by the dominated convergence theorem,

$$\int (f * u)^{\varepsilon}(x)\varphi(x)dx \to M(u) \int f(y)dy \int \varphi(x)dx$$

as $\varepsilon \to 0$. The proposition follows thereby. ■

In many classical cases the mean value M on $\mathbb{R}^N$ generated by the action $\mathcal{H}$ turns out to be an extension of the ones in use in mathematical analysis. We present below a few examples by way of illustration.

1.2.3.1 Mean Value for $\mathcal{B}_{\infty}(\mathbb{R}^N)$

Example 1.10 The mean value M generated by the action $\mathcal{H}$ on $\mathbb{R}^N$ is an extension of the mean value of the convergence at infinity on $\mathbb{R}^N$ (Example 1.7). Specifically, we have

$$D(m) \subset D(M) \text{ and } M(u) = m(u) \text{ for } u \in D(m), \tag{1.16}$$

where $D(m) = \mathcal{B}_{\infty}(\mathbb{R}^N)$ and $m(u) = \lim_{x \to \infty} u(x), u \in D(m)$. This follows immediately by Corollary 1.2 and use of the Lebesgue dominated convergence theorem.

1.2.3.2 Mean Value for $AP(\mathbb{R}^N)$

Example 1.11 We assume that $\mathcal{H}$ is compatible with addition in $\mathbb{R}^N$ and further, is self-adjoint, i.e., $\mathcal{H}$ satisfies

$$y \cdot H_{\varepsilon}(x) = x \cdot H_{\varepsilon}(y) \ \ (\varepsilon > 0) \text{ for } x, y \in \mathbb{R}^N, \tag{1.17}$$

where the dot represents the inner product in $\mathbb{R}^N$.

Under the above assumptions, the mean value M is an extension of the mean value for almost periodic functions on $\mathbb{R}^N$ (see Examples 1.4 and 1.9 with $G = \mathbb{R}^N$). In other words we have (1.16) with $D(m) = AP(\mathbb{R}^N)$ and $m(u)$ defined in Example 1.9. Let us show this. For each $k \in \mathbb{R}^N$, we denote by γ_k the complex function on $\mathbb{R}^N$ defined by

$$\gamma_k(y) = e^{2\pi i k \cdot y} \ \ (y \in \mathbb{R}^N), \tag{1.18}$$

where the dot stands for the inner product in $\mathbb{R}^N$, once again. We have here $\widehat{\mathbb{R}^N} = \{\gamma_k | k \in \mathbb{R}^N\}$ (see, e.g., Guichardet [57]). Hence, according to Example 1.4, the algebra $T(\mathbb{R}^N)$ of functions u of the form

$$u = \sum_{finite} a_k \gamma_k \ \ (a_k \in \mathbb{C}), \tag{1.19}$$

is dense in $AP(\mathbb{R}^N)$. Thus, we are through if we can check that $\gamma_k \in \Pi^\infty(\mathbb{R}^N)$ with $M(\gamma_k) = m(\gamma_k)$ for all $k \in \mathbb{R}^N$. If $k = \omega$ (the origin in $\mathbb{R}^N$) then $\gamma_k \in \Pi^\infty(\mathbb{R}^N)$ and $M(\gamma_k) = 1 = m(\gamma_{k=\omega})$ since γ_ω is the constant function 1. We assume $k \neq \omega$ in the sequel. Let $\varphi \in L^1(\mathbb{R}^N)$. By (1.17) one has

$$\int \gamma_k(H_\varepsilon(x))\varphi(x)dx = \mathcal{F}\varphi(H_\varepsilon(-k)) \ \ (\varepsilon > 0),$$

where $\mathcal{F}\varphi$ is the Fourier transform of φ. But, according to Corollary 1.2, $\lim_{\varepsilon\to 0} \mathcal{F}\varphi(H_\varepsilon(-k)) = 0$ (Riemann-Lebesgue lemma). Hence $\gamma_k \in \Pi^\infty(\mathbb{R}^N)$ with $M(\gamma_k) = 0 = m(\gamma_k)$. We deduce $T(\mathbb{R}^N) \subset \Pi^\infty(\mathbb{R}^N)$, hence $AP(\mathbb{R}^N) \subset \Pi^\infty(\mathbb{R}^N)$ with $M(u) = m(u)$ for $u \in AP(\mathbb{R}^N)$, as alleged.

1.2.3.3 Mean Value for $\mathcal{C}_{per}(Y)$

Example 1.12 We assume that $\mathcal{H}$ is self-adjoint and compatible with addition in $\mathbb{R}^N$. Then M is an extension of the mean value for Y-periodic functions on $\mathbb{R}^N$ with $Y = (-\frac{1}{2}, \frac{1}{2})^N$ (Example 1.8). In other words, we have to show (1.16) with $D(m) = \mathcal{C}_{per}(Y)$ and $m(u) = \int_Y u(y)dy \ \ (u \in D(m))$, Y specified above. Let $T_{\mathbb{Z}^N}(\mathbb{R}^N)$ denote the linear span of $\{\gamma_k | k \in \mathbb{Z}^N\}$ in $\mathcal{C}_{per}(Y)$, that is, the vector subspace of $\mathcal{C}_{per}(Y)$ consisting of functions u of the form (1.19). As is well known, $T_{\mathbb{Z}^N}(\mathbb{R}^N)$ is dense in $\mathcal{C}_{per}(Y)$. It therefore suffices to proceed exactly as in Example 1.11.

1.2.4 Some Specific Trace Results

Let Ω be an open set in $\mathbb{R}_x^N$ (the N-dimensional numerical space of variables $x = (x_1, \cdots, x_N)$). Let a real $\varepsilon > 0$ be fixed.

Our main purpose in the present section is to record a variety of functions $u \in L^1_{loc}(\Omega \times \mathbb{R}_y^N)$ behaving in such a way that it be possible to define a trace function such as $x \mapsto u(x, \frac{x}{\varepsilon})$ on Ω. Stress should be laid on the fact that such a trace is involved in an essential fashion in the mathematical analysis of homogenization problems.

Before we proceed any further, let us observe that such a trace does not arise in an evident fashion, for the set $D(\varepsilon) = \{(x, y) | y = \frac{x}{\varepsilon}, x \in \Omega\}$ is negligible for Lebesgue measure on $\mathbb{R}_x^N \times \mathbb{R}_y^N$ (see, e.g., Berger and Gostiaux [15, p. 33]). Therefore, if u is considered in $L_{loc}^1(\Omega \times \mathbb{R}_y^N)$ (or even in $L^p(\Omega \times \mathbb{R}_y^N)$ for $1 \leq p \leq \infty$), it will be meaningless to speak of «the function $x \mapsto u(x, \frac{x}{\varepsilon})$ $(x \in \Omega)$» unless u is required to lie in an appropriate subspace of $L_{loc}^1(\Omega \times \mathbb{R}_y^N)$.

Having made this point, let us return to the general setting based on the absorptive continuous action $\mathcal{H}$. Given a function $u \in L_{loc}^1(\Omega \times \mathbb{R}_y^N)$, the matter on hand is to define $u|_{D(\varepsilon)}$ in a reasonable sense; that is, to define the trace of u on the set $D(\varepsilon) = \{(x, y)|y = H_\varepsilon(x), x \in \Omega\}$. It is clear that the question is somewhat delicate because the set $D(\varepsilon)$ is in general negligible for the product measure $\lambda \otimes \lambda$; for example, that is the case if $\mathcal{H}$ is as in Examples 1.5 and 1.6, or more generally if $H_\varepsilon \in \mathcal{C}^1(\mathbb{R}^N; \mathbb{R}^N)$, or again if $(\mathbb{R}^N, \mathcal{H})$ is a contraction flow (see Nguetseng [99], Berger and Gostiaux [15, p. 33]). Nevertheless, as was implicitly pointed out above, this can be overcome provided one moves from the general context to various specific cases in which one should have to deal with functions behaving suitably.

Thus, we shall concentrate on four cases of practical interest, but also of frequent occurrence in homogenization theory.

1.2.4.1 $\mathcal{C}(\Omega \times \mathbb{R}_y^N)$

Starting with what may be qualified as the smoothest case, let us set

$$u^\varepsilon(x) = u(x, H_\varepsilon(x)) \quad (x \in \Omega) \tag{1.20}$$

for $u \in \mathcal{C}(\Omega \times \mathbb{R}_y^N)$. The right-hand side of (1.20) in the present case is defined for every $x \in \Omega$, and is nothing but the restriction, $u(x, y)|_{y=H_\varepsilon(x)}$ $(x \in \Omega)$, of u to $D(\varepsilon)$ in the most elementary understanding. This yields a continuous linear mapping $u \mapsto u^\varepsilon$ of $\mathcal{C}(\Omega \times \mathbb{R}_y^N)$ into $\mathcal{C}(\Omega)$ which, in particular, sends continuously $\mathcal{K}(\Omega; \mathcal{B}(\mathbb{R}_y^N))$ into $\mathcal{K}(\Omega)$.

1.2.4.2 $L^p(\Omega; \mathcal{B}(\mathbb{R}_y^N))$ $(1 \leq p < \infty)$

Let us prove the following

Proposition 1.6 *Viewed as a mapping of $\mathcal{K}(\Omega; \mathcal{B}(\mathbb{R}_y^N))$ into $\mathcal{K}(\Omega)$, the trace operator $u \mapsto u^\varepsilon$ (u^ε given by (1.20)) extends by continuity to a continuous linear mapping of $L^p(\Omega; \mathcal{B}(\mathbb{R}_y^N))$ into $L^p(\Omega)$ $(1 \leq p < \infty)$, still denoted by $u \mapsto u^\varepsilon$, with*

$$\|u^\varepsilon\|_{L^p(\Omega)} \leq \|u\|_{L^p(\Omega; \mathcal{B}(\mathbb{R}_y^N))} \tag{1.21}$$

for all $u \in L^p(\Omega; \mathcal{B}(\mathbb{R}_y^N))$.

Proof One clearly has (1.21) for all $u \in \mathcal{K}(\Omega; \mathcal{B}(\mathbb{R}_y^N))$. Hence the proposition follows by extension by continuity, $\mathcal{K}(\Omega; \mathcal{B}(\mathbb{R}_y^N))$ being dense in $L^p(\Omega; \mathcal{B}(\mathbb{R}_y^N))$. ∎

1.2.4.3 $L^\infty(\Omega; \mathcal{B}(\mathbb{R}_y^N))$

The present case is less straightforward.

Proposition 1.7 *For $u \in L^\infty(\Omega; \mathcal{B}(\mathbb{R}_y^N))$, the right-hand side of (1.20) is defined for almost all $x \in \Omega$, which determines a function u^ε belonging to $L^\infty(\Omega)$. Furthermore, the mapping $u \mapsto u^\varepsilon$ thus defined sends continuously and linearly $L^\infty(\Omega; \mathcal{B}(\mathbb{R}_y^N))$ into $L^\infty(\Omega)$ with*

$$\|u^\varepsilon\|_{L^\infty(\Omega)} \leq \|u\|_{L^\infty(\Omega;\mathcal{B}(\mathbb{R}_y^N))} \tag{1.22}$$

for all $u \in L^\infty(\Omega; \mathcal{B}(\mathbb{R}_y^N))$.

Proof Let $u \in L^\infty(\Omega; \mathcal{B}(\mathbb{R}_y^N))$. Let $\mathcal{N} \subset \Omega$ be a λ-negligible set such that for each $x \in \Omega \setminus \mathcal{N}$, the mapping $y \mapsto u(x, y)$ of $\mathbb{R}_y^N$ into $\mathbb{C}$ lies in $\mathcal{B}(\mathbb{R}_y^N)$. We may then consider the mapping $z \mapsto u(x, H_\varepsilon(z))$, which sends continuously Ω into $\mathbb{C}$. Hence, it follows that the complex number $u(x, H_\varepsilon(x))$ is well defined for each $x \in \Omega \setminus \mathcal{N}$, which yields a function $x \mapsto u^\varepsilon(x)$ from Ω to $\mathbb{C}$ defined almost everywhere in Ω by (1.20). The next step is to show that the function u^ε is measurable (for λ). For this purpose, fix freely some compact set $K \subset \Omega$ and some $\eta > 0$. Since u is measurable from Ω to $\mathcal{B}(\mathbb{R}_y^N)$, there is some compact set $K' \subset K$ such that $\lambda(K \setminus K') \leq \eta$ and further $u|_{K'}$ (the restriction of u to K') maps continuously K' into $\mathcal{B}(\mathbb{R}_y^N)$. Clearly the measurability of u^ε will have been established if we can verify that u^ε is continuous on K'. To show this, let $a \in K'$. Fix freely $\rho > 0$. Considering that the function $x \mapsto u(a, H_\varepsilon(x))$ is continuous on Ω and in particular at $x = a$, we can find a neighbourhood U of a in Ω such that $|u(a, H_\varepsilon(x)) - u(a, H_\varepsilon(a))| \leq \frac{\rho}{2}$ for $x \in U$. On the other hand, it is clear that there exists a neighbourhood V of a in Ω such that $\|u(x, \cdot) - u(a, \cdot)\|_\infty \leq \frac{\rho}{2}$ for all $x \in V \cap K'$. Hence $|u^\varepsilon(x) - u^\varepsilon(a)| \leq \rho$ for all $x \in U \cap V \cap K'$, and the measurability of u^ε thereby follows. Finally, it is clear that $|u(x, y)| \leq \|u(x, \cdot)\|_\infty$ for all $x \in \Omega \setminus \mathcal{N}$ and all $y \in \mathbb{R}_y^N$. We deduce $|u^\varepsilon(x)| \leq \|u(x, \cdot)\|_\infty$ for $x \in \Omega \setminus \mathcal{N}$. Hence $u^\varepsilon \in L^\infty(\Omega)$ and further (1.22) holds. The linearity of the operator $u \mapsto u^\varepsilon$ being evident, the proof is complete. ∎

1.2.4.4 $\mathcal{C}(\overline{\Omega}; L^\infty(\mathbb{R}_y^N))$ ($\overline{\Omega}$ the Closure of Ω in $\mathbb{R}_x^N$)

We assume here that Ω is bounded. We begin by proving a fundamental lemma.

Lemma 1.6 *Let* $u \in \mathcal{C}(\overline{\Omega}) \otimes L^\infty(\mathbb{R}^N_y)$*. There is a negligible set* $\mathcal{N} \subset \mathbb{R}^N_y$ *such that*

$$|u(x, y)| \le \sup_{z \in \overline{\Omega}} \|u(z, \cdot)\|_{L^\infty(\mathbb{R}^N_y)}$$

holds for all $x \in \overline{\Omega}$ *and all* $y \in \mathbb{R}^N \setminus \mathcal{N}$.

Proof The function u writes as

$$u = \sum_{i \in I} \varphi_i \otimes u_i,\ \varphi_i \in \mathcal{C}(\overline{\Omega}),\ u_i \in L^\infty(\mathbb{R}^N_y), \tag{1.23}$$

where I is a finite set (that depends on u). Now let D be a dense countable subset of $\overline{\Omega}$. For each $a \in D$, consider a negligible set $R_a \subset \mathbb{R}^N_y$ such that

$$|u(a, y)| \le \|u(a, \cdot)\|_{L^\infty(\mathbb{R}^N)}$$

for all $y \in \mathbb{R}^N_y \setminus R_a$. By letting $R = \bigcup_{a \in D} R_a$, it follows

$$|u(a, y)| \le \sup_{z \in \overline{\Omega}} \|u(z, \cdot)\|_{L^\infty(\mathbb{R}^N)} \tag{1.24}$$

for all $y \in \mathbb{R}^N_y \setminus R$ and for all $a \in D$. With this in mind, fix a constant $c > \sum_{i \in I} \|u_i\|_{L^\infty(\mathbb{R}^N_y)}$. On the other hand, let S be a negligible set in $\mathbb{R}^N_y$ such that $|u_i(y)| \le \|u_i\|_{L^\infty(\mathbb{R}^N)}$ for all $y \in \mathbb{R}^N \setminus S$ and all $i \in I$. Finally, put $\mathcal{N} = R \cup S$ (a negligible subset of $\mathbb{R}^N_y$). To conclude, fix freely $x \in \overline{\Omega}$. Consider independently some $\eta > 0$. In view of the density of D in $\overline{\Omega}$ there is an $a \in D$ such that

$$|\varphi_i(x) - \varphi_i(a)| \le \frac{\eta}{c}\ \ (i \in I). \tag{1.25}$$

Next, given any arbitrary $y \in \mathbb{R}^N \setminus \mathcal{N}$, write

$$u(x, y) = \sum_{i \in I} [\varphi_i(x) - \varphi_i(a)] u_i(y) + u(a, y)$$

and then use (1.24) and (1.25) to arrive at

$$|u(x, y)| \le \eta + \sup_{z \in \overline{\Omega}} \|u(z, \cdot)\|_{L^\infty(\mathbb{R}^N)}.$$

Hence the lemma follows by the arbitrariness of η. ■

We are now in a position to prove a trace result for the function space $\mathcal{C}(\overline{\Omega}; L^\infty(\mathbb{R}_y^N))$. Let us begin by noting that if u lies in $\mathcal{C}(\overline{\Omega}) \otimes L^\infty(\mathbb{R}_y^N)$, i.e., if u is given by (1.23), then clearly the trace of u on $D(\varepsilon)$ is quite naturally given by

$$u^\varepsilon = \sum_{i \in I} \varphi_i u_i^\varepsilon,$$

that is,

$$u^\varepsilon(x) = u(x, H_\varepsilon(x)) = \sum_{i \in I} \varphi_i(x) u_i(H_\varepsilon(x))$$

for almost all $x \in \Omega$. One of course has $u^\varepsilon \in L^\infty(\Omega)$.

Proposition 1.8 *The trace operator $u \mapsto u^\varepsilon$ of $\mathcal{C}(\overline{\Omega}) \otimes L^\infty(\mathbb{R}_y^N)$ into $L^\infty(\Omega)$ extends by continuity to a continuous linear mapping, still denoted by $u \mapsto u^\varepsilon$, of $\mathcal{C}(\overline{\Omega}; L^\infty(\mathbb{R}_y^N))$ into $L^\infty(\Omega)$ with*

$$\|u^\varepsilon\|_{L^\infty(\Omega)} \leq \sup_{z \in \overline{\Omega}} \|u(z, \cdot)\|_{L^\infty(\mathbb{R}_y^N)} \tag{1.26}$$

for all $u \in \mathcal{C}(\overline{\Omega}; L^\infty(\mathbb{R}_y^N))$.

Proof Let $u \in \mathcal{C}(\overline{\Omega}) \otimes L^\infty(\mathbb{R}_y^N)$. Thanks to Lemma 1.6 there is a negligible set $\mathcal{N} \subset \mathbb{R}_y^N$ such that

$$|u^\varepsilon(x)| \leq \sup_{z \in \overline{\Omega}} \|u(z, \cdot)\|_{L^\infty(\mathbb{R}^N)}$$

for all $x \in \Omega \setminus H_{\varepsilon^{-1}}(\mathcal{N})$. The next step is to check that $H_{\varepsilon^{-1}}(\mathcal{N})$ is a negligible subset of $\mathbb{R}_x^N$, which amounts to verifying that $H_{\varepsilon^{-1}}(\mathcal{N}) \cap K$ is negligible for any compact set $K \subset \mathbb{R}_x^N$. To this end, for each such K we have

$$\begin{aligned}
\lambda(H_{\varepsilon^{-1}}(\mathcal{N}) \cap K) &= \lambda(H_{\varepsilon^{-1}}[\mathcal{N} \cap H_\varepsilon(K)]) \\
&= H_\varepsilon(\lambda)(\mathcal{N} \cap H_\varepsilon(K)) \\
&= \varepsilon^s \lambda(\mathcal{N} \cap H_\varepsilon(K)) \text{ (Proposition 1.4)} \\
&= 0,
\end{aligned}$$

since $\mathcal{N} \cap H_\varepsilon(K)$ is a negligible subset of $\mathbb{R}_y^N$, $H_\varepsilon(K)$ being compact. Consequently, (1.26) holds for all $u \in \mathcal{C}(\overline{\Omega}) \otimes L^\infty(\mathbb{R}_y^N)$, and the proposition therefore follows by extension by continuity, $\mathcal{C}(\overline{\Omega}) \otimes L^\infty(\mathbb{R}_y^N)$ being dense in $\mathcal{C}(\overline{\Omega}; L^\infty(\mathbb{R}_y^N))$ (Bourbaki [22, p. 46]). ■

We shall conclude this section with three fundamental convergence results.

Proposition 1.9 *The following hold true as* $\varepsilon \to 0$*:*

(i) $u^\varepsilon \to \widetilde{u}$ *in* $L^\infty(\Omega)$*-weak∗ for* $u \in \mathcal{K}(\Omega; \Pi^\infty(\mathbb{R}^N_y))$,
(ii) $u^\varepsilon \to \widetilde{u}$ *in* $L^p(\Omega)$*-weak for* $u \in L^p(\Omega; \Pi^\infty(\mathbb{R}^N_y))$, $1 \leq p < \infty$,
(iii) $u^\varepsilon \to \widetilde{u}$ *in* $L^\infty(\Omega)$*-weak∗ for* $u \in \mathcal{C}(\overline{\Omega}; \Pi^\infty(\mathbb{R}^N_y))$*, provided* Ω *is bounded,*

where $\widetilde{u}(x) = M(u(x))$, $x \in \Omega$.

Proof Let us begin by noting some preliminary details. One has

$$|M(f(x,\cdot))| \leq \|f(x,\cdot)\|_\infty \quad (x \in \Omega) \tag{1.27}$$

for f lying in any of the preceding function spaces based on $\Pi^\infty(\mathbb{R}^N_y)$. From which we deduce $\widetilde{u} \in \mathcal{K}(\Omega)$ in case (i), $\widetilde{u} \in \mathcal{C}(\overline{\Omega})$ in case (iii), and $\widetilde{u} \in L^p(\Omega)$ in case (ii) (indeed, according to a classical result, $\widetilde{u} = M \circ u$ is measurable as the composition of the continuous mapping $M : \Pi^\infty(\mathbb{R}^N) \to \mathbb{C}$ and the measurable function $u : \Omega \to \Pi^\infty(\mathbb{R}^N)$. Hence the alleged statement follows by (1.27)).

Having done this, we observe that it is a simple exercise to check that (i) holds for any $u \in \mathcal{K}(\Omega) \otimes \Pi^\infty(\mathbb{R}^N_y)$ (recall that u writes as in (1.23) mutatis mutandis). Therefore the whole of (i) follows by the density of $\mathcal{K}(\Omega) \otimes \Pi^\infty(\mathbb{R}^N_y)$ in $\mathcal{K}(\Omega; \Pi^\infty(\mathbb{R}^N_y))$. More precisely, given $u \in \mathcal{K}(\Omega; \Pi^\infty(\mathbb{R}^N_y))$, let K be some compact set in Ω such that $u \in \mathcal{K}_K(\Omega; \Pi^\infty(\mathbb{R}^N_y))$ (the Banach space of those functions in $\mathcal{K}(\Omega; \Pi^\infty(\mathbb{R}^N_y))$ having supports in K). According to Bourbaki [22, p. 46, prop. 5], there is an ordinary sequence (u_n) in the tensor product $\mathcal{K}_K(\Omega) \otimes \Pi^\infty(\mathbb{R}^N_y)$ that converges to u in $\mathcal{K}_K(\Omega; \Pi^\infty(\mathbb{R}^N_y))$. Therefore, the whole of (i) follows by mere routine and use of (1.27) (with $f = u_n - u$) together with $\sup_{x\in K} |u^\varepsilon(x) - u^\varepsilon_n(x)| \leq \sup_{x\in K} \|u(x,\cdot) - u_n(x,\cdot)\|_\infty$.

In view of the density of $\mathcal{K}(\Omega; \Pi^\infty(\mathbb{R}^N_y))$ in $L^p(\Omega; \Pi^\infty(\mathbb{R}^N_y))$ $(1 \leq p < \infty)$, (ii) follows by (i) and use of (1.21) together with (1.27) valid almost everywhere for any $f \in L^p(\Omega; \Pi^\infty(\mathbb{R}^N_y))$. Finally, (iii) is obtained by the same line of proceeding as followed in (i), and use of (1.26),(1.27) and the density of $\mathcal{C}(\overline{\Omega}) \otimes \Pi^\infty(\mathbb{R}^N_y)$ in $\mathcal{C}(\overline{\Omega}; \Pi^\infty(\mathbb{R}^N_y))$. ■

Remark 1.12 For every $u \in L^1(\Omega; \Pi^\infty(\mathbb{R}^N_y))$, we have

$$\int_\Omega u(x, H_\varepsilon(x))dx \to \int_\Omega M(u(x,\cdot))dx$$

as $\varepsilon \to 0$. Indeed, this follows by part (ii) of Proposition 1.9.

1.2.5 A Fundamental Lemma

We conclude this chapter with a fundamental lemma (Lemma 1.9) whose usefulness will come to light later. In the sequel the action $\mathcal{H} = (H_\varepsilon)_{\varepsilon>0}$ is as in Sect. 1.2.3 and further is compatible with addition in $\mathbb{R}^N$. Finally, we assume that the following is satisfied.

$$(C)\qquad \text{For any fixed compact set } D \subset \mathbb{R}^N, \text{ one has} \quad \lim_{\varepsilon\to 0}\left(\sup_{y\in D}|H_{\varepsilon^{-1}}(y)|\right) = 0.$$

Remark 1.13 Let $(\mathcal{M}(\varepsilon))_{\varepsilon>0}$ be the family of $N \times N$ matrices attached to $\mathcal{H}$ as stated in Remark 1.11. Then $|H_{\varepsilon^{-1}}(y)| \leq \|\mathcal{M}(\varepsilon^{-1})\||y|$ for all $y \in \mathbb{R}^N$, where $|\cdot|$ denotes the Euclidean norm in $\mathbb{R}^N$, and $\|\mathcal{M}(\varepsilon^{-1})\|^2 = \sum_{1\leq i,j\leq N}|a_{ij}(\varepsilon^{-1})|^2$. Therefore, (C) is satisfied as soon as $\|\mathcal{M}(\varepsilon^{-1})\|$ tends to zero when $\varepsilon \to 0$. That is manifestly the case in Examples 1.5 and 1.6.

The aim now is to prove the following lemma.

Lemma 1.7 *Let $K \subset \mathbb{R}^N$ be a compact set, and Ω be a bounded open neighbourhood of K in $\mathbb{R}^N$. There is some real $\varepsilon_0 > 0$ depending solely on K, and for each $0 < \varepsilon \leq \varepsilon_0$, a finite set $F_\varepsilon \subset \mathbb{Z}^N$ such that*

$$K \subset \bigcup_{k\in F_\varepsilon} H_{\varepsilon^{-1}}(k + \overline{Y}) \subset \Omega \text{ with } Y = \left(-\frac{1}{2}, \frac{1}{2}\right)^N. \tag{1.28}$$

Proof Let $\varepsilon > 0$. First of all, let us observe that $\mathbb{R}^N$ is the union of the family $H_{\varepsilon^{-1}}(k + \overline{Y})$ $(k \in \mathbb{Z}^N)$. Next, let F_ε be the set of all those $k \in \mathbb{Z}^N$ for which $H_{\varepsilon^{-1}}(k+\overline{Y})$ intersects K. It is clear that K is contained in the union of all $H_{\varepsilon^{-1}}(k+\overline{Y})$ as k ranges over F_ε.

Now, consider some real $r > 0$ (e.g., $r = d(K, \mathbb{R}^N \setminus \Omega)$, d the Euclidian metric in $\mathbb{R}^N$) with the property that if x is in $\mathbb{R}^N$ and satisfies $d(x, K) < r$, then $x \in \Omega$. This being so, fix freely $k \in F_\varepsilon$. Let $x \in H_{\varepsilon^{-1}}(k + \overline{Y})$. Then $d(x, K) \leq d(x, K \cap H_{\varepsilon^{-1}}(k+\overline{Y})) \leq \operatorname{diam} H_{\varepsilon^{-1}}(k+\overline{Y}) = \operatorname{diam} H_{\varepsilon^{-1}}(\overline{Y})$. But one has $x-z \in H_{\varepsilon^{-1}}(\overline{Y}+\overline{Y})$ whenever $x, z \in H_{\varepsilon^{-1}}(\overline{Y})$. We deduce

$$\operatorname{diam} H_{\varepsilon^{-1}}(\overline{Y}) \leq \sup_{y\in\overline{Y}+\overline{Y}}|H_{\varepsilon^{-1}}(y)|.$$

Hence, by (C) we see that there is some $\varepsilon_0 > 0$ such that for $0 < \varepsilon \leq \varepsilon_0$, the union of all $H_{\varepsilon^{-1}}(k + \overline{Y})$ $(k \in F_\varepsilon)$ is contained in Ω. It remains to check that F_ε is finite. It is clear that $F_\varepsilon \subset H_\varepsilon(\overline{\Omega})$. Thus, assuming F_ε to be of infinite cardinality yields by the Bolzano-Weierstrass theorem ($H_\varepsilon(\overline{\Omega})$ being compact) some couple of points $h, k \in F_\varepsilon$ such that $0 < |k - h| < 1$, an absurdity, since $F_\varepsilon \subset \mathbb{Z}^N$. ■

The preceding lemma enables us to prove the next one which is besides of considerable interest by itself.

Lemma 1.8 *Let K be a compact set in $\mathbb{R}^N$. Let Ω be a bounded open neighbourhood of K in $\mathbb{R}^N$. There is some $\varepsilon_0 > 0$ such that*

$$\int_K |u(H_\varepsilon(x))|^p dx \leq \lambda(\Omega) \sup_{k\in\mathbb{Z}^N} \int_{k+Y} |u(y)|^p dy \quad (0 < \varepsilon \leq \varepsilon_0) \tag{1.29}$$

for all $u \in L^p_{loc}(\mathbb{R}^N)$ $(1 \leq p < \infty)$.

Remark 1.14 The right-hand side of (1.29) may or may not be finite.

Proof of Lemma 1.8 Let $u \in L^p_{loc}(\mathbb{R}^N)$ with $1 \leq p < \infty$. Let ε be arbitrary fixed with $0 < \varepsilon \leq \varepsilon_0$, ε_0 the constant in Lemma 1.7. From (1.28) one has

$$\int_K |u(H_\varepsilon(x))|^p dx \leq \sum_{k\in F_\varepsilon} \int_{H_{\varepsilon^{-1}}(k+Y)} |u(H_\varepsilon(x))|^p dx.$$

But, by Proposition 1.4,

$$\begin{aligned}\int_{H_{\varepsilon^{-1}}(k+Y)} |u(H_\varepsilon(x))|^p dx &= \int |((|u|^p \chi_{k+Y})(H_\varepsilon(x))| dx \\ &= \varepsilon^s \int_{k+Y} |u(y)|^p dy,\end{aligned}$$

where χ_{k+Y} stands for the characteristic function of $k + Y$ in $\mathbb{R}^N$. Hence

$$\int_K |u(H_\varepsilon(x))|^p dx \leq \varepsilon^s |F_\varepsilon| \sup_{k\in\mathbb{Z}^N} \int_{k+Y} |u(y)|^p dy,$$

where $|F_\varepsilon|$ denotes the (finite) cardinality of F_ε. Therefore, (1.29) follows by noticing that $\varepsilon^s |F_\varepsilon| = \lambda(\bigcup_{k\in F_\varepsilon} H_{\varepsilon^{-1}}(k + \overline{Y})) \leq \lambda(\Omega)$ (see Lemma 1.7). ∎

Finally comes the following

Lemma 1.9 *Let hypothesis (C) be satisfied. Let B be an open elementary set (Sect. 1.2.3). There is a constant $c > 0$ such that*

$$\int_B |u(H_\varepsilon(x))|^p dx \leq c \sup_{k\in\mathbb{Z}^N} \int_{k+Y} |u(y)|^p dy \quad (0 < \varepsilon \leq 1)$$

for all $u \in L^p_{loc}(\mathbb{R}^N)$ $(1 \leq p < \infty)$.

Proof For an arbitrary given $1 \leq p < \infty$, let $u \in L^p_{loc}(\mathbb{R}^N)$. Choosing $K = \overline{B}$ in Lemma 1.8 yields some $\alpha > 0$ (that depends solely on B) such that

$$\int_B |u(H_\varepsilon(x))|^p dx \leq \lambda(\Omega_B) \sup_{k \in \mathbb{Z}^N} \int_{k+Y} |u(y)|^p dy \tag{1.30}$$

for all $0 < \varepsilon \leq \alpha$, where Ω_B is any arbitrary given bounded open neighbourhood of $\overline{B}$ in $\mathbb{R}^N$. Two and only two cases are to be considered: (1) suppose $\alpha \geq 1$. Then (1.30) holds for $0 < \varepsilon \leq 1$, and the lemma is therefore proven. (2) Suppose $\alpha < 1$. Fix freely some $0 < \varepsilon \leq 1$. Then $0 < \alpha\varepsilon \leq \alpha$. Hence it follows that (1.30) holds for $\alpha\varepsilon$ in place of ε. But, $\chi_{H_\alpha(B)}$ denoting the characteristic function of $H_\alpha(B)$, we have

$$\begin{aligned}
\int_B |u(H_{\alpha\varepsilon}(x))|^p dx &= \int |u \circ H_\varepsilon(H_\alpha(x))|^p \chi_{H_\alpha(B)}(H_\alpha(x)) dx \\
&= \int (|u \circ H_\varepsilon|^p \chi_{H_\alpha(B)})(H_\alpha(x)) dx \\
&= \alpha^s \int_{H_\alpha(B)} |u(H_\varepsilon(x))|^p dx \text{ (Proposition 1.4)} \\
&\geq \alpha^s \int_B |u(H_\varepsilon(x))|^p dx,
\end{aligned}$$

since $B \subset H_\alpha(B)$. Therefore, the lemma follows with $c = \alpha^{-s}\lambda(\Omega_B)$. ■

Remark 1.15 If the unit cube $Y = (-\frac{1}{2}, \frac{1}{2})^N$ is substituted by the parallelepiped defined in (1.3), and $\mathbb{Z}^N$ by the network S made up of those $k \in \mathbb{R}^N$ given by (1.2), a quick survey of the whole Sect. 1.2.5 reveals that the preceding lemmas together with their proofs remain valid provided $\varepsilon^s |F_\varepsilon|$ in the proof of Lemma 1.8 is changed into $\varepsilon^s |F_\varepsilon| \lambda(Y)$, and $\lambda(\Omega)$ is changed into $\frac{\lambda(\Omega)}{\lambda(Y)}$.

1.3 Comments and Problems

The proof of Theorem 1.2 was given for the first time in Nguetseng [93]. It should be emphasized that this theorem (through (1.4)) is essential for the study, in the framework of deterministic homogenization, of evolution equations and of reiterated homogenization problems. The study of absorptive actions was introduced and thoroughly discussed by the author in the general context of locally compact spaces including locally compact Abelian groups (see [94, 99] for details).

For the sake of the cause our discussion of means is limited to $\mathbb{R}^N$. However, an analogous discussion in the general context of locally compact Abelian groups can be found in [94]. On this point the interested reader is referred to Rosenblatt

[118, 119] and to Rosenblatt and Yang [120], where many interesting questions about means are considered on various topological groups.

Let $\mathcal{H} = (H_\varepsilon)_{\varepsilon>0}$ be an absorptive continuous action of $\mathbb{R}_+^*$ on $\mathbb{R}^N$ with the further properties of being self-adjoint (i.e., of satisfying (1.17)) and compatible with addition in $\mathbb{R}^N$ (see Theorem 1.3). Let $u \in \Pi_{\mathcal{H}}^\infty(\mathbb{R}^N)$. In the present chapter and beyond one can find several examples of functions u providing different explicit expressions of the mean value $M(u)$ (see, e.g., Examples 1.10, 1.11, and 1.12). However, for the sake of coherency or for computation needs it may be useful to exhibit a simple and self-contained (i.e., depending solely on $\mathcal{H}$) expression of $M(u)$. A general answer is given by the equality

$$M(u) = \lim_{\varepsilon\to 0} \frac{1}{|H_\varepsilon(B)|} \int_{H_\varepsilon(B)} u(y)\, dy, \tag{1.31}$$

where B is, say an arbitrary given nonempty bounded open set in $\mathbb{R}^N$, and $|H_\varepsilon(B)| = \lambda(H_\varepsilon(B))$ (Lebesgue measure of $H_\varepsilon(B)$). Indeed, by noting that $\lambda(H_\varepsilon(B)) = \varepsilon^{-s}\lambda(B)$ (see Proposition 1.4), one readily sees that

$$\frac{1}{|B|} \int_B u(H_\varepsilon(x))\, dx = \frac{1}{|H_\varepsilon(B)|} \int_{H_\varepsilon(B)} u(y)\, dy.$$

Hence (1.31) follows, since by definition $\lim_{\varepsilon\to 0} \frac{1}{|B|} \int_B u(H_\varepsilon(x))\, dx = M(u)$.

Let us consider the particular action $H_\varepsilon(x) = \frac{x}{\varepsilon}$ ($\varepsilon > 0$) for $x \in \mathbb{R}^N$. One has here

$$M(u) = \lim_{R\to\infty} \frac{1}{|B(a, R)|} \int_{B(a,R)} u(y)\, dy, \tag{1.32}$$

where a is any arbitrary point in $\mathbb{R}^N$, and $B(a, R)$ is the open ball in $\mathbb{R}^N$ with center a and radius $R > 0$. Indeed, according to (1.31) one has

$$\begin{aligned} M(u) &= \lim_{\varepsilon\to 0} \frac{1}{|\frac{1}{\varepsilon}B(\omega, 1)|} \int_{\frac{1}{\varepsilon}B(\omega,1)} u(y)\, dy \\ &= \lim_{R\to\infty} \frac{1}{|RB(\omega, 1)|} \int_{RB(\omega,1)} u(y)\, dy \\ &= \lim_{R\to\infty} \frac{1}{|B(\omega, R)|} \int_{B(\omega,R)} u(y)\, dy \quad (\omega = (0, \cdots, 0) \text{ the origin in } \mathbb{R}^N). \end{aligned}$$

Hence (1.32) follows at once by the fact that $\tau_{-a}u \in \Pi_{\mathcal{H}}^\infty(\mathbb{R}^N)$ with $M(\tau_{-a}u) = M(u)$ (see Theorem 1.3).

Chapter 2
Homogenization Algebras on $\mathbb{R}^N$

This chapter is devoted to the study of a class of C^*-algebras, the so-called homogenization algebras, and their Gelfand representations. Various examples of homogenization algebras are provided and thoroughly discussed with a view to later considering deterministic homogenization problems beyond the periodic setting.

The family $\mathcal{H} = (H_\varepsilon)_{\varepsilon>0}$ throughout denotes an absorptive continuous action of $\mathbb{R}_+^*$ on $\mathbb{R}^N$ as stated in Sect. 1.2.3 (see in particular Remarks 1.10 and 1.11). Moreover it is assumed once for all that $\mathcal{H}$ is self-adjoint (i.e., $x \cdot H_\varepsilon(z) = z \cdot H_\varepsilon(x)$ for $x, z \in \mathbb{R}^N, \varepsilon > 0$) and compatible with addition in $\mathbb{R}^N$ (i.e., $H_\varepsilon(x + z) = H_\varepsilon(x) + H_\varepsilon(z)$ for $x, z \in \mathbb{R}^N, \varepsilon > 0$). Thus, the mean-value, M, generated by the action $\mathcal{H}$ on $\mathbb{R}^N$ is a mean value on $\mathbb{R}^N$ viewed as an Abelian topological group as well. Finally it is worth laying stress on the fact that all of the preceding hypotheses are satisfied by almost all actions involved in the mathematical analysis of homogenization problems.

2.1 Fundamentals of Homogenization Algebras

2.1.1 Definitions and Basic Results

Definition 2.1 By a *homogenization algebra* on $\mathbb{R}^N$ under the action $\mathcal{H}$ is meant a closed subalgebra A of $\mathcal{B}(\mathbb{R}_y^N)$ with the following properties:

(HA1) A with the supremum norm is separable
(HA2) A contains the constants
(HA3) A preserves complex conjugation (i.e., if $u \in A$, then $\overline{u} \in A$)
(HA4) $A \subset \Pi_{\mathcal{H}}^\infty(\mathbb{R}_y^N)$ (see Theorem 1.3).

For brevity we will most of the time write H-*algebra* in place of *homogenization algebra*. On the other hand, as stated in the preceding definition, the notion of

G. Nguetseng, *Homogenization Algebras and Applications*, Springer Monographs in Mathematics, https://doi.org/10.1007/978-3-031-84705-9_2

an H-algebra on $\mathbb{R}^N$ is intrinsically connected to the action considered on $\mathbb{R}^N$. However, whenever there is no risk of confusion, we will often speak of an H-algebra on $\mathbb{R}^N$ without further ado, thereby omitting the action considered on $\mathbb{R}^N$.

We shall always implicitly assume that an H-algebra A on $\mathbb{R}^N$ is equipped with the algebra structure induced by the natural one on $\mathcal{B}(\mathbb{R}^N)$ (see Chap. 1) and with the supremum norm on $\mathbb{R}^N$. Thus equipped, A is a commutative C^*-algebra with identity, and the Gelfand transformation $\mathcal{G}$ is an isometric C^*-isomorphism of A onto $\mathcal{C}(\Delta(A))$. We recall in passing that the spectrum, $\Delta(A)$, of A is provided with the Gelfand topology, which makes it into a compact space. Moreover, as a consequence of (HA1), $\Delta(A)$ is metrizable (see Chap. 1).

We turn now to some basic results about H-algebras. We first state a simple but fundamental lemma whose proof is left to the reader (see Nguetseng and Svanstedt [102] for details).

Lemma 2.1 *Let A be a closed subalgebra of $\mathcal{B}(\mathbb{R}^N)$ that contains the constants and preserves complex conjugation, so that A is a commutative C^*-algebra with identity. Let $\mathcal{G}$ be the Gelfand transformation on A. The following hold:*

(i) $\mathcal{G}$ and $\mathcal{G}^{-1}$ are order preserving, that is, if $u \in A$ and $u \geq 0$, then $\mathcal{G}(u) \geq 0$; if $f \in \mathcal{C}(\Delta(A))$ and $f \geq 0$, then $\mathcal{G}^{-1}(f) \geq 0$.
(ii) Given $p > 0$, if $u \in A$, then $|u|^p \in A$ and $\mathcal{G}(|u|^p) = |\mathcal{G}(u)|^p$.

This being so, we now wish to show that the mean value, M, on $\mathbb{R}^N$ generated by the action $\mathcal{H}$ admits an integral representation on any H-algebra A on $\mathbb{R}^N$ (under $\mathcal{H}$).

Proposition 2.1 *Let A be an H-algebra on $\mathbb{R}^N$. Then, there exists one and only one Radon measure β on $\Delta(A)$ such that*

$$M(u) = \int_{\Delta(A)} \mathcal{G}(u)(s)d\beta(s) \quad (u \in A). \tag{2.1}$$

Furthermore, β is positive and of total mass 1.

Proof The mapping $\varphi \mapsto M(\mathcal{G}^{-1}(\varphi))$ of $\mathcal{C}(\Delta(A))$ into $\mathbb{C}$ is a continuous linear form on $\mathcal{C}(\Delta(A))$ (see Theorem 1.3). In other words, there clearly exists a Radon measure β on $\Delta(A)$ such that (2.1) holds. The unicity of β follows by the fact that any function in $\mathcal{C}(\Delta(A))$ is of the form $\mathcal{G}(u)$, $u \in A$. Furthermore, β is positive by virtue of Lemma 2.1 and use of the fact that M is a positive linear form. Finally β is of total mass 1 because $\mathcal{G}(1) = 1$ and $M(1) = 1$. ■

The Radon measure β thus defined is referred to as the *M-measure on $\Delta(A)$* (or for A). Throughout the rest of the book, the compact space $\Delta(A)$ shall be considered implicitly equipped with the Radon measure β.

The following result is immediate by Proposition 2.1 and Lemma 2.1.

Proposition 2.2 *Let A be an H-algebra on $\mathbb{R}^N$. Let $p > 0$. Then*

$$M(|u|^p) = \int_{\Delta(A)} |\mathcal{G}(u)(s)|^p d\beta(s) \quad (u \in A).$$

Now, let $\mathbb{R}^N$ be factorized as in (1.12). For each i, $1 \leq i \leq m$, let $\mathcal{H}_i = (H_\varepsilon^i)_{\varepsilon>0}$ be an absorptive continuous action of $\mathbb{R}_+^*$ on $\mathbb{R}^{N_i}$. We assume moreover that $\mathcal{H}_i$ is self-adjoint and compatible with addition in $\mathbb{R}^{N_i}$. Put $\mathcal{H} = \mathcal{H}_1 \times \cdots \times \mathcal{H}_m$ (Sect. 1.2), that is, $\mathcal{H} = (H_\varepsilon)_{\varepsilon>0}$ denotes the action of $\mathbb{R}_+^*$ on $\mathbb{R}^N$ given by (1.13). This is an absorptive continuous action on $\mathbb{R}^N$ (Proposition 1.3). Moreover, $\mathcal{H}$ is self-adjoint and compatible with addition in $\mathbb{R}^N$, as is straightforward by a simple calculation. Now, for each $1 \leq i \leq m$, let A_i be an H-algebra on $\mathbb{R}^{N_i}$ under $\mathcal{H}_i$. Let A be the closure of $\bigotimes_{i=1}^m A_i$ in $\mathcal{B}(\mathbb{R}^N)$. Our main purpose is to check that A is an H-algebra on $\mathbb{R}^N$ under $\mathcal{H}$. The following lemma will be needed.

Lemma 2.2 *Under the preceding hypotheses and notation, we have*

$$\overline{\bigotimes_{i=1}^m \Pi_{\mathcal{H}_i}^\infty(\mathbb{R}^{N_i})} \subset \Pi_{\mathcal{H}}^\infty(\mathbb{R}^N),$$

where the left-hand side denotes the closure of $\bigotimes_{i=1}^m \Pi_{\mathcal{H}_i}^\infty(\mathbb{R}^{N_i})$ in $\mathcal{B}(\mathbb{R}^N)$.

Proof It clearly suffices to check that any function u of the form $u = \bigotimes_{i=1}^m u_i$ ($u_i \in \Pi_{\mathcal{H}_i}^\infty(\mathbb{R}^{N_i})$) lies in $\Pi_{\mathcal{H}}^\infty(\mathbb{R}^N)$. More precisely, let us show that each such u is in $\Pi_{\mathcal{H}}^\infty(\mathbb{R}^N)$ with

$$M(\bigotimes_{i=1}^m u_i) = \prod_{i=1}^m M(u_i). \tag{2.2}$$

But, since the tensor product $\bigotimes_{i=1}^m \mathcal{K}(\mathbb{R}^{N_i})$ is (sequentially) dense in the space $\mathcal{K}(\mathbb{R}^N)$ (see, e.g., Vo-Khac [138, pp. 151–152]) which in turn is dense in $L^1(\mathbb{R}^N)$, the whole problem reduces to showing that, given $u = \bigotimes_{i=1}^m u_i$ as above, we have

$$\lim_{\varepsilon\to 0} \int_{\mathbb{R}^N} u(H_\varepsilon(x)) f(x) dx = \left(\prod_{i=1}^m M(u_i)\right) \int_{\mathbb{R}^N} f(x) dx$$

for all $f = \bigotimes_{i=1}^m f_i$ with $f_i \in \mathcal{K}(\mathbb{R}^{N_i})$. However, this is straightforward by an easy calculation. ■

We are now in a position to show that A is an H-algebra on $\mathbb{R}^N$ (under the product action $\mathcal{H}$).

Proposition 2.3 *Under the preceding hypotheses and notation, the following assertions are true:*

(i) A *is an H-algebra on* $\mathbb{R}^N$ *under* $\mathcal{H}$.
(ii) $\Delta(A) = \Delta(A_1) \times \cdots \times \Delta(A_m)$.
(iii) Let β_i *be the M-measure on* $\Delta(A_i)$ $(1 \leq i \leq m)$, *and* β *be the M-measure on* $\Delta(A)$. *Then* $\beta = \bigotimes_{i=1}^{m} \beta_i$.

Proof A is evidently a closed subalgebra of $\mathcal{B}(\mathbb{R}^N)$ that contains the constants and preserves complex conjugation. Therefore (i) follows by Lemmas 1.4 and 2.2. Furthermore, (ii) results directly from Theorem 1.2. Finally, (iii) follows by combining (ii) with (2.2) and Corollary 1.1. Indeed, the result is that

$$\int_{\Delta(A)} \bigotimes_{i=1}^{m} \mathcal{G}(u_i) d\beta = \prod_{i=1}^{m} \int_{\Delta(A_i)} \mathcal{G}(u_i) d\beta_i$$

for $u_i \in A_i$ $(1 \leq i \leq m)$. The desired conclusion follows thereby (see, e.g., Bourbaki [22, Chap.III,§4]). ■

We will conclude this section by turning our attention to a concept of degeneracy.

Definition 2.2 An H-algebra A on $\mathbb{R}^N$ is said to be nondegenerate if the only function $u \in A$ verifying $u \geq 0$ and $M(u) = 0$ is the zero function in $\mathcal{B}(\mathbb{R}^N)$.

Proposition 2.4 *The following two assertions are equivalent for an H-algebra* A *on* $\mathbb{R}^N$.

(i) A *is nondegenerate.*
(ii) $Supp\beta = \Delta(A)$ *(*β *the M-measure for* A*).*

Proof Suppose (i) holds. Assuming (ii) to be wrong leads to some $r \in \Delta(A)$ lying off $Supp\beta$ (the support of β). By Urysohn's lemma we may consider some $\varphi \in \mathcal{C}(\Delta(A))$ such that $\varphi \geq 0, \varphi(r) = 1$ and $\varphi = 0$ on $Supp\beta$. Clearly $\beta(\varphi) \equiv \int_{\Delta(A)} \varphi(s) d\beta(s) = 0$. Therefore, letting $u = \mathcal{G}^{-1}(\varphi)$, it follows $M(u) = 0$. But $u \geq 0$ (Lemma 2.1). Therefore $u = 0$. Consequently $\varphi(s) = \mathcal{G}(u)(s) = 0$ for all $s \in \Delta(A)$, a contradiction and so (ii) is true. Reciprocally, assume (ii), and let $u \in A$ with $u \geq 0$ and $M(u) = 0$. Then $\varphi = \mathcal{G}(u) \geq 0$ and $\beta(\varphi) = 0$. It follows that φ vanishes on $Supp\beta$ (see, e.g., Bourbaki [22, p. 69]); hence $\varphi(s) = 0$ for all $s \in \Delta(A)$, according to (ii). Therefore $u = 0$ and so (i) follows. ■

2.1.2 *Examples of Homogenization Algebras*

2.1.2.1 The H-algebra $\mathcal{B}_\infty(\mathbb{R}^N)$

Example 2.1 Let $A = \mathcal{B}_\infty(\mathbb{R}^N)$. A quick survey of Examples 1.2 and 1.10 reveals that A is an H-algebra on $\mathbb{R}^N$ (under $\mathcal{H}$). Moreover, if $a\mathbb{R}^N$ denotes the Alexandroff compactification of $\mathbb{R}^N$, and ∞ its point at infinity, then the mapping $\Phi : A \to \mathcal{C}(a\mathbb{R}^N)$ defined for each $u \in A$ by $\Phi(u)(x) = u(x)$ if $x \in a\mathbb{R}^N \setminus \{\infty\}$, $\Phi(u)(\infty) = \lim_{y\to\infty} u(y)$, is an isometric C^*-isomorphism of A onto $\mathcal{C}(a\mathbb{R}^N)$. It follows at once that the mapping $\mathcal{G} \circ \Phi^{-1} : \mathcal{C}(a\mathbb{R}^N) \to \mathcal{C}(\Delta(A))$, where $\mathcal{G}$ denotes the Gelfand transformation on A, is an isometric C^*-isomorphism of $\mathcal{C}(a\mathbb{R}^N)$ onto $\mathcal{C}(\Delta(A))$.

Proposition 2.5 *There is some* $s_\infty \in \Delta(A)$ *such that*

$$\lim_{y\to\infty} u(y) = \mathcal{G}(u)(s_\infty) \tag{2.3}$$

for all $u \in \mathcal{B}_\infty(\mathbb{R}^N)$. *Moreover, the point* s_∞ *is unique.*

Proof Because $\mathcal{G} \circ \Phi^{-1}$ is an algebra isomorphism, the compact spaces $\Delta(A)$ and $a\mathbb{R}^N$ are homeomorphic. More precisely, there is a homeomorphism $\varphi : \Delta(A) \to a\mathbb{R}^N$ such that $(\mathcal{G} \circ \Phi^{-1})(f)(s) = f(\varphi(s))$ for $f \in \mathcal{C}(a\mathbb{R}^N)$ and $s \in \Delta(A)$ (see, e.g., Larsen [69, p. 90]). By letting $s_\infty = \varphi^{-1}(\infty)$ and taking in particular $s = s_\infty$ and $f = \Phi(u)$, $u \in A$, one is readily led to (2.3). This completes the proof, the unicity being evident. ■

Remark 2.1 Proposition 2.5 amounts to $\beta = \delta_{s_\infty}$ (Dirac measure on $\Delta(A)$ at s_∞), where β is the M-measure on $\Delta(A)$.

2.1.2.2 The H-algebra $\mathcal{C}_{per}(Y)$

Example 2.2 According to Examples 1.3 and 1.12, $\mathcal{C}_{per}(Y)$ is evidently an H-algebra in the case $Y = (-\frac{1}{2}, \frac{1}{2})^N$. Concerning the general case (see Sect. 1.1), (HA1) and (HA4) will be clarified in forthcoming Sect. 2.5.

Example 2.3 Let $A_i = \mathcal{C}_{per}(Y_i) \subset \mathcal{B}(\mathbb{R}^{N_i})$ with $Y_i = (-\frac{1}{2}, \frac{1}{2})^{N_i}$ where N_i is a positive integer, $1 \leq i \leq m$, and $m \geq 2$. Let $N = N_1 + \cdots + N_m$, $Y = (-\frac{1}{2}, \frac{1}{2})^N$. Then the H-algebra $\mathcal{C}_{per}(Y)$ is the closure in $\mathcal{B}(\mathbb{R}^N)$ of $\bigotimes_{i=1}^m \mathcal{C}_{per}(Y_i)$.

2.1.2.3 The H-algebra $\mathcal{B}_\infty(\mathbb{R}; A_1)$

Example 2.4 Let A_1 be an H-algebra on $\mathbb{R}^{N_1}$ under an absorptive continuous action $\mathcal{H}_1$ of $\mathbb{R}_+^*$ on $\mathbb{R}^{N_1}$ ($N_1 \geq 1$). Let $\mathcal{H}_0$ be an absorptive continuous action of $\mathbb{R}_+^*$ on $\mathbb{R}$. We assume that $\mathcal{H}_0$ and $\mathcal{H}_1$ are self-adjoint and compatible with addition in $\mathbb{R}$ and $\mathbb{R}^{N_1}$, respectively. Put $\mathcal{H}_2 = \mathcal{H}_1 \times \mathcal{H}_0$. This is an absorptive continuous action of $\mathbb{R}_+^*$ on $\mathbb{R}^{N_1+1}$ (Sect. 1.2) having in every respect the same properties with $\mathcal{H}_0$ (resp. $\mathcal{H}_1$). Let A be the closure of $A_1 \otimes \mathcal{B}_\infty(\mathbb{R})$ in $\mathcal{B}(\mathbb{R}^{N_1+1})$, $\mathcal{B}_\infty(\mathbb{R})$ being viewed here as an H-algebra on $\mathbb{R}$ under $\mathcal{H}_0$ (Example 2.1). According to Proposition 2.3, A is an H-algebra on $\mathbb{R}^{N_1+1}$ under the action $\mathcal{H}_2$. Our main purpose here is precisely to identify the H-algebra A. To this end, let $\mathcal{B}_\infty(\mathbb{R}; A_1)$ be the space of all $\psi \in \mathcal{B}(\mathbb{R}; A_1) \subset \mathcal{B}(\mathbb{R}^{N_1+1})$ such that

$$\lim_{|\tau|\to\infty} \|\psi(\cdot, \tau) - \widetilde{\psi}\|_\infty = 0,$$

where $\widetilde{\psi} \in A_1$. It is a classical fact that $\mathcal{B}_\infty(\mathbb{R}; A_1)$ is a Banach space under the $\mathcal{B}(\mathbb{R}^{N_1+1})$-norm. Furthermore, by letting $l_1(\psi) = \widetilde{\psi}$ for $\psi \in \mathcal{B}_\infty(\mathbb{R}; A_1)$, one defines a mapping $\psi \mapsto l_1(\psi)$ of $\mathcal{B}_\infty(\mathbb{R}; A_1)$ into A_1 denoted below by l_1, which is continuous linear. Let $\mathcal{B}_0(\mathbb{R}; A_1)$ be the space of those $\psi \in \mathcal{B}_\infty(\mathbb{R}; A_1)$ for which $l_1(\psi) = 0$. It should be kept in mind that $\mathcal{B}_0(\mathbb{R}; A_1)$ coincides with the closure of $A_1 \otimes \mathcal{K}(\mathbb{R})$ in $\mathcal{B}(\mathbb{R}^{N_1+1})$ (see, e.g., Bourbaki [22, pp. 45–56]).

Having made this point, let us next prove that $A = \mathcal{B}_\infty(\mathbb{R}; A_1)$. First, we clearly have $A_1 \otimes \mathcal{B}_\infty(\mathbb{R}) \subset \mathcal{B}_\infty(\mathbb{R}; A_1)$. This implies at once $A \subset \mathcal{B}_\infty(\mathbb{R}; A_1)$. To have the inverse inclusion, observe that each $\psi \in \mathcal{B}_\infty(\mathbb{R}; A_1)$ is (uniquely) expressible in the form

$$\psi = l_1(\psi) \otimes 1 + \psi_0, \tag{2.4}$$

where $\psi_0 \in \mathcal{B}_0(\mathbb{R}; A_1)$. Hence $\mathcal{B}_\infty(\mathbb{R}; A_1) \subset A$. This shows the claimed identity.

After identifying the H-algebra A, let us for further needs prove the following statement.

Proposition 2.6 *Let $A = \mathcal{B}_\infty(\mathbb{R}; A_1)$. The mapping*

$$L_1 : \mathcal{C}(\Delta(A)) \to \mathcal{C}(\Delta(A_1)),\ L_1(\mathcal{G}(\psi)) = \mathcal{G}(l_1(\psi)),\ \psi \in A$$

extends by continuity to an isometric isomorphism, still denoted by L_1, of $L^2(\Delta(A))$ onto $L^2(\Delta(A_1))$.

Proof Before embarking upon the proof, let us observe that the letter $\mathcal{G}$ above denotes the Gelfand transformation on A and A_1, as well. Now, let us prove the equality

$$\|L_1(\widehat{\psi})\|_{L^2(\Delta(A_1))} = \|\widehat{\psi}\|_{L^2(\Delta(A))} \quad (\psi \in A), \tag{2.5}$$

where $\widehat{\psi} = \mathcal{G}(\psi)$. For this purpose, fix freely $\psi \in A$. From (2.4) we have

$$M(|\psi|^2) = M(|l_1(\psi)|^2 \otimes 1) + 2\mathrm{Re} M(\overline{\psi}_0[l_1(\psi) \otimes 1]) + M(|\psi_0|^2),$$

where M denotes the mean value generated by $\mathcal{H}_2$ on $\mathbb{R}^{N_1+1}$. But the last two terms on the right reduce each to zero because M vanishes on $\mathcal{B}_0(\mathbb{R}; A_1)$. On the other hand, we have $M(|l_1(\psi)|^2 \otimes 1) = M(|l_1(\psi)|^2)$ (Lemma 2.2). Hence (2.5) follows by Proposition 2.2. Consequently, thanks to the density of $\mathcal{C}(\Delta(A))$ in $L^2(\Delta(A))$, the mapping L_1 extends by continuity to an isometry, still called L_1, of $L^2(\Delta(A))$ into

$L^2(\Delta(A_1))$. Thus, it only remains to check that this isometry is surjective. From the obvious equality $l_1(f \otimes 1) = f \ (f \in A_1)$ we see that $\mathcal{C}(\Delta(A_1)) \subset L_1(\mathcal{C}(\Delta(A))) \subset L_1(L^2(\Delta(A))) \subset L^2(\Delta(A_1))$. We deduce at once that $L_1(L^2(\Delta(A)))$ is dense in $L^2(\Delta(A_1))$. Hence the surjectivity of L_1 follows by the fact that $L_1(L^2(\Delta(A)))$ is closed in $L^2(\Delta(A_1))$. ■

2.1.3 Summable Families of H-algebras

A finite family $\{A_i\}_{1 \leq i \leq m}$ of H-algebras on $\mathbb{R}^N$ (under the action $\mathcal{H}$) is said to be *summable* if the vector space $A_1 + \cdots + A_m$ is stable under (pointwise) multiplication in $\mathcal{B}(\mathbb{R}^N)$. Let A be the closure of $A_1 + \cdots + A_m$ in $\mathcal{B}(\mathbb{R}^N)$. It is immediate that A is an H-algebra as soon as the family $\{A_i\}$ is summable.

Now, let $\{A_1, A_2\}$ be a pair of H-algebras on $\mathbb{R}^N$. Let

$$A_1/\mathbb{C} = \{\psi | \psi \in A_1, M(\psi) = 0\}.$$

We assume the following:

$$A_1/\mathbb{C} \text{ is stable under multiplication.} \tag{2.6}$$

$$\text{If } \varphi \in A_1/\mathbb{C} \text{ and } \psi \in A_2, \text{ then } \varphi\psi \in A_1/\mathbb{C}. \tag{2.7}$$

$$(A_1/\mathbb{C}) \cap A_2 = \{0\}. \tag{2.8}$$

With (2.8) in mind, let $\mathcal{V} = (A_1/\mathbb{C}) \oplus A_2$ (direct sum), and define the mapping $l_2 : \mathcal{V} \to \mathcal{V}$ to be the projection (of $\mathcal{V}$) on A_2 along $A_1/\mathbb{C}$. Thus, $A_1/\mathbb{C} = \{\psi | \psi \in \mathcal{V}, l_2(\psi) = 0\}$ and $A_2 = \{\psi | \psi \in \mathcal{V}, l_2(\psi) = \psi\}$ (range of l_2), and every $\psi \in \mathcal{V}$ is uniquely expressible in the form

$$\psi = \psi_0 + l_2(\psi) \text{ with } \psi_0 \in A_1/\mathbb{C}. \tag{2.9}$$

The aim is to establish the following

Proposition 2.7 *Let (2.6)–(2.8) be satisfied. Then the family $\{A_1, A_2\}$ is summable. Furthermore, if A is defined as the closure of $A_1 + A_2$ in $\mathcal{B}(\mathbb{R}^N)$, and if $\mathcal{G}$ denotes the Gelfand transformation on A_i $(i = 1, 2)$ and on A, as well, then the mapping*

$$L_2 : \mathcal{G}(\mathcal{V}) \to \mathcal{C}(\Delta(A_2)), L_2(\mathcal{G}(\psi)) = \mathcal{G}(l_2(\psi)) \ (\psi \in \mathcal{V})$$

extends by continuity to an isometric isomorphism L of $L^2(\Delta(A))$ onto $L^2(\Delta(A_2))$.

Proof By reason of (2.6)–(2.7), the space $\mathcal{V}$ is stable under multiplication. Hence the summability of $\{A_1, A_2\}$ follows by the trivial equality $A_1 + A_2 = (A_1/\mathbb{C}) + A_2$. Thus, A is an H-algebra. For the rest, the reasoning is the same as followed in proving Proposition 2.6: Let $\psi \in \mathcal{V}$. By (2.9) and use of the fact that the functions $|\psi_0|^2$, $\overline{\psi}_0 l_2(\psi)$ and $\psi_0\overline{l_2(\psi)}$ lie in $A_1/\mathbb{C}$ (thanks to (2.6) and (2.7)), one arrives at $M(|\psi|^2) = M(|l_2(\psi)|^2)$. Therefore $\|L_2(\widehat{\psi})\|_{L^2(\Delta(A_2))} = \|\widehat{\psi}\|_{L^2(\Delta(A))}$ for all $\psi \in \mathcal{V}$. Consequently, since $\mathcal{G}(\mathcal{V})$ is dense in $L^2(\Delta(A))$ (because $\mathcal{V}$ is dense in A), it follows that L_2 extends by continuity to an isometric mapping, denoted by L_2, of $L^2(\Delta(A))$ into $L^2(\Delta(A_2))$. Finally, by a simple argument similar to that followed in proving Proposition 2.6, one quickly sees that L_2 is surjective, thereby completing the proof. ∎

2.1.3.1 The H-algebra $\mathcal{B}_{\infty,\mathbb{Z}^N}(\mathbb{R}^N)$

Example 2.5 Let $\mathcal{B}_{\infty,\mathbb{Z}^N}(\mathbb{R}^N)$ be the closure of $\mathcal{B}_\infty(\mathbb{R}^N) + \mathcal{C}_{per}(Y)$ in $\mathcal{B}(\mathbb{R}^N)$, where $\mathcal{B}_\infty(\mathbb{R}^N)$ and $\mathcal{C}_{per}(Y)$ are the H-algebras of Examples 2.1 and 2.2, respectively. We claim that $\mathcal{B}_{\infty,\mathbb{Z}^N}(\mathbb{R}^N)$ is an H-algebra on $\mathbb{R}^N$. This follows immediately by Proposition 2.7 with $A_1 = \mathcal{B}_\infty(\mathbb{R}^N)$ and $A_2 = \mathcal{C}_{per}(Y)$. Indeed, it is an easy exercise to check that the pair thus considered satisfies (2.6)–(2.8) and is therefore summable, so that $\mathcal{B}_{\infty,\mathbb{Z}^N}(\mathbb{R}^N)$ is an H-algebra verifying moreover the conclusions of Proposition 2.7.

It is worth noting that $\mathcal{B}_{\infty,\mathbb{Z}^N}(\mathbb{R}^N)$ is identical with the closure in $\mathcal{B}(\mathbb{R}^N)$ of the set of functions of the form

$$u = \sum_{finite} \varphi_k \gamma_k,$$

where $\varphi_k \in \mathcal{B}_\infty(\mathbb{R}^N)$, γ_k given by (1.18) with $k \in \mathbb{Z}^N$.

2.1.4 *Representation of Smooth Functions in an H-algebra*

In what follows, A is an H-algebra on $\mathbb{R}^N$.

2.1.4.1 The Spaces A^m. Partial Derivatives on $\Delta(A)$

Let m be a positive integer. We put

$$A^m = \{u | u \in A \cap \mathcal{C}^m(\mathbb{R}^N),\ D_y^\alpha u \in A \text{ for } |\alpha| \le m\},$$

where

$$D_y^\alpha u = \frac{\partial^{|\alpha|} u}{\partial y_1^{\alpha_1} \cdots \partial y_N^{\alpha_N}}$$

with $\alpha = (\alpha_1, \cdots, \alpha_N) \in \mathbb{N}^N$, $|\alpha| = \alpha_1 + \cdots + \alpha_N$.

It is clear by the Leibniz formula that A^m is a subalgebra of A. More precisely, provided with the algebra structure on A, A^m is a commutative algebra with involution and with identity. In the sequel, A^m is considered equipped with the norm

$$\|u\|_m = \sup_{|\alpha|\leq m} \|D_y^\alpha u\|_\infty \quad (u \in A^m)$$

which makes it into a Banach space (the verification is a classical exercise left to the reader).

Now, let

$$A^\infty = \bigcap_{m\in\mathbb{N}} A^m,$$

where $A^0 = A$. Provided with the algebra operations in A, A^∞ is a commutative algebra with involution and with identity. The natural topology on A^∞ is the locally convex topology defined by the family of norms $\|\cdot\|_m$ $(m \geq 1)$. It is a routine exercise (left to the reader) to check that A^∞ thus topologized is a Fréchet space (for the construction of a locally convex topology by means of continuous seminorms see, e.g., Vo-Khac [138]).

At the present time, for $1 \leq m \leq \infty$, let

$$\mathcal{D}^m(\Delta(A)) = \{\varphi|\varphi \in \mathcal{C}(\Delta(A)), \mathcal{G}^{-1}(\varphi) \in A^m\},$$

or, equivalently, $\mathcal{D}^m(\Delta(A)) = \mathcal{G}(A^m)$. For convenience we put $\mathcal{D}(\Delta(A)) = \mathcal{D}^\infty(\Delta(A))$. It is quite obvious that $\mathcal{D}^m(\Delta(A))$ $(1 \leq m \leq \infty)$ is a subalgebra of $\mathcal{C}(\Delta(A))$ that contains the constants and preserves complex conjugation.

Having made this point, let us turn now to the definition of the notion of a partial derivative on $\Delta(A)$.

Definition 2.3 For any fixed i $(1 \leq i \leq N)$, let ∂_i be the unbounded linear operator from $\mathcal{C}(\Delta(A))$ to $\mathcal{C}(\Delta(A))$ defined as follows:

$$\begin{cases} D(\partial_i) = \mathcal{D}^1(\Delta(A)) \text{ (domain of } \partial_i) \\ \partial_i\varphi = (\mathcal{G} \circ \frac{\partial}{\partial y_i} \circ \mathcal{G}^{-1})(\varphi), \quad \varphi \in D(\partial_i). \end{cases}$$

The operator ∂_i is called the *partial derivative of index i* on $\Delta(A)$. For $\varphi \in \mathcal{D}^1(\Delta(A))$, the function $\partial_i\varphi \in \mathcal{C}(\Delta(A))$ is called the *partial derivative of index i of φ*.

We can next define higher-order partial derivatives on $\Delta(A)$. For fixed index i $(1 \leq i \leq N)$ and for any given integer $m \geq 2$, we set $\partial_i^m = \partial_i \circ \cdots \circ \partial_i$ (m times), defining thus an unbounded linear operator from $\mathcal{C}(\Delta(A))$ into itself with domain $D(\partial_i^m) = \{\varphi|\varphi \in D(\partial_i), \partial_i^j\varphi \in D(\partial_i), 1 \leq j \leq m-1\}$ and with $\partial_i^m\varphi$ (for $\varphi \in D(\partial_i^m)$) defined by iteration in an obvious fashion. The operator ∂_i^m is called the partial derivative of order m with respect to (or of) index i.

Finally, for each multi-index $\alpha = (\alpha_1, \cdots, \alpha_N) \in \mathbb{N}^N$, the partial derivative of index α is defined as

$$\partial^\alpha = \partial^{\alpha_1} \circ \cdots \circ \partial^{\alpha_N}$$

with the convention $\partial^{\alpha_i = 0} = Id$ for some $1 \leq i \leq N$, where $Id(\varphi) = \varphi$ for $\varphi \in \mathcal{C}(\Delta(A))$. Of course, the computation of $\partial^\alpha \varphi$ is an easy matter once the domain $D(\partial^\alpha)$ is specified, which is quite feasible, though a bit cumbersome.

On the other hand, an easy calculation reveals the important general relation

$$\partial^\alpha = \mathcal{G} \circ D_y^\alpha \circ \mathcal{G}^{-1} \quad (\alpha \in \mathbb{N}^N),$$

which furnishes a straightforward definition of ∂^α provided the domain $D(\partial^\alpha)$ is well defined, of course. So it should be noted that the derivative $\partial^\alpha \varphi \in \mathcal{C}(\Delta(A))$ is defined for any $\varphi \in \mathcal{D}^m(\Delta(A))$ and any multi-index α of length $|\alpha| \leq m$.

Let us now provide $\mathcal{D}^m(\Delta(A))$ with a suitable topology. For any given integer $m \geq 1$, let

$$\|\varphi\|_m = \sup_{|\alpha| \leq m} \|\partial^\alpha \varphi\|_\infty \quad (\varphi \in \mathcal{D}^m(\Delta(A)))$$

where $\|\partial^\alpha \varphi\|_\infty = \sup_{s \in \Delta(A)} |\partial^\alpha \varphi(s)|$. This defines a norm $\|\cdot\|_m$ on $\mathcal{D}^m(\Delta(A))$ (there is no risk of confusion with the norm on A^m). Furthermore, as m ranges over the positive integers, the norms $\|\cdot\|_m$ viewed as defined on $\mathcal{D}(\Delta(A))$ form a defining family of seminorms for a locally convex topology on $\mathcal{D}(\Delta(A))$. In the sequel we shall consider $\mathcal{D}(\Delta(A))$ endowed with this locally convex topology, and $\mathcal{D}^m(\Delta(A))$ (integer $m \geq 1$) equipped with the norm $\|\cdot\|_m$.

The preceding construction leads us quite naturally to the next proposition. The proof is a rather simple exercise left to the reader.

Proposition 2.8 *Provided with the induced algebra structure on $\mathcal{C}(\Delta(A))$, the space $\mathcal{D}^m(\Delta(A))$ $(1 \leq m \leq \infty)$ is a commutative algebra with involution and with identity. Furthermore, $\mathcal{D}^m(\Delta(A))$ $(1 \leq m < \infty)$ is a Banach space and the Gelfand transformation on A maps A^m isometrically over $\mathcal{D}^m(\Delta(A))$; $\mathcal{D}(\Delta(A)) = \mathcal{D}^\infty(\Delta(A))$ is a Fréchet space and the Gelfand transformation on A, considered as defined on A^∞, is a topological isomorphism of A^∞ onto $\mathcal{D}(\Delta(A))$.*

Remark 2.2 For any multi-index $\alpha \in \mathbb{N}^N$, the linear operator ∂^α maps continuously $\mathcal{D}(\Delta(A))$ into itself. Indeed this follows readily by the inequality $\|\partial^\alpha \varphi\|_m \leq \|\varphi\|_{m+|\alpha|}$ for all $\varphi \in \mathcal{D}(\Delta(A))$, where m is any arbitrary positive integer.

Let us prove one fundamental result.

Proposition 2.9 *Let* $\varphi \in \mathcal{D}^m(\Delta(A))$ $(m \geq 1)$. *For any multi-index* α *with* $1 \leq |\alpha| \leq m$, *we have*

$$\int_{\Delta(A)} \partial^\alpha \varphi(s) d\beta(s) = 0.$$

Proof It will be enough to show this for $m = 1$. So, we have to prove that

$$\int_{\Delta(A)} \partial_i \varphi(s) d\beta(s) = 0 \quad (1 \leq i \leq N).$$

By letting $\psi = \mathcal{G}^{-1}(\varphi) \in A^1$, and recalling Proposition 2.1, it amounts to showing that $M(\frac{\partial \psi}{\partial y_i}) = 0$ $(i = 1, \cdots, N)$. To this end, let $f \in \mathcal{D}(\mathbb{R}^N) = \mathcal{K}(\mathbb{R}^N) \cap \mathcal{C}^\infty(\mathbb{R}^N)$ with $\int_{\mathbb{R}^N} f(y)dy = 1$. By appealing to Proposition 1.5, we get

$$M\left(f * \frac{\partial \psi}{\partial y_i}\right) = M\left(\frac{\partial \psi}{\partial y_i}\right).$$

On the other hand

$$M\left(f * \frac{\partial \psi}{\partial y_i}\right) = M\left(\frac{\partial f}{\partial y_i} * \psi\right) = M(\psi) \int_{\mathbb{R}^N} \frac{\partial f}{\partial y_i}(y)dy = 0,$$

where we have made use of Proposition 1.5, once again. Hence $M(\frac{\partial \psi}{\partial y_i}) = 0$, and the proposition follows thereby. ∎

2.1.5 Distributions on $\Delta(A)$

Let A be an H-algebra on $\mathbb{R}^N$. By a distribution on $\Delta(A)$ is meant any continuous linear form on the Fréchet space $\mathcal{D}(\Delta(A))$. The set of all distributions on $\Delta(A)$ is denoted by $\mathcal{D}'(\Delta(A))$. The value of $T \in \mathcal{D}'(\Delta(A))$ on $\varphi \in \mathcal{D}(\Delta(A))$ is denoted by $\langle T, \varphi \rangle$. A distribution T on $\Delta(A)$ is said to be real (resp. positive) if $\langle T, \varphi \rangle$ is real (resp. nonnegative) for any $\varphi \in \mathcal{D}(\Delta(A))$ that assumes only real (resp. nonnegative real) values.

The topology on $\mathcal{D}'(\Delta(A))$ will be the strong dual topology, that is, the locally convex topology defined by the family of seminorms N_Φ given by

$$N_\Phi(T) = \sup_{\varphi \in \Phi} |\langle T, \varphi \rangle| \quad (T \in \mathcal{D}'(\Delta(A))),$$

where Φ ranges over the bounded subsets of $\mathcal{D}(\Delta(A))$ (for the benefit of the reader, we recall that a set $\Phi \subset \mathcal{D}(\Delta(A))$ is said to be bounded if $\sup_{\varphi\in\Phi} \|\varphi\|_m < \infty$ for any $m \geq 1$). Thus equipped, $\mathcal{D}'(\Delta(A))$ is a separated complete locally convex space (indeed, the strong dual of a bornological space, and in particular of a Fréchet space, is complete).

For $f \in L^1(\Delta(A))$, the mapping $\varphi \mapsto \int_{\Delta(A)} f(s)\varphi(s)d\beta(s)$ of $\mathcal{D}(\Delta(A))$ into $\mathbb{C}$ is a distribution on $\Delta(A)$ denoted by T_f. This yields a transformation $f \mapsto T_f$ which is a continuous linear mapping of $L^1(\Delta(A))$ into $\mathcal{D}'(\Delta(A))$. In case $\mathcal{D}(\Delta(A))$ is dense in $\mathcal{C}(\Delta(A))$ (this is fulfilled by all H-algebras involved in practical situations), this transformation is injective (see Bourbaki [23, p.46, Corol. 2]), and we may therefore identify T_f with f. Consequently, $L^1(\Delta(A)) \subset \mathcal{D}'(\Delta(A))$ with continuous embedding. Of course, this implies $L^p(\Delta(A)) \subset \mathcal{D}'(\Delta(A))$ $(1 \leq p \leq \infty)$ with continuous embedding.

The partial derivatives of a $T \in \mathcal{D}'(\Delta(A))$ are defined as expected. Let $\alpha \in \mathbb{N}^N$, $|\alpha| \neq 0$. Set $L_\alpha(\varphi) = (-1)^{|\alpha|}\partial^\alpha\varphi$ for $\varphi \in \mathcal{D}(\Delta(A))$, which defines a continuous linear mapping of $\mathcal{D}(\Delta(A))$ into itself (Remark 2.2). Therefore, its adjoint L_α^* maps continuously and linearly $\mathcal{D}'(\Delta(A))$ into itself. The partial derivative $\partial^\alpha T$ of a distribution T on $\Delta(A)$ is defined as $\partial^\alpha T = L_\alpha^* T$, i.e.,

$$\langle \partial^\alpha T, \varphi\rangle = (-1)^{|\alpha|}\langle T, \partial^\alpha \varphi\rangle, \quad \varphi \in \mathcal{D}(\Delta(A)).$$

Furthermore, the correspondance $T \mapsto \partial^\alpha T$ is a continuous linear mapping, denoted by ∂^α, of $\mathcal{D}'(\Delta(A))$ into itself.

Finally, it is a simple matter (use the Leibniz formula together with Proposition 2.9) to check that if $f \in \mathcal{D}^m(\Delta(A))$ and $|\alpha| \leq m$, then $\partial^\alpha T_f = T_{\partial^\alpha f}$. This shows that the distribution derivation is consistent with the derivation of functions introduced earlier (Definition 2.3).

2.2 Further Basic Spaces Attached to an H-algebra

2.2.1 *The Spaces* $\mathfrak{X}_A^p(\mathbb{R}^N)$ $(1 \leq p < \infty)$

Let $1 \leq p < \infty$. We shall first of all introduce the underlying space $\Xi^p(\mathbb{R}^N)$ consisting of those functions $u \in L^p_{loc}(\mathbb{R}^N)$ for which the sequence $(u \circ H_\varepsilon)_{0<\varepsilon\leq 1}$ is bounded in the Fréchet space $L^p_{loc}(\mathbb{R}^N)$. This is a vector subspace of $L^p_{loc}(\mathbb{R}^N)$. Now, let B be an open elementary set (Lemma 1.5). Let

$$\|u\|_{\Xi^p} = \sup_{0<\varepsilon\leq 1}\left(\int_B |u(H_\varepsilon(x))|^p dx\right)^{\frac{1}{p}} \tag{2.10}$$

for $u \in \Xi^p(\mathbb{R}^N)$, which defines a semi-norm on $\Xi^p(\mathbb{R}^N)$.

Lemma 2.3 *The above semi-norm is actually a norm under which $\Xi^p(\mathbb{R}^N)$ is a Banach space.*

Proof For each compact set $K \subset \mathbb{R}^N$, let

$$q_K(u) = \sup_{0<\varepsilon\leq 1} \left(\int_K |u(H_\varepsilon(x))|^p dx \right)^{\frac{1}{p}} \quad (u \in \Xi^p(\mathbb{R}^N)),$$

which provides $\Xi^p(\mathbb{R}^N)$ with a family of semi-norms $\{q_K\}$ (K ranging over the compact sets in $\mathbb{R}^N$). The natural topology on $\Xi^p(\mathbb{R}^N)$ is that defined by this family of semi-norms. So topologized, $\Xi^p(\mathbb{R}^N)$ is a separated locally convex space. Because $\mathbb{R}^N$ is a σ-compact locally compact space, this topology has a countable base of continuous seminorms, so that there is a countable base of neighbourhoods at the origin in $\Xi^p(\mathbb{R}^N)$. Consequently the proof of the completeness reduces to that of the sequential completeness of $\Xi^p(\mathbb{R}^N)$. But then this follows by a routine line of proceeding based on four facts, namely: (**1**) $L^p_{loc}(\mathbb{R}^N)$ is sequentially complete. (**2**) $\Xi^p(\mathbb{R}^N)$ is continuously embedded in $L^p_{loc}(\mathbb{R}^N)$. (**3**) The mapping $u \mapsto u \circ H_\varepsilon$ sends continuously $L^p_{loc}(\mathbb{R}^N)$ into itself. (**4**) Any given Cauchy ordinary sequence, say in $\Xi^p(\mathbb{R}^N)$ is bounded. Thus, $\Xi^p(\mathbb{R}^N)$ is a Fréchet space and the lemma is proved if we can check that the topology on $\Xi^p(\mathbb{R}^N)$ is identical with that defined by the single semi-norm defined in (2.10), in which case the said semi-norm will turn out to be a complete norm on $\Xi^p(\mathbb{R}^N)$. But then a little moment's reflection reveals that the whole problem reduces to showing that to each compact set $K \subset \mathbb{R}^N$ there is assigned some constant $c = c(K) > 0$ such that

$$\left(\int_K |u(H_\varepsilon(x))|^p dx \right)^{\frac{1}{p}} \leq c\|u\|_{\Xi^p} \quad (0 < \varepsilon \leq 1) \tag{2.11}$$

for all $u \in \Xi^p(\mathbb{R}^N)$. Let K be as above. Thanks to Proposition 1.2, there is some integer $n \geq 1$ such that $K \subset H_{n^{-1}}(B)$. Let u be freely fixed in $\Xi^p(\mathbb{R}^N)$. For $0 < \varepsilon \leq 1$, we have

$$\int_K |u(H_\varepsilon(x))|^p dx \leq \int_{H_{n^{-1}}(B)} |u(H_\varepsilon(x))|^p dx.$$

In the sequel, χ_S stands for the characteristic function of a set $S \subset \mathbb{R}^N$. It must be recalled that $\chi_S \circ H_\varepsilon = \chi_{H_{\varepsilon^{-1}}(S)}$. On the other hand, for the sake of convenience, f stands for the function in $L^1_{loc}(\mathbb{R}^N)$ given by $f(x) = |u(H_{n^{-1}\varepsilon}(x))|^p \quad (x \in \mathbb{R}^N)$. Then

$$\begin{aligned}\int_{H_{n^{-1}}(B)} |u(H_\varepsilon(x))|^p dx &= \int_{H_{n^{-1}}(B)} f(H_n(x))dx \\ &= \int f(H_n(x))\chi_{H_{n^{-1}}(B)}(x)dx\end{aligned}$$

$$= \int (f\chi_B)(H_n(x))dx$$

$$= n^s \int f(x)\chi_B(x)dx \text{ (Proposition 1.4)}$$

$$= n^s \int_B |u(H_{n^{-1}\varepsilon}(x))|^p dx \le n^s \|u\|^p_{\Xi^p}.$$

Hence (2.11) follows with $c = n^{\frac{s}{p}}$, which completes the proof. ■

Remark 2.3 We have $L^p(\mathbb{R}^N) \subset \Xi^p(\mathbb{R}^N)$ with $\|u\|_{\Xi^p} \le \|u\|_{L^p(\mathbb{R}^N)}$ for all $u \in L^p(\mathbb{R}^N)$.

Having made these preliminaries, let us now come to the heart of the matter.

Definition 2.4 Let A be an H-algebra on $\mathbb{R}^N$ (under $\mathcal{H}$). For any given $1 \le p < \infty$, we define $\mathfrak{X}^p_A(\mathbb{R}^N)$ as being the closure of A in the Banach space $\Xi^p(\mathbb{R}^N)$.

Throughout the remainder of the book we shall always consider $\mathfrak{X}^p_A(\mathbb{R}^N)$ tacitly equipped with the $\Xi^p(\mathbb{R}^N)$-norm defined in (2.10), which makes it into a Banach space. Attention is drawn to the fact that two distinct such norms are equivalent (see Proposition 1.2 and Remark 1.9). For short we will often write $\mathfrak{X}^p(\mathbb{R}^N)$ (or even $\mathfrak{X}^p$) in place of $\mathfrak{X}^p_A(\mathbb{R}^N)$ provided this involves no risk of confusion.

Remark 2.4 If $1 \le p < q < \infty$, then $\mathfrak{X}^q \subset \mathfrak{X}^p$ with continuous embedding.

Our next concern is to extend two basic notions to $\mathfrak{X}^p$, namely the mean value and the Gelfand transformation on an H-algebra . In this connection there are two simple propositions destined to play crucial roles in deterministic homogenization. Throughout the rest of this section, A denotes an H-algebra on $\mathbb{R}^N$ (for $\mathcal{H}$).

Proposition 2.10 *Let $1 \le p < \infty$. The mean value M viewed as defined on A, extends by continuity to a positive continuous linear form on $\mathfrak{X}^p$ still denoted by M. Furthermore, for any fixed bounded open set $\Omega \subset \mathbb{R}^N_x$, we have $u \circ H_\varepsilon \to M(u)$ in $L^p(\Omega)$-weak as $\varepsilon \to 0$ ($u \in \mathfrak{X}^p$), where for convenience we write $u \circ H_\varepsilon$ for $(u \circ H_\varepsilon)|_\Omega$.*

Proof Let $u \in A$. By Hölder's inequality,

$$\left| \int u(H_\varepsilon(x))\chi_B(x)dx \right| \le \lambda(B)^{\frac{1}{p'}} \|u\|_{\Xi^p} \quad (0 < \varepsilon \le 1),$$

where $\frac{1}{p} + \frac{1}{p'} = 1$. By letting $\varepsilon \to 0$, we get by (HA4) (Definition 2.1)

$$|M(u)| \le \lambda(B)^{-\frac{1}{p}} \|u\|_{\Xi^p} \quad (u \in A),$$

from which we deduce the first portion of the proposition except the positivity which is an obvious consequence of the second portion. But then the latter is achieved by the same standard argument mutatis mutandis as followed in proving part (i) of Theorem 1.3. ■

Proposition 2.11 *The Gelfand transformation $\mathcal{G} : A \to \mathcal{C}(\Delta(A))$ extends by continuity to a continuous linear mapping, still denoted by $\mathcal{G}$, of $\mathfrak{X}^p$ into $L^p(\Delta(A))$ $(1 \leq p < \infty)$.*

Proof For any arbitrary fixed $u \in A$, one has

$$\int |u(H_\varepsilon(x))|^p \chi_B(x)dx \leq \|u\|_{\Xi^p}^p \quad (0 < \varepsilon \leq 1).$$

By letting $\varepsilon \to 0$, it follows by (HA4) and use of Proposition 2.2 (see also Lemma 2.1) that

$$\|\mathcal{G}(u)\|_{L^p(\Delta(A))} \leq \lambda(B)^{-\frac{1}{p}} \|u\|_{\Xi^p} \quad (u \in A).$$

Hence the proposition follows in an obvious way. ■

The mapping $\mathcal{G} : \mathfrak{X}^p \to L^p(\Delta(A))$ thus constructed will be referred to as the *canonical mapping* of $\mathfrak{X}^p$ into $L^p(\Delta(A))$.

We now record a series of corollaries of the preceding proposition. We assume that $1 \leq p < \infty$ except where otherwise stated.

Corollary 2.1 *Let $u \in \mathfrak{X}^p$. Then*

(i) $M(u) = \int_{\Delta(A)} \mathcal{G}(u)d\beta$.
(ii) $\overline{u} \in \mathfrak{X}^p$ and $\overline{\mathcal{G}(u)} = \mathcal{G}(\overline{u})$.

Proof (i) follows in an obvious fashion by (2.1) and use of both Propositions 2.10 and 2.11. As regards (ii), we observe that the complex conjugation maps continuously $\mathfrak{X}^p$ into itself and $L^p(\Delta(A))$ into itself. Hence (ii) follows at once by Proposition 2.11 and use of the fact that the Gelfand transformation on A is a C^*-homomorphism of A into $\mathcal{C}(\Delta(A))$. ■

Corollary 2.2 *Let $1 < p, q < \infty$ with $\frac{1}{p} + \frac{1}{q} = \frac{1}{r} \leq 1$. If $u \in \mathfrak{X}^p$ and $v \in \mathfrak{X}^q$, then $uv \in \mathfrak{X}^r$ and $\mathcal{G}(uv) = \mathcal{G}(u)\mathcal{G}(v)$.*

Proof Given u and v as stated, it is a classical fact that uv lies in $L^r_{loc}(\mathbb{R}^N)$ and $\|uv\|_{L^r(\mathcal{O})} \leq \|u\|_{L^p(\mathcal{O})}\|v\|_{L^q(\mathcal{O})}$ (generalized Hölder's inequality) for any bounded open set $\mathcal{O} \subset \mathbb{R}^N$, where for convenience the same character is used to denote a function in $L^1_{loc}(\mathbb{R}^N)$ and its restriction to $\mathcal{O}$. Thus, given an open elementary set B in $\mathbb{R}^N$ (under the action $\mathcal{H}$), we have $\|(uv)^\varepsilon\|_{L^r(B)} \leq \|u^\varepsilon\|_{L^p(B)}\|v^\varepsilon\|_{L^q(B)}$ $(\varepsilon > 0)$ with as usual $f^\varepsilon = f \circ H_\varepsilon$ for $f \in L^1_{loc}(\mathbb{R}^N)$. Hence, it follows that uv lies in $\Xi^r(\mathbb{R}^N)$ and $\|uv\|_{\Xi^r} \leq \|u\|_{\Xi^p}\|v\|_{\Xi^q}$. One deduces by mere routine that $uv \in \mathfrak{X}^r(\mathbb{R}^N)$. The rest of the corollary follows readily by Proposition 2.11 in combination with the obvious facts that the desired identity holds for $u, v \in A$, and the bilinear mapping $(f, g) \mapsto fg$ sends continuously $\mathfrak{X}^p(\mathbb{R}^N) \times \mathfrak{X}^q(\mathbb{R}^N)$ into $\mathfrak{X}^r(\mathbb{R}^N)$ and $L^p(\Delta(A)) \times L^q(\Delta(A))$ into $L^r(\Delta(A))$. ∎

Corollary 2.3 *If $u \in \mathfrak{X}^p$, then $|u|^p \in \mathfrak{X}^1$ and $\mathcal{G}(|u|^p) = |\mathcal{G}(u)|^p$.*

Proof Our line of reasoning is based on one elementary but crucial inequality:

$$||a|^p - |b|^p| \leq p|a-b|(|a|+|b|)^{p-1} \quad (a, b \in \mathbb{C}). \tag{2.12}$$

This is easily deduced from the fact that for any couple of real numbers $0 < t_1 < t_2$, there is some $c \in (t_1, t_2)$ such that $\varphi(t_2) - \varphi(t_1) = \varphi'(c)(t_2 - t_1)$, where $\varphi(t) = t^p$ $(t > 0)$ (the details are left to the reader). Having done this, let $u \in \mathfrak{X}^p$. Let $(u_n)_{n\geq 0}$ be an ordinary sequence of functions in A such that, as $n \to \infty$, $u_n \to u$ in $\Xi^p(\mathbb{R}^N)$. By combining (2.12) (take $a = u_n(y)$ and $b = u(y)$, $y \in \mathbb{R}^N$) with an obvious Hölder's inequality we arrive at

$$\| |u_n|^p - |u|^p \|_{\Xi^1} \leq c \|u_n - u\|_{\Xi^p} \text{ (integers } n \geq 0),$$

where $c = p \sup_{n\geq 0} \| |u_n| + |u| \|_{\Xi^p} < \infty$. Hence $|u_n|^p \to |u|^p$ in $\Xi^1(\mathbb{R}^N)$ (as $n \to \infty$) with $|u_n|^p \in A$ (Lemma 2.1). This shows that $|u|^p \in \mathfrak{X}^1$. Furthermore, by Proposition 2.11, $\mathcal{G}(|u_n|^p) \to \mathcal{G}(|u|^p)$ in $L^1(\Delta(A))$. On the other hand, thanks to Proposition 2.11, once again, $|\mathcal{G}(u_n)|^p \to |\mathcal{G}(u)|^p$ in $L^1(\Delta(A))$. From all that we deduce by Lemma 2.1 the second portion of the corollary. ∎

The next corollary is immediate by Corollaries 2.1 and 2.3.

Corollary 2.4 *If $u \in \mathfrak{X}^p$, then $|u|^p \in \mathfrak{X}^1$ and*

$$M(|u|^p) = \int_{\Delta(A)} |\mathcal{G}(u)|^p d\beta.$$

Finally come the following three corollaries.

Corollary 2.5 *If $\psi \in A$ and $u \in \mathfrak{X}^p$, then $\psi u \in \mathfrak{X}^p$ and $\mathcal{G}(\psi)\mathcal{G}(u) = \mathcal{G}(\psi u)$.*

Proof This is a rather simple exercise left to the reader. ∎

Corollary 2.6 *If a function $u \in \mathfrak{X}^1$ assumes real values, then so also does $\mathcal{G}(u)$. If moreover $u \geq 0$ almost everywhere (for λ), then $\mathcal{G}(u) \geq 0$ almost everywhere (for β).*

Proof Let $u \in \mathfrak{X}^1$. Suppose u assumes real values. Immediately $\mathcal{G}(u)$ has the same property, as is apparent by Corollary 2.1(ii). Suppose moreover that $u \geq 0$ a.e. Let $\psi \in A$ with $\psi \geq 0$. Then $\psi u \in \mathfrak{X}^1$ (Corollary 2.5) with $\psi u \geq 0$, hence $M(\psi u) \geq 0$. We deduce, using Corollaries 2.1(i) and 2.5, that $\int_{\Delta(A)} \mathcal{G}(\psi)\mathcal{G}(u)d\beta \geq 0$. Consequently

$$\int f\mathcal{G}(u)d\beta \geq 0$$

for all $f \in \mathcal{C}(\Delta(A))$ with $f \geq 0$. Hence $\mathcal{G}(u) \geq 0$ (see, e.g., Bourbaki [23, p. 17, Corol. 3]). ∎

Corollary 2.7 *Let* $u \in \mathfrak{X}^1 \cap L^\infty$. *Then* $\mathcal{G}(u) \in L^\infty(\Delta(A))$ *and* $\|\mathcal{G}(u)\|_{L^\infty(\Delta(A))} \leq \|u\|_{L^\infty}$.

Proof Let us fix freely $\psi \in A$. From $|u| \leq \|u\|_{L^\infty}$ a.e. we deduce $|\psi u| \leq \|u\|_{L^\infty}|\psi|$ a.e. Therefore $M(|\psi u|) \leq \|u\|_{L^\infty} M(|\psi|)$. By combining Corollaries 2.1,2.3 and 2.5, it follows

$$\left|\int_{\Delta(A)} \mathcal{G}(u)\mathcal{G}(\psi)d\beta\right| \leq \|u\|_{L^\infty} \int_{\Delta(A)} |\mathcal{G}(\psi)|d\beta.$$

Hence

$$\left|\int_{\Delta(A)} \mathcal{G}(u) f d\beta\right| \leq \|u\|_{L^\infty} \|f\|_{L^1(\Delta(A))}$$

for all $f \in \mathcal{C}(\Delta(A))$. Whence the corollary follows. ∎

2.2.2 The Spaces $W^{m,p}(\Delta(A))$

Let A be an H-algebra. We assume that A^∞ is dense in A, which amounts to saying that $\mathcal{D}(\Delta(A))$ is dense in $\mathcal{C}(\Delta(A))$. Therefore $L^p(\Delta(A)) \subset \mathcal{D}'(\Delta(A))$ $(1 \leq p \leq \infty)$ with continuous embedding. Consequently, given $1 \leq p \leq \infty$ and an integer $m \geq 1$ one can define

$$W^{m,p}(\Delta(A)) = \{u | u \in L^p(\Delta(A)), \partial^\alpha u \in L^p(\Delta(A)) \text{ for } |\alpha| \leq m\}$$

where the partial derivatives $\partial^\alpha u$ are computed in the distribution sense on $\Delta(A)$, of course. For $1 \leq p < \infty$ we provide $W^{m,p}(\Delta(A))$ with the norm

$$\|u\|_{W^{m,p}(\Delta(A))} = \left(\sum_{|\alpha| \leq m} \|\partial^\alpha u\|^p_{L^p(\Delta(A))}\right)^{\frac{1}{p}} \quad (u \in W^{m,p}(\Delta(A)))$$

whereas $W^{m,\infty}(\Delta(A))$ is equipped with the norm

$$\|u\|_{W^{m,\infty}(\Delta(A))} = \sum_{|\alpha| \leq m} \|\partial^\alpha u\|_{L^\infty(\Delta(A))} \quad (u \in W^{m,\infty}(\Delta(A))).$$

In either case $W^{m,p}(\Delta(A))$ is a Banach space. In particular $W^{m,2}(\Delta(A))$, again denoted by $H^m(\Delta(A))$, is a Hilbert space. The proof of all that is a classical matter.

However, as will later become apparent, in practice the appropriate space is not the whole $W^{m,p}(\Delta(A))$ but its closed subspace

$$W^{m,p}(\Delta(A))/\mathbb{C} = \left\{ u | u \in W^{m,p}(\Delta(A)), \int_{\Delta(A)} u d\beta = 0 \right\}$$

instead, equipped with the seminorm

$$\|u\|_{W^{m,p}(\Delta(A))/\mathbb{C}} = \left(\sum_{|\alpha|=m} \|\partial^\alpha u\|^p_{L^p(\Delta(A))} \right)^{\frac{1}{p}}$$

with $1 \leq p < \infty$ (the case $p = \infty$ seems to have no interest for the present topic). In the sequel it is assumed that $1 \leq p < \infty$.

Throughout the rest of the book and except where otherwise stated, $W^{m,p}(\Delta(A))/\mathbb{C}$ shall be considered equipped with the above seminorm. Unfortunately, $W^{m,p}(\Delta(A))/\mathbb{C}$ so topologized is in general non-separated and non-complete.

Definition 2.5 We define $W^{m,p}_{\#}(\Delta(A))$ as the separated completion of $W^{m,p}(\Delta(A))/\mathbb{C}$, and J to be the canonical mapping of $W^{m,p}(\Delta(A))/\mathbb{C}$ into $W^{m,p}_{\#}(\Delta(A))$.

We refer the reader to Bourbaki [24, Chap. III, §3, $n^o 7$], Bourbaki [21, Chap. I, §1, $n^o 4$], and Edwards [49, pp. 61–62], for basic notions involved in Definition 2.5.

Remark 2.5 $W^{m,p}_{\#}(\Delta(A))$ is a Banach space and in particular a Hilbert space if $p = 2$. Furthermore, the following classical assertions hold true:

(1) J is linear.
(2) $J(W^{m,p}(\Delta(A))/\mathbb{C})$ is dense in $W^{m,p}_{\#}(\Delta(A))$.
(3) $\|J(v)\|_{W^{m,p}_{\#}(\Delta(A))} = \|v\|_{W^{m,p}(\Delta(A))/\mathbb{C}}$ $(v \in W^{m,p}(\Delta(A))/\mathbb{C})$.
(4) If F is a Banach space and L a continuous linear mapping of $W^{m,p}(\Delta(A))/\mathbb{C}$ into F, then there exists a unique continuous linear mapping L' : $W^{m,p}_{\#}(\Delta(A)) \to F$ such that $L = L' \circ J$.

The preceding remark leads us readily to the following proposition (the verification is an easy exercise left to the reader).

Proposition 2.12 *Let the distribution derivative ∂^α ($|\alpha| = m$) be viewed as a mapping of $W^{m,p}(\Delta(A))/\mathbb{C}$ into $L^p(\Delta(A))$. Then there exists a unique continuous linear mapping still denoted by ∂^α, of $W^{m,p}_{\#}(\Delta(A))$ into $L^p(\Delta(A))$ such that $\partial^\alpha J(v) = \partial^\alpha v$ for all $v \in W^{m,p}(\Delta(A))/\mathbb{C}$. Furthermore,*

$$\|u\|_{W^{m,p}_{\#}(\Delta(A))} = \left(\sum_{|\alpha|=m} \|\partial^\alpha u\|^p_{L^p(\Delta(A))} \right)^{\frac{1}{p}} \quad (u \in W^{m,p}_{\#}(\Delta(A))).$$

2.3 Almost Periodic Homogenization Algebras on $\mathbb{R}^N$

2.3.1 *Some Generalities on Almost Periodic Continuous Functions*

It is worth recalling that $AP(\mathbb{R}^N)$ is a commutative C^*-algebra with identity (Example 1.4). The mean value M generated by the action $\mathcal{H}$ is an extension of the classical mean value for almost periodic functions on $\mathbb{R}^N$ (Examples 1.9 and 1.11). Thus in particular, as $\varepsilon \to 0$ we have $u \circ H_\varepsilon \to M(u)$ in $L^\infty(\mathbb{R}^N)$-weak$*$ for any $u \in AP(\mathbb{R}^N)$.

However, $AP(\mathbb{R}^N)$ is not an H-algebra because it fails to satisfy property (HA1) in Definition 2.1. Contrary to all expectations, the algebra $AP(\mathbb{R}^N)$ nevertheless holds a major place in the present section because the almost periodicity is very likely the most pertinent generalization of the concept of periodicity.

The main purpose of this section is the bright notion of a spectrum. By the spectrum of a $u \in AP(\mathbb{R}^N)$ is meant the set

$$Sp(u) = \{k | k \in \mathbb{R}^N, M(\overline{\gamma}_k u) \neq 0\},$$

where γ_k is the complex exponential function defined before (see (1.18)). The set $Sp(u)$ is at most countable. Indeed, $Sp(u) \subset \bigcup_{n \in \mathbb{N}} Sp(u_n)$, where (u_n) is an ordinary sequence of elements of $T(\mathbb{R}^N)$ (the algebra of trigonometric polynomials on $\mathbb{R}^N$; see (1.19)) such that $u_n \to u$ in $\mathcal{B}(\mathbb{R}^N)$ as $n \to \infty$. In this connection there is the following classical fact: to any $u \in AP(\mathbb{R}^N)$ there is assigned a sequence $(u_n)_{n \in \mathbb{N}}$ such that

$$u_n \in T(\mathbb{R}^N),\ Sp(u_n) \subset Sp(u),\ u_n \to u \text{ in } \mathcal{B}(\mathbb{R}^N) \text{ as } n \to \infty. \tag{2.13}$$

On this point the reader is referred, e.g., to Guichardet [57, Chapter 5] for precise details.

2.3.2 *The H-algebra $AP_{\mathcal{R}}(\mathbb{R}^N)$*

Let $\mathcal{R}$ be a countable subgroup of $\mathbb{R}^N$. Let

$$AP_{\mathcal{R}}(\mathbb{R}^N) = \{u | u \in AP(\mathbb{R}^N), Sp(u) \subset \mathcal{R}\}.$$

Proposition 2.13 *Equipped with the algebra operations in $\mathcal{B}(\mathbb{R}^N)$ and the supremum norm, $AP_{\mathcal{R}}(\mathbb{R}^N)$ is an H-algebra on $\mathbb{R}^N$.*

Proof It is obvious by (2.13) (for $u \in AP(\mathbb{R}^N)$) that $AP_{\mathcal{R}}(\mathbb{R}^N)$ is a closed subalgebra of $\mathcal{B}(\mathbb{R}^N)$. Next, let $T_{\mathcal{R}}(\mathbb{R}^N) = \{u | u \in T(\mathbb{R}^N), Sp(u) \subset \mathcal{R}\}$. This is a dense countable subalgebra of $AP_{\mathcal{R}}(\mathbb{R}^N)$, according to (2.13); so that $AP_{\mathcal{R}}(\mathbb{R}^N)$ is separable. The rest of the required properties stated in Definition 2.1 being evident (see Example 1.11), the proof is complete. ■

Remark 2.6 $AP(\mathbb{R}^N) = \bigcup_{\mathcal{R}} AP_{\mathcal{R}}(\mathbb{R}^N)$, $\mathcal{R}$ ranging over the countable subgroups of $\mathbb{R}^N$.

Thus, $AP_{\mathcal{R}}(\mathbb{R}^N)$ is a commutative C^*-algebra with identity. Throughout the rest of the chapter we shall always consider the countable group $\mathcal{R}$ provided with the discrete topology. Consequently the dual group $\widehat{\mathcal{R}}$ is compact (see, e.g., Larsen [69, p. 122]). We will need the following lemma.

Lemma 2.4 *Let φ be the group homomorphism of $\mathbb{R}^N$ into $\widehat{\mathcal{R}}$ defined on each $y \in \mathbb{R}^N$ by $\langle \varphi(y), k \rangle = \gamma_k(y)$ $(k \in \mathcal{R})$. Then*

(i) φ is continuous.
(ii) $\varphi(\mathbb{R}^N)$ is dense in $\widehat{\mathcal{R}}$.

Proof First of all, it is clear that φ is a homomorphism of the additive group $\mathbb{R}^N$ into the multiplicative group $\widehat{\mathcal{R}}$.

(i) Let us show the continuity of φ. Clearly it is sufficient to show the continuity at ω (the origin). In other words we must check that for any $\varepsilon > 0$ and any finite set $F \subset \mathcal{R}$, there is some $\eta > 0$ such that $|y| \leq \eta$ implies $|\langle \varphi(y), k \rangle - 1| \leq \varepsilon$ for all $k \in F$. But for each such k, γ_k is a continuous homomorphism of $\mathbb{R}^N$ into $\mathbb{U}$ (the unit circle). Therefore there is some $\eta_k > 0$ such that the relation $|y| \leq \eta_k$ implies $|\gamma_k(y) - 1| \leq \varepsilon$. Hence the desired property follows by letting $\eta = \min_{k \in F} \eta_k$.
(ii) We now prove the density property. Let us consider the transposed homomorphism ${}^t\varphi : \mathcal{R} \to \widehat{\mathbb{R}^N}$ defined on each $k \in \mathcal{R}$ by $\langle {}^t\varphi(k), y \rangle = \langle \varphi(y), k \rangle$ $(y \in \mathbb{R}^N)$, where $\widehat{\widehat{\mathcal{R}}}$ is identified with $\mathcal{R}$, thanks to the Pontrjagin duality theorem (see, e.g., Guichardet [57, pp. 45–48]). According to Dieudonné [45, pp. 98–99], we are through if we can show that ${}^t\varphi$ is injective. But ${}^t\varphi(k) = \gamma_k$ $(k \in \mathcal{R})$. Hence if $k \in \mathcal{R}$ and ${}^t\varphi(k) = {}^t\varphi(\omega)$, then $k = \omega$. The density of $\varphi(\mathbb{R}^N)$ in $\widehat{\mathcal{R}}$ follows thereby.

■

The next lemma is also needed in the sequel.

Lemma 2.5 *There is a unique continuous linear mapping $L : AP_{\mathcal{R}}(\mathbb{R}^N) \to \mathcal{C}(\widehat{\mathcal{R}})$ such that*

$$Lu(\varphi(y)) = u(y) \ \ (y \in \mathbb{R}^N) \tag{2.14}$$

for all $u \in AP_{\mathcal{R}}(\mathbb{R}^N)$. Furthermore, L is an isometric C^*-isomorphism of $AP_{\mathcal{R}}(\mathbb{R}^N)$ onto $\mathcal{C}(\widehat{\mathcal{R}})$.

Proof Let $\psi : \mathcal{R} \to \widehat{\widehat{\mathcal{R}}}$ be the canonical homomorphism of $\mathcal{R}$ into $\widehat{\widehat{\mathcal{R}}}$, that is, $\langle \psi(k), \gamma \rangle = \langle \gamma, k \rangle$ for $k \in \mathcal{R}, \gamma \in \widehat{\mathcal{R}}$. It is worth recalling that ψ is actually a group isomorphism, according to the Pontrjagin duality theorem.

Now, let $u \in T_{\mathcal{R}}(\mathbb{R}^N)$, that is,

$$u = \sum_{k \in \mathcal{R}} a_k \gamma_k \quad (a_k \in \mathbb{C}),$$

where the sum is actually a finite sum, the non zero a_k's being finite in number. Put

$$Lu = \sum_{k \in \mathcal{R}} a_k \psi(k).$$

This defines a mapping $L : T_{\mathcal{R}}(\mathbb{R}^N) \to \mathcal{C}(\widehat{\mathcal{R}})$ which is linear and which satisfies (2.14). Hence, by Lemma 2.4 one readily arrives at $\|Lu\|_\infty = \|u\|_\infty$ for all $u \in T_{\mathcal{R}}(\mathbb{R}^N)$. It follows by the density of $T_{\mathcal{R}}(\mathbb{R}^N)$ in $AP_{\mathcal{R}}(\mathbb{R}^N)$ that L extends by continuity to a unique continuous linear mapping, still denoted by L, of $AP_{\mathcal{R}}(\mathbb{R}^N)$ into $\mathcal{C}(\widehat{\mathcal{R}})$, which is isometric and which satisfies (2.14) for all $u \in AP_{\mathcal{R}}(\mathbb{R}^N)$.

Our next purpose is to verify that L is surjective. To this end, let $f \in \mathcal{C}(\widehat{\mathcal{R}})$. Let $\eta > 0$. Thanks to the fact that $\widehat{\widehat{\mathcal{R}}}$ is total in $\mathcal{C}(\widehat{\mathcal{R}})$ (see, e.g., Guichardet [57, p. 38,Corollaire 2.3]), there is a family of complex numbers $(b_\gamma)_{\gamma \in \widehat{\widehat{\mathcal{R}}}}$ with $b_\gamma = 0$ for $\gamma \in \widehat{\widehat{\mathcal{R}}} \setminus F$, F a finite subset of $\widehat{\widehat{\mathcal{R}}}$, such that

$$\left\| \sum_{k \in \mathcal{R}} a_k \psi(k) - f \right\|_\infty \leq \eta \text{ with } a_k = b_{\psi(k)}.$$

We deduce at once that $L(AP_{\mathcal{R}}(\mathbb{R}^N))$ is dense in $\mathcal{C}(\widehat{\mathcal{R}})$. Hence $L(AP_{\mathcal{R}}(\mathbb{R}^N)) = \mathcal{C}(\widehat{\mathcal{R}})$, since L maps $AP_{\mathcal{R}}(\mathbb{R}^N)$ isometrically into $\mathcal{C}(\widehat{\mathcal{R}})$, and so L is surjective. In conclusion, L is an isometric isomorphism of the Banach space $AP_{\mathcal{R}}(\mathbb{R}^N)$ onto the Banach space $\mathcal{C}(\widehat{\mathcal{R}})$.

We shall finally show that L is a C^*-homomorphism. But since L is linear, we only need to check that $L(uv) = L(u)L(v)$ and $\overline{L(u)} = L(\overline{u})$ for $u, v \in AP_{\mathcal{R}}(\mathbb{R}^N)$. However it is an easy exercise to verify that this follows by (2.14) and use of Lemma 2.4. The unicity of L being a direct consequence of Lemma 2.4(ii), this completes the proof. ∎

We are now in a position to prove the following proposition.

Proposition 2.14 *Let $A = AP_{\mathcal{R}}(\mathbb{R}^N)$. The compact space $\Delta(A)$ can be provided with a group operation in such a way that it becomes a compact Abelian group and further, normalized Haar measure on $\Delta(A)$ is just the M-measure, β.*

Proof Let us consider the mapping $l = L \circ \mathcal{G}^{-1}$ of $\mathcal{C}(\Delta(A))$ into $\mathcal{C}(\widehat{\mathcal{R}})$, where L is the mapping in Lemma 2.5 and $\mathcal{G}$ the Gelfand transformation on A. It is clear that l is an isometric C^*-isomorphism of $\mathcal{C}(\Delta(A))$ onto $\mathcal{C}(\widehat{\mathcal{R}})$. Therefore, according to a classical result (see, e.g., Larsen [69, p. 90]), there exists a homeomorphism $h : \widehat{\mathcal{R}} \to \Delta(A)$ such that $l(f)(\gamma) = f(h(\gamma))$ $(\gamma \in \widehat{\mathcal{R}})$ for all $f \in \mathcal{C}(\Delta(A))$.

Now, if $s, s' \in \Delta(A)$, define $s + s' = h(\gamma\gamma')$, where $\gamma = h^{-1}(s)$ and $\gamma' = h^{-1}(s')$. This gives a binary operation $+$ that makes $\Delta(A)$ into an Abelian group additively written: the zero element is $0 = h(e)$, where e is the neutral element of $\widehat{\mathcal{R}}$, and for $s \in \Delta(A)$, we put $-s = h(\gamma^{-1})$ where $\gamma = h^{-1}(s)$. In other words, the group structure on $\widehat{\mathcal{R}}$ is carried over to $\Delta(A)$ by means of h.

Let us check that the mapping $(s, s') \mapsto s - s'$ sends continuously $\Delta(A) \times \Delta(A)$ into $\Delta(A)$. So, let $(a, b) \in \Delta(A) \times \Delta(A)$. Let V be a neighbourhood of $a - b$ in $\Delta(A)$. Let $\mathcal{V} = h^{-1}(V)$, $a_1 = h^{-1}(a)$, $b_1 = h^{-1}(b)$. Clearly $\mathcal{V}$ is a neighbourhood of $a_1 b_1^{-1}$ in $\widehat{\mathcal{R}}$. Since the mapping $(\sigma, \rho) \mapsto \sigma\rho^{-1}$ sends continuously $\widehat{\mathcal{R}} \times \widehat{\mathcal{R}}$ into $\widehat{\mathcal{R}}$, there is a neighbourhood $\mathcal{U}$ of (a_1, b_1) in $\widehat{\mathcal{R}} \times \widehat{\mathcal{R}}$ such that $\sigma\rho^{-1} \in \mathcal{V}$ for $(\sigma, \rho) \in \mathcal{U}$. But there certainly exist a neighbourhood $\mathcal{U}_{a_1}$ of a_1 and a neighbourhood $\mathcal{U}_{b_1}$ of b_1 such that $\mathcal{U}_{a_1} \times \mathcal{U}_{b_1} \subset \mathcal{U}$. Thus, $\sigma\rho^{-1} \in \mathcal{V}$ for all $(\sigma, \rho) \in \mathcal{U}_{a_1} \times \mathcal{U}_{b_1}$. One deduces at once that $s - r \in V$ for $s \in U_a = h(\mathcal{U}_{a_1})$ and $r \in U_b = h(\mathcal{U}_{b_1})$. Since $U = U_a \times U_b$ is a neighbourhood of (a, b), this shows the desired continuity. Consequently, $\Delta(A)$ thus equipped is a compact Abelian group and h is a group isomorphism of $\widehat{\mathcal{R}}$ onto $\Delta(A)$.

The next point is to show that normalized Haar measure on $\Delta(A)$ is just the M-measure β for A. It amounts to verifying that β is translation invariant, that is, $\beta(\tau_s f) = \beta(f)$ for $f \in \mathcal{C}(\Delta(A))$, $s \in \Delta(A)$, where it is well to recall that $\tau_s f$ stands for the function $r \mapsto f(r - s)$ of $\Delta(A)$ into $\mathbb{C}$. To accomplish this we need to verify the equality

$$\mathcal{G}(\tau_a u) = \tau_{j(a)} \mathcal{G}(u) \quad (u \in A, a \in \mathbb{R}^N), \tag{2.15}$$

where the mapping $j : \mathbb{R}^N \to \Delta(A)$ is defined by $j(y) = \delta_y$ $(y \in \mathbb{R}^N)$. However, let us first show that j is a group homomorphism. It is enough to verify that $j = h \circ \varphi$. Let $y \in \mathbb{R}^N$. Given $u \in A$, put for convenience $\widehat{u} = \mathcal{G}(u)$. Then $\widehat{u}(h(\varphi(y))) = l\widehat{u}(\varphi(y)) = Lu(\varphi(y)) = u(y)$ (Lemma 2.5) $= \widehat{u}(j(y))$. Hence $f(h(\varphi(y))) = f(j(y))$ and that for all $f \in \mathcal{C}(\Delta(A))$. Since $\mathcal{C}(\Delta(A))$ separates the points of $\Delta(A)$ and on the other hand y is arbitrary, the claimed statement is proven. Consequently

$$(\tau_{j(a)} \mathcal{G}(u))(j(y)) = \mathcal{G}(\tau_a u)(j(y))$$

for any $u \in A$, $a \in \mathbb{R}^N$. Hence (2.15) follows immediately by the density of $j(\mathbb{R}^N)$ in $\Delta(A)$ (see Lemma 1.1).

Finally, let $f \in \mathcal{C}(\Delta(A))$. Put $u = \mathcal{G}^{-1}(f) \in A$. For arbitrary given $a \in \mathbb{R}^N$, one has by (2.15)

$$\beta(\tau_{j(a)} \mathcal{G}(u)) = \beta(\mathcal{G}(\tau_a u)) = M(\tau_a u) = M(u) = \beta(\mathcal{G}(u)).$$

Thus $\beta(\tau_s f) = \beta(f)$ for all $s \in j(\mathbb{R}^N)$. Hence the claimed translation invariance follows by the density of $j(\mathbb{R}^N)$ in $\Delta(A)$ and use of the continuity of the mapping $s \mapsto \tau_s f$ of $\Delta(A)$ into $\mathcal{C}(\Delta(A))$. This completes the proof. ■

Remark 2.7 As seen in proving the preceding proposition, the groups $\widehat{\mathcal{R}}$ and $\Delta(A)$ are isomorphic. Consequently, if need be, one may write $\Delta(A) = \widehat{\mathcal{R}}$, thus generalizing the identity (1.1) (indeed, for $\mathcal{R} = \mathbb{Z}^N$ we have $AP_{\mathcal{R}}(\mathbb{R}^N) = \mathcal{C}_{per}(Y)$ and $\widehat{\mathcal{R}} = \mathbb{T}^N$).

Now, let us see what the dual of the compact Abelian group $\Delta(A)$ looks like.

Proposition 2.15 *Let $A = AP_{\mathcal{R}}(\mathbb{R}^N)$. We have*

$$\widehat{\Delta(A)} = \{\widehat{\gamma_k} | k \in \mathcal{R}\}$$

with $\widehat{\gamma_k} = \mathcal{G}(\gamma_k)$.

Proof To begin, let us fix freely some $k \in \mathcal{R}$, and show that $\widehat{\gamma_k} \in \widehat{\Delta(A)}$. Firstly, it is clear that $\widehat{\gamma_k}$ is continuous on $\Delta(A)$. Secondly, we have $\widehat{\gamma_k}(j(y)) = \gamma_k(y)$ for $y \in \mathbb{R}^N$, where j is the group homomorphism of $\mathbb{R}^N$ into $\Delta(A)$ given by $j(y) = \delta_y$ (see the proof of Proposition 2.14). Hence, by the density of $j(\mathbb{R}^N)$ in $\Delta(A)$, it follows that $|\widehat{\gamma_k}(s)| = 1$ for $s \in \Delta(A)$. So, $\widehat{\gamma_k}$ maps continuously $\Delta(A)$ into the unit circle. Therefore, it remains to verify that $\widehat{\gamma_k}$ is a group homomorphism of $\Delta(A)$ into the unit circle. This is straightforward. Indeed, by recalling that the mapping j is a homomorphism of the additive group $\mathbb{R}^N$ into the additive group $\Delta(A)$, we see that $\widehat{\gamma_k}(j(x)+j(y)) = \widehat{\gamma_k}(j(x+y)) = \langle j(x+y), \gamma_k\rangle = \gamma_k(x+y) = \gamma_k(x)\gamma_k(y) = \widehat{\gamma_k}(j(x))\widehat{\gamma_k}(j(y))$ for $x, y \in \mathbb{R}^N$. Hence the desired property follows by the density of $j(\mathbb{R}^N)$ in $\Delta(A)$.

To complete the proof we must show that any continuous character of $\Delta(A)$ is the Gelfand transform of some γ_k with $k \in \mathcal{R}$. So, let $g \in \widehat{\Delta(A)}$. Thus, $g = \mathcal{G}(f)$ with $f \in A$, and it is immediate that f is a continuous character of the additive group $\mathbb{R}^N$. Hence there is some $k \in \mathbb{R}^N$ such that $f = \gamma_k$, and k necessarily lies in $\mathcal{R}$, since $\gamma_k \in A = AP_{\mathcal{R}}(\mathbb{R}^N)$. ■

2.3.3 *The Spaces $H^m(\Delta(A))$*

Let $A = AP_{\mathcal{R}}(\mathbb{R}^N)$ as before. We start by noting that, according to Proposition 2.15, the dual group of $\Delta(A)$ may be identified with $\mathcal{R}$, so that one is justified in viewing the Fourier transform of a function $u \in L^1(\Delta(A))$ as a mapping

$$k \mapsto \mathcal{F}_k(u) = \int u(s)\overline{\widehat{\gamma_k}(s)}d\beta(s)$$

of $\mathcal{R}$ into $\mathbb{C}$. The complex numbers $\mathcal{F}_k(u)$ are the so-called *Fourier coefficients* of $u \in L^1(\Delta(A))$.

By recalling that the dual of the compact Abelian group $\Delta(A)$ is an orthonormal basis of the Hilbert space $L^2(\Delta(A))$ (see Guichardet [57, p. 56]), we have, for $u \in L^2(\Delta(A))$,

$$u = \sum_{k \in \mathcal{R}} \mathcal{F}_k(u)\widehat{\gamma}_k, \tag{2.16}$$

where the family on the right is summable in the $L^2(\Delta(A))$ norm. Hence

$$\|u\|^2_{L^2(\Delta(A))} = \sum_{k \in \mathcal{R}} |\mathcal{F}_k(u)|^2.$$

On the other hand, let $k = (k_1, \cdots, k_N) \in \mathcal{R}$ and $\alpha = (\alpha_1, \cdots, \alpha_N) \in \mathbb{N}^N$. We have $\gamma_k \in A^\infty$ and $D^\alpha \gamma_k = (2\pi i)^{|\alpha|} k^\alpha \gamma_k$ where $k^\alpha = \prod_{1 \leq j \leq N} k_j^{\alpha_j}$. Thus,

$$\partial^\alpha \widehat{\gamma}_k = (2\pi i)^{|\alpha|} k^\alpha \widehat{\gamma}_k. \tag{2.17}$$

Hence, if $u \in H^m(\Delta(A)) = W^{m,2}(\Delta(A))$, integer $m \geq 1$, then

$$\mathcal{F}_k(\partial^\alpha u) = (2\pi i)^{|\alpha|} k^\alpha \mathcal{F}_k(u) \quad (|\alpha| \leq m), \tag{2.18}$$

according to the equality

$$\int_{\Delta(A)} \overline{\widehat{\gamma}_k(s)} \partial^\alpha u(s) d\beta(s) = (-1)^{|\alpha|} \int_{\Delta(A)} u(s) \overline{\partial^\alpha \widehat{\gamma}_k(s)} d\beta(s).$$

Having made these preliminaries, we turn now to the proof of the following

Proposition 2.16 *Let (integer) $m \geq 1$. The following statements hold.*

(i) $\|\cdot\|_{H^m(\Delta(A))/\mathbb{C}}$ *is a norm on* $H^m(\Delta(A))/\mathbb{C}$.
(ii) $\mathcal{D}(\Delta(A))$ *is dense in* $H^m(\Delta(A))$.

Proof

(i) We only need to show that the seminorm under consideration is definite. Let $u \in H^m(\Delta(A))/\mathbb{C}$ with $\|u\|_{H^m(\Delta(A))/\mathbb{C}} = 0$. Then $\partial^\alpha u = 0$ for all $\alpha \in \mathbb{N}^N$ with $|\alpha| = m$. Fix freely any $k = (k_1, \cdots, k_N)$ with $k \neq \omega$ (the origin in $\mathbb{R}^N$). Consider an integer $1 \leq \nu \leq N$ such that $k_\nu \neq 0$, and let $\alpha = (\alpha_j)$ with $\alpha_\nu = m$, $\alpha_j = 0$ if $j \neq \nu$. Then $k^\alpha = k_\nu^m \neq 0$. Hence $\mathcal{F}_k(u) = 0$, according to (2.18). Therefore $u = 0$, which proves (i).

(ii) We introduce a sequence of nonempty finite sets $\mathcal{R}_n \subset \mathcal{R}$ (integers $n \geq 1$) such that $\mathcal{R}_n \subset \mathcal{R}_{n+1}$, $\mathcal{R} = \bigcup_{n\geq 1} \mathcal{R}_n$. Let $u \in H^m(\Delta(A))$. Set

$$u_n = \sum_{k\in\mathcal{R}_n} \mathcal{F}_k(u)\widehat{\gamma}_k \quad \text{(integers } n \geq 1).$$

We have $u_n \in \mathcal{D}(\Delta(A))$ and further, thanks to (2.17) and (2.18),

$$\partial^\alpha u_n = \sum_{k\in\mathcal{R}_n} \mathcal{F}_k(\partial^\alpha u)\widehat{\gamma}_k \quad (|\alpha| \leq m).$$

Therefore, by (2.16) (with $\partial^\alpha u$ in place of u), we see that $u_n \to u$ in $H^m(\Delta(A))$ as $n \to \infty$. This shows (ii).

■

The preceding proposition has one corollary worth mentioning.

Corollary 2.8 *The following statements hold.*

(i) $\mathcal{D}(\Delta(A))/\mathbb{C}$ *is dense in* $H^m(\Delta(A))/\mathbb{C}$.
(ii) $J(\mathcal{D}(\Delta(A))/\mathbb{C})$ *is dense in* $H^m_\#(\Delta(A)) = W^{m,2}_\#(\Delta(A))$.
(iii) For $u \in H^m_\#(\Delta(A))$, *we have* $\int_{\Delta(A)} \partial^\alpha u d\beta = 0 \ \ (|\alpha| = m)$.

Proof Part (i) is a direct consequence of Proposition 2.16(ii). We deduce at once (ii) by parts (1)–(3) of Remark 2.5. Finally, (iii) follows from Propositions 2.9 and 2.12 and use of part (ii) above. ■

Thus, $H^m(\Delta(A))/\mathbb{C}$ with the norm $\|\cdot\|_{H^m(\Delta(A))/\mathbb{C}}$ is a separated preHilbert space but not in general a Hilbert space, as will be illustrated in a while. However, this requires a preliminary result.

Lemma 2.6 *The following statements are equivalent:*

(i) There is a constant $c > 0$ *such that*

$$\begin{cases} \|u\|_{L^2(\Delta(A))} \leq c\|u\|_{H^1(\Delta(A))/\mathbb{C}} \\ \textit{for all } u \in H^1(\Delta(A))/\mathbb{C}. \end{cases} \tag{2.19}$$

(ii) $\mathcal{R}$ *is a discrete subgroup of* $\mathbb{R}^N$.

Proof Let $u \in H^1(\Delta(A))/\mathbb{C}$. Clearly

$$\|u\|^2_{L^2(\Delta(A))} = \sum_{k\in\mathcal{R}\backslash\{\omega\}} |\mathcal{F}_k(u)|^2$$

and

$$\|u\|^2_{H^1(\Delta(A))/\mathbb{C}} = 4\pi^2 \sum_{k\in\mathcal{R}\setminus\{\omega\}} |k|^2 |\mathcal{F}_k(u)|^2,$$

where $|k|$ denotes the Euclidian norm of k. This being so, suppose (i) holds. Then

$$\sum_{k\in\mathcal{R}\setminus\{\omega\}} |\mathcal{F}_k(u)|^2 \leq 4\pi^2 c^2 \sum_{k\in\mathcal{R}\setminus\{\omega\}} |k|^2 |\mathcal{F}_k(u)|^2 \tag{2.20}$$

and that for all $u \in H^1(\Delta(A))/\mathbb{C}$. By taking in particular $u = \widehat{\gamma}_l$ for arbitrary $l \in \mathcal{R} \setminus \{\omega\}$ we get

$$|k| \geq r > 0 \;\; (k \in \mathcal{R} \setminus \{\omega\}), \tag{2.21}$$

where $r = \frac{1}{2\pi c}$. This means precisely that $\mathcal{R}$ is a discrete subgroup of $\mathbb{R}^N$, and so (ii) is established. Conversely, let (ii) be satisfied. In other words, suppose a constant $r > 0$ exists such that (2.21) holds. Immediately we see that if $u \in H^1(\Delta(A))/\mathbb{C}$ then (2.20) holds with $c = \frac{1}{2\pi r}$, whence (2.19) follows at once. Consequently, (ii) implies (i). ■

We are now in a position to elucidate our allegation.

Proposition 2.17 *Let $H^1(\Delta(A))/\mathbb{C}$ be equipped with the norm $\|\cdot\|_{H^1(\Delta(A))/\mathbb{C}}$. Then, $H^1(\Delta(A))/\mathbb{C}$ is complete (thus, is a Hilbert space) if and only if $\mathcal{R}$ is a discrete subgroup of $\mathbb{R}^N$.*

Proof Since $H^1(\Delta(A))/\mathbb{C}$ is closed in $H^1(\Delta(A))$, appeal to the open mapping theorem (see, e.g., Brézis [27, p. 19]) reveals that the space $H^1(\Delta(A))/\mathbb{C}$ is complete if and only if it admits $\|\cdot\|_{H^1(\Delta(A))}$ and $\|\cdot\|_{H^1(\Delta(A))/\mathbb{C}}$ as two equivalent norms. But this happens if and only if condition (i) of Lemma 2.6 is fulfilled. Hence the proposition follows by the same lemma. ■

Thus, $H^1(\Delta(A))/\mathbb{C}$ is not complete if for example $\mathcal{R} = \mathbb{Q}^N$ ($\mathbb{Q}$ the rationals). The preceding proposition points out the fact that in general, as is usual in analysis, in number of cases one cannot reasonably do without passing to the completion.

2.3.4 A Fundamental Lemma

We conclude this section with a lemma that will prove crucial for almost periodic homogenization and beyond. Let $A = AP_{\mathcal{R}}(\mathbb{R}^N)$ be as before.

Let

$$V_{\mathcal{R}} = \mathcal{G}(\mathcal{V}_{\mathcal{R}}) \text{ with } \mathcal{V}_{\mathcal{R}} = \{\boldsymbol{\Phi} | \boldsymbol{\Phi} = (\Phi_j) \in (A^\infty)^N, \operatorname{div} \boldsymbol{\Phi} = 0\},$$

where $\mathcal{G}(\mathbf{\Phi}) = (\mathcal{G}(\Phi_1), \cdots, \mathcal{G}(\Phi_N))$ for $\mathbf{\Phi} = (\Phi_j) \in \mathcal{V}_{\mathcal{R}}$. Note that

$$V_{\mathcal{R}} = \{\mathbf{u}|\mathbf{u} = (u_j) \in \mathcal{D}(\Delta(A))^N, \widehat{\operatorname{div}}\,\mathbf{u} = 0\},$$

where $\widehat{\operatorname{div}}\,\mathbf{u} = \sum_{j=1}^{N} \partial_j u_j$. Finally, for $\mathbf{u} = (u_j)$ and $\mathbf{v} = (v_j)$ in $L^2(\Delta(A))^N$, we denote by $\mathbf{u} \cdot \mathbf{v}$ the function in $L^1(\Delta(A))$ given by

$$\mathbf{u} \cdot \mathbf{v} = \sum_{j=1}^{N} u_j v_j.$$

The result in which we are concerned states as follows.

Lemma 2.7 *The following two statements are equivalent for* $\boldsymbol{f} = (f_j) \in L^2(\Delta(A))^N$.

(i) For any $\boldsymbol{u} \in V_{\mathcal{R}}$, *we have*

$$\int_{\Delta(A)} \boldsymbol{f} \cdot \boldsymbol{u}\, d\beta = 0. \tag{2.22}$$

(ii) There is a unique $p \in H^1_\#(\Delta(A))$ *such that*

$$\partial_j p = f_j \ \ (1 \le j \le N). \tag{2.23}$$

Proof We begin by proving in two steps that (i) implies (ii). So we assume (i).

Step 1. The aim here is to show that

$$\mathcal{F}_\omega(f_j) \equiv \int_{\Delta(A)} f_j\, d\beta = 0 \ \ (1 \le j \le N) \tag{2.24}$$

and

$$k_i \mathcal{F}_k(f_j) - k_j \mathcal{F}_k(f_i) = 0 \ (1 \le i, j \le N) \text{ for } k = (k_l) \in \mathcal{R}. \tag{2.25}$$

For fixed index $1 \le \nu \le N$, take in (2.22) the particular $\mathbf{u} = (u_j)$ with $u_\nu = 1, u_j = 0$ if $j \ne \nu$. Hence (2.24) follows immediately by the arbitrariness of ν. As regards (2.25), fix any arbitrary two indices $1 \le i, j \le N$. Let $k \in \mathcal{R}$, and choose $\mathbf{u} = (u_l)$ in (2.22) such that $u_i = \overline{\partial_j \widehat{\gamma_k}}, u_j = -\overline{\partial_i \widehat{\gamma_k}}, u_l = 0$ if $l \ne i$ and $l \ne j$. Hence (2.25) follows.

Step 2. The goal is to prove the existence of a unique p in $H^1_\#(\Delta(A))$ satisfying (2.23). The unicity being evident, we shall only check the existence. We begin by considering the case of an $\mathbf{f}$ of the form

$$f_j = \sum_{k\in R} \mathcal{F}_k(f_j)\widehat{\gamma_k} \ \ (1 \leq j \leq N),$$

where R is an arbitrary given finite subset of $\mathcal{R}$. We next seek p (solution of (2.23)) of the form

$$p = J\left(\sum_{k\in R} a_k\widehat{\gamma_k}\right), a_k \in \mathbb{C}. \tag{2.26}$$

Then the problem reduces to finding a family of complex numbers $(a_k)_{k\in R}$ such that

$$2\pi i k_j a_k = \mathcal{F}_k(f_j) \ \ (k \in R, 1 \leq j \leq N).$$

This is a system of $N|R|$ ($|R|$ the cardinality of R) algebraic equations whose solution is possible thanks to the compatibility relations (2.24) and (2.25): Let us consider the mapping $k \mapsto a_k$ of R into $\mathbb{C}$ given by $a_\omega = 0$ if $\omega \in R$, $a_k = \frac{\mathcal{F}_k(f_\nu)}{2\pi i k_\nu}$ if $k \in R \setminus \{\omega\}$, where k_ν is any nonzero coordinate of k. The function p in (2.26) thus constructed satisfies (2.23).

Now, we consider the general case where we know merely that $\mathbf{f}$ lies in $L^2(\Delta(A))^N$ and satisfies (2.22) for all $\mathbf{u} \in V_\mathcal{R}$. As we did earlier in proving Proposition 2.16, let us consider a covering of $\mathcal{R}$ consisting of an increasing sequence $(\mathcal{R}_n)_{n\geq 1}$ of finite subsets. For each integer $n \geq 1$, let

$$\mathbf{f}_n = (f_{n,j}) \in L^2(\Delta(A))^N, \ f_{n,j} = \sum_{k\in\mathcal{R}_n} \mathcal{F}_k(f_j)\widehat{\gamma_k} \ \ (1 \leq j \leq N).$$

It is easy to verify that $\mathbf{f}_n$ satisfies conditions similar to (2.24) and (2.25), so that a simple copying of what we did before with R yields a function $p_n \in H^1_\#(\Delta(A))$ such that

$$\partial_j p_n = f_{n,j} \ \ (1 \leq j \leq N)$$

and that for each integer $n \geq 1$. On the other hand it is clear that $f_{n,j} \to f_j$ in $L^2(\Delta(A))$ as $n \to \infty$ $(1 \leq j \leq N)$. It follows that (p_n) is a Cauchy sequence in $H^1_\#(\Delta(A))$. Hence $p_n \to p$ in $H^1_\#(\Delta(A))$ as $n \to \infty$, and it is clear that p satisfies the system (2.23) of which it is the unique solution. This completes step 2.

Finally, we assume (ii) and try to prove (i). In view of Corollary 2.8(ii) there is a sequence of functions p_n in $\mathcal{D}(\Delta(A))/\mathbb{C}$ such that $J(p_n) \to p$ in $H^1_\#(\Delta(A))$ as $n \to \infty$. Therefore

$$\sum_{j=1}^{N} \int_{\Delta(A)} u_j \partial_j p_n \, d\beta \to \sum_{j=1}^{N} \int_{\Delta(A)} u_j \partial_j p \, d\beta \tag{2.27}$$

as $n \to \infty$, and that for all $\mathbf{u} = (u_j)_{1 \leq j \leq N} \in L^2(\Delta(A))^N$. Hence (i) follows by two facts: (1) The right-hand side of (2.27) is equal to $\int_{\Delta(A)} \mathbf{f} \cdot \mathbf{u} \, d\beta$. (2) As regards the left side we have

$$\sum_{j=1}^{N} \int_{\Delta(A)} u_j \partial_j p_n \, d\beta = -\int_{\Delta(A)} p_n \widehat{\operatorname{div}} \, \mathbf{u} \, d\beta$$

for any $\mathbf{u} \in \mathcal{D}(\Delta(A))^N$. Hence (i) follows readily. ■

2.4 Stepanoff Almost Periodic Functions

Stepanoff almost periodic functions are to $AP(\mathbb{R}^N)$ as $L^p(\mathbb{R}^N)$ functions are to $\mathcal{K}(\mathbb{R}^N)$. Before we can say more, however, let us introduce the appropriate underlying function spaces.

2.4.1 The Amalgam Space $(L^p, l^\infty)(\mathbb{R}^N)$

Let $1 \leq p < \infty$, $Y = (-\frac{1}{2}, \frac{1}{2})^N$. We define $(L^p, l^\infty)(\mathbb{R}^N)$ to be the space of all $u \in L^p_{loc}(\mathbb{R}^N)$ such that

$$\|u\|_{p,\infty} = \sup_{k \in \mathbb{Z}^N} \left[\int_{k+Y} |u(y)|^p dy \right]^{\frac{1}{p}} < \infty.$$

This is a vector subspace of $L^p_{loc}(\mathbb{R}^N)$, and $\|\cdot\|_{p,\infty}$ is a norm on $(L^p, l^\infty)(\mathbb{R}^N)$ that makes it into a Banach space (this can be proved by the same argument mutatis mutandis as used in proving Lemma 2.3; the details are left to the reader).

The Banach space $(L^p, l^\infty)(\mathbb{R}^N)$ is called the *amalgam* of L^p and l^∞ on $\mathbb{R}^N$. For further details on amalgams we refer the interested reader to Bertrandias et al. [16], Fournier and Stewart [53].

For further needs we wish to point out a useful characterization of functions in $(L^p, l^\infty)(\mathbb{R}^N)$. This requires a simple lemma that will be stated soon after a preliminary notation.

For any compact set $K \subset \mathbb{R}^N$, we denote by $j_1(K)$ the (finite) set of all those $k \in \mathbb{Z}^N$ for which $k + \overline{Y}$ intersects K. It is worth observing that K is then contained in the union of all $k+\overline{Y}$ as k ranges over $j_1(K)$. Finally, we will set $j_1(a) = j_1(\{a\})$, and denote the cardinality of any finite set S by $|S|$.

Lemma 2.8 *Let K be a compact set in $\mathbb{R}^N$. There exists a constant $c = c(K) > 0$ such that $|j_1(a + K)| \leq c$ for all $a \in \mathbb{R}^N$.*

Proof Let $a \in \mathbb{R}^N$. Fix freely any $k \in j_1(a + K)$. Let $x \in (a + K) \cap (k + \overline{Y})$. One has $x = a + z$ with $z \in K \cap (k - a + \overline{Y})$. Now, consider some $k_a \in j_1(a)$, i.e., $a \in k_a + \overline{Y}$. Clearly $z \in k - k_a + \overline{Y} + \overline{Y}$. Hence it follows $z \in k - k_a + k' + \overline{Y}$, where $k' \in j_1(\overline{Y} + \overline{Y})$. Finally, z lies in the set $K \cap (k - k_a + k' + \overline{Y})$ which is therefore nonempty. It amounts to saying that $k - k_a + k'$ lies in $j_1(K)$. Consequently $k = k_a - k' + l$ with $k_a \in j_1(a), -k' \in j_1(\overline{Y} + \overline{Y}), l \in j_1(K)$. Hence, since k is arbitrary, it follows that $j_1(a + K) \subset j_1(a) + j_1(\overline{Y} + \overline{Y}) + j_1(K)$. By the surjectivity of the mapping $(h, k, l) \mapsto h + k + l$ of $j_1(a) \times j_1(\overline{Y} + \overline{Y}) \times j_1(K)$ into $j_1(a) + j_1(\overline{Y} + \overline{Y}) + j_1(K)$, we deduce that $|j_1(a+K)| \leq |j_1(a)||j_1(\overline{Y}+\overline{Y})||j_1(K)|$. But $|j_1(a)| \leq 2^N$. Therefore the lemma follows. ■

Thanks to the preceding lemma we are now in a position to prove the following proposition.

Proposition 2.18 *The following two statements are equivalent for a function $u \in L^p_{loc}(\mathbb{R}^N)$ $(1 \leq p < \infty)$:*

(i) $u \in (L^p, l^\infty)(\mathbb{R}^N)$.
(ii) The set of all translates $\tau_a u$ $(a \in \mathbb{R}^N)$ is bounded in $L^p_{loc}(\mathbb{R}^N)$.

Proof That (ii) implies (i) is evident. Conversely, assuming (i), let K be any given compact set in $\mathbb{R}^N$. For any arbitrary $a \in \mathbb{R}^N$ we have $a + K \subset \bigcup_{k \in j_1(a+K)} (k + \overline{Y})$. Therefore

$$\int_K |u(y + a)|^p dy = \int_{a+K} |u(y)|^p dy \leq \sum_{k \in j_1(a+K)} \int_{k+Y} |u(y)|^p dy$$
$$\leq |j_1(a + K)| \|u\|^p_{p,\infty}.$$

Hence (ii) follows by Lemma 2.8. ■

This proposition has one corollary of great interest. To this end, let

$$N_{p,\infty}(u) = \sup_{a \in \mathbb{R}^N} \left[\int_Y |u(y + a)|^p dy \right]^{\frac{1}{p}}$$

for $u \in (L^p, l^\infty)(\mathbb{R}^N)$, which defines a norm $N_{p,\infty}$ on $(L^p, l^\infty)(\mathbb{R}^N)$.

Corollary 2.9 *The following hold.*

(i) The norms $\|\cdot\|_{p,\infty}$ and $N_{p,\infty}$ are equivalent.
(ii) For any given $a \in \mathbb{R}^N$, the translation operator $u \mapsto \tau_a u$ sends continuously $(L^p, l^\infty)(\mathbb{R}^N)$ into itself.

Proof

(i) We have

$$cN_{p,\infty}(u) \leq \|u\|_{p,\infty} \leq N_{p,\infty}(u) \tag{2.28}$$

for all $u \in (L^p, l^\infty)(\mathbb{R}^N)$, where the constant $c > 0$ depends only on Y. Indeed, the left inequality follows by Proposition 2.18 (see especially its proof) whereas the right one is trivial.
(ii) Proposition 2.18 implies directly that the translation τ_a maps the space $(L^p, l^\infty)(\mathbb{R}^N)$ into itself. Finally, the continuity follows easily by (2.28).

■

We will conclude the present section by drawing attention to one fundamental result established earlier in another context and which was still under an implicit form till now.

Proposition 2.19 *Suppose the action $\mathcal{H}$ on $\mathbb{R}^N$ satisfies hypothesis (C) of Chap. 1 (see Sect. 1.2.5). Then $(L^p, l^\infty)(\mathbb{R}^N) \subset \Xi^p(\mathbb{R}^N)$ with continuous embedding.*

Proof This is nothing else than Lemma 1.9. ■

2.4.2 The Spaces $L^p_{AP}(\mathbb{R}^N)$

Definition 2.6 Let $1 \leq p < \infty$. A function $u \in L^p_{loc}(\mathbb{R}^N)$ is said to be Stepanoff almost periodic if $u \in (L^p, l^\infty)(\mathbb{R}^N)$ and further if the set $\mathfrak{T}(u) = \{\tau_a u | a \in \mathbb{R}^N\}$ has compact closure in $(L^p, l^\infty)(\mathbb{R}^N)$.

We will denote by $L^p_{AP}(\mathbb{R}^N)$ the set of all functions $u \in L^p_{loc}(\mathbb{R}^N)$ that are Stepanoff almost periodic. $L^p_{AP}(\mathbb{R}^N)$ is a closed vector subspace of $(L^p, l^\infty)(\mathbb{R}^N)$. Indeed, as is classical (see, e.g., Bourbaki [24, II, pp. 29–30]), to say that the set $\mathfrak{T}(u)$ has compact closure in $(L^p, l^\infty)(\mathbb{R}^N)$ amounts to saying that it has the following property: To each $\eta > 0$ there is assigned some finite set $F \subset \mathbb{R}^N$ in such a manner that for any arbitrary $x \in \mathbb{R}^N$, there is some $a \in F$ such that $\|\tau_x u - \tau_a u\|_{p,\infty} < \eta$. Hence the claimed statement follows.

We provide $L^p_{AP}(\mathbb{R}^N)$ with the norm $\|\cdot\|_{p,\infty}$, which makes it into a Banach space.

Proposition 2.20 *The following assertions hold true.*

(i) *For each $u \in L^p_{AP}(\mathbb{R}^N)$, the mapping $x \mapsto \tau_x u$ sends continuously $\mathbb{R}^N$ into $(L^p, l^\infty)(\mathbb{R}^N)$.*
(ii) *$AP(\mathbb{R}^N)$ is dense in $L^p_{AP}(\mathbb{R}^N)$.*

Proof

(i) Given $u \in L^p_{AP}(\mathbb{R}^N)$, the aim is to check that if $(x_n)_{n\geq 1}$ is any arbitrary sequence of points in $\mathbb{R}^N$ that converges to $a \in \mathbb{R}^N$ as $n \to \infty$, then the sequence $(\tau_{x_n} u)_{n\geq 1}$ converges to $\tau_a u$ in $(L^p, l^\infty)(\mathbb{R}^N)$. But this follows readily by the Bolzano-Weierstrass theorem for the metrizable compact space $\overline{\mathfrak{T}(u)}$ (the closure of $\mathfrak{T}(u)$ in $(L^p, l^\infty)(\mathbb{R}^N)$) and use of two arguments: (1) The Banach space $(L^p, l^\infty)(\mathbb{R}^N)$ is continuously embedded in $L^p_{loc}(\mathbb{R}^N)$. (2) The mapping $x \mapsto \tau_x u$ sends continuously $\mathbb{R}^N$ into $L^p_{loc}(\mathbb{R}^N)$.
(ii) First of all, it goes without saying that $AP(\mathbb{R}^N)$ is a vector subspace of $L^p_{AP}(\mathbb{R}^N)$. Now, let $(\theta_n)_{n\geq 1}$ be a mollifier on $\mathbb{R}^N$ with $Supp\theta_n \subset \frac{1}{n}\overline{B}_N$, where $\overline{B}_N$ denotes the closed unit ball in $\mathbb{R}^N$. It is worth perhaps recalling that $\theta_n \in \mathcal{D}(\mathbb{R}^N) = \mathcal{C}^\infty_0(\mathbb{R}^N)$ with $\theta_n \geq 0$, $\int \theta_n(y)dy = 1$.

Let $u \in L^p_{AP}(\mathbb{R}^N)$. We have $u * \theta_n \in \mathcal{C}^\infty(\mathbb{R}^N)$ (a classical property) and, for $y \in \mathbb{R}^N$,

$$|(u * \theta_n)(y)| = \left| \int \theta_n(x)u(y-x)dx \right| \leq \|\theta_n\|_{L^{p'}} \left(\int_{y+\overline{B}_N} |u(x)|^p dx \right)^{\frac{1}{p}}$$

where $\frac{1}{p} + \frac{1}{p'} = 1$. By the same operation (based on Lemma 2.8) as made in proving Proposition 2.18 it follows

$$|(u * \theta_n)(y)| \leq c\|\theta_n\|_{L^{p'}(\mathbb{R}^N)}\|u\|_{p,\infty}$$

for all $y \in \mathbb{R}^N$, where the positive constant c depends only on B_N. Thus $u * \theta_n \in \mathcal{B}(\mathbb{R}^N)$ and moreover the mapping $u \mapsto u * \theta_n$ sends continuously $(L^p, l^\infty)(\mathbb{R}^N)$ into $\mathcal{B}(\mathbb{R}^N)$. Hence, by the property $\tau_x(u * \theta_n) = \tau_x u * \theta_n$ for all $x \in \mathbb{R}^N$, one quickly deduces that $u * \theta_n \in AP(\mathbb{R}^N)$. Finally, by applying Hölder's inequality to the functions $x \mapsto \theta_n(x)^{\frac{1}{p'}}$ and $x \mapsto \theta_n(x)^{\frac{1}{p}}[u(y-x) - u(y)]$ for fixed $y \in \mathbb{R}^N$, we get

$$\int_{k+Y} |(u * \theta_n)(y) - u(y)|^p dy \leq \int \theta_n(x) \left[\int_{k+Y} |u(y-x) - u(y)|^p dy \right] dx$$

and that for all $k \in \mathbb{Z}^N$. Therefore

$$\|u * \theta_n - u\|^p_{p,\infty} \leq \int \theta_n(x)\|u - \tau_x u\|^p_{p,\infty} dx$$

and that for any integer $n \geq 1$. Hence $u * \theta_n \to u$ in $(L^p, l^\infty)(\mathbb{R}^N)$ as $n \to \infty$, as easily seen by (i) and use of the fact that θ_n has support in $\frac{1}{n}\overline{B}_N$. This completes the proof. ■

Remark 2.8 Throughout the rest of this chapter, the action $\mathcal{H}$ is assumed to satisfy hypothesis (C) of Chap. 1 (Sect. 1.2.5).

As a consequence of Proposition 2.20, we have

Corollary 2.10 *The mean value M generated by $\mathcal{H}$, viewed as defined on $AP(\mathbb{R}^N)$ (Examples 1.9 and 1.11), extends by continuity to a positive continuous linear form on $L^p_{AP}(\mathbb{R}^N)$, again denoted by M. Furthermore, for a given $u \in L^p_{AP}(\mathbb{R}^N)$ and any bounded open set $\Omega \subset \mathbb{R}^N$, we have $u \circ H_\varepsilon \to M(u)$ in $L^p(\Omega)$-weak as $\varepsilon \to 0$.*

Proof The line of proceeding is the same mutatis mutandis as followed in proving Proposition 2.10, with taking account of the fact that $(L^p, l^\infty)(\mathbb{R}^N)$ is continuously embedded in $\Xi^p(\mathbb{R}^N)$ (Proposition 2.19) and $AP(\mathbb{R}^N)$ dense in $L^p_{AP}(\mathbb{R}^N)$ (Proposition 2.20). ■

This enables us to carry over to $L^p_{AP}(\mathbb{R}^N)$ the important notion of a spectrum:

$$Sp(u) = \{k | k \in \mathbb{R}^N, M(\overline{\gamma}_k u) \neq 0\} \quad (u \in L^p_{AP}(\mathbb{R}^N)),$$

where $\gamma_k(y) = e^{2\pi i k \cdot y}$ $(y \in \mathbb{R}^N)$.

In this matter it would not be out of interest to know more about the functions $u * \theta_n$ involved in the proof of Proposition 2.20. The answer is stated below.

Lemma 2.9 *Let $\theta \in \mathcal{D}(\mathbb{R}^N)$. For $u \in L^p_{AP}(\mathbb{R}^N)$, we have $u * \theta \in AP(\mathbb{R}^N)$ with $Sp(u * \theta) \subset Sp(u)$.*

Proof We only need to prove the second portion of the lemma (see the proof of the preceding proposition). The first step is to show the equality

$$M(u * \theta) = M(u) \int \theta(x) dx \tag{2.29}$$

for $u \in L^p_{AP}(\mathbb{R}^N)$. We already know from Proposition 1.5 that this equality holds for $u \in AP(\mathbb{R}^N)$. We deduce at once that it holds also for $u \in L^p_{AP}(\mathbb{R}^N)$, thanks to the continuity of the mapping $u \mapsto u * \theta$ of $L^p_{AP}(\mathbb{R}^N)$ into $AP(\mathbb{R}^N)$, of that of the mean value M on $AP(\mathbb{R}^N)$ as well as on $L^p_{AP}(\mathbb{R}^N)$, and use of the density of $AP(\mathbb{R}^N)$ in $L^p_{AP}(\mathbb{R}^N)$.

Having made this point, let $u \in L^p_{AP}(\mathbb{R}^N)$. Fix freely some $k \in \mathbb{R}^N$, and note that $\overline{\gamma}_k u \in L^p_{AP}(\mathbb{R}^N)$. Next, observe that

$$\overline{\gamma}_k(u * \theta) = (\overline{\gamma}_k u) * (\overline{\gamma}_k \theta)$$

and then use (2.29) to arrive at $M(\overline{\gamma}_k(u * \theta)) = \mathcal{F}\theta(k)M(\overline{\gamma}_k u)$, where $\mathcal{F}$ denotes the Fourier transformation on $\mathbb{R}^N$. Hence the lemma follows by the arbitrariness of $k \in \mathbb{R}^N$. ■

The preceding lemma enables us to complete Proposition 2.20 as follows.

Proposition 2.21 *Let* $u \in L^p_{AP}(\mathbb{R}^N)$ $(1 \leq p < \infty)$. *Let a sequence of functions* $\theta_n \in \mathcal{D}(\mathbb{R}^N)$ $(n \in \mathbb{N}^*)$ *be as in Proposition 2.20. Then*

(i) $u * \theta_n \in AP(\mathbb{R}^N) \cap \mathcal{C}^\infty(\mathbb{R}^N)$.
(ii) $u * \theta_n \to u$ *in* $(L^p, l^\infty)(\mathbb{R}^N)$ *as* $n \to \infty$.
(iii) $Sp(u) = \bigcup_{n \in \mathbb{N}^*} Sp(u * \theta_n)$.

Proof According to Proposition 2.20 together with its proof, we only need to check (iii). First, from (ii) we deduce using Corollary 2.10 that $M(\overline{\gamma}_k(u * \theta_n)) \to M(\overline{\gamma}_k u)$ as $n \to \infty$, where k is any arbitrary point in $\mathbb{R}^N$. Hence $Sp(u) \subset \bigcup_{n \in \mathbb{N}^*} Sp(u * \theta_n)$. Therefore (iii) follows readily by Lemma 2.9. ■

Remark 2.9 In view of Proposition 2.21(iii), the spectrum of u is at most countable, as a countable union of at most countable sets.

2.4.3 The Spaces $L^p_{\mathcal{R}}(\mathbb{R}^N)$

Let $\mathcal{R}$ be a countable subgroup of $\mathbb{R}^N$. We define

$$L^p_{\mathcal{R}}(\mathbb{R}^N) = \{u | u \in L^p_{AP}(\mathbb{R}^N), Sp(u) \subset \mathcal{R}\} \quad (1 \leq p < \infty).$$

In view of Remark 2.9 this definition makes sense. It is clear that this is a closed vector subspace of $L^p_{AP}(\mathbb{R}^N)$. We shall provide $L^p_{\mathcal{R}}(\mathbb{R}^N)$ with the (L^p, l^∞)-norm, which makes it into a Banach space. The following proposition is a direct consequence of Proposition 2.21.

Proposition 2.22 $AP_{\mathcal{R}}(\mathbb{R}^N)$ *is dense in* $L^p_{\mathcal{R}}(\mathbb{R}^N)$.

This has a corollary of considerable interest in practice.

Corollary 2.11 *We have* $L^p_{\mathcal{R}}(\mathbb{R}^N) \subset \mathfrak{X}^p_A(\mathbb{R}^N)$ $(A = AP_{\mathcal{R}}(\mathbb{R}^N))$ *with continuous embedding.*

Proof This follows at once by Propositions 2.19 and 2.22. ■

2.5 Basic Spaces of Periodic Functions

2.5.1 *Periodic Setting*

Let $\mathcal{B} = \{b_1, \cdots, b_N\}$ be a basis of the N-dimensional vector space $\mathbb{R}^N$. Prior to any discussion it should be stressed that $\mathcal{B}$ **may or may not be orthogonal**. To $\mathcal{B}$ there are attached

$$Y = \left\{ y | y = \sum_{i=1}^{N} t_i b_i, -\frac{1}{2} < t_i < \frac{1}{2}, \ (1 \le i \le N) \right\} \text{ (period of reference)} \tag{2.30}$$

and

$$S = \left\{ k | k = \sum_{i=1}^{N} n_i b_i, n_i \in \mathbb{Z} \ (1 \le i \le N) \right\} \text{ (group of periods).} \tag{2.31}$$

Y is referred to as the open parallelepiped centred at $\omega = (0, \cdots, 0)$ (the origin in $\mathbb{R}^N$), constructed on the vectors $\frac{1}{2}b_i \ \ (1 \le i \le N)$, and S as the subgroup of $\mathbb{R}^N$ spanned by $\mathcal{B}$. Since $\mathcal{B}$ is a basis of the vector space $\mathbb{R}^N$, there exists a bijective linear mapping φ of $\mathbb{R}^N$ onto itself such that $\varphi(b_i) = e_i \ \ (1 \le i \le N)$, where $\{e_1, \cdots, e_N\}$ is the canonical basis of $\mathbb{R}^N$. But φ being at the same time an automorphism of the topological group $\mathbb{R}^N$, it is clear that, considered as defined on S, φ is an isomorphism of the group S onto the subgroup of $\mathbb{R}^N$ spanned by the canonical basis. Hence S is a discrete subgroup of $\mathbb{R}^N$ isomorphic to $\mathbb{Z}^N$. Such an S, referred to as a *network* in $\mathbb{R}^N$, was considered earlier in Chap. 1. By passing to the quotient one readily deduces that the quotient groups $\mathbb{R}^N/S$ and $\mathbb{R}^N/\mathbb{Z}^N$ are isomorphic (see Bourbaki [24, III, p. 17, Remarque3]). The result is that $\mathbb{R}^N/S$ is a compact metrizable Abelian group (additively written).

Finally, we will need the so-called *subgroup associated to* S, namely

$$S^* = \{k | k \in \mathbb{R}^N, k \cdot l \in \mathbb{Z} \text{ for all } l \in S\}, \tag{2.32}$$

where the dot denotes the Euclidean inner product in $\mathbb{R}^N$. S^* is evidently a subgroup of $\mathbb{R}^N$. Better still, S^* is a network in $\mathbb{R}^N$ (Bourbaki [25, VII, p. 7, prop. 5]).

Remark 2.10 If $\mathcal{B}$ is the canonical basis of $\mathbb{R}^N$, then $Y = (-\frac{1}{2}, \frac{1}{2})^N$ and $S = S^* = \mathbb{Z}^N$. This particular case lies within the classical periodic setting.

Remark 2.11 It is worth noting that in a large measure, the use of the couple $\{S, Y\}$ (defined in (2.30) and (2.31)) in place of the usual one, $\{S = \mathbb{Z}^N, Y = (-\frac{1}{2}, \frac{1}{2})^N\}$, leads us to the same qualitative results. This was pointed out earlier in Remark 1.15 in a specific context. Likewise, Proposition 2.18 and Corollary 2.9 show that the amalgam space $(L^p, l^\infty)(\mathbb{R}^N)$ introduced (in Sect. 2.4) using the latter couple is identical (with equivalence of norms) with that which one can similarly define by means of the former couple.

2.5.2 The C^*-algebra $\mathcal{C}_{per}(Y)$ Revisited

Let $\mathcal{B}$, S and Y be as above.

It is well to recall that $\mathcal{C}_{per}(Y)$ denotes the vector subspace of $\mathcal{B}(\mathbb{R}^N)$ consisting of all Y-periodic functions, that is, those u in $\mathcal{B}(\mathbb{R}^N)$ satisfying $u(y+k) = u(y)$ for $y \in \mathbb{R}^N$ and $k \in S$. With the usual pointwise algebra operations and the supremum norm, $\mathcal{C}_{per}(Y)$ is a commutative C^*-algebra with identity.

Our first concern in the present section is to establish that $\mathcal{C}_{per}(Y)$ is an H-algebra. However, in view of Proposition 2.13, it suffices to prove the following assertion.

Proposition 2.23 *We have* $\mathcal{C}_{per}(Y) = AP_{S^*}(\mathbb{R}^N)$.

Proof Since $\gamma_k \in \mathcal{C}_{per}(Y)$ for $k \in S^*$, it is clear by (2.13) that $AP_{S^*}(\mathbb{R}^N) \subset \mathcal{C}_{per}(Y)$. It then remains to check the inverse inclusion. Let us consider the mapping $L : \mathcal{C}(\mathbb{R}^N/S) \to \mathcal{C}_{per}(Y)$ defined by $Lf(y) = f(\pi(y))$ for $y \in \mathbb{R}^N$, where π stands for the canonical homomorphism of $\mathbb{R}^N$ onto $\mathbb{R}^N/S$. It is an easy matter to verify that L is an isometric C^*-isomorphism of $\mathcal{C}(\mathbb{R}^N/S)$ onto $\mathcal{C}_{per}(Y)$ and moreover $L(\tau_{\pi(a)} f) = \tau_a Lf \quad (a \in \mathbb{R}^N)$ for all $f \in \mathcal{C}(\mathbb{R}^N/S)$. It follows readily that for $u \in \mathcal{C}_{per}(Y)$, the set of all translates $\tau_a u \quad (a \in \mathbb{R}^N)$ has compact closure in $\mathcal{B}(\mathbb{R}^N)$, as a consequence of the same property for $L^{-1}u$ in $\mathcal{C}(\mathbb{R}^N/S)$ (indeed, this space is identical with the space of almost periodic continuous complex functions on the Abelian compact group $\mathbb{R}^N/S$). Thus, $\mathcal{C}_{per}(Y) \subset AP(\mathbb{R}^N)$. Now, let $u \in \mathcal{C}_{per}(Y)$. Fix freely $k \in Sp(u)$. Let $\mathcal{U}$ be the linear span of the translates $\tau_x u \ (x \in \mathbb{R}^N)$, and $\overline{\mathcal{U}}$ the closure of $\mathcal{U}$ in $\mathcal{B}(\mathbb{R}^N)$. One has $\gamma_k \in \overline{\mathcal{U}}$ (see, e.g., Guichardet [57, p. 93]). Hence $\gamma_k \in \mathcal{C}_{per}(Y)$, and it is clear that this holds if and only if $k \in S^*$. Consequently $Sp(u) \subset S^*$, which completes the proof. ∎

Corollary 2.12 *The following statements hold true.*

(i) $\mathcal{C}_{per}(Y)$ *is an H-algebra on* $\mathbb{R}^N$ *(under* $\mathcal{H}$*)*
(ii) The linear span of the exponential functions $\gamma_k \quad (k \in S^*)$ *is dense in* $\mathcal{C}_{per}(Y)$.

Proof The whole corollary follows by Propositions 2.13 and 2.23 and use of (2.13). ∎

We conclude the present section with two remarks worth pointing out.

Remark 2.12 By combining Proposition 2.23 with Remark 2.7, one readily obtains $\Delta(\mathcal{C}_{per}(Y)) = \widehat{S^*}$ (up to an isomorphism), as stated by anticipation in Example 1.3. Observe that if $S = \mathbb{Z}^N$, then $\widehat{S^*} = \widehat{\mathbb{Z}^N} = \mathbb{T}^N$.

Remark 2.13 It is essential that Proposition 2.23 should be consistent with the notion of a mean value. More precisely, if m denotes the mean value for almost periodic functions on $\mathbb{R}^N$ (Chap. 1, Example 1.9), we must have $m(u) = \frac{1}{|Y|}\int_Y u(y)dy$ for $u \in \mathcal{C}_{per}(Y)$. It clearly suffices to show that $\int_Y \gamma_k(y)dy = 0$ for $k \in S^*, k \neq \omega$. To this end, let us consider the bijective linear mapping $\varphi : \mathbb{R}^N \to \mathbb{R}^N$ such that $\varphi(b_i) = e_i$, $1 \leq i \leq N$ (see the beginning of the present section). Let $\psi = \varphi^{-1}$. Put $Z = (-\frac{1}{2}, \frac{1}{2})^N$. We have $\psi(Z) = Y$, and ψ is a diffeomorphism of Z onto Y with a constant Jacobian determinant $J(\psi) \neq 0$. Therefore, for k as above, one classically has

$$\int_Y \gamma_k(y)dy = |J(\psi)| \int_Z \gamma_k(\psi(z))dz = |J(\psi)| \int_Z e^{2\pi i k\cdot\psi(z)}dz.$$

On the other hand, noticing that $b_i \in S$, so that $k \cdot b_i \in \mathbb{Z}$ for $i = 1, \cdots, N$, we introduce the point $k^* = (k_i^*) \in \mathbb{Z}^N$ with $k_i^* = k \cdot b_i$. Hence it follows that $k \cdot \psi(z) = k^* \cdot z$ for $z \in \mathbb{R}^N$. Consequently,

$$\int_Z e^{2\pi i k\cdot\psi(z)}dz = \int_Z e^{2\pi i k^*\cdot z}dz = 0,$$

as claimed.

2.5.3 The Spaces $W_{per}^{m,p}(Y)$

Let S, Y be as in (2.30) and (2.31). Let $1 \leq p \leq \infty$. We define

$$L_{per}^p(Y) = \{u | u \in L_{loc}^p(\mathbb{R}^N),\ u\ Y\text{-periodic}\},$$

where « u Y-periodic» means here $u(y+k) = u(y)$ for each $k \in S$ and for almost all $y \in \mathbb{R}^N$. This is of course a closed vector subspace of the Fréchet space $L_{loc}^p(\mathbb{R}^N)$.

$L_{per}^p(Y)$ is provided with the norm $\|\cdot\|_{L^p(Y)}$ given by

$$\|u\|_{L^p(Y)} = \left(\int_Y |u(y)|^p dy\right)^{\frac{1}{p}}$$

if $1 \leq p < \infty$, $\|u\|_{L^\infty(Y)} = \|u\|_{L^\infty(\mathbb{R}^N)}$ otherwise (note that $L_{per}^\infty(Y) \subset L^\infty(\mathbb{R}^N)$). Thus equipped, $L_{per}^p(Y)$ is a Banach space, $L_{per}^2(Y)$ being in particular a Hilbert space. Furthermore, $L_{per}^p(Y)$ is continuously embedded in $L_{loc}^p(\mathbb{R}^N)$.

Now, for $m \in \mathbb{N}^*$ and $1 \leq p < \infty$, set

$$W_{per}^{m,p}(Y) = \{u | u \in L_{per}^p(Y),\ D_y^\alpha u \in L_{per}^p(Y) \text{ for } |\alpha| \leq m\},$$

where the partial derivative $D_y^\alpha u$ is computed in the distribution sense on $\mathbb{R}^N$. Furthermore, we put

$$\|u\|_{W^{m,p}(Y)} = \left(\sum_{|\alpha| \leq m} \|D_y^\alpha u\|_{L^p(Y)}^p \right)^{\frac{1}{p}},$$

which defines a norm on $W_{per}^{m,p}(Y)$ that makes it into a Banach space, and in particular a Hilbert space when $p = 2$.

We will be mostly concerned with the space

$$W_{per}^{m,p}(Y)/\mathbb{C} = \left\{ u | u \in W_{per}^{m,p}(Y), \int_Y u(y)dy = 0 \right\}.$$

This is of course a closed subspace of $W_{per}^{m,p}(Y)$. However, we shall equip $W_{per}^{m,p}(Y)/\mathbb{C}$ with the seminorm

$$\|u\|_{W_{per}^{m,p}(Y)/\mathbb{C}} = \left(\sum_{|\alpha| = m} \|D_y^\alpha u\|_{L^p(Y)}^p \right)^{\frac{1}{p}},$$

which actually proves to be a norm on $W_{per}^{m,p}(Y)/\mathbb{C}$ equivalent to the $W_{per}^{m,p}(Y)$-norm. This classical result follows by a Poincaré type inequality the proof of which is immediate by a routine approach utilizing the compact embedding of $W^{m,p}(Y)$ in $L^p(Y)$ (see, e.g., Adams [3, p. 144]).

Let us prove one simple but basic result.

Proposition 2.24 *Let (θ_n) be a mollifier on $\mathbb{R}^N$, i.e., $\theta_n \in \mathcal{D}(\mathbb{R}^N)$, $\theta_n \geq 0$, $\int \theta_n dx = 1$, $Supp\theta_n \subset \varepsilon_n \overline{B}_N$ ($\overline{B}_N$ the closed unit ball in $\mathbb{R}^N$), $0 < \varepsilon_n \to 0$ as $n \to \infty$. For any $u \in L_{per}^p(Y)$ $(1 \leq p < \infty)$, one has $u * \theta_n \to u$ in $L_{per}^p(Y)$ as $n \to \infty$.*

Proof The line of reasoning is the same as followed in proving Proposition 2.20(ii). Let $u \in L_{per}^p(Y)$. By suitably using Hölder's inequality as was done in the said proof, one quickly arrives at

$$\|u * \theta_n - u\|_{L^p(Y)}^p \leq \int \theta_n(x) \|u - \tau_x u\|_{L^p(Y)}^p dx \quad (n \in \mathbb{N}).$$

Hence $u * \theta_n \to u$ in $L_{per}^p(Y)$ as $n \to \infty$ because θ_n has support in $\varepsilon_n \overline{B}_N$ and the mapping $x \mapsto \tau_x u$ sends continuously $\mathbb{R}^N$ into $L_{loc}^p(\mathbb{R}^N)$. ∎

This proposition has a few interesting corollaries.

Corollary 2.13 *Let $1 \leq p < \infty$. The following hold true.*

(i) $\mathcal{C}^{\infty}_{per}(Y) = \mathcal{C}^{\infty}(\mathbb{R}^N) \cap \mathcal{C}_{per}(Y)$ *is dense in* $L^p_{per}(Y)$.
(ii) $\mathcal{C}^{\infty}_{per}(Y)$ *is dense in* $W^{m,p}_{per}(Y)$ $(m \in \mathbb{N}^*)$.

Proof Part (i) is straightforward by Proposition 2.24 and use of the fact that $u * \theta_n \in \mathcal{C}^{\infty}_{per}(\mathbb{R}^N)$ for $u \in L^p_{per}(Y)$. Part (ii) follows by Proposition 2.24 and use of the equality $D^{\alpha}(u * \theta_n) = \theta_n * D^{\alpha}u$ for $u \in W^{m,p}_{per}(Y)$ together with the relation $u * \theta_n \in \mathcal{C}^{\infty}_{per}(Y)$, of course. ∎

Corollary 2.14 *Let $1 \leq p < \infty$. The following statements hold.*

(i) $L^p_{per}(Y) = L^p_{S^*}(\mathbb{R}^N)$ *with equality of norms.*
(ii) $L^p_{per}(Y) = \mathfrak{X}^p_{\mathcal{C}_{per}(Y)}(\mathbb{R}^N)$ *with equivalence of norms.*
(iii) The Gelfand transformation $\mathcal{G}$ on $A = \mathcal{C}_{per}(Y)$ extends by continuity to an isomorphism of $L^p_{per}(Y)$ onto $L^p(\Delta(A))$, still denoted by $\mathcal{G}$.

Proof (i) follows by the equality $\|u\|_{p,\infty} = \|u\|_{L^p(Y)}$ for $u \in L^p_{per}(Y)$, and use of Propositions 2.22,2.23 and Corollary 2.13. (ii) follows readily by Proposition 2.19 and Corollary 2.13, and use of the continuity of the embedding $\Xi^p(\mathbb{R}^N) \to L^p_{loc}(\mathbb{R}^N)$. Finally, (iii) follows by Proposition 2.11, Corollary 2.4 and part (ii) of the present corollary (the surjectivity is a direct consequence of the density of $\mathcal{C}(\Delta(A))$ in $L^p(\Delta(A))$). ∎

Remark 2.14 Let $A = \mathcal{C}_{per}(Y)$. One has

$$\frac{1}{|Y|}\int_Y |u(y)|^p dy = \int_{\Delta(A)} |\mathcal{G}(u)(s)|^p d\beta(s)$$

for $u \in L^p_{per}(Y)$. Therefore, in Corollary 2.14 the isomorphism $\mathcal{G}$ of $L^p_{per}(Y)$ onto $L^p(\Delta(A))$ is isometric provided $L^p_{per}(Y)$ is equipped with the equivalent norm $|Y|^{-\frac{1}{p}}\|\cdot\|_{L^p(Y)}$.

Corollary 2.15 *Let $1 < p < \infty$, $A = \mathcal{C}_{per}(Y)$, and let $L^p_{per}(Y)$ be equipped with the equivalent norm pointed out in Remark 2.14. The Gelfand transformation $\mathcal{G} : A \to \mathcal{C}(\Delta(A))$ extends by continuity to an isometric isomorphism of $L^p_{per}(Y)$ onto $L^p(\Delta(A))$ mapping isometrically $W^{m,p}_{per}(Y)$ over $W^{m,p}(\Delta(A))$ and $W^{m,p}_{per}(Y)/\mathbb{C}$ over $W^{m,p}(\Delta(A))/\mathbb{C}$.*

Proof We refer to Corollary 2.14 and Remark 2.14 for that part of the statement which is relevant to $L^p_{per}(Y)$. Next, still calling $\mathcal{G}$ the said isometric isomorphism of $L^p_{per}(Y)$ onto $L^p(\Delta(A))$, let us consider below $\mathcal{G}$ as a mapping of $W^{m,p}_{per}(Y)$ into $L^p(\Delta(A))$ (integer $m > 0$) with a view to showing that it is an isometric isomorphism of $W^{m,p}_{per}(Y)$ onto $W^{m,p}(\Delta(A))$. For that purpose, fix freely an integer

$m > 0$. To start with, let us note that $\partial^\alpha \mathcal{G}(u) = \mathcal{G}(D_y^\alpha u)$ $(|\alpha| \leq m)$ for $u \in W_{per}^{m,p}(Y)$. Indeed, this is initially true for $u \in A^\infty = \mathcal{C}_{per}^\infty(Y)$ and then its extends by continuity to $W_{per}^{m,p}(Y)$ by Corollaries 2.13 and 2.14(iii) and use of two facts: (1) $L^p(\Delta(A))$ is continuously embedded in $\mathcal{D}'(\Delta(A))$ thanks to the density of A^∞ in A (Corollary 2.12). (2) ∂^α maps continuously $\mathcal{D}'(\Delta(A))$ into itself. Having done this, it becomes quite apparent that $\mathcal{G}$ is an isometrical mapping of $W_{per}^{m,p}(Y)$ into $W^{m,p}(\Delta(A))$. Our next purpose is to prove that $\mathcal{G}$ is surjective. This however requires some preliminaries.

Let $\mathcal{O}$ be an arbitrary bounded open neighbourhood of the origin in $\mathbb{R}_y^N$. Let $S(\mathcal{O})$ denote the set of all $k \in S$ such that $k + \overline{Y}$ intersects $\mathcal{O}$, and Z the union of all $k + \overline{Y}$ as k ranges over $S(\mathcal{O})$. It is clear that $S(\mathcal{O})$ is finite (since $\mathcal{O}$ is bounded) and $\mathcal{O}$ is contained in Z (since $\mathbb{R}^N$ is the union of all $k + \overline{Y}$ as k ranges over S).

With this in mind, fix freely $f \in W^{m,p}(\Delta(A))$, and let $u \in L_{per}^p(Y)$, $\mathcal{G}(u) = f$. Our purpose now is to prove that $u \in W_{per}^{m,p}(Y)$. It clearly amounts to showing that $D_y^\alpha u \in L_{loc}^p(\mathbb{R}^N)$ for $0 < |\alpha| \leq m$. So, fix freely one such $\alpha \in \mathbb{N}^N$. Let $\varphi \in \mathcal{D}(\mathcal{O})$ (i.e., φ lies in $\mathcal{D}(\mathbb{R}^N)$ with support in $\mathcal{O}$). One has

$$\begin{aligned}
\langle D_y^\alpha u, \varphi \rangle &= (-1)^{|\alpha|} \int_Z u D^\alpha \varphi \, dy \\
&= (-1)^{|\alpha|} \sum_{k \in S(\mathcal{O})} \int_{k+Y} u D^\alpha \varphi \, dy \\
&= (-1)^{|\alpha|} \int_Y u D^\alpha \psi \, dy
\end{aligned}$$

with

$$\psi \in \mathcal{C}_{per}^\infty(Y), \ \psi = \sum_{k \in S} \tau_k \varphi \text{ (a locally finite sum)},$$

where it should be noted that $D^\gamma \psi(y) = \sum_{k \in S(\mathcal{O})} D^\gamma \varphi(k + y)$ for all $\gamma \in \mathbb{N}^N$ and all $y \in \overline{Y}$. But, by Corollaries 2.1, 2.5, and 2.14,

$$\frac{1}{|Y|} \int_Y u D^\alpha \psi \, dy = (-1)^{|\alpha|} \int_{\Delta(A)} \widehat{\psi} \partial^\alpha f \, d\beta \ \ (\widehat{\psi} = \mathcal{G}(\psi)).$$

We deduce by Hölder's inequality that

$$|\langle D_y^\alpha u, \varphi \rangle| \leq \|\partial^\alpha f\|_{L^p(\Delta(A))} \|\widehat{\psi}\|_{L^{p'}(\Delta(A))} |Y| \ \left(\frac{1}{p} + \frac{1}{p'} = 1\right).$$

Furthermore, recall that $\|\widehat{\psi}\|_{L^{p'}(\Delta(A))} = |Y|^{-\frac{1}{p'}}\|\psi\|_{L^{p'}(Y)}$, and use an elementary Hölder's inequality (for fixed $y \in Y$), namely

$$|\psi(y)| \le |S(\mathcal{O})|^{\frac{1}{p}} \left(\sum_{k \in S(\mathcal{O})} |\varphi(k+y)|^{p'} \right)^{\frac{1}{p'}},$$

to see that

$$\|\psi\|^{p'}_{L^{p'}(Y)} \le |S(\mathcal{O})|^{\frac{p'}{p}} \|\varphi\|^{p'}_{L^{p'}(\mathcal{O})},$$

where $|S(\mathcal{O})|$ stands for the (finite) cardinality of $S(\mathcal{O})$. Consequently

$$|\langle D^{\alpha}_{y} u, \varphi \rangle| \le c \|\partial^{\alpha} f\|_{L^{p}(\Delta(A))} \|\varphi\|_{L^{p'}(\mathcal{O})},$$

where $c = |S(\mathcal{O})|^{\frac{1}{p}}|Y|^{\frac{1}{p}}$. Hence, φ being arbitrary in $\mathcal{D}(\mathcal{O})$, it follows that the restriction of $D^{\alpha}_{y} u$ to $\mathcal{O}$ lies in $L^{p}(\mathcal{O})$. Since any compact set K is contained in one such $\mathcal{O}$, one deduces readily that $D^{\alpha}_{y} u$ lies in $L^{p}_{loc}(\mathbb{R}^{N}_{y})$. This proves the second part of the statement. Hence the last one follows at once (use Corollaries 2.1, 2.4, and 2.14 if necessary). ■

2.5.4 A Fundamental Lemma Revisited

This section deals with the periodic version of Lemma 2.7. We need beforehand some notation. Let

$$\mathcal{V}_{per}(Y) = \{\mathbf{\Phi} | \mathbf{\Phi} \in \mathcal{C}^{\infty}_{per}(Y)^{N}, \operatorname{div}_{y} \mathbf{\Phi} = 0\},$$

$$D_{y} = \left(\frac{\partial}{\partial y_{1}}, \dots, \frac{\partial}{\partial y_{N}} \right),$$

$$\mathbf{u} \cdot \mathbf{v} = \sum_{i=1}^{N} u_{i} v_{i} \text{ for } \mathbf{u} = (u_{i}) \in L^{p}_{per}(Y)^{N} \text{ and } \mathbf{v} = (v_{i}) \in L^{p'}_{per}(Y)^{N}$$

$$\text{with } \frac{1}{p} + \frac{1}{p'} = 1 \;\; (1 \le p < \infty).$$

Lemma 2.10 *The following statements are equivalent for a function* $f = (f_{i}) \in L^{2}_{per}(Y)^{N}$.

(i) f satisfies

$$\int_Y f \cdot \mathbf{\Phi}\, dy = 0 \tag{2.33}$$

for all $\mathbf{\Phi} \in \mathcal{V}_{per}(Y)$.
(ii) There is a unique $q \in H^1_{per}(Y)/\mathbb{C}$ such that

$$D_y q = f. \tag{2.34}$$

Proof It is evident that (ii) implies (i), whereas the converse follows readily by Lemma 2.7 with $\mathcal{R} = S^*$ and use of Proposition 2.23, Corollaries 2.5, 2.14, and 2.15 together with the equalities $V_{\mathcal{R}} = \mathcal{G}(\mathcal{V}_{per}(Y))$ and

$$\frac{1}{|Y|}\int_Y u(y)dy = \int_{\Delta(A)} \mathcal{G}(u)d\beta$$

for $u \in L^2_{per}(Y)$, where $A = \mathcal{C}_{per}(Y)$ and $|Y| = \lambda(Y)$. ■

We conclude with a useful generalization of this lemma.

Lemma 2.11 *Let $1 \le p < \infty$. The following two statements are equivalent for $f \in L^p_{per}(Y)^N$.*

(i) f satisfies (2.33) for all $\mathbf{\Phi} \in \mathcal{V}_{per}(Y)$.
(ii) There is a unique $q \in W^{1,p}_{per}(Y)/\mathbb{C}$ satisfying (2.34).

Proof Let $(\theta_n)_{n\in\mathbb{N}}$ be a mollifier in $\mathcal{D}(\mathbb{R}^N)$ (see, e.g., the proof of Proposition 2.24). Let $\mathbf{f}_n = \mathbf{f} * \theta_n$. For $\mathbf{\Phi} \in \mathcal{V}_{per}(Y)$, we have

$$\begin{aligned}\int_Y \mathbf{f}_n \cdot \mathbf{\Phi}\, dy &= \int \theta_n(x)\left(\int_{Y-x} \mathbf{f}(y)\cdot \mathbf{\Phi}_x(y)dy\right)dx \\ &= \int \theta_n(x)\left(\int_Y \mathbf{f}(y)\cdot \mathbf{\Phi}_x(y)dy\right)dx = 0 \text{ (see (2.33))},\end{aligned}$$

since the function $\mathbf{\Phi}_x = \tau_{-x}\mathbf{\Phi}$ lies in $\mathcal{V}_{per}(Y)$. Thus (2.33) holds with $\mathbf{f}_n$ in place of $\mathbf{f}$. Hence, appeal to Lemma 2.10 yields some $q_n \in H^1_{per}(Y)/\mathbb{C}$ such that $D_y q_n = \mathbf{f}_n$. But $\mathbf{f}_n \in \mathcal{C}^\infty_{per}(Y)^N$. Consequently $q_n \in \mathcal{C}^1_{per}(Y)$, thanks to a classical result from the theory of distributions (see Schwartz [126, p. 61]). Thus, q_n lies in $W^{1,p}_{per}(Y)/\mathbb{C}$, and that for any integer $n \ge 0$. But then it is clear that the sequence (q_n) is Cauchy in $W^{1,p}_{per}(Y)/\mathbb{C}$. Therefore it follows that $q_n \to q$ in $W^{1,p}_{per}(Y)/\mathbb{C}$ as $n \to \infty$. By combining this with the fact that $\mathbf{f}_n \to \mathbf{f}$ in $L^p_{per}(Y)^N$ as $n \to \infty$, we deduce readily that q satisfies Eq. (2.34) of which it is clearly the unique solution in $W^{1,p}_{per}(Y)/\mathbb{C}$. This shows that (i) implies (ii). The converse being evident, the lemma follows. ■

2.6 Discrete H-algebras

Our main purpose here is to discuss further examples of H-algebras of practical interest. Let us first of all introduce the underlying function spaces. In what follows, and unless otherwise stated, G denotes a locally compact Abelian group additively written. In practice G will be either $\mathbb{R}^N$ or the network $S = \mathbb{Z}^N$.

2.6.1 *Essential Functions on G*

We will start with some preliminaries. By a *finite net* in G is meant any finite family $A = (a_i)_{i\in I}$ of points in G. If $B = (b_j)_{j\in J}$ is one further finite net in G, one puts $A + B = (a_i + b_j)_{(i,j)\in I\times J}$, defining thus a new finite net in G referred to as the sum of A and B. Given some $b \in G$, it is clear that the translate $b+A = (b+a_i)_{i\in I}$ is again a finite net in G.

Now, for any finite net $A = (a_i)_{i\in I}$ in G, let

$$\mathfrak{M}_A(u) = \frac{1}{|I|}\sum_{i\in I} \tau_{a_i} u \quad (u \in \mathcal{B}(G)),$$

where $|I|$ denotes the cardinality of I, and $\tau_{a_i}u(y) = u(y-a_i)$ as usual. This defines a mapping $\mathfrak{M}_A : \mathcal{B}(G) \to \mathcal{B}(G)$ which is continuous linear of norm 1. Furthermore, one has $\tau_b\mathfrak{M}_A(u) = \mathfrak{M}_{b+A}(u)$ and $\mathfrak{M}_A(\mathfrak{M}_B(u)) = \mathfrak{M}_{A+B}(u) = \mathfrak{M}_B(\mathfrak{M}_A(u))$.

Let us now come to the heart of the matter.

Definition 2.7 A mapping $u : G \to \mathbb{C}$ is called an *essential function* on G if u lies in $\mathcal{B}(G)$ and further

$$(ES)\quad \begin{cases} \text{For any real number } \eta > 0, \text{ there is a finite net } A \text{ in } G \\ \text{such that } |\mathfrak{M}_A(u)(x) - \mathfrak{M}_A(u)(y)| \le \eta \text{ for all } x, y \in G. \end{cases}$$

We call an *η-net for u*, any A for which (ES) is satisfied. The set of all essential functions on G will be denoted by $ES(G)$.

It is an easy exercise to check that $ES(G)$ is a closed vector subspace of $\mathcal{B}(G)$. Equipped with the supremum norm, $ES(G)$ is therefore a Banach space. Furthermore, $ES(G)$ contains the constants and preserves translation and complex conjugation.

A useful characterization of essential functions is provided by the following theorem.

Theorem 2.1 *The following statements are equivalent for a function $u \in \mathcal{B}(G)$.*

(i) $u \in ES(G)$.
(ii) There exists a complex number $\mathfrak{M}(u)$ with the following property: Given $\eta > 0$, there is a finite net A in G such that

$$|\mathfrak{M}_A(u)(x) - \mathfrak{M}(u)| \leq \eta \;\; (x \in G). \tag{2.35}$$

Furthermore, $\mathfrak{M}(u)$ is unique.

Proof Part (ii) $\Rightarrow$ (i) follows by mere routine calculation. The unicity of $\mathfrak{M}(u)$ is easily obtained by noting that if A is as in (2.35) and if B is any further finite net in G, then (2.35) still holds with $A+B$ in place of A. It remains to prove part (i) $\Rightarrow$(ii).

Let $u \in ES(G)$. To begin, let us suppose that u assumes real values. For each finite net A in G let us define (following Jacobs [60, p. 379])

$$\overline{u}(A) = \sup_{x \in G} \mathfrak{M}_A(u)(x), \;\; \underline{u}(A) = \inf_{x \in G} \mathfrak{M}_A(u)(x)$$

and

$$I_A(u) = [\underline{u}(A), \overline{u}(A)] \;\; \text{(compact interval from } \underline{u}(A) \text{ to } \overline{u}(A)).$$

Next, we set $I(u) = \bigcap_A I_A(u)$, where A ranges over the finite nets in G. Let us begin by verifying that $I(u)$ is nonempty. It clearly suffices to check that the family $(I_A(u))$ has the finite intersection property. But this is straightforward by noticing that if A_i $(i = 1, \cdots, n)$ is any finite family of finite nets and if $A = A_1 + \cdots + A_n$, then $\mathfrak{M}_A(u)(\omega) \in \bigcap_{i=1}^n I_{A_i}(u)$, ω standing for the neutral element of G. With this in mind, given $\eta > 0$, if A is an η-net for u (Definition 2.7) then $0 \leq \overline{u}(A) - \underline{u}(A) \leq \eta$. One deduces by the arbitrariness of $\eta > 0$ that $I(u)$ has length zero. Hence it follows that $I(u)$ consists of exactly one point denoted by $m(u)$. Therefore, if $\eta > 0$ is given, and if A is an η-net for u, then one is quickly led to (2.35) with $m(u)$ in place of $\mathfrak{M}(u)$.

Finally, assuming that u is a complex function, we put $u = u_1 + iu_2$, where u_1 and u_2 are real functions in $ES(G)$. Let $\eta > 0$. As shown above, to each $k = 1, 2$ there is attached a finite net A_k in G such that $|\mathfrak{M}_{A_k}(u_k)(x) - m(u_k)| \leq \frac{\eta}{2}$ for all $x \in G$. Then, by letting $\mathfrak{M}(u) = m(u_1) + im(u_2)$ we are readily led to (2.35) with $A = A_1 + A_2$. This completes the proof. ∎

It is an easy exercise (left to the reader) to verify that the operator $\mathfrak{M}$ thus constructed is a mean value on G with domain $D(\mathfrak{M}) = ES(G)$ (see Definition 1.2). We refer to $\mathfrak{M}$ as the *essential mean value on* G.

2.6.2 Some Examples of Essential Functions

Example 2.6 Let m denote the mean value for $AP(G)$ (Example 1.9). The essential mean value on G is an extension of m. More precisely, one has $AP(G) \subset ES(G)$

and $\mathfrak{M}(u) = m(u)$ for $u \in AP(G)$. Indeed, fix freely some $u \in AP(G)$, and let $\eta > 0$. It is a classical fact that there is a finite family of points a_i $(1 \leq i \leq n)$ in G with the property that to any arbitrary given $x \in G$ there is assigned some index $1 \leq i \leq n$ such that

$$(1) \quad |u(x+y) - u(a_i + y)| < \frac{\eta}{2} \ (y \in G).$$

For each index i $(1 \leq i \leq n)$, let G_i denote the (non empty) set of all $x \in G$ satisfying (1). This yields a family $(G_i)_{1 \leq i \leq n}$ which is a covering of G, and which satisfies, for every $i = 1, \cdots, n$,

$$|u(x+y) - u(x'+y)| < \eta \ \ (y \in G) \text{ for all } x, x' \in G_i.$$

Hence, it follows by Jacobs [60, pp. 379–381] that there is a unique complex number $m(u)$, namely the mean value of u, with the property that to any $\eta > 0$ there is attached some finite net F in G such that $|\mathfrak{M}_F(u)(x) - m(u)| \leq \eta$ for all $x \in G$. Therefore, the claimed statement follows by Theorem 2.1.

Example 2.7 Here, G is either $\mathbb{R}^N$ or $\mathbb{Z}^N$. One has $\mathcal{B}_\infty(G) \subset ES(G)$ and $\mathfrak{M}(u) = m(u) = \lim_{|x| \to \infty} u(x)$ for $u \in \mathcal{B}_\infty(G)$ (see Example 1.7 and Remark 1.6). Indeed, since $\mathbb{C} + \mathcal{K}(G)$ is dense in $\mathcal{B}_\infty(G)$, our purpose reduces to showing that $\mathcal{K}(G) \subset ES(G)$ with $\mathfrak{M}(\varphi) = 0$ for $\varphi \in \mathcal{K}(G)$. To this end, consider a discrete subgroup H of G such that G/H is compact (take, e.g., $H = \mathbb{Z}^N$ if $G = \mathbb{R}^N$, $H = 2\mathbb{Z}^N$ if $G = \mathbb{Z}^N$). For $\psi \in \mathcal{K}(G)$, put

$$T(\psi) = \sum_{k \in H} \tau_k \psi \text{ (a locally finite sum).}$$

Clearly $T(\psi)$ is a continuous function on G. Moreover, $T(\psi)$ is H-periodic, i.e., $T(\psi)(y+k) = T(\psi)(y)$ for $y \in G$ and $k \in H$. Hence $T(\psi) \in \mathcal{B}(G)$. Having done this, let $(U_n)_{n \in \mathbb{N}}$ be an open covering of G with $\overline{U}_n \subset U_{n+1}$, $\overline{U}_n$ compact. Put $V_n = H \cap U_n$. This set is finite and $|V_n| \to \infty$ as $n \to \infty$ ($|V_n|$ stands for the cardinality of V_n). One may write $V_n = (h_i)_{i \in I_n}$ with $I_n = \{1, \cdots, \sigma(n)\}$ where $\sigma(n) = |V_n| = |I_n|$. Finally, let $\varphi \in \mathcal{K}(G)$. For fixed integer $n > 0$, let

$$\mathfrak{M}_{V_n}(\varphi) = \frac{1}{|I_n|} \sum_{i \in I_n} \tau_{h_i} \varphi.$$

Clearly

$$|\mathfrak{M}_{V_n}(\varphi)(x)| \leq \frac{1}{|I_n|} \sup_{y \in G} T(|\varphi|)(y) \quad (x \in G).$$

By letting $n \to \infty$, it follows $\|\mathfrak{M}_{V_n}(\varphi)\|_\infty \to 0$. Hence $\varphi \in ES(G)$ with $\mathfrak{M}(\varphi) = 0$, according to Theorem 2.1.

Example 2.8 Let $Y \subset \mathbb{R}^N$ be as in the general periodic setting of Sect. 2.5 (see in particular (2.30)). We have $\mathcal{C}_{per}(Y) \subset ES(\mathbb{R}^N)$ with $\mathfrak{M}(u) = \frac{1}{|Y|}\int_Y u(y)dy$ for $u \in \mathcal{C}_{per}(Y)$. Indeed, this follows by Proposition 2.23 (see also Remark 2.12) and Example 2.6.

Example 2.9 Let $\mathcal{R}$ be a network in $\mathbb{R}^N$ with $\mathcal{R} \subset \mathbb{Z}^N$. By an $\mathcal{R}$-periodic complex function on $\mathbb{Z}^N$ is meant any mapping $u : \mathbb{Z}^N \to \mathbb{C}$ such that $u(k+\nu) = u(k)$ ($k \in \mathbb{Z}^N$) for any $\nu \in \mathcal{R}$. We denote the space of all such functions by $P_{\mathcal{R}}(\mathbb{Z}^N)$. In order to point out the properties of $P_{\mathcal{R}}(\mathbb{Z}^N)$, let us first of all note that the quotient group $\mathbb{Z}^N/\mathcal{R}$ is finite (and therefore compact). In the sequel φ denotes the canonical homomorphism of $\mathbb{Z}^N$ onto $\mathbb{Z}^N/\mathcal{R}$, and F a finite subset of $\mathbb{Z}^N$ such that φ maps bijectively F onto $\mathbb{Z}^N/\mathcal{R}$. This being so, we have $P_{\mathcal{R}}(\mathbb{Z}^N) \subset \mathcal{B}(\mathbb{Z}^N) = l^\infty(\mathbb{Z}^N)$, since $u(\mathbb{Z}^N) = u(F)$ for $u \in P_{\mathcal{R}}(\mathbb{Z}^N)$. Furthermore, provided with the algebra operations in $\mathcal{B}(\mathbb{Z}^N)$ and the supremum norm, $P_{\mathcal{R}}(\mathbb{Z}^N)$ is a commutative C^*-algebra with identity. Now, as is classical, the mapping $f \mapsto f \circ \varphi$ is an isometric isomorphism of $\mathcal{C}(\mathbb{Z}^N/\mathcal{R})$ onto $P_{\mathcal{R}}(\mathbb{Z}^N)$. Hence it follows readily that $P_{\mathcal{R}}(\mathbb{Z}^N) \subset AP(\mathbb{Z}^N)$, from which we deduce by Example 2.6 that $P_{\mathcal{R}}(\mathbb{Z}^N) \subset ES(\mathbb{Z}^N)$. Furthermore, by noticing that the mapping $f \mapsto \mathfrak{M}(f \circ \varphi)$ of $\mathcal{C}(\mathbb{Z}^N/\mathcal{R})$ into $\mathbb{C}$ is the normalized Haar measure on $\mathbb{Z}^N/\mathcal{R}$, we have

$$\mathfrak{M}(f \circ \varphi) = \frac{1}{|\mathbb{Z}^N/\mathcal{R}|} \sum_{t \in \mathbb{Z}^N/\mathcal{R}} f(t) \quad (f \in \mathcal{C}(\mathbb{Z}^N/\mathcal{R}))$$

hence

$$\mathfrak{M}(u) = \frac{1}{|F|} \sum_{k \in F} u(k) \quad (u \in P_{\mathcal{R}}(\mathbb{Z}^N)),$$

where $|\Phi|$ denotes the cardinality of any finite set Φ.

Example 2.10 Let $S = \mathbb{Z}^N$. We have $l^1(S) \subset ES(S)$ and $\mathfrak{M}(u) = 0$ for $u \in l^1(S)$. Indeed, let $u \in l^1(S)$. First, we have $u \in l^\infty(S) = \mathcal{B}(S)$. Next, given a covering $(S_n)_{n \in \mathbb{N}}$ of S with S_n finite, $S_n \subset S_{n+1}$, we have $|\mathfrak{M}_{S_n}(u)(y)| \leq \frac{1}{|S_n|}\|u\|_1$ ($y \in S$) for all integers $n \geq 0$, where $\|u\|_1 = \sum_{k \in S} |u(k)|$. Hence the claimed statement follows by the fact that $|S_n| \to +\infty$ as $n \to \infty$ (proceed as in Example 2.7).

Example 2.11 Let $\mathcal{B}_{\infty,AP}(G)$ be the closure in $\mathcal{B}(G)$ of the subalgebra of $\mathcal{B}(G)$ made up of all finite sums

$$\sum_{finite} \varphi_i u_i \quad (\varphi_i \in \mathcal{B}_\infty(G), u_i \in AP(G)).$$

We have $\mathcal{B}_{\infty,AP}(G) \subset ES(G)$. Indeed, it is enough to check that $\varphi u \in ES(G)$ for all $\varphi \in \mathcal{B}_\infty(G)$ and $u \in AP(G)$. But we have $\varphi u = (\varphi - \xi)u + \xi u$ with $\xi = \lim_{|x| \to \infty} \varphi(x)$. Hence the desired result follows by Examples 2.6 and 2.7.

2.6.3 Consistency of the Essential Functions

For further needs we shall prove that the essential mean value is consistent with the mean value generated by the action $\mathcal{H}$ on $\mathbb{R}^N$. Before we can do this, however, we must establish one preliminary result.

Lemma 2.12 *Let* $u \in L^\infty(\mathbb{R}^N)$ *and* $a \in \mathbb{R}^N$. *Then* $(u - \tau_a u) \circ H_\varepsilon \to 0$ *in* $L^\infty(\mathbb{R}^N)$-*weak∗ as* $\varepsilon \to 0$.

Proof The aim is precisely to check that we have, for any given $\varphi \in L^1(\mathbb{R}^N)$,

$$\int (u - \tau_a u)(H_\varepsilon(x))\varphi(x)dx \to 0 \text{ as } \varepsilon \to 0. \tag{2.36}$$

However, in view of the density of $\mathcal{K}(\mathbb{R}^N)$ in $L^1(\mathbb{R}^N)$ it is enough to show this for $\varphi \in \mathcal{K}(\mathbb{R}^N)$. So, under this assumption, suppose $\varepsilon > 0$ is such that $H_{\varepsilon^{-1}}(a) \in \overline{B}_N$ (the closed unit ball in $\mathbb{R}^N$), and let $K = \overline{B}_N + Supp\varphi$. Then, it is worth bearing in mind that for such an $\varepsilon > 0$, the two functions $\varphi \circ H_{\varepsilon^{-1}}$ and $\tau_{-a}(\varphi \circ H_{\varepsilon^{-1}})$ have their supports in the compact set $H_\varepsilon(K)$. In the sequel $c > 0$ denotes a constant such that $\lambda(K)\|u\|_{L^\infty} \leq c$.

With this in mind, let $\eta > 0$. According to the uniform continuity of φ, we can find some $0 < r \leq 1$ such that $|\varphi(x) - \varphi(x')| \leq \frac{\eta}{c}$ for $x, x' \in \mathbb{R}^N, x - x' \in r\overline{B}_N$. Thus, given $\alpha > 0$ such that $H_{\varepsilon^{-1}}(a) \in rB_N$ for $0 < \varepsilon \leq \alpha$ (recall that such an α does exist by virtue of the absorptiveness of $\mathcal{H}$), one has

$$|\varphi(H_{\varepsilon^{-1}}(x)) - \varphi(H_{\varepsilon^{-1}}(x) + H_{\varepsilon^{-1}}(a))| \leq \frac{\eta}{c} \quad (x \in \mathbb{R}^N, 0 < \varepsilon \leq \alpha).$$

Hence, by the equality

$$\int (u - \tau_a u)(H_\varepsilon(x))\varphi(x)dx = \varepsilon^s \int u(x)[\varphi(H_{\varepsilon^{-1}}(x)) - \varphi(H_{\varepsilon^{-1}}(x) + H_{\varepsilon^{-1}}(a))]dx$$

and use of the fact that the right integration is actually over $H_\varepsilon(K)$, it follows

$$\left| \int (u - \tau_a u)(H_\varepsilon(x))\varphi(x)dx \right| \leq \eta$$

for $0 < \varepsilon \leq \alpha$. This shows (2.36), thereby proving the lemma. ■

We are now able to point out the consistency in connection with the mean value.

Proposition 2.25 *The mean value, M, generated by the action $\mathcal{H}$ on $\mathbb{R}^N$ is an extension of the essential mean value on $\mathbb{R}^N$. Specifically, we have* $ES(\mathbb{R}^N) \subset \Pi^\infty(\mathbb{R}^N)$ *and* $M(u) = \mathfrak{M}(u)$ *for* $u \in ES(\mathbb{R}^N)$.

Proof Let $u \in ES(\mathbb{R}^N)$. The whole problem amounts to showing that $u \circ H_\varepsilon \to \mathfrak{M}(u)$ in $L^\infty(\mathbb{R}^N)$-weak$*$ as $\varepsilon \to 0$. To do this, let $\varphi \in L^1(\mathbb{R}^N)$ be freely fixed. Let $\eta > 0$. There clearly exists a finite net A in $\mathbb{R}^N$ such that

$$\left| \int \mathfrak{M}_A(u)(H_\varepsilon(x))\varphi(x)dx - \mathfrak{M}(u) \int \varphi(x)dx \right| \leq \frac{\eta}{2} \;\; (\varepsilon > 0).$$

On the other hand, using Lemma 2.12 yields some $\alpha > 0$ such that

$$\left| \int u(H_\varepsilon(x))\varphi(x)dx - \int \mathfrak{M}_A(u)(H_\varepsilon(x))\varphi(x)dx \right| \leq \frac{\eta}{2} \;\; (0 < \varepsilon \leq \alpha).$$

Hence

$$\left| \int u(H_\varepsilon(x))\varphi(x)dx - \mathfrak{M}(u) \int \varphi(x)dx \right| \leq \eta \; (0 < \varepsilon \leq \alpha),$$

which completes the proof. ■

We conclude this section by discussing the other side of the said consistency: the notion of an essential algebra. We purposely restrict the discussion to essential functions on the discrete group $S = \mathbb{Z}^N$.

Definition 2.8 By an *essential algebra* on S is meant a closed subalgebra $\mathfrak{G}$ of $\mathcal{B}(S) = l^\infty(S)$ having the following properties:

(ESA1) $\mathfrak{G}$ with the supremum norm is separable.
(ESA2) $\mathfrak{G}$ contains the constants.
(ESA3) $\mathfrak{G}$ preserves complex conjugation.
(ESA4) $\mathfrak{G} \subset ES(S)$.

Some examples of essential algebras are collected below.

Example 2.12 The algebra $\mathcal{B}_\infty(S)$. This is an essential algebra on S. Indeed, in view of Example 1.2, we only need to check property (ESA4). But this was established in Example 2.7.

Example 2.13 The algebra $l_0^1(S)$. We define $l_0^1(S)$ to be the closure of $\mathbb{C} + l^1(S)$ in $l^\infty(S)$. Property (ESA1) for $\mathfrak{G} = l_0^1(S)$ results from the fact that $\mathbb{C} + l^1(S)$ is separable; properties (ESA2) and (ESA3) are evident. Finally, (ESA4) follows by Example 2.10. Thus, $l_0^1(S)$ is an essential algebra.

Example 2.14 The algebra $l_\infty^1(S)$. By this we denote the closure of $\mathcal{B}_\infty(S)+l^1(S)$ in $l^\infty(S)$. It is clear by Examples 2.10 and 2.12 that $l_\infty^1(S)$ is an essential algebra.

Example 2.15 The algebra $P_\mathcal{R}(S)$ introduced in Example 2.9 is an essential algebra on $S = \mathbb{Z}^N$.

As might be suspected, the essential algebras are destined to serve as a foundation of the forthcoming notion of a discrete H-algebra.

2.6.4 Discrete H-algebras on $\mathbb{R}^N$

Let

$$S = \mathbb{Z}^N, Y = \left(-\frac{1}{2}, \frac{1}{2}\right)^N.$$

Throughout this subsection, $\mathfrak{G}$ denotes a given essential algebra on S. Let $\mathcal{F}$ be the set of all functions $f : \mathbb{R}^N_y \to \mathbb{C}$ of the form

$$f = \sum_{k \in S} a(k)\tau_k\varphi \quad (a \in \mathfrak{G}, \varphi \in \mathcal{K}(Y)), \tag{2.37}$$

where, as usual, $\mathcal{K}(Y)$ is identified with the subspace of $\mathcal{K}(\mathbb{R}^N)$ formed of all functions having supports in Y. Attention must be drawn to the fact that the above sum is locally finite.

Remark 2.15 Let $y \in \mathbb{R}^N$. Then, either there is some $l \in S$ such that $y \in l + Y$, in which case $f(y) = a(l)\varphi(y - l)$, or such an l does not exist (which means that y lies on the boundary of some $k + Y$) and we have therefore $f(y) = 0$.

The following lemma is essential to the notion of a discrete homogenization algebra.

Lemma 2.13 *The following statements hold.*

(i) $\mathcal{F} \subset ES(\mathbb{R}^N)$.
(ii) $\mathcal{F}$ *is stable under multiplication and under complex conjugation.*

Proof

(i) Let $f \in \mathcal{F}$ with (2.37). The restriction of f to any compact set $K \subset \mathbb{R}^N$ is given by $f|_K = \sum_{k \in j_1(K)} a(k)(\tau_k\varphi)|_K$, where $j_1(K)$ stands for the (finite) set of those $k \in S$ for which $k + \overline{Y}$ intersects K. This shows that f is continuous on $\mathbb{R}^N$. Furthermore, Remark 2.15 shows that $|f(y)| \leq \|a\|_\infty \|\varphi\|_\infty$ for all $y \in \mathbb{R}^N$. Thus, $f \in \mathcal{B}(\mathbb{R}^N)$. To complete the proof of part (i), let $\theta = \sum_{k \in S} \tau_k |\varphi|$. Clearly $\theta \in \mathcal{B}(\mathbb{R}^N)$ and so we can consider a constant $c > 0$ such that $0 \leq \theta \leq c$. Now let $\eta > 0$. Let F be a finite net in S such that

$$|\mathfrak{M}_F(a)(k) - \mathfrak{M}(a)| \leq \frac{\eta}{2c} \quad (k \in S),$$

where $\mathfrak{M}$ denotes the essential mean value on S and on $\mathbb{R}^N_y$ as well. Finally, let $\psi = \mathfrak{M}(a) \sum_{k\in S} \tau_k \varphi$, and bear in mind that $\psi \in \mathcal{C}_{per}(Y)$, hence $\psi \in ES(\mathbb{R}^N_y)$ (Example 2.8). By the trivial equality

$$\mathfrak{M}_F(f) - \psi = \sum_{k\in S} [\mathfrak{M}_F(a)(k) - \mathfrak{M}(a)]\tau_k \varphi$$

we see that

$$|\mathfrak{M}_F(f)(y) - \psi(y)| \leq \frac{\eta}{2} \quad (y \in \mathbb{R}^N). \tag{2.38}$$

Finally, fix a finite net R in $\mathbb{R}^N_y$ such that $|\mathfrak{M}_R(\psi)(y) - \mathfrak{M}(\psi)| \leq \frac{\eta}{2}$ for all $y \in \mathbb{R}^N$ (Theorem 2.1), and observe that one then has by (2.38)

$$|\mathfrak{M}_{F+R}(f)(y) - \mathfrak{M}_R(\psi)(y)| \leq \frac{\eta}{2} \, (y \in \mathbb{R}^N).$$

Hence $|\mathfrak{M}_{F+R}(f)(y) - \mathfrak{M}(\psi)| \leq \eta$ for all $y \in \mathbb{R}^N$, which shows that $f \in ES(\mathbb{R}^N_y)$ with $\mathfrak{M}(f) = \mathfrak{M}(\psi)$. Therefore (i) follows.

(ii). It is obvious that $\mathcal{F}$ preserves complex conjugation. Finally if $f, g \in \mathcal{F}$, f given by (2.37) and $g = \sum_{k\in S} b(k)\tau_k \theta$, according to Remark 2.15 we have $fg = \sum_{k\in S} a(k)b(k)\tau_k(\varphi\theta)$. This completes the proof.

■

Remark 2.16 A quick survey of the preceding proof reveals that if f is of the form (2.37) with $\varphi \in \mathcal{K}(Y)$ and (more generally) $a \in ES(\mathbb{Z}^N)$, then $f \in ES(\mathbb{R}^N)$ and $\mathfrak{M}(f) = \mathfrak{M}(a) \int \varphi(y)dy$.

Having made this point, let A be the closure in $\mathcal{B}(\mathbb{R}^N_y)$ of the space of all functions of the form

$$\psi = c + \sum_{i\in I} f_i \quad (c \in \mathbb{C}, \, f_i \in \mathcal{F}), \tag{2.39}$$

where I is a finite set that depends on ψ.

Proposition 2.26 *A is an H-algebra on $\mathbb{R}^N_y$.*

Proof It is clear by Lemma 2.13 that the functions of the form (2.39) form a subalgebra of $\mathcal{B}(\mathbb{R}^N_y)$ that contains the constants, preserves complex conjugation and further, is a vector subspace of $ES(\mathbb{R}^N_y)$. Consequently, A is a closed subalgebra of $\mathcal{B}(\mathbb{R}^N_y)$ satisfying three of the four properties of Definition 2.1 namely (HA2)–(HA4), the last property, (HA4), resulting from Proposition 2.25. Therefore, only (HA1) remains to be verified. For this purpose, let $\mathfrak{G}_0$ be a dense countable subset of $\mathfrak{G}$ (according to (ESA1)) and $\mathcal{K}_0$ be a dense countable subset of $\mathcal{K}(Y)$ (exceptionally equipped with the supremum norm). Finally, let $\mathcal{F}_0$ be the set of all functions of the form (2.37) with $\mathfrak{G}_0$ in place of $\mathfrak{G}$ and $\mathcal{K}_0$ in that of $\mathcal{K}(Y)$, and let $\mathbb{C}_0 = \mathbb{Q} + i\mathbb{Q}$ ($\mathbb{Q}$ denotes the rationals). Then, it is clear that the functions of the form (2.39) with $\mathbb{C}_0$ in place of $\mathbb{C}$ and $\mathcal{F}_0$ in that of $\mathcal{F}$ constitute a dense countable subset of A. This completes the proof. ∎

Definition 2.9 The H-algebra A constructed above will be referred to as the *discrete homogenization algebra* on $\mathbb{R}^N$ associated to the essential algebra $\mathfrak{G}$ on $\mathbb{Z}^N$.

The term «discrete» is due to the fact that each function in A is generated by a complex function defined on the discrete group $S = \mathbb{Z}^N$. The so-called *discrete behaviour* of a function defined on $\mathbb{R}^N$ arises quite naturally. For example, it is easily seen that each function $f \in L^\infty(\mathbb{R}^N)$ is uniquely expressible in the form $f = u + f_0$ with

$$u = \sum_{k\in\mathbb{Z}^N} a(k)\chi_{k+Y} \quad \text{(the «discrete» part of } f\text{)}$$

and

$$f_0 \in L^\infty(\mathbb{R}^N), \int_{k+Y} f_0(y)dy = 0 \ (k \in \mathbb{Z}^N),$$

where χ_{k+Y} denotes the characteristic function of $k + Y$ in $\mathbb{R}^N$, and $a(k) = \int_{k+Y} f(y)dy$.

Thus, f may have different possible discrete behaviours, for example one may assume that $a \in \mathcal{B}_\infty(\mathbb{Z}^N)$, or more generally $a \in ES(\mathbb{Z}^N)$. This will later be revisited in a more practical context.

2.7 Comments and Problems

The basic problem leading to the theory of homogenization algebras states as follows.

Given a suitable absorptive continuous action $\mathcal{H} = (H_\varepsilon)_{\varepsilon>0}$ of $\mathbb{R}_+^*$ on $\mathbb{R}^N$ (e.g., $H_\varepsilon(x) = \frac{x}{\varepsilon}$),

$$\begin{cases} \text{Find a mean value } m \text{ on } \mathbb{R}^N \text{ such that} \\ \text{(1) The domain, } D(m), \text{ is a subalgebra of } \mathcal{B}(\mathbb{R}^N). \\ \text{(2) For each } u \in D(m),\, u \circ H_\varepsilon \to m(u) \text{ in } L^\infty(\mathbb{R}^N)\text{-weak}* \\ \text{as } \varepsilon \to 0. \end{cases} \tag{2.40}$$

The enormous success of the periodic homogenization theory is due not only to the availability of, say a «periodic solution» to (2.40), but also to the fact that such a solution admits an integral representation with respect to Lebesgue measure on $\mathbb{R}^N$. Indeed, as was discussed in Sect. 2.5, problem (2.40) has a «periodic» solution $m : D(m) \subset \mathcal{B}(\mathbb{R}^N) \to \mathbb{C}$ given by $D(m) = \mathcal{C}_{per}(Y)$ and $m(u) = \frac{1}{|Y|}\int_Y u(y)dy$, $u \in D(m)$, where Y is any N-dimensional (non-necessarily rectangular) open parallelepiped in $\mathbb{R}^N$ centred at the origin. It is of great interest to note that this is the only solution of (2.40) admitting an integral representation with respect to Lebesgue measure on $\mathbb{R}^N$. Therefore, any attempt at constructing a non-stochastic homogenization theory beyond the periodic setting requires a profound change of orientation. Hence, as far as the book is concerned, the Gelfand representation theory as the main tool needed to frame an extended setting including the periodic one as a particular case, and highly fitted for the mathematical analysis of real-life multiscale problems arising in science and engineering.

This being so, it is henceforth known from the present chapter that (2.40) admits a more general solution, that is to say, an unbounded linear form $m : D(m) \subset \mathcal{B}(\mathbb{R}^N) \to \mathbb{C}$, where $D(m)$ is a homogenization algebra on $\mathbb{R}^N$ under the action $\mathcal{H}$, and m is the restriction to $D(m)$ of the mean value M on $\mathbb{R}^N$ generated by $\mathcal{H}$. We have in particular

$$m(u) = \int_{\Delta(D(m))} \mathcal{G}(u)(s)d\beta(s) \quad (u \in D(m)), \tag{2.41}$$

where the metrizable compact space $\Delta(D(m))$ denotes the spectrum of the commutative C^*-algebra with identity, $D(m)$, β a positive Radon measure on $\Delta(D(m))$ of total mass 1, and $\mathcal{G}$ the Gelfand transformation on $D(m)$. This is illustrated by many concrete examples in the present chapter and in Chaps. 5, 7, and 8, as well.

Attention must be drawn to equality (2.41) which not only states that the general solution of (2.40) admits an integral representation but also constitutes a major foundation of the theory. Attention should also be paid to Sect. 2.5 where the coherency of the theory of homogenization algebras is pointed out in comparison with periodic homogenization algebras. It clearly appears that the former is a natural extension of the latter as illustrated below, where the relation $\mathcal{A} \simeq \mathcal{B}$ between two topological vector spaces $\mathcal{A}$ and $\mathcal{B}$ means that the latter are algebraically and topologically isomorphic. So, letting $A = \mathcal{C}_{per}(Y)$, where Y is as above, one has among other things:

1. $\mathcal{D}^m(\Delta(A)) \simeq \mathcal{C}^m_{per}(Y)$, $m \in \mathbb{N} \cup \{\infty\}$.
2. $L^p(\Delta(A)) \simeq L^p_{per}(Y)$, $1 \leq p < \infty$.
3. $W^{m,p}(\Delta(A)) \simeq W^{m,p}_{per}(Y)$, $1 < p < \infty$, $m \in \mathbb{N}^*$.
4. $\int_{\Delta(A)} \mathcal{G}(u)d\beta = \frac{1}{|Y|}\int_Y u(y)dy$, $u \in L^p_{per}(Y)$ $(1 \leq p < \infty)$.

Furthermore, it is worth anticipating that from the practical view-point in connection with multiscale problems, there is a total similarity between the results obtained in the H-algebras setting and those provided by the periodic setting, as will become apparent in the next chapters.

The use of function algebras in homogenization theory goes back to Jikov et al. [63] and Zhikov and Krivenko [151] under the names of algebras with mean values and ergodic algebras. Later on, the author introduced in [93, 95] the theory of homogenization algebras based on the Gelfand representation theory, as pointed out above. See also Nguetseng et al. [107] for the use of some specific function algebras in the framework of reiterated homogenization.

Closely connected to the notion of a homogenization algebra there is that of a homogenization structure. In short, the latter is to the former as a linear mapping is to a matrix in classical linear algebra. In fact, there is a bijective correspondance between homogenization structures and homogenization algebras. The interested reader is referred to [93, 95].

Let us finally attract attention to a few unsolved problems.

(1) Let $A = AP_{\mathcal{R}}(\mathbb{R}^N)$ (see Sect. 2.3). As established in Proposition 2.16, $\mathcal{D}(\Delta(A))$ is dense in $W^{m,2}(\Delta(A))$ $(m \geq 1)$. The somewhat delicate question to raise is: *For which real $p \geq 1$ (distinct from 2) and which integer $m \geq 1$ can one claim that $\mathcal{D}(\Delta(A))$ is dense in $W^{m,p}(\Delta(A))$? It goes without saying that Fourier analysis is helpless in the present case.*

(2) In the framework of Sect. 2.5, let $A = \mathcal{C}_{per}(Y)$. The canonical mapping $\mathcal{G} : L^p_{per}(Y) \to L^p(\Delta(A))$ defined for every $1 \leq p < \infty$ by Corollary 2.14 carries $W^{m,p}_{per}(Y)$ isometrically over $W^{m,p}(\Delta(A))$ provided $1 < p < \infty$ (Corollary 2.15). *The question is whether this result can be extended to the case $p = 1$.*

Chapter 3
Σ-Convergence: The Periodic Setting

In the general periodic setting introduced in Sect. 2.5.1 is developed the so-called Σ-convergence theory generalizing well-known Two-scale convergence and laying the foundations of a global approach to periodic homogenization.

Throughout this chapter the numerical space $\mathbb{R}^N$ is provided with an absorptive continuous action $\mathcal{H} = (H_\varepsilon)_{\varepsilon>0}$ of $\mathbb{R}_+^*$ satisfying the conditions stated in Chap. 2. We further assume here that condition (C) in Chap. 1 (Sect. 1.2.5) is satisfied. The character E denotes a set of positive real numbers admitting 0 as an accumulation point (for example $E = \mathbb{R}_+^*$). In the particular case when $E = (\varepsilon_n)_{n\in\mathbb{N}}$ with $0 < \varepsilon_n \leq 1$ and $\varepsilon_n \to 0$ as $n \to \infty$, we will refer to E as a *fundamental sequence.* Finally, Ω denotes an open set in $\mathbb{R}_x^N$ (the numerical space $\mathbb{R}^N$ of variables $x = (x_1, \cdots, x_N)$) whereas $Y \subset \mathbb{R}_y^N$ and $S \subset \mathbb{R}_y^N$ stand for the open parallelepiped and the network defined in (2.30) and (2.31), respectively. It is worth perhaps recalling that a function $u \in L^1_{loc}(\mathbb{R}_y^N)$ is termed Y*-periodic* if for every $k \in S$, one has $u(y+k) = u(y)$ for almost every $y \in \mathbb{R}^N$.

3.1 Σ-convergence in $L^p(\Omega)$

3.1.1 Definitions and Elementary Properties

Definition 3.1 Let $1 \leq p < \infty$. A sequence of functions $u_\varepsilon \in L^p(\Omega)$ ($\varepsilon \in E$) is said to weakly Σ-converge in $L^p(\Omega)$ to some $u_0 \in L^p(\Omega; L^p_{per}(Y))$ if as $E \ni \varepsilon \to 0$,

$$\int_\Omega u_\varepsilon(x)\psi(x, H_\varepsilon(x))dx \to \frac{1}{|Y|}\int\int_{\Omega\times Y} u_0(x,y)\psi(x,y)dxdy \tag{3.1}$$

for all $\psi \in L^{p'}(\Omega; \mathcal{C}_{per}(Y))$ $(\frac{1}{p'} + \frac{1}{p} = 1)$, where $|Y| = \lambda(Y)$.

G. Nguetseng, *Homogenization Algebras and Applications*, Springer Monographs in Mathematics, https://doi.org/10.1007/978-3-031-84705-9_3

We will briefly express *weak Σ-convergence* in $L^p(\Omega)$ by writing $u_\varepsilon \to u_0$ in $L^p(\Omega)$-weak Σ.

Remark 3.1 The trace function $x \mapsto \psi(x, H_\varepsilon(x))$ in (3.1) is understood in the sense of either Proposition 1.6 or Proposition 1.7 (according as $p > 1$ or $p = 1$). We will most of the time denote it by ψ^ε.

Remark 3.2 The unicity of u_0 in (3.1) is easily obtained by taking $\psi(x, y) = \varphi(x)w(y)$ with $\varphi \in \mathcal{K}(\Omega)$, $w = \sum_{k\in S} \tau_k\theta$ (a locally finite sum), where $\theta \in \mathcal{K}(Y)$. The function u_0 is referred to as the *weak Σ-limit* of $(u_\varepsilon)_{\varepsilon\in E}$.

Definition 3.2 Let the hypotheses be as in Definition 3.1. The sequence $(u_\varepsilon)_{\varepsilon\in E}$ is said to strongly Σ-converge in $L^p(\Omega)$ to some $u_0 \in L^p(\Omega; L^p_{per}(Y))$ if the following condition is fulfilled:

$$\begin{cases} \text{Given } \eta > 0 \text{ and } \psi \in L^p(\Omega; \mathcal{C}_{per}(Y)) \text{ satisfying} \\ |Y|^{-\frac{1}{p}} \|u_0 - \psi\|_{L^p(\Omega\times Y)} \le \dfrac{\eta}{2}, \text{ there is some } \alpha > 0 \\ \text{such that } \|u_\varepsilon - \psi^\varepsilon\|_{L^p(\Omega)} \le \eta \text{ for } \varepsilon \in E,\ \varepsilon \le \alpha. \end{cases} \tag{3.2}$$

We will write $u_\varepsilon \to u_0$ in $L^p(\Omega)$-strong Σ to shortly express *strong Σ-convergence* in $L^p(\Omega)$, and call u_0 a *strong Σ-limit* of $(u_\varepsilon)_{\varepsilon\in E}$.

Remark 3.3 A function such as ψ in (3.2) does exist, since $L^p(\Omega; \mathcal{C}_{per}(Y))$ is dense in $L^p(\Omega; L^p_{per}(Y))$.

We shall now establish the unicity of u_0 in Definition 3.2.

Proposition 3.1 *Suppose $u_\varepsilon \to u_0$ in $L^p(\Omega)$-strong Σ. Then u_0 is unique.*

Proof Suppose the sequence $(u_\varepsilon)_{\varepsilon\in E}$ has two strong Σ-limits $u_0, u_1 \in L^p(\Omega; L^p_{per}(Y))$. According to (3.2), let us choose some $\psi_i \in L^p(\Omega; \mathcal{C}_{per}(Y))$ such that $|Y|^{-\frac{1}{p}} \|u_i - \psi_i\|_{L^p(\Omega\times Y)} \le \frac{\eta}{6}$ $(i = 0, 1)$. This yields some $\alpha > 0$ such that $\|u_\varepsilon - \psi_i^\varepsilon\|_{L^p(\Omega)} \le \frac{\eta}{3}$ $(i = 0, 1)$ for $\varepsilon \in E$, $\varepsilon \le \alpha$. It follows $\|\psi_1^\varepsilon - \psi_0^\varepsilon\|_{L^p(\Omega)} \le \frac{2\eta}{3}$ for $\varepsilon \in E$, $\varepsilon \le \alpha$. By Remark 1.12, one deduces (by letting $E \ni \varepsilon \to 0$) $|Y|^{-\frac{1}{p}} \|\psi_1 - \psi_0\|_{L^p(\Omega\times Y)} \le \frac{2\eta}{3}$. Hence, by writing $u_1 - u_0 = u_1 - \psi_1 + \psi_1 - \psi_0 + \psi_0 - u_0$, we get $|Y|^{-\frac{1}{p}} \|u_1 - u_0\|_{L^p(\Omega\times Y)} \le \eta$. Since η is arbitrary, it follows that $u_1 = u_0$. ∎

Before we proceed any farther let us illustrate Σ-convergence in $L^p(\Omega)$ by one fundamental example.

Example 3.1 Let $\psi \in L^p(\Omega; \mathcal{C}_{per}(Y))$ $(1 \le p < +\infty)$. Then, as $\varepsilon \to 0$,

(1) $\psi^\varepsilon \to \psi$ in $L^p(\Omega)$-weak Σ.
(2) $\psi^\varepsilon \to \psi$ in $L^p(\Omega)$-strong Σ.

Indeed, this follows readily by Remark 1.12.

The next proposition gives a simple characterization of the strong Σ-limit when the latter is smooth enough in the periodicity variable.

Proposition 3.2 *Let $u_\varepsilon \in L^p(\Omega)$ ($\varepsilon \in E$). The following are equivalent for a function $u_0 \in L^p(\Omega; \mathcal{C}_{per}(Y))$.*

(i) $u_\varepsilon \to u_0$ in $L^p(\Omega)$-strong Σ.
(ii) $\|u_\varepsilon - u_0^\varepsilon\|_{L^p(\Omega)} \to 0$ as $E \ni \varepsilon \to 0$, where $u_0^\varepsilon(x) = u_0(x, H_\varepsilon(x))$.

Proof Assuming (i), we see readily that (ii) follows by choosing $\psi = u_0$ in (3.2). Assuming now (ii), let $\eta > 0$. Fix freely some $\psi \in L^p(\Omega; \mathcal{C}_{per}(Y))$ with $|Y|^{-\frac{1}{p}}\|u_0 - \psi\|_{L^p(\Omega\times Y)} \leq \frac{\eta}{2}$. By recalling that $\|u_0^\varepsilon - \psi^\varepsilon\|_{L^p(\Omega)}$ has limit $|Y|^{-\frac{1}{p}}\|u_0 - \psi\|_{L^p(\Omega\times Y)}$ when $E \ni \varepsilon \to 0$ (Remark 1.12), this yields some $\alpha_1 > 0$ such that $\|u_0^\varepsilon - \psi^\varepsilon\|_{L^p(\Omega)} \leq \frac{\eta}{2} + \frac{\eta}{3}$ for $\varepsilon \in E$, $\varepsilon \leq \alpha_1$. On the other hand, according to (ii), there is some $\alpha_2 > 0$ such that $\|u_\varepsilon - u_0^\varepsilon\|_{L^p(\Omega)} \leq \frac{\eta}{6}$ for $\varepsilon \in E$, $\varepsilon \leq \alpha_2$. Hence

$$\|u_\varepsilon - \psi^\varepsilon\|_{L^p(\Omega)} \leq \|u_\varepsilon - u_0^\varepsilon\|_{L^p(\Omega)} + \|u_0^\varepsilon - \psi^\varepsilon\|_{L^p(\Omega)} \leq \frac{\eta}{6} + \frac{5\eta}{6} = \eta$$

for $\varepsilon \in E$, $\varepsilon \leq \alpha = \min(\alpha_1, \alpha_2)$. This shows (i). ∎

The preceding proposition is likely to help us have a clear idea of the somewhat abstract concept of strong Σ-convergence. Likewise we have the following

Proposition 3.3 *Let $1 \leq p < \infty$. Suppose a sequence of functions $u_\varepsilon \in L^p(\Omega)$ ($\varepsilon \in E$) is strongly Σ-convergent to some $u_0 \in L^p(\Omega; L^p_{per}(Y))$. Then*

(i) $u_\varepsilon \to u_0$ in $L^p(\Omega)$-weak Σ.
(ii) $\|u_\varepsilon\|_{L^p(\Omega)} \to |Y|^{-\frac{1}{p}}\|u_0\|_{L^p(\Omega\times Y)}$ as $E \ni \varepsilon \to 0$.
Furthermore the converse is true if $p = 2$.

Proof Fix freely some $\psi \in L^{p'}(\Omega; \mathcal{C}_{per}(Y))$. Let $c > \|\psi\|_{L^{p'}(\Omega;\mathcal{B}(\mathbb{R}^N))}$. Let $\eta > 0$. Consider some $f \in L^p(\Omega; \mathcal{C}_{per}(Y))$ such that $|Y|^{-\frac{1}{p}}\|u_0 - f\|_{L^p(\Omega\times Y)} \leq \frac{\eta}{4c}$, and then, by hypothesis, some $\alpha_0 > 0$ such that $\|u_\varepsilon - f^\varepsilon\|_{L^p(\Omega)} \leq \frac{\eta}{2c}$ for $\varepsilon \in E$, $\varepsilon \leq \alpha_0$. On the other hand, since $f^\varepsilon \to f$ in $L^p(\Omega)$-weak Σ (Example 3.1), we may consider some $\alpha_1 > 0$ such that

$$\left|\int_\Omega f^\varepsilon\psi^\varepsilon\, dx - \frac{1}{|Y|}\int\int_{\Omega\times Y} f\psi\, dxdy\right| \leq \frac{\eta}{4}$$

provided $0 < \varepsilon \leq \alpha_1$. Hence, by writing

$$\int_\Omega u_\varepsilon\psi^\varepsilon\, dx - \frac{1}{|Y|}\int\int_{\Omega\times Y} u_0\psi\, dxdy = \frac{1}{|Y|}\int_{\Omega\times Y}(f - u_0)\psi\, dxdy$$
$$+ \int_\Omega (u_\varepsilon - f^\varepsilon)\psi^\varepsilon\, dx + \int_\Omega f^\varepsilon\psi^\varepsilon\, dx - \frac{1}{|Y|}\int\int_{\Omega\times Y} f\psi\, dxdy,$$

we readily see by Hölder's inequality and use of Propositions 1.6 or 1.7 (according as $p > 1$ or $p = 1$), that

$$\left|\int_{\Omega} u_{\varepsilon}\psi^{\varepsilon}\, dx - \frac{1}{|Y|}\int\int_{\Omega\times Y} u_0\psi\, dxdy\right| \leq \eta$$

for $\varepsilon \in E$, $\varepsilon \leq \alpha = \min(\alpha_0, \alpha_1)$. This shows (i).

We turn now to the proof of (ii). Let $\eta > 0$. Consider some f in $L^p(\Omega; \mathcal{C}_{per}(Y))$ such that $|Y|^{-\frac{1}{p}}\|u_0 - f\|_{L^p(\Omega\times Y)} \leq \frac{\eta}{6}$. By hypothesis this yields some $\alpha_0 > 0$ such that $\|u_\varepsilon - f^\varepsilon\|_{L^p(\Omega)} \leq \frac{\eta}{3}$ for $\varepsilon \in E$, $\varepsilon \leq \alpha_0$. Thus,

$$|Y|^{-\frac{1}{p}}\left|\|u_0\|_{L^p(\Omega\times Y)} - \|f\|_{L^p(\Omega\times Y)}\right| \leq \frac{\eta}{6}$$

and

$$\left|\|u_\varepsilon\|_{L^p(\Omega)} - \|f^\varepsilon\|_{L^p(\Omega)}\right| \leq \frac{\eta}{3} \quad (\varepsilon \in E,\ \varepsilon \leq \alpha_0).$$

On the other hand, according to Remark 1.12, there is some $\alpha_1 > 0$ such that $\left|\|f^\varepsilon\|_{L^p(\Omega)} - |Y|^{-\frac{1}{p}}\|f\|_{L^p(\Omega\times Y)}\right| \leq \frac{\eta}{2}$ for $0 < \varepsilon \leq \alpha_1$. Hence, by the obvious inequality

$$\begin{aligned}\left|\|u_\varepsilon\|_{L^p(\Omega)} - |Y|^{-\frac{1}{p}}\|u_0\|_{L^p(\Omega\times Y)}\right| &\leq \left|\|u_\varepsilon\|_{L^p(\Omega)} - \|f^\varepsilon\|_{L^p(\Omega)}\right| + \left|\|f^\varepsilon\|_{L^p(\Omega)}\right.\\ &\quad \left. - |Y|^{-\frac{1}{p}}\|f\|_{L^p(\Omega\times Y)}\right| + |Y|^{-\frac{1}{p}}\left|\|f\|_{L^p(\Omega\times Y)} - \|u_0\|_{L^p(\Omega\times Y)}\right|\end{aligned}$$

one arrives at once at

$$\left|\|u_\varepsilon\|_{L^p(\Omega)} - |Y|^{-\frac{1}{p}}\|u_0\|_{L^p(\Omega\times Y)}\right| \leq \eta$$

for $\varepsilon \in E$, $\varepsilon \leq \alpha = \min(\alpha_0, \alpha_1)$. Whence (ii) follows.

To finish, let us assume $p = 2$. For any arbitrary fixed $\psi \in L^2(\Omega; \mathcal{C}_{per}(Y))$, we have

$$\|u_\varepsilon - \psi^\varepsilon\|^2_{L^2(\Omega)} = \|u_\varepsilon\|^2_{L^2(\Omega)} - 2\mathrm{Re}\int_{\Omega} u_\varepsilon\overline{\psi^\varepsilon}\, dx + \|\psi^\varepsilon\|^2_{L^2(\Omega)}.$$

By letting $E \ni \varepsilon \to 0$ we see by (i)–(ii) that

$$\lim_{E\ni\varepsilon\to 0}\|u_\varepsilon - \psi^\varepsilon\|^2_{L^2(\Omega)} = \frac{1}{|Y|}\|u_0 - \psi\|^2_{L^2(\Omega\times Y)},$$

from which one readily deduces (3.2). This completes the proof. ■

The following remark points out an interesting relationship between Σ-convergence and classical weak and strong convergence in $L^p(\Omega)$.

Remark 3.4 Let $u_\varepsilon \in L^p(\Omega)$ $(\varepsilon \in E)$.

(1) If the sequence $(u_\varepsilon)_{\varepsilon\in E}$ is weakly Σ-convergent in $L^p(\Omega)$ to some $u_0 \in L^p(\Omega; L^p_{per}(Y))$, then it weakly converges in $L^p(\Omega)$ to $\widetilde{u}_0 \in L^p(\Omega)$ given by $\widetilde{u}_0(x) = \frac{1}{|Y|}\int_Y u_0(x, y)\, dy \quad (x \in \Omega)$. If moreover E is a fundamental sequence, then $(u_\varepsilon)_{\varepsilon\in E}$ is bounded in $L^p(\Omega)$.
(2) If $(u_\varepsilon)_{\varepsilon\in E}$ converges in the $L^p(\Omega)$ norm to some $u_0 \in L^p(\Omega)$, then $(u_\varepsilon)_{\varepsilon\in E}$ strongly Σ-converges in $L^p(\Omega)$ to u_0 (viewed as a function in, say $L^p(\Omega; \mathcal{C}_{per}(Y))$).

3.1.2 Basic Convergence Results

The present section is devoted to a series of basic results of practical interest in periodic homogenization.

Proposition 3.4 *Let $1 < p, q < \infty$ with $\frac{1}{p} + \frac{1}{q} = \frac{1}{r} \le 1$. Suppose a sequence $(u_\varepsilon)_{\varepsilon\in E}$ is strongly Σ-convergent in $L^p(\Omega)$ to some $u_0 \in L^p(\Omega; L^p_{per}(Y))$, and a sequence $(v_\varepsilon)_{\varepsilon\in E}$ is weakly Σ-convergent in $L^q(\Omega)$ to some $v_0 \in L^q(\Omega; L^q_{per}(Y))$. Then $u_\varepsilon v_\varepsilon \to u_0 v_0$ in $L^r(\Omega)$-weak Σ.*

Proof One may assume without loss of generality that E is a fundamental sequence. Consequently $(v_\varepsilon)_{\varepsilon\in E}$ is norm bounded in $L^q(\Omega)$ (Remark 3.4). This being so, fix freely some $\psi \in L^{r'}(\Omega; \mathcal{C}_{per}(Y))$ $(\frac{1}{r'} + \frac{1}{r} = 1)$, and let $c > 0$ with

$$|Y|^{-\frac{1}{q}} \|v_0\|_{L^q(\Omega\times Y)} \|\psi\|_{L^{r'}(\Omega;\mathcal{B}(\mathbb{R}^N))} \le c$$

and

$$\|\psi\|_{L^{r'}(\Omega;\mathcal{B}(\mathbb{R}^N))} \sup_{\varepsilon\in E} \|v_\varepsilon\|_{L^q(\Omega)} \le c.$$

Now, let $\eta > 0$. Based on the strong Σ-convergence of $(u_\varepsilon)_{\varepsilon\in E}$, introduce some $f \in L^p(\Omega; \mathcal{C}_{per}(Y))$ such that $|Y|^{-\frac{1}{p}} \|u_0 - f\|_{L^p(\Omega\times Y)} \le \frac{\eta}{6c}$, and then some real $\alpha_0 > 0$ such that $\|u_\varepsilon - f^\varepsilon\|_{L^p(\Omega)} \le \frac{\eta}{3c}$ for $\varepsilon \in E$, $\varepsilon \le \alpha_0$. Finally, since $f\psi \in L^{q'}(\Omega; \mathcal{C}_{per}(Y))$, by the weak Σ-convergence hypothesis on $(v_\varepsilon)_{\varepsilon\in E}$ we may consider some $\alpha_1 > 0$ such that

$$\left| \int_\Omega v_\varepsilon f^\varepsilon \psi^\varepsilon \, dx - \frac{1}{|Y|} \int\int_{\Omega\times Y} v_0 f \psi \, dxdy \right| \le \frac{\eta}{2}$$

for $\varepsilon \in E$, $\varepsilon \leq \alpha_1$. Hence, by writing

$$\begin{aligned}\int_\Omega u_\varepsilon v_\varepsilon \psi^\varepsilon\, dx - \frac{1}{|Y|}\int\int_{\Omega\times Y} u_0 v_0 \psi\, dxdy &= \int_\Omega (u_\varepsilon - f^\varepsilon) v_\varepsilon \psi^\varepsilon\, dx \\ &+ \frac{1}{|Y|}\int_{\Omega\times Y} (f - u_0) v_0 \psi\, dxdy + \int_\Omega v_\varepsilon f^\varepsilon \psi^\varepsilon\, dx \\ &- \frac{1}{|Y|}\int\int_{\Omega\times Y} v_0 f \psi\, dxdy\end{aligned}$$

one sees by suitable estimates and use of Propositions 1.6 or 1.7 according as $r > 1$ or $r = 1$ (the details are left to the reader), that

$$\left|\int_\Omega u_\varepsilon v_\varepsilon \psi^\varepsilon\, dx - \frac{1}{|Y|}\int\int_{\Omega\times Y} u_0 v_0 \psi\, dxdy\right| \leq \eta$$

provided $\varepsilon \in E$, $\varepsilon \leq \alpha = \min(\alpha_0, \alpha_1)$. The proof is complete. ■

The concept of Σ-convergence would be of no practical value if the following compactness theorem was not available.

Theorem 3.1 *Let $1 < p < \infty$. Suppose E is a fundamental sequence, and let a sequence $(u_\varepsilon)_{\varepsilon\in E}$ be norm bounded in $L^p(\Omega)$. Then, a subsequence E' can be extracted from E such that $(u_\varepsilon)_{\varepsilon\in E'}$ weakly Σ-converges in $L^p(\Omega)$ to some $u_0 \in L^p(\Omega; L^p_{per}(Y))$.*

Proof For each $\varepsilon \in E$, let

$$F_\varepsilon(\psi) = \int_\Omega u_\varepsilon(x)\psi(x, H_\varepsilon(x))dx, \quad \psi \in L^{p'}(\Omega; \mathcal{C}_{per}(Y)).$$

This yields a sequence $(F_\varepsilon)_{\varepsilon\in E}$ which is bounded in the Banach space $[L^{p'}(\Omega; \mathcal{C}_{per}(Y))]'$ (strong dual of $L^{p'}(\Omega; \mathcal{C}_{per}(Y))$), according to Proposition 1.6. Therefore, $L^{p'}(\Omega; \mathcal{C}_{per}(Y))$ being a separable Banach space, a subsequence E' can be extracted from E such that as $E' \ni \varepsilon \to 0$,

$$F_\varepsilon \to F_0 \text{ in } [L^{p'}(\Omega; \mathcal{C}_{per}(Y))]'\text{-weak} * .$$

The next point is to characterize the limit F_0. For that purpose, let $\psi \in L^{p'}(\Omega; \mathcal{C}_{per}(Y))$. One has

$$|F_\varepsilon(\psi)| \leq c\left(\int_\Omega |\psi(x, H_\varepsilon(x))|^{p'} dx\right)^{\frac{1}{p'}} \quad (\varepsilon \in E'),$$

with $\sup_{\varepsilon\in E}\|u_\varepsilon\|_{L^p(\Omega)} = c$. By letting $E' \ni \varepsilon \to 0$ and recalling Remark 1.12, it follows $|F_0(\psi)| \leq c|Y|^{-\frac{1}{p'}}\|\psi\|_{L^{p'}(\Omega\times Y)}$, and that for any $\psi \in L^{p'}(\Omega; \mathcal{C}_{per}(Y))$. Consequently, $L^{p'}(\Omega; \mathcal{C}_{per}(Y))$ being dense in $L^{p'}(\Omega; L^{p'}_{per}(Y))$, F_0 extends by continuity to an element of $[L^{p'}(\Omega; L^{p'}_{per}(Y))]' = L^p(\Omega; L^p_{per}(Y))$. More precisely, there exists $u_0 \in L^p(\Omega; L^p_{per}(Y))$ such that as $E' \ni \varepsilon \to 0$,

$$\int_\Omega u_\varepsilon(x)\psi(x, H_\varepsilon(x))dx \to \frac{1}{|Y|}\int\int_{\Omega\times Y} u_0(x, y)\psi(x, y)dxdy$$

for all $\psi \in L^{p'}(\Omega; \mathcal{C}_{per}(Y))$. This completes the proof. ■

The preceding theorem enables us to prove a useful result closely connected with Proposition 3.4.

Proposition 3.5 *Let $1 < p < \infty$. Let E be a fundamental sequence, $(u_\varepsilon)_{\varepsilon\in E}$ a sequence in $L^p(\Omega)$, and $(v_\varepsilon)_{\varepsilon\in E}$ a sequence in $L^{p'}(\Omega) \cap L^\infty(\Omega)$ $(\frac{1}{p'} + \frac{1}{p} = 1)$ such that*

(i) $u_\varepsilon \to u_0$ in $L^p(\Omega)$-weak Σ;
(ii) $v_\varepsilon \to v_0$ in $L^{p'}(\Omega)$-strong Σ;
(iii) $(v_\varepsilon)_{\varepsilon\in E}$ is bounded in $L^\infty(\Omega)$.
Then $u_\varepsilon v_\varepsilon \to u_0 v_0$ in $L^p(\Omega)$-weak Σ.

Proof By Proposition 3.4 we have $u_\varepsilon v_\varepsilon \to u_0 v_0$ in $L^1(\Omega)$-weak Σ. Thus, as $E \ni \varepsilon \to 0$, one has

$$\int_\Omega u_\varepsilon v_\varepsilon \psi^\varepsilon\, dx \to \frac{1}{|Y|}\int\int_{\Omega\times Y} u_0 v_0 \psi\, dxdy$$

for all $\psi \in \mathcal{K}(\Omega; \mathcal{C}_{per}(Y)))$. On the other hand, we observe that the sequence $(u_\varepsilon v_\varepsilon)_{\varepsilon\in E}$ is bounded in $L^p(\Omega)$. Therefore, thanks to Theorem 3.1 one can extract E' from E such that $(u_\varepsilon v_\varepsilon)_{\varepsilon\in E'}$ weakly Σ-converges in $L^p(\Omega)$ to some $z_0 \in L^p(\Omega; L^p_{per}(Y))$. This implies that, as $E' \ni \varepsilon \to 0$,

$$\int_\Omega u_\varepsilon v_\varepsilon \psi^\varepsilon\, dx \to \frac{1}{|Y|}\int\int_{\Omega\times Y} z_0 \psi\, dxdy$$

for all $\psi \in \mathcal{K}(\Omega; \mathcal{C}_{per}(Y)))$. Hence, it follows

$$\int\int_{\Omega\times Y} (z_0 - u_0 v_0)\psi\, dxdy = 0$$

for all $\psi \in \mathcal{K}(\Omega; \mathcal{C}_{per}(Y))$. By reasoning as in Remark 3.2 we deduce $z_0 = u_0 v_0$, thereby proving the proposition. ■

For further needs it is worth recalling that $L^{\infty}_{per}(Y)$ denotes the space of all functions in $L^{\infty}_{loc}(\mathbb{R}^N)$ that are Y-periodic (Sect. 2.5). We have $L^{\infty}_{per}(Y) \subset L^{\infty}(\mathbb{R}^N)$. Equipped with the $L^{\infty}(\mathbb{R}^N)$ norm, $L^{\infty}_{per}(Y)$ is a Banach space.

Proposition 3.6 *Let $1 < p < \infty$. Suppose a sequence $(u_\varepsilon)_{\varepsilon \in E}$ is weakly Σ-convergent in $L^p(\Omega)$ to some $u_0 \in L^p(\Omega; L^p_{per}(Y))$. Suppose further that Ω is bounded. Then (3.1) still holds for $\psi \in \mathcal{C}(\overline{\Omega}; L^{\infty}_{per}(Y))$.*

Remark 3.5 For $\psi \in \mathcal{C}(\overline{\Omega}; L^{\infty}_{per}(Y))$, the trace function $x \mapsto \psi^{\varepsilon}(x) = \psi(x, H_\varepsilon(x))$ is well defined, and it belongs to $L^{\infty}(\Omega)$ (Proposition 1.8).

Proof of Proposition 3.6 This proceeds in two steps.

Step 1. The first step is to show that (3.1) holds for any $\psi \in \mathcal{C}(\overline{\Omega}) \otimes L^{\infty}_{per}(Y)$. Clearly it is enough to prove this for the elementary functions

$$\psi = \varphi \otimes f, \ \ \varphi \in \mathcal{C}(\overline{\Omega}), \ \ f \in L^{\infty}_{per}(Y). \tag{3.3}$$

On the other hand, one may assume without loss of generality that E is a fundamental sequence, so that $(u_\varepsilon)_{\varepsilon \in E}$ is norm bounded in $L^p(\Omega)$.
Now, let us fix freely some ψ as in (3.3). Next, introduce a constant $c > 0$ satisfying

$$\begin{aligned} &\frac{1}{|Y|}\|\varphi\|_{L^{p'}(\Omega)}\|u_0\|_{L^p(\Omega\times Y)} \le c \ \ (\frac{1}{p} + \frac{1}{p'} = 1) \\ &c_0\|\varphi\|_\infty \sup_{\varepsilon\in E}\|u_\varepsilon\|_{L^p(\Omega)} \le c, \end{aligned} \tag{3.4}$$

where $c_0 > 0$ is a constant such that $\|w^{\varepsilon}\|_{L^{p'}(\Omega)} \le c_0\|w\|_{L^{p'}(Y)}$ $(0 < \varepsilon \le 1)$ for all $w \in L^{p'}_{per}(Y)$, according to Lemma 1.9 (see also Proposition 1.2 and Remarks 1.9 and 1.15). Having done this, let $\eta > 0$. By the relation $L^{\infty}_{per}(Y) \subset L^{p'}_{per}(Y)$ and use of the density of $\mathcal{C}_{per}(Y)$ in $L^{p'}_{per}(Y)$, we can find some $g \in \mathcal{C}_{per}(Y)$ such that

$$\|f - g\|_{L^{p'}(Y)} \le \frac{\eta}{3c}. \tag{3.5}$$

Put $\Phi = \varphi \otimes g$, and observe that $\Phi \in \mathcal{C}(\overline{\Omega}; \mathcal{C}_{per}(Y)) \subset L^{p'}(\Omega; \mathcal{C}_{per}(Y))$. Therefore, there is some $0 < \alpha \le 1$ such that

$$\left|\int_\Omega u_\varepsilon \Phi^{\varepsilon}\, dx - \frac{1}{|Y|}\int\int_{\Omega\times Y} u_0 \Phi\, dxdy\right| \le \frac{\eta}{3} \tag{3.6}$$

for $\varepsilon \in E,\ \varepsilon \leq \alpha$. On the other hand, let

$$\begin{aligned}\int_\Omega u_\varepsilon \psi^\varepsilon\, dx - \frac{1}{|Y|}\int\int_{\Omega\times Y} u_0\psi\, dxdy &= \int_\Omega u_\varepsilon(\psi^\varepsilon - \Phi^\varepsilon)\, dx \\ &+ \int_\Omega u_\varepsilon \Phi^\varepsilon\, dx - \frac{1}{|Y|}\int\int_{\Omega\times Y} u_0\Phi\, dxdy \\ &+ \frac{1}{|Y|}\int\int_{\Omega\times Y} u_0(\Phi - \psi)\, dxdy.\end{aligned} \tag{3.7}$$

The aim is to estimate the first and last integrals on the right. By Hölder's inequality we have

$$\begin{aligned}\left|\int_\Omega u_\varepsilon(\psi^\varepsilon - \Phi^\varepsilon)\, dx\right| &\leq \|u_\varepsilon\|_{L^p(\Omega)}\|\varphi\|_\infty\|f^\varepsilon - g^\varepsilon\|_{L^{p'}(\Omega)} \\ &\leq c_0 \sup_{\sigma\in E}\|u_\sigma\|_{L^p(\Omega)}\|\varphi\|_\infty\|f - g\|_{L^{p'}(Y)} \\ &\leq \frac{\eta}{3}\quad \text{for } \varepsilon \in E,\ \varepsilon \leq 1,\end{aligned}$$

thanks to (3.4) and (3.5). Likewise, the same line of argument leads us readily to

$$\begin{aligned}\left|\int\int_{\Omega\times Y} u_0(\Phi - \psi)\, dxdy\right| &\leq \|u_0\|_{L^p(\Omega\times Y)}\|\varphi\|_{L^{p'}(\Omega)}\|f - g\|_{L^{p'}(Y)} \\ &\leq |Y|\frac{\eta}{3}.\end{aligned}$$

We deduce, using (3.6),

$$\left|\int_\Omega u_\varepsilon \psi^\varepsilon\, dx - \frac{1}{|Y|}\int\int_{\Omega\times Y} u_0\psi\, dxdy\right| \leq \eta \tag{3.8}$$

for $\varepsilon \in E,\ \varepsilon \leq \alpha$. Therefore, (3.1) holds for ψ as in (3.3), hence by linearity for all $\psi \in \mathcal{C}(\overline{\Omega}) \otimes L^\infty_{per}(Y)$, as claimed.

Step 2. To start with, let us introduce a constant $c > 0$ such that

$$\frac{1}{|Y|}\|u_0\|_{L^1(\Omega\times Y)} \leq c \text{ and } \sup_{\varepsilon\in E}\|u_\varepsilon\|_{L^1(\Omega)} \leq c.$$

Fix freely some $\psi \in \mathcal{C}(\overline{\Omega}; L^\infty_{per}(Y))$ and let $\eta > 0$. By density (see e.g., Bourbaki [22, p. 46]), consider some $\Phi \in \mathcal{C}(\overline{\Omega}) \otimes L^\infty_{per}(Y)$ such that

$$\|\psi - \Phi\|_{\mathcal{C}(\overline{\Omega};L^\infty(\mathbb{R}^N))} \equiv \sup_{x\in\overline{\Omega}}\|\psi(x,\cdot) - \Phi(x,\cdot)\|_{L^\infty(\mathbb{R}^N)} \leq \frac{\eta}{3c}.$$

Furthermore, on account of Step 1, let $\alpha > 0$ be such that (3.6) holds for $\varepsilon \in E$, $\varepsilon \leq \alpha$. Finally, consider the obvious equality (3.7), and use the simple estimates

$$\begin{aligned}\left|\int_\Omega u_\varepsilon(\psi^\varepsilon - \Phi^\varepsilon)\,dx\right| &\leq \|u_\varepsilon\|_{L^1(\Omega)}\|\psi^\varepsilon - \Phi^\varepsilon\|_{L^\infty(\Omega)}\\ &\leq c\|\psi - \Phi\|_{\mathcal{C}(\overline{\Omega};L^\infty(\mathbb{R}^N))} \text{ (see Proposition 1.8)}\end{aligned}$$

and

$$\begin{aligned}\left|\iint_{\Omega\times Y} u_0(\Phi - \psi)\,dxdy\right| &\leq \|u_0\|_{L^1(\Omega\times Y)}\|\psi - \Phi\|_{\mathcal{C}(\overline{\Omega};L^\infty(\mathbb{R}^N))}\\ &\leq c|Y|\|\psi - \Phi\|_{\mathcal{C}(\overline{\Omega};L^\infty(\mathbb{R}^N))}\end{aligned}$$

to arrive at (3.8) for $\varepsilon \in E$, $\varepsilon \leq \alpha$. Hence (3.1) follows for all $\psi \in \mathcal{C}(\overline{\Omega}; L^\infty_{per}(Y))$. This completes the proof.

■

This proposition has two corollaries worth mentioning.

Corollary 3.1 *Let $u \in \mathcal{C}(\overline{\Omega}; L^\infty_{per}(Y))$. Suppose Ω is bounded. Then, as $0 < \varepsilon \to 0$, we have $u^\varepsilon \to u$ in $L^p(\Omega)$-weak Σ $(1 < p < \infty)$.*

Proof For arbitrary given $\psi \in L^{p'}(\Omega; \mathcal{C}_{per}(Y))$, Example 3.1 in combination with Proposition 3.6 yields

$$\int_\Omega u^\varepsilon \psi^\varepsilon\,dx \to \frac{1}{|Y|}\iint_{\Omega\times Y} u\psi\,dxdy$$

as $\varepsilon \to 0$. Hence the corollary follows by the arbitrariness of ψ. ■

Corollary 3.2 *Let $1 < p < \infty$. Suppose Ω is bounded. Let a sequence $(u_\varepsilon)_{\varepsilon\in E}$ be bounded in $L^p(\Omega)$. Then, the following are equivalent for $u_0 \in L^p(\Omega; L^p_{per}(Y))$.*

(i) $u_\varepsilon \to u_0$ in $L^p(\Omega)$-weak Σ.
(ii) As $E \ni \varepsilon \to 0$, one has (3.1) for all $\psi \in \mathcal{C}(\overline{\Omega}; L^\infty_{per}(Y))$.

Proof That (i) implies (ii) is nothing else than Proposition 3.6. Suppose now that (ii) holds. By noting that the space $\mathcal{K}(\Omega; \mathcal{C}_{per}(Y))$ is contained in $\mathcal{C}(\overline{\Omega}; L^\infty_{per}(Y))$ and at the same time is dense in $L^{p'}(\Omega; \mathcal{C}_{per}(Y))$, one sees readily that (i) follows by mere routine. ■

3.2 Σ-Convergence in the Sobolev Spaces $W^{m,p}(\Omega)$

For simplicity we assume throughout the present section that $\mathcal{H}$ is particularized as

$$H_\varepsilon(x) = \frac{x}{\varepsilon} \quad (x \in \mathbb{R}^N).$$

In this particular case, the Σ-convergence is commonly known as the two-scale convergence.

3.2.1 Σ-convergence in $W^{1,p}(\Omega)$

Let $1 < p < \infty$, and let E be a fundamental sequence.

Theorem 3.2 *Let $(u_\varepsilon)_{\varepsilon\in E}$ be a bounded sequence in $W^{1,p}(\Omega)$. There are a subsequence E' from E and a couple of functions $u_0 \in W^{1,p}(\Omega)$, $u_1 \in L^p(\Omega; W^{1,p}_{per}(Y)/\mathbb{C})$, such that, as $E' \ni \varepsilon \to 0$,*

$$u_\varepsilon \to u_0 \text{ in } W^{1,p}(\Omega)\text{-weak}, \tag{3.9}$$

$$u_\varepsilon \to u_0 \text{ in } L^p(\Omega)\text{-weak } \Sigma, \tag{3.10}$$

$$\frac{\partial u_\varepsilon}{\partial x_j} \to \frac{\partial u_0}{\partial x_j} + \frac{\partial u_1}{\partial y_j} \text{ in } L^p(\Omega)\text{-weak } \Sigma \ \ (1 \le j \le N). \tag{3.11}$$

Proof Theorem 3.1 in combination with the reflexivity of $W^{1,p}(\Omega)$ yields a subsequence E' from E and a family of functions $u \in W^{1,p}(\Omega)$ and $u_0, U_j \in L^p(\Omega; L^p_{per}(Y))$ $(1 \le j \le N)$ such that as $E' \ni \varepsilon \to 0$, we have (3.10),

$$\frac{\partial u_\varepsilon}{\partial x_j} \to U_j \text{ in } L^p(\Omega)\text{-weak } \Sigma \ \ (1 \le j \le N) \tag{3.12}$$

and

$$u_\varepsilon \to u \text{ in } W^{1,p}(\Omega)\text{-weak}. \tag{3.13}$$

The next point is to check the equality $u = u_0$. For that purpose, let $\varphi \in \mathcal{D}(\Omega) = C_0^\infty(\Omega)$ and $\psi \in C^\infty_{per}(Y)$. Fix freely some index $1 \le j \le N$. Clearly

$$\varepsilon \int_\Omega \left(u_\varepsilon \psi^\varepsilon \frac{\partial \varphi}{\partial x_j} + \frac{\partial u_\varepsilon}{\partial x_j} \psi^\varepsilon \varphi \right) dx = - \int_\Omega u_\varepsilon \left(\frac{\partial \psi}{\partial y_j} \right)^\varepsilon \varphi dx,$$

where, as usual, one sets $f^\varepsilon(x) = f(\frac{x}{\varepsilon})$ for $f \in L^1_{loc}(\mathbb{R}^N)$. By letting $E' \ni \varepsilon \to 0$, we get by (3.10)

$$\int_\Omega \left(\int_Y u_0(x,y) \frac{\partial \psi}{\partial y_j}(y) dy \right) \varphi(x) dx = 0.$$

Hence, by the arbitrariness of φ,

$$\int_Y u_0(x,y)\frac{\partial\psi}{\partial y_j}(y)dy = 0 \text{ (for a.e. } x \in \Omega).$$

By choosing ψ of the form $\psi = \sum_{k\in S} \tau_k\theta,\ \ \theta \in \mathcal{D}(Y)$, one deduces readily

$$\frac{\partial u_0(x,\cdot)}{\partial y_j} = 0 \ (1 \le j \le N) \text{ for a.e. } x \in \Omega,$$

where the partial derivatives are taken in the distribution sense on Y. Thus, for almost every $x \in \Omega$, the function $y \mapsto u_0(x, y)$ is constant in $\mathbb{R}^N$, since it is Y-periodic and constant in Y. According to Remark 3.4 it follows by comparison with (3.13) that $u_0 = u$. Therefore, (3.9) and (3.10) are established. It remains to prove (3.11) with u_1 as stated above.

To do this, let

$$F_j = U_j - \frac{\partial u_0}{\partial x_j} \ \ (1 \le j \le N).$$

This gives a function $\mathbf{F} = (F_j) \in L^p(\Omega; L^p_{per}(Y)^N)$. On the other hand, for $\varphi \in \mathcal{D}(\Omega)$ and $\mathbf{\Phi} = (\psi_j) \in \mathcal{V}_{per}(Y)$, we have

$$\sum_{j=1}^N \int_\Omega \left[\varphi\frac{\partial u_\varepsilon}{\partial x_j} + u_\varepsilon\frac{\partial\varphi}{\partial x_j}\right]\psi_j^\varepsilon\, dx = -\frac{1}{\varepsilon}\int_\Omega u_\varepsilon(div_y\mathbf{\Phi})^\varepsilon\varphi\, dx = 0.$$

By letting $E' \ni \varepsilon \to 0$ we obtain using (3.10), (3.12) and the arbitrariness of φ,

$$\int_Y \mathbf{F}(x,\cdot)\cdot\mathbf{\Phi}\, dy = 0 \quad \text{for a.e. } x \in \Omega,$$

and that for $\mathbf{\Phi}$ ranging over $\mathcal{V}_{per}(Y)$ (see Sect. 3.3 for clarification). Lemma 2.11 applies, and the result is that there exists a function $x \mapsto u_1(x)$ from Ω to $W^{1,p}_{per}(Y)/\mathbb{C}$ such that

$$F_j(x,\cdot) = \frac{\partial u_1(x)}{\partial y_j} \ \ (1 \le j \le N) \text{ for a.e. } x \in \Omega.$$

Finally, it is an easy task to check that the said function, denoted by u_1, lies in $L^p(\Omega; W^{1,p}_{per}(Y)/\mathbb{C})$ and so (3.11) follows at once. The theorem is proven. ■

As a consequence of this theorem there is the following corollary whose usefulness will come to light later on.

Corollary 3.3 *Let $(u_\varepsilon)_{\varepsilon\in E}$ be a bounded sequence in $W^{1,p}(\Omega)$. There are a subsequence E' from E and some $u_1 \in L^p(\Omega; W^{1,p}_{per}(Y)/\mathbb{C})$ such that as $E' \ni \varepsilon \to 0$,*

$$\int_\Omega \frac{u_\varepsilon(x)}{\varepsilon}\psi\left(x, \frac{x}{\varepsilon}\right)dx \to \frac{1}{|Y|}\int\int_{\Omega\times Y} u_1(x,y)\psi(x,y)dxdy$$

for all $\psi \in \mathcal{D}(\Omega)\otimes[\mathcal{C}^\infty_{per}(Y)/\mathbb{C}]$.

Proof Let $\Phi \in \mathcal{D}(\Omega)\otimes\mathcal{C}^\infty_{per}(Y)$. Clearly

$$\int_\Omega \frac{u_\varepsilon}{\varepsilon}\left(\frac{\partial\Phi}{\partial y_j}\right)^\varepsilon dx = -\int_\Omega u_\varepsilon\left(\frac{\partial\Phi}{\partial x_j}\right)^\varepsilon dx - \int_\Omega \frac{\partial u_\varepsilon}{\partial x_j}\Phi^\varepsilon\, dx.$$

Let E' and u_1 be as in Theorem 3.2. By letting $E' \ni \varepsilon \to 0$ above one easily arrives at

$$\int_\Omega \frac{u_\varepsilon}{\varepsilon}\left(\frac{\partial\Phi}{\partial y_j}\right)^\varepsilon dx \to \frac{1}{|Y|}\int\int_{\Omega\times Y} u_1(x,y)\frac{\partial\Phi}{\partial y_j}(x,y)dxdy \tag{3.14}$$

thanks, among other things, to the equality

$$-\int_Y \frac{\partial u_1}{\partial y_j}(x,y)\Phi(x,y)dy = \int_Y u_1(x,y)\frac{\partial\Phi}{\partial y_j}(x,y)dy.$$

By taking in (3.14) $\Phi = \frac{\partial\theta}{\partial y_j}$ with $\theta \in \mathcal{D}(\Omega)\otimes\mathcal{C}^\infty_{per}(Y)$ and then summing in $1 \leq j \leq N$, we see that

$$\int_\Omega \frac{u_\varepsilon}{\varepsilon}(\Delta_y\theta)^\varepsilon\, dx \to \frac{1}{|Y|}\int\int_{\Omega\times Y} u_1\Delta_y\theta\, dxdy$$

as $E' \ni \varepsilon \to 0$. With this in mind, let $\psi \in \mathcal{D}(\Omega)\otimes[\mathcal{C}^\infty_{per}(Y)/\mathbb{C}]$. Specifically, let

$$\psi = \sum_{i\in I}\varphi_i\otimes\psi_i,\quad \varphi_i \in \mathcal{D}(\Omega),\, \psi_i \in \mathcal{C}^\infty_{per}(Y)/\mathbb{C},$$

where the index set I is finite and depends on ψ. For each $i \in I$, let θ_i be the function in $W^{1,2}_{per}(Y)/\mathbb{C}$ uniquely determined by the equation $\Delta_y\theta_i = \psi_i$ in $\mathbb{R}^N_y$. Because of the hypoellipticity of the Laplace operator Δ_y in $\mathbb{R}^N_y$, the function θ_i actually lies in $\mathcal{C}^\infty_{per}(Y)$. Hence, by letting

$$\theta = \sum_{i\in I}\varphi_i\otimes\theta_i,$$

it follows that ψ is representable as $\psi = \Delta_y \theta, \theta \in \mathcal{D}(\Omega) \otimes [\mathcal{C}_{per}^{\infty}(Y)/\mathbb{C}]$. Therefore the corollary follows. ■

3.2.2 Σ-convergence in $H^m(\Omega)$

Our main purpose here is to prove the analogue of Theorem 3.2 for the Sobolev space $H^m(\Omega) = W^{m,2}(\Omega)$, where m is any arbitrary positive integer. To accomplish this we need to state and prove an appropriate lemma generalizing Lemma 2.10. Before we can do this, however, we require a few preliminaries.

Let $X = \mathbb{R}_y^N$ or Ω, and let V be a vector subspace of $L_{loc}^1(X)$. To start with, define $\mathcal{T}_0(V) = V$, $\mathcal{T}_1(V) = V^N = V \times \cdots \times V$ (N times), and for any integer $n \geq 2$, put $\mathcal{T}_n(V)$ for the space of all N^n-tuples $\mathbf{T} = (T_{i_1 \cdots i_n})_{1 \leq i_1, \cdots, i_n \leq N}$ with $T_{i_1 \cdots i_n} \in V$. In other words, each $\mathbf{T} \in \mathcal{T}_n(V)$ is a tensor function on X of order n with components $T_{i_1 \cdots i_n}$ in V.

For example, $\mathbf{T} \in \mathcal{T}_2(V)$ is of the form $\mathbf{T} = (T_{ij})_{1 \leq i,j \leq N}$, $\mathbf{T} \in \mathcal{T}_3(V)$ is of the form $\mathbf{T} = (T_{ijk})_{1 \leq i,j,k \leq N}$.

It is clear that $\mathcal{T}_n(V)$ is a vector space under the usual pointwise operations on functions.

We suppose in the sequel that $n > 1$. Let $\mathbf{T} = (T_{i_1 \cdots i_n}) \in \mathcal{T}_n(\mathcal{C}^1(\mathbb{R}_y^N))$, where it is understood that each of the indices $i_1, \cdots, i_n$ runs through $\mathbb{N}_N^* = \{i | i \in \mathbb{N}, 1 \leq i \leq N\}$. The divergence of $\mathbf{T}$ is defined to be the tensor function $\mathbf{R} = (R_{i_1 \cdots i_{n-1}}) \in \mathcal{T}_{n-1}(\mathcal{C}(\mathbb{R}_y^N))$ given by

$$R_{i_1 \cdots i_{n-1}} = \sum_{j=1}^{N} \frac{\partial T_{i_1 \cdots i_{n-1} j}}{\partial y_j} \quad (1 \leq i_1, \cdots, i_{n-1} \leq N)$$

and denoted by $\mathbf{R} = div_y \mathbf{T}$. If $\mathbf{T} \in \mathcal{T}_n(\mathcal{C}^1(\Omega))$, $div_x \mathbf{T}$ is defined analogously. We will set

$$\mathcal{V}_{per}^{(n)}(Y) = \{\mathbf{T} | \mathbf{T} \in \mathcal{T}_n(\mathcal{C}_{per}^{\infty}(Y)), div_y \mathbf{T} = \mathbf{0}\}$$

where $\mathbf{0}$ stands for the origin in the vector space $\mathcal{T}_{n-1}(\mathcal{C}_{per}^{\infty}(Y))$. It is clear that $\mathcal{V}_{per}^{(n)}(Y)$ is a vector subspace of $\mathcal{T}_n(\mathcal{C}_{per}^{\infty}(Y))$. On the other hand, notice that $\mathcal{V}_{per}^{(1)}(Y) = \mathcal{V}_{per}(Y)$ (Lemmas 2.10 and 2.11).

The use of Fourier coefficients will be essential in this section. It is worth recalling that the Fourier coefficients of a function $u \in L_{per}^1(Y)$ are given by

$$\mathcal{F}_k(u) = \int_Y e^{-2\pi i k \cdot y} u(y)\, dy \quad (k \in S^*)$$

where the network S^* is defined in (2.32), with of course $S^* = \mathbb{Z}^N$ if $S = \mathbb{Z}^N$.

We will need the following lemma whose proof is a simple adaptation to the periodic setting (Sect. 2.5) of the proof of Lemma 2.7. The details are left to the reader.

Lemma 3.1 *Let $\boldsymbol{F} = (F_i) \in L^2_{per}(Y)^N$. The following statements are equivalent:*

(i) ***F*** *satisfies*

$$\sum_{i=1}^{N} \int_Y F_i \psi_i \, dy = 0$$

for all $\boldsymbol{\Phi} = (\psi_i) \in \mathcal{V}_{per}(Y)$.

(ii) ***F*** *satisfies*

$$\mathcal{F}_{k=\omega}(F_i) = 0 \ \ (\omega \text{ the origin in } \mathbb{R}^N_y), \ \ 1 \le i \le N,$$

$$k_i \mathcal{F}_k(F_j) - k_j \mathcal{F}_k(F_i) = 0, \ \ k \in S^*, \ \ 1 \le i, j \le N.$$

(iii) There exists a unique $q \in H^1_{per}(Y)/\mathbb{C}$ such that

$$\frac{\partial q}{\partial y_i} = F_i \quad (1 \le i \le N).$$

We conclude these preliminaries with one simple but very useful remark.

Remark 3.6 For every multi-index $p = (p_1, \cdots, p_N) \in \mathbb{N}^N$ with $|p| = p_1 + \cdots + p_N = n$, there is an n-tuple $(i_1, \cdots, i_n)$, $1 \le i_1, \cdots, i_n \le N$ such that

$$D^p_y = \frac{\partial^n}{\partial y_{i_1} \cdots \partial y_{i_n}}, \tag{3.15}$$

where $D^p_y = \frac{\partial^{|p|}}{\partial y_1^{p_1} \cdots \partial y_N^{p_N}}$. Reciprocally, to each n-tuple $(i_1, \cdots, i_n)$ as above, there can be attached some $p \in \mathbb{N}^N$, $|p| = n$, such that (3.15) holds true.

We are now ready to generalize Lemma 2.10.

Lemma 3.2 *Let $m \in \mathbb{N}^*$. Let $\boldsymbol{F} = (F_{i_1 \cdots i_m}) \in \mathcal{T}_m(L^2_{per}(Y))$, $\boldsymbol{F}$ symmetric, i.e., $F_{i_{\sigma(1)} \cdots i_{\sigma(m)}} = F_{i_1 \cdots i_m}$ for every permutation σ of $\mathbb{N}^*_m = \{1, 2, \cdots, m-1, m\}$. Assume that*

$$\int_Y \boldsymbol{F} \cdot \boldsymbol{T} \, dy = 0 \quad (\boldsymbol{T} \in \mathcal{V}^{(m)}_{per}(Y)), \tag{3.16}$$

where

$$\boldsymbol{F} \cdot \boldsymbol{T} = \sum_{i_1, \cdots, i_m = 1}^{N} F_{i_1 \cdots i_m} T_{i_1 \cdots i_m}.$$

Then, there exists a unique $\theta \in H^m_{per}(Y)/\mathbb{C}$ *such that*

$$\frac{\partial^m \theta}{\partial y_{i_1} \cdots \partial y_{i_m}} = F_{i_1\cdots i_m} \ (1 \le i_1, \cdots, i_m \le N). \tag{3.17}$$

Proof The unicity of θ is evident and so we only have to prove the existence. This will be accomplished by induction on $m \ge 1$. The case $m = 1$ being achieved by Lemma 3.1, we assume $m > 1$ and make the hypothesis

$(\mathbf{H})_{m-1}$ *To any given* $\boldsymbol{R} = (R_{i_1\cdots i_{m-1}}) \in \mathcal{T}_{m-1}(L^2_{per}(Y))$ *which is symmetric and which satisfies*

$$\int_Y \mathbf{R} \cdot \boldsymbol{\Phi}\, dy = 0 \quad (\boldsymbol{\Phi} \in \mathcal{V}^{(m-1)}_{per}(Y)), \tag{3.18}$$

there can be attached some $\varphi \in H^{m-1}_{per}(Y)/\mathbb{C}$ *such that*

$$\frac{\partial^{m-1} \varphi}{\partial y_{i_1} \cdots \partial y_{i_{m-1}}} = R_{i_1\cdots i_{m-1}} \ (1 \le i_1, \cdots, i_{m-1} \le N).$$

Under this hypothesis, the aim is to show that there exists some $\theta \in H^m_{per}(Y)/\mathbb{C}$ such that (3.17) holds. To accomplish this, fix an arbitrary integer $1 \le l \le N$, and define $\mathbf{G}^l = (G^l_{i_1\cdots i_{m-1}})$ in $\mathcal{T}_{m-1}(L^2_{per}(Y))$ as

$$G^l_{i_1\cdots i_{m-1}} = F_{i_1\cdots i_{m-1}l} \quad (1 \le i_1, \cdots, i_{m-1} \le N).$$

On the other hand, let $\boldsymbol{\Phi} = (\psi_{i_1\cdots i_{m-1}}) \in \mathcal{V}^{(m-1)}_{per}(Y)$, $\boldsymbol{\Phi}$ arbitrary fixed. In (3.16), choose in particular $\mathbf{T}$ such that

$$T_{i_1\cdots i_{m-1}i_m} = \delta_{l i_{m-1}} \psi_{i_1\cdots i_{m-2}i_m} \quad (1 \le i_1, \cdots, i_m \le N),$$

where δ_{kh} denotes the usual Kronecker symbol. Then, by taking account of the fact that the two indices h and l may be permuted, it follows

$$\begin{aligned}\int_Y \mathbf{G}^l \cdot \boldsymbol{\Phi}\, dy &= \sum_{i_1,\cdots,i_{m-2}=1}^{N} \sum_{h=1}^{N} \int_Y F_{i_1\cdots i_{m-2}lh} \psi_{i_1\cdots i_{m-2}h}\, dy \\ &= \int_Y \mathbf{F} \cdot \mathbf{T}\, dy = 0.\end{aligned}$$

Thus, (3.18) holds with $\mathbf{R} = \mathbf{G}^l$. Therefore, according to hypothesis $(H)_{m-1}$, there is some $\varphi_l \in H^{m-1}_{per}(Y)/\mathbb{C}$ such that

$$\frac{\partial^{m-1} \varphi_l}{\partial y_{i_1} \cdots \partial y_{i_{m-1}}} = F_{i_1\cdots i_{m-1}l} \ (1 \le i_1, \cdots, i_{m-1} \le N). \tag{3.19}$$

Whence a family $(\varphi_l)_{1\leq l\leq N}$ in $H^{m-1}_{per}(Y)/\mathbb{C}$ such that (3.19) holds for each integer $1 \leq l \leq N$. Thus, the lemma is proved if we can verify that there is some $\theta \in H^1_{per}(Y)/\mathbb{C}$ such that $\varphi_l = \frac{\partial\theta}{\partial y_l}$ for $l = 1, \cdots, N$ (if it does exist, such a θ will turn out to lie in $H^m_{per}(Y)/\mathbb{C}$ and further, to satisfy (3.17)). However, as stated in Lemma 3.1, such a θ does exist if the vector function $\mathbf{f} = (\varphi_l)_{1\leq l\leq N} \in \mathcal{T}_1(L^2_{per}(Y))$ satisfies

$$\mathcal{F}_{k=\omega}(\varphi_l) = 0 \quad (1 \leq l \leq N) \tag{3.20}$$

and

$$k_l\mathcal{F}_k(\varphi j) - k_j\mathcal{F}_k(\varphi_l) = 0 \ \ (1 \leq j, l \leq N) \tag{3.21}$$

for every $k \in S^*$. Property (3.20) is trivial. As regards (3.21), let us first note that (3.19) is equivalent to

$$\begin{aligned} &(2\pi i)^{m-1}k_{i_1} \cdots k_{i_{m-1}}\mathcal{F}_k(\varphi_l) = \mathcal{F}_k(F_{i_1\cdots i_{m-1}l}) \\ &\text{for } k \in S^* \text{ and } 1 \leq i_1, \cdots, i_{m-1} \leq N. \end{aligned} \tag{3.22}$$

With this in mind, fix freely an $(m-1)$-tuple $(t_1, \cdots, t_{m-1})$ with $1 \leq t_1, \cdots, t_{m-1} \leq N$. Let $\mathbf{\Phi} = (\psi_l) \in \mathcal{V}^{(1)}_{per}(Y) = \mathcal{V}_{per}(Y)$. In (3.16) choose $\mathbf{T}$ such that

$$\begin{cases} T_{i_1\cdots i_{m-1}i_m} = 0 \text{ if } (i_1, \cdots, i_{m-1}) \neq (t_1, \cdots, t_{m-1}) \\ T_{t_1\cdots t_{m-1}l} = \psi_l \ \ (l = 1, \cdots, N). \end{cases}$$

Then

$$\sum_{l=1}^{N}\int_Y F_{t_1\cdots t_{m-1}l}\psi_l \, dy = 0$$

and that for any $\mathbf{\Phi} = (\psi_l) \in \mathcal{V}_{per}(Y)$. Appeal to Lemma 3.1 leads us to

$$k_l\mathcal{F}_k(F_{i_1\cdots i_{m-1}j}) - k_j\mathcal{F}_k(F_{i_1\cdots i_{m-1}l}) = 0 \ \ (k \in S^*, \ 1 \leq j, l \leq N)$$

and that for any $1 \leq i_1, \cdots, i_{m-1} \leq N$. Combining this with (3.22) yields

$$k_{i_1} \cdots k_{i_{m-1}}[k_l\mathcal{F}_k(\varphi j) - k_j\mathcal{F}_k(\varphi_l)] = 0 \ \ (1 \leq j, l \leq N)$$

for all $k \in S^*$ and all $1 \leq i_1, \cdots, i_{m-1} \leq N$. Finally, given freely $k \in S^*$ with $k \neq \omega$, let $1 \leq \nu \leq N$ be such that $k_\nu \neq 0$, and choose above (with arbitrary fixed j, l) the $(m-1)$-tuple $(i_1, \cdots, i_{m-1})$ such that $i_1 = \cdots = i_{m-1} = \nu$. Hence (3.21) follows. Consequently, there is a unique function $\theta \in H^m_{per}(Y)/\mathbb{C}$ satisfying (3.17). ■

Remark 3.7 It goes without saying that (3.16) is also necessary to the existence of a function $\theta \in H^m_{per}(Y)$ satisfying (3.17).

We are now in a position to establish the claimed theorem. Throughout the rest of this section, E denotes a fundamental sequence and m is any arbitrary given positive integer.

Theorem 3.3 *Let $(u_\varepsilon)_{\varepsilon\in E}$ be a bounded sequence in $H^m(\Omega)$. Then, there exist a subsequence E' extracted from E, and some functions $u_0 \in H^m(\Omega)$, $u_m \in L^2(\Omega; H^m_{per}(Y)/\mathbb{C})$, such that, as $E' \ni \varepsilon \to 0$,*

$$u_\varepsilon \to u_0 \text{ in } H^m(\Omega)\text{-weak}, \tag{3.23}$$

$$D^q u_\varepsilon \to D^q_x u_0 \text{ in } L^2(\Omega)\text{-weak } \Sigma \quad (0 \le |q| \le m-1), \tag{3.24}$$

$$D^q u_\varepsilon \to D^q_x u_0 + D^q_y u_m \text{ in } L^2(\Omega)\text{-weak } \Sigma \quad (|q| = m). \tag{3.25}$$

Proof Combining Theorem 3.1 with the weak compactness of the closed balls of $H^m(\Omega)$ yields a subsequence E' from E and some functions $u \in H^m(\Omega)$, u_0, $U_q \in L^2(\Omega; L^2_{per}(Y))$, $1 \le |q| \le m$, such that, as $E' \ni \varepsilon \to 0$,

$$u_\varepsilon \to u \text{ in } H^m(\Omega)\text{-weak}, \tag{3.26}$$

$$u_\varepsilon \to u_0 \text{ in } L^2(\Omega)\text{-weak } \Sigma, \tag{3.27}$$

$$D^q u_\varepsilon \to U_q \text{ in } L^2(\Omega)\text{-weak } \Sigma \quad (1 \le |q| \le m). \tag{3.28}$$

Now, observe that from (3.26) we have $u_\varepsilon \to u$ in $H^1(\Omega)$-weak as $E' \ni \varepsilon \to 0$. Hence, by taking (3.27) into account it follows by Theorem 3.2 (with $p = 2$) that $u_0 = u$. Therefore (3.23) and (3.24) for $|q| = 0$ are proved.

The next point is to show (3.24) for $1 \le |q| \le m-1$. Let such a q be fixed in $\mathbb{N}^N$. We have $D^q u_\varepsilon \in H^1(\Omega) = W^{1,2}(\Omega)$ and, as $E' \ni \varepsilon \to 0$, $D^q u_\varepsilon \to D^q_x u_0$ in $H^1(\Omega)$-weak, as is straightforward from (3.23) and (3.28). Therefore, by Theorem 3.2, once more, it follows that $U_q = D^q_x u_0$, which completes the proof of (3.24).

The last point is to prove (3.25). Turning (3.28) to account, the problem reduces to showing that some $u_m \in L^2(\Omega; H^m_{per}(Y)/\mathbb{C})$ exists such that $U_q = D^q_x u_0 + D^q_y u_m$ for $q \in \mathbb{N}^N$ with $|q| = m$. But then, by (3.28) and use of Remark 3.6 we have as $E' \ni \varepsilon \to 0$,

$$\frac{\partial^m u_\varepsilon}{\partial x_{i_1} \cdots \partial x_{i_m}} \to Z_{i_1 \cdots i_m} \text{ in } L^2(\Omega)\text{-weak } \Sigma, \tag{3.29}$$

where $Z_{i_1\cdots i_m} \in L^2(\Omega; L^2_{per}(Y))$, $1 \leq i_1, \cdots, i_m \leq N$. Therefore, we are through with (3.25) once we have found some u_m in $L^2(\Omega; H^m_{per}(Y)/\mathbb{C})$ such that

$$Z_{i_1\cdots i_m} = \frac{\partial^m u_0}{\partial x_{i_1} \cdots \partial x_{i_m}} + \frac{\partial^m u_m}{\partial y_{i_1} \cdots \partial y_{i_m}} \quad (1 \leq i_1, \cdots, i_m \leq N). \tag{3.30}$$

For that purpose, let

$$\begin{cases} \mathbf{F} = (F_{i_1\cdots i_m}) \\ F_{i_1\cdots i_m} = Z_{i_1\cdots i_m} - \frac{\partial^m u_0}{\partial x_{i_1}\cdots\partial x_{i_m}} \quad 1 \leq i_1, \cdots, i_m \leq N. \end{cases} \tag{3.31}$$

By this, one clearly defines a function

$$x \mapsto \mathbf{F}(x) = (F_{i_1\cdots i_m}(x))_{1 \leq i_1, \cdots, i_m \leq N}$$

of Ω into $\mathcal{T}_m(L^2_{per}(Y))$ with $\mathbf{F}(x)$ symmetric for almost all $x \in \Omega$. On the other hand, for $\mathbf{T} = (T_{i_1\cdots i_m}) \in \mathcal{V}^{(m)}_{per}(Y)$, let $\mathbf{T}^\varepsilon = (T^\varepsilon_{i_1\cdots i_m})$, where $T^\varepsilon_{i_1\cdots i_m}(x) = T_{i_1\cdots i_m}(\frac{x}{\varepsilon})$, $x \in \Omega$. We have $\mathbf{T}^\varepsilon \in \mathcal{T}_m(\mathcal{C}^\infty(\Omega))$ with $div_x \mathbf{T}^\varepsilon = \frac{1}{\varepsilon}(div_y \mathbf{T})^\varepsilon = 0$. Hence, if $\varphi \in \mathcal{D}(\Omega)$, then

$$\sum_{i_1, \cdots, i_m = 1}^{N} \int_\Omega \frac{\partial^m u_\varepsilon}{\partial x_{i_1} \cdots \partial x_{i_m}} \varphi T^\varepsilon_{i_1\cdots i_m} dx$$

$$= - \sum_{i_1, \cdots, i_m = 1}^{N} \int_\Omega \frac{\partial^{m-1} u_\varepsilon}{\partial x_{i_1} \cdots \partial x_{i_{m-1}}} T^\varepsilon_{i_1\cdots i_m} \frac{\partial \varphi}{\partial x_{i_m}} dx.$$

By passing to the limit (as $E' \ni \varepsilon \to 0$) using (3.24) (with Remark 3.6) and (3.29) together with the arbitrariness of φ, we arrive at

$$\int_Y \mathbf{F}(x) \cdot \mathbf{T} dy = 0 \quad (\mathbf{T} \in \mathcal{V}^{(m)}_{per}(Y))$$

for almost all $x \in \Omega$. By Lemma 3.2 this yields a function $x \mapsto u_m(x)$ of Ω into $H^m_{per}(Y)/\mathbb{C}$ such that

$$\frac{\partial^m u_m(x)}{\partial y_{i_1} \cdots \partial y_{i_m}} = F_{i_1\cdots i_m}(x) \quad (1 \leq i_1, \cdots, i_m \leq N)$$

for almost every $x \in \Omega$. Finally, it is an easy matter to show that the function $x \mapsto u_m(x)$, denoted below by u_m, belongs to the space $L^2(\Omega; H^m_{per}(Y)/\mathbb{C})$ and moreover

$$\frac{\partial^m u_m}{\partial y_{i_1} \cdots \partial y_{i_m}} = F_{i_1\cdots i_m} \quad (1 \leq i_1, \cdots, i_m \leq N).$$

By substituting in (3.31) one is at once led to (3.30) as claimed. Hence the theorem follows by (3.29) and use of Remark 3.6, once more. ∎

3.3 Comments and Problems

The reader who is familiar with periodic homogenization has certainly noticed that in the particular context of Sect. 3.2, what we term *Σ-convergence* is nothing else than two-scale convergence. The foundations of two-scale convergence were laid some three decades ago in the papers of Nguetseng [92] and Allaire [6]. Since then, a lot of works have been published on the subject, so that there is a considerable literature on two-scale convergence and its applications. To mention only a few, we refer to Zhikov [149, 150], Lukkassen et al. [77], Bourgeat et al. [26] (for a stochastic setting), Pak [113] (for a geometric approach), Clark and Showalter [39] (for the study of partially fissured media), Meirmanov [83, 84] (for the study of acoustic waves in porous media), Visintin [135, 136], Clark and Packer [37, 38], Zhang [148], Neuss-Radu [90], Casado-Diaz and Gayte [31], Amar [10], Marusic-Paloka and Marusic [81], Junca [64], Casado-Diaz [30], Miller [85].

Except for Sect. 3.2.2 the present chapter seems to be, say the periodic transcript of the general Σ-convergence theory developed in the next chapter. As a matter of fact, the latter is inspired by the former. The results of Sect. 3.2.2 were given for the first time in Nguetseng [97].

The following problem is worth pointing out: Given a sequence $(u_\varepsilon)_{\varepsilon\in E}$ in $L^p(\Omega)$ $(1 \leq p < \infty)$ and a function $u_0 \in L^p(\Omega; L^p_{per}(Y))$ such that $u_\varepsilon \to u_0$ in $L^p(\Omega)$-weak Σ (Definition 3.1) and $\|u_\varepsilon\|_{L^p(\Omega)} \to |Y|^{-\frac{1}{p}}\|u_0\|_{L^p(\Omega\times Y)}$ as $E \ni \varepsilon \to 0$, *the question is whether these two conditions suffice to state that the sequence $(u_\varepsilon)_{\varepsilon\in E}$ is strongly Σ-convergent in $L^p(\Omega)$ to u_0. For $p = 2$, the answer is positively given by Proposition 3.3. The case $p \neq 2$ seems to be out of reach. By remembering that the L^p spaces are uniformly convex for $1 < p < \infty$ one would expect that the problem will be in the natural order of events. Unfortunately the uniform convexity seems to be helpless here because the functions u_ε and u_0 have too different features.*

Finally, because of the crucial rôle of such results as Theorem 3.2 in homogenization theory, the reader's interest demands to clarify the derivation of the equation

$$\int_Y \mathbf{F}(x,\cdot)\cdot\mathbf{\Phi}\,dy = 0 \quad \text{for a.e. } x \in \Omega,$$

where $\mathbf{\Phi}$ is arbitrary in $\mathcal{V}_{per}(Y)$ (see Sect. 3.2.1). Actually, in proving Theorem 3.2, what one primarily obtains by letting $E' \ni \varepsilon \to 0$ for fixed $\mathbf{\Phi} \in \mathcal{V}_{per}(Y)$ and $\varphi \in \mathcal{D}(\Omega)$ is the equality

$$\int_\Omega \left(\int_Y \mathbf{F}(x,y)\cdot\mathbf{\Phi}(y)\,dy\right)\varphi(x)\,dx = 0$$

leading us readily, by the arbitrariness of φ, to some negligible set $\mathcal{N}_{\boldsymbol{\Phi}} \subset \Omega$ such that $\int_Y \mathbf{F}(x,\cdot)\cdot\boldsymbol{\Phi}\,dy = 0$ for all $x \in \Omega \setminus \mathcal{N}_{\boldsymbol{\Phi}}$. It follows $\int_Y \mathbf{F}(x,\cdot)\cdot\boldsymbol{\Phi}\,dy = 0$ for all $x \in \Omega \setminus \mathcal{N}$ and all $\boldsymbol{\Phi} \in D$, where D is a dense countable subset of $\mathcal{V}_{per}(Y)$ (observe that this is a separable space under the supremum norm on $\mathbb{R}^N$) and $\mathcal{N}$ the union of all $\mathcal{N}_{\boldsymbol{\Phi}}$ as $\boldsymbol{\Phi}$ ranges over D. One deduces at once that the latter equality stands with D substituted by $\mathcal{V}_{per}(Y)$.

Chapter 4
Σ-Convergence: The General Setting

Based on the theory of homogenization algebras, this chapter extends the concept of Σ-convergence from the periodic setting to a general framework built for the purpose of laying the foundations of a global deterministic homogenization theory.

Unless otherwise specified, $\mathcal{H} = (H_\varepsilon)_{\varepsilon>0}$ denotes an action of $\mathbb{R}_+^*$ on $\mathbb{R}^N$ satisfying the conditions stated in Chap. 3. The basic notation (such as Ω, $E, \cdots$) is as in Chap. 3 (see also Chap. 2). The letter A throughout denotes an H-algebra on $\mathbb{R}^N$ (under $\mathcal{H}$) admitting A^∞ as a dense subalgebra. It amounts to saying that $\mathcal{D}(\Delta(A))$ is dense in $\mathcal{C}(\Delta(A))$. Consequently, since $\mathcal{D}(\Delta(A))$ is continuously embedded in $\mathcal{C}(\Delta(A))$, one has $\mathcal{M}(\Delta(A)) \equiv \mathcal{C}'(\Delta(A)) \subset \mathcal{D}'(\Delta(A))$ with continuous embedding. Therefore, recalling that $L^p(\Delta(A)) \subset \mathcal{M}(\Delta(A))$ with continuous embedding, the result is that $L^p(\Delta(A))$ is continuously embedded in $\mathcal{D}'(\Delta(A))$ for $1 \leq p \leq \infty$, so that one may treat a function in $L^p(\Delta(A))$ like a distribution on $\Delta(A)$ if need be, and then consider Sobolev spaces on $\Delta(A)$. Finally, it should be recalled that the compact space $\Delta(A)$ is provided with the so-called M-measure for A denoted by β, and that the Gelfand transformation on A is denoted by $\mathcal{G}$, and the canonical mapping of $W^{m,p}(\Delta(A))/\mathbb{C}$ into $W_\#^{m,p}(\Delta(A))$ is denoted by J.

4.1 Σ-Convergence in $L^p(\Omega)$

4.1.1 Definitions and Elementary Properties

Definition 4.1 Let $1 \leq p < \infty$. A sequence of functions $u_\varepsilon \in L^p(\Omega)$ $(\varepsilon \in E)$ is said to weakly Σ-converge in $L^p(\Omega)$ to some $u_0 \in L^p(\Omega; L^p(\Delta(A))) = L^p(\Omega \times \Delta(A))$ if, as $E \ni \varepsilon \to 0$,

$$\int_\Omega u_\varepsilon(x)\psi(x, H_\varepsilon(x))dx \to \int\int_{\Omega\times\Delta(A)} u_0(x,s)\widehat{\psi}(x,s)dxd\beta(s) \tag{4.1}$$

G. Nguetseng, *Homogenization Algebras and Applications*, Springer Monographs in Mathematics, https://doi.org/10.1007/978-3-031-84705-9_4

for all $\psi \in L^{p'}(\Omega; A)$, where $\widehat{\psi} = \mathcal{G} \circ \psi$ (usual composition).

We will symbolically express this notion of convergence by writing $u_\varepsilon \to u_0$ in $L^p(\Omega)$-weak Σ. It is an easy exercise to check the unicity of u_0.

It should be anticipated that, as far as the present section is concerned, the whole of the forthcoming results together with their proofs are similar to those presented earlier in Sect. 3.1.1. Therefore, in most cases the proof is either sketched or merely omitted. In the sequel one often puts

$$\psi^\varepsilon(x) = \psi(x, H_\varepsilon(x)), \quad x \in \Omega \text{ (see Remark 3.1).}$$

Definition 4.2 Under the hypotheses of Definition 4.1, the sequence $(u_\varepsilon)_{\varepsilon \in E}$ is said to strongly Σ-converge in $L^p(\Omega)$ to $u_0 \in L^p(\Omega \times \Delta(A))$ if the following property is satisfied:

$$\begin{cases} \text{Given } \eta > 0 \text{ and } \psi \in L^p(\Omega; A) \text{ verifying } \|u_0 - \widehat{\psi}\|_{L^p(\Omega \times \Delta(A))} \leq \dfrac{\eta}{2}, \\ \text{there is some real number } \alpha > 0 \text{ such that } \|u_\varepsilon - \psi^\varepsilon\|_{L^p(\Omega)} \leq \eta \\ \text{for } \varepsilon \in E,\ \varepsilon \leq \alpha. \end{cases} \tag{4.2}$$

We then write $u_\varepsilon \to u_0$ in $L^p(\Omega)$-strong Σ.

Remark 4.1 Remark 3.3 carries over mutatis mutandis to the present setting. Also, the unicity of u_0 in (4.2) is established using the same argument mutatis mutandis as followed in proving Proposition 3.1.

Remark 4.2 Suppose $A = \mathcal{C}_{per}(Y)$. Then, Definitions 4.1 and 4.2 are equivalent to Definitions 3.1 and 3.2, respectively. This follows immediately by Corollaries 2.1, 2.5 and 2.14 (see also Remark 2.14).

We give below one typical example of a weakly as well as strongly Σ-convergent sequence in $L^p(\Omega)$ (see Remark 1.12).

Example 4.1 Let $\psi \in L^p(\Omega; A)$ $(1 \leq p < \infty)$. The sequence $(\psi^\varepsilon)_{\varepsilon>0}$ is both weakly and strongly Σ-convergent in $L^p(\Omega)$ to $\widehat{\psi}$.

The following proposition is likely to provide a clearer idea of strong Σ-convergence. The verification is a simple adaptation of the proof of Proposition 3.2. The details are left to the reader.

Proposition 4.1 *The following statements are equivalent for a sequence $(u_\varepsilon)_{\varepsilon \in E}$ in $L^p(\Omega)$ and a function $u_0 \in L^p(\Omega; A)$ $(1 \leq p < \infty)$:*

(i) $u_\varepsilon \to \widehat{u}_0$ *in* $L^p(\Omega)$*-strong* Σ
(ii) $\|u_\varepsilon - u_0^\varepsilon\|_{L^p(\Omega)} \to 0$ *as* $E \ni \varepsilon \to 0$.

The proof of the next result is a copy mutatis mutandis of that of Proposition 3.3 and is therefore omitted.

Proposition 4.2 *If a sequence $(u_\varepsilon)_{\varepsilon \in E}$ is strongly Σ-convergent in $L^p(\Omega)$ to some $u_0 \in L^p(\Omega \times \Delta(A))$ $(1 \leq p < \infty)$, then:*

(i) $u_\varepsilon \to u_0$ *in $L^p(\Omega)$-weak Σ*
(ii) $\|u_\varepsilon\|_{L^p(\Omega)} \to \|u_0\|_{L^p(\Omega \times \Delta(A))}$ *as* $E \ni \varepsilon \to 0$.
If $p = 2$, then the converse holds true.

Remark 4.3 Let $u_\varepsilon \in L^p(\Omega)$ $(\varepsilon \in E)$, $u_0 \in L^p(\Omega \times \Delta(A))$, $u \in L^p(\Omega)$.

(1) If $u_\varepsilon \to u_0$ in $L^p(\Omega)$-weak Σ, then $u_\varepsilon \to \widetilde{u}_0$ in $L^p(\Omega)$-weak as $E \ni \varepsilon \to 0$, where $\widetilde{u}_0(x) = \int_{\Delta(A)} u_0(x, s) d\beta(s)$ $(x \in \Omega)$. If moreover E is a fundamental sequence, then $(u_\varepsilon)_{\varepsilon \in E}$ is norm-bounded in $L^p(\Omega)$.
(2) If $u_\varepsilon \to u$ in $L^p(\Omega)$ as $E \ni \varepsilon \to 0$, then $u_\varepsilon \to u$ in $L^p(\Omega)$-strong Σ.

4.1.2 Basic Convergence Results

We start with a result of major interest, namely the so-called *Σ-compactness theorem.*

Theorem 4.1 *Let $1 < p < \infty$. $L^p(\Omega)$ is Σ-reflexive in the following sense: Given a fundamental sequence E and a bounded sequence $(u_\varepsilon)_{\varepsilon \in E}$ in $L^p(\Omega)$, a subsequence E' can be extracted from E such that the sequence $(u_\varepsilon)_{\varepsilon \in E'}$ weakly Σ-converges in $L^p(\Omega)$ to some $u_0 \in L^p(\Omega \times \Delta(A))$.*

Proof The line of proceeding is the same as followed in proving the analogous result in the periodic setting (Chap. 3). We shall nevertheless give a sketch of the proof not only because of the importance of the result, but also with a view to drawing attention to property (HA1) involved in the definition of an H-algebra (Chap. 2, Definition 2.1).

This being so, let us consider the functionals $F_\varepsilon \in [L^{p'}(\Omega; A)]'$ $(\varepsilon \in E)$ given by

$$F_\varepsilon(\psi) = \int_\Omega u_\varepsilon(x) \psi(x, H_\varepsilon(x)) dx, \quad \psi \in L^{p'}(\Omega; A).$$

Bearing in mind that the ordinary sequence $(F_\varepsilon)_{\varepsilon \in E}$ is bounded in $[L^{p'}(\Omega; A)]'$ (use Proposition 1.6), we next observe that the separability of A implies that of $L^{p'}(\Omega; A)$, so that the closed balls of the Banach space $[L^{p'}(\Omega; A)]'$, in addition to being compact under the relative weak $*$ topology on $[L^{p'}(\Omega; A)]'$, are metrizable under the same topology (this is a classical result from functional analysis). Hence a subsequence E' can be extracted from E such that $F_\varepsilon \to F_0$ in $[L^{p'}(\Omega; A)]'$-

weak∗ as $E' \ni \varepsilon \to 0$. By the same argument as followed in proving Theorem 3.1 one easily arrives at

$$|F_0(\psi)| \le c\|\widehat{\psi}\|_{L^{p'}(\Omega\times\Delta(A))}$$

for all $\psi \in L^{p'}(\Omega; A)$, or equivalently

$$|G_0(f)| \le c\|f\|_{L^{p'}(\Omega\times\Delta(A))}$$

for all $f \in L^{p'}(\Omega; \mathcal{C}(\Delta(A)))$, where c is a positive constant and G_0 is the continuous linear form on $L^{p'}(\Omega; \mathcal{C}(\Delta(A)))$ given by $G_0(f) = F_0(\mathcal{G}^{-1} \circ f)$, $f \in L^{p'}(\Omega; \mathcal{C}(\Delta(A)))$. Therefore, by the density of $L^{p'}(\Omega; \mathcal{C}(\Delta(A)))$ in $L^{p'}(\Omega; L^{p'}(\Delta(A))) = L^{p'}(\Omega \times \Delta(A))$, it follows that G_0 extends by continuity to a continuous linear form on $L^{p'}(\Omega \times \Delta(A))$. In other words, there is a function $u_0 \in L^p(\Omega \times \Delta(A))$ such that

$$F_0(\psi) = \int\int_{\Omega\times\Delta(A)} u_0(x,s)\widehat{\psi}(x,s)d\beta(s)$$

for all $\psi \in L^{p'}(\Omega; A)$. This completes the proof. ∎

The following proposition is of considerable interest in the mathematical analysis of multiscale problems. The proof is quite similar mutatis mutandis to that of Proposition 3.4. The details are left to the reader.

Proposition 4.3 *Let $1 < p, q < \infty$ with $\frac{1}{p} + \frac{1}{q} = \frac{1}{r} \le 1$. Suppose a sequence $(u_\varepsilon)_{\varepsilon\in E}$ is strongly Σ-convergent in $L^p(\Omega)$ to some $u_0 \in L^p(\Omega \times \Delta(A))$, and a sequence $(v_\varepsilon)_{\varepsilon\in E}$ is weakly Σ-convergent in $L^q(\Omega)$ to some $v_0 \in L^q(\Omega \times \Delta(A))$. Then $u_\varepsilon v_\varepsilon \to u_0 v_0$ in $L^r(\Omega)$-weak Σ.*

As a corollary of this proposition, we have the following statement whose proof is an easy adaptation of that of Proposition 3.5, and is not worth repeating.

Corollary 4.1 *Let $1 < p < \infty$. Let E be a fundamental sequence, $(u_\varepsilon)_{\varepsilon\in E}$ a sequence in $L^p(\Omega)$, and $(v_\varepsilon)_{\varepsilon\in E}$ a sequence in $L^{p'}(\Omega) \cap L^\infty(\Omega)$ $(\frac{1}{p'} = 1 - \frac{1}{p})$ such that:*

(i) $u_\varepsilon \to u_0$ in $L^p(\Omega)$-weak Σ
(ii) $v_\varepsilon \to v_0$ in $L^{p'}(\Omega)$-strong Σ
(iii) $(v_\varepsilon)_{\varepsilon\in E}$ is bounded in $L^\infty(\Omega)$.
Then $u_\varepsilon v_\varepsilon \to u_0 v_0$ in $L^p(\Omega)$-weak Σ.

Finally, for every $1 \le r < \infty$, let $\mathfrak{X}_A^{r,\infty}(\mathbb{R}_y^N) = \mathfrak{X}_A^r(\mathbb{R}_y^N) \cap L^\infty(\mathbb{R}_y^N)$ (Definition 2.4). We equip $\mathfrak{X}_A^{r,\infty}(\mathbb{R}_y^N)$ with the $L^\infty(\mathbb{R}^N)$-norm, which makes it into a Banach space.

Proposition 4.4 *Let $1 < p < \infty$. Let $(u_\varepsilon)_{\varepsilon\in E}$ be a weakly Σ-convergent sequence in $L^p(\Omega)$. Suppose Ω is bounded. Then (4.1) still holds for $\psi \in \mathcal{C}(\overline{\Omega}; \mathfrak{X}_A^{p',\infty}(\mathbb{R}^N))$.*

Proof The line of proceeding is the same as followed in proving Proposition 3.6. Let us observe beforehand that if ψ belongs to $\mathcal{C}(\overline{\Omega}; \mathfrak{X}_A^{p',\infty}(\mathbb{R}^N))$, then $\psi \in \mathcal{C}(\overline{\Omega}; L^\infty(\mathbb{R}^N))$. Hence, it follows by Proposition 1.8 that ψ^ε is well defined. More precisely, we have $\psi^\varepsilon \in L^\infty(\Omega) \subset L^{p'}(\Omega)$ (Ω being bounded). This being so, we assume in the sequel that E is a fundamental sequence. This assumption is of importance in that it ensures the boundedness of $(u_\varepsilon)_{\varepsilon\in E}$ in $L^p(\Omega)$ (see Remark 4.3) without however entailing loss of generality.

Having made this point, the aim now is to check that (4.1) holds true for ψ of the form

$$\psi = \varphi \otimes f, \quad \varphi \in \mathcal{C}(\overline{\Omega}), \quad f \in \mathfrak{X}_A^{p'}(\mathbb{R}^N). \tag{4.3}$$

One such ψ being freely fixed, let us introduce some real $c > 0$ such that

$$c_0 \|\varphi\|_\infty \sup_{\varepsilon\in E} \|u_\varepsilon\|_{L^p(\Omega)} \leq c$$

and

$$c_1 \|\varphi\|_{L^{p'}(\Omega)} \|u_0\|_{L^p(\Omega\times\Delta(A))} \leq c,$$

where c_0 and c_1 are two positive constants such that $\|w^\varepsilon\|_{L^{p'}(\Omega)} \leq c_0 \|w\|_{\Xi^{p'}}$ ($0 < \varepsilon \leq 1$) for all $w \in \Xi^{p'}(\mathbb{R}^N)$ (Sect. 2.2.1) and $\|\widehat{w}\|_{L^{p'}(\Delta(A))} \leq c_1 \|w\|_{\Xi^{p'}}$ for all $w \in \mathfrak{X}_A^{p'}(\mathbb{R}^N)$ (Proposition 2.11), with $w^\varepsilon = w \circ H_\varepsilon$ and $\widehat{w} = \mathcal{G}(w)$.

At the present time, let $\eta > 0$. Let $g \in A$ be such that

$$\|f - g\|_{\Xi^{p'}} \leq \frac{\eta}{3c}. \tag{4.4}$$

Let $\Phi = \varphi \otimes g$. Clearly $\Phi \in L^{p'}(\Omega; A)$. Consequently, there is by hypothesis a real number $0 < \alpha \leq 1$ such that

$$\left| \int_\Omega u_\varepsilon \Phi^\varepsilon \, dx - \iint_{\Omega\times\Delta(A)} u_0 \widehat{\Phi} \, dx d\beta \right| \leq \frac{\eta}{3} \tag{4.5}$$

for $\varepsilon \in E$, $\varepsilon \leq \alpha$. Hence, by the trivial identity

$$\begin{aligned}
\int_\Omega u_\varepsilon \psi^\varepsilon \, dx - \iint_{\Omega\times\Delta(A)} u_0 \widehat{\psi} \, dx d\beta &= \int_\Omega u_\varepsilon (\psi^\varepsilon - \Phi^\varepsilon) \, dx \\
&+ \int_\Omega u_\varepsilon \Phi^\varepsilon \, dx - \iint_{\Omega\times\Delta(A)} u_0 \widehat{\Phi} \, dx d\beta \\
&+ \iint_{\Omega\times\Delta(A)} u_0 (\widehat{\Phi} - \widehat{\Psi}) \, dx d\beta
\end{aligned} \tag{4.6}$$

and use of the estimates

$$\left|\int_{\Omega} u_\varepsilon(\psi^\varepsilon - \Phi^\varepsilon)\, dx\right| \leq c\|f - g\|_{\Xi^{p'}} \quad (\varepsilon \in E,\ \varepsilon \leq 1)$$

and

$$\left|\int\int_{\Omega\times\Delta(A)} u_0(\widehat{\Phi} - \widehat{\psi})\, dxd\beta\right| \leq c\|f - g\|_{\Xi^{p'}}$$

we see readily by (4.4) that

$$\left|\int_{\Omega} u_\varepsilon \psi^\varepsilon\, dx - \int\int_{\Omega\times\Delta(A)} u_0\widehat{\psi}\, dxd\beta\right| \leq \eta \tag{4.7}$$

for $\varepsilon \in E,\ \varepsilon \leq \alpha$. This shows that (4.1) holds for all ψ as in (4.3), and in particular for all ψ of the form $\psi = \varphi \otimes f$ with $\varphi \in \mathcal{C}(\overline{\Omega})$ and $f \in \mathfrak{X}_A^{p',\infty}(\mathbb{R}^N)$, hence by linearity for all $\psi \in \mathcal{C}(\overline{\Omega}) \otimes \mathfrak{X}_A^{p',\infty}(\mathbb{R}^N)$.

Our last purpose is to extend this to all $\psi \in \mathcal{C}(\overline{\Omega}; \mathfrak{X}_A^{p',\infty}(\mathbb{R}^N))$. To attain the end in view we need to know that the transformation $u \mapsto \widehat{u} = u \circ \mathcal{G}$ ($\mathcal{G}$ the canonical mapping of $\mathfrak{X}_A^{p'}(\mathbb{R}^N)$ into $L^{p'}(\Delta(A))$) of $\mathcal{C}(\overline{\Omega}; \mathfrak{X}_A^{p'}(\mathbb{R}^N))$ into $\mathcal{C}(\overline{\Omega}; L^{p'}(\Delta(A)))$ maps continuously $\mathcal{C}(\overline{\Omega}; \mathfrak{X}_A^{p',\infty}(\mathbb{R}^N))$ into $\mathcal{C}(\overline{\Omega}; L^\infty(\Delta(A)))$. Indeed, it clearly suffices to show that the canonical mapping $\mathcal{G} : \mathfrak{X}_A^{p'}(\mathbb{R}^N) \to L^{p'}(\Delta(A))$ sends continuously $\mathfrak{X}_A^{p',\infty}(\mathbb{R}^N)$ into $L^\infty(\Delta(A))$. But this follows at once by Corollary 2.7. With this in mind, let us introduce a constant $c > 0$ such that $\sup_{\varepsilon\in E} \|u_\varepsilon\|_{L^1(\Omega)} \leq c$ and $c_2\|u_0\|_{L^1(\Omega\times\Delta(A))} \leq c$, where the constant c_2 satisfies $\|\widehat{u}\|_{\mathcal{C}(\overline{\Omega};L^\infty(\Delta(A)))} \leq c_2\|u\|_{\mathcal{C}(\overline{\Omega};L^\infty(\mathbb{R}^N))}$ (actually, this holds with $c_2 = 1$) for all $u \in \mathcal{C}(\overline{\Omega}; \mathfrak{X}_A^{p',\infty}(\mathbb{R}^N))$.

Finally, fix freely some $\psi \in \mathcal{C}(\overline{\Omega}; \mathfrak{X}_A^{p',\infty}(\mathbb{R}^N))$. Let $\eta > 0$. By density, let $\Phi \in \mathcal{C}(\overline{\Omega}) \otimes \mathfrak{X}_A^{p',\infty}(\mathbb{R}^N)$ be such that $\|\psi - \Phi\|_{\mathcal{C}(\overline{\Omega};L^\infty(\mathbb{R}^N))} \leq \frac{\eta}{3c}$. By Proposition 1.8, one has

$$\left|\int_{\Omega} u_\varepsilon(\psi^\varepsilon - \Phi^\varepsilon)dx\right| \leq c\|\psi - \Phi\|_{\mathcal{C}(\overline{\Omega};L^\infty(\mathbb{R}^N))} \leq \frac{\eta}{3}.$$

On the other hand,

$$\begin{aligned}\left|\int\int_{\Omega\times\Delta(A)} u_0(\widehat{\Phi} - \widehat{\psi})dxd\beta\right| &\leq \|u_0\|_{L^1(\Omega\times\Delta(A))}\|\widehat{\Phi} - \widehat{\psi}\|_{\mathcal{C}(\overline{\Omega};L^\infty(\Delta(A)))}\\ &\leq c_2\|u_0\|_{L^1(\Omega\times\Delta(A))}\|\Phi - \psi\|_{\mathcal{C}(\overline{\Omega};L^\infty(\mathbb{R}^N))}\\ &\leq \frac{\eta}{3}.\end{aligned}$$

In the end, since (4.1) holds for Φ, as established above, there is some $\alpha > 0$ such that (4.5) holds true for $\varepsilon \in E,\ \varepsilon \leq \alpha$. From all that we deduce via (4.6) that (4.7) is satisfied for $\varepsilon \in E,\ \varepsilon \leq \alpha$. Hence, the proposition follows. ■

4.2 Σ-Convergence in the Sobolev Spaces $W^{m,p}(\Omega)$

Throughout the present section, the action $\mathcal{H} = (H_\varepsilon)_{\varepsilon>0}$ is specified as $H_\varepsilon(x) = \frac{x}{\varepsilon}$ $(x \in \mathbb{R}^N)$.

4.2.1 Generalities on Σ-Convergence in $W^{m,p}(\Omega)$

The notion of Σ-convergence in $W^{m,p}(\Omega)$ requires a few preliminaries. Let $1 < p < \infty$.

Following Schwartz [125], we denote by $\mathcal{D}'(\Omega; L^p(\Delta(A)))$ the space of all vector distributions on Ω with values in the Banach space $L^p(\Delta(A))$. For the benefit of the reader it is well to recall that each element of $\mathcal{D}'(\Omega; L^p(\Delta(A)))$ is nothing else than a continuous linear mapping $T : \mathcal{D}(\Omega) \to L^p(\Delta(A))$, $\mathcal{D}(\Omega)$ being provided with the usual inductive limit topology. The value of T on $\varphi \in \mathcal{D}(\Omega)$ is still denoted by $\langle T, \varphi\rangle \in L^p(\Delta(A))$. If B denotes any arbitrary bounded subset of $\mathcal{D}(\Omega)$ (see, e.g., Vo-Khac [138] for details), we set

$$N_B(T) = \sup_{\varphi\in B} \|\langle T, \varphi\rangle\|_{L^p(\Delta(A))},\quad T \in \mathcal{D}'(\Omega; L^p(\Delta(A)))$$

defining so a seminorm N_B on $\mathcal{D}'(\Omega; L^p(\Delta(A)))$. Equipped with the family of all such seminorms (B ranging over the bounded subsets of $\mathcal{D}(\Omega)$), $\mathcal{D}'(\Omega; L^p(\Delta(A)))$ is a separated complete locally convex space. The partial derivatives of $T \in \mathcal{D}'(\Omega; L^p(\Delta(A)))$, still denoted by $D_x^\alpha T = \frac{\partial^{|\alpha|}T}{\partial x_1^{\alpha_1}\cdots\partial x_N^{\alpha_N}}$, $\alpha = (\alpha_1, \cdots, \alpha_N) \in \mathbb{N}^N$, $|\alpha| = \alpha_1 + \cdots + \alpha_N$, are defined by

$$\langle D_x^\alpha T, \varphi\rangle = (-1)^{|\alpha|}\langle T, D_x^\alpha \varphi\rangle \ \ (\varphi \in \mathcal{D}(\Omega)).$$

One indeed has $D_x^\alpha T \in \mathcal{D}'(\Omega; L^p(\Delta(A)))$, and the operator D_x^α thus defined maps continuously and linearly $\mathcal{D}'(\Omega; L^p(\Delta(A)))$ into itself.

For any given $u \in L^p(\Omega; L^p(\Delta(A)))$, let $T_u \in \mathcal{D}'(\Omega; L^p(\Delta(A)))$,

$$\langle T_u, \varphi\rangle = \int_\Omega u(x)\varphi(x)dx \quad \varphi \in \mathcal{D}(\Omega).$$

The correspondance $u \mapsto T_u$ is an injective continuous linear mapping of $L^p(\Omega; L^p(\Delta(A)))$ into $\mathcal{D}'(\Omega; L^p(\Delta(A)))$. Consequently we may identify T_u with

u, so that

$$L^p(\Omega; L^p(\Delta(A))) \subset \mathcal{D}'(\Omega; L^p(\Delta(A)))$$

with continuous embedding. The result is that we can define, for any integer $m \geq 1$, the space

$$W^{m,p}(\Omega; L^p(\Delta(A))) = \{u | D_x^\alpha u \in L^p(\Omega; L^p(\Delta(A))),\ |\alpha| \leq m\}$$

with the usual convention that $D_x^\alpha u = u$ if $|\alpha| = 0$. This is of course a vector space. Equipped with the norm given by

$$\|u\|_{W^{m,p}(\Omega; L^p(\Delta(A)))} = \sum_{|\alpha| \leq m} \|D_x^\alpha u\|_{L^p(\Omega; L^p(\Delta(A)))},$$

$W^{m,p}(\Omega; L^p(\Delta(A)))$ is a Banach space.

We are henceforth in a position to clearly introduce the concept of weak Σ-convergence in $W^{m,p}(\Omega)$ for the H-algebra A on $\mathbb{R}^N$ under the action specified above.

Definition 4.3 A sequence $u_\varepsilon \in W^{m,p}(\Omega)$ $(\varepsilon \in E)$ is said to weakly Σ-converge in $W^{m,p}(\Omega)$ to some $u_0 \in W^{m,p}(\Omega; L^p(\Delta(A)))$ if there exists a function $u_m \in L^p(\Omega; W_\#^{m,p}(\Delta(A)))$ such that as $E \ni \varepsilon \to 0$, one has

(1) $D^\alpha u_\varepsilon \to D_x^\alpha u_0$ in $L^p(\Omega)$-weak Σ $(0 \leq |\alpha| \leq m-1)$
(2) $D^\alpha u_\varepsilon \to D_x^\alpha u_0 + \partial^\alpha u_m$ in $L^p(\Omega)$-weak Σ $(|\alpha| = m)$.

We then write $u_\varepsilon \to u_0$ in $W^{m,p}(\Omega)$-weak Σ.

It is clear that the couple $\{u_0, u_m\}$ is unique. The function u_m is referred to as a *corrector* for $(u_\varepsilon)_{\varepsilon \in E}$.

Remark 4.4 The derivative $\partial^\alpha u_m$ $(|\alpha| = m)$ needs some clarification. First, considered as a mapping of $W^{m,p}(\Delta(A))/\mathbb{C}$ into $L^p(\Delta(A))$, the partial derivative ∂^α (for $|\alpha| = m$) extends to a unique continuous linear mapping of $W_\#^{m,p}(\Delta(A))$ into $L^p(\Delta(A))$ still denoted by ∂^α (Chap. 2, Proposition 2.12). Thus, if $u \in L^p(\Omega; W_\#^{m,p}(\Delta(A)))$ then for almost every $x \in \Omega$, the derivative $\partial^\alpha u(x)$ is well defined as an element of $L^p(\Delta(A))$. Hence a function $x \mapsto \partial^\alpha u(x)$ of Ω into $L^p(\Delta(A))$ that proves to lie in $L^p(\Omega; L^p(\Delta(A)))$ and to be nothing else that $\partial^\alpha \circ u$. Furthermore, the correspondance $u \mapsto \partial^\alpha \circ u$ sends continuously and linearly $L^p(\Omega; W_\#^{m,p}(\Delta(A)))$ into $L^p(\Omega; L^p(\Delta(A)))$. Finally, for convenience we put $\partial^\alpha u = \partial^\alpha \circ u$ for $u \in L^p(\Omega; W_\#^{m,p}(\Delta(A)))$.

The following proposition is not out of interest.

Proposition 4.5 *Suppose $\mathcal{D}(\Delta(A))$ is dense in $W^{m,p}(\Delta(A))$. Let a sequence $u_\varepsilon \in W^{m,p}(\Omega)$ $(\varepsilon \in E)$ weakly Σ-converge in $W^{m,p}(\Omega)$ to some $u_0 \in$*

$W^{m,p}(\Omega; L^p(\Delta(A)))$. *Then* $u_\varepsilon \to \widetilde{u}_0$ *in* $W^{m,p}(\Omega)$*-weak as* $E \ni \varepsilon \to 0$, *where* $\widetilde{u}_0(x) = \int_{\Delta(A)} u_0(x,s)d\beta(s)$ $(x \in \Omega)$.

Proof By Definition 4.3 and use of Remark 4.3 we have, as $E \ni \varepsilon \to 0$,

$$D^\alpha u_\varepsilon \to \widetilde{D_x^\alpha u_0} \text{ in } L^p(\Omega)\text{-weak for } 0 \le |\alpha| \le m-1$$

and

$$D^\alpha u_\varepsilon \to \widetilde{D_x^\alpha u_0} + \widetilde{\partial^\alpha u_m} \text{ in } L^p(\Omega)\text{-weak for } |\alpha| = m.$$

But because $L^p(\Omega)$ under the weakened topology is continuously embedded in $\mathcal{D}'(\Omega)$ (with the weak $*$ topology) and further the differential operator D^α maps continuously $\mathcal{D}'(\Omega)$ into itself, we see readily by Remark 4.3, once again, that $\widetilde{D_x^\alpha u_0} = D^\alpha \widetilde{u}_0$ for $0 \le |\alpha| \le m-1$. On the other hand, combination of Proposition 2.9 and the density of $J(\mathcal{D}(\Delta(A))/\mathbb{C})$ in $W_\#^{m,p}(\Delta(A))$, as a consequence of that of $\mathcal{D}(\Delta(A))$ in $W^{m,p}(\Delta(A))$ (see Remark 2.5 and Proposition 2.12), yields $\widetilde{\partial^\alpha u_m} = 0$ (i.e., $(\partial^\alpha u_m(x))^\sim = 0$ a.e. in $x \in \Omega$) for $|\alpha| = m$. By the same argument as followed above we deduce that $\widetilde{D_x^\alpha u_0} = D^\alpha \widetilde{u}_0$ for $|\alpha| = m$. Hence the proposition follows at once. ■

In matters of Σ-convergence there is no gap between the usual periodic setting and the H-algebra setting based on the Gelfand representation theory. This is confirmed by the next proposition (see Remark 4.2).

Proposition 4.6 *Let* $1 < p < \infty$. *Assume that* $A = \mathcal{C}_{per}(Y)$. *Then, the following statements are equivalent for a sequence of functions* $u_\varepsilon \in W^{m,p}(\Omega)$ $(\varepsilon \in E)$.

(i) The sequence $(u_\varepsilon)_{\varepsilon \in E}$ *is weakly* Σ*-convergent in* $W^{m,p}(\Omega)$.
(ii) There exists a couple (u_0, u_m) *with* $u_0 \in W^{m,p}(\Omega; L^p_{per}(Y))$ *and* $u_m \in L^p(\Omega; W^{m,p}_{per}(Y)/\mathbb{C})$, *such that as* $E \ni \varepsilon \to 0$, *the following hold in the sense of Definition 3.1:*

(1) $D^\alpha u_\varepsilon \to D_x^\alpha u_0$ *in* $L^p(\Omega)$*-weak* Σ $(0 \le |\alpha| \le m-1)$.
(2) $D^\alpha u_\varepsilon \to D_x^\alpha u_0 + D_y^\alpha u_m$ *in* $L^p(\Omega)$*-weak* Σ $(|\alpha| = m)$.

Proof This follows readily by Corollaries 2.14 and 2.15 (see also Corollaries 2.1 and 2.2 together with Remark 2.14 if necessary). ■

Remark 4.5 Thus, as far as the periodic setting is concerned, we are fully justified in stating (as a definition) that a sequence of functions $u_\varepsilon \in W^{m,p}(\Omega)$ $(\varepsilon \in E)$ is said to be weakly Σ-convergent in $W^{m,p}(\Omega)$ if it satisfies condition (ii) of Proposition 4.6.

4.2.2 $W^{m,p}(\Omega)$-Proper Homogenization Algebras

The notion of a proper homogenization algebra requires a preliminary definition. Let $1 < p < +\infty$.

Definition 4.4 The Sobolev space $W^{m,p}(\Omega)$ is said to be Σ-reflexive (for the H-algebra A) if the following holds: Given a fundamental sequence E and a norm-bounded sequence of functions $u_\varepsilon \in W^{m,p}(\Omega)$ $(\varepsilon \in E)$, a subsequence E' can be extracted from E such that the corresponding sequence $(u_\varepsilon)_{\varepsilon\in E'}$ weakly Σ-converges in $W^{m,p}(\Omega)$.

We are now ready to introduce the notion of a proper H-algebra.

Definition 4.5 Let $1 < p < \infty$. The H-algebra A is termed $W^{m,p}(\Omega)$-proper if the following two conditions are fulfilled:

(WP1) $\mathcal{D}(\Delta(A))$ is dense in $W^{m,p}(\Delta(A))$.
(WP2) $W^{m,p}(\Omega)$ is Σ-reflexive (for A).

Proper H-algebras are destined to play an essential role in the mathematical analysis of homogenization problems beyond the usual periodic setting. Thus, the more proper H-algebras at our disposal, the wider the range of those multiscale problems that can be worked out beyond the classical periodic homogenization theory. Hence our quest of more and more proper H-algebras.

Before we proceed any farther let us make the following remark.

Remark 4.6 Suppose condition (WP1) in Definition 4.5 is satisfied. Then, the following assertions are true:

(1) $J(\mathcal{D}(\Delta(A))/\mathbb{C})$ is dense in $W^{m,p}_{\#}(\Delta(A))$,
(2) $\int_{\Delta(A)} \partial^\alpha u d\beta = 0 \quad (|\alpha| = m)$ for $u \in W^{m,p}_{\#}(\Delta(A))$.

This is straightforward by Propositions 2.9 and 2.12 together with Remark 2.5.

At the present time our main purpose is to state and prove a theorem that will in the sequel enable us to exhibit concrete examples of $W^{m,p}(\Omega)$-proper H-algebras on $\mathbb{R}^N$. Our line of proceeding will be to prove a series of lemmas culmunating in the desired result. Let us begin by setting the problem.

Let A_2 be one further H-algebra on $\mathbb{R}^N$ (under the action $\mathcal{H}$, as specified thereinbefore). We assume that A_2^∞ is dense in A_2. The M-measure (for A_2) on $\Delta(A_2)$ is denoted by β_2, and the canonical mapping of $W^{m,p}(\Delta(A_2))/\mathbb{C}$ into $W^{m,p}_{\#}(\Delta(A_2))$ is denoted by J_2, whereas we shall use the same $\mathcal{G}$ to denote the Gelfand transformation on A and on A_2, as well.

We assume that hypothesis (H) below is satisfied.

(H) *There exist an isometric isomorphism L of $L^p(\Delta(A))$ onto $L^p(\Delta(A_2))$, a dense vector subspace $\mathcal{V}$ of A, a surjective linear mapping $l : \mathcal{V} \to A_2$ and a vector subspace $\mathcal{V}_\infty$ of $A^\infty \cap \mathcal{V}$ such that:*

$$L(\mathcal{G}(v)) = \mathcal{G}(lv) \quad (v \in \mathcal{V}) \tag{4.8}$$

$$L(\widehat{v}u) = L(\widehat{v})L(u) \quad (u \in L^p(\Delta(A)),\ v \in \mathcal{V}) \tag{4.9}$$

$$(v - lv)^\varepsilon \to 0 \text{ in } L^p_{loc}(\mathbb{R}^N_x) \text{ as } \varepsilon \to 0 \quad (v \in \mathcal{V}) \tag{4.10}$$

$$\text{If } v \in \mathcal{V}_\infty \text{ then } D^\alpha_y v \in \mathcal{V} \quad (\alpha \in \mathbb{N}^N) \tag{4.11}$$

$$l \text{ maps } \mathcal{V}_\infty \text{ onto } A_2^\infty \tag{4.12}$$

$$D^\alpha_y(lv) = l(D^\alpha_y v) \quad (v \in \mathcal{V}_\infty,\ \alpha \in \mathbb{N}^N). \tag{4.13}$$

The aim is to prove that under this hypothesis, if A_2 is $W^{m,p}$-proper for some integer $m \geq 1$ and real $p > 1$, then so also is A. As mentioned above, this needs some lemmas. So, let (integer) $m \geq 1$ and (real) $p > 1$. Let (H) be satisfied.

Lemma 4.1 *Suppose $\mathcal{D}(\Delta(A_2))$ is dense in $W^{m,p}(\Delta(A_2))$. Then:*

(i) The mapping L sends $W^{m,p}(\Delta(A))$ isometrically over $W^{m,p}(\Delta(A_2))$.
(ii) $\mathcal{G}(\mathcal{V}_\infty)$ (hence $\mathcal{D}(\Delta(A))$) is dense in $W^{m,p}(\Delta(A))$.

Proof Let us begin by showing that the mapping L sends $W^{m,p}(\Delta(A))$ into $W^{m,p}(\Delta(A_2))$ with

$$\partial^\alpha(Lu) = L(\partial^\alpha u) \quad (1 \leq |\alpha| \leq m) \tag{4.14}$$

for all $u \in W^{m,p}(\Delta(A))$. To do this, we need the equality

$$\int_{\Delta(A_2)} Lu\, d\beta_2 = \int_{\Delta(A)} u\, d\beta \quad (u \in L^p(\Delta(A))) \tag{4.15}$$

which is easily obtained from (4.10) by use of (4.8) together with the density of $\mathcal{G}(\mathcal{V})$ in $L^p(\Delta(A))$.

This being so, let $u \in W^{m,p}(\Delta(A))$, $\alpha \in \mathbb{N}^N$ with $1 \leq |\alpha| \leq m$. Fix freely $\psi \in A_2^\infty$. By (4.12), consider $v \in \mathcal{V}_\infty$ such that $\psi = lv$. Then $\partial^\alpha \widehat{\psi} = L(\partial^\alpha \widehat{v})$ (this follows by (4.8), (4.11) and (4.13)), where $\widehat{\psi} = \mathcal{G}(\psi)$ and $\widehat{v} = \mathcal{G}(v)$. Bearing all that in mind, we have in the distribution sense on $\Delta(A_2)$,

$$\begin{aligned}
\langle \partial^\alpha Lu, \widehat{\psi} \rangle &= (-1)^{|\alpha|} \int_{\Delta(A_2)} Lu \partial^\alpha \widehat{\psi}\, d\beta_2 \\
&= (-1)^{|\alpha|} \int_{\Delta(A_2)} L(u \partial^\alpha \widehat{v})\, d\beta_2 \text{ (according to (4.9))} \\
&= (-1)^{|\alpha|} \int_{\Delta(A)} u \partial^\alpha \widehat{v}\, d\beta \text{ (thanks to (4.15))} \\
&= \int_{\Delta(A)} \widehat{v} \partial^\alpha u\, d\beta
\end{aligned}$$

$$= \int_{\Delta(A_2)} L(\widehat{v}\partial^\alpha u)\, d\beta_2 \text{ (by (4.15), once again)}$$

$$= \int_{\Delta(A_2)} \widehat{\psi} L(\partial^\alpha u)\, d\beta_2 \text{ (thanks to (4.8)–(4.9)).}$$

Hence (4.14) follows by the arbitrariness of ψ. One deduces immediately that L maps $W^{m,p}(\Delta(A))$ isometrically into $W^{m,p}(\Delta(A_2))$. Thus, in order to be through with part (i), we only need to check that L maps $W^{m,p}(\Delta(A))$ onto $W^{m,p}(\Delta(A_2))$. In view of what has just been established, it clearly amounts to showing that $L(W^{m,p}(\Delta(A)))$, which is then closed in $W^{m,p}(\Delta(A_2))$, is at the same time dense in the latter. But this follows by the relations

$$\mathcal{D}(\Delta(A_2)) = \mathcal{G}(A_2^\infty) = L(\mathcal{G}(\mathcal{V}_\infty)) \subset L(W^{m,p}(\Delta(A))) \subset W^{m,p}(\Delta(A_2))$$

and use of the fact that $\mathcal{D}(\Delta(A_2))$ is dense in $W^{m,p}(\Delta(A_2))$. This shows (i). Hence (ii) follows at once, since one has $\mathcal{D}(\Delta(A_2)) = L(\mathcal{G}(\mathcal{V}_\infty))$ and $\mathcal{G}(\mathcal{V}_\infty) \subset \mathcal{D}(\Delta(A))$. ■

Let us turn to the next lemma.

Lemma 4.2 *Assume that $\mathcal{D}(\Delta(A_2))$ is dense in $W^{m,p}(\Delta(A_2))$. Then, there exists an isometric isomorphism $L^\#$ of $W_\#^{m,p}(\Delta(A))$ onto $W_\#^{m,p}(\Delta(A_2))$ such that*

$$L^\#(Jf) = J_2(Lf) \text{ for } f \in W^{m,p}(\Delta(A))/\mathbb{C} \tag{4.16}$$

$$\partial^\alpha L^\# u = L(\partial^\alpha u) \quad (|\alpha| = m) \text{ for } u \in W_\#^{m,p}(\Delta(A)). \tag{4.17}$$

Proof For simplicity, let the restriction of L to $W^{m,p}(\Delta(A))/\mathbb{C}$ be still denoted by L. Then, thanks to Lemma 4.1 (see in particular (4.14) and (4.15)), L is an isometric isomorphism of $W^{m,p}(\Delta(A))/\mathbb{C}$ onto $W^{m,p}(\Delta(A_2))/\mathbb{C}$. With this in mind, we next turn to Chap. 2 and observe that Remark 2.5 therein enables us to write on the other hand

$$\|Lf\|_{W^{m,p}(\Delta(A_2))/\mathbb{C}} = \|J_2(Lf)\|_{W_\#^{m,p}(\Delta(A_2))}$$

for all $f \in W^{m,p}(\Delta(A))/\mathbb{C}$. Consequently, $J_2 \circ L$ is an isometry of $W^{m,p}(\Delta(A))/\mathbb{C}$ into $W_\#^{m,p}(\Delta(A_2))$. Hence, referring again to Remark 2.5 reveals the existence of a unique isometric mapping $L^\# : W_\#^{m,p}(\Delta(A)) \to W_\#^{m,p}(\Delta(A_2))$ such that (4.16) holds. Furthermore, the surjectivity of $L^\#$ follows by noting that

$$\begin{aligned} J_2[W^{m,p}(\Delta(A_2))/\mathbb{C}] &= J_2[L(W^{m,p}(\Delta(A))/\mathbb{C})] \\ &= L^\#[J(W^{m,p}(\Delta(A))/\mathbb{C})] \\ &\subset L^\#[W_\#^{m,p}(\Delta(A))] \subset W_\#^{m,p}(\Delta(A_2)) \end{aligned}$$

and using the density of $J_2[W^{m,p}(\Delta(A_2))/\mathbb{C}]$ in $W_{\#}^{m,p}(\Delta(A_2))$. Thus, only (4.17) remains to be shown. It clearly suffices to check this for $u = Jf$, $f \in W^{m,p}(\Delta(A))/\mathbb{C}$. But this results at once by (4.16) and use of (4.14), Remark 2.5 being the chief point of the argument, of course. ■

Before we can prove the main result in this section we need one further lemma and a few notation and remarks. To begin with, if $u \in L^p(\Omega; L^p(\Delta(A)))$, we set $(L_\Omega u)(x) = L(u(x))$ a.e. in $x \in \Omega$, which yields a mapping $L_\Omega : L^p(\Omega; L^p(\Delta(A))) \to L^p(\Omega; L^p(\Delta(A_2)))$ (this is easily shown). Furthermore, L_Ω is an isometric isomorphism of $L^p(\Omega; L^p(\Delta(A)))$ onto $L^p(\Omega; L^p(\Delta(A_2)))$. Likewise, provided $\mathcal{D}(\Delta(A_2))$ is dense in $W^{m,p}(\Delta(A_2))$, Lemma 4.2 reveals that the mapping

$$L_\Omega^\# : L^p(\Omega; W_\#^{m,p}(\Delta(A))) \to L^p(\Omega; W_\#^{m,p}(\Delta(A_2)))$$

defined by $L_\Omega^\# u = L^\# \circ u$ is an isometric isomorphism, with

$$\partial^\alpha(L_\Omega^\# u) = L_\Omega(\partial^\alpha u) \quad (|\alpha| = m) \tag{4.18}$$

for $u \in L^p(\Omega; W_\#^{m,p}(\Delta(A)))$ (see Remark 4.4).

Also, for $\psi \in \mathcal{K}(\Omega) \otimes \mathcal{V}$, we set $(l_\Omega \psi)(x) = l(\psi(x))$, $x \in \Omega$, which gives a function $l_\Omega \psi \in \mathcal{K}(\Omega) \otimes A_2$. For every $\psi \in \mathcal{K}(\Omega) \otimes \mathcal{V}$, we have, by (4.8) and (4.10),

$$L_\Omega(\mathcal{G}_\Omega(\psi)) = \mathcal{G}_\Omega(l_\Omega \psi) \tag{4.19}$$

$$(\psi - l_\Omega \psi)^\varepsilon \to 0 \text{ in } L^p(\Omega) \text{ as } \varepsilon \to 0, \tag{4.20}$$

where $\mathcal{G}_\Omega(f) = \mathcal{G} \circ f$ for any f in, say either $\mathcal{C}(\Omega; A)$ or $\mathcal{C}(\Omega; A_2)$.

Finally, in order to avoid some unfortunate confusion, as far as the H-algebra A_2 is concerned we shall speak of Σ_2-convergence (resp. Σ_2-reflexive) instead of Σ-convergence (resp. Σ-reflexive), the latter wording being reserved for the H-algebra A.

Lemma 4.3 *Let E be a fundamental sequence, and $(u_\varepsilon)_{\varepsilon \in E}$ be an ordinary sequence in $L^p(\Omega)$ $(1 < p < \infty)$ such that*

$$u_\varepsilon \to v_0 \text{ in } L^p(\Omega)\text{-weak } \Sigma_2, \tag{4.21}$$

where $v_0 \in L^p(\Omega \times \Delta(A_2))$. Then $u_\varepsilon \to L_\Omega^{-1} v_0$ in $L^p(\Omega)$-weak Σ.

Proof It is clear that the sequence $(u_\varepsilon)_{\varepsilon \in E}$ is bounded in $L^p(\Omega)$. On the other hand, $\mathcal{K}(\Omega) \otimes \mathcal{V}$ being dense in $L^{p'}(\Omega; A)$, by a routine argument already followed before, the problem reduces to showing that, as $E \ni \varepsilon \to 0$,

$$\int_\Omega u_\varepsilon \psi^\varepsilon \, dx \to \iint_{\Omega \times \Delta(A)} u_0(x, s) \widehat{\psi}(x, s) \, dx d\beta(s)$$

for all $\psi \in \mathcal{K}(\Omega) \otimes \mathcal{V}$, where $u_0 = L_\Omega^{-1} v_0 \in L^p(\Omega \times \Delta(A))$. But according to (4.20), it amounts to verifying that, as $E \ni \varepsilon \to 0$,

$$\int_\Omega u_\varepsilon (l_\Omega \psi)^\varepsilon \, dx \to \int \int_{\Omega \times \Delta(A)} u_0(x,s) \widehat{\psi}(x,s) \, dx d\beta(s) \tag{4.22}$$

for $\psi \in \mathcal{K}(\Omega) \otimes \mathcal{V}$. So, ψ being thus given, by (4.21) we have

$$\int_\Omega u_\varepsilon (l_\Omega \psi)^\varepsilon \, dx \to \int \int_{\Omega \times \Delta(A_2)} v_0(x,s) \mathcal{G}_\Omega (l_\Omega \psi)(x,s) \, dx d\beta(s)$$

as $E \ni \varepsilon \to 0$. Hence (4.22) follows by (4.19), (4.9) and use of (4.15). ■

Let us now come to the heart of the matter.

Theorem 4.2 *Let $1 < p < \infty$ and $m \in \mathbb{N}$ with $m \geq 1$. Suppose the H-algebra A_2 is $W^{m,p}(\Omega)$-proper. Then the H-algebra A is $W^{m,p}(\Omega)$-proper too.*

Proof In view of Lemma 4.1, $\mathcal{D}(\Delta(A))$ is dense in $W^{m,p}(\Delta(A))$ and so condition (WP1) of Definition 4.5 is fulfilled. It remains to check that (WP2) is also satisfied, namely that $W^{m,p}(\Omega)$ is Σ-reflexive (for A, of course). To do this, let E be any arbitrary given fundamental sequence, and $(u_\varepsilon)_{\varepsilon \in E}$ any arbitrary norm-bounded sequence in $W^{m,p}(\Omega)$. Since this space is Σ_2-reflexive (for A_2) by hypothesis, there are a subsequence E' extracted from E and a couple of functions $v_0 \in W^{m,p}(\Omega; L^p(\Delta(A_2)))$ and $v_m \in L^p(\Omega; W_\#^{m,p}(\Delta(A_2)))$ such that, as $E' \ni \varepsilon \to 0$,

$$\begin{cases} D^\alpha u_\varepsilon \to D_x^\alpha v_0 \text{ in } L^p(\Omega)\text{-weak } \Sigma_2 \quad (0 \leq |\alpha| \leq m-1) \\ D^\alpha u_\varepsilon \to D_x^\alpha v_0 + \partial^\alpha v_m \text{ in } L^p(\Omega)\text{-weak } \Sigma_2 \quad (|\alpha| = m). \end{cases} \tag{4.23}$$

Let $u_0 = L_\Omega^{-1} v_0$ and $u_m = (L_\Omega^\#)^{-1} v_m$. By recalling the preceding preliminaries we see that $u_m \in L^p(\Omega; W_\#^{m,p}(\Delta(A)))$ with further, thanks to (4.18), $\partial^\alpha u_m = L_\Omega^{-1}(\partial^\alpha v_m)$ for $|\alpha| = m$. On the other hand one clearly has $u_0 \in L^p(\Omega; L^p(\Delta(A)))$. Let us show that $u_0 \in W^{m,p}(\Omega; L^p(\Delta(A)))$ with

$$D_x^\alpha u_0 = L_\Omega^{-1} D_x^\alpha v_0 \quad (1 \leq |\alpha| \leq m). \tag{4.24}$$

To this end, fix freely $\varphi \in \mathcal{D}(\Omega)$. For any fixed multi-index α as above we have

$$\int_\Omega D_x^\alpha v_0(x) \varphi(x) dx = (-1)^{|\alpha|} \int_\Omega v_0(x) D^\alpha \varphi(x) dx$$

with $D_x^\alpha v_0 \in L^p(\Omega; L^p(\Delta(A_2)))$, of course. By a classical result from integration theory (see, e.g., Bourbaki [22, p. 142,Théorème1]) it follows

$$\int_\Omega L^{-1}(D_x^\alpha v_0(x)) \varphi(x) dx = (-1)^{|\alpha|} \int_\Omega L^{-1}(v_0(x)) D^\alpha \varphi(x) dx,$$

which equally writes as

$$\int_{\Omega}[L_{\Omega}^{-1}D_x^{\alpha}v_0](x)\varphi(x)dx = (-1)^{|\alpha|}\int_{\Omega}u_0(x)D^{\alpha}\varphi(x)dx$$
$$= \langle D_x^{\alpha}u_0, \varphi\rangle.$$

Hence (4.24) follows by the arbitrariness of φ, and so u_0 does lie in the space $W^{m,p}(\Omega; L^p(\Delta(A)))$ with explicitly known derivatives. Therefore the Σ-reflexivity of $W^{m,p}(\Omega)$ follows by applying Lemma 4.3 to (4.23). This completes the proof. ∎

Remark 4.7 As far as the reader can see, the present developments do not take care about real function spaces, which is not in favour of nonlinear analysis in homogenization theory. However, this gap can be filled. Indeed, suppose the following modifications are made in hypothesis (H):

(1) In addition to being an isometric isomorphism of $L^p(\Delta(A))$ onto $L^p(\Delta(A_2))$, the operator L maps $L^p_{\mathbb{R}}(\Delta(A)) = L^p(\Delta(A); \mathbb{R})$ onto $L^p_{\mathbb{R}}(\Delta(A_2))$.
(2) In addition to being a surjective linear mapping of $\mathcal{V}$ onto A_2, l maps ${}_{\mathbb{R}}\mathcal{V} = \mathcal{V} \cap \mathcal{C}_{\mathbb{R}}(\mathbb{R}^N)$ onto ${}_{\mathbb{R}}A_2 = A_2 \cap \mathcal{C}_{\mathbb{R}}(\mathbb{R}^N)$.
(3) It is required of $\mathcal{V}_\infty$ that it preserves complex conjugation.
Under these modifications, it is an easy exercise (left to the reader) to verify that in addition to the existing conclusions, the following are true.
(4) In Lemma 4.1 (with the same hypothesis): L carries $W^{m,p}_{\mathbb{R}}(\Delta(A)) = W^{m,p}(\Delta(A); \mathbb{R})$ isometrically over $W^{m,p}_{\mathbb{R}}(\Delta(A_2))$, and further $\mathcal{G}({}_{\mathbb{R}}\mathcal{V}_\infty)$ [hence $\mathcal{D}_{\mathbb{R}}(\Delta(A))$] is dense in $W^{m,p}_{\mathbb{R}}(\Delta(A))$.
(5) In Lemma 4.2 (with the same hypothesis): $L^\#$ carries

$$W^{m,p}_{\#}(\Delta(A); \mathbb{R}) = \{u|u \in W^{m,p}_{\#}(\Delta(A)), \partial^{\alpha}u \in L^p_{\mathbb{R}}(\Delta(A)) \ \ (|\alpha| = m)\}$$

isometrically over $W^{m,p}_{\#}(\Delta(A_2); \mathbb{R})$.

Finally, it is worth adding that all properties involved in the preceding modifications are verified in practice.

4.2.3 *Some Examples of $W^{1,p}(\Omega)$-Proper Homogenization Algebras*

We present below a few examples of the main $W^{1,p}(\Omega)$-proper homogenization algebras we will be dealing with in the next chapters.

4.2.3.1 The Periodic H-Algebras

Example 4.2 It is to be expected that the algebras $\mathcal{C}_{per}(Y)$ (Chap. 2, Sect. 2.5.2) constitute the typical family of proper H-algebras. Indeed, letting $A = \mathcal{C}_{per}(Y)$, let us first of all note that by Corollary 2.12, A^{∞} is dense in A. Next, Corollaries 2.13, 2.14 and 2.15 ensure that $\mathcal{D}(\Delta(A))$ is dense in $W^{1,p}(\Delta(A))$ for $1 < p < \infty$. Finally, Proposition 4.6 in combination with Theorem 3.2 shows that $W^{1,p}(\Omega)$ is Σ-reflexive. In conclusion, if $\{b_1, \cdots, b_N\}$ is any arbitrary given basis of the N-dimensional numerical space $\mathbb{R}^N$, and Y the associated open parallelepiped defined in Sect. 2.5.1, the H-algebra $\mathcal{C}_{per}(Y)$ is $W^{1,p}(\Omega)$-proper for any real $p > 1$.

4.2.3.2 The Almost Periodic H-Algebras

Example 4.3 We turn now to Chap. 2 (Sect. 2.3) where the almost periodic H-algebra $A = AP_{\mathcal{R}}(\mathbb{R}^N)$ is introduced from a given countable subgroup $\mathcal{R}$ of $\mathbb{R}^N_y$. First of all, we recall that $\mathcal{D}(\Delta(A))$ is dense in $\mathcal{C}(\Delta(A))$ (see the proof of Proposition 2.13). The aim is to check that $AP_{\mathcal{R}}(\mathbb{R}^N)$ is $H^1(\Omega)$-proper. It is already known that $\mathcal{D}(\Delta(A))$ is dense in $H^1(\Delta(A)) = W^{1,2}(\Delta(A))$ (see Proposition 2.16). Thus, it remains to check that $H^1(\Omega)$ is Σ-reflexive.

Proposition 4.7 *Let E be a fundamental sequence, $(u_\varepsilon)_{\varepsilon\in E}$ a bounded sequence in $H^1(\Omega)$. Then, there are a subsequence E' extracted from E, and some functions $u_0 \in H^1(\Omega)$, $u_1 \in L^2(\Omega; H^1_\#(\Delta(A)))$ such that, as $E' \ni \varepsilon \to 0$,*

$$u_\varepsilon \to u_0 \text{ in } H^1(\Omega)\text{-weak} \tag{4.25}$$

$$u_\varepsilon \to u_0 \text{ in } L^2(\Omega)\text{-weak } \Sigma \tag{4.26}$$

$$\frac{\partial u_\varepsilon}{\partial x_j} \to \frac{\partial u_0}{\partial x_j} + \partial_j u_1 \text{ in } L^2(\Omega)\text{-weak } \Sigma \quad (1 \le j \le N). \tag{4.27}$$

Proof The line of proceeding is the same as followed in proving Theorem 3.2. Appeal to both Theorem 4.1 and the reflexivity of $H^1(\Omega)$ yields a subsequence E' from E and functions $u \in H^1(\Omega)$, $u_0, U_j \in L^2(\Omega; L^2(\Delta(A)))$ $(1 \le j \le N)$ such that as $E' \ni \varepsilon \to 0$, (4.26) holds and further

$$u_\varepsilon \to u \text{ in } H^1(\Omega)\text{-weak}$$

and

$$\frac{\partial u_\varepsilon}{\partial x_j} \to U_j \text{ in } L^2(\Omega)\text{-weak } \Sigma. \tag{4.28}$$

Next, in order to prove that $u_0 = u$, let us consider $\varphi \in \mathcal{D}(\Omega)$ and $\psi \in A^\infty$. For fixed $1 \leq j \leq N$, the same argument as followed in proving Theorem 3.2 leads us to

$$\int_\Omega \left(\int_{\Delta(A)} u_0(x,s)\partial_j \widehat{\psi}(s) d\beta(s) \right) \varphi(x) dx = 0.$$

In view of the arbitrariness of φ, and then of that of ψ, we deduce $\partial_j u_0(x,\cdot) = 0$ in $\Delta(A)$ for almost every $x \in \Omega$, and that for $j = 1, \cdots, N$. For fixed x, this is equivalent to

$$k_j \mathcal{F}_k u_0(x,\cdot) = 0 \ \ (1 \leq j \leq N) \text{ for } k \in \mathcal{R}, \ k \neq \omega = (0,\cdots,0),$$

where $\mathcal{F}_k u_0(x,\cdot)$ denotes the Fourier coefficient at point $k \in \mathcal{R}$ of the function $u_0(x,\cdot) \in L^2(\Delta(A))$ (see Sect. 2.3.3). Hence

$$u_0(x,\cdot) = \widetilde{u}_0(x) = \int_{\Delta(A)} u_0(x,s) d\beta(s).$$

In other words $u_0(x,\cdot)$ is a constant function of the variable $s \in \Delta(A)$. By an obvious argument we deduce that $u_0 = u$, so that (4.25) and (4.26) are proved. Only (4.27) remains to be shown.

Let

$$\mathbf{F} = (F_j) \in L^2(\Omega; L^2(\Delta(A))^N),$$

$$F_j = U_j - \frac{\partial u_0}{\partial x_j} \quad (1 \leq j \leq N).$$

Let $\varphi \in \mathcal{D}(\Omega)$ and $\mathbf{\Phi} = (\psi_j) \in \mathcal{V}_\mathcal{R} = \{\mathbf{f} | \mathbf{f} \in (A^\infty)^N, div_y \mathbf{f} = 0\}$. Observe in passing that $\mathcal{G}(\mathcal{V}_\mathcal{R}) = V_\mathcal{R}$ (Sect. 2.3.4). We have

$$\sum_{j=1}^N \int_\Omega \left[\varphi \frac{\partial u_\varepsilon}{\partial x_j} + u_\varepsilon \frac{\partial \varphi}{\partial x_j} \right] \psi_j^\varepsilon \, dx = -\frac{1}{\varepsilon} \int_\Omega u_\varepsilon (div_y \mathbf{\Phi})^\varepsilon \varphi \, dx = 0.$$

By letting $E' \ni \varepsilon \to 0$ we see by (4.26), (4.28), and use of the arbitrariness of φ, that

$$\int_{\Delta(A)} \mathbf{F}(x,\cdot) \cdot \widehat{\mathbf{\Phi}} \, d\beta = 0 \text{ for a.e. } x \in \Omega \ \ (\text{with } \widehat{\mathbf{\Phi}} = (\mathcal{G}(\psi_j))_{1 \leq j \leq N})$$

and that for any arbitrary $\mathbf{\Phi} \in \mathcal{V}_\mathcal{R}$. Hence, appeal to Lemma 2.7 yields a function $u_1 \in L^2(\Omega; H^1_\#(\Delta(A)))$ such that

$$F_j(x,\cdot) = \partial_j u_1(x) \text{ for a.e. } x \in \Omega \quad (1 \leq j \leq N).$$

Therefore (4.27) follows. The proof is complete. ■

In conclusion, for any given countable subgroup $\mathcal{R}$ of $\mathbb{R}^N$, the almost periodic homogenization algebra $AP_{\mathcal{R}}(\mathbb{R}^N)$ is $H^1(\Omega)$-proper.

4.2.3.3 The H-Algebras $\mathcal{B}_{\infty,S^*}(\mathbb{R}^N)$

Example 4.4 It is worth perhaps reminding the reader of the definition of the algebra $\mathcal{B}_{\infty,S^*}(\mathbb{R}^N)$. The sets Y, S and S^* being introduced in (2.30)–(2.32), we define $\mathcal{B}_{\infty,S^*}(\mathbb{R}^N)$ as the closure in $\mathcal{B}(\mathbb{R}^N)$ of the set $\mathcal{B}_\infty(\mathbb{R}^N)+\mathcal{C}_{per}(Y)$. According to Example 2.5, $\mathcal{B}_{\infty,S^*}(\mathbb{R}^N)$ is an H-algebra when $S = \mathbb{Z}^N$ and $Y = (-\frac{1}{2}, \frac{1}{2})^N$. By the same argument (Proposition 2.7) as followed to establish this, we conclude immediately that $\mathcal{B}_{\infty,S^*}(\mathbb{R}^N)$ is still an H-algebra when the periodic H-algebra $\mathcal{C}_{per}(Y)$ is under the more general form discussed in Sect. 2.5. Our main purpose is to show that the H-algebra $A = \mathcal{B}_{\infty,S^*}(\mathbb{R}^N)$ is $W^{1,p}$-proper for $1 < p < \infty$.

In order to apply Theorem 4.2, let us put $A_2 = \mathcal{C}_{per}(Y)$, $\mathcal{V} = \mathcal{B}_0(\mathbb{R}^N)\oplus\mathcal{C}_{per}(Y)$, where $\mathcal{B}_0(\mathbb{R}^N)$ denotes the subalgebra of $\mathcal{B}_\infty(\mathbb{R}^N)$ consisting of functions that vanish at infinity. Observe that $\mathcal{B}_0(\mathbb{R}^N) \cap \mathcal{C}_{per}(Y) = \{0\}$, so that the above direct sum is a fact. On the other hand, it is clear that $\mathcal{V} = \mathcal{B}_\infty(\mathbb{R}^N) + \mathcal{C}_{per}(Y)$, so that $\mathcal{V}$ is a dense vector subspace of A.

Now, let $l : \mathcal{V} \to A_2$ be the projection of $\mathcal{V}$ on $A_2 = \mathcal{C}_{per}(Y)$ along $\mathcal{B}_0(\mathbb{R}^N)$. Thus, $l(\mathcal{B}_0(\mathbb{R}^N)) = \{0\}$, $l(\mathcal{V}) = A_2$, and every $\psi \in \mathcal{V}$ is uniquely expressible in the form

$$\psi = \psi_0 + l\psi \text{ with } \psi_0 \in \mathcal{B}_0(\mathbb{R}^N). \tag{4.29}$$

To l there is attached the linear mapping $L : \mathcal{G}(\mathcal{V}) \to \mathcal{C}(\Delta(A_2))$ defined by $L(\mathcal{G}(v)) = \mathcal{G}(lv)$ for $v \in \mathcal{V}$.

Proposition 4.8 *The mapping L extends by continuity to an isometric isomorphism, again called L, of $L^p(\Delta(A))$ onto $L^p(\Delta(A_2))$.*

Proof Let $\psi \in \mathcal{V}$. Let us consider again the inequality (2.12) in which we take $a = \psi(y)$ and $b = (l\psi)(y)$ for fixed $y \in \mathbb{R}^N$. By (4.29) we have

$$||\psi(y)|^p - |(l\psi)(y)|^p| \leq c|\psi_0(y)|,$$

where $c = p\||\psi| + |l\psi|\|_\infty$. For fixed $\varphi \in \mathcal{K}(\mathbb{R}^N)$ such that $\int \varphi(x)dx = 1$, it follows

$$\left|\int \left|\psi\left(\frac{x}{\varepsilon}\right)\right|^p \varphi(x)dx - \int \left|(l\psi)\left(\frac{x}{\varepsilon}\right)\right|^p \varphi(x)dx\right| \leq c\int \left|\psi_0\left(\frac{x}{\varepsilon}\right)\right| |\varphi(x)|dx$$

and that for all $\varepsilon > 0$. Letting $\varepsilon \to 0$ leads us readily to

$$M(|\psi|^p) = M(|l\psi|^p).$$

Hence, by Proposition 2.2 one deduces that

$$\|\mathcal{G}(\psi)\|_{L^p(\Delta(A))} = \|L(\mathcal{G}(\psi))\|_{L^p(\Delta(A_2))}.$$

Since this holds for any arbitrary $\psi \in \mathcal{V}$ and moreover $\mathcal{G}(\mathcal{V})$ is dense in $L^p(\Delta(A))$, it follows that the mapping L extends by continuity to an isometric mapping (still denoted by) L of $L^p(\Delta(A))$ into $L^p(\Delta(A_2))$. It remains to show that this isometric mapping is surjective. But this follows at once by the chain of relations

$$\mathcal{C}(\Delta(A_2)) = L(\mathcal{G}(\mathcal{V})) \subset L(L^p(\Delta(A))) \subset L^p(\Delta(A_2))$$

and use of the fact that $\mathcal{C}(\Delta(A_2))$ is dense in $L^p(\Delta(A_2))$. ■

So we have hypothesis (H) partly satisfied. The rest is somewhat easier to check. Indeed, the set $\mathcal{V}_\infty = \mathcal{D}(\mathbb{R}^N) + \mathcal{C}^\infty_{per}(Y)$ is a vector subspace of $\mathcal{V} \cap A^\infty$ satisfying manifestly properties (4.11)–(4.13). Concerning (4.8), comment is needless. Property (4.9) follows by the fact that l is multiplicative (i.e., $l(v\psi) = l(v)l(\psi)$ for $v, \psi \in \mathcal{V}$) and use of Proposition 4.8. Finally, (4.10) is a direct consequence of (4.29).

In conclusion the H-algebra $\mathcal{B}_{\infty,S^*}(\mathbb{R}^N)$ is $W^{1,p}(\Omega)$-proper for $1 < p < \infty$, as is immediate by Example 4.2 and application of Theorem 4.2.

Remark 4.8 $\mathcal{B}_{\infty,S^*}(\mathbb{R}^N)$ may also be defined as the closure in $\mathcal{B}(\mathbb{R}^N)$ of the space of all finite sums $\sum_{finite} \varphi_i u_i$ with $\varphi_i \in \mathcal{B}_\infty(\mathbb{R}^N)$, $u_i \in \mathcal{C}_{per}(Y)$.

Remark 4.9 A^∞ is dense in $A = \mathcal{B}_{\infty,S^*}(\mathbb{R}^N)$. Indeed, we have $\mathcal{D}(\mathbb{R}^N) + \mathcal{C}^\infty_{per}(Y) \subset A^\infty$ and $\mathcal{D}(\mathbb{R}^N) + \mathcal{C}^\infty_{per}(Y)$ is dense in A.

4.2.3.4 The H-Algebras $\mathcal{B}_{\infty,\mathcal{R}}(\mathbb{R}^N)$

Example 4.5 Given a countable subgroup $\mathcal{R}$ of $\mathbb{R}^N$, $\mathcal{B}_{\infty,\mathcal{R}}(\mathbb{R}^N)$ is defined as the closure in $\mathcal{B}(\mathbb{R}^N)$ of the algebra $\mathcal{B}_\infty(\mathbb{R}^N) + AP_{\mathcal{R}}(\mathbb{R}^N)$ (see Example 4.3). This is a generalization of Example 4.4 (see Sects. 2.3 and 2.5). Indeed, observe first that $\mathcal{B}_{\infty,\mathcal{R}}(\mathbb{R}^N)$ is an H-algebra, as is easily seen by noting that

$$\mathcal{B}_\infty(\mathbb{R}^N) + AP_{\mathcal{R}}(\mathbb{R}^N) = \mathcal{B}_0(\mathbb{R}^N) + AP_{\mathcal{R}}(\mathbb{R}^N)$$

and appealing to Proposition 2.7. Next $\mathcal{B}_{\infty,\mathcal{R}}(\mathbb{R}^N)$ is an $H^1(\Omega)$-proper H-algebra, as is readily seen by applying Theorem 4.2 with $A = \mathcal{B}_{\infty,\mathcal{R}}(\mathbb{R}^N)$, $A_2 = AP_{\mathcal{R}}(\mathbb{R}^N)$ (see Example 4.3), and reasoning as in Example 4.4.

Remark 4.10 A^{∞} is dense in $A = \mathcal{B}_{\infty,\mathcal{R}}(\mathbb{R}^N)$. This follows by the relation $\mathcal{D}(\mathbb{R}^N) + A_2^{\infty} \subset A^{\infty}$ with $A_2 = AP_{\mathcal{R}}(\mathbb{R}^N)$, and use of the fact that $\mathcal{D}(\mathbb{R}^N) + A_2^{\infty}$ is dense in A.

Remark 4.11 $\mathcal{B}_{\infty,\mathcal{R}}(\mathbb{R}^N)$ may be defined as the closure in $\mathcal{B}(\mathbb{R}^N)$ of the space of all finite sums $\sum_{finite} \varphi_i u_i$ with $\varphi_i \in \mathcal{B}_{\infty}(\mathbb{R}^N)$, $u_i \in AP_{\mathcal{R}}(\mathbb{R}^N)$. $\mathcal{B}_{\infty,\mathcal{R}}(\mathbb{R}^N)$ may also be defined as the closure in $\mathcal{B}(\mathbb{R}^N)$ of the space of all finite sums $\sum_{k\in F} \varphi_k \gamma_k$ with $\varphi_k \in \mathcal{B}_{\infty}(\mathbb{R}^N)$, $\gamma_k(y) = e^{2\pi i k\cdot y}$ $(y \in \mathbb{R}^N)$, where F ranges over the finite subsets of $\mathcal{R}$.

4.2.3.5 The H-Algebra $\mathcal{B}_{\infty}(\mathbb{R}; A_1)$

Let $\mathbb{R}^N = \mathbb{R}^{N-1} \times \mathbb{R}$ (integer $N > 1$). Points in $\mathbb{R}^N$ are denoted by either $y = (y', y_N)$ or $x = (x', x_N)$ with $x', y' \in \mathbb{R}^{N-1}$ and $x_N, y_N \in \mathbb{R}$. Let A_1 be an H-algebra on $\mathbb{R}^{N-1}$ under the action $\mathcal{H}_1 = (H_\varepsilon^1)_{\varepsilon>0}$ given by $H_\varepsilon^1(x') = \frac{x'}{\varepsilon}$ $(x' \in \mathbb{R}^{N-1})$. It is assumed that A_1^{∞} is dense in A_1. As shown earlier in Chap. 2 (Example 2.4), the space $A = \mathcal{B}_{\infty}(\mathbb{R}; A_1)$ of functions $\psi \in \mathcal{B}(\mathbb{R}^N)$ such that

$$\lim_{|y_N|\to\infty} \|\psi(\cdot, y_N) - \widetilde{\psi}\|_{\infty} = 0 \quad (\widetilde{\psi} \in A_1)$$

coincides with the closure of $A_1 \otimes \mathcal{B}_{\infty}(\mathbb{R})$ in $\mathcal{B}(\mathbb{R}^N)$, and is therefore an H-algebra on $\mathbb{R}^N$ under the action $\mathcal{H} = (H_\varepsilon)_{\varepsilon>0}$ given by $H_\varepsilon(x) = \frac{x}{\varepsilon}$ $(x \in \mathbb{R}^N)$. Indeed, this follows readily by Proposition 2.3, $\mathcal{B}_{\infty}(\mathbb{R})$ being considered here as an H-algebra on $\mathbb{R}$ under the action $\mathcal{H}_0 = (H_\varepsilon^0)_{\varepsilon>0}$, $H_\varepsilon^0(x_N) = \frac{x_N}{\varepsilon}$ $(x_N \in \mathbb{R})$, and the preceding action as the product $\mathcal{H} = \mathcal{H}_1 \times \mathcal{H}_0$. Furthermore, A^{∞} is dense in A, since $A_1^{\infty} \otimes (\mathcal{D}(\mathbb{R}) + \mathbb{C})$ is dense in A and at the same time is contained in A^{∞}. On the other hand, by letting $l_1(\psi) = \widetilde{\psi}$ for $\psi \in A = \mathcal{B}_{\infty}(\mathbb{R}; A_1)$, one defines a mapping $l_1 : A \to A_1$ which is linear, of course.

Now, let

$$A_2 = \{\psi | \psi \in A, \psi = \varphi \otimes 1, \varphi \in A_1\}.$$

Thus, each $\psi \in A_2$ is uniquely determined by some $\varphi \in A_1$ such that $\psi(y', y_N) = \varphi(y')$ for $(y', y_N) \in \mathbb{R}^{N-1} \times \mathbb{R}$. It is a rather simple exercise to check that A_2 is an H-algebra on $\mathbb{R}^N$ under $\mathcal{H}$, and further A_2^{∞} is dense in A_2 (indeed, the elements of A_2^{∞} are of the form $\varphi \otimes 1$, $\varphi \in A_1^{\infty}$). One has $A = A_2 \oplus \mathcal{B}_0(\mathbb{R}; A_1)$. In other words, each $\psi \in A$ is uniquely expressible in the form

$$\textbf{(D)} \quad \psi = l(\psi) + \psi_0 \quad \text{with } l(\psi) = l_1(\psi) \otimes 1, \psi_0 \in \mathcal{B}_0(\mathbb{R}; A_1);$$

and it is clear that $l(\psi) = \psi$ for $\psi \in A_2$, and $l(\psi) = 0$ for $\psi \in \mathcal{B}_0(\mathbb{R}; A_1)$. Thus, the mapping $l : A \to A_2$ defined above is the projection of A on A_2 along $\mathcal{B}_0(\mathbb{R}; A_1)$, and l is therefore linear and surjective.

We turn now to the proof of the following

Proposition 4.9 *Under the preceding hypotheses, if the H-algebra A_2 is $H^1(\Omega)$-proper, then so also is $A = \mathcal{B}_\infty(\mathbb{R}; A_1)$.*

Proof The machinery at our disposal by which to prove this is based on a suitable setting leading to hypothesis (H) (see Sect. 4.2.2). To begin with, let us introduce the mapping $L : \mathcal{C}(\Delta(A)) \to \mathcal{C}(\Delta(A_2))$ given by $L(f) = \mathcal{G}(l(\mathcal{G}^{-1}f))$ for $f \in \mathcal{C}(\Delta(A))$. One has $L(\mathcal{G}(v)) = \mathcal{G}(l(v))$ for $v \in A$. On the other hand, $|l(v)|^2 = |l_1(v)|^2 \otimes 1$. It follows $M(|l(v)|^2) = M(|l_1(v)|^2)$ (see (2.2)) for $v \in A$. By Propositions 2.2 and 2.6 one deduces that $\|Lu\|_{L^2(\Delta(A_2))} = \|L_1 u\|_{L^2(\Delta(A_1))} = \|u\|_{L^2(\Delta(A))}$ for all $u \in \mathcal{C}(\Delta(A))$. Hence, it follows that L extends by continuity to an isometric mapping, still called L, of $L^2(\Delta(A))$ into $L^2(\Delta(A_2))$. Now, by the property $Lu = u$ for $u \in \mathcal{C}(\Delta(A_2))$, we have

$$\mathcal{C}(\Delta(A_2)) \subset L(L^2(\Delta(A))) \subset L^2(\Delta(A_2)),$$

from which we deduce that $L(L^2(\Delta(A)))$ is dense in $L^2(\Delta(A_2))$. But this implies the surjectivity of L, since the former space is closed in the latter; and so the mapping L is an isometric isomorphism of $L^2(\Delta(A))$ onto $L^2(\Delta(A_2))$.

This being so, with a view to applying Theorem 4.2 we are led to put $\mathcal{V} = A$ and $\mathcal{V}_\infty = A_2^\infty + [A_1^\infty \otimes \mathcal{D}(\mathbb{R})]$, and then to verify that hypothesis (H) is satisfied. Clearly we only need to check that conditions (4.9) up to (4.13) are satisfied. But it is an easy matter to see that all these properties follow from the decomposition (D). In particular the latter yields $l(\varphi\psi) = l(\varphi)l(\psi)$ for $\varphi, \psi \in \mathcal{V}$, from which (4.9) follows in an obvious way. Likewise one has $v - l(v) \in \mathcal{B}_0(\mathbb{R}; A_1)$, whence (4.10) is readily obtained by dominated convergence. The verification of the rest is left to the reader. Therefore, the proposition follows by Theorem 4.2. ■

This is worth illustrating.

Example 4.6 The H-algebra $\mathcal{B}_\infty(\mathbb{R}; \mathcal{C}_{per}(Y'))$. We have here $A_1 = \mathcal{C}_{per}(Y')$ with $Y' = (-\frac{1}{2}, \frac{1}{2})^{N-1}$. More precisely, $\mathcal{C}_{per}(Y')$ is the algebra of continuous complex functions ψ on $\mathbb{R}^{N-1}$ such that $\psi(y+k) = \psi(y)$ for $k \in \mathbb{Z}^{N-1}$ and $y \in \mathbb{R}^{N-1}$ (see Example 4.2). As regards A_2, we have precisely $A_2 = AP_{\mathcal{R}}(\mathbb{R}^N)$ with $\mathcal{R} = \mathbb{Z}^{N-1} \times \{0\}$ (Example 4.3). This is an $H^1(\Omega)$-proper homogenization algebra admitting A_2^∞ as a dense subalgebra. Consequently, according to Proposition 4.9, $\mathcal{B}_\infty(\mathbb{R}; \mathcal{C}_{per}(Y'))$ is an H^1-proper homogenization algebra.

Example 4.7 The H-algebra $\mathcal{B}_\infty(\mathbb{R}; AP_{\mathcal{R}'}(\mathbb{R}^{N-1}))$. More generally, the H-algebra $A = \mathcal{B}_\infty(\mathbb{R}; AP_{\mathcal{R}'}(\mathbb{R}^{N-1}))$ attached to the countable subgroup $\mathcal{R}'$ of $\mathbb{R}^{N-1}$, of which A^∞ is a dense subalgebra, is $H^1(\Omega)$-proper. Indeed, the H-algebra $A_2 = AP_{\mathcal{R}}(\mathbb{R}^N)$ with $\mathcal{R} = \mathcal{R}' \times \{0\}$ is $H^1(\Omega)$-proper, whence the claimed property follows by Proposition 4.9.

4.2.4 *Some Examples of $W^{m,2}(\Omega)$-Proper Homogenization Algebras*

4.2.4.1 The H-Algebra $\mathcal{C}_{per}(Y)$

Example 4.8 The H-algebra $A = \mathcal{C}_{per}(Y)$ (Chap. 3) is $W^{m,2}(\Omega)$-proper for any integer $m \geq 1$. Indeed, the density of $\mathcal{D}(\Delta(A))$ in $H^m(\Delta(A)) = W^{m,2}(\Delta(A))$ is immediate by Propositions 2.16 and 2.23, the Σ-reflexivity of $H^m(\Omega)$ follows by combining Proposition 4.6 with Theorem 3.3.

4.2.4.2 The H-Algebra $\mathcal{B}_{\infty,S^*}(\mathbb{R}^N)$

Example 4.9 The H-algebra $A = \mathcal{B}_{\infty,S^*}(\mathbb{R}^N)$ (Example 4.4) is $H^m(\Omega)$-proper for any integer $m \geq 1$. Indeed, this is accomplished by applying Theorem 4.2 with $A_2 = \mathcal{C}_{per}(Y)$ (Example 4.8). The line of proceeding is a mere copy (with $p = 2$) of that followed in Example 4.4.

4.2.5 $W^{m,2}(\Omega)$*-Properness and Discrete H-Algebras*

The study of forthcoming so-called discrete homogenization problems requires a deeper discussion of the associated eponymous H-algebras (see Sect. 2.6), especially in the present context. Before we can accomplish this, however, we need to prove some preliminary results.

4.2.5.1 Preliminaries

Let $Y = (-\frac{1}{2}, \frac{1}{2})^N$ and $\varepsilon > 0$. For any nonempty set $B \subset \mathbb{R}^N$, put

$$i^\varepsilon(B) = \{k | k \in \mathbb{Z}^N, \varepsilon k \in B\},$$

$$j^\varepsilon(B) = \{k | k \in \mathbb{Z}^N, \varepsilon(k + \overline{Y}) \cap B \neq \emptyset\} \quad (\emptyset \text{ the void set}),$$

$$i_0^\varepsilon(B) = \{k | k \in \mathbb{Z}^N, \varepsilon(k + \overline{Y}) \subset B\},$$

$$I^\varepsilon(B) = \bigcup_{k \in i^\varepsilon(B)} \varepsilon(k + \overline{Y}),$$

$$J^\varepsilon(B) = \bigcup_{k \in j^\varepsilon(B)} \varepsilon(k + \overline{Y}),$$

$$I_0^\varepsilon(B) = \bigcup_{k \in i_0^\varepsilon(B)} \varepsilon(k + \overline{Y}).$$

Remark 4.12

1. $B \subset J^{\varepsilon}(B)$.
2. $i_0^{\varepsilon}(B) \subset i^{\varepsilon}(B) \subset j^{\varepsilon}(B)$.
3. If B has compact closure then $j^{\varepsilon}(B)$ is finite.

The following simple lemma is of considerable interest, as will presently become apparent.

Lemma 4.4 *We have $J^{\varepsilon}(\overline{B}) = I_0^{\varepsilon}(\mathring{B}) \cup J^{\varepsilon}(\partial B)$, where $\overline{B}$, $\mathring{B}$ and ∂B denote the closure, interior and boundary, respectively, of B.*

Proof This is a direct consequence of the equality $j^{\varepsilon}(\overline{B}) = i_0^{\varepsilon}(\mathring{B}) \cup j^{\varepsilon}(\partial B)$ (this follows by the connectedness of $\overline{Y}$ and use of Bourbaki [24, I, p. 81, Prop. 3]). ■

We also need the following lemma.

Lemma 4.5 *Let K be a compact set in $\mathbb{R}^N$. Then*

$$\lim_{\varepsilon \to 0} \lambda(J^{\varepsilon}(K)) = \lambda(K). \tag{4.30}$$

Proof For each integer $n \geq 1$, let

$$\Omega_n = \left\{ x | x \in \mathbb{R}^N, d(x, K) < \frac{1}{n} \right\},$$

where d stands for the Euclidean metric in $\mathbb{R}^N$. This yields a decreasing sequence of bounded open neighbourhoods of K such that $\bigcap_{n \geq 1} \Omega_n = K$. Hence, by a classical result from integration theory (see, e.g., Bourbaki [22, p. 150]), $\lim_{n \to +\infty} \lambda(\Omega_n) = \lambda(K)$.

This being so, let $\eta > 0$. Let us fix some integer $n \geq 1$ such that $\lambda(\Omega_n) - \lambda(K) < \eta$. According to Lemma 1.7 (see also the beginning of the proof of the said lemma), there is some $\alpha_n > 0$ such that $K \subset J^{\varepsilon}(K) \subset \Omega_n$ for $0 < \varepsilon \leq \alpha_n$. Hence

$$0 \leq \lambda(J^{\varepsilon}(K)) - \lambda(K) < \eta \quad \text{for } 0 < \varepsilon \leq \alpha_n$$

and so (4.30) is proved. ■

In the next statement, B denotes any given (nonempty) subset of $\mathbb{R}^N$ with compact closure and λ-negligible boundary ∂B. By recalling that the set of the discontinuities of its characteristic function χ_B is precisely the λ-negligible set ∂B, it follows that χ_B is λ-measurable, so that B is λ-integrable (see, e.g., Bourbaki [22, p. 173]).

Proposition 4.10 *Let $u \in \Pi^{\infty}(\mathbb{R}^N)$. Let $B \subset \mathbb{R}^N$ with compact closure and λ-negligible boundary. Then, as $\varepsilon \to 0$, we have*

$$\varepsilon^N \sum_{k \in i^{\varepsilon}(B)} \int_{k+Y} u(y)dy \to M(u)\lambda(B).$$

Proof Fix any arbitrary real $\eta > 0$. Since $\int_B u(\frac{x}{\varepsilon})dx \to M(u)\lambda(B)$ as $\varepsilon \to 0$, it is clear that we are through if we can find some real $\alpha > 0$ such that

$$\left| \int_B u^\varepsilon(x)dx - \varepsilon^N \sum_{k\in i^\varepsilon(B)} \int_{k+Y} u(y)dy \right| \leq \eta \tag{4.31}$$

for $0 < \varepsilon \leq \alpha$, where, as usual, $u^\varepsilon(x) = u(\frac{x}{\varepsilon})$. For this purpose, let $\varepsilon > 0$. We have

$$\int_B u^\varepsilon\, dx - \int_{I^\varepsilon(B)} u^\varepsilon\, dx = \int_{J^\varepsilon(B)\setminus I^\varepsilon(B)} u^\varepsilon\, dx - \int_{J^\varepsilon(B)\setminus B} u^\varepsilon\, dx.$$

It follows

$$\left|\int_B u^\varepsilon\, dx - \int_{I^\varepsilon(B)} u^\varepsilon\, dx\right| \leq \int_{J^\varepsilon(B)\setminus I^\varepsilon(B)} |u^\varepsilon|\, dx + \int_{J^\varepsilon(B)\setminus B} |u^\varepsilon|\, dx.$$

But a quick glance at Lemma 4.4 reveals that the sets $J^\varepsilon(B)\setminus B$ and $J^\varepsilon(B)\setminus I^\varepsilon(B)$ are each contained in $J^\varepsilon(\partial B)$. Therefore

$$\left|\int_B u^\varepsilon\, dx - \int_{I^\varepsilon(B)} u^\varepsilon\, dx\right| \leq 2\|u\|_\infty \lambda(J^\varepsilon(\partial B)).$$

Hence, by the equality

$$\int_{I^\varepsilon(B)} u^\varepsilon\, dx = \varepsilon^N \sum_{k\in i^\varepsilon(B)} \int_{k+Y} u(y)dy$$

and the fact that $\lambda(J^\varepsilon(\partial B)) \to 0$ (Lemma 4.5), we are at once led to some $\alpha > 0$ such that (4.31) holds for $0 < \varepsilon \leq \alpha$. The proposition follows thereby. ∎

4.2.5.2 Discrete H-Algebras Revisited

For the benefit of the reader it is perhaps worth recalling how a discrete H-algebra is framed. Given an essential algebra $\mathfrak{A}$ on $\mathbb{Z}^N$ (Sect. 2.6), we consider the set $\mathcal{F}_\mathfrak{A}$ of all complex functions f on $\mathbb{R}^N_y$ of the form

$$f = \sum_{k\in\mathbb{Z}^N} a(k)\tau_k\varphi \quad \text{with } a \in \mathfrak{A} \text{ and } \varphi \in \mathcal{K}(Y). \tag{4.32}$$

We denote by $\langle\mathcal{F}_\mathfrak{A}\rangle$ the linear span of $\mathcal{F}_\mathfrak{A}$. Then, by the discrete H-algebra attached to $\mathfrak{A}$ is meant the closure in $\mathcal{B}(\mathbb{R}^N_y)$ of the space $\mathbb{C}+\langle\mathcal{F}_\mathfrak{A}\rangle$, where $\mathbb{C}$ is here identified with the algebra of constant complex functions on $\mathbb{R}^N$. Unless otherwise specified, the discrete H-algebra attached to $\mathfrak{A}$ will be denoted by $A_{\mathfrak{A}}$. The following remark provides us with an alternative expression of $A_{\mathfrak{A}}$.

Remark 4.13 We have $A_{\mathfrak{A}} = \mathbb{C} + \overline{\langle \mathcal{F}_{\mathfrak{A}} \rangle}$, where $\overline{\langle \mathcal{F}_{\mathfrak{A}} \rangle}$ (the closed linear span of $\mathcal{F}_{\mathfrak{A}}$) stands for the closure of $\langle \mathcal{F}_{\mathfrak{A}} \rangle$ in $\mathcal{B}(\mathbb{R}^N)$. Let us check this. The relation $\mathbb{C} + \overline{\langle \mathcal{F}_{\mathfrak{A}} \rangle} \subset A_{\mathfrak{A}}$ is evident. Let us verify $A_{\mathfrak{A}} \subset \mathbb{C} + \overline{\langle \mathcal{F}_{\mathfrak{A}} \rangle}$. Let $u \in A_{\mathfrak{A}}$. By definition there is a sequence $(u_n)_{n \geq 1}$ with

$$u_n = c_n + v_n \quad (c_n \in \mathbb{C}, \, v_n \in \langle \mathcal{F}_{\mathfrak{A}} \rangle)$$

such that $u_n \to u$ in $\mathcal{B}(\mathbb{R}^N)$ as $n \to +\infty$. Noticing that each function v_n vanishes on ∂Y, we deduce that as $n \to +\infty$, $c_n \to c_0$ and further $v_n \to v_0$ in $\mathcal{B}(\mathbb{R}^N)$. Hence $u = c_0 + v_0 \in \mathbb{C} + \overline{\langle \mathcal{F}_{\mathfrak{A}} \rangle}$, as claimed.

Now, let $\mathcal{C}_{per}(Y)$ be the usual periodic H-algebras on $\mathbb{R}^N$, i.e., with $Y = (-\frac{1}{2}, \frac{1}{2})^N$ and $S = \mathbb{Z}^N$. Before we can proceed any further, let us check that the pair of H-algebras $A_{\mathfrak{A}}$ and $\mathcal{C}_{per}(Y)$ is summable. In other words, we have to check that if $u \in A_{\mathfrak{A}}$ and $\psi \in \mathcal{C}_{per}(Y)$, then $u\psi \in A_{\mathfrak{A}} + \mathcal{C}_{per}(Y)$. But in view of Remark 4.13, u writes as $u = c_0 + u_0$ with $c_0 \in \mathbb{C}$, $u_0 \in \overline{\langle \mathcal{F}_{\mathfrak{A}} \rangle}$; so that $u\psi = c_0\psi + u_0\psi$. It suffices then to show that $u_0\psi$ lies in $A_{\mathfrak{A}}$. But this is straightforward by noting that $\psi \mathcal{F}_{\mathfrak{A}} \subset \mathcal{F}_{\mathfrak{A}}$, hence $\psi \overline{\langle \mathcal{F}_{\mathfrak{A}} \rangle} \subset \overline{\langle \mathcal{F}_{\mathfrak{A}} \rangle}$. Therefore $u_0\psi \in A_{\mathfrak{A}}$ and so the pair of H-algebras under consideration is summable (see Sect. 2.1.3).

Thus, the closure of $A_{\mathfrak{A}} + \mathcal{C}_{per}(Y)$ in $\mathcal{B}(\mathbb{R}^N)$ is an H-algebra . Our main purpose is to prove that under some suitable hypothesis on $\mathfrak{A}$, the said H-algebra is $H^m(\Omega)$-proper for any integer $m \geq 1$. Before we can do that, however, a few preliminaries are needed. First, we set

$$\mathfrak{A}_0 = \{a | a \in \mathfrak{A}, \, \mathfrak{M}(a) = 0\} \quad (\mathfrak{M} \text{ the essential mean value on } \mathbb{Z}^N).$$

Thus, $\mathfrak{A} = \mathfrak{A}_0 \oplus \mathbb{C}$ (direct sum). On the other hand, for each vector subspace V of $\mathfrak{A}$, we put $\mathcal{F}_V$ for the set of functions f of the form (4.32) with V in place of $\mathfrak{A}$. However, for convenience we will put $\mathcal{F} = \mathcal{F}_{\mathfrak{A}}, \mathcal{F}_0 = \mathcal{F}_{\mathfrak{A}_0}$, provided this entails no confusion. It is clear that $\mathcal{F} = \mathcal{F}_0 \oplus \mathcal{F}_{\mathbb{C}}$. Next, in order to be in accordance with Sect. 4.2.2 we are led to put

$$A_2 = \mathcal{C}_{per}(Y)$$
$$A = \overline{A_{\mathfrak{A}} + \mathcal{C}_{per}(Y)} \quad (\text{ the bar denotes here the closure in } \mathcal{B}(\mathbb{R}^N)).$$

Finally, we assume that

$$\mathfrak{A}_0 \text{ is a subalgebra of } \mathfrak{A}, \tag{4.33}$$

which assumption stands throughout the rest of this section.

The following lemma is a basic one.

Lemma 4.6 *The following statements hold.*

(i) $\langle \mathcal{F}_0 \rangle$ *is a subalgebra of* $\mathcal{B}(\mathbb{R}^N)$.
(ii) $\langle \mathcal{F}_0 \rangle$ *is stable under complex conjugation.*
(iii) If $\psi \in \langle \mathcal{F}_0 \rangle$ *and* $g \in A_2$*, then* $\psi g \in \langle \mathcal{F}_0 \rangle$.
(iv) $\langle \mathcal{F}_0 \rangle \cap A_2 = \{0\}$.
(v) $\langle \mathcal{F}_0 \rangle + A_2$ *is dense in* A.
(vi) A^∞ *is dense in* A.

Proof It is obvious that $\langle \mathcal{F}_0 \rangle$ is a vector subspace of $\mathcal{B}(\mathbb{R}^N)$. Furthermore, $\langle \mathcal{F}_0 \rangle$ is stable under multiplication, as is easily seen by (4.33) and Lemma 2.13. This shows (i). Next, (ii) is an obvious fact, whereas (iii) follows by the property $g\mathcal{F}_0 \subset \mathcal{F}_0$. Let us check (iv): Let $\psi \in \langle \mathcal{F}_0 \rangle \cap A_2$. As pointed out in Remark 2.16, if f is given by (4.32), then $M(f) = \mathfrak{M}(f) = \mathfrak{M}(a) \int \varphi(y)dy$. Therefore $M(|\psi|^2) = 0$, since $|\psi|^2 = \psi\overline{\psi} \in \langle \mathcal{F}_0 \rangle \cap A_2$ (use (i) and (ii)). It follows $\psi = 0$ because A_2 is a nondegenerate H-algebra (Definition 2.2). Property (v) follows by the equality $\langle \mathcal{F}_0 \rangle + A_2 = \langle \mathcal{F} \rangle + A_2$ and use of the fact that the space on the right is dense in A. Finally, let us check (vi). In view of the fact that $A_2^\infty = \mathcal{C}_{per}^\infty(Y)$ is dense in A_2, the problem reduces to showing that if f is given by (4.32), then f is the limit in $\mathcal{B}(\mathbb{R}^N)$ of a sequence of functions of the form

$$g = \sum_{k \in \mathbb{Z}^N} a(k)\tau_k\theta \quad \text{with } a \in \mathfrak{A} \text{ and } \theta \in \mathcal{D}(Y).$$

This follows at once by two simple facts:

(1) $\mathcal{D}(Y)$ is dense in $\mathcal{K}(Y)$ (under the inductive limit topology), as is classical.
(2) Given f as in (4.32) and g as above (with the same $a \in \mathfrak{A}$), one has $\|f - g\|_\infty \leq \|a\|_\infty \|\varphi - \theta\|_\infty$.

■

Remark 4.14 We have $M(\psi) = 0$ for $\psi \in \langle \mathcal{F}_0 \rangle$.

We are now in a position to prove the claimed statement.

Proposition 4.11 *Suppose (4.33) holds. Then the H-algebra A is* $H^m(\Omega)$*-proper for any integer* $m \geq 1$.

Proof In view of Example 4.8 and Theorem 4.2, it suffices to show that the pair of H-algebras $\{A, A_2\}$ satisfies hypothesis (H) in Sect. 4.2.2. To begin, let $\mathcal{V} = \langle \mathcal{F}_0 \rangle \oplus A_2$. This is a dense subalgebra of A (Lemma 4.6). Let $l : \mathcal{V} \to A_2$ be the projection of $\mathcal{V}$ on A_2 along $\langle \mathcal{F}_0 \rangle$. Thus, $lv = v$ for $v \in A_2$ and $lv = 0$ for $v \in \langle \mathcal{F}_0 \rangle$. Moreover, each $v \in \mathcal{V}$ is uniquely expressible in the form $v = v_0 + lv$ with $v_0 \in \langle \mathcal{F}_0 \rangle$. It follows by Lemma 4.6 and use of Remark 4.14 that $M(|lv|^2) = M(|v|^2)$ for $v \in \mathcal{V}$. Consequently, the linear mapping $L : \mathcal{G}(\mathcal{V}) \to \mathcal{C}(\Delta(A_2))$ defined by $L(\mathcal{G}(v)) = \mathcal{G}(lv)$ for $v \in \mathcal{V}$ satisfies

$$\|Lu\|_{L^2(\Delta(A_2))} = \|u\|_{L^2(\Delta(A))}$$

for all $u \in \mathcal{G}(\mathcal{V})$. Therefore, thanks to the density of $\mathcal{G}(\mathcal{V})$ in $L^2(\Delta(A))$ (as a consequence of Lemma 4.6(v)), L extends by continuity to an isometric mapping, still denoted by L, of $L^2(\Delta(A))$ into $L^2(\Delta(A_2))$. But

$$\mathcal{C}(\Delta(A_2)) = L(\mathcal{G}(\mathcal{V})) \subset L[L^2(\Delta(A))] \subset L^2(\Delta(A_2)).$$

Hence we deduce that L is surjective in reason of the fact that $L[L^2(\Delta(A))]$ is then closed (since L is isometric) and dense in $L^2(\Delta(A_2))$ at the same time. Thus, the mapping L is an isometric isomorphism of $L^2(\Delta(A))$ onto $L^2(\Delta(A_2))$. Finally, let $\mathcal{F}_0^\infty$ be the set of all $f \in \mathcal{F}$ of the form

$$f = \sum_{k\in\mathbb{Z}^N} a(k)\tau_k\varphi \quad \text{with } a \in \mathfrak{A}_0 \text{ and } \varphi \in \mathcal{D}(Y).$$

Let $\mathcal{V}_\infty = \langle \mathcal{F}_0^\infty \rangle + A_2^\infty$. This is manifestly a vector subspace of $A^\infty \cap \mathcal{V}$. Clearly in view of the statement of hypothesis (H), we are through with the proof once it is shown that conditions (4.8)–(4.13) (with $p = 2$ as far as (4.9) and (4.10) are concerned) are fulfilled. This is however a simple exercise left to the reader except perhaps the verification of condition (4.10). Let us concentrate on this case.

Since $v - lv \in \langle \mathcal{F}_0 \rangle$ $(v \in \mathcal{V})$, it is quite sufficient to show the property for $f \in \mathcal{F}_0$ in place of $v - lv$. To this end, let a compact set $K \subset \mathbb{R}_x^N$ be fixed in an arbitrary fashion. Let B be an open neighbourhood of K with compact closure and λ-negligible boundary (B is for example an open ball containing K). Thanks to Lemma 1.7 we may consider some $\alpha > 0$ such that $J^\varepsilon(K) \subset B$ for $0 < \varepsilon \leq \alpha$. It is then clear that $K \subset J^\varepsilon(K) \subset I^\varepsilon(B)$ for $0 < \varepsilon \leq \alpha$. Hence, given $f \in \mathcal{F}_0$, we have

$$\int_K \left| f\left(\frac{x}{\varepsilon}\right)\right|^2 dx \leq \int_{I^\varepsilon(B)} \left| f\left(\frac{x}{\varepsilon}\right)\right|^2 dx = \varepsilon^N \sum_{k\in i^\varepsilon(B)} \int_{k+Y} |f(y)|^2 dy.$$

Therefore, since $|f|^2 \in \langle \mathcal{F}_0 \rangle$, thus $M(|f|^2) = 0$ (see Lemma 4.6 and Remark 4.14), the desired result follows by Proposition 4.10. Hence the proposition follows by Theorem 4.2 and use of Example 4.8. ■

Let us end the present section with some examples of those essential algebras on $\mathbb{Z}^N$ satisfying condition (4.33) (we refer the reader to Examples 2.10, 2.12-2.14).

Example 4.10 For $\mathfrak{A} = \mathcal{B}_\infty(\mathbb{Z}^N)$ we have $\mathfrak{A}_0 = \mathcal{B}_0(\mathbb{Z}^N)$ (the space of those $a \in \mathcal{B}_\infty(\mathbb{Z}^N)$ vanishing at infinity). This is of course a subalgebra of $\mathcal{B}_\infty(\mathbb{Z}^N)$.

Example 4.11 Let $\mathfrak{A} = l_0^1(\mathbb{Z}^N)$ (the closure of $\mathbb{C} + l^1(\mathbb{Z}^N)$ in $l^\infty(\mathbb{Z}^N)$). Here and in the next example, the bar denotes the closure in $l^\infty(\mathbb{Z}^N)$. We have here $\mathfrak{A}_0 = \overline{l^1(\mathbb{Z}^N)}$. Indeed, the relation $\overline{l^1(\mathbb{Z}^N)} \subset \mathfrak{A}_0$ being evident, we only need to check $\mathfrak{A}_0 \subset \overline{l^1(\mathbb{Z}^N)}$. So, given $a \in \mathfrak{A}_0$, let (a_n) be a sequence such that $a_n \to a$ in $l^\infty(\mathbb{Z}^N)$ as $n \to +\infty$, where $a_n = \xi_n + b_n$ with $\xi_n \in \mathbb{C}$ and $b_n \in l^1(\mathbb{Z}^N)$. Clearly $\xi_n = \mathfrak{M}(a_n) \to \mathfrak{M}(a) = 0$ as $n \to \infty$. Hence $\|a_n - b_n\|_\infty = |\xi_n| \to 0$ as $n \to \infty$.

We deduce that $b_n \to a$ in $l^\infty(\mathbb{Z}^N)$ as $n \to +\infty$. Hence $a \in \overline{l^1(\mathbb{Z}^N)}$, and the equality follows. Finally, it is then clear that (4.33) is satisfied.

Example 4.12 Let $\mathfrak{A} = l^1_\infty(\mathbb{Z}^N)$ (the closure of $\mathcal{B}_\infty(\mathbb{Z}^N) + l^1(\mathbb{Z}^N)$ in $l^\infty(\mathbb{Z}^N)$). We have $\mathfrak{A}_0 = \overline{\mathcal{B}_0(\mathbb{Z}^N) + l^1(\mathbb{Z}^N)}$. Let us check this. First, it is obvious that $\overline{\mathcal{B}_0(\mathbb{Z}^N) + l^1(\mathbb{Z}^N)} \subset \mathfrak{A}_0$ (observe that $\mathfrak{A}_0$ is closed in $l^\infty(\mathbb{Z}^N)$. Let us verify the inverse inclusion. Let $a \in \mathfrak{A}_0$. By definition, there is some sequence (a_n) with $a_n = s_n + b_n, s_n \in \mathcal{B}_\infty(\mathbb{Z}^N), b_n \in l^1(\mathbb{Z}^N)$, such that $a_n \to a$ in $l^\infty(\mathbb{Z}^N)$ when $n \to \infty$. Now, by writing $s_n = s_n^0 + \mathfrak{M}(s_n)$ with $s_n^0 \in \mathcal{B}_0(\mathbb{Z}^N)$ and $\mathfrak{M}(s_n) = \lim_{|k|\to+\infty} s_n(k)$, we see that $\|a_n - (s_n^0 + b_n)\|_\infty = |\mathfrak{M}(s_n)| = |\mathfrak{M}(a_n)|$. But $\mathfrak{M}(a_n) \to \mathfrak{M}(a) = 0$, hence $s_n^0 + b_n \to a$ in $l^\infty(\mathbb{Z}^N)$ as $n \to +\infty$. This shows that $a \in \overline{\mathcal{B}_0(\mathbb{Z}^N) + l^1(\mathbb{Z}^N)}$, which completes the proof of the claimed equality. One easily deduces that $\mathfrak{A}_0$ satisfies (4.33).

4.3 The Vague Σ-Convergence of Radon Measures

Let the basic notation and hypotheses be as stated in the beginning of the chapter save that the algebra A^∞ is not required to be dense in A.

Definition 4.6 A sequence $(\mu_\varepsilon)_{\varepsilon\in E}$ of Radon measures on Ω is said to be vaguely Σ-convergent to some $\mu_0 \in \mathcal{M}(\Omega \times \Delta(A)) = \mathcal{K}'(\Omega \times \Delta(A))$ if as $E \ni \varepsilon \to 0$, we have

$$\int_\Omega \psi(x, H_\varepsilon(x))d\mu_\varepsilon(x) \to \int\int_{\Omega\times\Delta(A)} \widehat{\psi}(x, s)d\mu_0(x, s)$$

for all $\psi \in \mathcal{K}(\Omega; A)$. We then write $\mu_\varepsilon \to \mu_0$ in $\mathcal{M}(\Omega)$-vague Σ.

Remark 4.15 It is an elementary exercise to check that the transformation $\psi \mapsto \widehat{\psi} = \mathcal{G} \circ \psi$ is a topological isomorphism of $\mathcal{K}(\Omega; A)$ onto $\mathcal{K}(\Omega \times \Delta(A)) = \mathcal{K}(\Omega; \mathcal{C}(\Delta(A)))$, each of the two spaces being provided with the corresponding inductive limit topology. Consequently, for fixed $\varepsilon \in E$, it is readily seen that to each $\mu \in \mathcal{M}(\Omega)$ there is attached a unique $T_\varepsilon(\mu) \in \mathcal{M}(\Omega \times \Delta(A))$ such that

$$\langle T_\varepsilon(\mu), \widehat{\psi}\rangle = \int_\Omega \psi(x, H_\varepsilon(x))d\mu(x)$$

for all $\psi \in \mathcal{K}(\Omega; A)$, where the brackets denote the duality pairing between $\mathcal{M}(\Omega \times \Delta(A))$ and $\mathcal{K}(\Omega \times \Delta(A))$. This yields a transformation $\mu \mapsto T_\varepsilon(\mu)$ that maps linearly $\mathcal{M}(\Omega)$ into $\mathcal{M}(\Omega \times \Delta(A))$. Thus, to say that a sequence of Radon measures $\mu_\varepsilon \in \mathcal{M}(\Omega)$ $(\varepsilon \in E)$ is vaguely Σ-convergent amounts to saying that as $E \ni \varepsilon \to 0$, the sequence of Radon measures $T_\varepsilon(\mu_\varepsilon)$ $(\varepsilon \in E)$ on $\Omega \times \Delta(A)$ is convergent in the weak $*$ topology on $\mathcal{M}(\Omega \times \Delta(A))$.

The following lemma is the keystone of the present section.

Lemma 4.7 *Let X be a locally compact space, and let $\mathcal{P} \subset \mathcal{M}(X)$. The following statements are equivalent.*

(i) $\mathcal{P}$ is bounded in the weak $*$ topology on $\mathcal{M}(X)$, i.e., $\sup_{\mu\in\mathcal{P}} |\mu(\varphi)| < \infty$ for each $\varphi \in \mathcal{K}(X)$.
(ii) $\mathcal{P}$ is locally bounded in norm, i.e., $\sup_{\mu\in\mathcal{P}} |\mu|(K) < \infty$ for each compact set $K \subset X$.

Proof As stated in Bourbaki [22, p. 60, Proposition15], assertion (i) is equivalent to the following:
(iii) For any compact set $H \subset X$, there exists a constant $c_H \geq 0$ such that $\sup_{\mu\in\mathcal{P}} |\mu(\varphi)| \leq c_H \|\varphi\|_\infty$ for all $\varphi \in \mathcal{K}_H(X) = \{f | f \in \mathcal{K}(X), Supp f \subset H\}$.

Thus, the problem reduces to proving the equivalence (ii) $\Leftrightarrow$ (iii). But the (ii) $\Rightarrow$ (iii) part being evident, we only need to concentrate on the (iii) $\Rightarrow$ (ii) part. So, assume (iii), and fix freely a compact set $K \subset X$. Let U be an open neighbourhood of K with compact closure $H = \overline{U}$. Then, according to (iii), we have $|\mu|(f) \leq c_H \|f\|_\infty$ for any $\mu \in \mathcal{P}$ and for all $f \in \mathcal{K}_H(X)$, $f \geq 0$, where c_H is a nonnegative constant. With this in mind, let $\mu \in \mathcal{P}$. Recalling that χ_U (the characteristic function of U) is lower semicontinuous on X, we have

$$|\mu|(U) = \sup_{f\in\mathcal{K}_+(X), f\leq\chi_U} |\mu|(f),$$

where $\mathcal{K}_+(X)$ denotes the set of all $\varphi \in \mathcal{K}(X)$ with $\varphi \geq 0$. But it is clear that each $f \in \mathcal{K}_+(X)$ with $f \leq \chi_U$ lies in $\mathcal{K}_H(X)$ and satisfies $\|f\|_\infty \leq 1$. Therefore $\sup_{\mu\in\mathcal{P}} |\mu|(U) \leq c_H$, hence $\sup_{\mu\in\mathcal{P}} |\mu|(K) \leq c_H$, which shows (ii). ∎

Based on the above lemma, we want to achieve a Σ-compactness theorem for Radon measures similar to Theorem 4.1. Actually, this will arise as a consequence of a more general result viz.

Theorem 4.3 *Let X be a metrizable σ-compact locally compact space. Let $(\mu_n)_{n\in\mathbb{N}}$ be an ordinary sequence of Radon measures on X. We assume that this sequence is bounded in the weak $*$ topology on $\mathcal{M}(X)$. Then one can extract a subsequence $(\mu_{k_n})_{n\in\mathbb{N}}$ from $(\mu_n)_{n\in\mathbb{N}}$ that converges in $\mathcal{M}(X)$-weak $*$ when $n \to \infty$.*

Proof We accomplish this in two steps.

Step 1 Let U be any arbitrary open set in X with compact closure. The aim here is to verify that any subsequence $(\mu_{t_n})_{n\in\mathbb{N}}$ extracted from $(\mu_n)_{n\in\mathbb{N}}$ admits a subsequence $(\mu_{t'_n})_{n\in\mathbb{N}}$ such that

$$\mu_{t'_n}|_U \to \nu' \text{ in } \mathcal{M}(U)\text{-weak } * \text{ as } n \to +\infty. \tag{4.34}$$

To show this, put $K = \overline{U}$ and call B' the closed unit ball of $\mathcal{M}(K)$ (strong dual of $\mathcal{C}(K)$). Provided with the relative weak $*$ topology on $\mathcal{M}(K)$, B' is a metrizable compact space (see, e.g., Bourbaki [21], Brézis [27]). Having made this point, let $(\mu_{t_n})_{n\in\mathbb{N}}$ be any arbitrary subsequence extracted from $(\mu_n)_{n\in\mathbb{N}}$. For each integer $n \geq 0$, put $\nu_n = \mu_{t_n}|_K$ (see Bourbaki [22, pp. 186,187]). Then $\nu_n \in \mathcal{M}(K)$ and further $\sup_n \|\nu_n\| = \sup_n |\nu_n|(K) = \sup_n |\mu_{t_n}|(K) < +\infty$ (Lemma 4.7), where n runs through $\mathbb{N}$. Thus, we may assume without loss of generality that the sequence $(\nu_n)_{n\in\mathbb{N}}$ is contained in B'. Hence we can extract a subsequence $(\nu_{r_n})_{n\in\mathbb{N}}$ from $(\nu_n)_{n\in\mathbb{N}}$ such that as $n \to +\infty$, $\nu_{r_n} \to \nu$ in $\mathcal{M}(K)$-weak $*$, whence $\nu_{r_n}|_U \to \nu|_U = \nu'$ in $\mathcal{M}(U)$-weak $*$. Therefore (4.34) follows with $t'_n = t_{r_n}$ by observing that $\nu_n|_U = \mu_{t_n}|_U$ (see Bourbaki [23, p. 85, Prop.4]).

Step 2 Let $(U_i)_{i\in\mathbb{N}}$ be a sequence of open sets in X with $\overline{U}_i$ compact, $\overline{U}_i \subset U_{i+1}$ and $\bigcup_{i\in\mathbb{N}} U_i = X$. By suitably applying the result of Step 1 we are readily led to two sequences $(\rho_i)_{i\in\mathbb{N}}$ with $\rho_i \in \mathcal{M}(U_i)$, and $(\mu_{t_n^{(i)}})_{(i,n)\in\mathbb{N}^2}$ with $\mu_{t_n^{(i)}} \in \mathcal{M}(X)$, framed as follows: $(\mu_{t_n^{(0)}})_{n\in\mathbb{N}}$ is a subsequence extracted from $(\mu_n)_{n\in\mathbb{N}}$ such that $\mu_{t_n^{(0)}}|_{U_0} \to \rho_0$ in $\mathcal{M}(U_0)$-weak $*$ as $n \to +\infty$; for $i \geq 1$, $(\mu_{t_n^{(i)}})_{n\in\mathbb{N}}$ is a subsequence extracted from $(\mu_{t_n^{(i-1)}})_{n\in\mathbb{N}}$ such that $\mu_{t_n^{(i)}}|_{U_i} \to \rho_i$ in $\mathcal{M}(U_i)$-weak $*$ as $n \to +\infty$. Following the Cantor diagonal process, let us consider the sequence $(\mu_{k_n})_{n\in\mathbb{N}}$ with $k_n = t_n^{(n)}$. It is an easy exercise to verify that this is a subsequence extracted from $(\mu_n)_{n\in\mathbb{N}}$ and further, for each $i \in \mathbb{N}$, we have $\mu_{k_n}|_{U_i} \to \rho_i$ in $\mathcal{M}(U_i)$-weak $*$ as $n \to +\infty$. On the other hand, it is clear that $\rho_i = \rho_{i+1}|_{U_i}$ $(i \in \mathbb{N})$, whence a (unique) Radon measure μ on X such that $\mu|_{U_i} = \rho_i$ for each $i \in \mathbb{N}$ (this is a classical property). Since each function $\varphi \in \mathcal{K}(X)$ lies in some $\mathcal{K}(U_i)$ (for a suitable integer i), we deduce that $\mu_{k_n} \to \mu$ in $\mathcal{M}(X)$-weak $*$ as $n \to +\infty$, thereby proving the theorem.

■

We can therefore prove the claimed Σ-compactness result for measures.

Corollary 4.2 *Let E be a fundamental sequence, and let $(\mu_\varepsilon)_{\varepsilon\in E}$ be a sequence of Radon measures on Ω. Suppose this sequence is bounded in the weak $*$ topology on $\mathcal{M}(\Omega)$. Then, there are a subsequence E' extracted from E and some Radon measure μ_0 on $\Omega \times \Delta(A)$ such that as $E' \ni \varepsilon \to 0$, we have $\mu_\varepsilon \to \mu_0$ in $\mathcal{M}(\Omega)$-vague Σ.*

Proof Since $\Omega \times \Delta(A)$ is a metrizable σ-compact locally compact space, it is clear by Remark 4.15 that the corollary follows readily from Theorem 4.3 if we can show that the sequence $(T_\varepsilon(\mu_\varepsilon))_{\varepsilon\in E}$ is bounded in the weak $*$ topology on $\mathcal{M}(\Omega \times \Delta(A))$. This is straightforward. Indeed, given any arbitrary $\psi \in \mathcal{K}(\Omega; A)$, let K be a compact set in Ω containing the support of ψ. Then

$$|\langle T_\varepsilon(\mu_\varepsilon), \widehat{\psi}\rangle| \leq c \sup_{x\in\Omega} \|\psi(x)\|_\infty \quad (\varepsilon \in E),$$

where $c = \sup_{r\in E} |\mu_r|(K) < +\infty$ (Lemma 4.7). This completes the proof. ■

The next corollary is nothing else than Theorem 4.1.

Corollary 4.3 *Let E be as above, and let $(u_\varepsilon)_{\varepsilon\in E}$ be a norm bounded sequence in $L^p(\Omega)$ $(1 < p < \infty)$. Then, a subsequence E' from E and a function $u_0 \in L^p(\Omega \times \Delta(A))$ exist such that $u_\varepsilon \to u_0$ in $L^p(\Omega)$-weak Σ as $E' \ni \varepsilon \to 0$.*

Proof By letting $\mu_\varepsilon = u_\varepsilon\, dx$ one gets a sequence $(\mu_\varepsilon)_{\varepsilon\in E}$ which is weakly $*$ bounded in $\mathcal{M}(\Omega)$. Therefore, according to Corollary 4.2, there is some $\mu_0 \in \mathcal{M}(\Omega \times \Delta(A))$ such that as $E' \ni \varepsilon \to 0$, one has

$$\int_\Omega u_\varepsilon(x)\psi(x, H_\varepsilon(x))dx \to \mu_0(\widehat{\psi})$$

for all $\psi \in \mathcal{K}(\Omega; A)$. By reasoning as in proving Theorems 3.1 and 4.1 one quickly arrives at $|\mu_0(\widehat{\psi})| \le c\|\widehat{\psi}\|_{L^{p'}(\Omega\times\Delta(A))}$ for all $\psi \in \mathcal{K}(\Omega; A)$, where $c = \sup_{\varepsilon\in E} \|u_\varepsilon\|_{L^p(\Omega)}$. Hence, $\mathcal{K}(\Omega \times \Delta(A)) = \mathcal{K}(\Omega; \mathcal{C}(\Delta(A)))$ being dense in $L^{p'}(\Omega \times \Delta(A))$, the corollary follows by extension by continuity and use of a routine argument. ∎

4.4 Comments and Problems

Σ-convergence is a relatively new theory. The first works on the subject go back to the early 2000s with Nguetseng [93, 95], Nguetseng and Nnang [100], Nguetseng and Svanstedt [102]. The eponymous homogenization method yielded by Σ-convergence theory is growing in popularity, proving day after day its capability of systematically tackling multiscale problems arising in periodic as well as nonperiodic physical processes. Concerning the applications of Σ-convergence to concrete multiscale problems in science and engineering we refer the interested reader to some recent papers such as Deugoue and Woukeng [43], Kenne et al. [66], Jäger and Woukeng [61].

The problem considered earlier in Sect. 3.3 is a particular aspect of a more general question: Given an H-algebra A on $\mathbb{R}^N$ (under an action $\mathcal{H}$), it is known from Proposition 4.2 that if a sequence $(u_\varepsilon)_{\varepsilon\in E}$ in $L^p(\Omega)$ $(1 \le p < \infty)$ and a function $u_0 \in L^p(\Omega \times \Delta(A))$ are such that $u_\varepsilon \to u_0$ in $L^p(\Omega)$-weak Σ and $\|u_\varepsilon\|_{L^p(\Omega)} \to \|u_0\|_{L^p(\Omega\times\Delta(A))}$ as $E \ni \varepsilon \to 0$, then $u_\varepsilon \to u_0$ in $L^p(\Omega)$-strong Σ provided $p = 2$. *The question is whether this result extends to the case where $p \neq 2$. This seems to be an unsolved problem.*

Attention must be drawn to what follows. Let $A = \mathcal{C}_{per}(Y)$ (see Chap. 3). To say (without further ado) that a sequence $(u_\varepsilon)_{\varepsilon\in E}$ is weakly Σ-convergent in $L^p(\Omega)$ has either of the following two meanings:

1. There is some $u_0 \in L^p(\Omega; L^p_{per}(Y))$ such that as $E \ni \varepsilon \to 0$ one has (3.1) for $\psi \in L^{p'}(\Omega; A)$. Here, A is viewed as a mere Banach algebra having no

connection with the Gelfand representation theory, and so this case falls within the self-contained periodic Σ-convergence theory introduced in Chap. 3.

2. There is some $u \in L^p(\Omega \times \Delta(A))$ such that as $E \ni \varepsilon \to 0$,

$$\int_{\Omega} u_\varepsilon \psi^\varepsilon \, dx \to \int\int_{\Omega \times \Delta(A)} u \widehat{\psi} \, dx d\beta$$

for $\psi \in L^{p'}(\Omega; A)$. A is considered here as an H-algebra, and this interpretation is relevant to the general setting of the present chapter. One has $u = \widehat{u}_0$.

The $W^{m,p}(\Omega)$-proper H-algebras play a crucial role in deterministic homogenization, as we will see in the next chapters.

The Σ-convergence of functions extends quite naturally to Radon measures, as seen in Sect. 4.3. Theorem 4.3 seems to be a classical result unknown to us. In default of a reference at hand, the author was in an obligation to give a proof of the latter theorem. Evidently the said proof becomes useless in case it would be available in the literature.

We conclude by attracting attention to a remark having a close connection with nonlinear analysis in homogenization. Let $u_\varepsilon \in W^{m,p}(\Omega)$ ($\varepsilon \in E$) with $u_\varepsilon \to u_0$ in $W^{m,p}(\Omega)$-weak Σ (Definition 4.3). Suppose each u_ε is real-valued, i.e., $u_\varepsilon \in W^{m,p}(\Omega; \mathbb{R})$ ($\varepsilon \in E$). Then $u_0 \in W^{m,p}(\Omega; L^p_{\mathbb{R}}(\Delta(A)))$ and $u_m \in L^p(\Omega; W^{m,p}_{\#}(\Delta(A); \mathbb{R}))$ (see Remark 4.7). In particular if $A = \mathcal{C}_{per}(Y)$, the conclusion reads $u_0 \in W^{m,p}(\Omega; L^p_{per}(Y; \mathbb{R}))$, $u_m \in L^p(\Omega; W^{1,p}_{per}(Y; \mathbb{R})/\mathbb{R})$.

Chapter 5
Homogenization of Elliptic Operators

Homogenization problems for linear as well as nonlinear elliptic PDEs are studied in two main directions: on one hand, in a self-contained periodic setting; on the other hand, in a general deterministic setting using homogenization algebras and including the periodic setting as a particular case. The basic approach in either direction is the Σ-convergence method. Higher-order PDEs, homogenization in perforated domains, quasilinear monotone operators, and Navier-Stokes type equations, are considered and discussed thoroughly.

Throughout this chapter the N-dimensional numerical space $\mathbb{R}^N$ is viewed in two ways: on one hand, as the so-called *macroscopic space* of variables $x = (x_1, \cdots, x_N)$ and denoted by $\mathbb{R}_x^N$; on the other hand, as the so-called *structure space* (this wording will come to light in the sequel) of variables $y = (y_1, \cdots, y_N)$ and denoted by $\mathbb{R}_y^N$. Thus, x is the *macroscopic variable* whereas y is the *structure variable*, again called *local* or *microscopic variable* in the periodic homogenization wording. Both variables are considered concomitantly throughout the mathematical analysis of a multiscale problem, generally as a couple (x, y). The spaces $\mathbb{R}_x^N$ and $\mathbb{R}_y^N$ are interconnected by means of an action $\mathcal{H} = (H_\varepsilon)_{\varepsilon>0}$ of the multiplicative group $\mathbb{R}_+^*$ on $\mathbb{R}^N$, each H_ε being viewed here as a bijection of $\mathbb{R}_x^N$ onto $\mathbb{R}_y^N$. It is precisely by means of such an action that the macroscopic and structure variables interact permanently.

However, the action $\mathcal{H}$ is specified here as $H_\varepsilon(x) = \frac{x}{\varepsilon}$, so that the whole chapter falls within the two-scale setting. On the other hand, Ω denotes a bounded open set in $\mathbb{R}_x^N$. Unless otherwise specified, Ω is assumed to have a smooth boundary, $\partial\Omega$.

G. Nguetseng, *Homogenization Algebras and Applications*, Springer Monographs in Mathematics, https://doi.org/10.1007/978-3-031-84705-9_5

5.1 Second Order Linear Elliptic Operators

5.1.1 Preliminaries

Let

$$a_{ij} \in \mathcal{C}(\overline{\Omega}; L^\infty(\mathbb{R}_y^N)) \ (1 \le i, j \le N) \tag{5.1}$$

with the symmetry property

$$a_{ji} = \overline{a}_{ij} \tag{5.2}$$

and the ellipticity condition:

$$\begin{cases} \text{There is a constant } \alpha > 0 \text{ such that} \\ \operatorname{Re} \sum_{i,j=1}^{N} a_{ij}(x, y)\xi_j \overline{\xi}_i \ge \alpha |\xi|^2, \ \xi = (\xi_i) \in \mathbb{C}^N, \\ \text{for all } x \in \overline{\Omega} \text{ and almost all } y \in \mathbb{R}_y^N, \end{cases} \tag{5.3}$$

where $|\xi|^2 = |\xi_1|^2 + \cdots + |\xi_N|^2$.

Remark 5.1 By (5.3) is meant precisely what follows: there is a constant $\alpha > 0$ such that, given $x \in \overline{\Omega}$, the inequality in (5.3) holds for any $\xi \in \mathbb{C}^N$ and for all $y \in \mathbb{R}^N$ lying outside some negligible set $T_x \subset \mathbb{R}^N$.

Next, for fixed $\varepsilon > 0$, let P^ε be the differential operator in Ω given by

$$P^\varepsilon = - \sum_{i,j=1}^{N} \frac{\partial}{\partial x_i} \left(a_{ij}^\varepsilon \frac{\partial}{\partial x_j} \right), \tag{5.4}$$

where the function $a_{ij}^\varepsilon \in L^\infty(\Omega)$ is defined in the sense of Proposition 1.8 by

$$a_{ij}^\varepsilon(x) = a_{ij}\left(x, \frac{x}{\varepsilon}\right) \quad (x \in \Omega). \tag{5.5}$$

For given $f \in H^{-1}(\Omega) = W^{-1,2}(\Omega)$, we consider the Dirichlet boundary value problem

$$\begin{cases} P^\varepsilon u_\varepsilon = f \text{ in } \Omega \\ u_\varepsilon = 0 \text{ on } \partial\Omega. \end{cases} \tag{5.6}$$

In order to get an existence and uniqueness result for (5.6) we need some preliminaries.

Lemma 5.1 *Let X be a separable metric space, and let $u \in \mathcal{C}(X) \otimes L^\infty(\mathbb{R}_y^N)$. Suppose for each $x \in X$ we have $Re\, u(x, y) \geq 0$ a.e. in $y \in \mathbb{R}^N$. Then, there is a negligible set $\mathcal{N} \subset \mathbb{R}_y^N$ such that $Re\, u(x, y) \geq 0$ for all $x \in X$ and for all $y \in \mathbb{R}^N \setminus \mathcal{N}$.*

Proof First of all, u writes as

$$u = \sum_{i \in I} \varphi_i \otimes u_i, \quad \varphi_i \in \mathcal{C}(X), \quad u_i \in L^\infty(\mathbb{R}_y^N),$$

where I is a finite set. Let D be a dense countable subset of X. For any fixed $a \in D$, there exists (by hypothesis) a negligible set $R_a \subset \mathbb{R}_y^N$ such that $\mathrm{Re}\, u(a, y) \geq 0$ for all $y \in \mathbb{R}^N \setminus R_a$. By letting $R = \bigcup_{a \in D} R_a$, we have therefore $\mathrm{Re}\, u(a, y) \geq 0$ for every $a \in D$ and all $y \in \mathbb{R}^N \setminus R$, and moreover R is negligible. Now, consider a negligible set $S \subset \mathbb{R}_y^N$ and a constant $c > 0$ as in the proof of Lemma 1.6, and put $\mathcal{N} = R \cup S$(a negligible subset of $\mathbb{R}_y^N$). Fix freely $x \in X$, and let $\eta > 0$. D being dense in X, there is some $a \in D$ such that inequality (1.25) in the preceding reference holds. Hence, if y is given in $\mathbb{R}^N \setminus \mathcal{N}$, then the equality under the same (1.25) leads us readily to $|u(x, y) - u(a, y)| \leq \eta$. Therefore, $\mathrm{Re}\, u(x, y) \geq \mathrm{Re}\, u(a, y) - \eta \geq -\eta$ and that for any given pair (x, y) with $x \in X$ and $y \in \mathbb{R}^N \setminus \mathcal{N}$. Hence the lemma follows by the arbitrariness of η. ∎

Based on the preceding lemma, we can now move towards our main purpose.

Lemma 5.2 *Under hypotheses (5.1) and (5.3), there exists a constant $0 < \alpha_0 < \alpha$ (that doesn't depend on ε) such that for any $\varepsilon > 0$, one has*

$$Re \sum_{i,j=1}^{N} a_{ij}^{\varepsilon}(x)\xi_j\overline{\xi}_i \geq \alpha_0|\xi|^2 \quad (\xi \in \mathcal{C}^N) \tag{5.7}$$

almost everywhere in $x \in \Omega$.

Proof For any fixed pair of indices $1 \leq i, j \leq N$, let $(b_{nij})_{n>0}$ be a sequence in $\mathcal{C}(\overline{\Omega}) \otimes L^\infty(\mathbb{R}_y^N)$ such that $b_{nij} \to a_{ij}$ in $\mathcal{C}(\overline{\Omega}; L^\infty(\mathbb{R}_y^N))$ as $n \to \infty$. Thus, letting

$$q_n = \sup_{x \in \overline{\Omega}} \sum_{i,j=1}^{N} \|b_{nij}(x, \cdot) - a_{ij}(x, \cdot)\|_{L^\infty(\mathbb{R}_y^N)},$$

we have $\lim_{n\to\infty} q_n = 0$. On the other hand, let us bear in mind that to each $x \in \overline{\Omega}$ there is attached a negligible set $S_x \subset \mathbb{R}^N_y$ such that

$$\left| \sum_{i,j=1}^{N} [b_{nij}(x, y) - a_{ij}(x, y)]\xi_j\overline{\xi}_i \right| \leq cq_n \quad \text{(integers } n > 0\text{)}$$

for all $y \in \mathbb{R}^N \setminus S_x$ and all $\xi = (\xi_i) \in \mathbb{C}^N$ with $|\xi| = 1$, where

$$c = \max_{1\leq i,j\leq N} \sup_{z\in\mathbb{C}^N, |z|=1} |z_j\overline{z}_i| < \infty.$$

Now, fix a real number $0 < \alpha_0 < \alpha$ (α the constant in (5.3)). Let $0 < \eta \leq \alpha - \alpha_0$. Fix an integer $m > 0$ such that $cq_n \leq \eta$ for all $n \geq m$. Then, for each such n we have

$$\begin{aligned} \operatorname{Re} \sum_{i,j=1}^{N} b_{nij}(x, y)\xi_j\overline{\xi}_i &\geq \operatorname{Re} \sum_{i,j=1}^{N} a_{ij}(x, y)\xi_j\overline{\xi}_i - \eta \\ &\geq \alpha - \eta \\ &\geq \alpha_0 \end{aligned}$$

for all $y \in \mathbb{R}^N \setminus Z_x$ and all $\xi \in K = \{z | z \in \mathbb{C}^N, |z| = 1\}$, where $Z_x = S_x \cup T_x$, T_x being as in Remark 5.1. By considering the mapping $v : \overline{\Omega} \times K \times \mathbb{R}^N \to \mathbb{C}$ given by

$$v(x, \xi, y) = \sum_{i,j=1}^{N} b_{nij}(x, y)\xi_j\overline{\xi}_i - \alpha_0 \quad (\text{ for fixed } n \geq m)$$

for $x \in \overline{\Omega}, \xi \in K, y \in \mathbb{R}^N$, and applying Lemma 5.1 with $X = \overline{\Omega} \times K$ and u replaced by v viewed as a function of the two variables $t = (x, \xi) \in X$ and $y \in \mathbb{R}^N$, we are readily led to

$$\operatorname{Re} \sum_{i,j=1}^{N} b_{nij}(x, y)\xi_j\overline{\xi}_i \geq \alpha_0$$

for all $x \in \overline{\Omega}$, for all $\xi \in \mathbb{C}^N, |\xi| = 1$, and all $y \in \mathbb{R}^N \setminus \mathcal{N}$, where $\mathcal{N}$ is a negligible subset of $\mathbb{R}^N$ depending only on m, and where n (with $n \geq m$) is arbitrary. Therefore, for arbitrary fixed $\varepsilon > 0$ one has

$$\operatorname{Re} \sum_{i,j=1}^{N} b^{\varepsilon}_{nij}(x)\xi_j\overline{\xi}_i \geq \alpha_0$$

for all $x \in \Omega \setminus \varepsilon\mathcal{N}$, where $n \geq m$ and $\xi \in \mathbb{C}^N$, $|\xi| = 1$, are fixed in an arbitrary fashion, and where the trace b^ε_{nij} is defined as in Proposition 1.8. Since $b^\varepsilon_{nij} \to a^\varepsilon_{ij}$ in $L^\infty(\Omega)$ as $n \to \infty$ (thanks to the latter reference), one finally deduces that a negligible set $\mathcal{R}_\varepsilon \subset \Omega$ exists such that

$$\operatorname{Re} \sum_{i,j=1}^{N} a^\varepsilon_{ij}(x)\xi_j\overline{\xi}_i \geq \alpha_0$$

for all $x \in \Omega \setminus \mathcal{R}_\varepsilon$ and all $\xi \in \mathbb{C}^N$ with $|\xi| = 1$. Hence (5.7) follows. This proves the lemma. ∎

Thanks to the preceding lemma we are henceforth justified in claiming the existence of a unique solution of the boundary value problem under consideration (the details are left to the reader):

Proposition 5.1 *Under hypotheses (5.1) and (5.3), to each real $\varepsilon > 0$ there is attached a function $u_\varepsilon \in H^1_0(\Omega)$ uniquely determined by (5.6).*

This being so, our main purpose is to investigate the behaviour of u_ε as $\varepsilon \to 0$. In all probability such an undertaking is hopeless without requiring the functions $y \mapsto a_{ij}(x, y)$ $(1 \leq i, j \leq N)$ to fulfil further suitable conditions characterizing their behaviours, referred to, as a whole, as a *structure hypothesis*.

The common structure hypothesis is the so-called *periodicity hypothesis* stating in the present case that the functions a_{ij} are periodic (with the same period) with respect to the structure variable $y = (y_1, \cdots, y_N)$. Section 5.1.2 below is indeed devoted to the periodic homogenization of (5.6).

Next, beyond the usual periodic theory, we shall carry out the homogenization of (5.6) in a very general deterministic setting characterized by an abstract assumption on $a_{ij}(x, y)$ (with respect to y) covering a wide range of concrete behaviours including the periodicity hypothesis as a particular case, of course. However, let us be through beforehand with the generalities.

As is classical, $H^1_0(\Omega)$ denotes the closure of $\mathcal{D}(\Omega)$ in $H^1(\Omega) = W^{1,2}(\Omega)$. Furthermore, if Ω has a smooth boundary, which is the case here, then $H^1_0(\Omega)$ is identical with the space of those $v \in H^1(\Omega)$ with $v = 0$ on $\partial\Omega$ (in the sense of traces; see Lions and Magenes [74]).

By letting

$$\|u\|_{H^1_0(\Omega)} = \left[\sum_{i=1}^{N} \left\|\frac{\partial u}{\partial x_i}\right\|^2_{L^2(\Omega)}\right]^{\frac{1}{2}} \quad (u \in H^1_0(\Omega)),$$

we define a norm on $H^1_0(\Omega)$ which is equivalent to the $H^1(\Omega)$ norm. The antiduality pairing between $H^1_0(\Omega)$ and its antidual, $H^{-1}(\Omega)$, is denoted by $\langle , \rangle$. The Hilbert space $L^2(\Omega)$ shall be identified with its antidual so that $H_0^1(\Omega) \subset L^2(\Omega) \subset H^{-1}(\Omega)$ with continuous embeddings, each space being dense in the next one.

For fixed $\varepsilon > 0$, let

$$a^\varepsilon(u, v) = \sum_{i,j=1}^{N} \int_\Omega a_{ij}^\varepsilon \frac{\partial u}{\partial x_j} \overline{\frac{\partial v}{\partial x_i}} dx \quad (u, v \in H_0^1(\Omega)).$$

This defines a Hermitian sesquilinear form $a^\varepsilon(\,,\,)$ on $H_0^1(\Omega) \times H_0^1(\Omega)$ satisfying

$$|a^\varepsilon(u, v)| \leq c\|u\|_{H_0^1(\Omega)} \|v\|_{H_0^1(\Omega)} \quad (u, v \in H_0^1(\Omega))$$

and

$$\operatorname{Re} a^\varepsilon(u, u) \geq \alpha_0 \|u\|^2_{H_0^1(\Omega)} \quad (u \in H_0^1(\Omega)), \tag{5.8}$$

where α_0 is the same as is (5.7) and $c = \max_{1\leq i,j\leq N} \sup_{x\in\overline{\Omega}} \|a_{ij}(x, \cdot)\|_{L^\infty}$. Indeed, (5.8) follows directly from Lemma 5.2 whereas the former inequality follows by Proposition 1.8.

We conclude these preliminaries with a classical but fundamental statement which is nothing else but the weak formulation of (5.6).

Proposition 5.2 *Let $\varepsilon > 0$. The following assertions are equivalent for a function $u_\varepsilon \in H^1(\Omega)$.*

(i) u_ε is a solution of (5.6).
(ii) $u_\varepsilon \in H_0^1(\Omega)$ and

$$a^\varepsilon(u_\varepsilon, v) = \langle f, v\rangle \tag{5.9}$$

for all $v \in H_0^1(\Omega)$.

5.1.2 Periodic Setting

5.1.2.1 Global Homogenization Theorem

We purposely restrict the study to the classical case where

$$Y = \left(-\frac{1}{2}, \frac{1}{2}\right)^N, \quad S = \mathbb{Z}^N,$$

though the whole dissertation remains rigorously valid with any arbitrary couple $\{Y, S\}$ underlying the general periodic setting fully described in Sect. 2.5. Indeed, as pointed out earlier, the only difference between the developments arising from either setting lies in the multiplicative factor $\frac{1}{|Y|}$ ($|Y|$ the measure of Y for λ).

We begin by collecting the basic notions and results necessary to our undertaking.

The Space $\mathbb{F}_0^1(\Omega \times Y)$ We define

$$\mathbb{F}_0^1(\Omega \times Y) = H_0^1(\Omega) \times L^2(\Omega; H_{per}^1(Y)/\mathbb{C}),$$

where $H_{per}^1(Y) = W_{per}^{1,2}(Y)$. This is a Hilbert space with norm

$$\|\mathbf{u}\|_{\mathbb{F}_0^1} = (\|u_0\|_{H_0^1(\Omega)}^2 + \|u_1\|_{L^2(\Omega; H_{per}^1(Y)/\mathbb{C})}^2)^{\frac{1}{2}}, \quad \mathbf{u} = (u_0, u_1) \in \mathbb{F}_0^1,$$

where we put for short $\mathbb{F}_0^1 = \mathbb{F}_0^1(\Omega \times Y)$. Furthermore

$$\mathcal{F}_0^\infty = \mathcal{D}(\Omega) \times (\mathcal{D}(\Omega) \otimes [\mathcal{C}_{per}^\infty(Y)/\mathbb{C}]) \text{ is dense in } \mathbb{F}_0^1, \tag{5.10}$$

as is straightforward from Corollary 2.13.

For fixed $1 \leq i \leq N$, we set

$$\mathbb{D}_i \mathbf{u} = \frac{\partial u_0}{\partial x_i} + \frac{\partial u_1}{\partial y_i}, \quad \mathbf{u} = (u_0, u_1) \in \mathbb{F}_0^1.$$

This defines a differential operator $\mathbb{D}_i$ that maps continuously and linearly $\mathbb{F}_0^1$ into $L^2(\Omega; L_{per}^2(Y))$. Furthermore

$$\|\mathbf{u}\|_{\mathbb{F}_0^1} = \left[\sum_{i=1}^N \|\mathbb{D}_i \mathbf{u}\|_{L^2(\Omega \times Y)}^2\right]^{\frac{1}{2}} \quad (\mathbf{u} \in \mathbb{F}_0^1),$$

as is easily verified by recalling that for $u \in H_{per}^1(Y)$, each partial derivative $\frac{\partial u}{\partial y_i}$ has mean value zero.

Except where otherwise stated, it is assumed from now that

$$a_{ij} \text{ is } Y\text{-periodic in } y \quad (1 \leq i, j \leq N), \tag{5.11}$$

that is, for fixed $x \in \overline{\Omega}$, we have $a_{ij}(x, y + k) = a_{ij}(x, y)$ $(1 \leq i, j \leq N)$ for all $k \in \mathbb{Z}^N$ and for almost every $y \in \mathbb{R}^N$. Thus,

$$a_{ij} \in \mathcal{C}(\overline{\Omega}; L_{per}^\infty(Y)) \quad (1 \leq i, j \leq N). \tag{5.12}$$

For $\mathbf{u}, \mathbf{v} \in \mathbb{F}_0^1$, let

$$a_\Omega(\mathbf{u}, \mathbf{v}) = \sum_{i,j=1}^{N} \int\int_{\Omega \times Y} a_{ij} \mathbb{D}_j \mathbf{u} \overline{\mathbb{D}_i \mathbf{v}}\, dxdy.$$

This defines a sesquilinear form $a_\Omega(,)$ on $\mathbb{F}_0^1 \times \mathbb{F}_0^1$ which, by reason of (5.1),(5.2) and (5.3), is continuous, Hermitian and $\mathbb{F}_0^1$-coercive. As regards the last property, we have precisely

$$\operatorname{Re} a_\Omega(\mathbf{v}, \mathbf{v}) \geq \alpha \|\mathbf{v}\|^2_{\mathbb{F}_0^1} \quad (\mathbf{v} \in \mathbb{F}_0^1). \tag{5.13}$$

Let $\mathbf{l}$ be the antilinear form on $\mathbb{F}_0^1$ defined by

$$\mathbf{l}(\mathbf{v}) = \langle f, v_0 \rangle, \quad \mathbf{v} = (v_0, v_1) \in \mathbb{F}_0^1.$$

Then, in view of (5.12) and (5.13), the variational problem

$$\begin{cases} \mathbf{u} = (u_0, u_1) \in \mathbb{F}_0^1 : \\ a_\Omega(\mathbf{u}, \mathbf{v}) = \mathbf{l}(\mathbf{v}) \quad (\mathbf{v} \in \mathbb{F}_0^1) \end{cases} \tag{5.14}$$

admits one and only one solution (this is straightforward by the Lax-Milgram lemma).

About the Forthcoming Homogenized Coefficients For fixed $x \in \overline{\Omega}$, let

$$a(x; u, v) = \sum_{i,j=1}^{N} \int_Y a_{ij}(x, y) \frac{\partial u}{\partial y_j}(y) \overline{\frac{\partial v}{\partial y_i}}(y) dy$$

for $u, v \in H^1_{per}(Y)/\mathbb{C}$. This defines a sesquilinear form $a(x; \cdot, \cdot)$ on $H^1_{per}(Y)/\mathbb{C} \times H^1_{per}(Y)/\mathbb{C}$. According to (5.1), (5.2) and (5.3), the said sesquilinear form is continuous, Hermitian and $(H^1_{per}(Y)/\mathbb{C})$-coercive, i.e.,

$$\operatorname{Re} a(x; u, u) \geq \alpha \|u\|^2_{H^1(Y)/\mathbb{C}} \quad (u \in H^1_{per}(Y)/\mathbb{C}). \tag{5.15}$$

Now, let $1 \leq j \leq N$ be fixed. Thanks to the preceding properties of $a(x; \cdot, \cdot)$, to each $x \in \overline{\Omega}$ there is assigned a unique function $\chi^j(x, \cdot) \in H^1_{per}(Y)/\mathbb{C}$ such that

$$a(x; \chi^j(x, \cdot), v) = \sum_{k=1}^{N} \int_Y a_{kj}(x, y) \overline{\frac{\partial v}{\partial y_k}}(y) dy \tag{5.16}$$

for all $v \in H^1_{per}(Y)/\mathbb{C}$. Furthermore, for any pair of indices $1 \le i, j \le N$ and any $x \in \overline{\Omega}$, let

$$q_{ij}(x) = \int_Y a_{ij}(x, y)dy - \sum_{l=1}^{N} \int_Y a_{il}(x, y)\frac{\partial \chi^j(x, y)}{\partial y_l}dy, \tag{5.17}$$

which yields a family of mappings $q_{ij} : \overline{\Omega} \to \mathbb{C}$ $(1 \le i, j \le N)$. Our goal is precisely to point out the main properties of this family. For that purpose, let us prove the following proposition.

Proposition 5.3 *The following statements hold true for* $1 \le i, j \le N$.

(i) $\chi^j \in \mathcal{C}(\overline{\Omega}; H^1_{per}(Y)/\mathbb{C})$.
(ii) $q_{ij} \in \mathcal{C}(\overline{\Omega})$.
(iii) $q_{ji} = \overline{q}_{ij}$.
(iv) There is a constant $\alpha_1 > 0$ *such that*

$$Re \sum_{i,j=1}^{N} q_{ij}(x)\xi_j\overline{\xi}_i \ge \alpha_1|\xi|^2$$

holds for all $x \in \overline{\Omega}$ *and all* $\xi \in \mathbb{C}^N$.

Proof

(i) For convenience we denote the right-hand side of (5.16) by $L_j(x; v)$ and we set on the other hand

$$r(x, x') = \max_{1\le k,l\le N} \|a_{kl}(x', \cdot) - a_{kl}(x, \cdot)\|_{L^\infty(\mathbb{R}^N)}, \quad (x, x' \in \overline{\Omega}).$$

Now, fix any arbitrary $x_0 \in \overline{\Omega}$. For $x \in \overline{\Omega}$ and $v \in H^1_{per}(Y)/\mathbb{C}$ we have

$$\begin{aligned} a(x; \chi^j(x) - \chi^j(x_0), v) = L_j(x; v) - L_j(x_0; v) + a(x_0; \chi^j(x_0), v) \\ - a(x; \chi^j(x_0), v), \end{aligned}$$

where, for short, one writes $\chi^j(x)$ in place of $\chi^j(x, \cdot)$. But an easy calculation leads us to some constant $c > 0$ such that

$$|L_j(x; v) - L_j(x_0; v)| \le cr(x, x_0)\|v\|_{H^1(Y)/\mathbb{C}}$$

and

$$|a(x; \chi^j(x_0), v) - a(x_0; \chi^j(x_0), v)| \le cr(x, x_0)\|v\|_{H^1(Y)/\mathbb{C}}.$$

Hence, by choosing $v = \chi^j(x) - \chi^j(x_0)$ in the preceding equality and taking the real part of either side of the resultant equality one obtains by (5.15)

$$\|\chi^j(x) - \chi^j(x_0)\|_{H^1(Y)/\mathbb{C}} \leq \frac{2c}{\alpha} r(x, x_0)$$

and that for any arbitrary $x \in \overline{\Omega}$. Therefore (i) follows at once, since $r(x, x_0) \to 0$ as $x \to x_0$ (use (5.1)).

(ii) According to (i) we have $\frac{\partial \chi^j}{\partial y_l} \in \mathcal{C}(\overline{\Omega}; L^2_{per}(Y))$ for $1 \leq j, l \leq N$. Hence (ii) follows readily by (5.17) and (5.1).

(iii) Let $x \in \overline{\Omega}$. Following Bensoussan, Lions and Papanicolaou [14], one can write

$$q_{ij}(x) = a(x; y_j - \chi^j(x), y_i).$$

On the other hand, (5.16) reads

$$a(x; \chi^j(x) - y_j, v) = 0,$$

where, since v is arbitrary in $H^1_{per}(Y)/\mathbb{C}$, we may choose $v = \chi^i(x)$. We deduce immediately that

$$q_{ij}(x) = a(x; \chi^j(x) - y_j, \chi^i(x) - y_i). \tag{5.18}$$

Hence (iii) follows readily, $a(x; \cdot, \cdot)$, being Hermitian.

(iv) Again following Bensoussan, Lions and Papanicolaou [14], let

$$q(x, \xi) = \operatorname{Re} \sum_{i,j=1}^{N} q_{ij}(x) \xi_j \overline{\xi}_i$$

for $x \in \overline{\Omega}$ and $\xi \in \mathbb{C}^N$, which defines a mapping q that sends continuously $\overline{\Omega} \times \mathbb{C}^N$ into $\mathbb{R}$ with further $q(x, t\xi) = t^2 q(x, \xi)$ for $x \in \overline{\Omega}$, $\xi \in \mathbb{C}^N$ and $t \in \mathbb{R}$. Thus, we shall have proved (iv) if we can check that there is some $\alpha_1 > 0$ such that $q(x, \xi) \geq \alpha_1$ for all $x \in \overline{\Omega}$ and $\xi \in K = \{z | z \in \mathbb{C}^N, |z| = 1\}$. To this end, fix freely $x \in \overline{\Omega}$, and let

$$u(\xi) = \sum_{j=1}^{N} [y_j - \chi^j(x)] \xi_j, \quad \xi = (\xi_i) \in \mathbb{C}^N.$$

By (5.18) and (5.3) we have

$$q(x, \xi) = \operatorname{Re} a(x; u(\xi), u(\xi)) \geq \alpha \sum_{i=1}^{N} \int_Y \left| \frac{\partial u(\xi)}{\partial y_i} \right|^2 dy.$$

This shows that $q(x,\xi) \geq 0$. Furthermore, if $q(x,\xi) = 0$ then there is some constant $c(x) \in \mathbb{C}$ such that $u(\xi) = c(x)$, i.e., $\sum_{j=1}^{N} \xi_j \chi^j(x) = -c(x) + \sum_{j=1}^{N} \xi_j y_j$, an equality between a polynomial function and a Y-periodic function! This entails $\xi_j = 0$ ($j = 1, \cdots, N$) and so $q(x,\cdot)$ is positive definite. One deduces that $q(x,\xi) > 0$ for $(x,\xi) \in \overline{\Omega} \times K$. But q being continuous on the compact set $\overline{\Omega} \times K$, there certainly exists some $(x_1,\xi_1) \in \overline{\Omega} \times K$ such that $q(x,\xi) \geq q(x_1,\xi_1) > 0$ for all $(x,\xi) \in \overline{\Omega} \times K$. Hence (iv) follows with $\alpha_1 = q(x_1,\xi_1)$. ■

Homogenization Results We are now in a position to prove the global homogenization theorem and its corollaries.

Theorem 5.1 *Under assumptions (5.1)–(5.3) and (5.11), let* $\mathbf{u} = (u_0, u_1)$ *be (uniquely) defined by (5.14). For each real* $\varepsilon > 0$*, let* $u_\varepsilon \in H_0^1(\Omega)$ *be the (unique) solution of (5.6) with (5.4)–(5.5). Then, as* $\varepsilon \to 0$*, we have*

$$u_\varepsilon \to u_0 \text{ in } H_0^1(\Omega)\text{-weak,} \tag{5.19}$$

$$u_\varepsilon \to u_0 \text{ in } L^2(\Omega), \tag{5.20}$$

$$\frac{\partial u_\varepsilon}{\partial x_j} \to \mathbb{D}_j \mathbf{u} = \frac{\partial u_0}{\partial x_j} + \frac{\partial u_1}{\partial y_j} \text{ in } L^2(\Omega)\text{-weak } \Sigma \quad (1 \leq j \leq N). \tag{5.21}$$

Remark 5.2 Here and throughout the rest of the book we implicitly use the following simple but useful fact: given $r_0 \in \mathbb{C}$ and $(r_\varepsilon)_{\varepsilon>0}$ with $r_\varepsilon \in \mathbb{C}$, we have $r_\varepsilon \to r_0$ as $\varepsilon \to 0$ if and only if for any fundamental sequence E (Chap. 3), we have $r_\varepsilon \to r_0$ as $E \ni \varepsilon \to 0$.

Proof of Theorem 5.1 Let $\varepsilon > 0$. To begin, let us turn to Proposition 5.2 and take in (5.9) the particular function $v = u_\varepsilon$. Then, by (5.8) and the inequality $\mathrm{Re}\langle f, u_\varepsilon\rangle \leq \|f\|_{H^{-1}(\Omega)} \|u_\varepsilon\|_{H_0^1(\Omega)}$ one quickly arrives at $\|u_\varepsilon\|_{H_0^1(\Omega)} \leq (\alpha^{-1}\|f\|_{H^{-1}(\Omega)})$, and that for any arbitrary real $\varepsilon > 0$. Therefore, given an arbitrary fundamental sequence E, Theorem 3.2 guarantees the existence of a subsequence E' from E and some couple $\mathbf{u} = (u_0, u_1) \in \mathbb{F}_0^1$ such that as $E' \ni \varepsilon \to 0$, we have (5.19) and (5.21). Immediately (5.20) follows from (5.19) by the Rellich-Kondrachov theorem. Thus, the theorem is proved if we can check that $\mathbf{u} = (u_0, u_1)$ satisfies (5.14). Indeed, thanks to the unicity in (5.14), the preceding convergence properties will then be valid for the whole E, hence merely for $\varepsilon \to 0$ (Remark 5.2).

Having made this point, let us fix freely $\mathbf{\Phi} = (\psi_0, \psi_1) \in \mathcal{F}_0^\infty$. For $\varepsilon > 0$, put

$$\Phi_\varepsilon = \psi_0 + \varepsilon \psi_1^\varepsilon, \tag{5.22}$$

where the notation $\psi_1^\varepsilon(x) = \psi_1(x, \frac{x}{\varepsilon})$ is worth recalling. It is clear that $\Phi_\varepsilon \in \mathcal{D}(\Omega)$; and so we may choose $v = \Phi_\varepsilon$ in (5.9) to get

$$a^\varepsilon(u_\varepsilon, \Phi_\varepsilon) = \langle f, \Phi_\varepsilon\rangle. \tag{5.23}$$

The aim now is to pass to the limit when $E' \ni \varepsilon \to 0$. To achieve this, let us begin by noting that for $1 \leq i \leq N$, we have

$$\frac{\partial \Phi_\varepsilon}{\partial x_i} = \frac{\partial \psi_0}{\partial x_i} + \left(\frac{\partial \psi_1}{\partial y_i}\right)^\varepsilon + \varepsilon \left(\frac{\partial \psi_1}{\partial x_i}\right)^\varepsilon.$$

Since $\varepsilon(\frac{\partial \psi_1}{\partial x_i})^\varepsilon \to 0$ in $L^p(\Omega)$ as $\varepsilon \to 0$, one deduces by Proposition 3.2 that

$$\frac{\partial \Phi_\varepsilon}{\partial x_i} \to \mathbb{D}_i \mathbf{\Phi} \text{ in } L^p(\Omega)\text{-strong } \Sigma \;\; (1 \leq p < \infty). \tag{5.24}$$

Hence, by taking Proposition 3.3 (i) and Remark 3.4 (1) into consideration, one readily sees that

$$\Phi_\varepsilon \to \psi_0 \text{ in } H_0^1(\Omega)\text{-weak as } \varepsilon \to 0. \tag{5.25}$$

We are now in a position to pass to the limit in (5.23). First, consider (5.24) with $p = 2$ and with the further property that the sequence of functions $\frac{\partial \Phi_\varepsilon}{\partial x_i}$ $(\varepsilon \in E)$ is norm-bounded in $L^\infty(\Omega)$. Recalling on the other hand that (5.21) holds when $E' \ni \varepsilon \to 0$, we deduce by Proposition 3.5 that

$$\frac{\partial u_\varepsilon}{\partial x_j} \overline{\frac{\partial \Phi_\varepsilon}{\partial x_i}} \to \mathbb{D}_j \mathbf{u} \overline{\mathbb{D}_i \mathbf{\Phi}} \text{ in } L^2(\Omega)\text{-weak } \Sigma \text{ as } E' \ni \varepsilon \to 0$$

for $1 \leq i, j \leq N$. But, thanks to Proposition 3.6 this may be expressed in terms of integrals using test functions ψ in $\mathcal{C}(\overline{\Omega}; L^\infty_{per}(Y))$ [instead of $L^2(\Omega; \mathcal{C}_{per}(Y))$] and therefore in particular $\psi = a_{ij}$ (see (5.12)). Hence

$$a^\varepsilon(u_\varepsilon, \Phi_\varepsilon) \to a_\Omega(\mathbf{u}, \mathbf{\Phi}) \text{ as } E' \ni \varepsilon \to 0. \tag{5.26}$$

Consequently, by passing to the limit in (5.23) using (5.25), one arrives at $a_\Omega(\mathbf{u}, \mathbf{\Phi}) = \mathbf{l}(\mathbf{\Phi})$ for all $\mathbf{\Phi} \in \mathcal{F}_0^\infty$. Hence (5.14) follows by (5.10). ■

The preceding theorem has three important corollaries.

Corollary 5.1 *Let* $\mathbf{u} = (u_0, u_1)$ *be the solution of (5.14). For almost every* $x \in \Omega$, $u_1(x)$ *is uniquely determined by the variational problem*

$$\begin{cases} u_1(x) \in H^1_{per}(Y)/\mathbb{C} : \\ a(x; u_1(x), w) = -\displaystyle\sum_{k,j=1}^N \frac{\partial u_0}{\partial x_j}(x) \int_Y a_{kj}(x, y) \overline{\frac{\partial w}{\partial y_k}}(y) dy \\ \textit{for all } w \in H^1_{per}(Y)/\mathbb{C}. \end{cases} \tag{5.27}$$

Proof Let $w \in H^1_{per}(Y)/\mathbb{C}$. For $\varphi \in \mathcal{D}(\Omega)$, take in (5.14) the test function $\mathbf{v} = (v_0, v_1)$ such that $v_0 = 0$ and $v_1(x) = \varphi(x)w$ for $x \in \Omega$. Then (5.27) (for almost all $x \in \Omega$) follows by the arbitrariness of φ. Furthermore, according to (5.15), $u_1(x)$ is the unique solution of (5.27). ■

Corollary 5.2 *Let u_1 be as above. Then*

$$u_1(x) = -\sum_{j=1}^{N} \frac{\partial u_0}{\partial x_j}(x)\chi^j(x) \quad (\text{ a.e. in } x \in \Omega), \tag{5.28}$$

where $\chi^j \in \mathcal{C}(\overline{\Omega}; H^1_{per}(Y)/\mathbb{C})$ (Proposition 5.3), $\chi^j(x)$ given by (5.16) ($1 \leq j \leq N$).

Proof Let $z(x)$ denote the right-hand side of (5.28). Multiplying both sides of (5.16) by $-\frac{\partial u_0}{\partial x_j}(x)$ and then summing over $1 \leq j \leq N$ we see that (5.27) still holds with $z(x)$ in place of $u_1(x)$. Hence (5.28) follows by unicity. ■

Corollary 5.3 *Let $\mathbf{u} = (u_0, u_1)$ be the solution of (5.14). Then u_0 is the weak solution of the Dirichlet boundary value problem*

$$\begin{cases} -\sum_{i,j=1}^{N} \dfrac{\partial}{\partial x_i}\left(q_{ij}\dfrac{\partial u_0}{\partial x_j}\right) = f \text{ in } \Omega, \\ \\ u_0 = 0 \text{ on } \partial\Omega, \end{cases} \tag{5.29}$$

where $q_{ij} \in \mathcal{C}(\overline{\Omega})$ (Proposition 5.3), $q_{ij}(x)$ defined by (5.17) ($1 \leq i, j \leq N$).

Proof By choosing in (5.14) the particular test functions $\mathbf{v} = (v_0, v_1)$ characterized by $v_1 = 0$ we are readily led to (5.29) which uniquely determines u_0 thanks to parts (ii) and (iv) of Proposition 5.3. ■

Problem (5.14) is referred to as the *global homogenized problem*. The term *global* is used here to stress the point that both the so-called *local problem*, namely (5.27), and *macroscopic homogenized problem*, viz. (5.29), are embodied in (5.14). The differential operator in (5.29) is referred to as the *homogenized operator associated to P^ε*. Its main properties are collected in Proposition 5.3.

5.1.2.2 Strong Approximation

Throughout the present section, u_ε denotes the solution of the Boundary value problem (5.6), and $\mathbf{u} = (u_0, u_1)$ that of the variational problem (5.14).

Proposition 5.4 *As $\varepsilon \to 0$, we have*

$$\frac{\partial u_\varepsilon}{\partial x_j} \to \mathbb{D}_j\mathbf{u} \text{ in } L^2(\Omega)\text{-strong } \Sigma \quad (1 \leq j \leq N). \tag{5.30}$$

Proof In view of (5.21), we know from Proposition 3.3 that (5.30) is achieved if we can show that as $\varepsilon \to 0$,

$$\left\| \frac{\partial u_\varepsilon}{\partial x_j} \right\|_{L^2(\Omega)} \to \|\mathbb{D}_j \mathbf{u}\|_{L^2(\Omega \times Y)} \ (1 \leq j \leq N). \tag{5.31}$$

For this purpose, let $\eta > 0$. Based on (5.10), fix some $\mathbf{\Phi} = (\psi_0, \psi_1) \in \mathcal{F}_0^\infty$ such that

$$\mathrm{Re}\, a_\Omega(\mathbf{u} - \mathbf{\Phi}, \mathbf{u} - \mathbf{\Phi}) \leq \frac{\alpha_0 \eta^2}{16} \ (\text{ same } \alpha_0 \text{ as in (5.7)}). \tag{5.32}$$

Next, fix freely $\varepsilon > 0$, and let Φ_ε be given by (5.22). We begin by proving that there is some $\varepsilon_1 > 0$ such that

$$\left\| \frac{\partial u_\varepsilon}{\partial x_j} - \frac{\partial \Phi_\varepsilon}{\partial x_j} \right\|_{L^2(\Omega)} \leq \frac{\eta}{2} \ (1 \leq j \leq N) \tag{5.33}$$

holds for $0 < \varepsilon < \varepsilon_1$. To do this, we consider the obvious equality

$$a^\varepsilon(u_\varepsilon - \Phi_\varepsilon, u_\varepsilon - \Phi_\varepsilon) = \langle f, u_\varepsilon \rangle - 2\mathrm{Re}\, a^\varepsilon(u_\varepsilon, \Phi_\varepsilon) + a^\varepsilon(\Phi_\varepsilon, \Phi_\varepsilon)$$

and we use various facts contained in Theorem 5.1 and its proof to deduce that

$$a^\varepsilon(u_\varepsilon - \Phi_\varepsilon, u_\varepsilon - \Phi_\varepsilon) \to a_\Omega(\mathbf{u} - \mathbf{\Phi}, \mathbf{u} - \mathbf{\Phi}) \text{ as } \varepsilon \to 0.$$

Hence there is some $\varepsilon_1 > 0$ such that

$$\mathrm{Re}\, a^\varepsilon(u_\varepsilon - \Phi_\varepsilon, u_\varepsilon - \Phi_\varepsilon) \leq \mathrm{Re}\, a_\Omega(\mathbf{u} - \mathbf{\Phi}, \mathbf{u} - \mathbf{\Phi}) + \frac{3\alpha_0 \eta^2}{16}$$

holds provided $0 < \varepsilon \leq \varepsilon_1$. But according to (5.8) (resp. (5.32)), the left (resp. right)-hand side is greater (resp. less) than or equal to $\alpha_0 \|u_\varepsilon - \Phi_\varepsilon\|^2_{H_0^1(\Omega)}$ (resp. $\frac{\alpha_0 \eta^2}{4}$). Therefore (5.33) follows provided $0 < \varepsilon \leq \varepsilon_1$. Consequently

$$\left| \left\| \frac{\partial u_\varepsilon}{\partial x_j} \right\|_{L^2(\Omega)} - \left\| \frac{\partial \Phi_\varepsilon}{\partial x_j} \right\|_{L^2(\Omega)} \right| \leq \frac{\eta}{2} \ (0 < \varepsilon \leq \varepsilon_1)$$

for $1 \leq j \leq N$. On the other hand, by (5.13) (with $\mathbf{v} = \mathbf{u} - \mathbf{\Phi}$) and (5.32) with taking account of the fact that $\alpha_0 < \alpha$, we have

$$\left| \|\mathbb{D}_j \mathbf{u}\|_{L^2(\Omega \times Y)} - \|\mathbb{D}_j \mathbf{\Phi}\|_{L^2(\Omega \times Y)} \right| \leq \frac{\eta}{4}.$$

Finally, appeal to (5.24) and use of Proposition 3.3 yields some $\varepsilon_2 > 0$ such that

$$\left| \left\| \frac{\partial \Phi_\varepsilon}{\partial x_j} \right\|_{L^2(\Omega)} - \|\mathbb{D}_j \mathbf{\Phi}\|_{L^2(\Omega\times Y)} \right| \le \frac{\eta}{4}$$

for $0 < \varepsilon \le \varepsilon_2$. Hence,by the trivial inequality

$$\begin{aligned} \left| \left\| \frac{\partial u_\varepsilon}{\partial x_j} \right\|_{L^2(\Omega)} - \|\mathbb{D}_j \mathbf{u}\|_{L^2(\Omega\times Y)} \right| &\le \left| \left\| \frac{\partial u_\varepsilon}{\partial x_j} \right\|_{L^2(\Omega)} - \left\| \frac{\partial \Phi_\varepsilon}{\partial x_j} \right\|_{L^2(\Omega)} \right| \\ &+ \left| \left\| \frac{\partial \Phi_\varepsilon}{\partial x_j} \right\|_{L^2(\Omega)} - \|\mathbb{D}_j \mathbf{\Phi}\|_{L^2(\Omega\times Y)} \right| \\ &+ \left| \|\mathbb{D}_j \mathbf{u}\|_{L^2(\Omega\times Y)} - \|\mathbb{D}_j \mathbf{\Phi}\|_{L^2(\Omega\times Y)} \right| \end{aligned}$$

one readily arrives at

$$\left| \left\| \frac{\partial u_\varepsilon}{\partial x_j} \right\|_{L^2(\Omega)} - \|\mathbb{D}_j \mathbf{u}\|_{L^2(\Omega\times Y)} \right| \le \eta \;\; (1 \le j \le N)$$

for $0 < \varepsilon \le \varepsilon_0 = \min(\varepsilon_1, \varepsilon_2)$. This shows (5.31) when $\varepsilon \to 0$. Hence the proposition follows. ■

Three corollaries of interest are worth pointing out.

Corollary 5.4 *Let $\eta > 0$. Then, to each function $\psi_1 \in H^1(\Omega) \otimes [C^1_{per}(Y)/\mathbb{C}]$ such that*

$$\|u_1 - \psi_1\|_{L^2(\Omega; H^1_{per}(Y)/\mathbb{C})} \le \frac{\eta}{4}, \tag{5.34}$$

there is assigned a real $\varepsilon_1 > 0$ such that

$$\left\| \frac{\partial}{\partial x_j}(u_\varepsilon - u_0 - \varepsilon \psi_1^\varepsilon) \right\|_{L^2(\Omega)} \le \eta \;\; (1 \le j \le N) \tag{5.35}$$

holds for $0 < \varepsilon \le \varepsilon_1$.

Proof Let us first observe that $H^1(\Omega) \otimes [C^1_{per}(Y)/\mathbb{C}]$ being dense in $L^2(\Omega; H^1_{per}(Y)/\mathbb{C})$, a function such as ψ_1 does exist. This being so, let $1 \le j \le N$. Let ψ_1 be as above. In view of (5.34), we have

$$\left\| \frac{\partial u_1}{\partial y_j} - \frac{\partial \psi_1}{\partial y_j} \right\|_{L^2(\Omega\times Y)} \le \frac{\eta}{4}.$$

According to Proposition 5.4 (see also Definition 3.2) this yields some $\varepsilon_2 > 0$ such that

$$\left\| \frac{\partial u_\varepsilon}{\partial x_j} - \frac{\partial u_0}{\partial x_j} - \left(\frac{\partial \psi_1}{\partial y_j}\right)^\varepsilon \right\|_{L^2(\Omega)} \leq \frac{\eta}{2}$$

for $0 < \varepsilon \leq \varepsilon_2$. But

$$\left(\frac{\partial \psi_1}{\partial y_j}\right)^\varepsilon = \varepsilon \frac{\partial \psi_1^\varepsilon}{\partial x_j} - \varepsilon \left(\frac{\partial \psi_1}{\partial x_j}\right)^\varepsilon .$$

On the other hand, by Proposition 1.6,

$$\left\| \left(\frac{\partial \psi_1}{\partial x_j}\right)^\varepsilon \right\|_{L^2(\Omega)} \leq \left\| \frac{\partial \psi_1}{\partial x_j} \right\|_{L^2(\Omega;\mathcal{B}(\mathbb{R}^N_y))} .$$

We deduce

$$\left\| \frac{\partial}{\partial x_j}(u_\varepsilon - u_0 - \varepsilon\psi_1^\varepsilon) \right\|_{L^2(\Omega)} \leq \frac{\eta}{2} + c\varepsilon \ \ (0 < \varepsilon \leq \varepsilon_2),$$

where c is a constant such that $c > \max_{1\leq i\leq N} \|\frac{\partial \psi_1}{\partial x_i}\|_{L^2(\Omega;\mathcal{B}(\mathbb{R}^N))}$. Hence (5.35) follows with $\varepsilon_1 = \min\{\varepsilon_2, \frac{\eta}{2c}\}$. ∎

Corollary 5.5 *Let $\eta > 0$. Then, to each function $\psi_1 \in H^1(\Omega) \otimes [\mathcal{C}^1_{per}(Y)/\mathbb{C}]$ verifying*

$$\|u_1 - \psi_1\|_{L^2(\Omega;H^1_{per}(Y)/\mathbb{C})} \leq \frac{\eta}{4\sqrt{N+1}},$$

there is attached some real $\varepsilon_0 > 0$ such that

$$\|u_\varepsilon - u_0 - \varepsilon\psi_1^\varepsilon\|_{H^1(\Omega)} \leq \eta \ \ (0 < \varepsilon \leq \varepsilon_0).$$

Proof Let ψ_1 be as in this statement. According to Corollary 5.4 there is some $\varepsilon_1 > 0$ such that

$$\sum_{j=1}^N \left\| \frac{\partial}{\partial x_j}(u_\varepsilon - u_0 - \varepsilon\psi_1^\varepsilon) \right\|^2_{L^2(\Omega)} \leq \frac{N}{N+1}\eta^2$$

for $0 < \varepsilon \leq \varepsilon_1$. On the other hand, by (5.20) and use of the fact that $\varepsilon\psi_1^\varepsilon \to 0$ in $L^2(\Omega)$, there is some $\varepsilon_1' > 0$ such that $\|u_\varepsilon - u_0 - \varepsilon\psi_1^\varepsilon\|^2_{L^2(\Omega)} \leq \frac{\eta^2}{N+1}$ for $0 < \varepsilon \leq \varepsilon_1'$. Hence the corollary follows with $\varepsilon_0 = \min\{\varepsilon_1, \varepsilon_1'\}$. ∎

The next corollary is an obvious consequence of Corollary 5.5.

Corollary 5.6 *Suppose* $u_1 \in H^1(\Omega) \otimes [\mathcal{C}^1_{per}(Y)/\mathbb{C}]$. *Then, as* $\varepsilon \to 0$, *we have* $u_\varepsilon - u_0 - \varepsilon u_1^\varepsilon \to 0$ *in* $H^1(\Omega)$.

The function $\varepsilon u_1^\varepsilon$ is called a *first order corrector*, whereas a function such as $\varepsilon \psi_1^\varepsilon$ in Corollary 5.5 is a *first order semi-corrector*.

Remark 5.3 There is a sequence $(\psi_n, \varepsilon_n)_{n\in\mathbb{N}^*}$ with $\varepsilon_n > 0$ and $\psi_n \in H^1(\Omega) \otimes [\mathcal{C}^1_{per}(Y)/\mathbb{C}]$ such that

(i) $\psi_n \to u_1$ in $L^2(\Omega; H^1_{per}(Y)/\mathbb{C})$ as $n \to \infty$
(ii) $\|u_\varepsilon - u_0 - \varepsilon\psi_n^\varepsilon\|_{H^1(\Omega)} \leq \frac{1}{n}$ for $0 < \varepsilon \leq \varepsilon_n$ $(n \in \mathbb{N}^*)$.

This follows readily from Corollary 5.5.

We conclude the present section with the description of some conditions that are likely to lead us to the hypothesis of Corollary 5.6.

Example 5.1 We assume here that each a_{ij} depends on the sole structure variable y. Then the functions χ^j do not depend on x, so that the homogenized coefficients q_{ij} are reduced to complex numbers. Thus, the homogenized operator is a strongly elliptic differential operator in Ω with constant coefficients. Consequently, if Ω is smooth enough, say of class C^2, and if $f \in L^2(\Omega)$, then by a classical regularity result (see, e.g., Agmon et al. [4]), the function u_0, as the solution of (5.29), lies in $H^1_0(\Omega) \cap H^2(\Omega)$. Furthermore, if the coefficients a_{ij} are smooth enough then by the same argument as followed before we arrive at $\chi^j \in \mathcal{C}^1_{per}(Y)/\mathbb{C}$. Hence it follows by (5.28) that the hypothesis of Corollary 5.6 is satisfied.

5.1.3 *Abstract Setting*

5.1.3.1 Some Concrete Examples of Structure Hypotheses

Our next purpose is to investigate the homogenization of the boundary value problem (5.6), this time beyond the usual periodic setting. As was said earlier, this is not possible without requiring the family $\{a_{ij}\}$ to satisfy a structure hypothesis giving specific information on the way the said family is framed from an algebraic, a topological or geometric point of view. The present section is devoted to presenting some concrete examples of such structure hypotheses. As will presently become apparent, a great unity arises from those examples in so far as they seem to be but different interpretations of a so-called *abstract structure hypothesis*.

Example 5.2 (Almost Periodicity Hypothesis) We assume here that

$$a_{ij}(x, \cdot) \in L^2_{AP}(\mathbb{R}^N_y)\ (1 \leq i, j \leq N) \text{ for fixed } x \in \overline{\Omega}, \tag{5.36}$$

where $L^2_{AP}(\mathbb{R}^N_y)$ denotes the space of all functions in $L^2_{loc}(\mathbb{R}^N_y)$ that are almost periodic in the sense of Stepanoff (see Sect. 2.4). It is of interest to observe that (5.36) may cover various concrete cases. By way of illustration, it is easily seen that (5.36) is satisfied in each of the following cases:

(1) To each $x \in \overline{\Omega}$ there is attached a network $S(x)$ in $\mathbb{R}^N_y$ (Sect. 2.5) such that $a_{ij}(x,\cdot)$ is $S(x)$-periodic $(1 \leq i, j \leq N)$, i.e., $a(x, y+k) = a(x, y)$ for all $k \in S(x)$ and almost every $y \in \mathbb{R}^N$.
(2) There is a family of networks S_{ij} $(1 \leq i, j \leq N)$ in $\mathbb{R}^N_y$ such that for any fixed $x \in \overline{\Omega}$, $a_{ij}(x,\cdot)$ is S_{ij}-periodic.

Example 5.3 Let $Y' = (-\frac{1}{2}, \frac{1}{2})^{N-1} \subset \mathbb{R}^{N-1}_{y'} (N > 1)$. We suppose here that

$$a_{ij}(x,\cdot) \in \mathcal{B}_\infty(\mathbb{R}; L^2_{per}(Y'))(1 \leq i, j \leq N) \text{ for fixed } x \in \overline{\Omega}, \tag{5.37}$$

where $\mathcal{B}_\infty(\mathbb{R}; L^2_{per}(Y'))$ denotes the space of functions $u \in \mathcal{B}(\mathbb{R}; L^2_{per}(Y'))$ such that $u(\cdot, y_N)$ converges in $L^2_{per}(Y')$ when $|y_N| \to \infty$.

Example 5.4 The structure hypothesis is here

$$a_{ij}(x,\cdot) \in \mathcal{B}_\infty(\mathbb{R}; L^2_{\mathcal{R}'}(\mathbb{R}^{N-1}))(1 \leq i, j \leq N) \text{ for fixed } x \in \overline{\Omega}, \tag{5.38}$$

where $\mathcal{R}'$ is a countable subgroup of $\mathbb{R}^{N-1}$ (see Sect. 2.4.3).

Example 5.5 As a structure hypothesis we have also

$$a_{ij}(x,\cdot) \in L^2(\mathbb{R}^N_y) + L^2_{per}(Y) \text{ for fixed } x \in \overline{\Omega} \ \ (1 \leq i, j \leq N), \tag{5.39}$$

where $Y = (-\frac{1}{2}, \frac{1}{2})^N$.

Example 5.6 (Discrete and Semidiscrete Structure Hypotheses) First of all, let us note that, under hypotheses (5.1)–(5.3), a_{ij} is uniquely expressible under the form

$$a_{ij} = c_{ij} + d_{ij} \ \ (1 \leq i, j \leq N) \tag{5.40}$$

with

$$(1) \quad c_{ij} \in \mathcal{C}(\overline{\Omega}; L^\infty(\mathbb{R}^N_y)), \int_{k+Y} c_{ij}(x, y)dy = 0 \ \ (x \in \overline{\Omega}, k \in \mathbb{Z}^N)$$

$$(2) \quad d_{ij} \in \mathcal{C}(\overline{\Omega}; L^\infty(\mathbb{R}^N_y)), d_{ij}(x,\cdot) = \sum_{k \in \mathbb{Z}^N} r_{ij}(x,k)\chi_{k+Y} \ \ (x \in \overline{\Omega}),$$

where χ_{k+Y} is the characteristic function of $k+Y$ (the same Y as in Example 5.5) and r_{ij} is a mapping of $\overline{\Omega} \times \mathbb{Z}^N$ into $\mathbb{C}$ such that

$$r_{ij} \in \mathcal{C}(\overline{\Omega}; l^\infty(\mathbb{Z}^N)) \ (1 \leq i, j \leq N) \tag{5.41}$$

$$\operatorname{Re} \sum_{i,j=1}^{N} r_{ij}(x,k)\xi_j\overline{\xi}_i \geq \alpha|\xi|^2 \text{ for } \xi = (\xi_i) \in \mathbb{C}^N \quad (x \in \overline{\Omega}, k \in \mathbb{Z}^N) \tag{5.42}$$

$$r_{ji} = \overline{r}_{ij} \ (1 \leq i, j \leq N). \tag{5.43}$$

Indeed, the trivial decomposition (5.40) was pointed out earlier in Sect. 2.6, whereas (5.41)–(5.43) follow readily by the equality $r_{ij}(x,k) = \int_{k+Y} a_{ij}(x,y)dy$.

Remark 5.4 (5.42) and (5.43) imply similar properties for d_{ij} $(1 \leq i, j \leq N)$.

Having made this point, it is not out of interest to anticipate that each of the following cases constitutes a structure hypothesis under which the behaviour, as $\varepsilon \to 0$, of u_ε (the solution of (5.6)) can be successfully investigated. More precisely, a_{ij} being naturally decomposed as stated in (5.40), we have the following different structure hypotheses, where $x \in \overline{\Omega}$ and $1 \leq i, j \leq N$ are fixed.

$$c_{ij}(x,\cdot) \in L^2_{per}(Y) \text{ and } r_{ij}(x,\cdot) \in \mathcal{B}_\infty(\mathbb{Z}^N) \tag{5.44}$$

$$c_{ij}(x,\cdot) \in L^2_{per}(Y) \text{ and } r_{ij}(x,\cdot) \in \mathcal{B}_\infty(\mathbb{Z}^N) + l^1(\mathbb{Z}^N) \tag{5.45}$$

$$c_{ij}(x,\cdot) \in L^2_{AP}(\mathbb{R}^N) \text{ and } r_{ij}(x,\cdot) \in \mathcal{B}_\infty(\mathbb{Z}^N). \tag{5.46}$$

Each of these structure hypotheses is said to be *semidiscrete*. In the particular case where $c_{ij} = 0$ $(1 \leq i, j \leq N)$, it is clear that $a_{ij}(x,\cdot)$ $(1 \leq i, j \leq N)$ is constant on each cell $k+Y$ $(k \in \mathbb{Z}^N)$. The corresponding structure hypothesis is then termed *discrete* in so far as the behaviour of $a_{ij}(x,\cdot)$ is entirely determined by that of the «discrete» function $r_{ij}(x,\cdot)$.

5.1.3.2 Global Homogenization Theorem

The basic notation being as in Chap. 4, let A be an H-algebra on $\mathbb{R}^N_y$ (under the prescribed action) admitting A^∞ as a dense subalgebra. Moreover, we assume that A is $H^1(\Omega)$-proper (Definition 4.5), where $H^1(\Omega) = W^{1,2}(\Omega)$. The first step of our line of proceeding (which is the same mutatis mutandis as followed in Sect. 5.1.2) is to attach to A the basic tools we shall be using throughout.

The Spaces $A^m/\mathbb{C}$ We recall the notation (introduced earlier in Chap. 2) $A/\mathbb{C} = \{\psi | \psi \in A, M(\psi) = 0\}$ (M the mean value generated by the action $\mathcal{H}$ on $\mathbb{R}^N$). For any $m \in \overline{\mathbb{N}} = \mathbb{N} \cup \{\infty\}$, we set $A^m/\mathbb{C} = A^m \cap (A/\mathbb{C})$. In particular, the space $A^\infty/\mathbb{C}$ will be of frequent use. Of course, we have $\mathcal{G}(A/\mathbb{C}) = \mathcal{C}(\Delta(A))/\mathbb{C} =$

$\{\varphi|\varphi \in \mathcal{C}(\Delta(A)), \int_{\Delta(A)} \varphi d\beta = 0\}$. More generally, $\mathcal{G}(A^m/\mathbb{C}) = \mathcal{D}^m(\Delta(A))/\mathbb{C} \equiv \mathcal{D}^m(\Delta(A)) \cap [\mathcal{C}(\Delta(A))/\mathbb{C}]$.

The Space $\mathbb{F}_0^1(\Omega \times \Delta(A))$ Let

$$\mathbb{F}_0^1(\Omega \times \Delta(A)) = H_0^1(\Omega) \times L^2(\Omega; H_\#^1(\Delta(A))).$$

The natural norm on $\mathbb{F}_0^1 = \mathbb{F}_0^1(\Omega \times \Delta(A))$ making it into a Hilbert space is given by

$$\|\mathbf{u}\|_{\mathbb{F}_0^1} = (\|u_0\|^2_{H_0^1(\Omega)} + \|u_1\|^2_{L^2(\Omega; H_\#^1(\Delta(A)))})^{\frac{1}{2}}, \quad \mathbf{u} = (u_0, u_1) \in \mathbb{F}_0^1.$$

Furthermore, it is clear by Remark 4.6 that

$$\mathcal{F}_0^\infty = \mathcal{D}(\Omega) \times [\mathcal{D}(\Omega) \otimes J(\mathcal{D}(\Delta(A))/\mathbb{C})] \text{ is dense in } \mathbb{F}_0^1. \tag{5.47}$$

For fixed $1 \leq i \leq N$, let

$$\mathbb{D}_i \mathbf{u} = \frac{\partial u_0}{\partial x_i} + \partial_i u_1, \quad \mathbf{u} = (u_0, u_1) \in \mathbb{F}_0^1.$$

This defines a differential operator on $\Omega \times \Delta(A)$ that maps continuously and linearly $\mathbb{F}_0^1$ into $L^2(\Omega; L^2(\Delta(A)))$. Furthermore, thanks to Remark 4.6, once again, we have

$$\|\mathbf{u}\|_{\mathbb{F}_0^1} = \left(\sum_{i=1}^N \|\mathbb{D}_i \mathbf{u}\|^2_{L^2(\Omega \times \Delta(A))} \right)^{\frac{1}{2}} \quad (\mathbf{u} \in \mathbb{F}_0^1).$$

The Sesquilinear Form $a_\Omega(,)$ We assume here and throughout that the family $\{a_{ij}\}$ satisfies

$$a_{ij}(x, \cdot) \in \mathfrak{X}^2 \ \ (1 \leq i, j \leq N) \text{ for fixed } x \in \overline{\Omega}, \tag{5.48}$$

where $\mathfrak{X}^2 = \mathfrak{X}_A^2(\mathbb{R}_y^N)$ (Chap. 2, Sect. 2.2). This is referred to as *the abstract structure hypothesis on the coefficients a_{ij}*. Recalling (5.1), we have therefore

$$a_{ij} \in \mathcal{C}(\overline{\Omega}; \mathfrak{X}^{2,\infty}) \ \ (1 \leq i, j \leq N), \tag{5.49}$$

where $\mathfrak{X}^{2,\infty} = \mathfrak{X}^2 \cap L^\infty(\mathbb{R}_y^N)$ is equipped with the L^∞ norm. We can next define $\widehat{a}_{ij} = \mathcal{G}_\Omega(a_{ij}) = \mathcal{G} \circ a_{ij}$, and then consider the family $\{\widehat{a}_{ij}\}(1 \leq i, j \leq N)$ whose main properties are collected below.

Proposition 5.5 *The following statements hold for $1 \leq i, j \leq N$.*

$$\widehat{a}_{ij} \in \mathcal{C}(\overline{\Omega}; L^\infty(\Delta(A))) \tag{5.50}$$

$$\widehat{a}_{ji} = \overline{\widehat{a}}_{ij} \tag{5.51}$$

$$Re \sum_{i,j=1}^{N} \widehat{a}_{ij}(x,s)\xi_j\overline{\xi}_i \geq \alpha|\xi|^2 \tag{5.52}$$

for all $x \in \overline{\Omega}$, *for almost all* $s \in \Delta(A)$, *and for all* $\xi = (\xi_i) \in \mathbb{C}^N$.

Proof (5.50) follows readily by (5.49) and use of Corollary 2.7 (notice that $\mathfrak{X}^2 \subset \mathfrak{X}^1$), (5.51) is a direct consequence of (5.2) and use of Corollary 2.1; and finally, (5.52) follows by (5.3) and use of Corollary 2.6. ■

This being so, for $\mathbf{u}, \mathbf{v} \in \mathbb{F}_0^1$, put

$$a_\Omega(\mathbf{u}, \mathbf{v}) = \sum_{i,j=1}^{N} \int\int_{\Omega\times\Delta(A)} \widehat{a}_{ij}(x,s)\mathbb{D}_j\mathbf{u}(x,s)\overline{\mathbb{D}_i\mathbf{v}}(x,s)dxd\beta(s).$$

This defines a sesquilinear form $a_\Omega(,)$ on $\mathbb{F}_0^1 \times \mathbb{F}_0^1$ which, according to Proposition 5.5, is continuous, Hermitian and $\mathbb{F}_0^1$-coercive in the sense that

$$\text{Re}\, a_\Omega(\mathbf{v}, \mathbf{v}) \geq \alpha\|\mathbf{v}\|^2_{\mathbb{F}_0^1} \quad (\mathbf{v} \in \mathbb{F}_0^1). \tag{5.53}$$

Hence, denoting by $\mathbf{l}$ the antilinear form on $\mathbb{F}_0^1$ defined by

$$\mathbf{l}(\mathbf{v}) = \langle f, v_0\rangle, \quad \mathbf{v} = (u_0, v_1) \in \mathbb{F}_0^1,$$

it follows that the variational problem

$$\begin{cases} \mathbf{u} = (u_0, u_1) \in \mathbb{F}_0^1 : \\ a_\Omega(\mathbf{u}, \mathbf{v}) = \mathbf{l}(\mathbf{v}) \quad (\mathbf{v} \in \mathbb{F}_0^1) \end{cases} \tag{5.54}$$

admits a unique solution (the role of (5.53) is essential, of course).

About the Forthcoming Homogenized Coefficients Let $x \in \overline{\Omega}$. We introduce the sesquilinear form $a(x; \cdot, \cdot)$ on $H^1_\#(\Delta(A)) \times H^1_\#(\Delta(A))$ given by

$$a(x; u, v) = \sum_{i,j=1}^{N} \int_{\Delta(A)} \widehat{a}_{ij}(x,s)\partial_j u(s)\overline{\partial_i v}(s)d\beta(s)$$

for $u, v \in H^1_\#(\Delta(A))$. According to Proposition 5.5, this sesquilinear form is continuous, Hermitian and $H^1_\#(\Delta(A))$-coercive, i.e.,

$$\mathrm{Re}\, a(x; v, v) \geq \alpha \|v\|^2_{H^1_\#(\Delta(A))} \quad (v \in H^1_\#(\Delta(A)). \tag{5.55}$$

Likewise, by letting

$$\mathcal{A}(x; \mathbf{U}, \mathbf{V}) = \sum_{i,j=1}^{N} \int_{\Delta(A)} \widehat{a}_{ij}(x, s) U_j(s) \overline{V}_i(s) d\beta(s)$$

for $\mathbf{U} = (U_i)$ and $\mathbf{V} = (V_i)$ in $L^2(\Delta(A))^N$, we define a sesquilinear form $\mathcal{A}(x; \cdot, \cdot)$ on $L^2(\Delta(A))^N \times L^2(\Delta(A))^N$ which is continuous, Hermitian and $L^2(\Delta(A))^N$-coercive, i.e.,

$$\mathrm{Re}\, \mathcal{A}(x; \mathbf{V}, \mathbf{V}) \geq \alpha \|\mathbf{V}\|^2_{L^2(\Delta(A))^N} \quad (\mathbf{V} \in L^2(\Delta(A))^N). \tag{5.56}$$

Furthermore

$$a(x; u, v) = \mathcal{A}(x; \nabla u, \nabla v) \quad (u, v \in H^1_\#(\Delta(A))) \tag{5.57}$$

where

$$\nabla v = (\partial_1 v, \cdots, \partial_N v) \quad (v \in H^1_\#(\Delta(A))).$$

Now, fix freely $1 \leq j \leq N$. Thanks to the properties (see in particular (5.55)) of the form $a(x; \cdot, \cdot)$, we can consider the mapping χ^j of $\overline{\Omega}$ into $H^1_\#(\Delta(A))$ defined for each $x \in \overline{\Omega}$ by the variational equation

$$a(x; \chi^j(x), v) = \sum_{k=1}^{N} \int_{\Delta(A)} \widehat{a}_{kj}(x, s) \overline{\partial_k v}(s) d\beta(s) \tag{5.58}$$

for all $v \in H^1_\#(\Delta(A))$. This yields a family $\chi^j (1 \leq j \leq N)$, to which we attach the family of functions $q_{ij} : \overline{\Omega} \to \mathbb{C}$ $(1 \leq i, j \leq N)$ given by

$$q_{ij}(x) = \int_{\Delta(A)} \widehat{a}_{ij}(x, s) d\beta(s) - \sum_{l=1}^{N} \int_{\Delta(A)} \widehat{a}_{il}(x, s) \partial_l \chi^j(x, s) d\beta(s) \tag{5.59}$$

for $x \in \overline{\Omega}$. As one might expect, this leads us to a statement which is in all respects similar to Proposition 5.3. At the risk of annoying the reader about repeating, we however think it is opportune to state word for word the analogue of the latter proposition.

Proposition 5.6 *The following statements hold for* $1 \le i, j \le N$:

(i) $\chi^j \in \mathcal{C}(\overline{\Omega}; H^1_\#(\Delta(A)))$
(ii) $q_{ij} \in \mathcal{C}(\overline{\Omega})$
(iii) $q_{ji} = \overline{q}_{ij}$
(iv) *There is a constant* $\alpha_1 > 0$ *such that*

$$Re \sum_{i,j=1}^{N} q_{ij}(x)\xi_j\overline{\xi}_i \ge \alpha_1 |\xi|^2$$

holds for all $x \in \overline{\Omega}$ *and all* $\xi \in \mathbb{C}^N$.

Proof First of all, let us note that the chief point of the argument is here Proposition 5.3. Now, in a straightforward manner, part (i) results from the proof of Proposition 5.3(i) by appropriate substitutions (the details are left to the reader). From (i) we deduce at once that $\partial_l \chi^j \in \mathcal{C}(\overline{\Omega}; L^2(\Delta(A)))$. Hence (ii) follows by (5.59) and use of (5.50). Turning now to the proof of (iii), let us introduce the family $(p^i)_{1\le i\le N}$ defined by $p^i = (\delta_{i1}, \cdots, \delta_{iN})$ where δ_{ij} denotes the Kronecker symbol. We shall view each p^i as a (constant) function in $L^2(\Delta(A))^N$. A simple calculation using (5.57) gives $q_{ij}(x) = \mathcal{A}(x; p^j - \nabla\chi^j(x), p^i)$. On the other hand, by (5.58) we have $\mathcal{A}(x; \nabla\chi^j(x) - p^j, \nabla\chi^i(x)) = 0$. Therefore $q_{ij}(x) = \mathcal{A}(x; \nabla\chi^j(x) - p^j, \nabla\chi^i(x) - p^i)$. Hence (iii) follows by the fact that $\mathcal{A}(x; \cdot, \cdot)$ is Hermitian. Finally, as regards (iv), the whole of the proof of Proposition 5.3(iv) is adopted provided two modifications are made:

(1) We have here $\mathbf{u}(\xi) = \sum_{j=1}^{N} [p^j - \nabla\chi^j(x)]\xi_j \in L^2(\Delta(A))^N$.
(2) By (5.56) we have

$$q(x, \xi) = \mathrm{Re}\, \mathcal{A}(x; u(\xi), u(\xi)) \ge \alpha \|u(\xi)\|^2_{L^2(\Delta(A))}.$$

The rest proceeds without change. However, attention must be drawn to the property $\int_{\Delta(A)} \partial_l \chi^j(x, s) d\beta(s) = 0$ thanks to which $u(\xi) = 0$ implies $\xi_j = 0$ $(j = 1, \cdots, N)$.

■

Global Homogenization Theorem We will now turn our attention to the statement and proof of the global homogenization theorem and its corollaries. For any function $v : \Omega \to H^1(\Delta(A))/\mathbb{C}$, we set $J_\Omega(v) = J \circ v$, and we keep in mind that

$$\mathcal{F}_0^\infty = \mathcal{D}(\Omega) \times J_\Omega(\mathcal{D}(\Omega) \otimes [\mathcal{D}(\Delta(A))/\mathbb{C}]). \tag{5.60}$$

Theorem 5.2 *Under the assumptions (5.1)–(5.3) and (5.48), let* $\mathbf{u} = (u_0, u_1)$ *be defined by (5.54). For any real number* $\varepsilon > 0$, *let* $u_\varepsilon \in H_0^1(\Omega)$ *be the unique solution of (5.6) (with (5.4) and (5.5)). Then, as* $\varepsilon \to 0$, *we have*

$$u_\varepsilon \to u_0 \text{ in } H_0^1(\Omega)\text{-weak} \tag{5.61}$$

$$u_\varepsilon \to u_0 \text{ in } L^2(\Omega) \tag{5.62}$$

$$\frac{\partial u_\varepsilon}{\partial x_j} \to \mathbb{D}_j \mathbf{u} = \frac{\partial u_0}{\partial x_j} + \partial_j u_1 \text{ in } L^2(\Omega)\text{-weak } \Sigma \ (1 \leq j \leq N). \tag{5.63}$$

Proof The line of proceeding is exactly the same as followed in proving Theorem 5.1, and we will not go too deeply into details.

Starting at (5.9) with $v = u_\varepsilon$ reveals at once that the (generalized) sequence $(u_\varepsilon)_{\varepsilon>0}$ is norm-bounded in $H_0^1(\Omega)$. Hence, given any arbitrary fundamental sequence E, combination of the $H^1(\Omega)$-properness of A, Proposition 4.5 and the Rellich-Kondrachov theorem, leads us readily to (5.61)–(5.63) as $E' \ni \varepsilon \to 0$, where E' is a subsequence from E, and where the couple $\mathbf{u} = (u_0, u_1)$ is an element of $\mathbb{F}_0^1$. Clearly we are through once we have shown that $\mathbf{u}$ is the solution of (5.54).

To this end, fix freely $\mathbf{\Phi} = (\psi_0, J_\Omega(\widehat{\psi}_1)) \in \mathcal{F}_0^\infty$ (see (5.60)), where we recall that $\psi_0 \in \mathcal{D}(\Omega)$, $\psi_1 \in \mathcal{D}(\Omega) \otimes (A^\infty/\mathbb{C})$ and $\widehat{\psi}_1 = \mathcal{G}_\Omega(\psi_1) \equiv \mathcal{G} \circ \psi_1$. The next step is to pass to the limit in (5.23) when $E' \ni \varepsilon \to 0$, where Φ_ε is given by (5.22). The way to the desired result is quite apparent. As $E' \ni \varepsilon \to 0$ we have, according to Corollary 4.1,

$$\frac{\partial u_\varepsilon}{\partial x_j} \overline{\frac{\partial \Phi_\varepsilon}{\partial x_i}} \to \mathbb{D}_j \mathbf{u} \overline{\mathbb{D}_i \mathbf{\Phi}} \text{ in } L^2(\Omega)\text{-weak } \Sigma \ (1 \leq i, j \leq N). \tag{5.64}$$

Indeed, this results on one hand from (5.63) (as $E' \ni \varepsilon \to 0$), on the other hand from (5.24) plus the fact that the sequence of functions $\frac{\partial \Phi_\varepsilon}{\partial x_i}$ $(\varepsilon \in E)$ is norm-bounded in $L^\infty(\Omega)$, so that Corollary 4.1 applies.

Now, Proposition 4.4 allows us to express (5.64) as

$$\int_\Omega \frac{\partial u_\varepsilon}{\partial x_j} \overline{\frac{\partial \Phi_\varepsilon}{\partial x_i}} \psi^\varepsilon dx \to \int_{\Omega \times \Delta(A)} \mathbb{D}_j \mathbf{u} \overline{\mathbb{D}_i \mathbf{\Phi}} \widehat{\psi} \, dx d\beta \ (1 \leq i, j \leq N)$$

for all $\psi \in \mathcal{C}(\overline{\Omega}; \mathfrak{X}^{2,\infty})$, hence in particular for $\psi = a_{ij}$ by reason of (5.49). We deduce at once (5.26), hence $a_\Omega(\mathbf{u}, \mathbf{\Phi}) = \mathbf{l}(\mathbf{\Phi})$ by passing to the limit in (5.23) using (5.25). Therefore (5.54) follows by the arbitrariness of $\mathbf{\Phi}$ and use of (5.47). The theorem follows thereby. ∎

Corollaries 5.1, 5.2 and 5.3 together with their proofs carry over mutatis mutandis to the present setting. We will therefore omit the proofs, contenting ourselves with summarizing the appropriate statements.

Corollary 5.7 *Let* $\mathbf{u} = (u_0, u_1)$ *be defined by (5.54). Then:*

(i) u_0 *is the weak solution of the boundary value problem (5.29), where the* q_{ij}*'s are defined by (5.59) and have the properties collected in Proposition 5.6,*
(ii) u_1 *is uniquely defined for almost every* $x \in \Omega$ *by the so-called local problem, namely:*

$$\begin{cases} u_1(x) \in H^1_\#(\Delta(A)) : \\ a(x; u_1(x), w) = -\displaystyle\sum_{k,j=1}^{N} \frac{\partial u_0}{\partial x_j}(x) \int_{\Delta(A)} \widehat{a}_{kj}(x,s)\overline{\partial_k w}(s) d\beta(s) \\ \text{for all } w \in H^1_\#(\Delta(A)). \end{cases}$$

Furthermore, u_1 *is explicitly given by (5.28) with* $\chi^j \in \mathcal{C}(\overline{\Omega}; H^1_\#(\Delta(A)))$, χ^j *defined on each point* $x \in \overline{\Omega}$ *by the variational Eq. (5.58).*

5.1.3.3 Strong Approximation

The proof of the following proposition is a copy mutatis mutandis of that of Proposition 5.4, and is therefore not worth repeating.

Proposition 5.7 *Let* $\mathbf{u} = (u_0, u_1)$ *be the solution of (5.54). Then, as* $\varepsilon \to 0$, *we have*

$$\frac{\partial u_\varepsilon}{\partial x_j} \to \mathbb{D}_j \mathbf{u} = \frac{\partial u_0}{\partial x_j} + \partial_j u_1 \text{ in } L^2(\Omega)\text{-strong } \Sigma \ (1 \le j \le N).$$

By using the fact that

$$J_\Omega(H^1(\Omega) \otimes [\mathcal{D}^1(\Delta(A))/\mathbb{C}]) \text{ is dense in } L^2(\Omega; H^1_\#(\Delta(A))),$$

we deduce the following corollary by the same argument as used in proving Corollary 5.4. The details are left to the reader.

Corollary 5.8 *Let* $\eta > 0$. *Then, to each* $\psi_1 \in H^1(\Omega) \otimes (A^1/\mathbb{C})$ *such that*

$$\|u_1 - J_\Omega(\widehat{\psi}_1)\|_{L^2(\Omega; H^1_\#(\Delta(A)))} \le \frac{\eta}{4},$$

there is assigned some $\varepsilon_1 > 0$ *such that*

$$\left\| \frac{\partial}{\partial x_j}(u_\varepsilon - u_0 - \varepsilon \psi_1^\varepsilon) \right\|_{L^2(\Omega)} \le \eta \ (0 < \varepsilon \le \varepsilon_1) \ 1 \le j \le N.$$

By the same line of reasoning as followed to establish Corollary 5.5, we have likewise

Corollary 5.9 *Let* $\eta > 0$. *To each* $\psi_1 \in H^1(\Omega) \otimes (A^1/\mathbb{C})$ *verifying*

$$\|u_1 - J_\Omega(\widehat{\psi}_1)\|_{L^2(\Omega; H^1_\#(\Delta(A)))} \leq \frac{\eta}{4\sqrt{N+1}},$$

there is attached some $\varepsilon_0 > 0$ *such that* $\|u_\varepsilon - u_0 - \varepsilon\psi_1^\varepsilon\|_{H^1(\Omega)} \leq \eta$ $(0 < \varepsilon \leq \varepsilon_0)$.

One easily deduces the following corollaries.

Corollary 5.10 *There is a sequence* $(\psi_n, \varepsilon_n)_{n \in \mathbb{N}^*}$ *with* $\varepsilon_n > 0$ *and* $\psi_n \in H^1(\Omega) \otimes (A^1/\mathbb{C})$ $(n \in \mathbb{N}^*)$ *such that*

(i) $J_\Omega(\widehat{\psi}_n) \to u_1$ *in* $L^2(\Omega; H^1_\#(\Delta(A)))$ *as* $n \to \infty$
(ii) $\|u_\varepsilon - u_0 - \varepsilon\psi_n^\varepsilon\|_{H^1(\Omega)} \leq \frac{1}{n}$ *for* $0 < \varepsilon \leq \varepsilon_n$ $(n \in \mathbb{N}^*)$

Corollary 5.11 *Suppose* $u_1 \in J_\Omega(H^1(\Omega) \otimes [\mathcal{D}^1(\Delta(A))/\mathbb{C}])$, *and let* $u_1 = J_\Omega(\widehat{\psi}_1)$ *with* $\psi_1 \in H^1(\Omega) \otimes (A^1/\mathbb{C})$. *Then* $u_\varepsilon - u_0 - \varepsilon\psi_1^\varepsilon \to 0$ *in* $H^1(\Omega)$ *as* $\varepsilon \to 0$.

5.1.4 Some Concrete Deterministic Homogenization Problems

First of all, it is worth recalling that a homogenization problem is posed as soon as one has on one hand, say a suitable boundary value problem, on the other hand a so-called structure hypothesis. Thus, each of the structure hypotheses presented in Sect. 5.1.3 determines a specific homogenization problem for the boundary value problem (5.6). Such a problem is qualified as a concrete homogenization problem because the associated structure hypothesis (e.g., the periodicity hypothesis) appears under a natural form, as opposed to the abstract hypothesis (5.48).

The present section deals with the study of a few concrete homogenization problems for (5.6). Thanks to the preceding developments (see in particular Theorem 5.2), the whole problem in each case will reduce to showing that the associated concrete hypothesis can write as (5.48) for a suitable $H^1(\Omega)$-proper homogenization algebra (to be determined).

5.1.4.1 Problem I

Let us begin by revisiting the periodicity hypothesis (5.11) under which the homogenization problem for (5.6) was carried out in a self-contained fashion in Sect. 5.1.2. It should however be noted that the same periodicity hypothesis may be viewed in terms of the abstract theory of Sect. 5.1.3 as well. Indeed, one has $a_{ij}(x, \cdot) \in L^\infty_{per}(Y) \subset L^2_{per}(Y) = \mathfrak{X}^2_{\mathcal{C}_{per}(Y)}(\mathbb{R}^N)$ (see Sect. 2.5) for $x \in \overline{\Omega}$, $1 \leq i, j \leq N$. Hence, $\mathcal{C}_{per}(Y)$ being an $H^1(\Omega)$-proper homogenization algebra on

$\mathbb{R}^N$ (under $\mathcal{H}$) admitting $\mathcal{C}^{\infty}_{per}(Y)$ as a dense subalgebra (see Example 4.2), an alternative solution to the homogenization problem for (5.6) under hypothesis (5.11) is provided by Theorem 5.2 with $A = \mathcal{C}_{per}(Y)$. It is a rather simple exercise to check using appropriate results of Sect. 2.5 that both of the homogenization approaches thus pointed out are consistent in that they provide the same result expressed in different terms.

5.1.4.2 Problem II

Our goal here is to work out the homogenization problem for (5.6) under the almost periodicity hypothesis (5.36). We refer the reader to Sects. 2.3 and 2.4 (see also Sect. 1.1) for all notions and notation on almost periodic functions in use here.

Let us prove a fundamental lemma.

Lemma 5.3 *Under hypotheses (5.1) and (5.36), there exists a countable subgroup $\mathcal{R}$ of $\mathbb{R}^N_y$ such that*

$$a_{ij}(x,\cdot) \in L^2_{\mathcal{R}}(\mathbb{R}^N_y) \ \ (1 \leq i, j \leq N) \text{ for all } x \in \overline{\Omega}. \tag{5.65}$$

Proof Let D be a dense countable set in $\overline{\Omega}$. Put

$$\mathcal{S} = \bigcup_{1 \leq i,j \leq N} \bigcup_{\xi \in D} Sp(a_{ij}(\xi)),$$

where $a_{ij}(x) = a_{ij}(x,\cdot)$, $x \in \overline{\Omega}$, as usual. $\mathcal{S}$ is a countable subset of $\mathbb{R}^N$. Let $\mathcal{R} = Gr(\mathcal{S})$ denote the set of all finite sums

$$\sum_{finite} n_i k_i \text{ with } n_i \in \mathbb{Z}, k_i \in \mathcal{S}.$$

It is clear that $\mathcal{R}$ is a countable subgroup of $\mathbb{R}^N_y$. Moreover $Sp(a_{ij}(\xi)) \subset \mathcal{R}$ for $\xi \in D$, $1 \leq i, j \leq N$. So that each $a_{ij}(\xi)$ lies in $L^2_{\mathcal{R}}(\mathbb{R}^N_y)$. To finish, let us show that this implies (5.65). So, fix freely $1 \leq i, j \leq N$ and $x \in \overline{\Omega}$. Since $L^{\infty}(\mathbb{R}^N)$ is continuously embedded in the amalgam space $(L^2, l^{\infty})(\mathbb{R}^N)$, the function a_{ij} maps continuously $\overline{\Omega}$ into $(L^2, l^{\infty})(\mathbb{R}^N)$. Therefore, given an arbitrary $\eta > 0$, by density there is some $\xi \in D$ such that $\|a_{ij}(x) - a_{ij}(\xi)\|_{2,\infty} \leq \frac{\eta}{c}$, where c is a positive constant such that $|M(u)| \leq c\|u\|_{2,\infty}$ for all $u \in L^2_{AP}(\mathbb{R}^N)$ (Corollary 2.10). It follows

$$|M(\overline{\gamma}_k a_{ij}(x)) - M(\overline{\gamma}_k a_{ij}(\xi))| \leq \eta$$

for all $k \in \mathbb{R}^N$, where $\gamma_k(y) = e^{2\pi i k \cdot y}$ $(y \in \mathbb{R}^N)$. Since $M(\overline{\gamma}_k a_{ij}(\xi)) = 0$ for $k \in \mathbb{R}^N \setminus \mathcal{R}$, we deduce by the arbitrariness of η that $M(\overline{\gamma}_k a_{ij}(x)) = 0$ for $k \in \mathbb{R}^N \setminus \mathcal{R}$. Hence (5.65) follows. ∎

Now, let $A = AP_{\mathcal{R}}(\mathbb{R}^N_y)$ (Sect. 2.3). This is an $H^1(\Omega)$-proper homogenization algebra on $\mathbb{R}^N$ (Example 4.3) admitting A^∞ as a dense subalgebra. Furthermore, according to Corollary 2.11, we have $L^2_{\mathcal{R}}(\mathbb{R}^N) \subset \mathfrak{X}^2 \equiv \mathfrak{X}^2_A(\mathbb{R}^N)$. Therefore (5.48) follows by Lemma 5.3. Hence, it follows by Theorem 5.2 and the subsequent developments that we are through with the almost periodic homogenization of the boundary value problem (5.6).

5.1.4.3 Problem III

Our purpose is to study the homogenization of (5.6) under the structure hypothesis (5.37). It is useful to observe that $\mathcal{B}_\infty(\mathbb{R}; L^2_{per}(Y'))$, naturally equipped with the $\mathcal{B}(\mathbb{R}; L^2_{per}(Y'))$ norm, is continuously embedded in $\Xi^2(\mathbb{R}^N)$. Indeed, this follows readily by the facts that $\mathcal{B}_\infty(\mathbb{R}; L^2_{per}(Y'))$ is continuously embedded in the amalgam space $(L^2, l^\infty)(\mathbb{R}^N)$ (this is easy to check) which in turn is continuously embedded in $\Xi^2(\mathbb{R}^N)$ (Proposition 2.19).

With this in mind, let $A = \mathcal{B}_\infty(\mathbb{R}; \mathcal{C}_{per}(Y'))$ (see Examples 2.4 and 4.6). This is an $H^1(\Omega)$-proper H-algebra on $\mathbb{R}^N$ admitting A^∞ as a dense subalgebra. Moreover it is quite clear that A is dense in $\mathcal{B}_\infty(\mathbb{R}; L^2_{per}(Y'))$. Hence, recalling the previous result, it follows at once $\mathcal{B}_\infty(\mathbb{R}; L^2_{per}(Y')) \subset \mathfrak{X}^2 \equiv \mathfrak{X}^2_A(\mathbb{R}^N)$ and therefore (5.48). Thanks to Theorem 5.2, this leads us to the achievement of the problem under consideration.

5.1.4.4 Problem IV

Generalizing Problem III, we consider the homogenization of (5.6) under the structure hypothesis (5.38). Proceeding as above, one is led to the H-algebra $A = \mathcal{B}_\infty(\mathbb{R}; AP_{\mathcal{R}'}(\mathbb{R}^{N-1}))$ which is $H^1(\Omega)$-proper, admitting moreover A^∞ as a dense subalgebra (Example 4.7). On the other hand it is clear by Proposition 2.22 that A is dense in the space $\mathcal{B}_\infty(\mathbb{R}; L^2_{\mathcal{R}'}(\mathbb{R}^{N-1}))$. Since $\mathcal{B}(\mathbb{R}; L^2_{\mathcal{R}'}(\mathbb{R}^{N-1}))$ is continuously embedded in $\Xi^2(\mathbb{R}^N)$, it follows $\mathcal{B}_\infty(\mathbb{R}; L^2_{\mathcal{R}'}(\mathbb{R}^{N-1})) \subset \mathfrak{X}^2 \equiv \mathfrak{X}^2_A(\mathbb{R}^N)$, hence (5.48) follows. We arrive at the same conclusion as before.

Remark 5.5 The structure hypothesis (5.38) can be released from the constraint on the spectra if the a_{ij}'s do not depend on $x \in \overline{\Omega}$. Specifically, assuming the coefficients a_{ij} to be independent of x, we consider the homogenization of (5.6) under the structure hypothesis

$$a_{ij} \in \mathcal{B}_\infty(\mathbb{R}; L^2_{AP}(\mathbb{R}^{N-1})) \ (1 \le i, j \le N).$$

Then, by the same argument as followed in proving Lemma 5.3 one quickly arrives at a countable subgroup $\mathcal{R}'$ of $\mathbb{R}^{N-1}$ such that $a_{ij} \in \mathcal{B}_\infty(\mathbb{R}; L^2_{\mathcal{R}'}(\mathbb{R}^{N-1}))$ for $1 \leq i, j \leq N$. Hence (5.48) follows thereby.

5.1.4.5 Problem V

The matter on hand is about the investigation of the behaviour, as $\varepsilon \to 0$, of the solution, u_ε, of (5.6) under the structure hypothesis (5.39).

Let $A = \mathcal{B}_{\infty,\mathbb{Z}^N}(\mathbb{R}^N)$ (Example 2.5). This is an H-algebra on $\mathbb{R}^N$ of which A^∞ is a dense subalgebra. Moreover A is $H^1(\Omega)$-proper, as shown in Example 4.4. On the other hand, we have $\mathcal{K}(\mathbb{R}^N) \subset A, \mathcal{K}(\mathbb{R}^N)$ dense in $L^2(\mathbb{R}^N)$ and $L^2(\mathbb{R}^N)$ continuously embedded in $\Xi^2(\mathbb{R}^N)$ (Remark 2.3). We deduce at once $L^2(\mathbb{R}^N) \subset \mathfrak{X}^2_A$. Since $L^2_{per}(Y) = \mathfrak{X}^2_{\mathcal{C}_{per}(Y)}$ (Corollary 2.14(ii)) and further $\mathfrak{X}^2_{\mathcal{C}_{per}(Y)} \subset \mathfrak{X}^2_A$ because $\mathcal{C}_{per}(Y) \subset A$, we deduce $L^2(\mathbb{R}^N) + L^2_{per}(Y) \subset \mathfrak{X}^2_A$, hence (5.48) follows. Therefore, the solution of the problem is provided by Theorem 5.2 and the subsequent developments.

5.1.4.6 Problem VI

We assume here that for fixed $x \in \overline{\Omega}$, the function $a_{ij}(x, \cdot)$ is constant on each cell $k + Y$ $(1 \leq i, j \leq N)$, where $Y = (-\frac{1}{2}, \frac{1}{2})^N$. In other words, let $(r_{ij})_{1\leq i,j\leq N}$ be a family of functions of $\overline{\Omega} \times \mathbb{Z}^N$ into $\mathbb{C}$ satisfying (5.41)–(5.43), where α is a given positive real number. Let $(a_{ij})_{1\leq i,j\leq N}$ be the family of functions of $\overline{\Omega} \times \mathbb{R}^N$ into $\mathbb{C}$ defined by

$$a_{ij}(x, \cdot) = \sum_{k\in\mathbb{Z}^N} r_{ij}(x, k)\chi_{k+Y} \text{ (locally finite sum)}, \tag{5.66}$$

where χ_{k+Y} denotes the characteristic function of $k+Y$. It is an elementary exercise to verify that the a_{ij}'s satisfy (5.1)–(5.3).

Remark 5.6 For any fixed $k \in \mathbb{Z}^N$, we have $a_{ij}(x, y) = r_{ij}(x, k)$ for all $y \in k+Y$.

Our purpose is to investigate the behaviour, as $\varepsilon \to 0$, of the solution, u_ε, of the boundary value problem (5.6) under the structure hypothesis

$$r_{ij}(x, \cdot) \in \mathcal{B}_\infty(\mathbb{Z}^N) \ (1 \leq i, j \leq N) \text{ for fixed } x \in \overline{\Omega}. \tag{5.67}$$

Let us note in passing what follows: (5.67) says that to each $x \in \overline{\Omega}$ there is attached some $\xi_{ij}(x) \in \mathbb{C}$ such that $r_{ij}(x, k) \to \xi_{ij}(x)$ as $|k| \to \infty$. Such a structure hypothesis is justifiably termed *discrete*.

This being so, let A_σ denote the discrete H-algebra on $\mathbb{R}^N$ associated to $\mathfrak{G} = \mathcal{B}_\infty(\mathbb{Z}^N)$ (see Sect. 2.6; see especially Sect. 4.2.5). Let A be the closure of $A_\sigma +$

$\mathcal{C}_{per}(Y)$ in $\mathcal{B}(\mathbb{R}^N)$. A is an $H^1(\Omega)$-proper homogenization algebra on $\mathbb{R}^N$ admitting A^∞ as a dense subalgebra (Proposition 4.11). Thus, we are through if we show that $a_{ij}(x,\cdot) \in \mathfrak{X}^2_A(\mathbb{R}^N)$ for $x \in \overline{\Omega}$ and $1 \leq i, j \leq N$. In the sequel c stands for a positive constant such that $|r_{pq}(z,k)| \leq c$ for all $z \in \overline{\Omega}$, $k \in \mathbb{Z}^N$ and $1 \leq p, q \leq N$.

For arbitrary fixed $x \in \overline{\Omega}$ and $1 \leq i, j \leq N$, let $\eta > 0$. In view of the density of $\mathcal{K}(Y)$ in $L^2(Y)$, let $\varphi \in \mathcal{K}(Y)$ be such that $\|1 - \varphi\|_{L^2(Y)} \leq \frac{\eta}{c}$, where 1 actually denotes the constant function $u \in L^2(Y)$ assuming the value 1. Put

$$\psi = \sum_{k\in\mathbb{Z}^N} r_{ij}(x,k)\tau_k\varphi.$$

We have $\psi \in A_\sigma$, therefore $\psi \in A$. On the other hand, in view of (5.66), $\|a_{ij}(x,\cdot) - \psi\|_{2,\infty} \leq \eta$. It follows that $a_{ij}(x,\cdot)$ lies in the closure of A in the amalgam space $(L^2, l^\infty)(\mathbb{R}^N)$. But since the latter is continuously embedded in $\Xi^2(\mathbb{R}^N)$ (this argument was used before), we deduce $a_{ij}(x,\cdot) \in \mathfrak{X}^2_A(\mathbb{R}^N)$, as claimed. Thus, (5.48) being satisfied, the matter on hand is achieved thanks to Theorem 5.2 and the subsequent results.

Remark 5.7 The above development carries over, without the slightest change, to the case where $\mathcal{B}_\infty(\mathbb{Z}^N)$ is substituted in (5.67) by $\mathcal{B}_\infty(\mathbb{Z}^N) + l^1(\mathbb{Z}^N)$.

5.1.4.7 Problem VII

Let a family (a_{ij}) be given with (5.1)–(5.3). For each real $\varepsilon > 0$, let u_ε be the solution of the boundary value problem (5.6). The task before us is to investigate the behaviour of u_ε as $\varepsilon \to 0$ under the structure hypothesis:

$$\begin{cases} \text{For fixed } x \in \overline{\Omega} \text{ and } 1 \leq i, j \leq N, \\ \text{the limit } \lim\limits_{|k|\to\infty} \int_{k+Y} a_{ij}(x,y)dy \text{ exists} \\ \text{and the function } a_{ij}(x,\cdot) - d_{ij}(x,\cdot) \text{ is } Y\text{periodic,} \end{cases} \tag{5.68}$$

where $d_{ij}(x,\cdot)$ denotes that function in $L^\infty(\mathbb{R}^N_y)$ assuming the constant value $\int_{k+Y} a_{ij}(x,y)dy$ throughout $k+Y$ $(k \in \mathbb{Z}^N)$, $Y = (-\frac{1}{2}, \frac{1}{2})^N$.

To begin, let

$$r_{ij}(x,k) = \int_{k+Y} a_{ij}(x,y)dy \quad (k \in \mathbb{Z}^N)$$

and

$$a_{ij}(x,\cdot) - d_{ij}(x,\cdot) = c_{ij}(x,\cdot)$$

for all $x \in \overline{\Omega}$. Then we have (5.40)–(5.43). Therefore (5.68) becomes

$$r_{ij}(x,\cdot) \in \mathcal{B}_\infty(\mathbb{Z}^N) \text{ and } c_{ij}(x,\cdot) \text{ is } Y\text{-periodic } (1 \leq i, j \leq N) \text{ for fixed } x \in \overline{\Omega}.$$

The aim is to check that under this hypothesis, we have (5.48) for an $H^1(\Omega)$-proper homogenization algebra A of which A^∞ is a dense subalgebra. But this is somewhat straightforward. Let A be the $H^1(\Omega)$-proper homogenization algebra considered in Problem VI. According to the result achieved in the said problem, we have $d_{ij}(x,\cdot) \in \mathfrak{X}^2_A$. On the other hand, we have $c_{ij}(x,\cdot) \in L^\infty_{per}(Y) \subset L^2_{per}(Y) = \mathfrak{X}^2_{\mathcal{C}_{per}(Y)}$ (Corollary 2.14). Therefore $c_{ij}(x,\cdot) \in \mathfrak{X}^2_A$. Hence (5.48) follows by (5.40).

Remark 5.8 In Problem VI the basic data are the r_{ij}'s whereas in Problem VII the starting-coefficients are the a_{ij}'s. Attention is drawn to the fact that in either problem the integrals $\int_{k+Y} a_{ij}(x,y)dy$ constitute the foundation-stone for the framing of the structure hypothesis.

5.2 Second Order Semilinear Elliptic Operators

5.2.1 *Setting of the Abstract Homogenization Problem*

Let $(a_{ij})_{1\leq i,j\leq N}$ be a family of **real-valued functions** on $\overline{\Omega} \times \mathbb{R}^N_y$ satisfying (5.1)–(5.3), where the complex conjugate and real part operations are to be disregarded and $\mathbb{C}$ replaced by $\mathbb{R}$, of course. On the other hand, let h be a **real-valued** continuous function on $\mathbb{R}^N \times \mathbb{R}$ with the properties:

$$h \in \mathcal{C}(\mathbb{R}; \mathcal{B}_{\mathbb{R}}(\mathbb{R}^N_y)), \tag{5.69}$$

where $\mathcal{B}_{\mathbb{R}}(\mathbb{R}^N_y) \equiv \mathcal{B}(\mathbb{R}^N_y; \mathbb{R})$,

$$[h(y,r) - h(y,t)](r-t) \geq 0 \ (y \in \mathbb{R}^N, r, t \in \mathbb{R}), \tag{5.70}$$

$$\begin{cases} \text{There exist two constants } c > 0 \text{ and } \dfrac{1}{2} \leq \sigma \leq 1 \text{ such that} \\ |h(y,r)| \leq c|r|^\sigma \quad (y \in \mathbb{R}^N, r \in \mathbb{R}). \end{cases} \tag{5.71}$$

Remark 5.9 For example $h(y,r) = g(y)\theta(r)$ with $g \in \mathcal{B}_{\mathbb{R}}(\mathbb{R}^N)$, $g \geq 0$, and θ being any of the following functions:

1. $\theta(r) = r$
2. $\theta(r) = r^+ \equiv \frac{r+|r|}{2}$
3. $\theta(r) = |r|^{-\frac{1}{2}} r$ if $r \neq 0$, $\theta(0) = 0$.

For any $u \in L^2_{\mathbb{R}}(\Omega) \equiv L^2(\Omega; \mathbb{R})$, we define

$$h[u](x, y) = h(y, u(x)) \ \ (x \in \Omega, y \in \mathbb{R}^N),$$

which yields a measurable function $h[u]$ of Ω into $\mathcal{B}_{\mathbb{R}}(\mathbb{R}^N_y)$ (this is classical; see,e.g., Bourbaki [22, p.174,Theo.1]). Hence, by (5.71) it follows that

$$h[u] \in L^2(\Omega; \mathcal{B}_{\mathbb{R}}(\mathbb{R}^N_y)) \ \ (u \in L^2_{\mathbb{R}}(\Omega)). \tag{5.72}$$

Therefore, for any $u \in L^2_{\mathbb{R}}(\Omega)$, the trace $h^\varepsilon[u]$, i.e.,

$$h^\varepsilon[u](x) = h\left(\frac{x}{\varepsilon}, u(x)\right) \ \ (x \in \Omega),$$

is well defined according to Proposition 1.6, and we have $h^\varepsilon[u] \in L^2_{\mathbb{R}}(\Omega)$.

This being so, let $f \in H^{-1}(\Omega; \mathbb{R})$ (dual of $H^1_0(\Omega; \mathbb{R})$). Our main purpose in this section is to work out the homogenization of the boundary value problem

$$\begin{cases} h^\varepsilon[u] + P^\varepsilon u_\varepsilon = f \text{ in } \Omega, \\ u_\varepsilon = 0 \text{ on } \partial\Omega, \end{cases} \tag{5.73}$$

where the differential operator P^ε is given by (5.4)–(5.5). However we shall point out beforehand an existence and uniqueness result for (5.73).

For that purpose, let

$$I_\varepsilon(v) = \frac{1}{2}a^\varepsilon(v, v) - \langle f, v\rangle + \int_\Omega G^\varepsilon(x, v(x))dx, \quad v \in H^1_0(\Omega; \mathbb{R}),$$

where the bilinear form $a^\varepsilon(\cdot, \cdot)$ is defined in Sect. 5.1 (of course, with the slight modifications indicated above), $\langle , \rangle$ denotes the duality pairing between $H^{-1}(\Omega; \mathbb{R})$ and $H^1_0(\Omega; \mathbb{R})$, and

$$G^\varepsilon(x, r) = \int_0^r h\left(\frac{x}{\varepsilon}, t\right) dt \ \ (x \in \Omega, r \in \mathbb{R}).$$

This defines a functional $I_\varepsilon : H^1_0(\Omega; \mathbb{R}) \to \mathbb{R}$ which is continuous and Gâteaux-differentiable with Gâteaux derivative, $I'_\varepsilon(u) \in H^{-1}(\Omega; \mathbb{R})$, on each $u \in H^1_0(\Omega)$ given by

$$\langle I'_\varepsilon(u), v\rangle = a^\varepsilon(u, v) - \langle f, v\rangle + \int_\Omega h^\varepsilon[u]v dx \ \ (v \in H^1_0(\Omega; \mathbb{R})).$$

Furthermore, I_ε is strictly convex and $I_\varepsilon(v) \to \infty$ as $\|v\|_{H^1(\Omega)} \to \infty$. Consequently, for any given $\varepsilon > 0$ there is a unique $u_\varepsilon \in H^1_0(\Omega; \mathbb{R})$ such that

$I_\varepsilon(u_\varepsilon) \leq I_\varepsilon(v)$ for all $v \in H_0^1(\Omega; \mathbb{R})$; or equivalently such that $I'_\varepsilon(u_\varepsilon) = 0$, and so u_ε is the unique (weak) solution of the boundary value problem (5.73) (see, e.g., Lions [71] for details).

Having made this point, let us come to the heart of the matter. As was emphasized earlier in Sect. 5.1, in order to be «well posed» the problem on hand still needs a structure hypothesis.

Let A be an H-algebra on $\mathbb{R}^N$ of which A^∞ is a dense subalgebra. We assume moreover that A is $H^1(\Omega)$-proper. Finally, our goal is to investigate the behaviour of the solution, u_ε, of (5.73) as $\varepsilon \to 0$ under the abstract structure hypothesis

$$a_{ij}(x, \cdot) \in \mathfrak{X}_A^2(\mathbb{R}_y^N) \;\; (1 \leq i, j \leq N) \text{ for fixed } x \in \overline{\Omega} \tag{5.74}$$

$$h(\cdot, r) \in A \text{ for fixed } r \in \mathbb{R}. \tag{5.75}$$

5.2.2 *Preliminaries*

The next point deals with the function spaces that fit the problem before us. Let

$$H^1(\Delta(A); \mathbb{R}) = \{u | u, \partial_i u \in L^2(\Delta(A); \mathbb{R}), 1 \leq i \leq N\}.$$

This is a closed vector subspace of $H^1(\Delta(A))$ (the verification is an easy matter). We provide $H_{\mathbb{R}}^1(\Delta(A)) \equiv H^1(\Delta(A); \mathbb{R})$ with the $H^1(\Delta(A))$-norm, which makes it into a Hilbert space. Furthermore, $\mathcal{D}_{\mathbb{R}}(\Delta(A)) \equiv \mathcal{D}(\Delta(A); \mathbb{R})$ is dense in $H_{\mathbb{R}}^1(\Delta(A))$ (this follows from the $H^1(\Omega)$-properness of A). Let

$$H_{\mathbb{R}}^1(\Delta(A))/\mathbb{R} = \left\{u | u \in H_{\mathbb{R}}^1(\Delta(A)), \int_{\Delta(A)} u(s) d\beta(s) = 0\right\}.$$

This is a closed vector subspace of $H^1(\Delta(A))$. We equip it not with the corresponding norm but with the $H^1(\Delta(A))/\mathbb{C}$ seminorm, making it into a real preHilbert space in general non-separated and non-complete. We denote the separated completion of $H_{\mathbb{R}}^1(\Delta(A))/\mathbb{R}$ by $H_\#^1(\Delta(A); \mathbb{R})$.

We shall see presently that $H_\#^1(\Delta(A); \mathbb{R})$ is, in a sense, a closed vector subspace of $H_\#^1(\Delta(A))$. Let

$${}_{\mathbb{R}}H_\#^1(\Delta(A)) = \{u | u \in H_\#^1(\Delta(A)), \partial_i u \in L_{\mathbb{R}}^2(\Delta(A)) \; (1 \leq i \leq N)\}.$$

This is manifestly a closed vector subspace of $H_\#^1(\Delta(A))$ and therefore a Hilbert space with the $H_\#^1(\Delta(A))$ norm. However, more can be said about this space.

Proposition 5.8 *The following statements hold true:*

(i) *${}_{\mathbb{R}}H_\#^1(\Delta(A))$ is identical with the closure of $J(H_{\mathbb{R}}^1(\Delta(A))/\mathbb{R})$ in $H_\#^1(\Delta(A))$.*
(ii) *${}_{\mathbb{R}}H_\#^1(\Delta(A))$ is isometrically isomorphic to $H_\#^1(\Delta(A); \mathbb{R})$.*

Proof Let F be the closure of $J(H^1_{\mathbb{R}}(\Delta(A))/\mathbb{R})$ in $H^1_{\#}(\Delta(A))$, J the canonical mapping of $H^1(\Delta(A))/\mathbb{C}$ into its separated completion. Since $L^2_{\mathbb{R}}(\Delta(A))$ is closed in $L^2(\Delta(A))$, one clearly has $F \subset {}_{\mathbb{R}}H^1_{\#}(\Delta(A))$. It thus remains to show ${}_{\mathbb{R}}H^1_{\#}(\Delta(A)) \subset F$. To this end, let u be arbitrary given in ${}_{\mathbb{R}}H^1_{\#}(\Delta(A))$. Let (u_n) be an ordinary sequence in $H^1(\Delta(A))/\mathbb{C}$ such that $J(u_n) \to u$ in $H^1_{\#}(\Delta(A))$ as $n \to \infty$ (Remark 2.5). Thus, for $1 \leq i \leq N$, we have $\partial_i u_n \to \partial_i u$ in $L^2(\Delta(A))$ as $n \to \infty$. It follows $\partial_i(\text{Re } (u_n)) \to \partial_i u$ in $L^2(\Delta(A))$ as $n \to \infty$, since $\text{Re } (\partial_i u) = \partial_i u$ and $\partial_i(\text{Re } (u_n)) = \text{Re } (\partial_i u_n)$ (the verification is immediate). Letting $v_n = \text{Re } (u_n)$, we have $v_n \in H^1_{\mathbb{R}}(\Delta(A))/\mathbb{R}$ and $J(v_n) \to u$ in $H^1_{\#}(\Delta(A))$ as $n \to \infty$. Hence $u \in F$. This shows part (i). As regards part (ii), let i be the canonical embedding of $H^1_{\mathbb{R}}(\Delta(A))/\mathbb{R}$ into $H^1(\Delta(A))/\mathbb{C}$, and j be the canonical mapping of $H^1_{\mathbb{R}}(\Delta(A))/\mathbb{R}$ into its separated completion, $H^1_{\#}(\Delta(A); \mathbb{R})$. Then, according to Bourbaki [24, II,p.24,Prop.15] there exists a unique continuous linear mapping $\hat{\imath} : H^1_{\#}(\Delta(A); \mathbb{R}) \to H^1_{\#}(\Delta(A))$ such that $J \circ \mathrm{i} = \hat{\imath} \circ j$. It follows at once that $\|\hat{\imath}(j(u))\|_{H^1_{\#}(\Delta(A))} = \|j(u)\|_{H^1_{\#}(\Delta(A);\mathbb{R})}$ for all $u \in H^1_{\mathbb{R}}(\Delta(A))/\mathbb{R}$, since $\|J(u)\|_{H^1_{\#}(\Delta(A))} = \|j(u)\|_{H^1_{\#}(\Delta(A);\mathbb{R})}$ for such a u. One deduces by density that $\hat{\imath}$ is an isometry of $H^1_{\#}(\Delta(A); \mathbb{R})$ into $H^1_{\#}(\Delta(A))$. On the other hand, clearly

$$J(H^1_{\mathbb{R}}(\Delta(A))/\mathbb{R}) = \hat{\imath}(j(H^1_{\mathbb{R}}(\Delta(A))/\mathbb{R})) \subset \hat{\imath}(H^1_{\#}(\Delta(A); \mathbb{R})) \subset {}_{\mathbb{R}}H^1_{\#}(\Delta(A)).$$

Hence $\hat{\imath}(H^1_{\#}(\Delta(A); \mathbb{R})) = {}_{\mathbb{R}}H^1_{\#}(\Delta(A))$, as readily seen by part (i) and use of the fact that $\hat{\imath}(H^1_{\#}(\Delta(A); \mathbb{R}))$ is closed in $H^1_{\#}(\Delta(A))$. We conclude that $\hat{\imath}$ is an isometric isomorphism of $H^1_{\#}(\Delta(A); \mathbb{R})$ onto ${}_{\mathbb{R}}H^1_{\#}(\Delta(A))$, as claimed. This completes the proof. ∎

Remark 5.10 Let $H^1_{\#}(\Delta(A); \mathbb{R})$ be identified with ${}_{\mathbb{R}}H^1_{\#}(\Delta(A))$ by means of the isomorphism $\hat{\imath}$. Then

1. $J(H^1_{\mathbb{R}}(\Delta(A))/\mathbb{R})$ is dense in $H^1_{\#}(\Delta(A); \mathbb{R})$.
2. The restriction of J to $H^1_{\mathbb{R}}(\Delta(A))/\mathbb{R}$ is just the canonical mapping $j : H^1_{\mathbb{R}}(\Delta(A))/\mathbb{R} \to H^1_{\#}(\Delta(A); \mathbb{R})$.

Throughout the rest of this section, $H^1_{\#}(\Delta(A); \mathbb{R})$ is identified (by means of $\hat{\imath}$) with ${}_{\mathbb{R}}H^1_{\#}(\Delta(A))$.

Now, we define

$$\mathbb{F}^1_0(\Omega \times \Delta(A); \mathbb{R}) = H^1_0(\Omega; \mathbb{R}) \times L^2(\Omega; H^1_{\#}(\Delta(A); \mathbb{R})).$$

This is manifestly a closed vector subspace of $\mathbb{F}^1_0 \equiv \mathbb{F}^1_0(\Omega \times \Delta(A))$, hence a Hilbert space under the $\mathbb{F}^1_0$-norm (see Sect. 5.1). Furthermore, let

$$\mathcal{F}^\infty_0 = \mathcal{D}(\Omega; \mathbb{R}) \times [\mathcal{D}(\Omega; \mathbb{R}) \otimes J(\mathcal{D}(\Delta(A); \mathbb{R})/\mathbb{R})],$$

where $\mathcal{D}(\Delta(A); \mathbb{R})/\mathbb{R}$ denotes the space of those functions in $\mathcal{D}(\Delta(A); \mathbb{R})$ with zero integral (for β). Recalling that $\mathcal{D}(\Delta(A); \mathbb{R})$ is dense in $H^1(\Delta(A); \mathbb{R})$, we see immediately by Remark 5.10(1) that

$$\mathcal{F}_0^\infty \text{ is dense in } \mathbb{F}_0^1(\Omega \times \Delta(A); \mathbb{R}). \tag{5.76}$$

Finally, let $\widehat{h} = \mathcal{G} o h$ ($\mathcal{G}$ the Gelfand transformation on A), that is, $\widehat{h}(r) = \mathcal{G}(h(r))$ for $r \in \mathbb{R}$, where $h(r) = h(\cdot, r)$. In view of (5.69)–(5.71) we have

$$\begin{aligned}
&\widehat{h} \in \mathcal{C}(\mathbb{R}; \mathcal{C}_{\mathbb{R}}(\Delta(A))),\\
&[\widehat{h}(s,r) - \widehat{h}(s,t)](r-t) \geq 0,\\
&|\widehat{h}(s,r)| \leq c|r|^\sigma,
\end{aligned}$$

for $s \in \Delta(A)$, $r, t \in \mathbb{R}$. Hence, letting

$$\widetilde{h}(r) = \int_{\Delta(A)} \widehat{h}(s,r)d\beta(s) \ \ (r \in \mathbb{R})$$

yields a function $\widetilde{h} \in \mathcal{C}(\mathbb{R}; \mathbb{R})$ satisfying

$$[\widetilde{h}(r) - \widetilde{h}(t)](r-t) \geq 0, \tag{5.77}$$

$$|\widetilde{h}(r)| \leq c|r|^\sigma \tag{5.78}$$

for all $r, t \in \mathbb{R}$.

We will conclude these preliminaries with one simple but useful lemma. First of all, let $u \in L^2_{\mathbb{R}}(\Omega)$. According to (5.72) and (5.75) we have $h[u] \in L^2(\Omega; A)$. On the other hand, by (5.78) one has $\widetilde{h}[u] \in L^2_{\mathbb{R}}(\Omega)$, where $\widetilde{h}[u](x) = \widetilde{h}(u(x))$ for $x \in \Omega$.

Lemma 5.4 *Let $u \in L^2_{\mathbb{R}}(\Omega)$, u independent of $\varepsilon > 0$. Then, when $\varepsilon \to 0$, we have $h^\varepsilon[u] \to \widetilde{h}[u]$ in $L^2(\Omega)$-weak.*

Proof Let $\widehat{h}[u](x,s) = \widehat{h}(s, u(x))$ for $x \in \Omega, s \in \Delta(A)$. This defines a real function $\widehat{h}[u]$ on $\Omega \times \Delta(A)$ that lies in $L^2(\Omega; \mathcal{C}(\Delta(A)))$, as is easily seen by the properties of $\widehat{h}$. Furthermore, as $\varepsilon \to 0$, we have $h^\varepsilon[u] \to \widehat{h}[u]$ in $L^2(\Omega)$-weak Σ, since $\widehat{h}[u] = \mathcal{G} \circ h[u]$ and $h^\varepsilon[u](x) = h[u](x, \frac{x}{\varepsilon})$. Hence the lemma follows by a well known fact (see part (1) of Remark 4.3) and use of the equality $\widetilde{h}[u](x) = \int_{\Delta(A)} \widehat{h}(s, u(x))d\beta(s)$. ∎

5.2.3 Homogenization Results

We consider the variational problem

$$\begin{cases}
\mathbf{u} = (u_0, u_1) \in \mathbb{F}_0^1(\Omega \times \Delta(A); \mathbb{R}) : \\
\int_\Omega \widetilde{h}[u_0](x)v_0(x)dx + a_\Omega(\mathbf{u}, \mathbf{v}) = \mathbf{l}(\mathbf{v}) \\
\text{for all } \mathbf{v} = (v_0, v_1) \in \mathbb{F}_0^1(\Omega \times \Delta(A); \mathbb{R}),
\end{cases} \tag{5.79}$$

where $\mathbf{l}$ denotes the continuous linear form on $\mathbb{F}_0^1(\Omega \times \Delta(A); \mathbb{R})$ given by $\mathbf{l}(\mathbf{v}) = \langle f, v_0 \rangle$ for $\mathbf{v} = (v_0, v_1) \in \mathbb{F}_0^1(\Omega \times \Delta(A); \mathbb{R})$, and $a_\Omega(\cdot, \cdot)$ is the continuous, symmetric, $\mathbb{F}_0^1(\Omega \times \Delta(A); \mathbb{R})$-coercive (a property similar to (5.53)) bilinear form on $\mathbb{F}_0^1(\Omega \times \Delta(A); \mathbb{R}) \times \mathbb{F}_0^1(\Omega \times \Delta(A); \mathbb{R})$ defined in Sect. 5.1.3 minus the bar of complex conjugation and real part operations, respectively.

An existence and uniqueness result for (5.79) is obtained by the same line of argument as followed earlier in showing the similar result for (5.73). Indeed, the chief point of the argument is here the functional

$$I(\mathbf{v}) = \frac{1}{2} a_\Omega(\mathbf{v}, \mathbf{v}) - \mathbf{l}(\mathbf{v}) + \int_\Omega G(v_0(x))dx$$

for $\mathbf{v} = (v_0, v_1) \in \mathbb{F}_0^1(\Omega \times \Delta(A); \mathbb{R})$, where $G(r) = \int_0^r \widetilde{h}(t)dt$ $(r \in \mathbb{R})$. It is a rather easy exercise (left to the reader) to check that all of the required conditions (see in particular (5.77)–(5.78)) of existence and uniqueness result in (5.79) are fulfilled.

We are now in a position to prove the main homogenization result in the present abstract setting . We will need to refer to the equality

$$\mathcal{F}_0^\infty = \mathcal{D}_\mathbb{R}(\Omega) \times J_\Omega(\mathcal{D}_\mathbb{R}(\Omega) \otimes [\mathcal{D}_\mathbb{R}(\Delta(A))/\mathbb{R}]) \tag{5.80}$$

which is the «real version» of (5.60).

Theorem 5.3 *Let the basic assumptions on the functions* h, a_{ij} $(1 \le i, j \le N)$ *be those stated earlier in Sect. 5.2.1. For each* $\varepsilon > 0$*, let* $u_\varepsilon \in H_0^1(\Omega; \mathbb{R})$ *be the (unique) weak solution of (5.73). Under the structure hypothesis (5.74)–(5.75), let* $\mathbf{u} = (u_0, u_1)$ *be (uniquely) defined by (5.79). Then, as* $\varepsilon \to 0$*, we have (5.61)–(5.63).*

Proof Let $\varepsilon > 0$ be fixed. We have the following weak formulation of (5.73):

$$\int_\Omega h^\varepsilon[u_\varepsilon]v\, dx + a^\varepsilon(u_\varepsilon, v) = \langle f, v \rangle \quad (v \in H_0^1(\Omega; \mathbb{R})). \tag{5.81}$$

By taking in particular $v = u_\varepsilon$ and using $h^\varepsilon[u_\varepsilon]u_\varepsilon \ge 0$ (from (5.70)–(5.71)) together with (5.8) (where Re is disregarded) we see readily that the sequence $(u_\varepsilon)_{\varepsilon>0}$ is norm-bounded in $H^1(\Omega)$. Consequently, the sequence $(h^\varepsilon[u_\varepsilon])_{\varepsilon>0}$ is bounded in $L^2(\Omega)$, as is immediate by (5.71).

Thus, given any arbitrary fundamental sequence E, the same argument as followed in proving Theorem 5.2 yields a subsequence E' from E and some $\mathbf{u} = (u_0, u_1) \in \mathbb{F}_0^1(\Omega \times \Delta(A); \mathbb{R})$ and $g \in L^2_\mathbb{R}(\Omega)$ such that as $E' \ni \varepsilon \to 0$, we have (5.61)–(5.63) and

$$h^\varepsilon[u_\varepsilon] \to g \text{ in } L^2(\Omega)\text{-weak}. \tag{5.82}$$

It is clear that the theorem is proved if we can check that $\mathbf{u} = (u_0, u_1)$ satisfies the variational equation in (5.79). For this purpose, fix freely $\mathbf{\Phi} = (\psi_0, J_\Omega(\widehat{\psi}_1)) \in \mathcal{F}_0^\infty$ (see (5.80)). It is worth pointing out that $\psi_0 \in \mathcal{D}_\mathbb{R}(\Omega)$, $\psi_1 \in \mathcal{D}_\mathbb{R}(\Omega) \otimes (A_\mathbb{R}^\infty/\mathbb{R})$, where we denote by $A_\mathbb{R}^\infty/\mathbb{R}$ the subspace of $A^\infty/\mathbb{C}$ consisting of those functions that assume real values. Take in (5.81) the particular $v = \Phi_\varepsilon$ given by (5.22). The aim is to pass to the limit (when $E' \ni \varepsilon \to 0$) in the corresponding equation. This is without doubt a routine exercise. Indeed, by referring to the proof of Theorem 5.2 one quickly arrives at

$$\int_\Omega g\psi_0\, dx + a_\Omega(\mathbf{u}, \mathbf{\Phi}) = \mathbf{l}(\mathbf{\Phi}),$$

where we have made use of (5.82), of course. In view of the arbitrariness of $\mathbf{\Phi}$ we deduce by (5.76) that

$$\int_\Omega g v_0\, dx + a_\Omega(\mathbf{u}, \mathbf{v}) = \mathbf{l}(\mathbf{v})$$

for all $\mathbf{v} = (v_0, v_1) \in \mathbb{F}_0^1(\Omega \times \Delta(A); \mathbb{R})$. Thus, it only remains to show that $g = \widetilde{h}[u_0]$. To this end let $\varphi \in \mathcal{D}_\mathbb{R}(\Omega)$. Fix freely $\eta > 0$. We have by (5.70)

$$(h^\varepsilon[u_\varepsilon] - h^\varepsilon[u_0 + \eta\varphi])(u_\varepsilon - u_0 - \eta\varphi) \geq 0.$$

It follows

$$\int_\Omega (h^\varepsilon[u_\varepsilon] - h^\varepsilon[u_0 + \eta\varphi])(u_\varepsilon - u_0)\, dx + \int_\Omega h^\varepsilon[u_0 + \eta\varphi]\eta\varphi\, dx - \int_\Omega h^\varepsilon[u_\varepsilon]\eta\varphi\, dx \geq 0.$$

By letting $E' \ni \varepsilon \to 0$ we deduce using (5.62), (5.82) and Lemma 5.4 that

$$\int_\Omega \widetilde{h}[u_0 + \eta\varphi]\varphi\, dx - \int_\Omega g\varphi\, dx \geq 0.$$

By passing to the limit as $\eta \to 0$, it follows by dominated convergence that

$$\int_\Omega \widetilde{h}[u_0]\varphi\, dx - \int_\Omega g\varphi\, dx \geq 0.$$

Hence $g = \widetilde{h}[u_0]$ by reason of the arbitrariness of φ. ∎

The variational problem (5.79) is the so-called *global homogenized problem* for (5.73). The local and macroscopic homogenized problems derive from (5.79) in the same way as in Sect. 5.1.3. Corollary 5.7 carries over mutatis mutandis to the present case. The macroscopic homogenized problem is here

$$\widetilde{h}[u_0] - \sum_{i,j=1}^{N} \frac{\partial}{\partial x_i}\left(q_{ij}\frac{\partial u_0}{\partial x_j}\right) = f \text{ in } \Omega,$$

$$u_0 = 0 \text{ on } \partial\Omega,$$

where the coefficients q_{ij} are continuous real functions on $\overline{\Omega}$ with the symmetry and ellipticity properties stated in Proposition 5.6 (with obvious slight modifications).

Finally, concerning strong approximation, it is not hard to verify that Sect. 5.1.3.3 carries over mutatis mutandis to the present context.

5.2.4 *Some Concrete Examples*

By way of illustration we present below some concrete homogenization problems whose study leads to the application of the preceding theory.

5.2.4.1 Problem I (Periodic Setting)

Let us consider the homogenization problem for (5.73) under the periodicity hypothesis

$$\begin{cases} a_{ij}(x,\cdot) \text{ is } Y\text{-periodic } (1 \le i, j \le N) \text{ for fixed } x \in \overline{\Omega}, \\ h(\cdot, r) \text{ is } Y\text{-periodic for fixed } r \in \mathbb{R}, \end{cases} \tag{5.83}$$

where $Y = (-\frac{1}{2}, \frac{1}{2})^N$. To begin, we see immediately that in view of (5.12) and (5.69), hypothesis (5.83) amounts to

$$\begin{cases} a_{ij}(x,\cdot) \in L^\infty_{per}(Y) \ (1 \le i, j \le N) \text{ for fixed } x \in \overline{\Omega}, \\ h(\cdot, r) \in \mathcal{C}_{per}(Y) \text{ for fixed } r \in \mathbb{R}. \end{cases}$$

But $L^\infty_{per}(Y) \subset L^2_{per}(Y) = \mathfrak{X}^2_{\mathcal{C}_{per}(Y)}$ (Sect. 2.5). So that, the H-algebra $\mathcal{C}_{per}(Y)$ being $H^1(\Omega)$-proper and $\mathcal{C}^\infty_{per}(Y)$ dense in $\mathcal{C}_{per}(Y)$, the problem before us reduces to the abstract setting with $A = \mathcal{C}_{per}(Y)$. In conclusion, not only the H-algebra $\mathcal{C}_{per}(Y)$ meets the requirements underlying the abstract setting so that the above theorem applies to the present case, but also all the homogenization results achieved in Sect. 5.2.3 carry over mutatis mutandis to the periodic setting (see Proposition 4.6).

5.2.4.2 Problem II (Almost Periodic Setting)

The question before us is as above except (5.83) which is substituted by

$$\begin{cases} a_{ij}(x,\cdot) \in L^2_{AP}(\mathbb{R}^N_y)\ (1 \le i,j \le N) \text{ for fixed } x \in \overline{\Omega} \\ h(\cdot,r) \in AP(\mathbb{R}^N_y) \text{ for fixed } r \in \mathbb{R}. \end{cases} \tag{5.84}$$

According to Lemma 5.3, let $\mathcal{R}_1$ be a countable subgroup of $\mathbb{R}^N_y$ such that $a_{ij}(x,\cdot) \in L^2_{\mathcal{R}_1}(\mathbb{R}^N_y)$ $(1 \le i,j \le N)$ for any $x \in \overline{\Omega}$ (actually, the $\mathcal{R}$ of Lemma 5.3 is here called $\mathcal{R}_1$ for mere convenience). On the other hand, by letting $\mathcal{S}_2 = \bigcup_{r\in\mathbb{Q}} Sp(h(r))$, where $\mathbb{Q}$ denotes the usual rational numbers, the same line of proceeding as followed in proving Lemma 5.3 yields a countable subgroup $\mathcal{R}_2 = Gr(\mathcal{S}_2)$ of $\mathbb{R}^N$ such that $h(\cdot,r) \in AP_{\mathcal{R}_2}(\mathbb{R}^N_y)$ for all $r \in \mathbb{R}$. Then, the group $\mathcal{R} = Gr(\mathcal{R}_1 \cup \mathcal{R}_2)$ generated by $\mathcal{R}_1 \cup \mathcal{R}_2$ is manifestly a countable subgroup of $\mathbb{R}^N$ such that

$$\begin{cases} a_{ij}(x,\cdot) \in L^2_{\mathcal{R}}(\mathbb{R}^N_y)\ (1 \le i,j \le N) \text{ for fixed } x \in \overline{\Omega} \\ h(\cdot,r) \in AP_{\mathcal{R}}(\mathbb{R}^N_y) \text{ for fixed } r \in \mathbb{R}, \end{cases} \tag{5.85}$$

hence (5.74)–(5.75) follows with $A = AP_{\mathcal{R}}(\mathbb{R}^N_y)$ (see Corollary 2.11). Since $AP_{\mathcal{R}}(\mathbb{R}^N_y)$ has the required properties (see Example 4.3), the homogenization of (5.73) under hypothesis (5.84) follows by Theorem 5.3.

Let us conclude Problem II with a still more concrete example.

Example 5.7 Let $\mathcal{B}_i (i = 1,2)$ be two bases of the N-dimensional vector space $\mathbb{R}^N_y$. Let Y_i be the open parallelepiped centred at $\omega = (0,\cdots,0)$ constructed on $\frac{1}{2}\mathcal{B}_i$, and S_i the network in $\mathbb{R}^N_y$ attached to $\mathcal{B}_i$ (or Y_i), $i = 1,2$, as specified in Sect. 2.5. Finally, let S_i^* be the subgroup associated to $S_i (i = 1,2)$, which turns out to be a network (see Chap. 2, in particular (2.32), for the definition of S_i^*).

This being so, let the matter on hand be the homogenization of (5.73) under the structure hypothesis

$$\begin{cases} a_{ij}(x,\cdot) \text{ is } Y_1\text{-periodic } (1 \le i,j \le N) \text{ for fixed } x \in \overline{\Omega}, \\ h(\cdot,r) \text{ is } Y_2\text{-periodic for fixed } r \in \mathbb{R}. \end{cases} \tag{5.86}$$

Before proceeding any farther let Problem I be revisited with the purpose of laying emphasis on the fact that the whole development made there remains rigorously valid if the standard unit cube $Y = (-\frac{1}{2},\frac{1}{2})^N$ is substituted by any conformable parallelepiped such as $Y_i (i = 1,2)$. Having made this point, let us note that the problem thus stated does not in general fall within the framework of periodic homogenization. However, it is an almost periodic homogenization problem, as will presently become apparent. Indeed, in view of (5.1) and (5.69), hypothesis (5.86) amounts to

$$\begin{cases} a_{ij}(x,\cdot) \in L^\infty_{per}(Y_1)\ (1 \leq i, j \leq N) \text{ for fixed } x \in \overline{\Omega}, \\ h(\cdot, r) \in \mathcal{C}_{per}(Y_2) \text{ for fixed } r \in \mathbb{R}. \end{cases}$$

But $L^\infty_{per}(Y_1) \subset L^2_{per}(Y_1) = L^2_{S_1^*}(\mathbb{R}^N)$ (Corollary 2.14 (i)) and $\mathcal{C}_{per}(Y_2) = AP_{S_2^*}(\mathbb{R}^N)$ (Proposition 2.23). Hence (5.85) follows with $\mathcal{R} = Gr(S_1^* \cup S_2^*)$.

5.2.4.3 Problem III

Let the problem on hand be the homogenization of (5.73) under the hypothesis

$$a_{ij}(x,\cdot) \text{ is } Y\text{-periodic } (1 \leq i, j \leq N) \text{ for fixed } x \in \overline{\Omega}, \tag{5.87}$$

$$\lim_{|y|\to\infty} h(y, r) \text{ exists for fixed } r \in \mathbb{R}. \tag{5.88}$$

Assumptions (5.1) and (5.69) allow us to reduce (5.87)–(5.88) to

$$\begin{cases} a_{ij}(x,\cdot) \in \mathfrak{X}^2_{\mathcal{C}_{per}(Y)}\ (1 \leq i, j \leq N) \text{ for fixed } x \in \overline{\Omega}, \\ h(\cdot, r) \in \mathcal{B}_\infty(\mathbb{R}^N_y) \text{ for fixed } r \in \mathbb{R}. \end{cases}$$

Hence hypothesis (5.74)–(5.75) follows with $A = \mathcal{B}_{\infty,\mathbb{Z}^N}(\mathbb{R}^N)$ (Example 2.5). We deduce that the problem under consideration falls within the abstract setting , since A^∞ is dense in A and A is H^1-proper (Example 4.4).

5.2.4.4 Problem IV

More generally, it is possible to work out the preceding homogenization problem under the structure hypothesis derived from (5.87)–(5.88) by replacing (5.87) by

$$a_{ij}(x,\cdot) \in L^2_{AP}(\mathbb{R}^N_y)\ (1 \leq i, j \leq N) \text{ for fixed } x \in \overline{\Omega}.$$

This leads us to the same conclusion with $A = \mathcal{B}_{\infty,\mathcal{R}}(\mathbb{R}^N)$ (Example 4.5).

5.2.4.5 Problem V

For the sake of clearness we assume here that the coefficients a_{ij} do not depend on $x \in \overline{\Omega}$. Thus $a_{ij} \in L^\infty_{\mathbb{R}}(\mathbb{R}^N_y)$. The question before us is the homogenization of (5.73) under the structure hypothesis

$$\begin{cases} a_{ij} \text{ is constant on } k + Y (k \in \mathbb{Z}^N) \ (1 \leq i, j \leq N), \\ \lim_{|k|\to\infty} \int_{k+Y} a_{ij}(y)dy \in \mathbb{R} \text{ exists } (1 \leq i, j \leq N), \\ h(\cdot, r) \text{ is } Y\text{-periodic for fixed } r \in \mathbb{R}. \end{cases} \tag{5.89}$$

By letting $r_{ij}(k) = \int_{k+Y} a_{ij}(y)dy$, we see at once that the hypothesis on the a_{ij}'s amounts to

$$\begin{cases} a_{ij} = \sum_{k\in\mathbb{Z}^N} r_{ij}(k)\chi_{k+Y}, \\ r_{ij} \in \mathcal{B}_\infty(\mathbb{Z}^N). \end{cases}$$

On the other hand it is clear that

$$h(\cdot, r) \in \mathcal{C}_{per}(Y) \text{ for fixed } r \in \mathbb{R}.$$

Therefore, by reasoning as in Problem VI of Sect. 5.1.4 we see readily that (5.89) amounts to (5.74)–(5.75) where A is the same H-algebra as in the latter reference.

5.3 Semilinear Elliptic Operators of Order 2m

5.3.1 Setting of the Homogenization Problem

Given an arbitrary integer $m \geq 1$, let

$$a_{pq} \in \mathcal{C}(\overline{\Omega}; L^\infty_{\mathbb{R}}(\mathbb{R}^N_y)) \tag{5.90}$$

with

$$a_{qp} = a_{pq}, \tag{5.91}$$

where p and q range each separately over all multi-indices in $\mathbb{N}^N$ of length m (we recall that the length of $p = (p_1, \cdots, p_N) \in \mathbb{N}^N$ is $|p| = p_1 + \cdots + p_N$).

Let $\varepsilon > 0$. We consider the differential operator P^ε in Ω given by

$$P^\varepsilon u = (-1)^m \sum_{|p|=|q|=m} D^p(a^\varepsilon_{pq} D^q u) \tag{5.92}$$

with

$$D^p = \frac{\partial^{|p|}}{\partial x_1^{p_1} \cdots \partial x_N^{p_N}},$$

where the function $a^\varepsilon_{pq} \in L^\infty_{\mathbb{R}}(\Omega)$ is defined by

$$a^\varepsilon_{pq}(x) = a_{pq}\left(x, \frac{x}{\varepsilon}\right) \quad (x \in \Omega)$$

in the sense of Proposition 1.8.

In the sequel $H^m_0(\Omega; \mathbb{R})$, the closure of $\mathcal{D}_{\mathbb{R}}(\Omega)$ in $H^m(\Omega)$, is equipped with the norm

$$\|u\|_m = \left(\sum_{|p|=m} \|D^p u\|^2_{L^2(\Omega)} \right)^{\frac{1}{2}} \quad (u \in H^m_0(\Omega; \mathbb{R}))$$

equivalent to the $H^m(\Omega)$-norm. For $u, v \in H^m_0(\Omega; \mathbb{R})$, we set

$$a^\varepsilon(u, v) = \sum_{|p|=|q|=m} \int_\Omega a^\varepsilon_{pq} D^q u D^p v \, dx,$$

which defines a continuous, symmetric bilinear form $a^\varepsilon(\cdot, \cdot)$ on $H^m_0(\Omega; \mathbb{R}) \times H^m_0(\Omega; \mathbb{R})$. We require the family $\{a^\varepsilon(\cdot, \cdot)\}(\varepsilon > 0)$ to satisfy:

$$\begin{cases} \text{There exists some real } \alpha > 0 \text{ such that} \\ a^\varepsilon(u, u) \geq \alpha \|u\|^2_m \ \ (u \in H^m_0(\Omega; \mathbb{R})) \text{ for all } \varepsilon > 0. \end{cases} \tag{5.93}$$

On the other hand, let h be a continuous mapping of $\mathbb{R}^N_y \times \mathbb{R}$ into $\mathbb{R}$ with properties (5.69), (5.70), and further

$$\begin{cases} \text{There exist constants } c > 0, d \geq 0 \text{ and } \dfrac{1}{2} \leq \sigma \leq 1 \text{ such that} \\ |h(y, r)| \leq c|r|^\sigma + d \ \ (y \in \mathbb{R}^N, r \in \mathbb{R}). \end{cases} \tag{5.94}$$

Remark 5.9 is worth pointing out, of course. Also, the notation $h[u]$ and $h^\varepsilon[u]$ introduced earlier in Sect. 5.2 need not be reset.

Finally, for given $f \in H^{-m}(\Omega; \mathbb{R})$ (dual of $H^m_0(\Omega; \mathbb{R})$) and for fixed $\varepsilon > 0$ we consider the semilinear Dirichlet boundary value problem

$$h^\varepsilon[u_\varepsilon] + P^\varepsilon u_\varepsilon = f \text{ in } \Omega, u_\varepsilon \in H^m_0(\Omega; \mathbb{R}), \tag{5.95}$$

which uniquely determines u_ε (see Sect. 5.2). Our purpose is to investigate the behaviour, as $\varepsilon \to 0$, of u_ε, first in the periodic setting, next in an abstract deterministic setting intended for the mathematical analysis of non-periodic multiscale problems.

5.3.2 *Periodic Setting*

5.3.2.1 Preliminaries

Let $Y \subset \mathbb{R}_y^N$ be the standard unit cube we dealt with earlier in Sect. 5.1.1. Unless otherwise stated, the study proceeds under the structure hypothesis:

$$\begin{cases} a_{pq}(x, \cdot) \text{ is } Y\text{-periodic } (|p| = |q| = m) \text{ for fixed } x \in \overline{\Omega} \\ h(\cdot, r) \text{ is } Y\text{-periodic for fixed } r \in \mathbb{R}. \end{cases} \tag{5.96}$$

Having made this point, we collect below some preliminary results we shall use in the sequel. To begin, let

$$\mathbb{F}_0^m(\Omega \times Y; \mathbb{R}) = H_0^m(\Omega; \mathbb{R}) \times L^2(\Omega; H_{per}^m(Y; \mathbb{R})/\mathbb{R}),$$

where $H_{per}^m(Y; \mathbb{R})/\mathbb{R}$ denotes the closed vector subspace of $H_{per}^m(Y)$ made up of those of its elements that are real-valued functions with mean value zero. Provided with the norm

$$\|\mathbf{u}\|_{\mathbb{F}_0^m} = (\|u_0\|_m^2 + \|u_m\|^2_{L^2(\Omega; H_{per}^m(Y)/\mathbb{C})})^{\frac{1}{2}}, \mathbf{u} = (u_0, u_m) \in \mathbb{F}_0^m,$$

$\mathbb{F}_0^m \equiv \mathbb{F}_0^m(\Omega \times Y; \mathbb{R})$ is a Hilbert space.

For fixed $q \in \mathbb{N}^N$, $|q| = m$, let

$$\mathbb{D}^q \mathbf{u} = D_x^q u_0 + D_y^q u_m, \mathbf{u} = (u_0, u_m) \in \mathbb{F}_0^m(\Omega \times Y; \mathbb{R}).$$

This defines a differential operator $\mathbb{D}^q$ on $\Omega \times \mathbb{R}_y^N$ that maps continuously and linearly $\mathbb{F}_0^m(\Omega \times Y; \mathbb{R})$ into $L^2(\Omega; L_{per}^2(Y; \mathbb{R}))$. Furthermore, it is rather obvious that

$$\|\mathbf{u}\|_{\mathbb{F}_0^m} = \left(\sum_{|q|=m} \|\mathbb{D}^q \mathbf{u}\|^2_{L^2(\Omega \times Y)} \right)^{\frac{1}{2}}, \mathbf{u} \in \mathbb{F}_0^m(\Omega \times Y; \mathbb{R}).$$

Another basic function space is

$$\mathcal{F}_0^\infty = \mathcal{D}_\mathbb{R}(\Omega) \times (\mathcal{D}_\mathbb{R}(\Omega) \otimes [\mathcal{C}_{per}^\infty(Y; \mathbb{R})/\mathbb{R}]).$$

It is immediate by Corollary 2.13 that

$$\mathcal{F}_0^\infty \text{ is dense in } \mathbb{F}_0^m(\Omega \times Y; \mathbb{R}). \tag{5.97}$$

Now, for any $\mathbf{u}, \mathbf{v} \in \mathbb{F}_0^m(\Omega \times Y; \mathbb{R})$, let

$$a_\Omega(\mathbf{u}, \mathbf{v}) = \sum_{|p|=|q|=m} \int\int_{\Omega\times Y} a_{pq} \mathbb{D}^q \mathbf{u} \mathbb{D}^p \mathbf{v}\, dxdy.$$

This defines a bilinear form $a_\Omega(\cdot,\cdot)$ on $\mathbb{F}_0^m(\Omega \times Y; \mathbb{R}) \times \mathbb{F}_0^m(\Omega \times Y; \mathbb{R})$ which is continuous symmetric (see (5.90)–(5.91)). We will in a little while establish the $\mathbb{F}_0^m$-coerciveness of $a_\Omega(\cdot,\cdot)$. Let us prove beforehand one basic result.

To any given

$$\mathbf{\Phi} = (\varphi, \psi) \in \mathcal{D}_\mathbb{R}(\Omega) \times [\mathcal{D}_\mathbb{R}(\Omega) \otimes \mathcal{C}_{per}^\infty(Y, \mathbb{R})] \tag{5.98}$$

let us attach the sequence $(\Phi_\varepsilon)_{\varepsilon>0}$ defined as

$$\Phi_\varepsilon = \varphi + \varepsilon^m \psi^\varepsilon, \tag{5.99}$$

where $\psi^\varepsilon(x) = \psi(x, \frac{x}{\varepsilon})$, $x \in \Omega$. We have $\Phi_\varepsilon \in \mathcal{D}_\mathbb{R}(\Omega)$, of course.

Proposition 5.9 *Let* $1 \le r < \infty$. *The following hold for any integer* $m \ge 1$.

(i) For any $\psi \in \mathcal{D}_\mathbb{R}(\Omega) \otimes \mathcal{C}_{per}^\infty(Y, \mathbb{R})$, *we have*

$$\sup_{0<\varepsilon\le 1} \varepsilon^m \|D^q \psi^\varepsilon\|_\infty < \infty \text{ and } \sup_{0<\varepsilon\le 1} \varepsilon^m \|D^q \psi^\varepsilon\|_{L^r(\Omega)} < \infty \tag{5.100}$$

where q is any multi-index of length m.

(ii) For any $\mathbf{\Phi}$ *as in (5.98) we have, as* $\varepsilon \to 0$,

$$D^q \Phi_\varepsilon \to \mathbb{D}^q \mathbf{\Phi} = D_x^q \varphi + D_y^q \psi \text{ in } L^r(\Omega)\text{-strong } \Sigma \ (|q| = m). \tag{5.101}$$

Proof Let $\psi \in \mathcal{D}_\mathbb{R}(\Omega) \otimes \mathcal{C}_{per}^\infty(Y, \mathbb{R})$. One has

$$\psi = \sum_{i\in I} \varphi_i \otimes \theta_i \text{ with } \varphi_i \in \mathcal{D}_\mathbb{R}(\Omega), \theta_i \in \mathcal{C}_{per}^\infty(Y, \mathbb{R}),$$

where I is a (nonempty) finite set. Fix freely $\varepsilon > 0$. For $q \in \mathbb{N}^N$, $|q| = m$, we have by the Leibniz formula

$$\begin{aligned} \varepsilon^m D^q \psi^\varepsilon &= \sum_{\gamma\le q} \frac{q!\varepsilon^{m-|\gamma|}}{(q-\gamma)!\gamma!} \sum_{i\in I} (D_y^\gamma \theta_i)^\varepsilon D_x^{q-\gamma} \varphi_i \\ &= \sum_{\gamma\le q} \varepsilon^{m-|\gamma|} G_\gamma^\varepsilon, \end{aligned}$$

where

$$G_\gamma = \frac{q!}{(q-\gamma)!\gamma!} \sum_{i \in I} (D_x^{q-\gamma} \varphi_i) \otimes (D_y^\gamma \theta_i).$$

But

$$|G_\gamma^\varepsilon(x)| \le \|G_\gamma\|_\infty \ (x \in \Omega)$$

and

$$\|G_\gamma^\varepsilon\|_{L^r(\Omega)} \le \|G_\gamma\|_{L^r(\Omega;\mathcal{B}(\mathbb{R}_y^N))} \text{ (Proposition 1.6).}$$

Hence (5.100) follows by the fact that $\varepsilon^{m-|\gamma|} \le 1$ for any $0 < \varepsilon \le 1$ and any $\gamma \in \mathbb{N}^N$ with $\gamma \le q$ (i.e., with $\gamma_i \le q_i$ for $i = 1, \cdots, N$). This proves (i).

We turn next to the proof of (ii). We proceed by induction on $m \ge 1$. Let $\mathbf{\Phi}$ be arbitrary given by (5.98). We begin by proving (ii) for $m = 1$. Though we had to deal with this case before, we reconsider it here for the sake of clarification. So, let Φ_ε be attached to $\mathbf{\Phi}$ as in (5.99). For $1 \le i \le N$, we have

$$\frac{\partial \Phi_\varepsilon}{\partial x_i} = \frac{\partial \varphi}{\partial x_i} + \left(\frac{\partial \psi}{\partial y_i}\right)^\varepsilon + \varepsilon \left(\frac{\partial \psi}{\partial x_i}\right)^\varepsilon.$$

But, as $\varepsilon \to 0$,

$$\varepsilon \left(\frac{\partial \psi}{\partial x_i}\right)^\varepsilon \to 0 \text{ in } L^r(\Omega).$$

Therefore, for $m = 1$, (5.101) follows by Proposition 3.2.

Now, given $m > 1$, we assume that (5.101) holds for any function as in (5.98), and any multi-index q of length $m-1$. The aim is then to deduce that the same is true of length m (in place of $m-1$). So, fix freely some $q \in \mathbb{N}^N$, $q = (q_1, \cdots, q_N)$, $|q| = m$. Let q_ν be some non-zero component of q. Thus, $q_\nu = 1 + q'_\nu$ with $q'_\nu \ge 0$, of course. Consequently, q writes as $q = e + q'$, where e is that element of $\mathbb{N}^N$ whose $\nu - th$ component is 1 and the others are zero, and $|q'| = m - 1$.

With this in mind, let $\mathbf{\Phi} = (\varphi, \psi)$ be as above. Considering the corresponding Φ_ε with $0 < \varepsilon \le 1$ arbitrary fixed, we have

$$D^q \Phi_\varepsilon = D^{q'} \frac{\partial \Phi_\varepsilon}{\partial x_\nu} = D^{q'} \left(\frac{\partial \varphi}{\partial x_\nu} + \varepsilon^m \left[\left(\frac{\partial \psi}{\partial x_\nu}\right)^\varepsilon + \frac{1}{\varepsilon} \left(\frac{\partial \psi}{\partial y_\nu}\right)^\varepsilon \right] \right).$$

Therefore

$$D^q \Phi_\varepsilon = \varepsilon \left[\varepsilon^{m-1} D^{q'} \left(\frac{\partial \psi}{\partial x_\nu}\right)^\varepsilon \right] + D^{q'} \phi_\varepsilon, \tag{5.102}$$

where

$$\phi_\varepsilon = \frac{\partial \varphi}{\partial x_\nu} + \varepsilon^{m-1} \left(\frac{\partial \psi}{\partial y_\nu} \right)^\varepsilon .$$

In view of part (i), the term between brackets is bounded in $L^r(\Omega)$ as ε ranges over $(0, 1]$. Consequently, as $\varepsilon \to 0$, the first term of the sum on the right of (5.102) converges to zero in $L^r(\Omega)$-strong Σ. Finally, by letting

$$\mathbf{\Psi} = \left(\frac{\partial \varphi}{\partial x_\nu}, \frac{\partial \psi}{\partial y_\nu} \right)$$

we see readily that, according to the induction hypothesis we have, as $\varepsilon \to 0$,

$$D^{q'} \phi_\varepsilon \to \mathbb{D}^{q'} \mathbf{\Psi} = D_x^{q'} \frac{\partial \varphi}{\partial x_\nu} + D_y^{q'} \frac{\partial \psi}{\partial y_\nu} \text{ in } L^r(\Omega)\text{-strong } \Sigma.$$

Hence, since $\mathbb{D}^{q'} \mathbf{\Psi} = \mathbb{D}^q \mathbf{\Phi}$, (5.101) follows at once. This completes the proof. ■

Remark 5.11 Proposition 5.9 as well as its proof remains valid if:

(1) Ω is unbounded.
(2) The function spaces involved in the statements are considered over $\mathbb{C}$.

The preceding proposition has some useful corollaries.

Corollary 5.12 *Let $\mathbf{\Phi}$ and Φ_ε be as in (5.98)–(5.99). Then $\Phi_\varepsilon \to \varphi$ in $H_0^m(\Omega; \mathbb{R})$-weak as $\varepsilon \to 0$.*

Proof As a consequence of Proposition 5.9, for any multi-index q of length m we have, as $\varepsilon \to 0$, $D^q \Phi_\varepsilon \to D_x^q \varphi + D_y^q \psi$ in $L^2(\Omega)$-weak Σ. Since $\int_Y D_y^q \psi dy = 0$, it follows by Remark 3.4 that $D^q \Phi_\varepsilon \to D^q \varphi$ in $L^2(\Omega)$-weak as $\varepsilon \to 0$, and that for all $q \in \mathbb{N}^N$, $|q| = m$. Hence the corollary follows. ■

Corollary 5.13 *We have*

$$a_\Omega(\mathbf{v}, \mathbf{v}) \geq \alpha \|\mathbf{v}\|^2_{\mathbb{F}_0^m} \quad \textit{(same } \alpha \textit{ as in (5.93))} \tag{5.103}$$

for all $\mathbf{v} \in \mathbb{F}_0^m(\Omega \times Y; \mathbb{R})$.

Proof We begin by showing (5.103) for $\mathbf{v} = \mathbf{\Phi} \in \mathcal{F}_0^\infty$. For any couple of multi-indices p, q with $|p| = |q| = m$, we have, as $\varepsilon \to 0$,

$$D^q \Phi_\varepsilon D^p \Phi_\varepsilon \to \mathbb{D}^q \mathbf{\Phi} \mathbb{D}^p \mathbf{\Phi} \text{ in } L^2(\Omega)\text{-weak } \Sigma,$$

as is obvious by Proposition 5.9 and use of Proposition 3.5. We deduce at once by Proposition 3.6 (as we did in proving Theorem 5.1) that

$$a^{\varepsilon}(\Phi_{\varepsilon}, \Phi_{\varepsilon}) \to a_{\Omega}(\mathbf{\Phi}, \mathbf{\Phi}) \text{ as } \varepsilon \to 0.$$

On the other hand, it is clear that

$$\|\Phi_{\varepsilon}\|_{m}^{2} \to \|\mathbf{\Phi}\|_{\mathbb{F}_{0}^{m}}^{2} \text{ as } \varepsilon \to 0.$$

Hence (5.103) with $\mathbf{v} = \mathbf{\Phi}$ follows by (5.93). One deduces readily the whole corollary by (5.97) and use of the continuity of $a_{\Omega}(\cdot, \cdot)$. ■

In order to be through with the preliminaries, let us introduce the mapping $\widetilde{h} : \mathbb{R} \to \mathbb{R}$ given by

$$\widetilde{h}(r) = \int_{Y} h(y, r)dy \quad (r \in \mathbb{R}).$$

From (5.69),(5.70) and (5.94), one has

$$\begin{cases} \widetilde{h} \in \mathcal{C}(\mathbb{R}; \mathbb{R}), \\ [\widetilde{h}(r) - \widetilde{h}(t)](r - t) \geq 0, \quad (r, t \in \mathbb{R}), \\ |\widetilde{h}(r)| \leq c|r|^{\sigma} + d \quad (r \in \mathbb{R}). \end{cases} \tag{5.104}$$

For $u \in L^{2}(\Omega; \mathbb{R})$, we put $\widetilde{h}[u] = \widetilde{h} \circ u$, i.e., $\widetilde{h}[u](x) = \widetilde{h}(u(x)), x \in \Omega$. As is classical, $\widetilde{h}[u]$ is a measurable function of Ω into $\mathbb{R}$. Hence it follows by the third property in (5.104) that $\widetilde{h}[u] \in L^{2}(\Omega; \mathbb{R})$. Furthermore, $h^{\varepsilon}[u] \to \widetilde{h}[u]$ in $L^{2}(\Omega)$-weak as $\varepsilon \to 0$ (see Lemma 5.4).

Finally, let $\mathbf{l}$ denote the continuous linear form on $\mathbb{F}_{0}^{m}(\Omega \times Y; \mathbb{R})$ given by

$$\mathbf{l}(\mathbf{v}) = \langle f, v_{0} \rangle \text{ for } \mathbf{v} = (v_{0}, v_{m}) \in \mathbb{F}_{0}^{m}(\Omega \times Y; \mathbb{R}),$$

where $\langle , \rangle$ stands for the duality pairing between $H^{-m}(\Omega; \mathbb{R})$ and $H_{0}^{m}(\Omega, \mathbb{R})$. We will need the following variational problem

$$\begin{cases} \mathbf{u} = (u_{0}, u_{m}) \in \mathbb{F}_{0}^{m}(\Omega \times Y; \mathbb{R}) : \\ \displaystyle\int_{\Omega} \widetilde{h}[u_{0}](x) v_{0}(x) dx + a_{\Omega}(\mathbf{u}, \mathbf{v}) = \mathbf{l}(\mathbf{v}) \\ \text{for all } \mathbf{v} = (v_{0}, v_{m}) \in \mathbb{F}_{0}^{m}(\Omega \times Y; \mathbb{R}), \end{cases} \tag{5.105}$$

which uniquely determines $\mathbf{u} = (u_{0}, u_{m})$, thanks to (5.90),(5.91), (5.103) and (5.104) (see Sect. 5.2).

5.3.2.2 Homogenization Results

The Global Homogenization Theorem

Theorem 5.4 *For each* $\varepsilon > 0$*, let* u_ε *be the solution of the boundary value problem (5.95). Under the periodicity hypothesis (5.96), let* $\mathbf{u} = (u_0, u_m)$ *be the solution of (5.105). Then, as* $\varepsilon \to 0$,

$$u_\varepsilon \to u_0 \text{ in } H_0^m(\Omega; \mathbb{R})\text{-weak}, \tag{5.106}$$

$$u_\varepsilon \to u_0 \text{ in } H_0^{m-1}(\Omega; \mathbb{R}), \tag{5.107}$$

$$D^q u_\varepsilon \to \mathbb{D}^q \mathbf{u} = D_x^q u_0 + D_y^q u_m \text{ in } L^2(\Omega)\text{-weak } \Sigma \ (|q| = m). \tag{5.108}$$

Proof Let $\varepsilon > 0$. Obviously,

$$\int_\Omega h^\varepsilon[u_\varepsilon] v \, dx + a^\varepsilon(u_\varepsilon, v) = \langle f, v \rangle \tag{5.109}$$

for all $v \in H_0^m(\Omega; \mathbb{R})$. Proceeding as we did before (see the proof of Theorem 5.3) leads to the norm boundedness of the sequences $(u_\varepsilon)_{\varepsilon>0}$ in $H_0^m(\Omega, \mathbb{R})$, and $(h^\varepsilon[u_\varepsilon])_{\varepsilon>0}$ in $L^2_{\mathbb{R}}(\Omega)$. Hence, given a fundamental sequence E, appeal to Theorem 3.3, to the compactness of the embedding $H_0^m(\Omega; \mathbb{R}) \hookrightarrow H_0^{m-1}(\Omega; \mathbb{R})$, and finally to a usual compactness argument for $L^2_{\mathbb{R}}(\Omega)$, yields a subsequence E' from E, a vector function $\mathbf{u} = (u_0, u_m) \in \mathbb{F}_0^m(\Omega \times Y; \mathbb{R})$, and a scalar function $g \in L^2_{\mathbb{R}}(\Omega)$ such that, as $E' \ni \varepsilon \to 0$, we have (5.106)–(5.108) and (5.82). Thus, we are through if we can check that $\mathbf{u} = (u_0, u_m)$ satisfies (5.105). But then, thanks to the density result (5.97) it is enough that (5.105) be satisfied by $\mathbf{u} = (u_0, u_m)$ for the sole smooth functions $\mathbf{v} = \mathbf{\Phi} = (\varphi, \psi) \in \mathcal{F}_0^\infty$. This is rather straightforward. Bearing in mind that $a_{pq} \in \mathcal{C}(\overline{\Omega}; L^\infty_{per}(Y))$ for $|p| = |q| = m$ (from (5.90) and (5.96)), let $\mathbf{\Phi}$ and thereafter Φ_ε be as in (5.98)–(5.99). As $E' \ni \varepsilon \to 0$, we have by (5.101) (for $r = 2$) and (5.108), and use of Propositions 3.5 and 3.6,

$$a^\varepsilon(u_\varepsilon, \Phi_\varepsilon) \to a_\Omega(\mathbf{u}, \mathbf{\Phi}).$$

On the other hand, by the same routine line of reasoning as followed in proving Theorem 5.3 we get $g = \widetilde{h}[u_0]$. Hence, by passing to the limit (as $E' \ni \varepsilon \to 0$) in (5.109) using Corollary 5.12 one quickly arrives at (5.105) for any arbitrary $\mathbf{v} = \mathbf{\Phi} \in \mathcal{F}_0^\infty$. The theorem follows thereby. ■

Local Problem For fixed $x \in \overline{\Omega}$, let

$$a(x; z, w) = \sum_{|p|=|q|=m} \int_Y a_{pq}(x, \cdot) D_y^q z D_y^p w dy$$

for $z, w \in H^m_{per}(Y; \mathbb{R})/\mathbb{R}$. This defines a continuous symmetric bilinear form $a(x; \cdot, \cdot)$ on $H^m_{per}(Y; \mathbb{R})/\mathbb{R} \times H^m_{per}(Y; \mathbb{R})/\mathbb{R}$. Moreover this bilinear form is $H^m_{per}(Y; \mathbb{R})/\mathbb{R}$-coercive. Specifically, we have

$$a(x; w, w) \geq \alpha \|w\|^2_{H^m(Y)/\mathbb{C}} \quad (w \in H^m_{per}(Y; \mathbb{R})/\mathbb{R}). \tag{5.110}$$

Indeed, by choosing $\mathbf{v} = (v_0, v_m)$ in (5.103) such that $v_0 = 0$ and $v_m(x, y) = \varphi(x)w(y)$ ($x \in \Omega$, $y \in \mathbb{R}^N$) with $\varphi \in \mathcal{K}_{\mathbb{R}}(\Omega)$ and $w \in H^m_{per}(Y; \mathbb{R})/\mathbb{R}$, one readily arrives at

$$\int_\Omega \varphi(x)^2 [a(x; w, w) - \alpha \|w\|^2_{H^m(Y)/\mathbb{C}}] dx \geq 0$$

for any $\varphi \in \mathcal{K}_{\mathbb{R}}(\Omega)$; from which (5.110) follows at once by Bourbaki [23, p.47,Cor.3].

Having done this, let us introduce the so-called *cell (or structure) functions*. More precisely, let $q \in \mathbb{N}^N$ with $|q| = m$. For fixed $x \in \overline{\Omega}$, we define $\chi^q(x)$ as the solution of the variational problem

$$\begin{cases} \chi^q(x) \in H^m_{per}(Y; \mathbb{R})/\mathbb{R} : \\ a(x; \chi^q(x), w) = \sum_{|\nu|=m} \int_Y a_{\nu q}(x, y) D^\nu w(y) dy \\ \text{for all } w \in H^m_{per}(Y; \mathbb{R})/\mathbb{R}. \end{cases} \tag{5.111}$$

The existence and unicity of $\chi^q(x)$ can be obtained by the Lax-Milgram lemma thanks in particular to (5.110). This yields a mapping $\chi^q : \overline{\Omega} \to H^m_{per}(Y; \mathbb{R})/\mathbb{R}$. It is not hard to check that χ^q lies in $\mathcal{C}(\overline{\Omega}; H^m_{per}(Y; \mathbb{R})/\mathbb{R})$ (proceed as in the proof of Proposition 5.6).

Now, by choosing in (5.105) $\mathbf{v} = (v_0, v_m)$ such that $v_0 = 0$ and $v_m(x) = \varphi(x)w$ with $\varphi \in \mathcal{D}_{\mathbb{R}}(\Omega)$ and $w \in H^m_{per}(Y; \mathbb{R})/\mathbb{R}$, we get the *local problem* (at $x \in \Omega$) in the variational form, namely

$$\begin{cases} u_m(x) \in H^m_{per}(Y; \mathbb{R})/\mathbb{R} : \\ a(x; u_m(x), w) = -\sum_{|q|=m} D^q u_0(x) \sum_{|p|=m} \int_Y a_{pq}(x, y) D^p w(y) dy \\ \text{for all } w \in H^m_{per}(Y; \mathbb{R})/\mathbb{R}, \end{cases} \tag{5.112}$$

of which $u_m(x)$ is the unique solution by virtue of (5.110). Hence, by an obvious combination with (5.111), it follows

$$u_m(x) = -\sum_{|\nu|=m} D^\nu u_0(x) \chi^\nu(x). \tag{5.113}$$

Macroscopic Homogenized Problem Our main purpose here is to derive the boundary value problem for u_0. First of all, for fixed $x \in \overline{\Omega}$, let

$$c_{pq}(x) = \int_Y a_{pq}(x, y)dy - \sum_{|\nu|=m} \int_Y a_{p\nu}(x, y)D^\nu \chi^q(x, y)dy \tag{5.114}$$

where p and q are any two multi-indices of length m. This defines a function $c_{pq} \in \mathcal{C}_\mathbb{R}(\overline{\Omega})$ and therefore a family $(c_{pq})_{|p|=|q|=m}$ of so-called *homogenized coefficients* involved in the framing of the homogenized operator. On this point, let $\widetilde{P}$ be the differential operator in Ω defined by

$$\widetilde{P}u = (-1)^m \sum_{|p|=|q|=m} D^p(c_{pq}D^q u).$$

The function u_0 (in Theorem 5.4) is a solution of the macroscopic homogenized problem, viz.

$$\widetilde{h}[u_0] + \widetilde{P}u_0 = f \text{ in } \Omega, u_0 \in H_0^m(\Omega; \mathbb{R}). \tag{5.115}$$

Indeed, by choosing in (5.105) the particular $\mathbf{v} = (v_0, v_m)$ with $v_m = 0$ and then substituting (5.113) we obtain

$$\int_\Omega \widetilde{h}[u_0]v_0\, dx + \sum_{|p|=|q|=m} \int_\Omega c_{pq}D^q u_0 D^p v_0\, dx = (f, v_0) \tag{5.116}$$

for all $v_0 \in H_0^m(\Omega; \mathbb{R})$. Hence (5.115) follows.

The unicity of u_0 will follow by the next proposition.

Proposition 5.10 *The following assertions are equivalent for a real-valued function u_0 on Ω:*

(i) u_0 satisfies (5.115).
(ii) The couple (u_0, u_m), where u_m is given by (5.113), satisfies (5.105).

Proof It is quite clear that we only need to check that (i) implies (ii). So, suppose (i) holds. First, it is obvious that $u_m \in L^2(\Omega; H^m_{per}(Y; \mathbb{R})/\mathbb{R})$. So that $\mathbf{u} = (u_0, u_m) \in \mathbb{F}_0^m(\Omega \times Y; \mathbb{R})$. Next, by inserting (5.114) in (5.116) (which is equivalent to (5.115)), and then introducing u_m by means of (5.113), we see readily that (5.116) becomes

$$\sum_{|p|=|q|=m} \iint_{\Omega\times Y} a_{pq}(D^q u_0 + D_y^q u_m)D^p v_0\, dxdy = (f, v_0) - \int_\Omega \widetilde{h}[u_0]v_0\, dx$$

where v_0 is arbitrary fixed in $H_0^m(\Omega; \mathbb{R})$. On the other hand, let us turn to (5.112) and take there $w = v_m(x)$, where $v_m \in L^2(\Omega; H^m_{per}(Y; \mathbb{R})/\mathbb{R})$. After integrating over Ω we arrive at

$$\sum_{|p|=|q|=m} \int\int_{\Omega\times Y} a_{pq}(D^q u_0 + D_y^q u_m) D_y^p v_m \, dxdy = 0$$

Hence (ii) follows by summing these two equations. ■

Corollary 5.14 *The function u_0 is uniquely determined by (5.115).*

Proof Indeed, this follows by Proposition 5.10 and use of the unicity of the solution of (5.105). ■

We will conclude the present section with a discussion of some basic properties of the operator $\widetilde{P}$.

Proposition 5.11 *We have $c_{qp} = c_{pq}$.*

Proof For $\nu \in \mathbb{N}^N$, $|\nu| = m$, let $\pi_\nu(y) = \frac{y^\nu}{\nu!}$ $(y \in \mathbb{R}^N)$, where we recall that $y^\nu = y_1^{\nu_1} \cdots y_N^{\nu_N}$ and $\nu! = \nu_1! \cdots \nu_N!$ for $y = (y_1, \cdots, y_N)$ and $\nu = (\nu_1, \cdots, \nu_N)$. We observe that for $p \in \mathbb{N}^N$ with $|p| = m$, $D_y^p \pi_\nu = 0$ if $p \neq \nu$, $D_y^\nu \pi_\nu = 1$.

This being so the proposition is certainly proved if we can check that

$$c_{pq}(x) = a(x; \chi^q(x) - \pi_q, \chi^p(x) - \pi_p) \;\; (x \in \overline{\Omega}). \tag{5.117}$$

But this is straightforward once we have noticed that on one hand, $c_{pq}(x) = a(x; \pi_q - \chi^q, \pi_p)$, on the other hand, the equation in (5.111) reads

$$a(x; \pi_q - \chi^q(x), w) = 0 \text{ for any } w \in H_{per}^m(Y; \mathbb{R})/\mathbb{R};$$

hence $a(x; \pi_q - \chi^q(x), \chi^p(x)) = 0$. Equality (5.117) follows thereby. ■

Proposition 5.12 *We assume that*

$$a(x; w, w) \geq \alpha \sum_{|p|=m} \int_Y |D_y^p w|^2 dy \tag{5.118}$$

for all $x \in \overline{\Omega}$ and $w \in H_{loc}^m(\mathbb{R}^N; \mathbb{R})$. Then, there is some $\alpha_0 > 0$ such that

$$\sum_{|p|=|q|=m} c_{pq}(x)\xi^{p+q} \geq \alpha_0 |\xi|^{2m}$$

for all $x \in \overline{\Omega}$ and all $\xi \in \mathbb{R}^N$.

Proof We introduce the mapping $\mathcal{Q} : \overline{\Omega} \times \mathbb{R}^N \to \mathbb{R}$ given by

$$\mathcal{Q}(x, \xi) = \sum_{|p|=|q|=m} c_{pq}(x)\xi^{p+q} \quad (x \in \overline{\Omega}, \xi \in \mathbb{R}^N).$$

This mapping is continuous and further $\mathcal{Q}(x, t\xi) = t^{2m}\mathcal{Q}(x, \xi)$ for $x \in \overline{\Omega}, \xi \in \mathbb{R}^N$ and $t \in \mathbb{R}$. Thus, the proposition is established if we can show that there is some $\alpha_0 > 0$ such that $\mathcal{Q}(x, \xi) \geq \alpha_0$ for all $x \in \overline{\Omega}$ and $\xi \in K = \{z|z \in \mathbb{R}^N, |z| = 1\}$ (this is a routine argument; see Sect. 5.1).

For this purpose, let $x \in \overline{\Omega}$ be fixed. Put

$$u(\xi) = \sum_{|\nu|=m} \xi^\nu[\chi^\nu(x) - \pi_\nu] \quad (\xi \in \mathbb{R}^N).$$

By (5.117)–(5.118) one has

$$\mathcal{Q}(x, \xi) = a(x; u(\xi), u(\xi)) \geq \alpha \sum_{|q|=m} \int_Y |D_y^q u(\xi)|^2 dy$$

for all $\xi \in \mathbb{R}^N$. Therefore, we not only have $\mathcal{Q}(x, \xi) \geq 0$ but also $\mathcal{Q}(x, \xi) > 0$ for $\xi \neq \omega$ (ω the origin in $\mathbb{R}^N$). Indeed, assuming $\mathcal{Q}(x, \xi) = 0$ for some ξ implies $D_y^q u(\xi) = 0$, hence

$$\xi^q = \sum_{|\nu|=m} \xi^\nu D_y^q \chi^\nu(x) \quad (|q| = m).$$

Since $\int_Y D_y^q \chi^\nu(x)dy = 0$, we deduce $\xi^q = 0$ for all $q \in \mathbb{N}^N$ with $|q| = m$, hence $\xi = \omega$. As a consequence of this, we have $\mathcal{Q}(x, \xi) > 0$ on the compact set $\overline{\Omega} \times K$. Whence some $(x_0, \xi_0) \in \overline{\Omega} \times K$ such that $\mathcal{Q}(x, \xi) \geq \mathcal{Q}(x_0, \xi_0) = \alpha_0 > 0$ for all $(x, \xi) \in \overline{\Omega} \times K$. This completes the proof. ■

5.3.3 *Abstract Setting*

5.3.3.1 Preliminaries

Let A be an H-algebra on $\mathbb{R}^N$ of which A^∞ is a dense subalgebra. We assume moreover that A is $H^m(\Omega)$-proper. Our goal is to discuss the homogenization of (5.95) under the structure hypothesis:

$$a_{pq}(x, \cdot) \in \mathfrak{X}_A^2(\mathbb{R}_y^N) \; (|p| = |q| = m) \text{ for fixed } x \in \overline{\Omega} \tag{5.119}$$

$$h(\cdot, r) \in A \text{ for fixed } r \in \mathbb{R}. \tag{5.120}$$

It is clear that this setting generalizes that of Sect. 5.2. Consequently, except in case of necessity, we shall not carry out the details but instead refer the reader to the preceding section.

We start with some preliminaries. Let

$$H^m(\Delta(A);\mathbb{R}) = \{u|\partial^p u \in L^2(\Delta(A);\mathbb{R}), |p| \le m\}.$$

Equipped with the $H^m(\Delta(A))$ norm, $H^m_{\mathbb{R}}(\Delta(A)) \equiv H^m(\Delta(A);\mathbb{R})$ is a real Hilbert space admitting $\mathcal{D}_{\mathbb{R}}(\Delta(A)) \equiv \mathcal{D}(\Delta(A);\mathbb{R})$ as a dense subspace. We shall be mostly interested in the space

$$H^m_{\mathbb{R}}(\Delta(A))/\mathbb{R} = \left\{ u|u \in H^m_{\mathbb{R}}(\Delta(A)), \int_{\Delta(A)} u(s)d\beta(s) = 0 \right\}$$

equipped with the $H^m(\Delta(A))/\mathbb{C}$ seminorm under which it is in general non-separated and non-complete. The separated completion of $H^m_{\mathbb{R}}(\Delta(A))/\mathbb{R}$ is denoted by $H^m_{\#}(\Delta(A);\mathbb{R})$ (see Sect. 2.2). Proposition 5.8 carries over to the present general case (this is an easy exercise left to the reader), so that we may identify $H^m_{\#}(\Delta(A);\mathbb{R})$ (both algebraically and topologically) with a closed vector subspace of $H^m_{\#}(\Delta(A))$. The result is that:

(R1) $H^m_{\#}(\Delta(A);\mathbb{R})$ coincides with the space of all $u \in H^m_{\#}(\Delta(A))$ such that $\partial^p u \in L^2_{\mathbb{R}}(\Delta(A))$ for $|p| = m$.

(R2) Equipped with the $H^m_{\#}(\Delta(A))$ norm, $H^m_{\#}(\Delta(A);\mathbb{R})$ is a real Hilbert space admitting $J(H^m_{\mathbb{R}}(\Delta(A))/\mathbb{R})$ as a dense vector subspace (J is the canonical mapping of $H^m(\Delta(A))/\mathbb{C}$ into its separated completion, $H^m_{\#}(\Delta(A))$).

(R3) The restriction of J to $H^m_{\mathbb{R}}(\Delta(A))/\mathbb{R}$ is nothing but the canonical mapping $j : H^m_{\mathbb{R}}(\Delta(A))/\mathbb{R} \to H^m_{\#}(\Delta(A);\mathbb{R})$.

Having made this point, let

$$\mathbb{F}^m_0(\Omega \times \Delta(A);\mathbb{R}) = H^m_0(\Omega;\mathbb{R}) \times L^2(\Omega; H^m_{\#}(\Delta(A);\mathbb{R})).$$

Next, put

$$\|\mathbf{u}\|_{\mathbb{F}^m_0} = \left(\sum_{|q|=m} \|\mathbb{D}^q \mathbf{u}\|^2_{L^2(\Omega\times\Delta(A))} \right)^{\frac{1}{2}}, \quad \mathbf{u} = (u_0, u_m) \in \mathbb{F}^m_0(\Omega \times \Delta(A);\mathbb{R}),$$

where

$$\mathbb{D}^q \mathbf{u} = D^q_x u_0 + \partial^q u_m.$$

This defines a norm on $\mathbb{F}_0^m(\Omega \times \Delta(A); \mathbb{R})$ making it into a real Hilbert space. Furthermore,

$$\mathcal{F}_0^\infty \text{ is dense in } \mathbb{F}_0^m(\Omega \times \Delta(A); \mathbb{R}), \tag{5.121}$$

as is straightforward from the density of $\mathcal{D}(\Delta(A); \mathbb{R})$ in $H^m(\Delta(A); \mathbb{R})$ and use of (R3). We recall that the space $\mathcal{F}_0^\infty$ was introduced in Sect. 5.2.2 and that an expression of it using the mapping J_Ω instead of J is given in (5.80). On this point it is worth perhaps recalling that for any function $u : \Omega \to H^m(\Delta(A))/\mathbb{C}$, in particular for $u : \Omega \to H_{\mathbb{R}}^m(\Delta(A))/\mathbb{R}$, $J_\Omega(u)$ is the function of Ω into $H_\#^m(\Delta(A))$ defined by $J_\Omega(u) = J \circ u$, i.e., $J_\Omega(u)(x) = J(u(x)), x \in \Omega$. This holds in particular for $u = \varphi \otimes w$ with $\varphi \in \mathcal{D}(\Omega)$ and $w \in \mathcal{D}(\Delta(A))/\mathbb{C}$, u in this case being viewed as a function not of $\Omega \times \Delta(A)$ into $\mathbb{C}$ but of Ω into $\mathcal{D}(\Delta(A))$, i.e., $u(x) = \varphi(x)w, x \in \Omega$.

The above remark is valid mutatis mutandis with $\mathcal{G}_\Omega$ in place of J_Ω. We recall that for a function, say $u : \Omega \to A$, $\mathcal{G}_\Omega(u)$ denotes the function of Ω into $\mathcal{C}(\Delta(A))$ given by $\mathcal{G}_\Omega(u)(x) = \mathcal{G}(u(x)), x \in \Omega$.

The Bilinear Form $a_\Omega(\cdot, \cdot)$ First of all, from (5.90) and (5.119) we deduce that

$$a_{pq} \in \mathcal{C}(\overline{\Omega}; \mathfrak{X}_A^{2,\infty}(\mathbb{R}_y^N)) \quad (|p| = |q| = m). \tag{5.122}$$

In the sequel we put $\widehat{a}_{pq} = \mathcal{G}_\Omega(a_{pq})$. By proceeding as we did in proving Proposition 5.5 we obtain readily the following properties:

$$\widehat{a}_{pq} \in \mathcal{C}(\Omega; L_{\mathbb{R}}^\infty(\Delta(A))) \tag{5.123}$$

$$\widehat{a}_{qp} = \widehat{a}_{pq} \tag{5.124}$$

for $|p| = |q| = m$. We can now introduce the bilinear form $a_\Omega(\cdot, \cdot)$ on $\mathbb{F}_0^m(\Omega \times \Delta(A); \mathbb{R}) \times \mathbb{F}_0^m(\Omega \times \Delta(A); \mathbb{R})$ given by

$$a_\Omega(\mathbf{u}, \mathbf{v}) = \sum_{|p|=|q|=m} \int\int_{\Omega\times\Delta(A)} \widehat{a}_{pq}(x, s)\mathbb{D}^q\mathbf{u}(x, s)\mathbb{D}^p\mathbf{v}(x, s)dxd\beta(s)$$

for $\mathbf{u}, \mathbf{v} \in \mathbb{F}_0^m(\Omega \times \Delta(A); \mathbb{R})$. In reason of (5.123)–(5.124), this is a continuous symmetric bilinear form.

The $\mathbb{F}_0^m$-coerciveness of $a_\Omega(\cdot, \cdot)$ will follow by the next proposition.

Let

$$\mathbf{\Phi} = (\varphi, J_\Omega(\widehat{\psi})) \in \mathcal{F}_0^\infty = \mathcal{D}_{\mathbb{R}}(\Omega) \times J_\Omega(\mathcal{D}_{\mathbb{R}}(\Omega) \otimes [\mathcal{D}_{\mathbb{R}}(\Delta(A))/\mathbb{R}]). \tag{5.125}$$

Thus

$$\psi \in \mathcal{D}_{\mathbb{R}}(\Omega) \otimes (A_{\mathbb{R}}^\infty/\mathbb{R}).$$

We attach to $\mathbf{\Phi}$ the sequence of functions $\Phi_\varepsilon \in \mathcal{D}_{\mathbb{R}}(\Omega)$ given by (5.99).

Proposition 5.13 *Let $1 \leq r < \infty$. The following hold for any integer $m \geq 1$.*

(i) For any $\psi \in \mathcal{D}(\Omega) \otimes A^\infty$, the boundedness properties (5.100) hold true for any multi-index q of length m.
(ii) For any $\mathbf{\Phi} = (\varphi, J_\Omega(\widehat{\psi})) \in \mathcal{F}_0^\infty$, we have, as $\varepsilon \to 0$,

$$D^q \Phi_\varepsilon \to \mathbb{D}^q \mathbf{\Phi} = D^q \varphi + \partial^q \widehat{\psi} \text{ in } L^r(\Omega)\text{-strong } \Sigma \ (|q| = m). \tag{5.126}$$

Proof The proof of part (i) is a mere copy of that of the analogous part of Proposition 5.9. As regards part (ii), the line of reasoning is exactly the same mutatis mutandis as followed before. For the benefit of the reader we will however give a summary of the matter. By induction on $m \geq 1$, it is clear that the verification of the case $m = 1$ leads us readily by Proposition 4.1 to

$$\frac{\partial \Phi_\varepsilon}{\partial x_i} \to \mathbb{D}_i \mathbf{\Phi} = \frac{\partial \varphi}{\partial x_i} + \partial_i \widehat{\psi} \text{ in } L^r(\Omega)\text{-strong } \Sigma \text{ as } \varepsilon \to 0 \ (1 \leq i \leq N)$$

for all $\mathbf{\Phi}$ as in the statement. Next, under the induction hypothesis, we fix freely some integer $m > 1$, and we consider in an arbitrary fashion $\mathbf{\Phi} = (\varphi, J_\Omega(\widehat{\psi})) \in \mathcal{F}_0^\infty$ and subsequently the corresponding $\Phi_\varepsilon (0 < \varepsilon \leq 1)$. Given an arbitrary multi-index $q = (q_1, \cdots, q_N)$ of length m, we consider an integer $1 \leq \nu \leq N$ such that $q_\nu \neq 0$. This enables us to write $q = e + q'$, where e is that multi-index whose $\nu - th$ component is 1 and the others are zero, and $|q'| = m - 1$. This being so, by the same argument as followed in proving part (ii) of Proposition 5.9 one readily arrives at

$$D^q \Phi_\varepsilon - D^{q'} \phi_\varepsilon \to 0 \text{ in } L^r(\Omega)\text{-strong } \Sigma \text{ as } \varepsilon \to 0,$$

where

$$\phi_\varepsilon = \frac{\partial \varphi}{\partial x_\nu} + \varepsilon^{m-1} \left(\frac{\partial \psi}{\partial y_\nu} \right)^\varepsilon .$$

But, by letting

$$\mathbf{\Psi} = \left(\frac{\partial \varphi}{\partial x_\nu}, J_\Omega \left(\widehat{\frac{\partial \psi}{\partial y_\nu}} \right) \right) \in \mathcal{F}_0^\infty$$

and using the induction hypothesis, we have, as $\varepsilon \to 0$,

$$D^{q'} \phi_\varepsilon \to \mathbb{D}^{q'} \mathbf{\Psi} = D^{q'} \frac{\partial \varphi}{\partial x_\nu} + \partial^{q'} \widehat{\frac{\partial \psi}{\partial y_\nu}} \text{ in } L^r(\Omega)\text{-strong } \Sigma.$$

Hence, since $\mathbb{D}^q \mathbf{\Phi} = \mathbb{D}^{q'} \mathbf{\Psi}$, (5.126) follows. This shows (ii). ∎

Two useful corollaries must be mentioned.

Corollary 5.15 *Let* $\mathbf{\Phi}$ *and* Φ_ε *be as in (5.125) and (5.99), respectively. Then* $\Phi_\varepsilon \to \varphi$ *in* $H_0^m(\Omega; \mathbb{R})$*-weak as* $\varepsilon \to 0$.

Proof The proof of Corollary 5.12 carries over mutatis mutandis to the present setting. The details are left to the reader. ■

Corollary 5.16 *We have*

$$a_\Omega(\mathbf{v}, \mathbf{v}) \geq \alpha \|\mathbf{v}\|^2_{\mathbb{F}_0^m} \tag{5.127}$$

for all $\mathbf{v} \in \mathbb{F}_0^m(\Omega \times \Delta(A); \mathbb{R})$.

Proof The line of proceeding is the same mutatis mutandis as followed in proving Corollary 5.13, each result utilized in the latter corollary having an analogue in the present setting. The details are left to the reader. ■

Finally, we reintroduce the function $\widetilde{h} : \mathbb{R} \to \mathbb{R}$ defined by

$$\widetilde{h}(r) = \int_{\Delta(A)} \widehat{h}(s, r) d\beta(s) \quad (r \in \mathbb{R}).$$

The function $\widetilde{h}$ satisfies (5.104), and further we have Lemma 5.4.

5.3.3.2 Homogenization Results

The Global Homogenization Theorem

Let $\mathbf{l}$ denote the continuous linear form on $\mathbb{F}_0^m(\Omega \times \Delta(A); \mathbb{R})$ defined by

$$\mathbf{l}(\mathbf{v}) = (f, v_0) \text{ for } \mathbf{v} = (v_0, v_m) \in \mathbb{F}_0^m(\Omega \times \Delta(A); \mathbb{R}).$$

We consider the variational problem

$$\begin{cases} \mathbf{u} = (u_0, u_m) \in \mathbb{F}_0^m(\Omega \times \Delta(A); \mathbb{R}) : \\ \displaystyle\int_\Omega \widetilde{h}[u_0](x) v_0(x) dx + a_\Omega(\mathbf{u}, \mathbf{v}) = \mathbf{l}(\mathbf{v}) \\ \text{for all } \mathbf{v} = (v_0, v_m) \in \mathbb{F}_0^m(\Omega \times \Delta(A); \mathbb{R}) \end{cases} \tag{5.128}$$

which uniquely determines $\mathbf{u} = (u_0, u_m)$, thanks to (5.104), (5.123), (5.124) and (5.127). Next comes the global homogenization theorem.

Theorem 5.5 *For each $\varepsilon > 0$, let u_ε be the solution of the boundary value problem (5.95). Under the abstract structure hypothesis (5.119)–(5.120), let $\mathbf{u} = (u_0, u_m)$ be the solution of (5.128). Then, as $\varepsilon \to 0$,*

$$u_\varepsilon \to u_0 \text{ in } H_0^m(\Omega; \mathbb{R})\text{-weak}, \tag{5.129}$$

$$u_\varepsilon \to u_0 \text{ in } H_0^{m-1}(\Omega; \mathbb{R}), \tag{5.130}$$

$$D^q u_\varepsilon \to \mathbb{D}^q \mathbf{u} = D_x^q u_0 + \partial^q u_m \text{ in } L^2(\Omega)\text{-weak } \Sigma \ \ (|q| = m). \tag{5.131}$$

Proof The proof of Theorem 5.4 carries over mutatis mutandis to the present case and is therefore omitted. The reader can easily realize that everything necessary to carry over the proof of Theorem 5.4 has been done, so that any result utilized in the said proof has a counterpart in this setting. Thus in particular, the $H^m(\Omega)$-properness of the H-algebra A yields, on one hand, the density property (5.121), on the other hand, the Σ-reflexivity of $H^m(\Omega)$ (for A). Property (5.122) enables Proposition 4.4 to be used for passing to the limit in (5.109) as $\varepsilon \to 0$. Last but not least, Corollary 5.15 and Lemma 5.4 are turned to full account. The details are left to the reader. ■

As a consequence of this, there is the following

Corollary 5.17 *Let the hypotheses and notation be as above. Then, as $\varepsilon \to 0$,*

$$D^q u_\varepsilon \to \mathbb{D}^q \mathbf{u} = D_x^q u_0 + \partial^q u_m \text{ in } L^2(\Omega)\text{-strong } \Sigma \ \ (|q| = m).$$

Proof In view of (5.131), we only need to prove (according to Proposition 4.2) that as $\varepsilon \to 0$,

$$\|D^q u_\varepsilon\|_{L^2(\Omega)} \to \|\mathbb{D}^q \mathbf{u}\|_{L^2(\Omega \times \Delta(A))} \ \ (|q| = m). \tag{5.132}$$

The line of proceeding is the same as followed in proving Proposition 5.4. Let $\eta > 0$. In view of (5.121), we can find $\mathbf{\Phi} \in \mathcal{F}_0^\infty$ such that

$$a_\Omega(\mathbf{u} - \mathbf{\Phi}, \mathbf{u} - \mathbf{\Phi}) \leq \frac{\alpha \eta^2}{16}, \tag{5.133}$$

where α is the same as in (5.93). On the other hand, let $\Phi_\varepsilon (\varepsilon > 0)$ be attached to $\mathbf{\Phi}$ as in (5.125) and (5.99). Our purpose is to check that there is some $\varepsilon_1 > 0$ such that

$$\|D^q u_\varepsilon - D^q \Phi_\varepsilon\|_{L^2(\Omega)} \leq \frac{\eta}{2} \ \ (|q| = m) \ \text{ for } 0 < \varepsilon \leq \varepsilon_1. \tag{5.134}$$

To begin, (5.109) shows that

$$a^\varepsilon(u_\varepsilon - \Phi_\varepsilon, u_\varepsilon - \Phi_\varepsilon) = (f, u_\varepsilon) - \int_\Omega h^\varepsilon[u_\varepsilon] u_\varepsilon dx + a^\varepsilon(\Phi_\varepsilon, \Phi_\varepsilon) + 2a^\varepsilon(u_\varepsilon, \Phi_\varepsilon).$$

A quick survey of the proofs of Corollary 5.13 and Theorem 5.5 or Theorem 5.4 (use in particular (5.129)–(5.131)) reveals that as $\varepsilon \to 0$, the right-hand side of the above equation tends to $a_\Omega(\mathbf{u} - \mathbf{\Phi}, \mathbf{u} - \mathbf{\Phi})$. Hence there exists some $\varepsilon_1 > 0$ such that

$$a^\varepsilon(u_\varepsilon - \Phi_\varepsilon, u_\varepsilon - \Phi_\varepsilon) \leq a_\Omega(\mathbf{u} - \mathbf{\Phi}, \mathbf{u} - \mathbf{\Phi}) + \frac{3\alpha\eta^2}{16}$$

for $0 < \varepsilon \leq \varepsilon_1$. Therefore, (5.134) follows by (5.93) and (5.133).

Now, let $q \in \mathbb{N}^N$ be fixed with $|q| = m$. By (5.134) we have

$$\left| \|D^q u_\varepsilon\|_{L^2(\Omega)} - \|D^q \Phi_\varepsilon\|_{L^2(\Omega)} \right| \leq \frac{\eta}{2} \quad (0 < \varepsilon \leq \varepsilon_1).$$

On the other hand, (5.133) combines with Corollary 5.16 to give

$$\left| \|\mathbb{D}^q \mathbf{u}\|_{L^2(\Omega \times \Delta(A))} - \|\mathbb{D}^q \mathbf{\Phi}\|_{L^2(\Omega \times \Delta(A))} \right| \leq \frac{\eta}{4} \quad (0 < \varepsilon \leq \varepsilon_2).$$

Hence it follows by mere routine (see the proof of Proposition 5.4) that

$$\left| \|D^q u_\varepsilon\|_{L^2(\Omega)} - \|\mathbb{D}^q \mathbf{u}\|_{L^2(\Omega \times \Delta(A))} \right| \leq \eta$$

for $0 < \varepsilon \leq \varepsilon_0 = \min(\varepsilon_1, \varepsilon_2)$, i.e., (5.132) is proved. This completes the proof. ■

Our next purpose is to derive the local and macroscopic homogenized problems.

Local Problem Let $x \in \overline{\Omega}$. For $z, w \in H^m_\#(\Delta(A); \mathbb{R})$, we put

$$a(x; z, w) = \sum_{|p|=|q|=m} \int_{\Delta(A)} \widehat{a}_{pq}(x, \cdot)\partial^q z \partial^p w \, d\beta,$$

defining thus a bilinear form $a(x; \cdot, \cdot)$ on $H^m_\#(\Delta(A); \mathbb{R}) \times H^m_\#(\Delta(A); \mathbb{R})$ which is continuous symmetric with further

$$a(x, w, w) \geq \alpha \|w\|^2_{H^m_\#(\Delta(A))} \quad (w \in H^m_\#(\Delta(A); \mathbb{R})). \tag{5.135}$$

Indeed, this is immediate from (5.127) by the same argument as followed earlier to obtain (5.110). Thanks to (5.135) we can define the so-called *structure function* of index q at point x as being the (unique) solution of the variational problem

$$\begin{cases} \chi^q(x) \in H^m_\#(\Delta(A); \mathbb{R}) : \\ a(x; \chi^q(x), w) = \sum_{|\nu|=m} \int_{\Delta(A)} \widehat{a}_{\nu q}(x, s)\partial^\nu w(s) d\beta(s) \\ \text{for all } w \in H^m_\#(\Delta(A); \mathbb{R}). \end{cases}$$

The mapping $x \mapsto \chi^q(x)$ thus defined sends continuously $\overline{\Omega}$ into $H^m_\#(\Delta(A);\mathbb{R})$, as is immediate by an easy adaptation of the proof of Proposition 5.3(i).

On the other hand, the same reasoning as followed in Sect. 5.3.3 leads us to the local problem at $x \in \Omega$, namely

$$\begin{cases} u_m(x) \in H^m_\#(\Delta(A);\mathbb{R}) : \\ a(x; u_m(x), w) = -\sum_{|q|=m} D^q u_0(x) \sum_{|p|=m} \int_{\Delta(A)} \widehat{a}_{pq}(x,s)\partial^p w(s) d\beta(s) \\ \text{for all } w \in H^m_\#(\Delta(A);\mathbb{R}), \end{cases}$$

which uniquely determines $u_m(x)$, thanks to (5.135), once again. Furthermore one has (5.113).

Macroscopic Homogenized Problem The same routine method as followed before leads us to the macroscopic homogenized problem, namely (5.115), where the differential operator $\widetilde{P}$ is framed as in Sect. 5.3.2 with $c_{pq} \in \mathcal{C}_\mathbb{R}(\overline{\Omega})$ ($|p| = |q| = m$) given by

$$c_{pq}(x) = \int_{\Delta(A)} \widehat{a}_{pq}(x,s) d\beta(s) - \sum_{|\nu|=m} \int_{\Delta(A)} \widehat{a}_{p\nu}(x,s)\partial^\nu \chi^q(x,s) d\beta(s)$$

for all $x \in \overline{\Omega}$.

As one might expect, Proposition 5.10 carries over mutatis mutandis to the present setting. Consequently, the function u_0 in Theorem 5.5 is the unique solution of (5.115).

5.3.3.3 Strong Approximation

The aim here is to generalize the main approximation results of Sects. 5.1.2 and 5.1.3 to higher order differential operators. Specifically, we achieve in the present section a strong approximation result requiring no further regularity hypothesis, contrary to the custom. Our approach is based on the obvious fact that

$$J_\Omega(H^m_\mathbb{R}(\Omega) \otimes [\mathcal{D}_\mathbb{R}(\Delta(A))/\mathbb{R}]) \text{ is dense in } L^2(\Omega; H^m_\#(\Delta(A);\mathbb{R})). \tag{5.136}$$

Theorem 5.6 *Let the hypotheses and notation be as in Theorem 5.5. Let $[m]$ be the number of all $q \in \mathbb{N}^N$ with $|q| \leq m$. Then, for $\eta > 0$ and $\psi_m \in H^m_\mathbb{R}(\Omega) \otimes (A^m_\mathbb{R}/\mathbb{R})$ such that*

$$\|u_m - J_\Omega(\widehat{\psi}_m)\|_{L^2(\Omega; H^m_\#(\Delta(A)))} \leq \frac{\eta}{4[m]^{\frac{1}{2}}}, \tag{5.137}$$

there is some $\varepsilon_0 > 0$ *such that*

$$\|u_\varepsilon - u_0 - \varepsilon^m \psi_m^\varepsilon\|_{H^m(\Omega)} \leq \eta \tag{5.138}$$

provided $0 < \varepsilon \leq \varepsilon_0$.

Proof Thanks to (5.136) we can fix some ψ_m as above such that (5.137) holds true. Let $q \in \mathbb{N}^N$ with $|q| = m$. Then

$$\|\partial^q u_m - \partial^q \widehat{\psi}_m\|_{L^2(\Omega\times\Delta(A))} \leq \frac{\eta}{4[m]^{\frac{1}{2}}}.$$

By Corollary 5.17, this yields some $0 < \varepsilon_2 \leq 1$ such that

$$\|D^q u_\varepsilon - D^q u_0 - (D_y^q \psi_m)^\varepsilon\|_{L^2(\Omega)} \leq \frac{\eta}{2[m]^{\frac{1}{2}}} \tag{5.139}$$

for $0 < \varepsilon \leq \varepsilon_2$. Now, we have specifically

$$\psi_m = \sum_{i\in I} \varphi_i \otimes \theta_i \text{ with } \varphi_i \in H_{\mathbb{R}}^m(\Omega), \theta_i \in A_{\mathbb{R}}^m/\mathbb{R}, I \text{ finite.}$$

By the same Leibniz formula as used in proving Proposition 5.9, we have

$$\varepsilon^m D^q \psi_m^\varepsilon - (D_y^q \psi_m)^\varepsilon = \sum_{\gamma\leq q, \gamma\neq q} \varepsilon^{m-|\gamma|} G_\gamma^\varepsilon,$$

where $G_\gamma \in L^2(\Omega; A)$, G_γ defined in the proof of Proposition 5.9. But $\|G_\gamma^\varepsilon\|_{L^2(\Omega)} \leq \|G_\gamma\|_{L^2(\Omega;A)}$ (Proposition 1.6). On the other hand, $\varepsilon^{m-|\gamma|} \leq \varepsilon$ for $0 < \varepsilon \leq 1$, since $\gamma \leq q$ with $\gamma \neq q$ implies $|\gamma| < m$. We deduce

$$\left\| \sum_{\gamma\leq q, \gamma\neq q} \varepsilon^{m-|\gamma|} G_\gamma^\varepsilon \right\|_{L^2(\Omega)} \leq \varepsilon \sum_{\gamma\leq q, \gamma\neq q} \|G_\gamma\|_{L^2(\Omega;A)}$$

for $0 < \varepsilon \leq 1$. By taking account of (5.139), it follows

$$\|D^q(u_\varepsilon - u_0 - \varepsilon^m \psi_m^\varepsilon)\|_{L^2(\Omega)} \leq \frac{\eta}{2[m]^{\frac{1}{2}}} + \varepsilon c,$$

where c is a positive constant (independent of ε) such that

$$c \geq \sum_{\gamma\leq p, \gamma\neq p} \|G_\gamma\|_{L^2(\Omega;A)}$$

for all $p \in \mathbb{N}^N$ with $|p| = m$. Hence

$$\|D^q(u_\varepsilon - u_0 - \varepsilon^m \psi_m^\varepsilon)\|_{L^2(\Omega)} \leq \frac{\eta}{[m]^{\frac{1}{2}}} \quad (|q| = m)$$

for $0 < \varepsilon \leq \varepsilon_1$, where $\varepsilon_1 = \min(\varepsilon_2, \frac{\eta}{2c[m]^{\frac{1}{2}}})$.

Finally, if $|q| < m$ then, as $\varepsilon \to 0$, on one hand it is clear that $\varepsilon^m D^q \psi_m^\varepsilon \to 0$ in $L^2(\Omega)$, on the other hand, $D^q(u_\varepsilon - u_0) \to 0$ in $L^2(\Omega)$ as a consequence of (5.130) which results itself from (5.129) by the compactness of the embedding $H_0^m(\Omega; \mathbb{R}) \hookrightarrow H_0^{m-1}(\Omega; \mathbb{R})$.

The result is that there is some $\varepsilon_0 > 0$ such that

$$\|D^q(u_\varepsilon - u_0 - \varepsilon^m \psi_m^\varepsilon)\|_{L^2(\Omega)} \leq \frac{\eta}{[m]^{\frac{1}{2}}} \quad (|q| \leq m)$$

for $0 < \varepsilon \leq \varepsilon_0$. Hence the theorem follows. ■

The following corollary results directly from Theorem 5.6. The verification is left to the reader.

Corollary 5.18 *Under the notation and hypotheses of Theorem 5.5, suppose moreover that* $u_m \in J_\Omega(H_{\mathbb{R}}^m(\Omega) \otimes [\mathcal{D}_{\mathbb{R}}^m(\Delta(A))/\mathbb{R}])$, *i.e., there is some* $\psi_m \in H_{\mathbb{R}}^m(\Omega) \otimes (A_{\mathbb{R}}^\infty/\mathbb{R})$ *such that* $u_m(x) = J(\psi_m(x))$ *a.e. in* $x \in \Omega$. *Then* $u_\varepsilon - u_0 - \varepsilon^m \psi_m^\varepsilon \to 0$ *in* $H^m(\Omega)$ *as* $\varepsilon \to 0$.

Remark 5.12 We have precisely $[m] = \frac{(m+N)!}{m!N!}$.

5.3.4 Some Concrete Deterministic Homogenization Problems

We discuss here a few concrete homogenization problems for (5.95). As before, our concern in each case under consideration will be to reduce the underlying concrete structure hypothesis to the abstract hypothesis (5.119)–(5.120) so as to make it possible to apply the results achieved earlier (see in particular Theorems 5.5 and 5.6).

5.3.4.1 Problem I (Periodic Setting)

The homogenization problem for (5.95) under the periodicity hypothesis (5.96) was worked out in a self-contained manner in Sect. 5.3.2. Considered now from the angle of Sect. 5.3.3 this same problem is readily solved by Theorem 5.5 with $A = \mathcal{C}_{per}(Y)$ (see Example 4.8 and Problem I of Sect. 5.1.4). In fact, it is a rather simple matter to check using appropriate tools from Chap. 2 (see in particular Corollary 2.15) that the whole Sect. 5.3.3 carries over mutatis mutandis to the periodic setting. In particular the strong approximation theory (purposely not discussed in Sect. 5.3.2) developed above yields the following result which is nothing else than the periodic version of Theorem 5.6.

Theorem 5.7 *Let the hypotheses and notation be as in Theorem 5.4. Let* $[m]$ *be as above. Then, given* $\eta > 0$ *and* $\psi_m \in H^m_{\mathbb{R}}(\Omega) \otimes [C^m_{per}(Y;\mathbb{R})/\mathbb{R}]$ *such that*

$$\|u_m - \psi_m\|_{L^2(\Omega;H^m_{per}(Y)/\mathbb{C})} \leq \frac{\eta}{4[m]^{\frac{1}{2}}},$$

there is some $\varepsilon_0 > 0$ *such that* $\|u_\varepsilon - u_0 - \varepsilon^m \psi^\varepsilon_m\|_{H^m(\Omega)} \leq \eta$ *for* $0 < \varepsilon \leq \varepsilon_0$.

5.3.4.2 Problem II

We consider here the homogenization of (5.95) under the structure hypothesis

$$\begin{cases} a_{pq}(x,\cdot) \text{ is } Y\text{-periodic } (|p| = |q| = m) \text{ for fixed } x \in \overline{\Omega} \\ h(\cdot, r) \in \mathcal{B}_\infty(\mathbb{R}^N_y) \text{ for fixed } r \in \mathbb{R}, \end{cases} \tag{5.140}$$

where $Y = (-\frac{1}{2}, \frac{1}{2})^N$. Let $A = \mathcal{B}_{\infty,\mathbb{Z}^N}(\mathbb{R}^N)$ be the H-algebra discussed in Example 4.4. Recalling that A is precisely the closure of $\mathcal{B}_\infty(\mathbb{R}^N_y) + \mathcal{C}_{per}(Y)$ in $\mathcal{B}(\mathbb{R}^N_y)$, we see readily that $h(\cdot, r) \in A$ for fixed $r \in \mathbb{R}$. On the other hand, it is clear that $a_{pq}(x,\cdot) \in L^\infty_{per}(Y) \subset L^2_{per}(Y)$. But $L^2_{per}(Y) = \mathfrak{X}^2_{\mathcal{C}_{per}(Y)} \subset \mathfrak{X}^2_A$ (see Chap. 2). Moreover, A is an $H^m(\Omega)$-proper H-algebra admitting A^∞ as a dense subalgebra (see Examples 4.4 and 4.9). Therefore the problem under consideration reduces to the abstract setting based on (5.119)–(5.120) with $A = \mathcal{B}_{\infty,\mathbb{Z}^N}(\mathbb{R}^N)$.

5.3.4.3 Problem III

The problem before us is as above save that (5.140) is here substituted by the structure hypothesis

$$\begin{cases} a_{pq}(x,\cdot) \in L^2(\mathbb{R}^N_y) + L^2_{per}(Y) \ \ (|p| = |q| = m) \text{ for fixed } x \in \overline{\Omega} \\ h(\cdot, r) \in \mathcal{C}_{per}(Y) \text{ for fixed } r \in \mathbb{R}. \end{cases}$$

It is not hard to see that this structure hypothesis reduces to (5.119)–(5.120) with the same A as in Problem II. The details are left to the reader.

5.3.4.4 Problem IV

We assume for simplicity that the coefficients a_{pq} do not depend on $x \in \overline{\Omega}$. The problem on hand is the homogenization of the boundary value problem (5.95) under the structure hypothesis:

$$\begin{cases} a_{pq} \text{ is constant on } k + Y \ (k \in \mathbb{Z}^N) \quad (|p| = |q| = m), \\ \lim\limits_{|k|\to\infty} \displaystyle\int_{k+Y} a_{pq}(y)dy \in \mathbb{R} \text{ exists} \quad (|p| = |q| = m), \\ h(\cdot, r) \text{ is } Y\text{-periodic for fixed } r \in \mathbb{R}. \end{cases}$$

This problem generalizes Problem V of Sect. 5.2.4. Moreover, the H-algebra underlying the argument followed in the latter problem turns out to be $H^m(\Omega)$-proper (see Problem VI of Sect. 5.1.4; see also Sect. 4.2.5). Consequently, the solution of the present case can be achieved by the same line of argument.

5.4 Deterministic Homogenization in Perforated Domains

5.4.1 Introduction

Let S be an **infinite** subset of $\mathbb{Z}^N$, and T be a connected compact set in $\mathbb{R}^N_y$ with smooth boundary and nonempty interior with

$$T \subset Y = \left(-\frac{1}{2}, \frac{1}{2}\right)^N.$$

For fixed $\varepsilon > 0$, let

$$t^\varepsilon = \{k | k \in S, \varepsilon(k + T) \subset \Omega\}$$

$$T^\varepsilon = \bigcup_{k \in t^\varepsilon} \varepsilon(k + T)$$

and

$$\Omega^\varepsilon = \Omega \setminus T^\varepsilon,$$

where S and Ω are so chosen that t^ε is nonempty. Observe that t^ε is finite, since Ω is bounded. It clearly follows that Ω^ε is an open set in $\mathbb{R}^N_x$. This is commonly referred to as a *perforated domain* whereas T is the *reference hole* and $\varepsilon(k + T)$ the *hole of size* $\varepsilon > 0$.

Now, for arbitrary given $\varepsilon > 0$, let

$$\begin{cases} a_0^\varepsilon u_\varepsilon - \sum_{i,j=1}^{N} \frac{\partial}{\partial x_i}\left(a_{ij}^\varepsilon \frac{\partial u_\varepsilon}{\partial x_j}\right) = f \text{ in } \Omega^\varepsilon, \\ \sum_{i,j=1}^{N} a_{ij}^\varepsilon \frac{\partial u_\varepsilon}{\partial x_j} \nu_i = 0 \text{ on } \partial T^\varepsilon \quad (\text{boundary of } T^\varepsilon), \\ u_\varepsilon = 0 \text{ on } \partial\Omega \quad (\text{boundary of } \Omega), \end{cases} \tag{5.141}$$

where $f \in L^2(\Omega)$, $\boldsymbol{\nu} = (\nu_i)_{1 \le i \le N}$ the outer unit normal to ∂T^ε with respect to Ω^ε, $a_0, a_{ij} \in L^\infty(\mathbb{R}_y^N)$ with $a_{ji} = \overline{a}_{ij}$ $(1 \le i, j \le N)$ and with the property that there are two constants $\alpha_0 > 0$ and $\alpha > 0$ such that

$$\operatorname{Re} a_0(y) \ge \alpha_0 \text{ a.e. in } \mathbb{R}_y^N$$

and

$$\operatorname{Re} \sum_{i,j=1}^{N} a_{ij}(y)\xi_j\overline{\xi}_i \ge \alpha|\xi|^2 \quad (\xi \in \mathbb{C}^N) \text{ a.e. in } \mathbb{R}_y^N. \tag{5.142}$$

Thanks in particular to (5.142), it is a classical matter to show that the Neumann-type boundary value problem (5.141) admits a unique solution $u_\varepsilon \in H^1(\Omega^\varepsilon)$.

The present section is devoted to the mathematical analysis of the limiting behaviour of u_ε as $\varepsilon \to 0$ in a general deterministic setting based on an abstract structure hypothesis on the coefficients a_0, a_{ij} as well as on the manner in which the holes are distributed.

5.4.2 *Preliminaries on the Perforation*

Let

$$\Theta = \bigcup_{k \in S} (k + T).$$

Based on the compactness of T one can easily verify that this is a closed subset of $\mathbb{R}_y^N$. Next, for arbitrary fixed $\varepsilon > 0$, let $Q^\varepsilon = \Omega \setminus \varepsilon\Theta$. This is an open set in $\mathbb{R}_x^N$ with further $Q^\varepsilon \subset \Omega^\varepsilon$.

Lemma 5.5 *Let K be a fixed compact set in Ω (thus, K does not depend on $\varepsilon > 0$). There is some $\varepsilon_0 > 0$ such that $(\Omega^\varepsilon \setminus Q^\varepsilon) \subset (\Omega \setminus K)$ for $0 < \varepsilon \le \varepsilon_0$.*

Proof Let U be an open neighbourhood of $\partial\Omega$ (a compact set) in $\mathbb{R}_x^N$ such that $U \subset (\mathbb{R}_x^N \setminus K)$ (such a U does exist). It is an evident fact that one may assume without loss of generality that U is bounded. Hence, by Lemma 1.7 (see also its proof) there is some $\varepsilon_0 > 0$ such that

$$J^\varepsilon(\partial\Omega) \subset U \text{ for } 0 < \varepsilon \leq \varepsilon_0,$$

where $J^\varepsilon(\partial\Omega)$ is defined in Sect. 4.2.5. On the other hand, we have

$$(\Omega^\varepsilon \setminus \mathcal{Q}^\varepsilon) \subset J^\varepsilon(\partial\Omega) \quad (\varepsilon > 0), \tag{5.143}$$

as is easily seen by the equality $j^\varepsilon(\overline{\Omega}) = i_0^\varepsilon(\Omega) \cup j^\varepsilon(\partial\Omega)$ (see Lemma 4.4). Hence the lemma follows. ∎

Remark 5.13

(1) Each point in $\Omega^\varepsilon \setminus \mathcal{Q}^\varepsilon$ lies in some $\varepsilon(k+T)$ that meets $\partial\Omega$. Indeed, such a hole, $\varepsilon(k+T)$, meets both Ω and $\mathbb{R}_x^N \setminus \Omega$, and therefore $\partial\Omega$ also, since it is connected (see, e.g., Bourbaki [24, I,p.81,Proposition 3]).
(2) According to (5.143) we have $\lim_{\varepsilon\to 0} \lambda(\Omega^\varepsilon \setminus \mathcal{Q}^\varepsilon) = 0$, since by Lemma 4.5 one has $\lim_{\varepsilon\to 0} \lambda(J^\varepsilon(\partial\Omega)) = \lambda(\partial\Omega) = 0$.

Now, for fixed $\varepsilon > 0$, let

$$V_\varepsilon = \{v | v \in H^1(\Omega^\varepsilon), v = 0 \text{ on } \partial\Omega\}.$$

We equip V_ε with the $H^1(\Omega^\varepsilon)$-norm, which makes it into a Hilbert space. Our next purpose is to construct a suitable extension operator sending continuously and linearly V_ε into $H_0^1(\Omega)$. Specifically, we want to prove the following

Proposition 5.14 *There exists a mapping $P_\varepsilon : V_\varepsilon \to H_0^1(\Omega)$ with the following properties:*

$$P_\varepsilon \text{ sends continuously and linearly } V_\varepsilon \text{ into } H_0^1(\Omega). \tag{5.144}$$

$$(P_\varepsilon v)|_{\Omega^\varepsilon} = v \text{ for all } v \in V_\varepsilon. \tag{5.145}$$

$$\|D(P_\varepsilon v)\|_{L^2(\Omega)^N} \leq c\|Dv\|_{L^2(\Omega^\varepsilon)^N} \text{ for all } v \in V_\varepsilon, \tag{5.146}$$

where the constant $c > 0$ depends only on Y and T, and D denotes the gradient operator in Ω, i.e., $D = (\frac{\partial}{\partial x_1}, \cdots, \frac{\partial}{\partial x_N})$.

Remark 5.14 When $S = \mathbb{Z}^N$, Proposition 5.14 is a classical result (see, e.g., Cioranescu and Saint Jean Paulin [35]). However, in the present case the holes are distributed in an arbitrary fashion, which we express by merely assuming S to be an infinite subset of $\mathbb{Z}^N$. Though resembling enough, the present situation seems unusual and therefore it requires some care.

The proof of Proposition 5.14 will use one fundamental lemma (whose proof is to be found in, e.g., Cioranescu and Saint Jean Paulin [35]), viz.

Lemma 5.6 *Let* $Y^* = Y \setminus T$. *Fix freely* $k \in \mathbb{Z}^N$. *There exists a continuous linear mapping* $p_k : H^1(k+Y^*) \to H^1(k+Y)$ *such that*

$$(i)\ (p_k w)|_{k+Y^*} = w \quad (w \in H^1(k+Y^*))$$

$$(ii)\ \|D_y(p_k w)\|_{L^2(k+Y)^N} \le c_0 \|D_y w\|_{L^2(k+Y^*)^N} \quad (w \in H^1(k+Y^*)),$$

where the constant $c_0 > 0$ *depends only on* Y *and* T, *and* D_y *is the gradient operator in* $\mathbb{R}^N_y$.

We turn now to the proof of the preceding proposition.

Proof of Proposition 5.14 Let us start with a useful remark. Let $v \in V_\varepsilon$. Suppose for some $k \in t^\varepsilon$, the set $\varepsilon(k+Y^*)$ is not contained in Ω. Then the extension of $v|_{\Omega\cap\varepsilon(k+Y^*)}$ by 0 in the complement of $\Omega \cap \varepsilon(k+Y^*)$ in $\varepsilon(k+Y^*)$ belongs to $H^1(\varepsilon(k+Y^*))$. Hence it follows that we may always view each $v \in V_\varepsilon$ as defined in the whole of $\varepsilon(k+Y^*)$, $k \in t^\varepsilon$.

With this in mind, let us now fix any arbitrary $v \in V_\varepsilon$, and try to define $P_\varepsilon v$. Since

$$\Omega = \bigcup_{k\in j^\varepsilon(\overline{\Omega})} [\Omega \cap \varepsilon(k+\overline{Y})] \text{ (see Sect. 4.2.5)},$$

it amounts to defining $P_\varepsilon v$ on each $\Omega\cap\varepsilon(k+Y)$ $(k \in j^\varepsilon(\overline{\Omega}))$. So, given $k \in j^\varepsilon(\overline{\Omega})$, let us first assume that k lies outside t^ε. Then we define $P_\varepsilon v$ on $\Omega \cap \varepsilon(k+Y)$ as

$$P_\varepsilon v(x) = v(x) \text{ a.e. in } \Omega \cap \varepsilon(k+Y).$$

We next suppose $k \in t^\varepsilon$. Let us for a while still call v the restriction of the latter to $\Omega\cap\varepsilon(k+Y^*)$. Let v_ε in $H^1(k+Y^*)$ be defined by $v_\varepsilon(y) = v(\varepsilon y)$ a.e. in $y \in k+Y^*$. Then, we define $P_\varepsilon v$ in $\Omega \cap \varepsilon(k+Y)$ as

$$P_\varepsilon v(x) = (p_k v_\varepsilon)\left(\frac{x}{\varepsilon}\right) \text{ a.e. in } x \in \Omega \cap \varepsilon(k+Y),$$

where $p_k v_\varepsilon$ is given by Lemma 5.6. It is clear that (5.145) holds true, and further the transformation $v \mapsto P_\varepsilon v$ maps linearly V_ε into $H^1_0(\Omega)$. Thus, the proof is complete once (5.146) is established (indeed, the continuity of P_ε will then follow by a well-known inequality). To achieve this, let $v \in V_\varepsilon$ be identified with its extension by 0 outside Ω. We may write

$$\begin{aligned}\|D(P_\varepsilon v)\|^2_{L^2(\Omega)^N} &= \sum_{k\in j^\varepsilon(\overline{\Omega})} \|D(P_\varepsilon v)\|^2_{L^2(\Omega\cap\varepsilon(k+Y))^N} \\ &= \sum_{k\in t^\varepsilon} \|D(P_\varepsilon v)\|^2_{L^2(\Omega\cap\varepsilon(k+Y))^N} \\ &+ \sum_{k\in j^\varepsilon(\overline{\Omega})\backslash t^\varepsilon} \|D(P_\varepsilon v)\|^2_{L^2(\Omega\cap\varepsilon(k+Y))^N}.\end{aligned}$$

Next, an elementary calculation utilizing Lemma 5.6 reveals that we have, for $k \in t^\varepsilon$,

$$\begin{aligned}\|D(P_\varepsilon v)\|^2_{L^2(\Omega\cap\varepsilon(k+Y))^N} &\le \|D(P_\varepsilon v)\|^2_{L^2(\varepsilon(k+Y))^N} \\ &\le c_0^2\|Dv\|^2_{L^2(\varepsilon(k+Y^*))^N} \\ &= c_0^2\|Dv\|^2_{L^2(\Omega^\varepsilon\cap\varepsilon(k+Y^*))^N} \\ &= c_0^2\|Dv\|^2_{L^2(\Omega^\varepsilon\cap\varepsilon(k+Y))^N}.\end{aligned}$$

On the other hand, it is obvious that for $k \in j^\varepsilon(\overline{\Omega}) \setminus t^\varepsilon$ we have

$$\begin{aligned}\|D(P_\varepsilon v)\|^2_{L^2(\Omega\cap\varepsilon(k+Y))^N} &= \|Dv\|^2_{L^2(\Omega\cap\varepsilon(k+Y))^N} \\ &= \|Dv\|^2_{L^2(\Omega^\varepsilon\cap\varepsilon(k+Y))^N}.\end{aligned}$$

Hence (5.146) follows with $c = \max\{c_0, 1\}$. The proof is complete. ■

5.4.3 *The Abstract Homogenization Problem for (5.141)*

5.4.3.1 Setting of the Problem and Preliminaries

Let $G = \mathbb{R}^N_y \setminus \Theta$. This is an open subset of $\mathbb{R}^N_y$, of course. In the sequel χ_G denotes the characteristic function of G in $\mathbb{R}^N_y$.

Now, let A be an $H^1(\Omega)$-proper homogenization algebra on $\mathbb{R}^N$ admitting A^∞ as a dense subalgebra. Then, *the abstract homogenization problem* for (5.141) consists in investigating the behaviour, as $\varepsilon \to 0$, of the solution, u_ε, of (5.141) under the assumptions

$$\chi_G \in \mathfrak{X}^p_A(\mathbb{R}^N_y), \tag{5.147}$$

$$M(\chi_G) > 0, \tag{5.148}$$

$$a_0, a_{ij} \in \mathfrak{X}_A^q(\mathbb{R}_y^N) \;\; (1 \leq i, j \leq N), \tag{5.149}$$

where the given positive real numbers p and q are such that

$$\frac{1}{p} + \frac{1}{q} \leq \frac{1}{2}. \tag{5.150}$$

The assumptions (5.147)–(5.150) constitute the so-called *abstract structure hypothesis* in the present context.

Before we can concentrate on the problem as such, however, we need some preliminary results.

Lemma 5.7 *Under assumption (5.147), there exists a β-measurable set $\widehat{G} \subset \Delta(A)$ such that $\widehat{\chi}_G = \chi_{\widehat{G}}$ (the characteristic function of $\widehat{G}$ in $\Delta(A)$), where $\widehat{\chi}_G = \mathcal{G}(\chi_G)$.*

Proof Clearly $\chi_G \in \mathfrak{X}_A^2 = \mathfrak{X}_A^2(\mathbb{R}_y^N)$, since $p \geq 2$. Therefore, it follows $\chi_G\chi_G \in \mathfrak{X}_A^1$ with $\mathcal{G}(\chi_G\chi_G) = \mathcal{G}(\chi_G)\mathcal{G}(\chi_G)$. In view of the obvious equality $\widehat{\chi}_G\widehat{\chi}_G = \widehat{\chi}_G$, one deduces that $\widehat{\chi}_G(s) \in \{0, 1\}$ for a.e. $s \in \Delta(A)$. More precisely, there is a β-negligible set $\mathcal{N} \subset \Delta(A)$ such that $\widehat{\chi}_G(s) \in \{0, 1\}$ for all $s \in \Delta(A) \setminus \mathcal{N}$. By letting $\widehat{G} = \{s | s \in \Delta(A) \setminus \mathcal{N}, \widehat{\chi}_G(s) = 1\}$, we see readily that $\widehat{\chi}_G = \chi_{\widehat{G}}$ a.e. in $\Delta(A)$. Hence, it follows by Corollary 2.7 that $\chi_{\widehat{G}} \in L^\infty(\Delta(A))$. Consequently, $\widehat{G}$ is β-measurable and in particular β-integrable. ■

Remark 5.15 As a consequence of Lemma 5.7 we have $\chi_G^\varepsilon \to \chi_{\widehat{G}}$ in $L^p(\Omega)$-weak Σ as $\varepsilon \to 0$. Furthermore, $\beta(\widehat{G}) = \int_{\Delta(A)} \widehat{\chi}_G(s)d\beta(s) = M(\chi_G)$. Finally, (5.148) says that $\widehat{G}$ is not β-negligible.

At the present time, we observe that the functions $\widehat{a}_0 = \mathcal{G}(a_0)$ and $\widehat{a}_{ij} = \mathcal{G}(a_{ij})$ lie in $L^\infty(\Delta(A))$ (this follows by joining (5.149)–(5.150) with Corollaries 2.6 and 2.7) and further

$$\operatorname{Re} \widehat{a}_0(s) \geq \alpha_0 > 0 \text{ for a.e. } s \in \Delta(A) \tag{5.151}$$

and

$$\operatorname{Re} \sum_{i,j=1}^N \widehat{a}_{ij}(s)\xi_j\overline{\xi}_i \geq \alpha|\xi|^2 (\xi \in \mathbb{C}^N) \text{ for a.e. } s \in \Delta(A), \tag{5.152}$$

with $\widehat{a}_{ji} = \overline{\widehat{a}}_{ij}$.

Having made this point, with the basic notation of Sect. 5.1.3, let

$$b_\Omega(\mathbf{u}, \mathbf{v}) = \sum_{i,j=1}^N \int\int_{\Omega\times\widehat{G}} \widehat{a}_{ij}(s)\mathbb{D}_j\mathbf{u}(x, s)\overline{\mathbb{D}_i\mathbf{v}}(x, s)dxd\beta(s)$$

$$+ \int\int_{\Omega\times\widehat{G}} \widehat{a}_0(s)u_0(x)\overline{v}_0(x)dxd\beta(s)$$

for $\mathbf{u} = (u_0, u_1)$ and $\mathbf{v} = (v_0, v_1)$ in $\mathbb{F}_0^1 = H_0^1(\Omega) \times L^2(\Omega; H_\#^1(\Delta(A)))$. This defines a sesquilinear form $b_\Omega(\cdot, \cdot)$ on $\mathbb{F}_0^1 \times \mathbb{F}_0^1$ which is continuous, Hermitian and noncoercive (the lack of coercivity can be established without difficulty; see Nguetseng [96]).

Let us nevertheless point out a partial (but enough for our purpose)uniqueness result for the noncoercive variational problem

$$\begin{cases} \mathbf{u} = (u_0, u_1) \in \mathbb{F}_0^1 : \\ b_\Omega(\mathbf{u}, \mathbf{v}) = \mathbf{l}(\mathbf{v}) \text{ for every } \mathbf{v} = (v_0, v_1) \in \mathbb{F}_0^1, \end{cases} \tag{5.153}$$

where $\mathbf{l}$ is the continuous antilinear form on $\mathbb{F}_0^1$ given by

$$\mathbf{l}(\mathbf{v}) = M(\chi_G) \int_\Omega f(x)\overline{v}_0(x)dx \text{ for } \mathbf{v} = (v_0, v_1) \in \mathbb{F}_0^1.$$

Lemma 5.8 *Let (5.147)–(5.150) be satisfied. Then, there is at most one function* $u_0 \in H_0^1(\Omega)$ *with the property*

$$(\mathcal{P}) \qquad \begin{cases} \textit{There is some } u_1 \in L^2(\Omega; H_\#^1(\Delta(A))) \textit{ such that} \\ \textit{the couple } \mathbf{u} = (u_0, u_1) \textit{ is a solution of (5.153)} \end{cases}$$

Proof Suppose there are two functions $u_0, z_0 \in H_0^1(\Omega)$ satisfying $(\mathcal{P})$. Then, there are some $u_1, z_1 \in L^2(\Omega; H_\#^1(\Delta(A)))$ such that $\mathbf{u} = (u_0, u_1)$ and $\mathbf{z} = (z_0, z_1)$ are solutions of (5.153). By (5.151)–(5.152) it follows $\mathbf{N}(\mathbf{z} - \mathbf{u}) = 0$, where

$$\mathbf{N}(\mathbf{v}) = \left[\int_\Omega |v_0(x)|^2 dx + \sum_{j=1}^N \iint_{\Omega \times \widehat{G}} |\mathbb{D}_j \mathbf{v}(x, s)|^2 dx d\beta(s) \right]^{\frac{1}{2}}$$

for $\mathbf{v} = (v_0, v_1) \in \mathbb{F}_0^1$. Hence $z_0 = u_0$, which completes the proof. ∎

Remark 5.16 Let us name $\mathcal{W}$ the vector space $\mathbb{F}_0^1$ equipped with the seminorm $\mathbf{N}$ defined above. This is clearly a non-separated, noncomplete preHilbert space. Let $\mathbb{W}$ be the separated completion of $\mathcal{W}$, and $\mathbf{i}$ be the canonical mapping of $\mathcal{W}$ into $\mathbb{W}$. It can be shown (see Nguetseng [96]) that there exists a continuous, Hermitian coercive sesquilinear form B on $\mathbb{W} \times \mathbb{W}$ such that $B(\mathbf{i}(\mathbf{u}), \mathbf{i}(\mathbf{v})) = b_\Omega(\mathbf{u}, \mathbf{v})$ for $\mathbf{u}, \mathbf{v} \in \mathcal{W}$, and further if L denotes the antilinear form on $\mathbb{W}$ such that $L(\mathbf{i}(\mathbf{v})) = \mathbf{l}(\mathbf{v})$ for $\mathbf{v} \in \mathcal{W}$, then a vector function $\mathbf{u} = (u_0, u_1)$ is a solution of (5.153) if and only if $\mathbf{i}(\mathbf{u})$ is the (unique) solution of the coercive variational problem

$$\begin{cases} \mathbf{w} \in \mathbb{W} : \\ B(\mathbf{w}, \mathbf{v}) = L(\mathbf{v}) \text{ for all } \mathbf{v} \in \mathbb{W}. \end{cases}$$

From which one deduces readily that (5.153) admits a (non unique) solution.

5.4.3.2 The Abstract Homogenization Theorem

We are now in a position to prove the following

Theorem 5.8 *Let the H-algebra A be as stated above. For fixed* $\varepsilon > 0$*, let* $u_\varepsilon \in H^1(\Omega^\varepsilon)$ *be uniquely given by (5.141), and* P_ε *be the extension operator introduced in Proposition 5.14. Then, under the abstract structure hypothesis (5.147)–(5.150), as* $\varepsilon \to 0$ *one has*

$$P_\varepsilon u_\varepsilon \to u_0 \text{ in } H_0^1(\Omega)\text{-weak}, \tag{5.154}$$

where u_0 *is the unique function in* $H_0^1(\Omega)$ *with property* $(\mathcal{P})$*.*

Proof To begin, u_ε (for arbitrary fixed $\varepsilon > 0$) lies in V_ε and

$$\int_{\Omega^\varepsilon} a_0^\varepsilon u_\varepsilon \overline{v}\, dx + \sum_{i;j=1}^{N} \int_{\Omega^\varepsilon} a_{ij}^\varepsilon \frac{\partial u_\varepsilon}{\partial x_j} \overline{\frac{\partial v}{\partial x_i}} dx = \int_{\Omega^\varepsilon} f \overline{v}\, dx \tag{5.155}$$

for all $v \in V_\varepsilon$. It follows by mere routine that

$$\sup_{\varepsilon > 0} \|u_\varepsilon\|_{H^1(\Omega^\varepsilon)} < \infty.$$

By Proposition 5.14 one deduces that the sequence $(P_\varepsilon u_\varepsilon)_{\varepsilon>0}$ is norm bounded in $H_0^1(\Omega)$. This being so, let us fix freely a fundamental sequence E. Combination of the $H^1(\Omega)$-properness of A, Proposition 4.5 and the Rellich-Kondrachov theorem, yields a subsequence E' from E and two functions $u_0 \in H_0^1(\Omega)$ and $u_1 \in L^2(\Omega; H_\#^1(\Delta(A)))$ such that as $E' \ni \varepsilon \to 0$, one has (5.154), $P_\varepsilon u_\varepsilon \to u_0$ in $L^2(\Omega)$, and

$$\frac{\partial P_\varepsilon u_\varepsilon}{\partial x_j} \to \mathbb{D}_j \mathbf{u} = \frac{\partial u_0}{\partial x_j} + \partial_j u_1 \text{ in } L^2(\Omega)\text{-weak } \Sigma \ (1 \le j \le N) \tag{5.156}$$

with $\mathbf{u} = (u_0, u_1)$. Thus, it is clear that, thanks to Lemma 5.8, the theorem will have been proved if we can check that $\mathbf{u} = (u_0, u_1)$, which lies in $\mathbb{F}_0^1$, satisfies the variational equation in (5.153). So, let

$$\boldsymbol{\Phi} = (\psi_0, J_\Omega(\widehat{\psi}_1)) \in \mathcal{F}_0^\infty \text{ (see (5.47) and (5.60), as well)}$$

and $\Phi_\varepsilon = \psi_0 + \varepsilon\psi_1^\varepsilon$, $\psi_1^\varepsilon(x) = \psi_1\,(x, \frac{x}{\varepsilon})$ $(\varepsilon > 0, x \in \Omega)$. It is worth perhaps recalling that $\mathcal{F}_0^\infty = \mathcal{D}(\Omega) \times [\mathcal{D}(\Omega) \otimes J(\mathcal{D}(\Delta(A))/\mathbb{C})]$, and further $\mathcal{F}_0^\infty$ is dense in $\mathbb{F}_0^1$. Also, one has $\psi_0 \in \mathcal{D}(\Omega)$ and $\psi_1 \in \mathcal{D}(\Omega) \otimes (A^\infty/\mathbb{C})$, so that $\Phi_\varepsilon \in \mathcal{D}(\Omega)$

and further all the functions $\Phi_\varepsilon(\varepsilon > 0)$ have their supports in a fixed compact set $K \subset \Omega$. Therefore, appeal to Lemma 5.5 yields some real $\varepsilon_0 > 0$ such that

$$\Phi_\varepsilon = 0 \text{ in } \Omega^\varepsilon \setminus Q^\varepsilon \quad (0 < \varepsilon \leq \varepsilon_0).$$

Now, choose $v = \Phi_\varepsilon|_{\Omega^\varepsilon}$ in (5.155) (with $0 < \varepsilon \leq \varepsilon_0$), and then use the decomposition $\Omega^\varepsilon = Q^\varepsilon \cup (\Omega^\varepsilon \setminus Q^\varepsilon)$ together with $Q^\varepsilon = \Omega \cap \varepsilon G$ to arrive at

$$\int_\Omega a_0^\varepsilon \chi_G^\varepsilon \overline{\Phi_\varepsilon} P_\varepsilon u_\varepsilon \, dx + \sum_{i,j=1}^N \int_\Omega a_{ij}^\varepsilon \frac{\partial P_\varepsilon u_\varepsilon}{\partial x_j} \overline{\frac{\partial \Phi_\varepsilon}{\partial x_i}} \chi_G^\varepsilon dx = \int_\Omega f \overline{\Phi_\varepsilon} \chi_G^\varepsilon \, dx \tag{5.157}$$

for $0 < \varepsilon \leq \varepsilon_0$. We shall now try to pass to the limit above when $E' \ni \varepsilon \to 0$. For that purpose it is useful to note that for $w \in \mathfrak{X}_A^r$ (real $r \geq 1$), we have $w^\varepsilon \to \widehat{w}$ in $L^r(\Omega)$-weak Σ as $\varepsilon \to 0$. This is easily deduced from Example 4.1 (observe that, Ω being bounded, A may be identified with constant functions in $L^r(\Omega; A)$) by a routine approach using the density of A in $\mathfrak{X}_A^r$, Proposition 2.11, and the continuity of the embedding $\Xi^r \hookrightarrow L_{loc}^r$.

With this in mind, if we put $\frac{1}{r} = \frac{1}{p} + \frac{1}{q}$, then by (5.147), (5.149), (5.150) and use of Corollary 2.2 together with $\mathfrak{X}_A^r \subset \mathfrak{X}_A^2$ (note that $r > 2$) we see readily that $a_0^\varepsilon \chi_G^\varepsilon \to \widehat{a}_0 \chi_{\widehat{G}}$ in $L^2(\Omega)$-weak Σ as $\varepsilon \to 0$, therefore

$$a_0^\varepsilon \chi_G^\varepsilon \to \int_{\Delta(A)} \widehat{a}_0 \chi_{\widehat{G}} \, d\beta \text{ in } L^2(\Omega)\text{-weak as } \varepsilon \to 0.$$

On the other hand, it is clear that $\overline{\Phi_\varepsilon} P_\varepsilon u_\varepsilon \to \overline{\psi}_0 u_0$ in $L^2(\Omega)$ as $E' \ni \varepsilon \to 0$. We deduce that, as $E' \ni \varepsilon \to 0$,

$$\int_\Omega a_0^\varepsilon \chi_G^\varepsilon \overline{\Phi_\varepsilon} P_\varepsilon u_\varepsilon \, dx \to \int\int_{\Omega \times \widehat{G}} \widehat{a}_0(s) u_0(x) \overline{\psi}_0(x) dx d\beta(s).$$

Now, fix $1 \leq i \leq N$. The sequence $(\frac{\partial \Phi_\varepsilon}{\partial x_i})_{0<\varepsilon\leq 1}$ is bounded in $L^\infty(\Omega)$ and further, as $\varepsilon \to 0$,

$$\frac{\partial \Phi_\varepsilon}{\partial x_i} \to \mathbb{D}_i \mathbf{\Phi} = \frac{\partial \psi_0}{\partial x_i} + \partial_i \widehat{\psi}_1 \text{ in } L^2(\Omega)\text{-strong } \Sigma.$$

Hence, by (5.156) and use of Corollary 4.1 it follows that

$$\frac{\partial P_\varepsilon u_\varepsilon}{\partial x_j} \overline{\frac{\partial \Phi_\varepsilon}{\partial x_i}} \to \mathbb{D}_j \mathbf{u} \overline{\mathbb{D}_i \mathbf{\Phi}} \text{ in } L^2(\Omega)\text{-weak } \Sigma \quad (1 \leq i, j \leq N).$$

We deduce by Proposition 4.4 that when $E' \ni \varepsilon \to 0$, one has

$$\sum_{i,j=1}^{N} \int_{\Omega} a_{ij}^{\varepsilon} \frac{\partial P_{\varepsilon} u_{\varepsilon}}{\partial x_j} \overline{\frac{\partial \Phi_{\varepsilon}}{\partial x_i}} \chi_G^{\varepsilon} \, dx \to \sum_{i,j=1}^{N} \int\int_{\Omega \times \widehat{G}} \widehat{a}_{ij}(s) \mathbb{D}_j \mathbf{u}(x,s) \overline{\mathbb{D}_i \mathbf{\Phi}}(x,s) dx d\beta(s),$$

as can be readily seen by noticing that $\chi_G a_{ij} \in \mathcal{C}(\overline{\Omega}; \mathfrak{X}_A^{2,\infty})$. Finally by an argument followed before one has, as $\varepsilon \to 0$,

$$\int_{\Omega} f \overline{\Phi}_{\varepsilon} \chi_G^{\varepsilon} \, dx \to M(\chi_G) \int_{\Omega} f(x) \overline{\psi}_0(x) \, dx.$$

Consequently, by passing to the limit in (5.157) when $E' \ni \varepsilon \to 0$, one is easily led to (5.153) by the density of $\mathcal{F}_0^{\infty}$ in $\mathbb{F}_0^1$. ■

5.4.4 Some Concrete Homogenization Problems for (5.141)

This section is intended to work out a few concrete homogenization problems for (5.141). More precisely, we consider here the boundary value problem (5.141) with various concrete data, namely

1. a specific distribution of holes,
2. a concrete structure hypothesis on the coefficients $\{a_0, a_{ij}\}$,

and we show that in each specific setting the concrete homogenization problem under consideration reduces to the abstract problem of Sect. 5.4.3.

5.4.4.1 Preliminaries

Let the basic notation and hypotheses be as before. By the *distribution function* of *the holes* is understood the characteristic function of S in $\mathbb{Z}^N$ denoted below by θ, i.e., $\theta = \chi_S$. Thus,

$$\chi_{\Theta} = \sum_{k \in \mathbb{Z}^N} \theta(k) \chi_{k+T} \text{ (a locally finite sum).} \tag{5.158}$$

One generally has $\theta \in ES(\mathbb{Z}^N)$ (see Sect. 2.6), as will presently become apparent. In this connection let us prove a result of fundamental interest.

Proposition 5.15 *Let $\mathfrak{A}$ denote an H-algebra on $\mathbb{R}^N$. We assume that $\theta \in ES(\mathbb{Z}^N)$ and further $\sum_{k\in\mathbb{Z}^N} \theta(k)\tau_k\varphi \in \mathfrak{A}$ for all $\varphi \in \mathcal{K}(Y)$. Then, $\chi_\Theta \in \mathfrak{X}^p_{\mathfrak{A}}$ $(1 \leq p < \infty)$ and further*

$$M(\chi_\Theta) = \mathfrak{M}(\theta)\lambda(T), \tag{5.159}$$

where λ is Lebesgue measure on $\mathbb{R}^N$ and $\mathfrak{M}(\theta)$ the essential mean of θ.

Before we can prove this proposition let us point out a preliminary lemma which is nothing else than another version of an existing result. For any $B \subset \mathbb{R}^N_x$, $i^\varepsilon(B)$ stands for the set of all $k \in \mathbb{Z}^N$ such that $\varepsilon k \in B$.

Lemma 5.9 *Let $u \in \mathfrak{X}^{p,\infty}_{\mathfrak{A}}(1 \leq p < \infty)$. Let $B \subset \mathbb{R}^N_x$ with compact closure and λ-negligible boundary. Then, as $\varepsilon \to 0$,*

$$\varepsilon^N \sum_{k\in i^\varepsilon(B)} \int_{k+Y} u(y)dy \to M(u)\lambda(B). \tag{5.160}$$

Proof More generally, if $u \in \mathfrak{X}^p_{\mathfrak{A}}$, for any bounded open set $\mathcal{O} \subset \mathbb{R}^N_x$ one has $u^\varepsilon \to M(u)$ in $L^p(\mathcal{O})$-weak when $\varepsilon \to 0$, as is straightforward by the density of $\mathfrak{A}$ in $\mathfrak{X}^p_{\mathfrak{A}}$ and use of Proposition 2.10. By choosing in particular $\mathcal{O}$ such that $B \subset \mathcal{O}$, one sees at once that $\int_B u^\varepsilon dx \to M(u)\lambda(B)$ as $\varepsilon \to 0$. With this in mind, given $u \in \mathfrak{X}^{p,\infty}_{\mathfrak{A}}$, we see readily that the rest of the proof is a copy of that of Proposition 4.10 (with $\|u\|_\infty$ changed into $\|u\|_{L^\infty}$) and is not worth repeating. ■

We are now in a position to prove Proposition 5.15.

Proof of Proposition 5.15 We begin by verifying that $\chi_\Theta \in \mathfrak{X}^p_{\mathfrak{A}}$. Let $\eta > 0$. χ_T being regarded as a function in $L^p(Y)$, there certainly exists some $\varphi \in \mathcal{K}(Y)$ such that $\|\chi_T - \varphi\|_{L^p(Y)} \leq \eta$. It follows readily that $\|\chi_\Theta - \psi\|_{p,\infty} \leq \eta$, where

$$\psi = \sum_{k\in\mathbb{Z}^N} \theta(k)\tau_k\varphi.$$

But by hypothesis $\psi \in \mathfrak{A}$. Hence $\chi_\Theta \in \mathfrak{X}^p_{\mathfrak{A}}$ as claimed, since the space $(L^p, l^\infty)(\mathbb{R}^N)$ is continuously embedded in Ξ^p (see Proposition 2.19).

We next prove (5.159). By Lemma 5.9 one has

$$M(\chi_\Theta)\lambda(T) = \lim_{\varepsilon\to 0} \varepsilon^N \sum_{k\in i^\varepsilon(T)} \int_{k+Y} \chi_\Theta(y)dy.$$

By (5.158) we deduce

$$M(\chi_\Theta) = \lim_{\varepsilon\to 0} \varepsilon^N \sum_{k\in i^\varepsilon(T)} \theta(k). \tag{5.161}$$

Therefore, we only need to compute the right-side. This will proceed in two steps. Let us put for convenience

$$M_\varepsilon(a) = \varepsilon^N \sum_{k \in i^\varepsilon(T)} a(k) \ \ (a \in ES(\mathbb{Z}^N)).$$

Step 1. Fix freely $l \in \mathbb{Z}^N$. Let $\eta > 0$. We want to show that one can find some $\varepsilon_0 > 0$ such that

$$|M_\varepsilon(\tau_l\theta) - M_\varepsilon(\theta)| \leq \eta \ \ (0 < \varepsilon \leq \varepsilon_0). \tag{5.162}$$

But

$$\theta(k) = \frac{1}{\lambda(T)} \int_{k+Y} \chi_\Theta(y)\, dy \ \ (k \in \mathbb{Z}^N).$$

Whence

$$M_\varepsilon(\tau_l\theta) - M_\varepsilon(\theta) = \int_{I^\varepsilon(T)} u^\varepsilon dx$$

where

$$u(y) = \frac{1}{\lambda(T)}[\chi_\Theta(y-l) - \chi_\Theta(y)] \ \ (y \in \mathbb{R}^N)$$

and

$$I^\varepsilon(T) = \bigcup_{k \in i^\varepsilon(T)} \varepsilon(k + \overline{Y}) \ \ \text{(see Sect. 4.2.5)}.$$

One deduces by mere routine

$$|M_\varepsilon(\tau_l\theta) - M_\varepsilon(\theta)| \leq \|u\|_{L^\infty}[\lambda(J^\varepsilon(T)\setminus I^\varepsilon(T)) + \lambda(J^\varepsilon(T)\setminus T)] + \left|\int u^\varepsilon \chi_T \, dx\right|.$$

But, according to Lemma 4.4 the sets $J^\varepsilon(T)\setminus I^\varepsilon(T)$ and $J^\varepsilon(T)\setminus T$ are contained in $J^\varepsilon(\partial T)$ and further $\lambda(J^\varepsilon(\partial T)) \to \lambda(\partial T) = 0$ as $\varepsilon \to 0$ (Lemma 4.5). On the other hand, by Lemma 2.12 one has $\int u^\varepsilon \chi_T \, dx \to 0$ as $\varepsilon \to 0$. Hence (5.162) follows for some suitable $\varepsilon_0 > 0$.

Step 2. Let $\eta > 0$. Appeal to Theorem 2.1 yields a finite family of points $l_i \in \mathbb{Z}^N$ $(1 \leq i \leq m)$ such that

$$\left|\frac{1}{m}\sum_{i=1}^m \theta(k - l_i) - \mathfrak{M}(\theta)\right| \leq \eta \ \ (k \in \mathbb{Z}^N).$$

It follows

$$\left| \frac{1}{m} \sum_{i=1}^{m} M_\varepsilon(\tau_{l_i}\theta) - \mathfrak{M}(\theta)\varepsilon^N |i^\varepsilon(T)| \right| \leq \eta\varepsilon^N |i^\varepsilon(T)| \ \ (\varepsilon > 0),$$

where $|i^\varepsilon(T)|$ denotes the cardinality of $i^\varepsilon(T)$. On the other hand, according to Step 1, there is some $\varepsilon_0 > 0$ such that (5.162) holds for $l = l_i$ $(i = 1, \cdots, m)$. We deduce that

$$\left| \frac{1}{m} \sum_{i=1}^{m} M_\varepsilon(\tau_{l_i}\theta) - M_\varepsilon(\theta) \right| \leq \frac{1}{m} \sum_{i=1}^{m} |M_\varepsilon(\tau_{l_i}\theta) - M_\varepsilon(\theta)| \leq \eta$$

for $0 < \varepsilon \leq \varepsilon_0$. Hence

$$|M_\varepsilon(\theta) - \mathfrak{M}(\theta)\varepsilon^N |i^\varepsilon(T)|| \leq \eta(1 + \varepsilon^N |i^\varepsilon(T)|) \ \ (0 < \varepsilon \leq \varepsilon_0).$$

But, by (5.160) with $u = 1$ and $B = T$ we have $\varepsilon^N |i^\varepsilon(T)| = \lambda(I^\varepsilon(T)) \to \lambda(T)$ as $\varepsilon \to 0$. Hence (5.159) follows by letting $\varepsilon \to 0$ and recalling (5.161). This completes the proof.

■

Corollary 5.19 *Under the hypotheses of Proposition 5.15, assumption (5.148) is satisfied. If moreover $\mathfrak{A} \subset A$, then (5.147) is also satisfied.*

Proof The verification of (5.147) being evident, we shall only check (5.148). But, in view of the equality $\chi_G = 1 - \chi_\Theta$, this follows by (5.159) and use of the obvious inequalities $\mathfrak{M}(\theta) \leq 1$ and $\lambda(T) < \lambda(Y) = 1$. ■

5.4.4.2 The Holes Are Equidistributed

Throughout this section, the configuration of the holes is characterized by the assumption that $\theta(k) = 1$ for all $k \in \mathbb{Z}^N$. This means concretely that $S = \mathbb{Z}^N$, what we appropriately express by saying that the *holes are equidistributed*. In this setting we consider the homogenization problem for (5.141) under various concrete structure hypotheses on the coefficients $\{a_0, a_{ij}\}$.

Let us point out beforehand the intrinsic behaviour of the function χ_G in this case.

Lemma 5.10 *Under the hypothesis that the holes are equidistributed, the following are true:*

(i) $\chi_G \in \mathfrak{X}^p_{C_{per}(Y)}$ $(1 \leq p < \infty)$.
(ii) $M(\chi_G) > 0$.

Proof This is straightforward by Proposition 5.15 and Corollary 5.19 with $\mathfrak{A} = \mathcal{C}_{per}(Y)$. Indeed, the constant function $\theta = 1$ lies in $ES(\mathbb{Z}^N)$ and further $\sum_{k\in\mathbb{Z}^N} \tau_k\varphi \in \mathcal{C}_{per}(Y)$ for all φ in $\mathcal{K}(\mathbb{R}^N)$ and in particular for $\varphi \in \mathcal{K}(Y)$. ■

Let us now give a few concrete examples of problems leading to the application of the theory developed in Sect. 5.4.3.

Example 5.8 (Periodic Setting) In the present context we consider the homogenization problem for (5.141) under the structure hypothesis

$$a_0, a_{ij} \in L^q_{per}(Y)\ (1 \leq i, j \leq N) \text{ for some given } 2 < q < \infty. \tag{5.163}$$

According to Lemma 5.10 this typical example reduces readily to the abstract setting of Sect. 5.4.3. Indeed, by choosing some $2 < p < \infty$ such that $\frac{1}{p} + \frac{1}{q} \leq \frac{1}{2}$, we see by Lemma 5.10 and Corollary 2.14 (ii) via (5.163) that the abstract structure hypothesis (5.147)–(5.150) is satisfied with $A = \mathcal{C}_{per}(Y)$, which H-algebra is $H^1(\Omega)$-proper and admits A^∞ as a dense subalgebra. Finally, it is of interest to observe that the developments of Sect. 5.4.3 carry over mutatis mutandis to the periodic setting (see Remark 4.2 and Proposition 4.6).

Example 5.9 (Almost Periodic Setting) In the same context, the question on hand is the homogenization problem for (5.141) under the structure hypothesis

$$a_0, a_{ij} \in L^q_{AP}(\mathbb{R}^N_y)\ (1 \leq i, j \leq N) \text{ for some given } 2 < q < \infty. \tag{5.164}$$

Let $\mathfrak{S}$ denote the union of the family $\{Sp(a_0), Sp(a_{ij})\}(1 \leq i, j \leq N)$. As is classical (see Sects. 2.3 and 2.4), the set $\mathfrak{S}$ is at most countable. Let R_0 be the subgroup of $\mathbb{R}^N_y$ spanned by $\mathfrak{S}$, i.e., the smallest subgroup of $\mathbb{R}^N_y$, denoted by $Gr(\mathfrak{S})$, containing $\mathfrak{S}$ (see the proof of Lemma 5.3). $R_0 = Gr(\mathfrak{S})$ is a countable subgroup of $\mathbb{R}^N$ and we have

$$a_0, a_{ij} \in L^q_{R_0}(\mathbb{R}^N), 1 \leq i, j \leq N, \tag{5.165}$$

where

$$L^q_{R_0}(\mathbb{R}^N_y) = \{v | v \in L^q_{AP}(\mathbb{R}^N), Sp(v) \subset R_0\}.$$

Now, recalling that $L^p_{per}(Y) = L^p_{\mathbb{Z}^N}(\mathbb{R}^N)$, we next set $\mathcal{R} = Gr(R_0 \cup \mathbb{Z}^N)$ i.e., $\mathcal{R}$ denotes the (countable) subgroup of $\mathbb{R}^N_y$ spanned by $R_0 \cup \mathbb{Z}^N$. We have

$$a_0, a_{ij} \in L^q_{\mathcal{R}}(\mathbb{R}^N)\ \ (1 \leq i, j \leq N)$$

and

$$\chi_G \in L^p_{\mathcal{R}}(\mathbb{R}^N)\ \ (1 \leq p < \infty)\ \ \text{(see Lemma 5.10).}$$

Finally, let $A = AP_{\mathcal{R}}(\mathbb{R}^N)$. This is an $H^1(\Omega)$-proper homogenization algebra on $\mathbb{R}^N$ admitting A^∞ as a dense subalgebra. Hence, since $L^p_{\mathcal{R}}(\mathbb{R}^N) \subset \mathfrak{X}^p_A$ $(1 \leq p < \infty)$ (Corollary 2.11), we see that by choosing some $p > 2$ such that (5.150) holds, and taking Lemma 5.10 into account, the problem under consideration reduces to the abstract setting with $A = AP_{\mathcal{R}}(\mathbb{R}^N)$.

Example 5.10 The problem to be worked out here states as in Example 5.8 with however this difference that (5.163) is replaced by the structure hypothesis

$$\begin{aligned} &a_0, a_{ij} \in L^q(\mathbb{R}^N_y) + L^q_{per}(Y) \ (1 \leq i, j \leq N) \\ &\text{for some given } 2 < q < \infty. \end{aligned} \tag{5.166}$$

Fixing a real $p > 2$ such that (5.150) holds, and proceeding as we did in solving Problem V of Sect. 5.1.4, we are readily led to the abstract setting based on (5.147)–(5.150) with $A = \mathcal{B}_{\infty,\mathbb{Z}^N}(\mathbb{R}^N_y)$ (see Example 4.9).

Example 5.11 The problem we consider here states as in Example 5.10 save that (5.166) is substituted by the structure hypothesis

$$\begin{cases} a_0 \in \mathcal{B}_\infty(\mathbb{R}^N_y), \\ a_{ij} \in L^q_{per}(Y) \ \ (1 \leq i, j \leq N) \text{ for some given } 2 < q < \infty. \end{cases}$$

To achieve this, we reconsider the $H^1(\Omega)$-proper H-algebra $A = \mathcal{B}_{\infty,\mathbb{Z}^N}(\mathbb{R}^N_y)$ utilized above. Recalling that A is precisely the closure, in $\mathcal{B}(\mathbb{R}^N)$, of the algebra $\mathcal{B}_\infty(\mathbb{R}^N) + \mathcal{C}_{per}(Y)$, we see immediately that (5.149) follows by the relations $a_0 \in A \subset \mathfrak{X}^q_A$ and $a_{ij} \in L^q_{per}(Y) = \mathfrak{X}^q_{\mathcal{C}_{per}(Y)} \subset \mathfrak{X}^q_A$ whereas (5.147) and (5.148) follow by Lemma 5.10 and use of the latter argument for a_{ij}. It only remains to choose $p > 2$ so as to have (5.150) fulfilled.

Example 5.12 We assume here that the coefficients $\{a_0, a_{ij}\}$ are constant in each cell $k + Y$, that is, they are of the form

$$\begin{cases} a_0 = \sum_{k \in \mathbb{Z}^N} r_0(k)\chi_{k+Y} \text{ (a locally finite sum)}, \\ a_{ij} = \sum_{k \in \mathbb{Z}^N} r_{ij}(k)\chi_{k+Y} \ \ (1 \leq i, j \leq N), \end{cases}$$

where the complex functions $k \mapsto r_0(k)$ and $k \mapsto r_{ij}(k)$ (defined on $\mathbb{Z}^N$) are given. Our goal is then to investigate the behaviour, as $\varepsilon \to 0$, of u_ε (the solution of (5.141)) under the structure hypothesis

$$\begin{cases} \lim_{|k|\to\infty} \int_{k+Y} a_0(y)dy = \xi_0 \in \mathbb{C}, \\ \lim_{|k|\to\infty} \int_{k+Y} a_{ij}(y)dy = \xi_{ij} \in \mathbb{C} \ \ (1 \leq i, j \leq N), \end{cases}$$

which structure hypothesis amounts to

$$r_0, r_{ij} \in \mathcal{B}_\infty(\mathbb{Z}^N) \ \ (1 \leq i, j \leq N).$$

We are thus led to introduce the discrete H-algebra (on $\mathbb{R}^N$), A_σ, associated to $\mathfrak{S} = \mathcal{B}_\infty(\mathbb{Z}^N)$ (see Sects. 2.6 and 4.2.5) and thereafter the closure, A, of $A_\sigma + \mathcal{C}_{per}(Y)$ in $\mathcal{B}(\mathbb{R}^N)$. A is an $H^1(\Omega)$-proper homogenization algebra on $\mathbb{R}^N$ admitting A^∞ as a dense subalgebra (see Proposition 4.11). By the same argument as followed in solving Problem VI of Sect. 5.1.4 one quickly arrives at

$$a_0, a_{ij} \in \mathfrak{X}_A^q \ \ (1 \leq i, j \leq N) \text{ for } 1 \leq q < \infty,$$

whence (5.147)–(5.150) follows by Lemma 5.10, the reals p and q being suitably chosen, of course.

5.4.4.3 The Holes Are Distributed in an Almost Periodic Fashion

We have need introducing beforehand the notion of an almost periodic essential algebra on $\mathbb{Z}^N$. Let us begin by recalling that the dual of the discrete group $\mathbb{Z}^N$ is the N-torus $\mathbb{T}^N = (\mathbb{R}/\mathbb{Z})^N$. However, for the sake of convenience $\mathbb{T}^N$ is identified with $\mathbb{R}^N/\mathbb{Z}^N$ (this is classical, see, e.g., Bourbaki [24, III,p.18,Corollaire]), so that to any $\rho \in \mathbb{T}^N$, there corresponds bijectively a continuous character ω_ρ of $\mathbb{Z}^N$ given by $\omega_\rho(k) = e^{2\pi i \nu \cdot k}$ $(k \in \mathbb{Z}^N)$, where ν is any point in $\mathbb{R}^N$ such that $\rho = \nu + \mathbb{Z}^N$, and $\nu \cdot k$ is the Euclidian scalar product in $\mathbb{R}^N$.

It is also worth recalling that the algebra of almost periodic (continuous) complex functions on $\mathbb{Z}^N$ is denoted by $AP(\mathbb{Z}^N)$ (see Example 1.4). Provided with the algebra structure induced by that of $\mathcal{B}(\mathbb{Z}^N) = l^\infty(\mathbb{Z}^N)$ and with the supremum norm, $AP(\mathbb{Z}^N)$ is a commutative C^*-algebra with identity. Furthermore $AP(\mathbb{Z}^N) \subset ES(\mathbb{Z}^N)$ with $m(a) = \mathfrak{M}(a)$ for $a \in AP(\mathbb{Z}^N)$, where m denotes the mean value for almost periodic functions on $\mathbb{Z}^N$ (see Examples 1.9 and 2.6).

This being so, let R be a countable subgroup of $\mathbb{R}^N$ such that $\mathbb{Z}^N \subset R$. Let $\dot{R} = \pi(R)$, where π is the canonical homomorphism of $\mathbb{R}^N$ onto $\mathbb{T}^N = \mathbb{R}^N/\mathbb{Z}^N$. It is clear that $\dot{R}$ is a countable subgroup of $\mathbb{T}^N$. In the sequel we put $\pi(\nu) = \dot{\nu}$ for $\nu \in R$. Let

$$AP_{\dot{R}}(\mathbb{Z}^N) = \{a | a \in AP(\mathbb{Z}^N), Sp(a) \subset \dot{R}\},$$

where

$$Sp(a) = \{\rho | \rho \in \mathbb{T}^N, \mathfrak{M}(\overline{\omega}_\rho a) \neq 0\} \ \ (\text{ the spectrum of } a).$$

$AP_{\dot{R}}(\mathbb{Z}^N)$ has all the properties of an essential algebra on $\mathbb{Z}^N$ (see Definition 2.8). We will refer to $AP_{\dot{R}}(\mathbb{Z}^N)$ as an *almost periodic essential algebra* on $\mathbb{Z}^N$.

Now, let $\mathfrak{A}_R$ be the discrete H-algebra on $\mathbb{R}^N$ associated to $AP_{\dot{R}}(\mathbb{Z}^N)$ (see Proposition 2.26, Definition 2.9 and Sect. 4.2.5). Our goal is to prove the following proposition.

Proposition 5.16 *We have* $\mathfrak{A}_R \subset AP_R(\mathbb{R}^N)$.

Proof Let $\mathcal{F}$ be the set of all complex functions f on $\mathbb{R}^N$ of the form

$$f = \sum_{k \in \mathbb{Z}^N} a(k)\tau_k\varphi \text{ with } a \in AP_{\dot{R}}(\mathbb{Z}^N) \text{ and } \varphi \in \mathcal{K}(Y), \tag{5.167}$$

and $\langle\mathcal{F}\rangle$ the linear span of $\mathcal{F}$. Recalling that $\mathfrak{A}_R$ is precisely the closure of $\mathbb{C} + \langle\mathcal{F}\rangle$ in $\mathcal{B}(\mathbb{R}^N_y)$, we see at once that the whole problem reduces to proving that $\mathcal{F} \subset AP_R(\mathbb{R}^N)$.

We begin by showing that $\mathcal{F} \subset AP(\mathbb{R}^N)$. This will proceed in two steps. Let $\varphi \in \mathcal{K}(Y)$, φ assumed to be nontrivial, i.e., $\|\varphi\|_\infty \neq 0$.

(1) Given $\nu \in \mathbb{R}^N$, the aim here is to check that the function

$$g_\nu = \sum_{k \in \mathbb{Z}^N} e^{2\pi i\nu\cdot k}\tau_k\varphi$$

lies in $AP(\mathbb{R}^N)$. To this end, let $u = \sum_{k\in\mathbb{Z}^N} \tau_k(\overline{\gamma}_\nu\varphi)$, where $\gamma_\nu(y) = e^{2\pi i\nu\cdot y}$ $(y \in \mathbb{R}^N)$. Clearly $u \in \mathcal{C}_{per}(Y) \subset AP(\mathbb{R}^N)$. Hence $g_\nu \in AP(\mathbb{R}^N)$ since $g_\nu = \gamma_\nu u$.

(2) Now, let f be given by (5.167) but, more generally, with $a \in AP(\mathbb{Z}^N)$. Let $\eta > 0$. Since the trigonometric polynomials on $\mathbb{Z}^N$ form a dense subalgebra of $AP(\mathbb{Z}^N)$, one can find some $z \in AP(\mathbb{Z}^N)$, $z(k) = \sum_{\nu\in F} c_\nu e^{2\pi i\nu\cdot k}$ $(k \in \mathbb{Z}^N)$ with $c_\nu \in \mathbb{C}$ and F a finite subset of $\mathbb{R}^N$ (that depends on z), such that $\|a - z\|_\infty \le \frac{\eta}{\|\varphi\|_\infty}$. Next, we set $g = \sum_{k\in\mathbb{Z}^N} z(k)\tau_k\varphi$. We have $g = \sum_{\nu\in F} c_\nu g_\nu$ (g_ν given above). Hence $g \in AP(\mathbb{R}^N)$, according to step (1). But $\|f - g\|_\infty \le \eta$. Therefore $f \in AP(\mathbb{R}^N)$, in view of the arbitrariness of η.

Finally, let f be given by (5.167) as such. As shown above, f belongs to $AP(\mathbb{R}^N)$. Thus, it only remains to show that the spectrum of f is contained in R. To achieve this, let us fix freely some $\nu \in \mathbb{R}^N$. Then, by mere routine

$$\overline{\gamma}_\nu f = \sum_{k\in\mathbb{Z}^N} a(k)e^{-2\pi i\nu\cdot k}\tau_k(\gamma_\nu\varphi).$$

Hence, by Remark 2.16 (see also Example 2.6 and Proposition 2.25), it follows

$$M(\overline{\gamma}_\nu f) = \mathfrak{M}(\overline{\omega}_{\dot{\nu}}a)\mathcal{F}\varphi(-\nu),$$

where $\mathcal{F}\varphi$ denotes the Fourier transform of φ (viewed as lying in $\mathcal{K}(\mathbb{R}^N)$). Consequently, if ν lies outside R, since this implies that $\dot{\nu}$ lies off $\dot{R}$, then $\mathfrak{M}(\overline{\omega}_{\dot{\nu}}a) = 0$. Therefore $M(\overline{\gamma}_\nu f) = 0$. Hence $f \in AP_R(\mathbb{R}^N)$. The proposition follows. ∎

The following corollary enables us to discuss the homogenization of (5.141) when θ is an almost periodic function on $\mathbb{Z}^N$.

Corollary 5.20 *Let the notation and hypotheses be as above. Suppose $\theta \in AP_{\dot{R}}(\mathbb{Z}^N)$. Then the following hold:*

(i) $\chi_G \in \mathfrak{X}^p_{AP_R(\mathbb{R}^N)}$ $(1 \leq p < \infty)$,
(ii) $M(\chi_G) > 0$.

Proof Combination of Proposition 5.15 and Corollary 5.19 yields at once $\chi_G \in \mathfrak{X}^p_{\mathfrak{A}_R}$ $(1 \leq p < \infty)$ and $M(\chi_G) > 0$. Therefore the corollary follows by Proposition 5.16. ∎

Thanks to the preceding corollary it is possible to work out the homogenization of (5.141) under various structure hypotheses on the coefficients $\{a_0, a_{ij}\}$. Let us illustrate this by some examples.

In what follows it is assumed that $\theta \in AP_{\dot{R}}(\mathbb{Z}^N)$, where R is a countable subgroup of $\mathbb{R}^N$ containing $\mathbb{Z}^N$ (e.g., $R = \mathbb{Q}^N$).

Example 5.13 If we reconsider the structure hypothesis (5.163) then by corollary 5.20 (with a suitable choice of p) we are immediately led to (5.147)–(5.150) with $A = AP_R(\mathbb{R}^N)$. Indeed $\mathcal{C}_{per}(Y) \subset AP_R(\mathbb{R}^N)$, since $\mathbb{Z}^N \subset R$.

Example 5.14 Returning to structure hypothesis (5.164), let $\mathcal{R} = Gr(R_0 \cup R)$ where R_0 is the same as in (5.165). Then, thanks to Corollary 5.20 (with p suitably chosen), we have (5.147)–(5.150) with $A = AP_{\mathcal{R}}(\mathbb{R}^N)$.

Example 5.15 If we reconsider the structure hypothesis (5.166), then the same line of argument as followed in Example 5.10 combines with Corollary 5.20 to yield (5.147)–(5.150) with $A = \mathcal{B}_{\infty,R}(\mathbb{R}^N)$ (see Example 4.5). It is worth mentioning that substituting $L^q_{per}(Y)$ by $L^q_{AP}(\mathbb{R}^N)$ in (5.166) leads to a similar result with a suitable $\mathcal{R}$ in place of R.

Example 5.16 Likewise the structure hypothesis of Example 5.11 leads readily, by Corollary 5.20, to the abstract setting with $A = \mathcal{B}_{\infty,R}(\mathbb{R}^N)$.

5.4.4.4 The Holes Are Periodically Distributed

We assume here that the function θ is periodic, that is, there is a network $\mathcal{R}_0$ in $\mathbb{R}^N$ with $\mathcal{R}_0 \subset \mathbb{Z}^N$ (e.g., $\mathcal{R}_0 = \prod_{i=1}^N m_i\mathbb{Z}$ with $m_i \in \mathbb{N}^*$) such that $\theta(k+\nu) = \theta(k)$ $(k \in \mathbb{Z}^N)$ for any $\nu \in \mathcal{R}_0$.

Let us recall that $\mathcal{R}_0$ is spanned by a basis $b_1, \cdots, b_N$ of the N-dimensional vector space $\mathbb{R}^N$, and that to $\mathcal{R}_0$ there is attached the parallelepiped

$$Y_0 = \left\{ y | y = \sum_{i=1}^{N} t_i b_i, -\frac{1}{2} < t_i < \frac{1}{2} \ (1 \leq i \leq N) \right\}$$

such that the couple $\mathcal{R}_0$, Y_0 constitutes the so-called Y_0-periodic setting discussed in the fullest detail in Sect. 2.5.

So, let $\mathcal{C}_{per}(Y_0)$ denote the H-algebra of all Y_0-periodic continuous complex functions on $\mathbb{R}^N$, that is, precisely those functions $\psi \in \mathcal{B}(\mathbb{R}^N)$ such that $\psi(y+\nu) = \psi(y)$ $(y \in \mathbb{R}^N)$ for all $\nu \in \mathcal{R}_0$. It is worth recalling that $\mathcal{C}_{per}(Y_0)$ is an $H^1(\Omega)$-proper H-algebra with $\mathcal{C}^\infty_{per}(Y_0) = \mathcal{C}_{per}(Y_0) \cap \mathcal{C}^\infty(\mathbb{R}^N)$ as a dense subalgebra (see Chaps. 3 and 4).

The following proposition enables to discuss the homogenization of (5.141) when the holes are periodically distributed. So we assume in the sequel that θ is $\mathcal{R}_0$-periodic as stated above.

Proposition 5.17 *We assume that θ is $\mathcal{R}_0$-periodic. Then, the following hold:*

(i) $\chi_G \in \mathfrak{X}^p_{\mathcal{C}_{per}(Y_0)}$ $(1 \leq p < \infty)$,
(ii) $M(\chi_G) > 0$.

Proof Let us first of all recall that $\theta \in ES(\mathbb{Z}^N)$ (Example 2.9). This being so, it is an elementary exercise to show that the functions $\sum_{k \in \mathbb{Z}^N} \theta(k) \tau_k \varphi$ $(\varphi \in \mathcal{K}(Y))$ lie in $\mathcal{C}_{per}(Y_0)$. Therefore the proposition follows by Proposition 5.15 and Corollary 5.19. ∎

Let us give in this configuration a few examples of structure hypotheses on $\{a_0, a_{ij}\}$ that reduce to the abstract setting (5.147)–(5.150).

Example 5.17 The simplest case corresponds without doubt to the structure hypothesis

$$a_0, a_{ij} \in L^q_{per}(Y_0) \ (1 \leq i, j \leq N) \text{ for some } 2 < q < \infty.$$

Indeed, thanks to Proposition 5.17 (with a suitable choice of p), this leads readily to (5.147)–(5.150) with $A = \mathcal{C}_{per}(Y_0)$. The same result is obtained for (5.163), since $\mathcal{C}_{per}(Y) \subset \mathcal{C}_{per}(Y_0)$.

Example 5.18 Suppose we are concerned in the homogenization of (5.141) under the structure hypothesis (5.164). By the relation $\mathcal{C}_{per}(Y_0) = AP_{\mathcal{R}_0^*}(\mathbb{R}^N)$ (Proposition 2.23) and use of Proposition 5.17 one sees at once that the problem reduces to the abstract setting (5.147)–(5.150) with $A = AP_{\mathcal{R}}(\mathbb{R}^N)$, where $\mathcal{R} = Gr(R_0 \cup \mathcal{R}_0^*)$ with the same R_0 as in (5.165).

Example 5.19 Let $\mathcal{R}_1$ be a network in $\mathbb{R}^N$ with $\mathcal{R}_1 \subset \mathbb{Z}^N$. Let Y_1 be the attached open parallelepiped analogous to Y_0 (i.e., Y_1 is to $\mathcal{R}_1$ as Y_0 is to $\mathcal{R}_0$). The Y_1-periodic structure hypothesis

$$a_0, a_{ij} \in L^q_{per}(Y_1) \;\; (1 \leq i, j \leq N) \text{ for some } 2 < q < \infty$$

associated with the fact that θ is $\mathcal{R}_0$-periodic, of course, reduces to (5.147)–(5.150) with $A = AP_{\mathcal{R}}(\mathbb{R}^N)$, where $\mathcal{R} = Gr(\mathcal{R}_0^* \cup \mathcal{R}_1^*)$. The details are left to the reader. Thus, the present example falls within almost periodic homogenization though the data are periodic.

Likewise it is an easy matter to check that, in the present configuration of the holes, each of the structure hypotheses in Examples 5.10 and 5.11 reduces to the abstract setting (5.147)–(5.150) with $A = \mathcal{B}_{\infty,\mathcal{R}_0^*}(\mathbb{R}^N)$. Finally, the same examples with Y substituted by Y_0 lead to the same result.

5.4.4.5 The Holes Are Concentrated in a Neighbourhood of the Hyperplane $\{x_N = 0\}$

We introduce the factorization $\mathbb{R}^N = \mathbb{R}^{N-1} \times \mathbb{R}$ $(N > 1)$. Thus, each $y \in \mathbb{R}^N$ writes as $y = (y', y_N)$ with $y' = (y_1, \cdots, y_{N-1}) \in \mathbb{R}^{N-1}$. Accordingly $\mathbb{Z}^N = \mathbb{Z}^{N-1} \times \mathbb{Z}$ and so each $k \in \mathbb{Z}^N$ writes as $k = (k', k_N)$.

This being so, we discuss below one concrete example in the present context.

Example 5.20 (Equidistributed Concentration) We take here

$$S = \mathbb{Z}^{N-1} \times \{0\}. \tag{5.168}$$

Thus, the distribution function of the holes, $\theta : \mathbb{Z}^N \to \mathbb{R}$, is given by

$$\theta(k', k_N) = \begin{cases} 1 \;\; (k' \in \mathbb{Z}^{N-1}) & \text{if } k_N = 0 \\ 0 \;\; (k' \in \mathbb{Z}^{N-1}) & \text{if } k_N \neq 0. \end{cases}$$

Let

$$A = \mathcal{B}_\infty(\mathbb{R}; \mathcal{C}_{per}(Y')) \text{ with } Y' = \left(-\frac{1}{2}, \frac{1}{2}\right)^{N-1} \quad (N > 1).$$

A is an $H^1(\Omega)$-proper homogenization algebra admitting A^∞ as a dense subalgebra. Our first purpose is to check that (5.147) and (5.148) hold in the present case. To this end, let $\varphi \in \mathcal{K}(Y)$ be freely fixed, and let

$$\psi = \sum_{k \in \mathbb{Z}^N} \theta(k) \tau_k \varphi.$$

It is clear that

$$\psi = \sum_{k' \in \mathbb{Z}^{N-1}} \tau_{(k',0)}\varphi.$$

One readily deduces that $\psi \in \mathcal{B}(\mathbb{R}; \mathcal{C}_{per}(Y'))$. Furthermore, by reasoning as in Remark 2.15 we see at once that to each $y = (y', y_N) \in \mathbb{R}^N$ there is assigned some $l' \in \mathbb{Z}^{N-1}$ such that $\psi(y', y_N) = \varphi(y' - l', y_N)$. It follows that $\sup_{z \in \mathbb{R}^{N-1}} |\psi(z, y_N)| \leq \sup_{z \in \mathbb{R}^{N-1}} |\varphi(z, y_N)|$ and that for all $y_N \in \mathbb{R}^N$, where φ is justifiably viewed as a function in $\mathcal{K}(\mathbb{R}; \mathcal{B}(\mathbb{R}^{N-1}))$ [with support in $(-\frac{1}{2}, \frac{1}{2})$]. One deduces that

$$\psi \in \mathcal{K}(\mathbb{R}; \mathcal{C}_{per}(Y')) \subset A.$$

Hence, by the same argument as followed in proving the first statement of Proposition 5.15 one is easily led to $\chi_\Theta \in \mathfrak{X}^p_A$ $(1 \leq p < \infty)$ and therefore to (5.147), where the real $p \geq 1$ is arbitrary for the moment. However, it is of interest to keep in mind that χ_Θ lies more precisely in the closure of $\mathcal{K}(\mathbb{R}; \mathcal{C}_{per}(Y'))$ in $\Xi^p(\mathbb{R}^N)$. For the purpose of verifying (5.148), let us show that $M(\psi) = 0$. This is straightforward. Indeed, for arbitrary fixed $x = (x', x_N)$, the relation $|\psi(\frac{x}{\varepsilon})| \leq \|\psi(\cdot, \frac{x_N}{\varepsilon})\|_\infty$ shows that as $\varepsilon \to 0$, we have $\psi(\frac{x}{\varepsilon}) \to 0$ for any x lying outside the hyperplane $\{x_N = 0\}$ (which is λ-negligible, of course). Consequently, given any compact set $K \subset \mathbb{R}^N$, we have by dominated convergence

$$\int_K \psi\left(\frac{x}{\varepsilon}\right) dx \to 0$$

as $\varepsilon \to 0$, which shows that $M(\psi) = 0$. Since this holds for any $\varphi \in \mathcal{K}(Y)$, we conclude that $M(\chi_\Theta) = 0$, hence $M(\chi_G) = 1$. Therefore (5.148) follows.

We illustrate this by considering two concrete problems in which it is assumed that the open set Ω intersects the hyperplane $\{x_N = 0\}$.

Problem I We consider here the homogenization problem for (5.141) under the perforation hypothesis (5.168) and the structure hypothesis

$$\begin{aligned} &a_0, a_{ij} \in L^q(\mathbb{R}; L^q_{per}(Y')) \ (1 \leq i, j \leq N) \\ &\text{for some given } 2 < q < \infty. \end{aligned} \tag{5.169}$$

Clearly the problem reduces to verifying that (5.169) implies (5.149). This is straightforward. Indeed, as pointed out above, $\mathcal{K}(\mathbb{R}; \mathcal{C}_{per}(Y'))$ is contained in A, and at the same time is a dense subspace of $L^q(\mathbb{R}; L^q_{per}(Y'))$. But the latter is continuously embedded in $\Xi^q(\mathbb{R}^N)$ via $(L^q, l^\infty)(\mathbb{R}^N)$. Hence (5.149) follows and so Theorem 5.8 applies.

Problem II The statement is as above save that (5.169) is substituted by

$$\begin{aligned} & a_0, a_{ij} \in \mathcal{B}_\infty(\mathbb{R}; L^q_{per}(Y')) \ \ (1 \leq i, j \leq N) \\ & \text{for some given } 2 < q < \infty. \end{aligned} \tag{5.170}$$

The same argument mutatis mutandis as followed above leads to the same conclusion.

5.5 Quasilinear Monotone Operators

5.5.1 Introduction

The basic notation and hypotheses are as before, except that Ω denotes here a bounded open set in $\mathbb{R}^N_x$ without any requirement of smoothness. For given real $q \geq 1$ and integer $d > 1$, we set

$$L^q(\Omega)^d = L^q(\Omega) \times \cdots \times L^q(\Omega) \ \ (d \text{ times}),$$

and we provide $L^q(\Omega)^d$ with the norm

$$\|\mathbf{u}\|_{L^q(\Omega)^d} = \left(\sum_{i=1}^d \|u_i\|^q_{L^q(\Omega)}\right)^{\frac{1}{q}}, \mathbf{u} = (u_i) \in L^q(\Omega)^d,$$

which makes it into a Banach space, and in particular into a Hilbert space if $q = 2$. On the other hand, let

$$|\mathbf{u}| = \left(\sum_{i=1}^d |u_i|^2\right)^{\frac{1}{2}}, \quad \mathbf{u} = (u_i) \in L^q(\Omega)^d.$$

Here and beyond it should be kept in mind that $|\mathbf{u}| \in L^q(\Omega)$ and further

$$d^{-\frac{1}{q}} \|\mathbf{u}\|_{L^q(\Omega)^d} \leq \| |\mathbf{u}| \|_{L^q(\Omega)} \leq d^{\frac{1}{2}-\frac{1}{q}} \|\mathbf{u}\|_{L^q(\Omega)^d}.$$

Attention must be drawn to the fact that all this remains valid if Ω is unbounded, or even if Ω is any locally compact space equipped with a positive Radon measure.

This being so, let $1 < p \leq 2$. Let $\mathbf{a} = (a_i)_{1 \leq i \leq N}$ be a function of $\mathbb{R}^N_y \times \mathbb{R}^N_\mu$ into $\mathbb{R}^N$ with the following properties:

$$\begin{aligned} & \text{For fixed } \mu \in \mathbb{R}^N, \text{ the function } y \mapsto \mathbf{a}(y, \mu) \text{ of } \mathbb{R}^N \text{ into } \mathbb{R}^N \\ & \text{denoted by } \mathbf{a}(\cdot, \mu), \text{ is measurable.} \end{aligned} \tag{5.171}$$

$$\mathbf{a}(y, \omega) = \omega \text{ (the origin in } \mathbb{R}^N) \text{ for a.e. } y \in \mathbb{R}^N. \tag{5.172}$$

There are two constants $\alpha, c > 0$ such that

$$\begin{aligned} &(1)\ (\mathbf{a}(y, \mu) - \mathbf{a}(y, \nu)) \cdot (\mu - \nu) \geq \alpha |\mu - \nu|^p \\ &(2)\ |\mathbf{a}(y, \mu) - \mathbf{a}(y, \nu)| \leq c |\mu - \nu|^{p-1} \end{aligned} \tag{5.173}$$

for any couple of points $\mu, \nu \in \mathbb{R}^N$ and for almost every $y \in \mathbb{R}^N$, where the dot denotes the usual Euclidian inner product in $\mathbb{R}^N$, and $|\cdot|$ the corresponding norm.

Now, let $f \in W^{-1,p'}(\Omega; \mathbb{R})$, $p' = \frac{p}{p-1}$. For fixed $\varepsilon > 0$ we consider the boundary value problem

$$-\operatorname{div}\mathbf{a}\left(\frac{x}{\varepsilon}, Du_\varepsilon\right) = f \text{ in } \Omega,\ u_\varepsilon \in W_0^{1,p}(\Omega; \mathbb{R}), \tag{5.174}$$

which uniquely determines u_ε, as will presently become apparent in the next section. Here $D = (\frac{\partial}{\partial x_1}, \cdots, \frac{\partial}{\partial x_N})$ stands for the usual gradient operator in Ω.

We investigate here the behaviour, as $\varepsilon \to 0$, of u_ε under an abstract assumption on a_i $(1 \leq i \leq N)$ covering a variety of concrete structure hypotheses beyond the classical periodic setting. A few examples of such structure hypotheses are next worked out in detail in order to help the reader have a clear idea of what a non periodic setting of the homogenization problem for (5.174) may look like.

5.5.2 Preliminaries

Lemma 5.11 *Let $u \in \mathcal{C}(\overline{\Omega}; L^\infty(\mathbb{R}_y^N; \mathbb{R}))$. Suppose for each $x \in \overline{\Omega}$ we have $u(x, y) \geq 0$ for a.e. $y \in \mathbb{R}^N$. Then, for fixed $\varepsilon > 0$, one has $u^\varepsilon(x) \geq 0$ a.e. in Ω, where u^ε is defined in Proposition 1.8.*

Proof Let us consider a sequence of functions $u_n \in \mathcal{C}(\overline{\Omega}) \otimes L^\infty(\mathbb{R}_y^N; \mathbb{R})$ (integers $n \geq 0$) such that $u_n \to u$ in $\mathcal{C}(\overline{\Omega}; L^\infty(\mathbb{R}_y^N; \mathbb{R}))$ as $n \to \infty$. Set $q_n = sup_{x\in\overline{\Omega}} \|u_n(x, \cdot) - u(x, \cdot)\|_{L^\infty(\mathbb{R}^N)}$. Fix freely $n \geq 0$, and let $x \in \overline{\Omega}$. Then $u_n(x, y) + q_n \geq 0$ for a.e. $y \in \mathbb{R}^N$. By Lemma 5.1 we deduce that there is a negligible set $\mathcal{N}_n \subset \mathbb{R}_y^N$ such that $u_n(x, y) + q_n \geq 0$ for all $x \in \overline{\Omega}$ and all $y \in \mathbb{R}^N \setminus \mathcal{N}_n$. It follows readily that $u_n(x, y) + q_n \geq 0$ for all $x \in \overline{\Omega}, n \geq 0$ and $y \in \mathbb{R}^N \setminus \mathcal{N}$, where $\mathcal{N} = \bigcup_{n\in\mathbb{N}} \mathcal{N}_n$ (a λ-negligible set). Therefore, for any fixed $\varepsilon > 0$ we have $u_n^\varepsilon(x) + q_n \geq 0$ for all $x \in \Omega \setminus \varepsilon\mathcal{N}$ and for any integer $n \geq 0$. But, as $n \to \infty$, we have $u_n^\varepsilon \to u^\varepsilon$ in $L^\infty(\Omega)$ (Proposition 1.8) and $q_n \to 0$. Hence, by extraction of a suitable subsequence, it follows that $u^\varepsilon(x) \geq 0$ for a.e. $x \in \Omega$, as claimed. ■

Now, let $G : \mathbb{R}^N \times \mathbb{R}^N \to \mathbb{R}$ be a function with the following properties:

$$\text{The function } G(\cdot, \mu) \text{ (fixed } \mu) \text{ is measurable from } \mathbb{R}^N \text{ to } \mathbb{R}. \tag{5.175}$$

$$G(y, \omega) = 0 \text{ for a.e. } y \in \mathbb{R}^N. \tag{5.176}$$

$$\begin{aligned} &|G(y, \mu) - G(y, \nu)| \leq c|\mu - \nu|^{p-1} \quad (\mu, \nu \in \mathbb{R}^N) \\ &\text{for a.e. } y \in \mathbb{R}^N, \text{ where } c \text{ is that constant in (5.173).} \end{aligned} \tag{5.177}$$

Given $\boldsymbol{\Psi} \in \mathcal{C}(\overline{\Omega}; \mathbb{R})^N = \mathcal{C}(\overline{\Omega}; \mathbb{R}) \times \cdots \times \mathcal{C}(\overline{\Omega}; \mathbb{R})$ (N times), it is an easy exercise to check that the function $(x, y) \mapsto G(y, \boldsymbol{\Psi}(x))$ of $\overline{\Omega} \times \mathbb{R}^N_y$ into $\mathbb{R}$ belongs to $\mathcal{C}(\overline{\Omega}; L^\infty(\mathbb{R}^N_y))$. Hence, the function $x \mapsto G(\frac{x}{\varepsilon}, \boldsymbol{\Psi}(x))$ of Ω into $\mathbb{R}$, denoted by $G^\varepsilon(\cdot, \boldsymbol{\Psi})$, is well defined as a function in $L^\infty(\Omega; \mathbb{R})$ (Proposition 1.8). This leads us to the following proposition, where $1 < p \leq 2$ and $p' = \frac{p}{p-1}$.

Proposition 5.18 *The transformation $\boldsymbol{\Psi} \mapsto G^\varepsilon(\cdot, \boldsymbol{\Psi})$ of $\mathcal{C}(\overline{\Omega}; \mathbb{R})^N$ into $L^\infty(\Omega; \mathbb{R})$ extends by continuity to a mapping, still denoted by $\boldsymbol{\Psi} \mapsto G^\varepsilon(\cdot, \boldsymbol{\Psi})$, of $L^p(\Omega; \mathbb{R})^N$ into $L^{p'}(\Omega; \mathbb{R})$ with the property*

$$\|G^\varepsilon(\cdot, \boldsymbol{\Psi}) - G^\varepsilon(\cdot, \boldsymbol{\Phi})\|_{L^{p'}(\Omega)} \leq cN^{(\frac{1}{2} - \frac{1}{p})(p-1)} \|\boldsymbol{\Psi} - \boldsymbol{\Phi}\|^{p-1}_{L^p(\Omega)^N} \tag{5.178}$$

for all $\boldsymbol{\Psi}, \boldsymbol{\Phi} \in L^p(\Omega; \mathbb{R})^N$.

Proof Fix freely some $\boldsymbol{\Psi}, \boldsymbol{\Phi} \in \mathcal{C}(\overline{\Omega}; \mathbb{R})^N$. By Lemma 5.11 with

$$u(x, y) = c|\boldsymbol{\Psi}(x) - \boldsymbol{\Phi}(x)|^{p-1} - |G(y, \boldsymbol{\Psi}(x)) - G(y, \boldsymbol{\Phi}(x))|$$

and use of (5.177), we have

$$\left|G\left(\frac{x}{\varepsilon}, \boldsymbol{\Psi}(x)\right) - G\left(\frac{x}{\varepsilon}, \boldsymbol{\Phi}(x)\right)\right| \leq c|\boldsymbol{\Psi}(x) - \boldsymbol{\Phi}(x)|^{p-1} \tag{5.179}$$

for a.e. $x \in \Omega$. Hence (5.178) follows at once for all $\boldsymbol{\Psi}, \boldsymbol{\Phi} \in \mathcal{C}(\overline{\Omega}; \mathbb{R})^N$. Therefore, $\mathcal{C}(\overline{\Omega}; \mathbb{R})^N$ being dense in $L^p(\Omega; \mathbb{R})^N$, the proposition follows in an obvious fashion. ■

So, by taking $G = a_i$ in Proposition 5.18 we see that for $u \in W^{1,p}(\Omega; \mathbb{R})$, the function $x \mapsto a_i(\frac{x}{\varepsilon}, Du(x))$ of Ω into $\mathbb{R}$ denoted by $a_i^\varepsilon(\cdot, Du)$, lies in $L^{p'}(\Omega; \mathbb{R})$. Consequently the left-hand side of (5.174) is fully justified. Furthermore, by letting

$$\mathbf{a}^\varepsilon(\cdot, Du) = (a_i^\varepsilon(\cdot, Du))_{1 \leq i \leq N}$$

we get a mapping $u \mapsto \mathbf{a}^\varepsilon(\cdot, Du)$ of $W^{1,p}(\Omega; \mathbb{R})$ into $L^{p'}(\Omega; \mathbb{R})^N$ that attains the value $\mathbf{0}$ (the origin in $L^{p'}(\Omega; \mathbb{R})^N$) on the function $u = 0$ in $W^{1,p}(\Omega; \mathbb{R})$ (this follows by (5.176), of course). Moreover, the following hold for $u, v \in W^{1,p}(\Omega; \mathbb{R})$:

(1) $[\mathbf{a}(\frac{x}{\varepsilon}, Du(x)) - \mathbf{a}(\frac{x}{\varepsilon}, Dv(x))] \cdot (Du(x) - Dv(x)) \geq \alpha|Du(x) - Dv(x)|^p$ a.e. in $x \in \Omega$,

(2) $\|\mathbf{a}^{\varepsilon}(\cdot, Du) - \mathbf{a}^{\varepsilon}(\cdot, Dv)\|_{L^{p'}(\Omega)^N} \leq cN^{\frac{p-1}{2}} \|Du - Dv\|^{p-1}_{L^p(\Omega)^N}$.

Therefore, it is a classical exercise to show that the boundary value problem (5.174) (where $\varepsilon > 0$ is arbitrary fixed) admits one and only one solution. Indeed, let us first observe that by letting

$$\|u\|_{W_0^{1,p}(\Omega)} = \|Du\|_{L^p(\Omega)^N} \quad (u \in W_0^{1,p}(\Omega; \mathbb{R}))$$

we define a norm on $W_0^{1,p}(\Omega; \mathbb{R})$ which is equivalent to the $W^{1,p}(\Omega)$ norm (thanks to Friedrichs inequality). With this in mind, it is an easy matter to check using the representation

$$\langle T_\varepsilon u, v\rangle = \int_\Omega \mathbf{a}^{\varepsilon}(\cdot, Du) \cdot Dv dx \quad (u, v \in W_0^{1,p}(\Omega; \mathbb{R}))$$

that the operator

$$T_\varepsilon : W_0^{1,p}(\Omega; \mathbb{R}) \to W^{-1,p'}(\Omega; \mathbb{R})$$

$$T_\varepsilon u = -\text{div}\, \mathbf{a}^{\varepsilon}(\cdot, Du),\, u \in W_0^{1,p}(\Omega; \mathbb{R}),$$

satisfies the hypotheses of the well-known Browder theorem ensuring the existence of a unique $u_\varepsilon \in W_0^{1,p}(\Omega; \mathbb{R})$ such that $T_\varepsilon u_\varepsilon = f$ (see, e.g., Zeidler [147]).

Now, according to (5.175) and (5.177), G is a so-called Carathéodory function (see, e.g., Fučik and Kufner [55]). Consequently, for any N-tuple of measurable functions $\psi_i : \mathbb{R}^N_y \to \mathbb{R}$ $(1 \leq i \leq N)$, the function

$$y \mapsto G(y, \psi_1(y), \cdots, \psi_N(y))$$

of $\mathbb{R}^N$ into $\mathbb{R}$ is measurable (for the proof see Vajnberg [134], Krasnosel'skii [68]).

In the sequel a function of the preceding form is denoted by $y \mapsto G(y, \boldsymbol{\Psi}(y))$, or more simply by $G(\cdot, \boldsymbol{\Psi})$, where $\boldsymbol{\Psi} = (\psi_i)_{1\leq i\leq N}$. By (5.176)–(5.177) we see readily that $G(\cdot, \boldsymbol{\Psi})$ lies in $L^{p'}(\mathbb{R}^N_y; \mathbb{R})$ for $\boldsymbol{\Psi} \in L^p(\mathbb{R}^N_y; \mathbb{R})^N$ $(1 < p \leq 2)$. On the other hand, $G(\cdot, \boldsymbol{\Psi}) \in L^\infty(\mathbb{R}^N_y; \mathbb{R})$ for $\boldsymbol{\Psi} \in \mathcal{B}(\mathbb{R}^N_y; \mathbb{R})^N$.

This enables us to state and prove the following lemma.

Lemma 5.12 *Let* $\mathbf{w} = (w_i)_{1\leq i\leq N} \in \mathcal{C}(\overline{\Omega}; \mathcal{B}(\mathbb{R}^N_y; \mathbb{R})^N)$. *Let* $G(\cdot, \mathbf{w})$ *stand for the function* $(x, y) \mapsto G(y, \mathbf{w}(x, y))$ *of* $\overline{\Omega} \times \mathbb{R}^N$ *into* $\mathbb{R}$. *Then* $G(\cdot, \mathbf{w}) \in \mathcal{C}(\overline{\Omega}; L^\infty(\mathbb{R}^N_y))$, *so that the function* $x \mapsto G(\frac{x}{\varepsilon}, \mathbf{w}(x, \frac{x}{\varepsilon}))$ *of* Ω *into* $\mathbb{R}$, *denoted by* $G(\cdot, \mathbf{w})^\varepsilon$, *is well defined and belongs to* $L^\infty(\Omega)$. *Furthermore, one has* $G(\cdot, \mathbf{w})^\varepsilon = G^\varepsilon(\cdot, \mathbf{w}^\varepsilon)$ *with* $\mathbf{w}^\varepsilon = (w_i^\varepsilon)_{1\leq i\leq N}$.

Proof It is immediate by (5.175)–(5.177) that the function $G(\cdot, \mathbf{w})$ lies in $\mathcal{C}(\overline{\Omega}; L^\infty(\mathbb{R}^N_y))$. Hence the first portion of the lemma follows by Proposition 1.8. Finally, fix freely a couple of points $x \in \overline{\Omega}$, $y \in \mathbb{R}^N$ (y lying off some negligible subset of $\mathbb{R}^N$, as stated in (5.177)) and for arbitrary fixed $\varepsilon > 0$, choose $\mu = \mathbf{w}^\varepsilon(x)$ and $\nu = \mathbf{w}(x, y)$ in (5.177) to obtain

$$|G(y, \mathbf{w}^\varepsilon(x)) - G(y, \mathbf{w}(x, y))| \leq c|\mathbf{w}^\varepsilon(x) - \mathbf{w}(x, y)|^{p-1};$$

that is,

$$-c|\mathbf{w}^\varepsilon(x)-\mathbf{w}(x, y)|^{p-1} \leq G(y, \mathbf{w}^\varepsilon(x))-G(y, \mathbf{w}(x, y)) \leq c|\mathbf{w}^\varepsilon(x)-\mathbf{w}(x, y)|^{p-1},$$

and that for any fixed $x \in \overline{\Omega}$ and a.e. $y \in \mathbb{R}^N$. By Lemma 5.11 with

$$u(x, y) = c|\mathbf{w}^\varepsilon(x) - \mathbf{w}(x, y)|^{p-1} - G(y, \mathbf{w}^\varepsilon(x)) + G(y, \mathbf{w}(x, y))$$

it follows $G(\cdot, \mathbf{w})^\varepsilon \geq G^\varepsilon(\cdot, \mathbf{w}^\varepsilon)$ a.e. in Ω. In an analogous way one sees readily that $G(\cdot, \mathbf{w})^\varepsilon \leq G^\varepsilon(\cdot, \mathbf{w}^\varepsilon)$ a.e. in Ω. Hence the second portion of the lemma follows thereby. ∎

5.5.3 *Abstract Setting*

5.5.3.1 Setting of the Problem and Preliminaries

Throughout the present section, A denotes a $W^{1,p}(\Omega)$-proper homogenization algebra on $\mathbb{R}^N$ admitting A^∞ as a dense subalgebra.

Our main purpose here is to discuss the homogenization of (5.174) under the abstract hypothesis

$$a_i(\cdot, \boldsymbol{\Psi}) \in \mathfrak{X}_A^{p'}(\mathbb{R}_y^N) \ \ (1 \leq i \leq N) \text{ for all } \boldsymbol{\Psi} \in (A_\mathbb{R})^N, \tag{5.180}$$

where $A_\mathbb{R} = A \cap \mathcal{C}(\mathbb{R}^N; \mathbb{R})$ and $\frac{1}{p} + \frac{1}{p'} = 1$, $1 < p \leq 2$.

This problem is referred to as the *abstract homogenization problem for* (5.174). As we shall see in the next section, the hypothesis (5.180) covers a wide range of concrete behaviours such as the periodicity, the almost periodicity, the convergence at infinity, and many more besides. However, before we can embark upon the analysis of the homogenization problem as such, we need further notation and preliminary results.

The first point deals with the function spaces suitable to the problem on hand. Let

$$W^{1,p}(\Delta(A); \mathbb{R}) = \{u | u, \partial_i u \in L^p(\Delta(A); \mathbb{R}), 1 \leq i \leq N\}.$$

This is a closed vector subspace of $W^{1,p}(\Delta(A))$. Thus, the natural norm on $W^{1,p}(\Delta(A); \mathbb{R})$ is obviously the $W^{1,p}(\Delta(A))$ norm. We are especially interested in the subspace

$$W^{1,p}(\Delta(A);\mathbb{R})/\mathbb{R} = \left\{u|u \in W_{\mathbb{R}}^{1,p}(\Delta(A)), \int_{\Delta(A)} u\, d\beta = 0\right\},$$

where $W_{\mathbb{R}}^{1,p}(\Delta(A)) = W^{1,p}(\Delta(A);\mathbb{R})$. As pointed out before (see Sect. 2.2.2), the appropriate topology on $W^{1,p}(\Delta(A);\mathbb{R})/\mathbb{R}$ is that defined by the $W^{1,p}(\Delta(A))/\mathbb{C}$ seminorm (see Chap. 2), which is in general non-separated and non-complete. Hence the passage to the separated completion of $W_{\mathbb{R}}^{1,p}(\Delta(A))/\mathbb{R}$ denoted by $W_{\#}^{1,p}(\Delta(A);\mathbb{R})$. A quick survey of Sect. 5.2.2 reveals that Proposition 5.8 and Remark 5.10 remain rigorously valid when $H^1(\Delta(A))$ is substituted there by $W^{1,p}(\Delta(A))$. It follows

$$W_{\#}^{1,p}(\Delta(A);\mathbb{R}) = \{u|u \in W_{\#}^{1,p}(\Delta(A)), \partial_i u \in L_{\mathbb{R}}^p(\Delta(A))\ (1 \le i \le N)\}$$

where $L_{\mathbb{R}}^p(\Delta(A)) = L^p(\Delta(A);\mathbb{R})$. It is worth recalling that the norm on $W_{\#}^{1,p}(\Delta(A);\mathbb{R})$ is given by

$$\|u\|_{W_{\#}^{1,p}(\Delta(A))} = \left[\sum_{i=1}^{N} \|\partial_i u\|_{L^p(\Delta(A))}^p\right]^{\frac{1}{p}} \quad (u \in W_{\#}^{1,p}(\Delta(A);\mathbb{R})).$$

On the other hand, among other things it is fundamental to note that $J(W_{\mathbb{R}}^{1,p}(\Delta(A))/\mathbb{R})$ is dense in $W_{\#}^{1,p}(\Delta(A);\mathbb{R})$, J being the canonical mapping of $W^{1,p}(\Delta(A))/\mathbb{C}$ into its separated completion, $W_{\#}^{1,p}(\Delta(A))$.

Now, let

$$\mathbb{F}_0^{1,p}(\Omega \times \Delta(A);\mathbb{R}) = W_0^{1,p}(\Omega;\mathbb{R}) \times L^p(\Omega; W_{\#}^{1,p}(\Delta(A);\mathbb{R})).$$

We equip this space with the norm

$$\|\mathbf{v}\|_{\mathbb{F}_0^{1,p}} = \|v_0\|_{W_0^{1,p}(\Omega)} + \|v_1\|_{L^p(\Omega; W_{\#}^{1,p}(\Delta(A)))}$$

with $\mathbf{v} = (v_0, v_1) \in \mathbb{F}_0^{1,p}(\Omega \times \Delta(A);\mathbb{R})$, which makes it into a Banach space. Furthermore, the space

$$\mathcal{F}_0^{\infty} = \mathcal{D}_{\mathbb{R}}(\Omega) \times [\mathcal{D}_{\mathbb{R}}(\Omega) \otimes J(\mathcal{D}_{\mathbb{R}}(\Delta(A))/\mathbb{R})]$$

is dense in $\mathbb{F}_0^{1,p}(\Omega \times \Delta(A);\mathbb{R})$ (this is straightforward by the density of $\mathcal{D}(\Delta(A))$ in $W^{1,p}(\Delta(A))$).

Remark 5.17 Attention is drawn to equality (5.80) where $J_\Omega(v) = J \circ v$ for any function $v : \Omega \to W^{1,p}(\Delta(A))/\mathbb{C}$.

The following lemma will prove fundamental.

Lemma 5.13 *Let* $1 < r < \infty$. *For* $u \in \mathcal{C}(\overline{\Omega}; \mathfrak{X}_A^{r,\infty})$, *we have, as* $\varepsilon \to 0$, $u^\varepsilon \to \widehat{u} = \mathcal{G} \circ u$ *in* $L^r(\Omega)$*-weak* Σ.

Proof This is straightforward by Proposition 4.4 (proceed as in the proof of Corollary 3.1). ■

The next lemma is a simple but useful tool.

Lemma 5.14 *Let* $(u_\varepsilon)_{\varepsilon\in E}$, $u_\varepsilon \in L^r(\Omega)$ $(1 \leq r < \infty)$. *Suppose* $u_\varepsilon(x) \geq 0$ *for a.e.* $x \in \Omega$ *and further* $u_\varepsilon \to u_0$ *in* $L^r(\Omega)$*-weak* Σ *as* $E \ni \varepsilon \to 0$. *Then* $u_0(x, s) \geq 0$ *for a.e.* $(x, s) \in \Omega \times \Delta(A)$.

Proof Let $\varphi \in \mathcal{K}(\Omega \times \Delta(A))$ with $\varphi \geq 0$. By identifying $\mathcal{K}(\Omega \times \Delta(A))$ with $\mathcal{K}(\Omega; \mathcal{C}(\Delta(A)))$ we can find some $\psi \in \mathcal{K}(\Omega; A)$ such that $\widehat{\psi} = \varphi$ i.e., $\mathcal{G}(\psi(x)) = \varphi(x)$ for all $x \in \Omega$; and it is clear that $\psi \geq 0$, hence $\psi^\varepsilon(x) = \psi(x, \frac{x}{\varepsilon}) \geq 0$ for every $x \in \Omega$. It follows $\int_\Omega u_\varepsilon \psi^\varepsilon dx \geq 0$. By letting $E \ni \varepsilon \to 0$, we deduce

$$\int\int_{\Omega\times\Delta(A)} u_0(x, s)\varphi(x, s)dxd\beta(s) \geq 0.$$

Therefore the lemma follows by the arbitrariness of φ and use of Bourbaki [23, p.47,Cor.3]. ■

Remark 5.18 Lemmas 5.13 and 5.14 require neither the $W^{1,p}(\Omega)$-properness of A nor that A^∞ should be a dense subalgebra of A.

At the present time we suppose (5.180) holds. Let the index $1 \leq i \leq N$ be fixed. For $\boldsymbol{\varphi} = (\varphi_j)_{1\leq j\leq N} \in \mathcal{C}_\mathbb{R}(\Delta(A))^N = \mathcal{C}(\Delta(A); \mathbb{R})^N$, let

$$b_i(\boldsymbol{\varphi}) = \mathcal{G}(a_i(\cdot, \mathcal{G}^{-1}\boldsymbol{\varphi})), \tag{5.181}$$

where $\mathcal{G}^{-1}\boldsymbol{\varphi} = (\mathcal{G}^{-1}\varphi_j)_{1\leq j\leq N}$. As pointed out in Sect. 5.5.2, $a_i(\cdot, \mathcal{G}^{-1}\boldsymbol{\varphi})$ lies in $L^\infty_\mathbb{R}(\mathbb{R}^N_y)$. By joining this with (5.180) it follows that $a_i(\cdot, \mathcal{G}^{-1}\boldsymbol{\varphi}) \in \mathfrak{X}_A^{p',\infty} = \mathfrak{X}_A^{p'}(\mathbb{R}^N_y) \cap L^\infty(\mathbb{R}^N_y)$. Therefore $b_i(\boldsymbol{\varphi}) \in L^\infty(\Delta(A))$ (Corollary 2.7). Thus, by (5.181) there is defined a mapping $\boldsymbol{\varphi} \mapsto b_i(\boldsymbol{\varphi})$ of $\mathcal{C}_\mathbb{R}(\Delta(A))^N$ into $L^\infty(\Delta(A))$ denoted by b_i. Let us note that

$$b_i \circ \widehat{\boldsymbol{\Psi}} = \mathcal{G}(a_i(\cdot, \boldsymbol{\Psi})) \text{ for } \boldsymbol{\Psi} = (\psi_j)_{1\leq j\leq N} \in \mathcal{C}(\overline{\Omega}; (A_\mathbb{R})^N), \tag{5.182}$$

i.e., $b_i(\widehat{\boldsymbol{\Psi}}(x)) = \mathcal{G}(a_i(\cdot, \boldsymbol{\Psi}(x)))$ for $x \in \overline{\Omega}$, where $\widehat{\boldsymbol{\Psi}} = (\mathcal{G} \circ \psi_j)_{1\leq j\leq N}$.

In the sequel we will set

$$\mathbf{b}(\boldsymbol{\Phi}) = (b_i \circ \boldsymbol{\Phi})_{1\leq i\leq N} \tag{5.183}$$

for $\boldsymbol{\Phi} \in \mathcal{C}(\overline{\Omega}; \mathcal{C}_\mathbb{R}(\Delta(A))^N)$.

Let us turn now to the statement and proof of the following proposition.

Proposition 5.19 *Let $1 < p \leq 2$. Suppose (5.180) holds. Let the notation be as above. Then the following statements hold.*

(i) For $\boldsymbol{\Psi} \in \mathcal{C}(\overline{\Omega}; (A_{\mathbb{R}})^N)$, we have $b_i \circ \widehat{\boldsymbol{\Psi}} \in \mathcal{C}(\overline{\Omega}; L^\infty(\Delta(A)))$ and

$$a_i^\varepsilon(\cdot, \boldsymbol{\Psi}^\varepsilon) \to b_i \circ \widehat{\boldsymbol{\Psi}} \text{ in } L^{p'}(\Omega)\text{-weak } \Sigma \text{ as } \varepsilon \to 0 \ (1 \leq i \leq N). \tag{5.184}$$

(ii) The mapping $\boldsymbol{\Phi} \mapsto \mathbf{b}(\boldsymbol{\Phi})$ (defined by (5.183)) of $\mathcal{C}(\overline{\Omega}; \mathcal{C}_{\mathbb{R}}(\Delta(A))^N)$ into $L^\infty(\Omega \times \Delta(A))^N$ extends by continuity to a mapping still denoted by $\mathbf{b}$, of $L^p(\Omega; L^p_{\mathbb{R}}(\Delta(A))^N)$ into $L^{p'}(\Omega \times \Delta(A))^N$ such that

$$\|\mathbf{b}(\mathbf{u}) - \mathbf{b}(\mathbf{v})\|_{L^{p'}(\Omega \times \Delta(A))^N} \leq c N^{\frac{p-1}{2}} \|\mathbf{u} - \mathbf{v}\|^{p-1}_{L^p(\Omega; L^p(\Delta(A))^N)} \tag{5.185}$$

and

$$(\mathbf{b}(\mathbf{u}) - \mathbf{b}(\mathbf{v})) \cdot (\mathbf{u} - \mathbf{v}) \geq \alpha |\mathbf{u} - \mathbf{v}|^p \text{ a.e. in } \Omega \times \Delta(A) \tag{5.186}$$

for all $\mathbf{u}, \mathbf{v} \in L^p(\Omega; L^p_{\mathbb{R}}(\Delta(A))^N)$.

Proof Let $\boldsymbol{\Psi} \in \mathcal{C}(\overline{\Omega}; (A_{\mathbb{R}})^N)$. Combination of (5.180) and Lemma 5.12 yields at once $a_i(\cdot, \boldsymbol{\Psi}) \in \mathcal{C}(\overline{\Omega}; \mathfrak{X}_A^{p',\infty})$, where $\mathfrak{X}_A^{p',\infty}$ is equipped with the L^∞ norm. Hence, by (5.182) and use of Corollary 2.7 it follows readily that $b_i \circ \widehat{\boldsymbol{\Psi}} \in \mathcal{C}(\overline{\Omega}; L^\infty(\Delta(A)))$. Furthermore, recalling the equality $a_i^\varepsilon(\cdot, \boldsymbol{\Psi}^\varepsilon) = a_i(\cdot, \boldsymbol{\Psi})^\varepsilon$ (choose $G = a_i$ in Lemma 5.12), we see immediately that (5.184) follows by Lemma 5.13.

With this in mind, we now concentrate on the next part of the proposition. Let $\mathbf{u} = (u_i), \mathbf{v} = (v_i) \in \mathcal{C}(\overline{\Omega}; (A_{\mathbb{R}})^N)$. Let $x \in \overline{\Omega}$ be fixed. According to part (1) of (5.173) we have

$$(\mathbf{a}(\cdot, \mathbf{u}(x, \cdot)) - \mathbf{a}(\cdot, \mathbf{v}(x, \cdot))) \cdot (\mathbf{u}(x, \cdot) - \mathbf{v}(x, \cdot)) \geq \alpha |\mathbf{u}(x, \cdot) - \mathbf{v}(x, \cdot)|^p$$

a.e. in $\mathbb{R}^N_y$, where $\mathbf{a}(\cdot, \mathbf{u}(x, \cdot)) = (a_i(\cdot, \mathbf{u}(x, \cdot)))_{1 \leq i \leq N}$ (idem with $\mathbf{v}$), the real function $a_i(\cdot, \mathbf{u}(x, \cdot))$ being defined as in Lemma 5.12. We deduce by Lemma 5.11 that

$$(\mathbf{a}^\varepsilon(\cdot, \mathbf{u}^\varepsilon) - \mathbf{a}^\varepsilon(\cdot, \mathbf{v}^\varepsilon)) \cdot (\mathbf{u}^\varepsilon - \mathbf{v}^\varepsilon) \geq \alpha |\mathbf{u}^\varepsilon - \mathbf{v}^\varepsilon|^p \text{ a.e. in } \Omega, \tag{5.187}$$

where $\mathbf{a}^\varepsilon(\cdot, \mathbf{u}^\varepsilon) = (a_i^\varepsilon(\cdot, \mathbf{u}^\varepsilon))_{1 \leq i \leq N}$ (idem with $\mathbf{v}^\varepsilon$). On the other hand, by taking in (5.179) $G = a_i$, $\boldsymbol{\Psi} = \mathbf{u}^\varepsilon$ and $\boldsymbol{\Phi} = \mathbf{v}^\varepsilon$, we obtain

$$|a_i^\varepsilon(\cdot, \mathbf{u}^\varepsilon) - a_i^\varepsilon(\cdot, \mathbf{v}^\varepsilon)| \leq c |\mathbf{u}^\varepsilon - \mathbf{v}^\varepsilon|^{p-1} \text{ a.e. in } \Omega \ (1 \leq i \leq N). \tag{5.188}$$

The next step is to pass to the limit above when $\varepsilon \to 0$. First of all, it is clear that $u_i^\varepsilon - v_i^\varepsilon \to \widehat{u}_i - \widehat{v}_i$ in $L^p(\Omega)$-strong Σ as $\varepsilon \to 0$. It follows readily by (5.184) that the left-hand side of (5.187) weakly Σ-converges in $L^1(\Omega)$ to $(b(\widehat{\mathbf{u}})-b(\widehat{\mathbf{v}}))\cdot(\widehat{\mathbf{u}}-\widehat{\mathbf{v}})$ (see Proposition 4.3). By an obvious argument (see Chap. 4) one has on the other hand $|\mathbf{u}^\varepsilon - \mathbf{v}^\varepsilon|^p \to |\widehat{\mathbf{u}}-\widehat{\mathbf{v}}|^p$ in $L^1(\Omega)$-weak Σ as $\varepsilon \to 0$. Hence, by letting $\varepsilon \to 0$ in (5.187) one obtains by Lemma 5.14

$$(\mathbf{b}(\widehat{\mathbf{u}}) - \mathbf{b}(\widehat{\mathbf{v}})) \cdot (\widehat{\mathbf{u}} - \widehat{\mathbf{v}}) \geq \alpha |\widehat{\mathbf{u}} - \widehat{\mathbf{v}}|^p \text{ a.e. in } \Omega \times \Delta(A)$$

and that for all $\mathbf{u}, \mathbf{v} \in \mathcal{C}(\overline{\Omega}; (A_\mathbb{R})^N)$. We will later deduce (5.186) from this inequality. Meanwhile, let us turn to (5.188), which reads

$$-c|\mathbf{u}^\varepsilon - \mathbf{v}^\varepsilon|^{p-1} \leq a_i^\varepsilon(\cdot, \mathbf{u}^\varepsilon) - a_i^\varepsilon(\cdot, \mathbf{v}^\varepsilon) \leq c|\mathbf{u}^\varepsilon - \mathbf{v}^\varepsilon|^{p-1}$$

a.e. in Ω $(1 \leq i \leq N)$. Since $|\mathbf{u}^\varepsilon - \mathbf{v}^\varepsilon|^{p-1} \to |\widehat{\mathbf{u}}-\widehat{\mathbf{v}}|^{p-1}$ in $L^{p'}(\Omega)$-weak as $\varepsilon \to 0$, by (5.184) and use of Lemma 5.14 we are readily led to

$$|b_i \circ \widehat{\mathbf{u}} - b_i \circ \widehat{\mathbf{v}}| \leq c|\widehat{\mathbf{u}} - \widehat{\mathbf{v}}|^{p-1} \text{ a.e. in } \Omega \times \Delta(A) \ \ (1 \leq i \leq N).$$

Whence

$$\|\mathbf{b}(\widehat{\mathbf{u}}) - \mathbf{b}(\widehat{\mathbf{v}})\|_{L^{p'}(\Omega\times\Delta(A))} \leq cN^{\frac{p-1}{2}} \|\widehat{\mathbf{u}} - \widehat{\mathbf{v}}\|^{p-1}_{L^p(\Omega;L^p(\Delta(A))^N)}.$$

Therefore, since the mapping $\mathbf{u} \mapsto \widehat{\mathbf{u}}$ sends $\mathcal{C}(\overline{\Omega}; (A_\mathbb{R})^N)$ isomorphically over $\mathcal{C}(\overline{\Omega}; \mathcal{C}_\mathbb{R}(\Delta(A))^N)$, we see at once that (5.185) and (5.186) have just been proven for all $\mathbf{u}, \mathbf{v} \in \mathcal{C}(\overline{\Omega}; \mathcal{C}_\mathbb{R}(\Delta(A))^N)$. Hence, thanks to the density of $\mathcal{C}(\overline{\Omega}; \mathcal{C}_\mathbb{R}(\Delta(A))^N)$ in $L^p(\Omega; L^p_\mathbb{R}(\Delta(A))^N)$, it is clear that the mapping $\mathbf{u} \mapsto \mathbf{b}(\mathbf{u})$ of $\mathcal{C}(\overline{\Omega}; \mathcal{C}_\mathbb{R}(\Delta(A))^N)$ into $L^\infty(\Omega \times \Delta(A))^N$ extends by continuity to a mapping, still called $\mathbf{b}$, of $L^p(\Omega; L^p_\mathbb{R}(\Delta(A))^N)$ into $L^{p'}(\Omega \times \Delta(A))^N$ satisfying (5.185) for all $\mathbf{u}, \mathbf{v} \in L^p(\Omega; L^p_\mathbb{R}(\Delta(A))^N)$. Finally, (5.186) follows in an obvious way by the same density result and use of a classical argument from integration theory (see, e.g., Bourbaki [22, p.131]). ■

Attention must be drawn to an important corollary of this proposition.

Corollary 5.21 *Let $\Phi_\varepsilon = \psi_0 + \varepsilon\psi_1^\varepsilon$ with $\psi_0 \in \mathcal{D}_\mathbb{R}(\Omega)$ and $\psi_1 \in \mathcal{D}_\mathbb{R}(\Omega) \otimes A_\mathbb{R}^\infty$ with $A_\mathbb{R}^\infty = A^\infty \cap \mathcal{C}_\mathbb{R}(\mathbb{R}^N)$. Then, as $\varepsilon \to 0$,*

$$a_i^\varepsilon(\cdot, D\Phi_\varepsilon) \to b_i \circ (D_x\psi_0 + \nabla\widehat{\psi}_1) \text{ in } L^{p'}(\Omega)\text{-weak } \Sigma \ \ (1 \leq i \leq N), \tag{5.189}$$

where $\nabla\widehat{\psi}_1 = (\partial_j\widehat{\psi}_1)_{1\leq j\leq N}$. Furthermore, if E is a fundamental sequence, and $(v_\varepsilon)_{\varepsilon\in E}$ is a sequence in $L^p(\Omega)$ such that $v_\varepsilon \to v_0$ in $L^p(\Omega)$-weak Σ, then

$$\int_\Omega a_i^\varepsilon(\cdot, D\Phi_\varepsilon)v_\varepsilon\, dx \to \int\int_{\Omega\times\Delta(A)} b_i \circ (D_x\psi_0 + \nabla\widehat{\psi}_1)v_0\, dxd\beta \ \ (1 \leq i \leq N)$$

as $E \ni \varepsilon \to 0$.

Proof It is clear that

$$D\Phi_\varepsilon = D_x\psi_0 + (D_y\psi_1)^\varepsilon + \varepsilon(D_x\psi_1)^\varepsilon.$$

Therefore, bearing in mind that $1 < p \leq 2$, it follows

$$\begin{aligned}&\|a_i^\varepsilon(\cdot, D\Phi_\varepsilon) - a_i^\varepsilon(\cdot, D_x\psi_0 + (D_y\psi_1)^\varepsilon)\|_{L^{p'}(\Omega)} \leq \\ &c\varepsilon^{p-1}\|D_x\psi_1\|^{p-1}_{L^p(\Omega;\mathcal{B}(\mathbb{R}^N))^N}\end{aligned} \tag{5.190}$$

for $1 \leq i \leq N$. This being so, by Proposition 5.19 we have, as $\varepsilon \to 0$, $a_i^\varepsilon(\cdot, D_x\psi_0 + (D_y\psi_1)^\varepsilon) \to b_i \circ (D_x\psi_0 + \nabla\widehat{\psi}_1)$ in $L^{p'}(\Omega)$-weak Σ. Hence (5.189) follows by (5.190). Similarly, since the function $a_i(\cdot, D_x\psi_0 + D_y\psi_1)$ lies in $\mathcal{C}(\overline{\Omega}; \mathfrak{X}_A^{p',\infty})$, we see readily that the last part of the corollary follows by Proposition 4.4 and use of Lemma 5.12 together with (5.190), once again. ■

5.5.3.2 The Abstract Homogenization Theorem

Our purpose here is to state and prove the main homogenization result for the boundary value problem (5.174) under the abstract deterministic hypothesis (5.180). Before we can do this, however, we need further preliminaries.

Let us begin by reintroducing the «global differential operator» $\mathbb{D}$ on $\Omega \times \Delta(A)$ which was being dealt with throughout. More precisely, for $\mathbf{v} = (v_0, v_1) \in \mathbb{F}_0^{1,p}(\Omega \times \Delta(A); \mathbb{R})$, let

$$\mathbb{D}_j\mathbf{v} = \frac{\partial v_0}{\partial x_j} + \partial_j v_1 \quad (1 \leq j \leq N).$$

It is clear that the operator $\mathbb{D}_j$ sends continuously and linearly $\mathbb{F}_0^{1,p}(\Omega \times \Delta(A); \mathbb{R})$ into $L^p(\Omega \times \Delta(A); \mathbb{R})$. In the sequel we let $\mathbb{D} = (\mathbb{D}_j)_{1\leq j\leq N}$, i.e., $\mathbb{D}\mathbf{v} = (\mathbb{D}_j\mathbf{v})_{1\leq j\leq N}$, $\mathbf{v} \in \mathbb{F}_0^{1,p}(\Omega \times \Delta(A); \mathbb{R})$. Then

$$\mathbb{D}\mathbf{v} = D_x v_0 + \nabla v_1 \text{ for } \mathbf{v} = (v_0, v_1) \in \mathbb{F}_0^{1,p}(\Omega \times \Delta(A); \mathbb{R}).$$

Having done this, we next consider the variational problem

$$\begin{cases}\mathbf{u} = (u_0, u_1) \in \mathbb{F}_0^{1,p}(\Omega \times \Delta(A); \mathbb{R}) : \\ \displaystyle\int\int_{\Omega\times\Delta(A)} \mathbf{b}(\mathbb{D}\mathbf{u}) \cdot \mathbb{D}\mathbf{v}\, dxd\beta = \langle f, v_0\rangle \\ \text{for all } \mathbf{v} = (v_0, v_1) \in \mathbb{F}_0^{1,p}(\Omega \times \Delta(A); \mathbb{R}),\end{cases} \tag{5.191}$$

where $\langle , \rangle$ denotes the duality pairing between $W^{-1,p'}(\Omega;\mathbb{R})$ and $W_0^{1,p}(\Omega;\mathbb{R})$. Let us establish the following lemma.

Lemma 5.15 *For $1 < p \leq 2$, the variational problem (5.191) admits at most one solution. If in particular $p = 2$, then there exists one and only one vector function $\mathbf{u} = (u_0, u_1)$ satisfying (5.191).*

Proof Assuming that (5.191) admits two solutions $\mathbf{u} = (u_0, u_1)$ and $\mathbf{u}' = (u_0', u_1')$ leads by (5.186) to

$$\alpha \|\mathbb{D}(\mathbf{u}' - \mathbf{u})\|^p_{L^p(\Omega\times\Delta(A))^N} \leq \int\int_{\Omega\times\Delta(A)} [\mathbf{b}(\mathbb{D}\mathbf{u}') - \mathbf{b}(\mathbb{D}\mathbf{u})]\cdot\mathbb{D}(\mathbf{u}' - \mathbf{u})dxd\beta = 0.$$

We deduce $\mathbb{D}\mathbf{z} = 0$, where $\mathbf{z} = \mathbf{u}' - \mathbf{u}$, $\mathbf{z} = (z_0, z_1)$. Thus,

$$\frac{\partial z_0}{\partial x_j} + \partial_j z_1 = 0 \text{ a.e. in } \Omega \times \Delta(A) \ \ (1 \leq j \leq N).$$

But, for a.e. $x \in \Omega$, we have $\int_{\Delta(A)} \partial_j z_1(x,\cdot)d\beta = 0$ (see Remark 4.6). We deduce at once $\frac{\partial z_0}{\partial x_j} = 0$ $(1 \leq j \leq N)$, hence $z_0 = 0$. Consequently $z_1 = 0$. Finally $\mathbf{u} = \mathbf{u}'$, which shows the first part of the lemma. As regards the next part, let T be the operator of $\mathbb{F}_0^1 \equiv \mathbb{F}_0^{1,2}(\Omega \times \Delta(A);\mathbb{R})$ into $(\mathbb{F}_0^1)'$ (dual of $\mathbb{F}_0^1$) defined by

$$\langle T\mathbf{z}, \mathbf{v}\rangle = \int\int_{\Omega\times\Delta(A)} \mathbf{b}(\mathbb{D}\mathbf{z})\cdot\mathbb{D}\mathbf{v}\, dxd\beta \ \ (\mathbf{z}, \mathbf{v} \in \mathbb{F}_0^1),$$

where the brackets denote the duality pairing between $(\mathbb{F}_0^1)'$ and $\mathbb{F}_0^1$. Let us first of all observe that $\mathbb{F}_0^1$ (as a Hilbert space with an obvious equivalent norm) is a reflexive Banach space. Now, by (5.185) it is an easy matter to show that the operator T is bounded and demicontinuous. Next, by (5.186) and the obvious property $\mathbf{b}(\mathbf{0}) = \mathbf{0}$, we have

$$\frac{\langle T\mathbf{v}, \mathbf{v}\rangle}{\|\mathbf{v}\|_{\mathbb{F}_0^1}} \geq \alpha\|\mathbf{v}\|_{\mathbb{F}_0^1} \ \ (\mathbf{v} \in \mathbb{F}_0^1, \mathbf{v} \neq \mathbf{0}),$$

from which it follows that T is coercive. Finally, it is clear by (5.186) that T is strictly monotone. Consequently, if we consider $\mathbf{l} \in (\mathbb{F}_0^1)'$ given by $\langle \mathbf{l}, \mathbf{v}\rangle = \langle f, v_0\rangle$ for $\mathbf{v} = (v_0, v_1) \in \mathbb{F}_0^1$, then, thanks to Browder theorem there is a unique $\mathbf{u} \in \mathbb{F}_0^1$ such that $T\mathbf{u} = \mathbf{l}$, that is, such that (5.191) holds. This completes the proof. ∎

We are now in a position to achieve the desired result, namely

Theorem 5.9 *Let $1 < p \leq 2$. Suppose (5.180) holds. For each $\varepsilon > 0$, let u_ε be the solution of (5.174). Then, as $\varepsilon \to 0$, we have*

$$u_\varepsilon \to u_0 \text{ in } W_0^{1,p}(\Omega)\text{-weak,} \tag{5.192}$$

$$u_\varepsilon \to u_0 \text{ in } L^p(\Omega), \tag{5.193}$$

$$\frac{\partial u_\varepsilon}{\partial x_j} \to \mathbb{D}_j \mathbf{u} = \frac{\partial u_0}{\partial x_j} + \partial_j u_1 \text{ in } L^p(\Omega)\text{-weak } \Sigma \ (1 \leq j \leq N), \tag{5.194}$$

where $\mathbf{u} = (u_0, u_1)$ is the unique solution of (5.191).

Proof One has

$$\int_\Omega \mathbf{a}^\varepsilon(\cdot, Du_\varepsilon) \cdot Dv\, dx = \langle f, v \rangle \tag{5.195}$$

for all $v \in W_0^{1,p}(\Omega; \mathbb{R})$. Taking in particular $v = u_\varepsilon$ and using the properties of the mapping $v \mapsto \mathbf{a}^\varepsilon(\cdot, Dv)$ (see Sect. 5.5.2) gives $\|u_\varepsilon\|_{W_0^{1,p}(\Omega)} \leq (\alpha^{-1}\|f\|_{W^{-1,p'}(\Omega)})^{\frac{1}{p-1}}$ for all $\varepsilon > 0$. Therefore, given an arbitrary fundamental sequence E, the $W^{1,p}(\Omega)$-properness of A (see also Proposition 4.5) and the compactness of the embedding $W_0^{1,p}(\Omega; \mathbb{R}) \hookrightarrow L^p_{\mathbb{R}}(\Omega)$ yield a subsequence E' from E and some $\mathbf{u} = (u_0, u_1) \in \mathbb{F}_0^{1,p}(\Omega \times \Delta(A); \mathbb{R})$ such that (5.192), (5.193) and (5.194) hold when $E' \ni \varepsilon \to 0$. Clearly, thanks to Lemma 5.15, the theorem will readily follow by the arbitrariness of E if we show that $\mathbf{u} = (u_0, u_1)$ satisfies (5.191). For that purpose, let $\mathbf{\Phi} \in \mathcal{F}_0^\infty$, i.e., $\mathbf{\Phi} = (\psi_0, J_\Omega(\widehat{\psi}_1))$ with $\psi_0 \in \mathcal{D}_{\mathbb{R}}(\Omega)$, $\psi_1 \in \mathcal{D}_{\mathbb{R}}(\Omega) \otimes (A^\infty_{\mathbb{R}}/\mathbb{R})$ (see Remark 5.17). Define Φ_ε as in Corollary 5.21. Next take in (5.195) the particular function $v = u_\varepsilon - \Phi_\varepsilon$ and use the inequality

$$\int_\Omega [\mathbf{a}^\varepsilon(\cdot, Du_\varepsilon) - \mathbf{a}^\varepsilon(\cdot, D\Phi_\varepsilon)] \cdot (Du_\varepsilon - D\Phi_\varepsilon) dx \geq 0$$

to arrive at

$$0 \leq \langle f, u_\varepsilon - \Phi_\varepsilon \rangle - \int_\Omega \mathbf{a}^\varepsilon(\cdot, D\Phi_\varepsilon) \cdot (Du_\varepsilon - D\Phi_\varepsilon) dx. \tag{5.196}$$

Now, let us observe that $\Phi_\varepsilon \to \psi_0$ in $W^{1,p}(\Omega)$-weak Σ as $\varepsilon \to 0$ (see Definition 4.3), that is, as $\varepsilon \to 0$, we have $\Phi_\varepsilon \to \psi_0$ in $L^p(\Omega)$-weak Σ (this is obvious) and

$$\frac{\partial \Phi_\varepsilon}{\partial x_j} \to \mathbb{D}_j \mathbf{\Phi} = \frac{\partial \psi_0}{\partial x_j} + \partial_j \widehat{\psi}_1 \text{ in } L^p(\Omega)\text{-weak } \Sigma \ (1 \leq j \leq N)$$

as is immediate by Proposition 5.13 (see especially (5.126)) and use of Proposition 4.2. Noting that this implies $\Phi_\varepsilon \to \psi_0$ in $W_0^{1,p}(\Omega)$-weak as $\varepsilon \to 0$, we next pass to the limit (as $E' \ni \varepsilon \to 0$) in (5.196) using (5.192), (5.194) and Corollary 5.21 with $v_\varepsilon = u_\varepsilon - \Phi_\varepsilon$, and we obtain

$$0 \leq \langle f, u_0 - \psi_0 \rangle - \int\int_{\Omega \times \Delta(A)} \mathbf{b}(\mathbb{D}\mathbf{\Phi}) \cdot \mathbb{D}(\mathbf{u} - \mathbf{\Phi}) dx d\beta. \tag{5.197}$$

Thanks to the density of $\mathcal{F}_0^\infty$ in $\mathbb{F}_0^{1,p}(\Omega \times \Delta(A); \mathbb{R})$ together with (5.185), this still holds for any $\mathbf{\Phi} \in \mathbb{F}_0^{1,p}(\Omega \times \Delta(A); \mathbb{R})$. Finally, take in (5.197) the particular $\mathbf{\Phi} = \mathbf{u} - t\mathbf{v}$ with $t > 0$ and $\mathbf{v} = (v_0, v_1) \in \mathbb{F}_0^{1,p}(\Omega \times \Delta(A); \mathbb{R})$, then divide both sides of the resultant inequality by t and, letting $t \to 0$, pass to the limit using (5.185). Hence (5.191) follows at once. ■

The variational problem (5.191) is referred to as the *global homogenized problem for* (5.174) under the abstract structure hypothesis (5.180). It is a rather simple exercise to verify that (5.191) is equivalent to looking for $\mathbf{u} = (u_0, u_1) \in \mathbb{F}_0^{1,p}(\Omega \times \Delta(A); \mathbb{R})$ satisfying

$$\int\int_{\Omega \times \Delta(A)} \mathbf{b}(\mathbb{D}\mathbf{u}) \cdot \nabla v_1 \, dx d\beta = 0 \tag{5.198}$$

and

$$\int_\Omega \left[\int_{\Delta(A)} \mathbf{b}(\mathbb{D}\mathbf{u}) d\beta \right] \cdot D_x v_0 \, dx = \langle f, v_0 \rangle \tag{5.199}$$

for all $v_1 \in L^p(\Omega; W_\#^{1,p}(\Delta(A); \mathbb{R}))$ and all $v_0 \in W_0^{1,p}(\Omega; \mathbb{R})$, respectively.

The next point is to derive the *macroscopic homogenized problem for* (5.174). To this end, let $r \in \mathbb{R}^N$ be freely fixed, and let us introduce the so-called *cell problem at point* r, viz.

$$\begin{cases} \pi(r) \in W_\#^{1,p}(\Delta(A); \mathbb{R}) : \\ \displaystyle\int_{\Delta(A)} \mathbf{b}(r + \nabla\pi(r)) \cdot \nabla\theta \, d\beta = 0 \\ \text{for all } \theta \in W_\#^{1,p}(\Delta(A); \mathbb{R}). \end{cases} \tag{5.200}$$

In default of the availability of an existence and uniqueness result one can, however, at least state that (5.200) admits at most one solution, as is straightforward by (5.186). In the sequel, $\mathfrak{S}_A$ denotes the set of those $r \in \mathbb{R}^N$ for which (5.200) has a solution (which solution is unique!). Thus, $r \mapsto \pi(r)$ may be viewed as a mapping of $\mathfrak{S}_A$ into $W_{\#}^{1,p}(\Delta(A); \mathbb{R})$. This yields a mapping

$$\mathbf{q} : \mathfrak{S}_A \to \mathbb{R}^N \quad \text{given by}$$

$$\mathbf{q}(r) = \int_{\Delta(A)} \mathbf{b}(r + \nabla\pi(r))d\beta \quad (r \in \mathfrak{S}_A).$$

Now, choose v_1 in (5.198) of the form $v_1(x) = \varphi(x)\theta \ (x \in \Omega)$ with $\varphi \in \mathcal{D}_{\mathbb{R}}(\Omega)$ and $\theta \in W_{\#}^{1,p}(\Delta(A); \mathbb{R})$. Then, for a.e. $x \in \Omega$,

$$\int_{\Delta(A)} \mathbf{b}(Du_0(x) + \nabla u_1(x)) \cdot \nabla\theta \, d\beta = 0 \text{ for all } \theta \in W_{\#}^{1,p}(\Delta(A); \mathbb{R}).$$

It follows readily that $Du_0(x) \in \mathfrak{S}_A$ for almost every $x \in \Omega$ and further $u_1 = \pi \circ Du_0$, where the right-hand side stands for the function $x \mapsto \pi(Du_0(x))$ of Ω into $W_{\#}^{1,p}(\Delta(A); \mathbb{R})$. Hence we see at once by (5.199) that u_0 is the (unique) solution of the boundary value problem

$$-\operatorname{div}(\mathbf{q} \circ Du_0) = f \text{ in } \Omega, \quad u_0 \in W_0^{1,p}(\Omega; \mathbb{R}).$$

5.5.4 Some Concrete Homogenization Problems for (5.174)

It is worth recalling that a homogenization problem is posed as soon as one has on one hand a suitable boundary value problem, on the other hand a so-called structure hypothesis. The term *concrete* is used here to lay emphasis on the fact that the structure hypothesis appears under a natural form, as opposed to the abstract hypothesis (5.180).

This section deals with the study of a few concrete homogenization problems for (5.174). Thanks to the results achieved in the preceding sections (see especially Theorem 5.9), the whole problem in each case will reduce to showing that the associated concrete structure hypothesis can write as (5.180) for a suitable $W^{1,p}(\Omega)$-proper homogenization algebra A on $\mathbb{R}^N$. However, some preliminaries are necessary.

5.5.4.1 Preliminaries

In what follows, A denotes a closed subalgebra of $\mathcal{B}(\mathbb{R}^N)$. We equip A with the supremum norm on $\mathbb{R}^N$, which makes it into a Banach algebra. Let $A_{\mathbb{R}} = A \cap \mathcal{C}(\mathbb{R}^N; \mathbb{R})$. Finally, let G be a function on $\mathbb{R}^N \times \mathbb{R}^N$ into $\mathbb{R}$ satisfying (5.177). We start with a lemma of great interest.

Lemma 5.16 *Suppose we have*

$$G(\cdot,\mu) \in A \text{ for fixed } \mu \in \mathbb{R}^N. \tag{5.201}$$

Then

$$G(\cdot,\boldsymbol{\Psi}) \in A \text{ for all } \boldsymbol{\Psi} \in (A_{\mathbb{R}})^N. \tag{5.202}$$

Proof To begin, we note that according to (5.177) and (5.201), the function $\mu \mapsto G(\cdot,\mu)$ lies in $\mathcal{C}(\mathbb{R}^N; A)$. With this in mind, fix freely some $\boldsymbol{\Psi}$ as in (5.202). Let K be a compact set in $\mathbb{R}^N$ such that $\boldsymbol{\Psi}(y) \in K$ for all $y \in \mathbb{R}^N$. Let G_K denote the function of $\mathbb{R}^N \times K$ into $\mathbb{R}$ defined by $G_K(y,\mu) = G(y,\mu)$ $(y \in \mathbb{R}^N, \mu \in K)$. Clearly $G_K \in \mathcal{C}(K; A_{\mathbb{R}})$, $A_{\mathbb{R}}$ being equipped with the supremum norm, which makes it into a real Banach algebra. Thus, the aim is to prove that $G_K(\cdot,\boldsymbol{\Psi}) \in A_{\mathbb{R}}$, where $G_K(\cdot,\boldsymbol{\Psi})$ stands for the function $y \mapsto G_K(y,\boldsymbol{\Psi}(y))$ of $\mathbb{R}^N$ into $\mathbb{R}$. Actually, the problem before us reduces to showing that $f(\cdot,\boldsymbol{\Psi}) \in A_{\mathbb{R}}$ for any $f \in \mathcal{C}(K; A_{\mathbb{R}})$. Let us start with the particular case of functions $f \in \mathcal{C}_{\mathbb{R}}(K) \otimes A_{\mathbb{R}}$. But then, it clearly suffices to establish the property for the elementary functions $f = \chi \otimes \varphi$, i.e., $f(y,\mu) = \chi(\mu)\varphi(y)$ $(y \in \mathbb{R}^N, \mu \in K)$ with $\chi \in \mathcal{C}_{\mathbb{R}}(K)$ and $\varphi \in A_{\mathbb{R}}$. Given freely one such f, by the Stone-Weierstrass theorem (see, e.g., Schwartz [127, p.376]) there is a sequence $(\zeta_n)_{n\in\mathbb{N}}$ of polynomial functions (with real coefficients) in $\mu = (\mu_1, \cdots, \mu_N) \in K$ such that $\zeta_n \to \chi$ in $\mathcal{C}(K)$ as $n \to \infty$. We deduce $\zeta_n \circ \boldsymbol{\Psi} \to \chi \circ \boldsymbol{\Psi}$ in $\mathcal{B}(\mathbb{R}^N)$ as $n \to \infty$, whence $\chi \circ \boldsymbol{\Psi} \in A_{\mathbb{R}}$, since $\zeta_n \circ \boldsymbol{\Psi} \in A_{\mathbb{R}}$. Hence $f(\cdot,\boldsymbol{\Psi}) \in A_{\mathbb{R}}$, and that for any $f \in \mathcal{C}_{\mathbb{R}}(K) \otimes A_{\mathbb{R}}$.

Bearing this in mind, let us turn now to the general case of a function $f \in \mathcal{C}(K; A_{\mathbb{R}})$. Thanks to a classical density result, there is a sequence of functions $f_n \in \mathcal{C}_{\mathbb{R}}(K) \otimes A_{\mathbb{R}}$ $(n \in \mathbb{N})$ such that one has $\sup_{\mu\in K} \|f_n(\cdot,\mu) - f(\cdot,\mu)\|_\infty \to 0$ as $n \to \infty$. It follows readily that $f_n(\cdot,\boldsymbol{\Psi}) \to f(\cdot,\boldsymbol{\Psi})$ in $\mathcal{B}(\mathbb{R}^N)$ as $n \to \infty$, hence $f(\cdot,\boldsymbol{\Psi}) \in A_{\mathbb{R}}$. Therefore the desired result follows by the choice $f = G_K$. ■

This lemma has an immediate consequence in connection with the homogenization of (5.174). Specifically, for $1 < p \leq 2$ one has the following

Proposition 5.20 *Suppose A is a $W^{1,p}(\Omega)$-proper H-algebra with A^∞ as a dense subalgebra. Assume that $a_i(\cdot,\mu) \in A$ $(1 \leq i \leq N)$ for fixed $\mu \in \mathbb{R}^N$. Then (5.180) holds, and so Theorem 5.9 is valid.*

Proof This is straightforward by Lemma 5.16 (with $G = a_i$) and use of the fact that $A \subset \mathfrak{X}_A^{p'}(\mathbb{R}^N)$. ■

We also wish to draw attention to a property of great interest as regards the applications. Let $Y = (-\frac{1}{2}, \frac{1}{2})^N$.

Definition 5.1 For given $1 \leq q < \infty$, the function $\mathbf{a} = (a_i) : \mathbb{R}^N \times \mathbb{R}^N \to \mathbb{R}$ is said to amalgamate the spaces $\mathcal{B}(\mathbb{R}^N)$ and $(L^q, l^\infty)(\mathbb{R}^N)$ if for any $\boldsymbol{\Psi} \in \mathcal{B}(\mathbb{R}^N; \mathbb{R})^N$, we have

$$\sup_{k \in \mathbb{Z}^N} \int_{k+Y} |\mathbf{a}(y - r, \boldsymbol{\Psi}(y)) - \mathbf{a}(y, \boldsymbol{\Psi}(y))|^q dy \to 0 \text{ as } |r| \to 0.$$

Remark 5.19 For brevity this convergence property also reads

$$\lim_{|r| \to 0} \|\tau_r a_i(\cdot, \boldsymbol{\Psi}) - a_i(\cdot, \boldsymbol{\Psi})\|_{q,\infty} = 0 \ \ (1 \leq i \leq N),$$

where $\tau_r a_i$ stands for the function $(y, \mu) \mapsto \tau_r a_i(y, \mu) = a_i(y - r, \mu)$ of $\mathbb{R}^N \times \mathbb{R}^N$ into $\mathbb{R}$, and where $\|\cdot\|_{q,\infty}$ denotes the norm in $(L^q, l^\infty)(\mathbb{R}^N)$ (see Sect. 2.4).

This is worth illustrating.

Example 5.21 Suppose the family $\{\mathbf{a}(\cdot, \mu)\}_{\mu \in \mathbb{R}^N}$ is uniformly equicontinuous in the following classical sense: To any given $\eta > 0$ there is attached some $\rho > 0$ such that $|\mathbf{a}(y - r, \mu) - \mathbf{a}(y, \mu)| \leq \eta$ for all $y, \mu \in \mathbb{R}^N$ provided $|r| \leq \rho$. Then the function $\mathbf{a}$ amalgamates $\mathcal{B}(\mathbb{R}^N)$ and $(L^q, l^\infty)(\mathbb{R}^N)$ for all $1 \leq q < \infty$.

Example 5.22 Suppose we have $a_i \in \mathcal{C}_\mathbb{R}(\mathbb{R}^N) \otimes L^q_{AP}(\mathbb{R}^N; \mathbb{R})$ $(1 \leq i \leq N)$, i.e., each a_i is of the form $a_i(y, \mu) = \sum_{\text{finite}} \chi_{ij}(\mu)\varphi_{ij}(y)$ $(\mu, y \in \mathbb{R}^N$, fixed $i)$, where $\chi_{ij} \in \mathcal{C}_\mathbb{R}(\mathbb{R}^N)$ and $\varphi_{ij} \in L^q_{AP}(\mathbb{R}^N; \mathbb{R})$. Then $\mathbf{a}$ amalgamates $\mathcal{B}(\mathbb{R}^N)$ and $(L^q, l^\infty)(\mathbb{R}^N)$. Indeed, it is enough to verify this for a_i of the form $a_i(y, \mu) = \chi(\mu)\varphi(y)$ $(y, \mu \in \mathbb{R}^N)$ with $\chi \in \mathcal{C}_\mathbb{R}(\mathbb{R}^N)$ and $\varphi \in L^q_{AP}(\mathbb{R}^N; \mathbb{R})$. Fix freely $\boldsymbol{\Psi} \in \mathcal{B}(\mathbb{R}^N; \mathbb{R})^N$, and let K be some compact set in $\mathbb{R}^N$ such that $\boldsymbol{\Psi}(y) \in K$ for all $y \in \mathbb{R}^N$. Then

$$\sup_{k \in \mathbb{Z}^N} \int_{k+Y} |a_i(y - r, \boldsymbol{\Psi}(y)) - a_i(y, \boldsymbol{\Psi}(y))|^q dy \leq c_K \|\varphi - \tau_r \varphi\|^q_{q,\infty}$$

with $c_K = sup_{\mu \in K} |\chi(\mu)|^q$. Hence the desired property follows by the fact that $\tau_r \varphi \to \varphi$ in $(L^q, l^\infty)(\mathbb{R}^N)$ as $|r| \to 0$ (see Proposition 2.20).

Example 5.23 Assume that $a_i \in \mathcal{C}_\mathbb{R}(\mathbb{R}^N) \otimes L^q(\mathbb{R}^N; \mathbb{R})$ $(1 \leq i \leq N)$. Then the function $\mathbf{a} = (a_i)$ amalgamates $\mathcal{B}(\mathbb{R}^N)$ and $(L^q, l^\infty)(\mathbb{R}^N)$. Indeed, for $u \in L^q(\mathbb{R}^N)$ with $1 \leq q < \infty$, we have $\tau_r u \to u$ in $L^q(\mathbb{R}^N)$ as $|r| \to 0$ and further $L^q(\mathbb{R}^N)$ is continuously embedded in $(L^q, l^\infty)(\mathbb{R}^N)$. Therefore the claimed property is obtained by the same argument as followed in Example 5.22.

5.5.4.2 Problem I (Periodic Setting)

We consider the homogenization problem for (5.174) under the classical periodicity hypothesis

$$a_i(\cdot,\mu) \text{ is } Y\text{-periodic } (1 \le i \le N) \text{ for fixed } \mu \in \mathbb{R}^N,$$

where Y is the same as before. Let $A = \mathcal{C}_{per}(Y)$. Let $1 \le i \le N$. For $\boldsymbol{\Psi} \in (A_{\mathbb{R}})^N$, one clearly has $a_i(\cdot,\boldsymbol{\Psi}) \in L^\infty_{per}(Y;\mathbb{R})$ (see Sect. 5.5.2). Therefore $a_i(\cdot,\boldsymbol{\Psi}) \in L^{p'}_{per}(Y) = \mathfrak{X}^{p'}_A$ and so the problem is solved by applying Theorem 5.9 (valid here for all $1 < p \le 2$), since the H-algebra A has all the required properties. Concretely, the periodic homogenization of (5.174) can be easily deduced from the abstract formulation by simply rewriting the text in periodic terms (see Chaps. 2 and 3).

Remark 5.20 This development is valid without the slightest change if Y is any parallelepiped as introduced in Sect. 2.5.

5.5.4.3 Problem II (Almost Periodic Setting)

We assume here that the function $\mathbf{a} = (a_i)$ amalgamates the spaces $\mathcal{B}(\mathbb{R}^N)$ and $(L^2, l^\infty)(\mathbb{R}^N)$. Our purpose is then to study the homogenization of (5.174) for $p = 2$, under the structure hypothesis

$$a_i(\cdot,\mu) \in L^1_{AP}(\mathbb{R}^N)\ (1 \le i \le N) \text{ for fixed } \mu \in \mathbb{R}^N. \tag{5.203}$$

It is worth perhaps recalling that $L^1_{AP}(\mathbb{R}^N)$ is the space of all those functions in $L^1_{loc}(\mathbb{R}^N)$ that are termed almost periodic in the sense of Stepanoff (Sect. 2.4).

This being so, a trivial adaptation of Lemma 5.3 leads us to a countable subgroup $\mathcal{R}$ of $\mathbb{R}^N$ such that

$$a_i(\cdot,\mu) \in L^1_{\mathcal{R}}(\mathbb{R}^N)\ (1 \le i \le N) \text{ for fixed } \mu \in \mathbb{R}^N, \tag{5.204}$$

where $L^1_{\mathcal{R}}(\mathbb{R}^N) = \{v | v \in L^1_{AP}(\mathbb{R}^N), Sp(v) \subset \mathcal{R}\}$ (Chap. 2). This suggests that we should consider the H-algebra $A = AP_{\mathcal{R}}(\mathbb{R}^N)$ which is $W^{1,2}$-proper and which admits A^∞ as a dense subalgebra.

Now, let $(\theta_n)_{n\ge 1}$ be a mollifier on $\mathbb{R}^N$ with $Supp\theta_n \subset \frac{1}{n}\overline{B}_N$ (see the proof of Proposition 2.20). Let

$$\xi^i_n(y,\mu) = \int \theta_n(r) a_i(y - r,\mu)dr \quad (1 \le i \le N) \text{ for } y,\mu \in \mathbb{R}^N. \tag{5.205}$$

For fixed $1 \leq i \leq N$ and $n \in \mathbb{N}^*$, the family $\{\xi_n^i(\cdot,\mu)\}_{\mu\in\mathbb{R}^N}$ satisfies

$$\xi_n^i(\cdot,\mu) \in A_{\mathbb{R}} \text{ for fixed } \mu \in \mathbb{R}^N \tag{5.206}$$

and

$$|\xi_n^i(y,\mu) - \xi_n^i(y,\nu)| \leq c|\mu - \nu|^{p-1} \quad (y,\mu,\nu \in \mathbb{R}^N) \tag{5.207}$$

for $p = 2$. Indeed (5.206) follows by (5.204) and Lemma 2.9, whereas (5.207) results from (5.173)(2), of course. Therefore, thanks to Lemma 5.16, it follows that

$$\xi_n^i(\cdot,\boldsymbol{\Psi}) \in A \text{ for all } \boldsymbol{\Psi} \in (A_{\mathbb{R}})^N. \tag{5.208}$$

On the other hand, by the obvious equality

$$\xi_n^i(y,\mu) - a_i(y,\mu) = \int \theta_n(r)[a_i(y-r,\mu) - a_i(y,\mu)]dr$$

and use of Hölder's inequality (see the proof of Proposition 2.20) one arrives at

$$|\xi_n^i(y,\mu) - a_i(y,\mu)|^2 \leq \int \theta_n(r)|a_i(y-r,\mu) - a_i(y,\mu)|^2 dr.$$

For arbitrary given $\boldsymbol{\Psi} \in (A_{\mathbb{R}})^N$, it follows

$$\|\xi_n^i(\cdot,\boldsymbol{\Psi}) - a_i(\cdot,\boldsymbol{\Psi})\|_{2,\infty}^2 \leq \int \theta_n(r)\|\tau_r a_i(\cdot,\boldsymbol{\Psi}) - a_i(\cdot,\boldsymbol{\Psi})\|_{2,\infty}^2 dr,$$

where the integral is actually computed over $\frac{1}{n}B_N$. Bearing the latter observation in mind, let $\eta > 0$. According to Remark 5.19, consider some $\rho > 0$ such that $\|\tau_r a_i(\cdot,\boldsymbol{\Psi}) - a_i(\cdot,\boldsymbol{\Psi})\|_{2,\infty} \leq \eta$ for $|r| \leq \rho$. Hence we see readily that $\|\xi_n^i(\cdot,\boldsymbol{\Psi}) - a_i(\cdot,\boldsymbol{\Psi})\|_{2,\infty} \leq \eta$ for any integer $n \geq \frac{1}{\rho}$ and so $\xi_n^i(\cdot,\boldsymbol{\Psi}) \to a_i(\cdot,\boldsymbol{\Psi})$ in $(L^2,l^\infty)(\mathbb{R}^N)$ as $n \to \infty$. We deduce by (5.208) and use of Proposition 2.22 that $a_i(\cdot,\boldsymbol{\Psi}) \in L^2_{\mathcal{R}}(\mathbb{R}^N)$. Since the latter space is continuously embedded in $\mathfrak{X}^2_A(\mathbb{R}^N)$ (see Corollary 2.11), we see that (5.180) follows at once with $p = 2$ and $A = AP_{\mathcal{R}}(\mathbb{R}^N)$. Hence the desired solution follows by Theorem 5.9.

Remark 5.21 Let the question before us be the homogenization problem for (5.174) under the structure hypothesis

$$a_i(\cdot,\mu) \in AP(\mathbb{R}^N) \quad (1 \leq i \leq N) \text{ for fixed } \mu \in \mathbb{R}^N,$$

where we recall that $AP(\mathbb{R}^N)$ denotes the algebra of almost periodic continuous functions on $\mathbb{R}^N$ (see Chap. 1). Since $AP(\mathbb{R}^N) \subset L^1_{AP}(\mathbb{R}^N)$, we are naturally led to the structure hypothesis (5.203), under which we know that a solution is available. It is however possible to proceed otherwise. Indeed, by Lemma 5.3 there is a countable subgroup $\mathcal{R}$ of $\mathbb{R}^N$ such that $a_i(\cdot, \mu) \in AP_{\mathcal{R}}(\mathbb{R}^N)$ ($\mu \in \mathbb{R}^N, 1 \leq i \leq N$). Hence the desired result follows by Proposition 5.20. This approach releases us from the amalgamation hypothesis (Definition 5.1).

Example 5.24 For the sake of coherence we should exhibit an example of a function $\mathbf{a} = (a_i) : \mathbb{R}^N_y \times \mathbb{R}^N_\mu \to \mathbb{R}^N$ satisfying (5.171)–(5.173), (5.203), and amalgamating the spaces $\mathcal{B}(\mathbb{R}^N)$ and $(L^q, l^\infty)(\mathbb{R}^N)$ for $1 < q < \infty$ (and in particular for $q = 2$ as far as Problem II is concerned). For that purpose let

$$\textbf{(1)}\quad \mathbf{a}(y, \mu) = \sum_{j=1}^{m} \theta_j(y)\mathbf{u}_j(\mu) \ \ (y, \mu \in \mathbb{R}^N), \text{ integer } m > 0,$$

where the families (θ_j) and $(\mathbf{u}_j)$ satisfy:

$$\textbf{(2)}\quad \theta_j \in L^q_{AP}(\mathbb{R}^N; \mathbb{R}) \cap L^\infty(\mathbb{R}^N),\ \theta_j(y) \geq \gamma_j > 0 \text{ for a.e. } y \in \mathbb{R}^N$$

$$\textbf{(3)}\quad \mathbf{u}_j \in \mathcal{C}(\mathbb{R}^N; \mathbb{R})^N,\ \mathbf{u}_j(\omega) = \omega \text{ (the origin in } \mathbb{R}^N)$$

$$\textbf{(4)}\quad (\mathbf{u}_j(\mu) - \mathbf{u}_j(\nu)) \cdot (\mu - \nu) \geq \alpha_j |\mu - \nu|^p$$

$$\textbf{(5)}\quad |\mathbf{u}_j(\mu) - \mathbf{u}_j(\nu)| \leq c_j |\mu - \nu|^{p-1},\quad \mu, \nu \in \mathbb{R}^N,$$

with $1 < p \leq 2$ and $\alpha_j, c_j > 0$. It is clear that the function $\mathbf{a} = (a_i)$ thus defined satisfies (5.203). Furthermore, there is no difficulty in verifying that conditions (5.171), (5.172) and (5.173) are fulfilled. Finally, $\mathbf{a}$ amalgamates the spaces $\mathcal{B}(\mathbb{R}^N; \mathbb{R})$ and $(L^q, l^\infty)(\mathbb{R}^N)$ for $1 < q < \infty$, as is readily seen by Example 5.22.

For still more illustration, let us consider one further example.

Example 5.25 Let $\mathbf{a} = (a_i)$ be given by **(1)** with **(3)**, **(4)** and **(5)** (see Example 5.24), but with **(2)** substituted by

$$\textbf{(6)}\quad \theta_j \in L^\infty_{per}(Y_j; \mathbb{R}) = \{v | v \in L^\infty_{loc}(\mathbb{R}^N; \mathbb{R}),\ v\ Y_j\text{-periodic}\},$$

where $Y_1, \cdots, Y_m$ is a family of parallelepipeds constructed on different (orthogonal as well as non-orthogonal) bases of the N-dimensional vector space $\mathbb{R}^N$, and where θ_j is bounded from below as in **(2)**. It is worth recalling that there is a unique network $S_j \subset \mathbb{R}^N$ attached to Y_j as described in Sect. 2.5, so that the property v *Y_j-periodic* is understood as $v(y + k) = v(y)$ for all $k \in S_j$ and for a.e. $y \in \mathbb{R}^N$. Now, let $S_j^* = \{k | k \in \mathbb{R}^N, k \cdot l \in \mathbb{Z} \text{ for all } l \in S_j\}$. This is a network in $\mathbb{R}^N$, and $L^\infty_{per}(Y_j) \subset L^q_{per}(Y_j) = L^q_{S_j^*}(\mathbb{R}^N)$ (see Corollary 2.14 and Sect. 2.4.3). Thus, since $L^\infty_{per}(Y_j) \subset L^\infty(\mathbb{R}^N)$, we see that **(6)** reduces to **(2)**. More precisely, put $R = \bigcup_{1 \leq j \leq m} S_j^*$ and define $\mathcal{R} = \text{Gr}(R)$ as in Sect. 5.1 (see the proof of Lemma 5.3), that is, $\mathcal{R}$ stands for the subgroup of $\mathbb{R}^N$ generated by R. It is clear that

$\mathcal{R}$ is a countable set and further $a_i(\cdot, \mu) \in L^q_{\mathcal{R}}(\mathbb{R}^N)$ $(1 \le i \le N)$ for fixed $\mu \in \mathbb{R}^N$. This implies (5.204), of course.

5.5.4.4 Problem III (Hybrid Setting)

We consider here an example of a problem in a so-called *hybrid setting*. The amalgamation hypothesis is not required.

Example 5.26 Let the homogenization problem under consideration be the same as before, but under a different structure hypothesis, namely

$$\begin{cases} a_i(\cdot, \mu) \in \mathcal{B}_\infty(\mathbb{R}^N) \ (1 \le i \le m-1) \text{ for fixed } \mu \in \mathbb{R}^N \\ a_i(\cdot, \mu) \text{ is } Y\text{-periodic continuous } (m \le i \le N) \text{ for fixed } \mu \in \mathbb{R}^N, \end{cases} \tag{5.209}$$

where m is a given integer with $2 \le m \le N$. This problem falls within the preceding abstract setting . More precisely, the hypothesis (5.209) reduces to (5.180) with $A = \mathcal{B}_{\infty,\mathbb{Z}^N}(\mathbb{R}^N$. Indeed, by the relation $\mathcal{B}_\infty(\mathbb{R}^N) \subset A$ and use of Lemma 5.16 we see that for $\boldsymbol{\Psi} \in (A_\mathbb{R})^N$, we have $a_i(\cdot, \boldsymbol{\Psi}) \in A$ $(1 \le i \le m-1)$. On the other hand, let $m \le i \le N$. For fixed $\mu \in \mathbb{R}^N$, we have by hypothesis $a_i(\cdot, \mu) \in \mathcal{C}_{per}(Y)$. Therefore $a_i(\cdot, \mu) \in A$. We deduce by Lemma 5.16, once again, that $a_i(\cdot, \boldsymbol{\Psi}) \in A$ for $\boldsymbol{\Psi}$ as above. As a result we have $a_i(\cdot, \boldsymbol{\Psi}) \in A$ for $\boldsymbol{\Psi} \in (A_\mathbb{R})^N$, and that for $1 \le i \le N$. Hence (5.180) follows at once.

5.5.4.5 Problem IV (Anisotropic Setting)

We assume here that the function $\mathbf{a} = (a_i)$ amalgamates the spaces $\mathcal{B}(\mathbb{R}^N)$ and $(L^2, l^\infty)(\mathbb{R}^N)$, and we consider the homogenization problem for (5.174) with $p = 2$, under the structure hypothesis

$$a_i(\cdot, \mu) \in \mathcal{B}_\infty(\mathbb{R}; L^\infty_{per}(Y')) \ (1 \le i \le N) \text{ for fixed } \mu \in \mathbb{R}^N, \tag{5.210}$$

where $Y' = (-\frac{1}{2}, \frac{1}{2})^{N-1}$ $(N > 1)$. In the sequel, points in $\mathbb{R}^N$ are sometimes written as $y = (y', y_N)$ with $y' = (y_1, \cdots, y_{N-1}) \in \mathbb{R}^{N-1}$. If v denotes a function of $\mathbb{R}^N$ into $\mathbb{R}$, for fixed $y_N \in \mathbb{R}$ we will denote by $v(\cdot, y_N)$ the function $y' \mapsto v(y', y_N)$ of $\mathbb{R}^{N-1}$ into $\mathbb{R}$.

Thus, μ being fixed, (5.210) means precisely that for fixed $y_N \in \mathbb{R}$, the function $a_i(\cdot, y_N, \mu)$ is Y'-periodic, therefore $a_i(\cdot, y_N, \mu) \in L^\infty_{per}(Y')$ and further the function $y_N \mapsto a_i(\cdot, y_N, \mu)$ maps continuously $\mathbb{R}$ into the Banach space $L^\infty_{per}(Y')$, with the additional property of converging in the $L^\infty_{per}(Y')$ norm to some $l_i(\cdot, \mu) \in L^\infty_{per}(Y')$ when $|y_N| \to \infty$. So, the behaviour of $a_i(\cdot, \mu)$ is not the same with respect to either of the variables y' and y_N, for which reason the present setting is termed *anisotropic*.

Now, let $A = \mathcal{B}_\infty(\mathbb{R}; \mathcal{C}_{per}(Y'))$. This is a $W^{1,2}$-proper homogenization algebra on $\mathbb{R}^N$, with A^∞ as a dense subalgebra (see Sect. 4.2.3 and in particular Example 4.6). Consequently, we are certainly through with the matter on hand if we can show that (5.180) holds with $p = 2$ and A as above. To begin, let $\mu \in \mathbb{R}^N$ be fixed. Clearly

$$a_i(y, \mu) = l_i(y', \mu) + a_i^0(y, \mu) \text{ for } y = (y', y_N) \in \mathbb{R}^N \quad (1 \leq i \leq N)$$

where $l_i(\cdot, \mu) \in L^\infty_{per}(Y')$ and $a_i^0(\cdot, \mu) \in \mathcal{B}_0(\mathbb{R}; L^\infty_{per}(Y'))$. Let $\xi_n^i(\cdot, \mu) = a_i(\cdot, \mu) * \theta_n$, i.e., $\xi_n^i(y, \mu)$ given by (5.205). We have

$$\xi_n^i(y, \mu) = f_n^i(y, \mu) + g_n^i(y, \mu) \ (1 \leq i \leq N) \text{ for } y = (y', y_N) \in \mathbb{R}^N$$

with

$$f_n^i(y, \mu) = \int \theta_n(r) a_i^1(y - r, \mu) dr$$

and

$$g_n^i(y, \mu) = \int \theta_n(r) a_i^0(y - r, \mu) dr, \tag{5.211}$$

where $a_i^1(\cdot, \mu) \in L^\infty(\mathbb{R}^N)$, $a_i^1(y', y_N, \mu) = l_i(y', \mu)$ for $(y', y_N) \in \mathbb{R}^N$. It is a rather simple exercise (left to the reader) to show that $g_n^i(\cdot, \mu) \in \mathcal{B}_0(\mathbb{R}; \mathcal{C}_{per}(Y'))$ and $f_n^i(\cdot, \mu) \in \mathcal{C}_{per}(Y')$. Consequently $\xi_n^i(\cdot, \mu) \in A = \mathcal{B}_\infty(\mathbb{R}; \mathcal{C}_{per}(Y'))$. Hence, since (5.207) is fulfilled (for $p = 2$), (5.208) follows by Lemma 5.16. Proceeding by the same argument as followed in solving Problem II one quickly arrives at $\xi_n^i(\cdot, \Psi) \to a_i(\cdot, \Psi)$ in $(L^2, l^\infty)(\mathbb{R}^N)$ as $n \to \infty$, and that for any $\Psi \in (A_\mathbb{R})^N$. Hence (5.180) (with $p = 2$) follows by (5.208) and use of the fact that $(L^2, l^\infty)(\mathbb{R}^N)$ is continuously embedded in $\Xi^2(\mathbb{R}^N)$ (see Problem III).

5.6 Homogenization of Stationary Navier-Stokes Type Equations

5.6.1 Introduction

In this section we discuss the homogenization of stationary Navier-Stokes type equations in Ω. To start with, let $0 < \varepsilon < 1$. Let P^ε be the differential operator in Ω given by

$$P^\varepsilon = -\sum_{i,j=1}^N \frac{\partial}{\partial x_i}\left(a_{ij}^\varepsilon \frac{\partial}{\partial x_j}\right), \quad a_{ij}^\varepsilon(x) = a_{ij}\left(\frac{x}{\varepsilon}\right) \ (x \in \Omega)$$

with $a_{ij} \in L^\infty_\mathbb{R}(\mathbb{R}^N_y)$, $a_{ji} = a_{ij}$ $(1 \le i, j \le N)$,

$$\sum_{i,j=1}^{N} a_{ij}(y)\xi_j\xi_i \ge \alpha|\xi|^2 \text{ for all } \xi = (\xi_i) \in \mathbb{R}^N \text{ and for a.e. } y \in \mathbb{R}^N. \tag{5.212}$$

In the sequel P^ε is considered acting on scalar functions, say on $H^1(\Omega)$, as well as on vector functions $\mathbf{u} = (u^i) \in H^1(\Omega)^N$ in a *diagonal way*, i.e.,

$$(P^\varepsilon \mathbf{u})^i = P^\varepsilon u^i \quad (1 \le i \le N).$$

This being so, given $\mathbf{f} = (f^i) \in H^{-1}(\Omega; \mathbb{R})^N$, let

$$P^\varepsilon \mathbf{u}_\varepsilon + \sum_{j=1}^{N} u^j_\varepsilon \frac{\partial \mathbf{u}_\varepsilon}{\partial x_j} + \operatorname{grad} p_\varepsilon = \mathbf{f} \text{ in } \Omega, \tag{5.213}$$

$$\operatorname{div} \mathbf{u}_\varepsilon = 0 \text{ in } \Omega, \tag{5.214}$$

$$\mathbf{u}_\varepsilon = 0 \text{ on } \partial\Omega, \tag{5.215}$$

with $\mathbf{u}_\varepsilon = (u^i_\varepsilon)$, where

$$\frac{\partial \mathbf{u}_\varepsilon}{\partial x_j} = \left(\frac{\partial u^1_\varepsilon}{\partial x_j}, \ldots, \frac{\partial u^N_\varepsilon}{\partial x_j} \right).$$

Prior to any discussion let us point out an existence and uniqueness result for this boundary value problem. To this end we introduce, for fixed $0 < \varepsilon < 1$, the bilinear form a^ε on $H^1_0(\Omega; \mathbb{R})^N \times H^1_0(\Omega; \mathbb{R})^N$ given by

$$a^\varepsilon(\mathbf{u}, \mathbf{v}) = \sum_{k=1}^{N} \sum_{i,j=1}^{N} \int_\Omega a^\varepsilon_{ij} \frac{\partial u^k}{\partial x_j} \frac{\partial v^k}{\partial x_i} dx$$

for $\mathbf{u} = (u^k)$ and $\mathbf{v} = (v^k) \in H^1_0(\Omega; \mathbb{R})^N$. The form a^ε is symmetric and further, by (5.212), we have

$$a^\varepsilon(\mathbf{v}, \mathbf{v}) \ge \alpha \|\mathbf{v}\|^2_{H^1_0(\Omega)^N} \quad (0 < \varepsilon < 1) \tag{5.216}$$

for every $\mathbf{v} = (v^k) \in H^1_0(\Omega; \mathbb{R})^N$, where

$$\|\mathbf{v}\|_{H^1_0(\Omega)^N} = \left(\sum_{k=1}^{N} \int_\Omega |Dv^k|^2 dx \right)^{\frac{1}{2}}$$

with $Dv^k = (\frac{\partial v^k}{\partial x_1}, \cdots, \frac{\partial v^k}{\partial x_N})$. On the other hand,

$$|a^\varepsilon(\mathbf{u}, \mathbf{v}) \le c_0 \|\mathbf{u}\|_{H_0^1(\Omega)^N} \|\mathbf{v}\|_{H_0^1(\Omega)^N} \quad (0 < \varepsilon < 1) \tag{5.217}$$

for $\mathbf{u}, \mathbf{v} \in H_0^1(\Omega; \mathbb{R})^N$, where c_0 is a positive constant. Another tool we shall be dealing with throughout is the trilinear form b on $H_0^1(\Omega; \mathbb{R})^N \times H_0^1(\Omega; \mathbb{R})^N \times H_0^1(\Omega; \mathbb{R})^N$ given by

$$b(\mathbf{u}, \mathbf{v}, \mathbf{w}) = \sum_{k=1}^N \sum_{j=1}^N \int_\Omega u^j \frac{\partial v^k}{\partial x_j} w^k \, dx$$

for $\mathbf{u} = (u^k), \mathbf{v} = (v^k), \mathbf{w} = (w^k) \in H_0^1(\Omega; \mathbb{R})^N$. One has

$$b(\mathbf{u}, \mathbf{v}, \mathbf{v}) = 0 \text{ for } \mathbf{u} \in V, \mathbf{v} \in H_0^1(\Omega; \mathbb{R})^N, \tag{5.218}$$

$$|b(\mathbf{u}, \mathbf{v}, \mathbf{w})| \le c(N) \|\mathbf{u}\|_{H_0^1(\Omega)^N} \|\mathbf{v}\|_{H_0^1(\Omega)^N} \|\mathbf{w}\|_{H_0^1(\Omega)^N} \tag{5.219}$$

for all $\mathbf{u}, \mathbf{v}, \mathbf{w} \in H_0^1(\Omega; \mathbb{R})^N$, where $c(N)$ is a positive constant that depends solely on N, and

$$V = \{\mathbf{u} | \mathbf{u} \in H_0^1(\Omega; \mathbb{R})^N, \ \operatorname{div} \mathbf{u} = 0\}.$$

See Lions [72], Temam [133], for precise details. Note also in passing that V is a Hilbert space with the $H_0^1(\Omega; \mathbb{R})$-norm.

Proposition 5.21 *Suppose N is either 2 or 3, and further*

$$c(N) \|\mathbf{f}\|_{H^{-1}(\Omega; \mathbb{R})^N} < \alpha^2, \tag{5.220}$$

where α and $c(N)$ are those constants in (5.216) and (5.219), respectively. Then, the boundary value problem (5.213)–(5.215) determines a unique pair $(\mathbf{u}_\varepsilon, p_\varepsilon)$ with $\mathbf{u}_\varepsilon \in H_0^1(\Omega; \mathbb{R})^N$, $p_\varepsilon \in L^2_{\mathbb{R}}(\Omega)/\mathbb{R}$.

Proof Let

$$\begin{cases} \mathbf{u}_\varepsilon \in V : \\ a^\varepsilon(\mathbf{u}_\varepsilon, \mathbf{v}) + b(\mathbf{u}_\varepsilon, \mathbf{u}_\varepsilon, \mathbf{v}) = \langle \mathbf{f}, \mathbf{v} \rangle \text{ for all } \mathbf{v} = (v^k) \in V \end{cases} \tag{5.221}$$

with

$$\langle \mathbf{f}, \mathbf{v} \rangle = \sum_{k=1}^N \langle f^k, v^k \rangle,$$

where the brackets denote the duality pairing between $H^{-1}(\Omega;\mathbb{R})$ and $H_0^1(\Omega;\mathbb{R})$ as well as between $H^{-1}(\Omega;\mathbb{R})^N$ and $H_0^1(\Omega;\mathbb{R})^N$. The proposition is proved if one can show that (5.221) has a unique solution and further is equivalent to (5.213)–(5.215). Following, e.g., Lions [72, p.99] one can check that this variational problem has at least one solution, so that we only need to verify that (5.221) admits at most one solution. To see this, let us begin by noting that, thanks to (5.216), any solution $\mathbf{u}^*$ of (5.221) satisfies

$$\|\mathbf{u}^*\|_{H_0^1(\Omega)^N} \le \frac{1}{\alpha}\|\mathbf{f}\|_{H^{-1}(\Omega;\mathbb{R})^N}. \tag{5.222}$$

Next, by assuming that (5.221) admits $\mathbf{u}^*$ and $\mathbf{u}^{**}$ as solutions and letting $\mathbf{u} = \mathbf{u}^* - \mathbf{u}^{**}$ one easily arrives at

$$a^\varepsilon(\mathbf{u},\mathbf{v}) + b(\mathbf{u}^*,\mathbf{u},\mathbf{v}) + b(\mathbf{u},\mathbf{u}^*,\mathbf{v}) - b(\mathbf{u},\mathbf{u},\mathbf{v}) = 0$$

for all $\mathbf{v} \in V$. By the particular choice $\mathbf{v} = \mathbf{u}$ and use of (5.216) and (5.218) one is led to

$$\alpha\|\mathbf{u}\|^2_{H_0^1(\Omega)^N} + b(\mathbf{u},\mathbf{u}^*,\mathbf{u}) \le 0$$

from which one deduces by (5.219) that

$$\alpha\|\mathbf{u}\|^2_{H_0^1(\Omega)^N} \le c(N)\|\mathbf{u}\|^2_{H_0^1(\Omega)^N}\|\mathbf{u}^*\|_{H_0^1(\Omega)^N}.$$

Hence, by (5.222),

$$\left(\alpha - \frac{c(N)}{\alpha}\|\mathbf{f}\|_{H^{-1}(\Omega;\mathbb{R})^N}\right)\|\mathbf{u}\|^2_{H_0^1(\Omega)^N} \le 0.$$

Whence $\mathbf{u} = 0$ thanks of (5.220). This shows the unicity in (5.221). Now, by following a classical line of argument (see, e.g., Temam [133, p.14]), one sees readily that there is $p_\varepsilon \in \mathcal{D}'(\Omega)$ such that (5.213) holds with in addition (5.214)–(5.215). Let us verify that p_ε is actually a function. To start with, one has $H_0^1(\Omega;\mathbb{R}) \subset L^4_{\mathbb{R}}(\Omega)$, since $N = 2$ or 3 (see, e.g., Temam [133, pp.291–296]). Consequently $u_\varepsilon^i u_\varepsilon^j \in L^2_{\mathbb{R}}(\Omega)$ $(1 \le i, j \le N)$. By noting that

$$\sum_{j=1}^N u_\varepsilon^j \frac{\partial \mathbf{u}_\varepsilon}{\partial x_j} = \sum_{j=1}^N \frac{\partial}{\partial x_j}\left(u_\varepsilon^j \mathbf{u}_\varepsilon\right) \quad \text{(see (5.214))},$$

it follows $\sum_{j=1}^N u_\varepsilon^j \frac{\partial \mathbf{u}_\varepsilon}{\partial x_j} \in H^{-1}(\Omega;\mathbb{R})^N$, from which we deduce using (5.213) that $grad p_\varepsilon$ lies in $H^{-1}(\Omega;\mathbb{R})^N$. Hence $p_\varepsilon \in L^2_{\mathbb{R}}(\Omega)/\mathbb{R}$, according to a classical result (see the preceding reference, once again). Conversely, if a couple $(\mathbf{u}_\varepsilon, p_\varepsilon) \in$

$H_0^1(\Omega;\mathbb{R})^N \times L^2_{\mathbb{R}}(\Omega)/\mathbb{R}$ is a solution of (5.213)–(5.215), then $\mathbf{u}_\varepsilon$ (uniquely) solves (5.221) (the verification is an elementary exercise left to the reader). The proposition follows thereby. ■

Our purpose is to study the homogenization of the boundary value problem (5.213)–(5.215) in two different settings: first, in the periodic setting based on the structure hypothesis

$$a_{ij} \text{ is } Y\text{-periodic } (1 \le i, j \le N), \tag{5.223}$$

where Y is the same as in Sect. 5.1.1; next, in an abstract setting on the lines of Sect. 5.1.3.

It is not out of interest to note that the homogenization of such problems as (5.213)–(5.215) has a close connection with the modelling of heterogeneous fluid flows (for more details see Voke and Collins [140]).

Throughout the rest of the present Sect. 5.6 it is assumed that N is either 2 or 3.

5.6.2 *Periodic Setting*

5.6.2.1 Global Homogenization Theorem

Some preliminaries are needed. Let

$$\mathcal{V}_Y = \{\mathbf{\Phi} | \mathbf{\Phi} \in \mathcal{C}^\infty_{per}(Y;\mathbb{R})^N,\ M(\mathbf{\Phi}) = 0 \text{ and } \mathrm{div}_y \mathbf{\Phi} = 0\}$$

$$V_Y = \{\mathbf{w} | \mathbf{w} \in [H^1_{per}(Y;\mathbb{R})/\mathbb{R}]^N,\ \mathrm{div}_y \mathbf{w} = 0\},$$

where div_y denotes the divergence operator in $\mathbb{R}^N_y$, and $M(\mathbf{\Phi}) = \int_Y \mathbf{\Phi}(y)dy$. Equipped with the $[H^1_{per}(Y)/\mathbb{C}]^N$-norm, V_Y is a Hilbert space. Furthermore, $\mathcal{V}_Y$ is dense in V_Y. Indeed, given a mollifier $(\theta_n)_{n>0}$ on $\mathbb{R}^N$, if $\mathbf{w} = (w^k) \in V_Y$ and if we set $\Phi_n^k = \theta_n * w^k$ $(1 \le k \le N)$, it is an elementary exercise to verify that for each integer $n > 0$, the vector function $\mathbf{\Phi}_n = (\Phi_n^k)$ lies in $\mathcal{V}_Y$ and further $\Phi_n^k \to w^k$ in $H^1_{per}(Y)/\mathbb{C}$ when $n \to \infty$ (see Proposition 2.24 and Corollary 2.13).

With this in mind, let

$$\mathbb{F}_0^1 = V \times L^2(\Omega; V_Y).$$

Equipped with the norm

$$\|\mathbf{v}\|_{\mathbb{F}_0^1} = (\|\mathbf{v}_0\|^2_{H_0^1(\Omega)^N} + \|\mathbf{v}_1\|^2_{L^2(\Omega;V_Y)})^{\frac{1}{2}},\ \ \mathbf{v} = (\mathbf{v}_0, \mathbf{v}_1) \in \mathbb{F}_0^1,$$

$\mathbb{F}_0^1$ is a Hilbert space and further the space

$$\mathcal{F}_0^\infty = \mathcal{V} \times [\mathcal{D}(\Omega;\mathbb{R}) \otimes \mathcal{V}_Y]$$

with

$$\mathcal{V} = \{\mathbf{v} | \mathbf{v} \in \mathcal{D}(\Omega; \mathbb{R})^N, \operatorname{div} \mathbf{v} = 0\},$$

is dense in $\mathbb{F}_0^1$, as is easily seen by the density of $\mathcal{V}_Y$ in V_Y, and by that of $\mathcal{V}$ in V (see Temam [133, p.18]). Now, let

$$a_\Omega(\mathbf{u}, \mathbf{v}) = \sum_{i,j,k=1}^{N} \int\int_{\Omega \times Y} a_{ij} \left(\frac{\partial u_0^k}{\partial x_j} + \frac{\partial u_1^k}{\partial y_j} \right) \left(\frac{\partial v_0^k}{\partial x_i} + \frac{\partial v_1^k}{\partial y_i} \right) dxdy$$

for $\mathbf{u} = (\mathbf{u}_0, \mathbf{u}_1)$ and $\mathbf{v} = (\mathbf{v}_0, \mathbf{v}_1)$ in $\mathbb{F}_0^1$. This defines a symmetric continuous bilinear form a_Ω on $\mathbb{F}_0^1 \times \mathbb{F}_0^1$. Furthermore, a_Ω is coercive, i.e.,

$$a_\Omega(\mathbf{v}, \mathbf{v}) \geq \alpha \|\mathbf{v}\|^2_{\mathbb{F}_0^1} \quad (\mathbf{v} \in \mathbb{F}_0^1),$$

as is easily checked using (5.212) and the fact that $\int_Y \frac{\partial v_1^k}{\partial y_j} dy = 0$.

In the sequel we put

$$b_\Omega(\mathbf{u}, \mathbf{v}, \mathbf{w}) = b(\mathbf{u}_0, \mathbf{v}_0, \mathbf{w}_0),$$

$$L(\mathbf{v}) = \langle \mathbf{f}, \mathbf{v}_0 \rangle,$$

for $\mathbf{u} = (\mathbf{u}_0, \mathbf{u}_1)$, $\mathbf{v} = (\mathbf{v}_0, \mathbf{v}_1)$ and $\mathbf{w} = (\mathbf{w}_0, \mathbf{w}_1)$ in $\mathbb{F}_0^1$, which defines a continuous trilinear form b_Ω on $\mathbb{F}_0^1 \times \mathbb{F}_0^1 \times \mathbb{F}_0^1$ and a continuous linear form L on $\mathbb{F}_0^1$, respectively, with further $b_\Omega(\mathbf{u}, \mathbf{v}, \mathbf{v}) = 0$ for $\mathbf{u}, \mathbf{v} \in \mathbb{F}_0^1$.

Throughout the remainder of this section it is assumed that (5.223) holds.

Lemma 5.17 *Suppose (5.220) holds. Then the variational problem*

$$\begin{cases} \mathbf{u} \in \mathbb{F}_0^1 : \\ a_\Omega(\mathbf{u}, \mathbf{v}) + b_\Omega(\mathbf{u}, \mathbf{u}, \mathbf{v}) = L(\mathbf{v}) \text{ for all } \mathbf{v} \in \mathbb{F}_0^1 \end{cases} \tag{5.224}$$

admits at most one solution.

Proof This proceeds by the same line of argument as followed in proving the unicity of u_ε in (5.221). ∎

Theorem 5.10 *Let (5.220) be satisfied. For each $0 < \varepsilon < 1$, let $\mathbf{u}_\varepsilon = (u_\varepsilon^k) \in H^1(\Omega; \mathbb{R})^N$ be defined by (5.213)–(5.215) (or equivalently by (5.221)). Then, as $\varepsilon \to 0$,*

$$\mathbf{u}_\varepsilon \to \mathbf{u}_0 \text{ in } H_0^1(\Omega)^N\text{-weak}, \tag{5.225}$$

$$\frac{\partial u_\varepsilon^k}{\partial x_j} \to \frac{\partial u_0^k}{\partial x_j} + \frac{\partial u_1^k}{\partial y_j} \ \textit{in } L^2(\Omega)\textit{-weak } \Sigma \ (1 \le j, k \le N), \tag{5.226}$$

where $\mathbf{u} = (\mathbf{u}_0, \mathbf{u}_1)$ *is the unique solution of (5.224).*

Proof Let $0 < \varepsilon < 1$. To start with, we have

$$a^\varepsilon(\mathbf{u}_\varepsilon, \mathbf{v}) + b(\mathbf{u}_\varepsilon, \mathbf{u}_\varepsilon, \mathbf{v}) - \int_\Omega p_\varepsilon \mathrm{div}\, \mathbf{v}\, dx = (\mathbf{f}, \mathbf{v}) \tag{5.227}$$

for all $\mathbf{v} = (v^k) \in H_0^1(\Omega; \mathbb{R})^N$. It follows at once by mere routine that the sequence $(\mathbf{u}_\varepsilon)_{0<\varepsilon<1}$ is norm bounded in $H_0^1(\Omega; \mathbb{R})^N$. On the other hand, by (5.217) and (5.219) one can see readily that

$$|(\mathrm{grad}\ p_\varepsilon, \mathbf{v})| \le (\|\mathbf{f}\|_{H^{-1}(\Omega;\mathbb{R})^N} + c(N)\|\mathbf{u}_\varepsilon\|^2_{H_0^1(\Omega)^N} + c_0\|\mathbf{u}_\varepsilon\|_{H_0^1(\Omega)^N})\|\mathbf{v}\|_{H_0^1(\Omega)^N}$$

for all $\mathbf{v} \in H_0^1(\Omega; \mathbb{R})^N$. Hence the boundedness of $(\mathrm{grad} p_\varepsilon)_{0<\varepsilon<1}$ in the space $H^{-1}(\Omega; \mathbb{R})^N$ follows at once from that of $(\mathbf{u}_\varepsilon)_{0<\varepsilon<1}$. By a classical result used before one deduces that the sequence $(p_\varepsilon)_{0<\varepsilon<1}$ is bounded in $L^2_\mathbb{R}(\Omega)$. As a consequence of all that, if a fundamental sequence E is freely fixed, then by Theorems 3.1 and 3.2 there are a subsequence E' from E and functions $\mathbf{u}_0 = (u_0^k) \in H_0^1(\Omega; \mathbb{R})^N$, $\mathbf{u}_1 = (u_1^k) \in L^2(\Omega; [H^1_{per}(Y)/\mathbb{R}]^N)$, $p \in L^2(\Omega; L^2_{per}(Y; \mathbb{R}))$ such that as $E' \ni \varepsilon \to 0$, one has (5.225)–(5.226) and

$$p_\varepsilon \to p \text{ in } L^2(\Omega)\text{-weak } \Sigma. \tag{5.228}$$

Let us bear in mind that $\mathbf{u} = (u_0, u_1) \in \mathbb{F}_0^1$, as is easily deduced from (5.214). Now, for $0 < \varepsilon < 1$, let

$$\mathbf{\Phi}_\varepsilon = \mathbf{\Phi}_0 + \varepsilon \mathbf{\Phi}_1^\varepsilon, \text{ i.e., } \mathbf{\Phi}_\varepsilon(x) = \mathbf{\Phi}_0(x) + \varepsilon \mathbf{\Phi}_1\left(x, \frac{x}{\varepsilon}\right) \text{ for } x \in \Omega \tag{5.229}$$

with $\mathbf{\Phi}_0 \in \mathcal{D}_\mathbb{R}(\Omega)^N$ and $\mathbf{\Phi}_1 \in \mathcal{D}_\mathbb{R}(\Omega) \otimes \mathcal{V}_Y$. Clearly $\mathbf{\Phi}_\varepsilon \in \mathcal{D}_\mathbb{R}(\Omega)^N$. Thus, by (5.227),

$$a^\varepsilon(\mathbf{u}_\varepsilon, \mathbf{\Phi}_\varepsilon) + b(\mathbf{u}_\varepsilon, \mathbf{u}_\varepsilon, \mathbf{\Phi}_\varepsilon) - \int_\Omega p_\varepsilon \mathrm{div}\, \mathbf{\Phi}_\varepsilon\, dx = (\mathbf{f}, \mathbf{\Phi}_\varepsilon), \tag{5.230}$$

and the aim is then to pass to the limit when $E' \ni \varepsilon \to 0$. First, by the same argument as followed in proving (5.26) one is led to

$$a^\varepsilon(\mathbf{u}_\varepsilon, \mathbf{\Phi}_\varepsilon) \to a_\Omega(\mathbf{u}, \mathbf{\Phi}) \text{ as } E' \ni \varepsilon \to 0,$$

where $\mathbf{\Phi} = (\mathbf{\Phi}_0, \mathbf{\Phi}_1)$. On the other hand, thanks to the Rellich theorem we see by (5.225) that $\mathbf{u}_\varepsilon \to \mathbf{u}_0$ in $L^2(\Omega)^N$, hence $u_\varepsilon^j \Phi_\varepsilon^k \to u_0^j \Phi_0^k$ in $L^2(\Omega)$ $(1 \le j, k \le$

N) as $E' \ni \varepsilon \to 0$. By joining this with (5.226) it follows by Proposition 3.4 that

$$b(\mathbf{u}_\varepsilon, \mathbf{u}_\varepsilon, \boldsymbol{\Phi}_\varepsilon) \to b_\Omega(\mathbf{u}, \mathbf{u}, \boldsymbol{\Phi}) \text{ as } E' \ni \varepsilon \to 0.$$

Now, from (5.228) we see that, as $E' \ni \varepsilon \to 0$,

$$\int_\Omega p_\varepsilon \mathrm{div}\, \boldsymbol{\Phi}_\varepsilon \, dx \to \int\int_{\Omega \times Y} p \mathrm{div}\, \boldsymbol{\Phi}_0 \, dxdy.$$

On the other hand it is clear that $\boldsymbol{\Phi}_\varepsilon \to \boldsymbol{\Phi}_0$ in $H_0^1(\Omega; \mathbb{R})^N$-weak as $E' \ni \varepsilon \to 0$.

Thus is collected what is necessary to pass to the limit in (5.230) when $E' \ni \varepsilon \to 0$. The result is

$$a_\Omega(\mathbf{u}, \boldsymbol{\Phi}) + b_\Omega(\mathbf{u}, \mathbf{u}, \boldsymbol{\Phi}) - \int_\Omega p_0 \mathrm{div}\, \boldsymbol{\Phi}_0 \, dx = (\mathbf{f}, \boldsymbol{\Phi}_0) \tag{5.231}$$

where $p_0(x) = \int_Y p(x, y)dy$ $(x \in \Omega)$, and that for all $\boldsymbol{\Phi} = (\boldsymbol{\Phi}_0, \boldsymbol{\Phi}_1) \in \mathcal{D}_\mathbb{R}(\Omega)^N \times [\mathcal{D}_\mathbb{R}(\Omega) \otimes \mathcal{V}_Y]$, hence for $\boldsymbol{\Phi} \in \mathcal{F}_0^\infty$. Therefore, (5.224) follows readily by the density of $\mathcal{F}_0^\infty$ in $\mathbb{F}_0^1$. In view of Lemma 5.17 this completes the proof. ■

As is usual in homogenization, it is possible to provide an explicit representation of the function $\mathbf{u}_1$. For that purpose we introduce the bilinear form a on $V_Y \times V_Y$ given by

$$a(\mathbf{v}, \mathbf{w}) = \sum_{i,j,k=1}^{N} \int_Y a_{ij} \frac{\partial v^k}{\partial y_j} \frac{\partial w^k}{\partial y_i} dy$$

for $\mathbf{v} = (v^k)$ and $\mathbf{w} = (w^k)$ in V_Y. Next, for each pair of indices $1 \leq i, k \leq N$, we consider the variational problem

$$\begin{cases} \chi_{ik} \in V_Y : \\ a(\chi_{ik}, \mathbf{w}) = \displaystyle\sum_{l=1}^{N} \int_Y a_{li} \frac{\partial w^k}{\partial y_l} dy \text{ for all } \mathbf{w} = (w^j) \in V_Y, \end{cases} \tag{5.232}$$

which determines χ_{ik} in a unique manner.

Lemma 5.18 *Under the hypotheses and notation of Theorem 5.10, we have*

$$\mathbf{u}_1(x, y) = - \sum_{j,k=1}^{N} \frac{\partial u_0^k}{\partial x_j}(x) \chi_{jk}(y) \tag{5.233}$$

for a.e. $(x, y) \in \Omega \times \mathbb{R}^N$.

Proof The line of proceeding is the same mutatis mutandis as followed in proving Corollary 5.2 (the details are left to the reader). ■

5.6.2.2 Macroscopic Homogenized Operator

Let

$$q_{ijkh} = \delta_{kh} \int_Y a_{ij}(y)dy - \sum_{l=1}^{N} \int_Y a_{il}(y) \frac{\partial \chi_{jh}^k}{\partial y_l}(y)dy,$$

where $\chi_{jh} = (\chi_{jh}^k)$ is defined exactly as in (5.232). To the coefficients q_{ijkh} we attach the differential operator $\mathfrak{L}$ on Ω mapping $\mathcal{D}'(\Omega)^N$ into $\mathcal{D}'(\Omega)^N$ as

$$(\mathfrak{L}\mathbf{z})^k = - \sum_{i,j,h=1}^{N} q_{ijkh} \frac{\partial^2 z^h}{\partial x_i \partial x_j} \quad (1 \leq k \leq N) \tag{5.234}$$

for $\mathbf{z} = (z^k)$, $z^k \in \mathcal{D}'(\Omega)$. $\mathfrak{L}$ is the so-called *macroscopic homogenized operator* associated to $P^\varepsilon (0 < \varepsilon < 1)$.

At the present time, we consider the boundary value problem

$$\mathfrak{L}\mathbf{u}_0 + \sum_{j=1}^{N} u_0^j \frac{\partial \mathbf{u}_0}{\partial x_j} + \text{grad } p_0 = \mathbf{f} \text{ in } \Omega, \tag{5.235}$$

$$\text{div } \mathbf{u}_0 = 0, \tag{5.236}$$

$$\mathbf{u}_0 = 0. \tag{5.237}$$

Lemma 5.19 *Suppose (5.220) holds. Then, the boundary value problem (5.235)–(5.237) admits at most one weak solution* $(\mathbf{u}_0, p_0)$ *with* $\mathbf{u}_0 \in H_0^1(\Omega; \mathbb{R})^N$ *and* $p_0 \in L^2_{\mathbb{R}}(\Omega)/\mathbb{R}$.

Proof By noting that the function $u_1 \in L^2(\Omega; V_Y)$ (in Theorem 5.10) is given by (5.233) and satisfies

$$\sum_{i,j,k=1}^{N} \int_Y a_{ij}(y) \left(\frac{\partial u_0^k}{\partial x_j}(x) + \frac{\partial u_1^k}{\partial y_j}(x, y) \right) \frac{\partial v^k}{\partial y_i}(y)\, dy = 0$$

for all $\mathbf{v} = (v^k) \in V_Y$ and for almost all $x \in \Omega$, it can be shown without difficulty that if a couple $(\mathbf{u}_0, p_0) \in H_0^1(\Omega; \mathbb{R})^N \times L^2_{\mathbb{R}}(\Omega)/\mathbb{R}$ satisfies (5.235)–(5.237), then the vector function $\mathbf{u} = (\mathbf{u}_0, \mathbf{u}_1)$ is the solution of (5.224). Hence, the unicity in (5.235)–(5.237) follows by Lemma 5.17. ∎

This leads us to the following theorem.

Theorem 5.11 *Suppose (5.220) holds. For each real* $0 < \varepsilon < 1$*, let* $(\mathbf{u}_\varepsilon, p_\varepsilon) \in H_0^1(\Omega; \mathbb{R})^N \times [L^2_{\mathbb{R}}(\Omega)/\mathbb{R}]$ *be defined by (5.213)–(5.215). Then, as* $\varepsilon \to 0$*, we have*

$\mathbf{u}_\varepsilon \to \mathbf{u}_0$ in $H_0^1(\Omega)^N$-weak and $p_\varepsilon \to p_0$ in $L^2(\Omega)$-weak, where the pair $(\mathbf{u}_0, p_0)$ lies in $H_0^1(\Omega;\mathbb{R})^N \times [L^2_{\mathbb{R}}(\Omega)/\mathbb{R}]$ and is the unique weak solution of (5.235)–(5.237).

Proof By the proof of Theorem 5.10 we know that from any arbitrary fundamental sequence E one can extract a subsequence E' such that as $E' \ni \varepsilon \to 0$, we have (5.225)–(5.226) and $p_\varepsilon \to p_0$ in $L^2(\Omega)$-weak (this follows by (5.228)), and further (5.231) holds for $\mathbf{\Phi} = (\mathbf{\Phi}_0, \mathbf{\Phi}_1) \in \mathcal{D}_{\mathbb{R}}(\Omega)^N \times [\mathcal{D}_{\mathbb{R}}(\Omega) \otimes \mathcal{V}_Y]$, where $\mathbf{u} = (\mathbf{u}_0, \mathbf{u}_1) \in \mathbb{F}_0^1$. Next, by substituting (5.233) in (5.231) and then choosing the $\mathbf{\Phi}$'s such that $\mathbf{\Phi}_1 = \mathbf{0}$, a simple computation yields (5.235) with evidently (5.236) and (5.237). Hence, the theorem follows by Lemma 5.19. ∎

Remark 5.22 The operator $\mathfrak{L}$ is strongly elliptic, i.e., there is a constant $\alpha_0 > 0$ such that

$$\sum_{i,j,k,h=1}^{N} q_{ijkh}\xi_{ik}\xi_{jh} \geq \alpha_0 \sum_{k,h=1}^{N} |\xi_{kh}|^2 \tag{5.238}$$

for all $(\xi_{kh})_{1\leq k,h\leq N} \in \mathbb{R}^{N^2}$. Indeed, it can be shown as in Sect. 5.1.2.1 that $q_{ijkh} = a(\chi_{ik} - \pi_{ik}, \chi_{jh} - \pi_{jh})$, where the function $\pi_{ik} = (\pi_{ik}^1, \cdots, \pi_{ik}^N) : \mathbb{R}_y^N \to \mathbb{R}^N$ is given by $\pi_{ik}^r(y) = y_i\delta_{kr}$ $(r = 1, \cdots, N)$ for $y \in \mathbb{R}^N$. Hence (5.238) follows as usual.

5.6.3 Abstract Setting

5.6.3.1 Global Homogenization Theorem

The letter A throughout denotes an H^1-proper H-algebra on $\mathbb{R}^N$ admitting A^∞ as a dense subalgebra. For the benefit of the reader, we recall that J denotes the canonical mapping of $H^1(\Delta(A))/\mathbb{C}$ into its separated completion, $H^1_\#(\Delta(A))$. In the sequel we put $\mathbf{J}(\mathbf{u}) = (J(u^i))_{1\leq i\leq N}$ for $\mathbf{u} = (u^i)$, with $u^i \in H^1(\Delta(A))/\mathbb{C}$. We have $\mathbf{J}(\mathbf{u}) \in H^1_\#(\Delta(A))^N$, and the transformation $\mathbf{u} \mapsto \mathbf{J}(\mathbf{u})$ maps $[H^1(\Delta(A))/\mathbb{C}]^N$ isometrically into $H^1_\#(\Delta(A))^N$, and $[H^1_{\mathbb{R}}(\Delta(A))/\mathbb{R}]$ isometrically into $H^1_\#(\Delta(A);\mathbb{R})$, where

$H^1_\#(\Delta(A);\mathbb{R}) = \{u|u \in H^1_\#(\Delta(A)), \partial_i u \in L^2_{\mathbb{R}}(\Delta(A))\ (1 \leq i \leq N)\}$ (see Sect. 5.2.2 for details).

We will need some appropriate function spaces. We set

$\mathbb{E}_0^1 = H_0^1(\Omega;\mathbb{R})^N \times L^2(\Omega; H^1_\#(\Delta(A);\mathbb{R})^N)$ (equipped with an obvious norm)

$\mathcal{E}_0^\infty = \mathcal{D}_{\mathbb{R}}(\Omega)^N \times (\mathcal{D}_{\mathbb{R}}(\Omega) \otimes \mathbf{J}[\mathcal{D}_{\mathbb{R}}(\Delta(A))/\mathbb{R}]^N)$,

$V_A = \{\mathbf{u}|\mathbf{u} = (u^i) \in H^1_\#(\Delta(A);\mathbb{R})^N, \widehat{\operatorname{div}}\,\mathbf{u} = 0\}$,

where

$$\widehat{\operatorname{div}}\,\mathbf{u} = \sum_{i=1}^{N} \partial_i u^i .$$

Equipped with the $H^1_{\#}(\Delta(A); \mathbb{R})^N$-norm, V_A is a Hilbert space. We finally put

$$\mathbb{F}^1_0 = V \times L^2(\Omega; V_A) \text{ (equipped with an obvious norm).}$$

Now, in all that follows we assume that

$$a_{ij} \in \mathfrak{X}^2_A \quad (1 \leq i, j \leq N). \tag{5.239}$$

We will need the bilinear form a_Ω on $\mathbb{E}^1_0 \times \mathbb{E}^1_0$ given by

$$a_\Omega(\mathbf{u}, \mathbf{v}) = \sum_{i,j,k=1}^{N} \int\int_{\Omega\times\Delta(A)} \widehat{a}_{ij} \left(\frac{\partial u^k_0}{\partial x_j} + \partial_j u^k_1\right)\left(\frac{\partial v^k_0}{\partial x_i} + \partial_i v^k_1\right) dx d\beta$$

for $\mathbf{u} = (\mathbf{u}_0, \mathbf{u}_1)$ and $\mathbf{v} = (\mathbf{v}_0, \mathbf{v}_1)$ in $\mathbb{E}^1_0$ with $\mathbf{u}_0 = (u^k_0)$, $\mathbf{u}_1 = (u^k_1)$ (and analogous expressions for $\mathbf{v}_0$ and $\mathbf{v}_1$), where $\widehat{a}_{ij} = \mathcal{G}(a_{ij})$, and where β is the M-measure on $\Delta(A)$. It is clear that the bilinear form a_Ω is symmetric, continuous and coercive.

We also define b_Ω and L as in Sect. 5.6.2. It is an easy exercise to check that Lemma 5.17 together with its proof carries over mutatis mutandis to the present setting. Therefore, the variational problem

$$\begin{cases} \mathbf{u} \in \mathbb{F}^1_0 = V \times L^2(\Omega; V_A) : \\ a_\Omega(\mathbf{u}, \mathbf{v}) + b_\Omega(\mathbf{u}, \mathbf{u}, \mathbf{v}) = L(\mathbf{v}) \text{ for all } \mathbf{v} \in \mathbb{F}^1_0 \end{cases} \tag{5.240}$$

admits at most one solution. We are thus led to the analogue of Theorem 5.10.

Theorem 5.12 *Let (5.220) be satisfied. Suppose (5.239) holds. For each $0 < \varepsilon < 1$, let $\mathbf{u}_\varepsilon = (u^k_\varepsilon) \in H^1_0(\Omega; \mathbb{R})^N$ be defined by (5.213)–(5.215) (or equivalently by (5.221)). Then, as $\varepsilon \to 0$,*

$$\mathbf{u}_\varepsilon \to \mathbf{u}_0 \text{ in } H^1_0(\Omega)^N\text{-weak}, \tag{5.241}$$

$$\frac{\partial u^k_\varepsilon}{\partial x_j} \to \frac{\partial u^k_0}{\partial x_j} + \partial_j u^k_1 \text{ in } L^2(\Omega)\text{-weak } \Sigma \quad (1 \leq j \leq N), \tag{5.242}$$

where $\mathbf{u} = (\mathbf{u}_0, \mathbf{u}_1)$ (with $\mathbf{u}_0 = (u^k_0)$ and $\mathbf{u}_1 = (u^k_1)$) is the solution of (5.240).

Proof By the same line of argument as followed in proving Theorem 5.10 one sees readily that the sequences $(\mathbf{u}_\varepsilon)_{0<\varepsilon<1}$ and $(p_\varepsilon)_{0<\varepsilon<1}$ are bounded in $H^1_0(\Omega; \mathbb{R})^N$ and

$L^2_{\mathbb{R}}(\Omega)$, respectively. Thus, given an arbitrary fundamental sequence E, there are a subsequence E' from E and some $\mathbf{u} = (\mathbf{u}_0, \mathbf{u}_1) \in \mathbb{F}^1_0$, $p \in L^2(\Omega; L^2_{\mathbb{R}}(\Delta(A)))$ such that as $E' \ni \varepsilon \to 0$ one has (5.228), (5.241) and (5.242).

This being so, for fixed $0 < \varepsilon < 1$, let

$$\mathbf{\Phi}_\varepsilon = \mathbf{\Phi}_0 + \varepsilon \mathbf{\Phi}_1^\varepsilon \text{ with } \mathbf{\Phi}_0 \in \mathcal{D}_{\mathbb{R}}(\Omega)^N, \ \mathbf{\Phi}_1 \in \mathcal{D}_{\mathbb{R}}(\Omega) \otimes (A^\infty_{\mathbb{R}}/\mathbb{R})^N$$

and

$$\mathbf{\Phi} = (\mathbf{\Phi}_0, \mathbf{J}_\Omega(\widehat{\mathbf{\Phi}}_1)).$$

It is clear that $\mathbf{\Phi}$ lies in $\mathcal{E}^\infty_0$. The aim is to pass to the limit in (5.230) (with $\mathbf{\Phi}_\varepsilon$ as above) as $E' \ni \varepsilon \to 0$. By proceeding as in the proof of Theorem 5.10 and turning the density of $\mathcal{E}^\infty_0$ in $\mathbb{E}^1_0$ to account it follows

$$a_\Omega(\mathbf{u}, \mathbf{v}) + b_\Omega(\mathbf{u}, \mathbf{u}, \mathbf{v}) - \int\int_{\Omega \times \Delta(A)} p(\operatorname{div} \mathbf{v}_0 + \widehat{\operatorname{div}}\, \mathbf{v}_1) dx d\beta = \langle \mathbf{f}, \mathbf{v}_0 \rangle, \tag{5.243}$$

for all $\mathbf{v} = (\mathbf{v}_0, \mathbf{v}_1) \in \mathbb{E}^1_0$, hence for all $\mathbf{v} \in \mathbb{F}^1_0$. The theorem follows at once by the unicity in (5.240). ■

With a view to providing an explicit representation of the function $\mathbf{u}_1$, we introduce the bilinear form $\mathbf{a}$ on $H^1_\#(\Delta(A); \mathbb{R})^N \times H^1_\#(\Delta(A); \mathbb{R})^N$ given by

$$\mathbf{a}(\mathbf{v}, \mathbf{w}) = \sum_{i,j,k=1}^N \int_{\Delta(A)} \widehat{a}_{ij} \partial_j v^k \partial_i w^k \, d\beta$$

for $\mathbf{v} = (v^k)$ and $\mathbf{w} = (w^k)$ in $H^1_\#(\Delta(A); \mathbb{R})^N$. It is a rather simple exercise (left to the reader) to check that this bilinear form has the same properties with its homologue of Sect. 5.6.2. We next consider the variational problem

$$\begin{cases} \chi_{ik} \in V_A : \\ \mathbf{a}(\chi_{ik}, \mathbf{w}) = \displaystyle\sum_{l=1}^N \int_{\Delta(A)} \widehat{a}_{li} \partial_l w^k \, d\beta \text{ for all } \mathbf{w} = (w^j) \in V_A, \end{cases} \tag{5.244}$$

which uniquely determines χ_{ik}.

Lemma 5.20 *Under the assumptions and notation of Theorem 5.12, we have*

$$u_1(x, s) = -\sum_{i,k=1}^N \frac{\partial u_0^k}{\partial x_i}(x) \chi_{ik}(s) \tag{5.245}$$

for a.e. $(x, s) \in \Omega \times \Delta(A)$.

Proof This is quite similar to the proof of Lemma 5.18 and is not worth repeating. ∎

5.6.3.2 Macroscopic Homogenized Equations

For $1 \leq i, j, k, h \leq N$, let

$$q_{ijkh} = \delta_{kh} \int_{\Delta(A)} \widehat{a}_{ij}(s) d\beta(s) - \sum_{l=1}^{N} \int_{\Delta(A)} \widehat{a}_{il}(s) \partial_l \chi_{jh}^k(s) d\beta(s),$$

where $\chi_{jh} = (\chi_{jh}^k)$ is defined as in (5.244). To these coefficients let us associate the differential operator $\mathfrak{L}$ on Ω given by (5.234). Finally, let us consider the boundary value problem (5.235)–(5.237).

Lemma 5.21 *Under the hypotheses of Theorem 5.12, the boundary value problem (5.235)–(5.237) admits at most one weak solution* $(\mathbf{u}_0, p_0)$ *with* $\mathbf{u}_0 \in H_0^1(\Omega; \mathbb{R})^N$, $p_0 \in L^2_{\mathbb{R}}(\Omega)/\mathbb{R}$.

Proof This is a copy mutatis mutandis of the proof of Lemma 5.19 and is worth omitting. ∎

This leads us to the following

Theorem 5.13 *Let the hypotheses of Theorem 5.12 be satisfied. For each real* $0 < \varepsilon < 1$, *let* $(\mathbf{u}_\varepsilon, p_\varepsilon) \in H_0^1(\Omega; \mathbb{R})^N \times [L^2_{\mathbb{R}}(\Omega)/\mathbb{R}]$ *be defined by (5.213)–(5.215). Then, as* $\varepsilon \to 0$, *we have* $\mathbf{u}_\varepsilon \to \mathbf{u}_0$ *in* $H_0^1(\Omega)^N$*-weak and* $p_\varepsilon \to p_0$ *in* $L^2(\Omega)$*-weak, where the pair* $(\mathbf{u}_0, p_0)$ *lies in* $H_0^1(\Omega; \mathbb{R})^N \times [L^2_{\mathbb{R}}(\Omega)/\mathbb{R}]$ *and is the weak solution of (5.235)–(5.237).*

Proof We proceed by the same line of argument mutatis mutandis as followed earlier in proving Theorem 5.11. The boundary value problem for $(\mathbf{u}_0, p_0)$ is derived from (5.243) by substituting (5.245) therein and then choosing in particular $\mathbf{v}_1 = \mathbf{0}$; finally, the unicity of the solution of the said boundary value problem, viz. (5.235)–(5.237), is ensured by Lemma 5.21. ∎

It is possible to suitably express q_{ijkh} as we did in Sect. 5.6.2. To this end, let $\mathcal{M}$ be the space of all $\mathbf{F} = (F^{ij})_{1 \leq i,j \leq N}$ with $F^{ij} \in L^2_{\mathbb{R}}(\Delta(A))$. Equipped with the norm

$$\|\mathbf{F}\|_{\mathcal{M}} = \left(\sum_{i,j=1}^{N} \|F^{ij}\|^2_{L^2(\Delta(A))} \right)^{\frac{1}{2}}, \mathbf{F} = (F^{ij}) \in \mathcal{M},$$

$\mathcal{M}$ is a Hilbert space. Now, we introduce the bilinear form $\mathcal{A}$ on $\mathcal{M} \times \mathcal{M}$ given by

$$\mathcal{A}(\mathbf{F}, \mathbf{G}) = \sum_{i,j,k=1}^{N} \int_{\Delta(A)} \widehat{a}_{ij}(s) F^{jk}(s) G^{ik}(s) d\beta(s)$$

for $\mathbf{F}, \mathbf{G} \in \mathcal{M}$. Clearly

$$\mathbf{a}(\mathbf{u}, \mathbf{v}) = \mathcal{A}(\nabla \mathbf{u}, \nabla \mathbf{v}), \quad \mathbf{u}, \mathbf{v} \in H^1_{\#}(\Delta(A); \mathbb{R})^N,$$

where $\nabla \mathbf{u} = (\partial_j u^k)_{1 \leq j,k \leq N}$ for $\mathbf{u} = (u^k) \in H^1_{\#}(\Delta(A); \mathbb{R})^N$. By a routine line of proceeding one gets readily

$$q_{ijkh} = \mathcal{A}(\nabla \chi_{ik} - \theta_{ik}, \nabla \chi_{jh} - \theta_{jh}),$$

where χ_{ik} is given by (5.244), and $\theta_{ik} = (\theta_{ik}^{lm})_{1 \leq l,m \leq N}$ with $\theta_{ik}^{lm} = \delta_{il}\delta_{km}$. This being so, it is a simple exercise to verify that Remark 5.22 carries over mutatis mutandis to the present setting.

5.6.4 Some Concrete Examples

This section deals with a few concrete examples of homogenization problems for (5.213)–(5.215) that fall within the scope of Sect. 5.6.3. In fact, it clearly amounts to exhibiting some examples of concrete structure hypotheses on the a_{ij}'s that reduce to the abstract hypothesis (5.239). Let us observe in passing that (5.239) includes the case of the periodicity hypothesis. Indeed, (5.223) implies (5.239) with $A = \mathcal{C}_{per}(Y)$,as was pointed out in several sections thereinbefore.

Example 5.27 Suppose we have $a_{ij} \in L^2_{AP}(\mathbb{R}^N)$ $(1 \leq i, j \leq N)$, i.e., each a_{ij} is almost periodic in the sense of Stepanov (Sect. 2.4). Then (5.239) is satisfied with $A = AP_{\mathcal{R}}(\mathbb{R}^N)$, where $\mathcal{R}$ is a suitable countable subgroup of $\mathbb{R}^N$ (see Sect. 5.1.3).

Example 5.28 We consider here the homogenization problem for (5.213)–(5.215) under the structure hypothesis $a_{ij} \in \mathcal{B}_\infty(\mathbb{R}; L^\infty_{per}(Y'))$, $1 \leq i, j \leq N$ (see Problem IV of Sect. 5.5). Thanks to the relation $\mathcal{B}_\infty(\mathbb{R}; L^\infty_{per}(Y')) \subset \mathcal{B}_\infty(\mathbb{R}; L^2_{per}(Y'))$, the object in view can be accomplished by the same argument as followed in solving Problem II of Sect. 5.1.4.

Example 5.29 The structure hypothesis $a_{ij} \in \mathcal{B}_\infty(\mathbb{R}; L^2_{AP}(\mathbb{R}^{N-1}))$ $(1 \leq i, j \leq N)$ generalizes that of Example 5.28, and it reduces to (5.239) with $A = \mathcal{B}_\infty(\mathbb{R}; AP_{\mathcal{R}'}(\mathbb{R}^{N-1}))$, $\mathcal{R}'$ a suitable countable subgroup of $\mathbb{R}^{N-1}_{y'}$ (see Remark 5.5).

Example 5.30 Suppose the given structure hypothesis is not (5.223), but $a_{ij} \in L^2(\mathbb{R}^N) + L^2_{per}(Y)$ $(1 \leq i, j \leq N)$ instead. This reduces to (5.239) with $A = \mathcal{B}_{\infty,\mathbb{Z}^N}(\mathbb{R}^N)$ (see Sect. 5.1.3).

5.7 Comments and Problems

The approach to homogenization which is followed here is the so-called Σ-convergence method. Let us give a summary description of it.

Broadly speaking, the statement of a deterministic homogenization problem for a boundary value problem consists of two main data. On one hand, a well-posed boundary value problem depending on, say a «small parameter» $\varepsilon > 0$, the main purpose being to investigate the behaviour, as $\varepsilon \to 0$, of the solution (for fixed $\varepsilon > 0$) of the said boundary value problem. On the other hand, a so-called *structure hypothesis* on the coefficients of the differential operator associated to the boundary value problem under consideration, which structure hypothesis exhibits in a concrete fashion the behaviours, with respect to the structure variables $y = (y_1, \cdots, y_N) \in \mathbb{R}^N$, of the coefficients of the differential operator. The most popular example of a structure hypothesis is without doubt the periodicity hypothesis (see, e.g., (5.11)). Several other structure hypotheses are given in Sect. 5.1.3 and thoroughly studied in Sect. 5.1.4. However, as seen thereinbefore, a structure hypothesis may be more complex than in the study of a problem such as (5.6). For example, a deterministic homogenization problem in a perforated domain requires two kinds of structure hypotheses: on one hand, a structure hypothesis on the coefficients of the differential operator under consideration, on the other hand, a structure hypothesis on the manner in which the holes are distributed (see Sect. 5.4). It is a matter of course that this is not apparent when the study is confined to the classical case of «periodically» perforated domains.

The preliminary step of the Σ-convergence method aims at determining which H-algebra , A, fits the problem. It is required that in addition to having A^∞ as a dense subalgebra, such an A should be $W^{m,p}(\Omega)$-proper for some suitable (integer) $m \geq 1$ and (real) $p > 1$. Throughout Sects. 5.1,5.2 and 5.6 this is satisfied with $m = 1$ and $p = 2$. It is worth perhaps recalling that the density of A^∞ in A implies $L^p(\Delta(A)) \subset \mathcal{D}'(\Delta(A))$ $(1 \leq p \leq \infty)$ with continuous embedding, which enables us to develop within the Gelfand representation theory an appropriate tool analogous to that in use in periodic homogenization (see Chaps. 2, 3 and 4).

The next step of the Σ-convergence method is to determine the appropriate function spaces $\mathfrak{X}^p_A$ $(1 \leq p < \infty)$ by means of which the different data should be embedded in the Gelfand representation theory. Whereas this is straightforward in Sects. 5.1, 5.2, 5.3 and 5.6, it is no easy matter when dealing with topics such as *perforated domains* (Sect. 5.4) or *nonlinear homogenization* (Sect. 5.5) to manage the transition between the concrete statement and the abstract setting of a deterministic homogenization problem. Anyhow, once the said transition is effected, the study proceeds with the same ease as in periodic homogenization.

The results of Sects. 5.2 and 5.3 were partly announced in Nguetseng [97].

Likewise the strong approximation approach followed in Sects. 5.1 and 5.3 goes back to Nguetseng [97].

Attention must be called to the misuse of language referring to $\varepsilon u_1^\varepsilon$ (in Corollary 5.6) as a first order corrector. Indeed, the latter function in the present context does not belong to the space $H_0^1(\Omega)$ containing the weak solution of (5.6) (see Bensoussam, Lions and Papanicolaou [14, Chapter 1, Section 5] for precise details). Anyhow it should be stressed that contrary to the classical notion of a first order corrector, the concept of a first order semi-corrector, $\varepsilon\psi_1^\varepsilon$, introduced in Sect. 5.1.2 (see in particular Corollary 5.5) requires no regularity assumptions. Moreover it is readily seen, e.g., in the setting of Sect. 5.1.2 (see Corollary 5.5, once again, together with its proof) that $\varepsilon\psi_1^\varepsilon$ may be chosen in $H_0^1(\Omega) \otimes [C_{per}^1(Y)/\mathbb{C}]$ or even in $\mathcal{D}^m(\Omega) \otimes [C_{per}^m(Y)/\mathbb{C}]$ for arbitrary $m \in \mathbb{N}^* \cup \{\infty\}$, if need be. Of course the preceding comments and remarks carry over mutatis mutandis to Sects. 5.1.3 and 5.3.

There are many settings within which one can carry out the homogenization of the boundary value problem (5.141), each setting corresponding to both the structure hypothesis on the coefficients $\{a_0, a_{ij}\}$ and the fashion in which the holes are distributed. The classical setting corresponds to the assumption that the coefficients are periodic while the holes are equidistributed. This case was first studied by Cioranescu and Saint Jean Paulin [35]. Since then homogenization in perforated domains has been attracting the attention of an increasing number of researchers. For a detailed bibliography on the subject we refer for example to Cioranescu and Donato [34], Chiado Piat and De Franceschi [33], Chechkin and Piatnitski [32], Jikov et al. [63].

The study of homogenization problems in domains perforated in a neighbourhood of the hyperplane $\{x_N = 0\}$ actually goes back to the early eighties of the last century (see, e.g., Lions [73], Nguetseng [91]). Later on, Oleinik and Shaposhnikova [112] investigated the more general case in which the hyperplane $\{x_N = 0\}$ is substituted by a manifold of small dimension $d < N$. However, in either case only differential operators with constant coefficients were considered till then. The author's work in [96] seems to be the first study in which the distribution of holes of Example 5.20 is associated to a differential operator with variable coefficients. Let us point out a substantial generalization in Sango [122], where the hyperplane $\{x_N = 0\}$ is replaced by a manifold of dimension $N - 1$ and the differential operator is quasilinear with variable coefficients.

Some generalizations are possible as far as Sect. 5.4 is concerned.

1. For example, concerning the manner in which the holes are concentrated in a neighbourhood of the hyperplane of equation $x_N = 0$ (see Sect. 5.4.4), one can consider a periodic concentration instead of the equiconcentration of the holes stated in (5.168).

More precisely, let $\mathcal{R}$ be a network in $\mathbb{R}^{N-1}$ with $\mathcal{R} \subset \mathbb{Z}^{N-1}$. Let r be a mapping of $\mathbb{Z}^{N-1}$ into $\{0, 1\}$ such that $r(k' + \nu) = r(k')$ for all $\nu \in \mathcal{R}$ and all $k' \in \mathbb{Z}^{N-1}$ (see Example 2.9). To this mapping there is attached the distribution function of the holes, namely the characteristic function of S (see Sect. 5.4.1) denoted by $\theta = \chi_S$, and given here by

$$\theta(k', k_N) = \begin{cases} r(k') & (k' \in \mathbb{Z}^{N-1}) \text{ if } k_N = 0 \\ 0 & (k' \in \mathbb{Z}^{N-1}) \text{ if } k_N \neq 0, \end{cases}$$

where points in $\mathbb{Z}^N$ read $k = (k', k_N)$ according to the factorization $\mathbb{Z}^N = \mathbb{Z}^{N-1} \times \mathbb{Z}$. It is worth recalling that the network $\mathcal{R}$ is spanned by a basis $b_1, \cdots, b_{N-1}$ of the $(N-1)$-dimensional vector space $\mathbb{R}^{N-1}$, and is closely connected to the open parallelepiped

$$Z = \left\{ y' | y' = t_1 b_1 + \cdots + t_{N-1} b_{N-1}, -\frac{1}{2} < t_i < \frac{1}{2} \ (1 \leq i \leq N-1) \right\}$$

within the couple $\{\mathcal{R}, Z\}$ underlying a specific periodic setting in $\mathbb{R}^{N-1}$ (see Sect. 2.5).

Finally, put $A = \mathcal{B}_\infty(\mathbb{R}; \mathcal{C}_{per}(Z))$. This is an $H^1(\Omega)$-proper homogenization algebra on $\mathbb{R}^N$ (under the action $\mathcal{H}$) admitting A^∞ as a dense subalgebra. By the same argument mutatis mutandis as followed in Example 5.20 one sees readily that (5.147) and (5.148) are satisfied. One can then consider the analogs of Problem 1 and 2 of Sect. 5.4.4 with Z in place of Y', and obtain the same results mutatis mutandis.

2. The study of Sect. 5.4 extends in a large measure to nonlinear elliptic differential operators. For example, in the notation of Sect. 5.4.1, for fixed $\varepsilon > 0$ let

$$\begin{aligned} a_0\left(\frac{x}{\varepsilon}\right) u_\varepsilon - \operatorname{div} \mathbf{a}\left(\frac{x}{\varepsilon}, Du_\varepsilon\right) &= f \text{ in } \Omega^\varepsilon \\ \mathbf{a}\left(\frac{x}{\varepsilon}, Du_\varepsilon\right) \cdot \nu &= 0 \text{ on } \partial T^\varepsilon \\ u_\varepsilon &= 0 \text{ on } \partial\Omega, \end{aligned} \tag{5.246}$$

where the functions f and a_0 are as in (5.141) save that they assume real values in the present setting, and the function $\mathbf{a} = (a_i)_{1 \leq i \leq N}$ from $\mathbb{R}^N \times \mathbb{R}^N$ to $\mathbb{R}^N$ is the same as in Sect. 5.5 with however the difference that one takes here $p = 2$. Finally, let A be an H-algebra as in Sect. 5.4.3. We assume that

$$\begin{aligned} &\chi_G \in \mathfrak{X}^2_A(\mathbb{R}^N) \quad \text{with (5.148)} \\ &a_0 \in \mathfrak{X}^2_A(\mathbb{R}^N) \\ &a_i(\cdot, \boldsymbol{\Psi}) \in \mathfrak{X}^2_A(\mathbb{R}^N) \quad (1 \leq i \leq N) \text{ for } \boldsymbol{\Psi} \in (A_{\mathbb{R}})^N. \end{aligned} \tag{5.247}$$

Under the abstract hypothesis (5.247) one can achieve for (5.246) a result analogous to Theorem 5.8.

The pioneering work in the direction of Sect. 5.5 seems to be the paper of Nguetseng and Nnang [100]. For the sake of clearness we have chosen a rather simple differential operator to illustrate the Σ-convergence method in Sect. 5.5. For the study of more «sophisticated» nonlinear elliptic operators in this context see, e.g., Woukeng [142, 143]. One can also fit the Σ-convergence theory to the study, in the present topic, of differential operators in the Orlicz-Sobolev spaces setting, as opposed to the usual Sobolev spaces setting. See Nnang [111] for a study in this direction.

The results of Sect. 5.6 were first given in Nguetseng and Signing [101] and further presented in lectures of the author.

Chapter 6
Homogenization of Parabolic Operators I

The periodic homogenization in both space and time of linear as well as nonlinear parabolic differential operators is discussed. Particular attention is paid to the couple of quasilinear monotone operators $u \mapsto \frac{\partial u}{\partial t} - \mathrm{div}\mathbf{a}(\frac{x}{\varepsilon}, \frac{t}{\varepsilon^k}, Du)$ $(k = 1, 2)$ corresponding to a classical case and a somewhat singular case, respectively. A comparative analysis of both cases points out great differences of behaviour arising from the third scale inherent in the case $k = 2$.

6.1 Linear Parabolic Operators

6.1.1 Introduction

Let $\Omega \subset \mathbb{R}^N_x$ (integer $N > 1$) be a smooth bounded open set. Let

$$Q = \Omega \times (0, T), \quad 0 < T < \infty.$$

We consider a family of functions a_{ij} $(1 \leq i, j \leq N)$ with

$$a_{ij} \in \mathcal{C}(\overline{Q}; L^{\infty}(\mathbb{R}^N_y \times \mathbb{R}_z)) \tag{6.1}$$

$$a_{ji} = \overline{a}_{ij} \tag{6.2}$$

G. Nguetseng, *Homogenization Algebras and Applications*, Springer Monographs in Mathematics, https://doi.org/10.1007/978-3-031-84705-9_6

and the ellipticity condition:

$$\begin{cases} \text{There exists a constant } \alpha > 0 \text{ such that} \\ \mathrm{Re} \displaystyle\sum_{i,j=1}^{N} a_{ij}(x, t, y, z)\xi_j\overline{\xi}_i \geq \alpha|\xi|^2, \ \xi = (\xi_i) \in \mathbb{C}^N, \\ \text{for all } (x, t) \in \overline{Q} \text{ and almost all } (y, z) \in \mathbb{R}^N \times \mathbb{R}. \end{cases} \tag{6.3}$$

Before we proceed any farther let us observe that Remark 5.1 carries over mutatis mutandis to the present setting.

Now, for fixed $\varepsilon > 0$, let $a_{ij}^{\varepsilon} \in L^{\infty}(\Omega)$ with

$$a_{ij}^{\varepsilon}(x, t) = a_{ij}\left(x, t, \frac{x}{\varepsilon}, \frac{t}{\varepsilon}\right), \quad (x, t) \in Q,$$

in the sense of Proposition 1.8. The main purpose of this section is to study the periodic homogenization of the initial-boundary value problem

$$\begin{cases} \dfrac{\partial u_\varepsilon}{\partial t} - \displaystyle\sum_{i,j=1}^{N} \frac{\partial}{\partial x_i}\left(a_{ij}^{\varepsilon}\frac{\partial u_\varepsilon}{\partial x_j}\right) = f \text{ in } Q, \\ u_\varepsilon = 0 \text{ on } \partial\Omega \times (0, T), \\ u_\varepsilon(x, 0) = 0 \text{ in } \Omega, \end{cases} \tag{6.4}$$

where f is given in $L^2(Q)$. However, prior to any discussion we need to make sure that an existence and unicity result for (6.4) is available. In this connection we have the following lemma which is but an evident adaptation of Lemma 5.2.

Lemma 6.1 *Under hypotheses (6.1) and (6.3), there is a constant $0 < \alpha_0 < \alpha$ such that for any $\varepsilon > 0$, one has*

$$Re \sum_{i,j=1}^{N} a_{ij}^{\varepsilon}(x, t)\xi_j\overline{\xi}_i \geq \alpha_0|\xi|^2 \quad (\xi \in \mathbb{C}^N) \tag{6.5}$$

for a.e. $(x, t) \in Q$.

Thanks to this lemma, the existence of a unique solution of (6.4) becomes a classical fact (see, e.g., Lions and Magenes [74, p. 264]):

Proposition 6.1 *Under hypotheses (6.1) and (6.3), for each $\varepsilon > 0$ there is a unique u_ε in $L^2(0, T; H_0^1(\Omega)) \cap \mathcal{C}([0, T]; L^2(\Omega))$ satisfying (6.4).*

Remark 6.1 Here and throughout the rest of the chapter we put $L^p(0, T; B) = L^p(X; B)$ for $X = (0, T)$, $1 \leq p \leq \infty$, and for any Banach space B.

For further needs we introduce

$$V^2 = \left\{ u | u \in L^2(0, T; H_0^1(\Omega)), u' = \frac{\partial u}{\partial t} \in L^2(0, T; H^{-1}(\Omega)) \right\}$$

equipped with the norm

$$\|u\|_{V^2} = (\|u\|^2_{L^2(0,T;H_0^1(\Omega))} + \|u'\|^2_{L^2(0,T;H^{-1}(\Omega))})^{\frac{1}{2}} \quad (u \in V^2),$$

which makes it into a Hilbert space. Furthermore, assuming $L^2(\Omega)$ to be identified with its anti-dual so that $H_0^1(\Omega) \subset L^2(\Omega) \subset H^{-1}(\Omega)$ (with continuous embeddings), we have $V^2 \subset \mathcal{C}([0, T]; L^2(\Omega))$ with continuous embedding. Hence, given any $0 \leq t \leq T$, we can define $u(t) \in L^2(\Omega)$ for any $u \in V^2$, and further the mapping $u \mapsto u(t)$ sends continuously V^2 into $L^2(\Omega)$. These are classical facts.

This being so, it is an easy matter to check that the solution u_ε of (6.4) lies in V^2 and further $u_\varepsilon(0) = 0$.

6.1.2 Preliminaries

We set

$$Y = \left(-\frac{1}{2}, \frac{1}{2}\right)^N \text{ and } Z = \left(-\frac{1}{2}, \frac{1}{2}\right),$$

and we assume that for fixed $(x, t) \in \overline{Q}$, the function $(y, z) \mapsto a_{ij}(x, t, y, z)$ of $\mathbb{R}^N \times \mathbb{R}$ into $\mathbb{C}$ is $Y \times Z$-periodic, i.e., $a_{ij}(x, t, y + k, z + l) = a_{ij}(x, t, y, z)$ for all $(k, l) \in \mathbb{Z}^N \times \mathbb{Z}$ and for a.e. $(y, z) \in \mathbb{R}^N \times \mathbb{R}$.

Thus, in view of (6.1), we have

$$a_{ij} \in \mathcal{C}(\overline{Q}; L^\infty_{per}(Y \times Z)) \quad (1 \leq i, j \leq N). \tag{6.6}$$

Now, let

$$\mathcal{H} = L^2_{per}(Z; H^1_{per}(Y)/\mathbb{C}),$$

where the right-hand side denotes the space of Z-periodic functions in $L^2_{loc}(\mathbb{R}; H^1_{per}(Y)/\mathbb{C})$, that is, those u in the latter space satisfying $u(z + l) = u(z)$ for all $l \in \mathbb{Z}$ and for a.e. $z \in \mathbb{R}$. Equipped with the norm

$$\|u\|_{\mathcal{H}} = \left(\int_Z \|u(z)\|^2_{H^1(Y)/\mathbb{C}} dz\right)^{\frac{1}{2}} \quad (u \in \mathcal{H}),$$

$\mathcal{H}$ is a Hilbert space. For $u \in \mathcal{H}$, we will most of the time write $u(\cdot, z)$ instead of $u(z)$. We need a few preliminary results.

Lemma 6.2 *Given any bounded sequence $(u_\varepsilon)_{\varepsilon\in E}$ in V^2, where E is a fundamental sequence, there are some subsequence E' from E and some $\mathbf{u} = (u_0, u_1) \in V^2 \times L^2(Q; \mathcal{H})$ such that as $E' \ni \varepsilon \to 0$,*

$$u_\varepsilon \to u_0 \textit{ in } V^2\textit{-weak}$$

and

$$\frac{\partial u_\varepsilon}{\partial x_j} \to \frac{\partial u_0}{\partial x_j} + \frac{\partial u_1}{\partial y_j} \textit{ in } L^2(Q)\textit{-weak } \Sigma \quad (1 \le j \le N).$$

Proof Let us note beforehand that the notion of Σ-convergence is here understood in the sense of Chap. 3, and is relevant to the algebra $A = \mathcal{C}_{per}(Y \times Z)$. Having made this point, one readily sees that the above boundedness hypothesis yields, via the reflexivity of V^2 and Theorem 3.1, some subsequence E' from E, some function $u_0 \in V^2$ and a family of functions $f_j \in L^2(Q; L^2_{per}(Y \times Z))$ $(1 \le j \le N)$ such that as $E' \ni \varepsilon \to 0$, we have $u_\varepsilon \to u_0$ in V^2-weak and $\frac{\partial u_\varepsilon}{\partial x_j} \to f_j$ in $L^2(Q)$-weak Σ $(1 \le j \le N)$. Thus, we are through if we can show that there is some $u_1 \in L^2(Q; \mathcal{H})$ such that

$$f_j = \frac{\partial u_0}{\partial x_j} + \frac{\partial u_1}{\partial y_j} \quad (1 \le j \le N).$$

To this end, let $\mathbf{\Phi} = (\psi_j)_{1\le j\le N} \in \mathcal{V}_{per}(Y)$ (Sect. 2.5), $\varphi \in \mathcal{D}(Q)$ and $w \in \mathcal{C}^\infty_{per}(Z)$. Clearly

$$\sum_{j=1}^N \int_Q \frac{\partial u_\varepsilon}{\partial x_j} \psi_j^\varepsilon w^\varepsilon \varphi \, dxdt = -\sum_{j=1}^N \int_Q u_\varepsilon \psi_j^\varepsilon w^\varepsilon \frac{\partial \varphi}{\partial x_j} dxdt,$$

where $\psi_j^\varepsilon(x) = \psi_j(\frac{x}{\varepsilon})$ and $w^\varepsilon(t) = w(\frac{t}{\varepsilon})$ for $x \in \Omega$ and $0 < t < T$. The aim now is to pass to the limit (as $E' \ni \varepsilon \to 0$) on both sides. This is an easy matter as far as the left-hand side is concerned. As regards the right-hand side, we accomplish this thanks to the compactness of the embedding $V^2 \hookrightarrow L^2(Q)$ (see Lions [72, p. 58,Théorème 5.1]), the latter implying that $u_\varepsilon \to u_0$ in $L^2(Q)$ as $E' \ni \varepsilon \to 0$. Finally, as $E' \ni \varepsilon \to 0$, the result is that

$$\sum_{j=1}^N \int\int\int_{Q\times Y\times Z} f_j \psi_j w \varphi \, dxdtdydz$$
$$= \sum_{j=1}^N \int\int\int_{Q\times Y\times Z} \frac{\partial u_0}{\partial x_j} \psi_j w \varphi \, dxdtdydz.$$

Using first the arbitrariness of φ and then that of w, one easily arrives at

$$\sum_{j=1}^{N} \int_Y F_j(x,t,y,z)\psi_j(y)dy = 0$$

for almost every $(x,t) \in Q$ and almost every $z \in \mathbb{R}$, where $\boldsymbol{\Phi} = (\psi_j)$ is arbitrary in $\mathcal{V}_{per}(Y)$, and

$$F_j(x,t,y,z) = f_j(x,t,y,z) - \frac{\partial u_0}{\partial x_j}(x,t).$$

Hence, by Lemma 2.10, there is a function $(x,t,z) \mapsto u_1(x,t,\cdot,z)$ of $Q \times \mathbb{R}$ into $H^1_{per}(Y)/\mathbb{C}$ such that

$$F_j(x,t,\cdot,z) = \frac{\partial u_1(x,t,\cdot,z)}{\partial y_j} \quad (1 \le j \le N)$$

for a.e. $(x,t) \in Q$ and for a.e. $z \in \mathbb{R}$. But $F_j \in L^2(Q; L^2_{per}(Y \times Z))$ $(1 \le j \le N)$. One readily infers from that the following facts whose verification is left to the reader:

(1) For a.e. $(x,t) \in Q$, the function $z \mapsto u_1(x,t,\cdot,z)$ is measurable from $\mathbb{R}$ to $H^1_{per}(Y)/\mathbb{C}$ (use Bourbaki [23, p. 89]). One easily deduces that the said function lies in $L^2_{loc}(\mathbb{R}; H^1_{per}(Y)/\mathbb{C})$.
(2) For a.e. $(x,t) \in Q$, the function $z \mapsto u_1(x,t,\cdot,z)$ is Z-periodic.

From which we obtain a function $(x,t) \mapsto u_1(x,t,\cdot,\cdot)$ from Q to $\mathcal{H}$, denoted by u_1, such that

$$F_j(x,t) = \frac{\partial}{\partial y_j} u_1(x,t,\cdot,\cdot) \quad (1 \le j \le N)$$

for a.e. $(x,t) \in Q$. Finally, it is a routine exercise to check that $u_1 \in L^2(Q; \mathcal{H})$, thereby completing the proof. ■

Remark 6.2 As $E \ni \varepsilon \to 0$, one has $u_\varepsilon \to u_0$ in V^2-weak if and only if $u_\varepsilon \to u_0$ in $L^2(0,T; H^1_0(\Omega))$-weak and $\frac{\partial u_\varepsilon}{\partial t} \to \frac{\partial u_0}{\partial t}$ in $L^2(0,T; H^{-1}(\Omega))$-weak.

Now, for any fixed $\varepsilon > 0$, define $\Phi_\varepsilon \in \mathcal{D}(Q)$ as

$$\Phi_\varepsilon = \psi_0 + \varepsilon \psi_1^\varepsilon, \tag{6.7}$$

i.e., $\Phi_\varepsilon(x,t) = \psi_0(x,t) + \varepsilon\psi_1(x,t,\frac{x}{\varepsilon},\frac{t}{\varepsilon})$, where

$$\psi_0 \in \mathcal{D}(Q),\ \psi_1 \in \mathcal{D}(Q) \otimes \mathcal{C}^\infty_{per}(Y \times Z). \tag{6.8}$$

Some simple but useful properties are gathered in the next lemma.

Lemma 6.3 *Let* $1 \le p < \infty$. *We have as* $\varepsilon \to 0$,

$$(i)\ \Phi_\varepsilon \to \psi_0 \ in\ L^p(Q),$$

$$(ii)\ \frac{\partial \Phi_\varepsilon}{\partial x_i} \to \frac{\partial \psi_0}{\partial x_i} + \frac{\partial \psi_1}{\partial y_i}\ in\ L^p(Q)\text{-}strong\ \Sigma \quad (1 \le i \le N),$$

$$(iii)\ \frac{\partial \Phi_\varepsilon}{\partial t} \to \frac{\partial \psi_0}{\partial t} + \frac{\partial \psi_1}{\partial z}\ in\ L^p(Q)\text{-}strong\ \Sigma,$$

$$(iv)\ \Phi_\varepsilon \to \psi_0\ in\ L^p(0, T; W_0^{1,p}(\Omega))\text{-}weak.$$

Proof The proof of (i)–(iii) is a simple exercise left to the reader. As regards (iv), one sees readily by (i)–(iii) and use of elementary properties (see Chap. 3) that $\Phi_\varepsilon \to \psi_0$ in $W_0^{1,p}(Q)$-weak. Hence the claimed property follows at once by the fact that $W_0^{1,p}(Q)$ is continuously embedded in $L^p(0, T; W_0^{1,p}(\Omega))$. ■

Now, let $\varepsilon > 0$. For fixed $0 < t < T$, we consider the sesquilinear form $a^\varepsilon(t; \cdot, \cdot)$ on $H_0^1(\Omega) \times H_0^1(\Omega)$ given by

$$a^\varepsilon(t; u, v) = \sum_{i,j=1}^{N} \int_\Omega a_{ij}^\varepsilon(x, t) \frac{\partial u}{\partial x_j}(x) \overline{\frac{\partial v}{\partial x_i}}(x) dx, \quad u, v \in H_0^1(\Omega).$$

We have

$$|a^\varepsilon(t; u, v)| \le c \|u\|_{H_0^1(\Omega)} \|v\|_{H_0^1(\Omega)} \quad (u, v \in H_0^1(\Omega)) \tag{6.9}$$

and

$$\operatorname{Re} a^\varepsilon(t; u, u) \ge \alpha_0 \|u\|_{H_0^1(\Omega)}^2 \quad (u \in H_0^1(\Omega)), \tag{6.10}$$

where $c = \max_{1 \le i,j \le N} \sup_{(x,t) \in \overline{Q}} \|a_{ij}(x, t, \cdot, \cdot)\|_{L^\infty(\mathbb{R}^{N+1})}$, and where α_0 is that constant in (6.5). Indeed, (6.9) follows by a routine argument and use of Proposition 1.8, whereas (6.10) is a direct consequence of Lemma 6.1.

On the other hand there is no difficulty in verifying that for $u, v \in L^2(0, T; H_0^1(\Omega))$, the function $t \mapsto a^\varepsilon(t; u(t), v(t))$ is integrable on $(0, T)$.

At the present time we put

$$\mathbb{D}_i \mathbf{v} = \frac{\partial v_0}{\partial x_i} + \frac{\partial v_1}{\partial y_i} \quad (1 \le i \le N), \mathbf{v} = (v_0, v_1) \in H_0^1(\Omega) \times L^2(Q; \mathcal{H})$$

and for fixed $0 \leq t \leq T$ we consider the sesquilinear form $a_\Omega(t; \cdot, \cdot)$ on $[H_0^1(\Omega) \times L^2(\Omega; \mathcal{H})]^2$ given by

$$a_\Omega(t; \mathbf{u}, \mathbf{v}) = \sum_{i,j=1}^{N} \int_\Omega \int_Y \int_Z a_{ij}(x, t, y, z) \mathbb{D}_j \mathbf{u}(x, y, z) \overline{\mathbb{D}_i \mathbf{v}}(x, y, z) dx dy dz$$

for $\mathbf{u}, \mathbf{v} \in H_0^1(\Omega) \times L^2(\Omega; \mathcal{H})$.

One further preliminary result is the following lemma.

Lemma 6.4 *Let a sequence $(u_\varepsilon)_{\varepsilon \in E}$ be as in Lemma 6.2, and let E' be a subsequence from E such that the conclusion of Lemma 6.2 holds. Then, as $E' \ni \varepsilon \to 0$,*

$$\int_0^T a^\varepsilon(t; u_\varepsilon(t), \Phi_\varepsilon(t)) dt \to \int_0^T a_\Omega(t; \boldsymbol{u}(t), \boldsymbol{\Phi}(t)) dt,$$

where $\Phi_\varepsilon(t) = \Phi_\varepsilon(\cdot, t)$, Φ_ε as in (6.7) and (6.8), and $\boldsymbol{\Phi} = (\psi_0, \psi_1)$.

Proof Lemma 6.2 in combination with Lemma 6.3 (for $p = 2$) and use of Proposition 3.5 yield

$$\frac{\partial u_\varepsilon}{\partial x_j} \frac{\partial \Phi_\varepsilon}{\partial x_i} \to \left(\frac{\partial u_0}{\partial x_j} + \frac{\partial u_1}{\partial y_j} \right) \left(\frac{\partial \psi_0}{\partial x_i} + \frac{\partial \psi_1}{\partial y_i} \right) \text{ in } L^2(Q)\text{-weak } \Sigma \text{ as } E' \ni \varepsilon \to 0,$$

and that for $1 \leq i, j \leq N$. Hence, in view of (6.6), the Lemma follows by Proposition 3.6 and use of the Fubini theorem. ■

6.1.3 Global Homogenization Theorem

In what follows, it is assumed that $L^2(\Omega)$ is identified with its antidual so that $H_0^1(\Omega) \subset L^2(\Omega) \subset H^{-1}(\Omega)$ with continuous embeddings, $H^{-1}(\Omega)$ being viewed here as the antidual of $H_0^1(\Omega)$. Therefore we may use the same symbol, $[\cdot, \cdot]$, to denote the inner product in $L^2(\Omega)$ and the antiduality pairing between $H^{-1}(\Omega)$ and $H_0^1(\Omega)$.

Before we can state and prove the main result in the present section we require further preliminaries. First, it is worth recalling that the natural norm on

$$\mathbb{V} = H_0^1(\Omega) \times L^2(\Omega; \mathcal{H})$$

is given by

$$\|\mathbf{v}\| = (\|v_0\|^2_{H_0^1(\Omega)} + \|v_1\|^2_{L^2(\Omega;\mathcal{H})})^{\frac{1}{2}}, \quad \mathbf{v} = (v_0, v_1) \in \mathbb{V},$$

which makes it into a Hilbert space. However, by letting

$$\|\mathbf{v}\|_{\mathbb{V}} = \left(\sum_{i=1}^{N} \int_{\Omega} \int_{Y} \int_{Z} |\mathbb{D}_i \mathbf{v}(x, y, z)|^2 dx dy dz \right)^{\frac{1}{2}}$$

for $\mathbf{v} = (v_0, v_1) \in \mathbb{V}$, one defines a norm, $\|\cdot\|_{\mathbb{V}}$, on $\mathbb{V}$ which is equal to the preceding one (the verification is an easy exercise left to the reader) and which proves more convenient. Indeed, one has

$$\operatorname{Re} a_{\Omega}(t; \mathbf{v}, \mathbf{v}) \geq \alpha \|\mathbf{v}\|_{\mathbb{V}}^2 \quad (\mathbf{v} \in \mathbb{V}) \tag{6.11}$$

for all $0 \leq t \leq T$, as is immediate by (6.3).

Attention is also drawn to the equality

$$L^2(0, T; \mathbb{V}) = L^2(0, T; H_0^1(\Omega)) \times L^2(Q; \mathcal{H}) \tag{6.12}$$

which will be used in the sequel without further ado.

Finally, we put

$$\mathbb{F}_{\mathcal{H}}^{1,2} = V_0^2 \times L^2(Q; \mathcal{H})$$

where

$$V_0^2 = \{v | v \in V^2, v(0) = 0\},$$

the latter space being equipped with the V^2 norm making it into a Hilbert space; and we consider the variational problem:

$$\begin{cases} \mathbf{u} = (u_0, u_1) \in \mathbb{F}_{\mathcal{H}}^{1,2} : \\ \int_0^T [u_0'(t), v_0(t)] dt + \int_0^T a_{\Omega}(t; \mathbf{u}(t), \mathbf{v}(t)) dt = \int_0^T [f(t), v_0(t)] dt \\ \text{for all } \mathbf{v} = (v_0, v_1) \in \mathbb{F}_{\mathcal{H}}^{1,2}. \end{cases} \tag{6.13}$$

Lemma 6.5 *The variational problem (6.13) has at most one solution.*

Proof It is enough to show that $f = 0$ implies $\mathbf{u} = \mathbf{0}$. To do this, choose in (6.13) (with $f = 0$) $\mathbf{v} = \mathbf{u}$ and use (6.11) and the classical formula

$$\int_0^T ([u'(t), v(t)] + [v'(t), u(t)]) dt = [u(T), v(T)] - [u(0), v(0)] \tag{6.14}$$

for $u, v \in V^2$, to arrive at

$$\|u_0(T)\|^2_{L^2(\Omega)} + 2\alpha \int_0^T \|\mathbf{u}(t)\|^2_{\mathbb{V}} dt = 0.$$

Hence $\mathbf{u} = \mathbf{0}$. ∎

Remark 6.3 In fact an existence and uniqueness result for (6.13) is available (see, e.g., Lions and Magenes [74, pp. 253–258]). However, the reader who is familiar with homogenization theory certainly knows that one only needs to prove the unicity of the solution of the so-called global homogenized problem, viz. (6.13). Indeed, the proof of the existence of such a solution is intrinsically involved in the homogenization approach followed in the book.

We will also need the following

Lemma 6.6 *The space* $\mathcal{F}_0^\infty = \mathcal{D}(Q) \times (\mathcal{D}(Q) \otimes [\mathcal{C}^\infty_{per}(Y)/\mathbb{C}] \otimes \mathcal{C}^\infty_{per}(Z))$ *is dense in* $L^2(0, T; \mathbb{V})$.

Proof Bearing (6.12) in mind, let us begin by noting that $\mathcal{D}(Q)$ is dense in $L^2(0, T; H^1_0(\Omega))$ (the verification is an easy exercise left to the reader). On the other hand, $\mathcal{K}(Q) \otimes \mathcal{H}$ is dense in $\mathcal{K}(Q; \mathcal{H})$ and the latter is a dense subspace of $L^2(Q; \mathcal{H})$, of course. Therefore, since $\mathcal{D}(Q)$ is dense in $\mathcal{K}(Q)$, we see that we are through if we can show that $\mathfrak{A} = [\mathcal{C}^\infty_{per}(Y)/\mathbb{C}] \otimes \mathcal{C}^\infty_{per}(Z)$ is dense in $\mathcal{H}$. But it amounts to showing that if $u \in \mathcal{H}$ and $(u, \psi)_\mathcal{H} = 0$ for all $\psi \in \mathfrak{A}$, where $(,)_\mathcal{H}$ denotes the inner product in $\mathcal{H}$, then $u = 0$. So, fix freely such a $u \in \mathcal{H}$. Then

$$\int_Z \left(\sum_{j=1}^N \int_Y \frac{\partial u(\cdot, z)}{\partial y_j} \overline{\frac{\partial \psi(\cdot, z)}{\partial y_j}} dy \right) dz = 0$$

for all $\psi \in \mathfrak{A}$. Let $k = (k_1, \cdots, k_N) \in \mathbb{Z}^N$ with $|k| \neq 0$ be freely fixed. Let $\varphi \in \mathcal{D}(Z)$ be arbitrary given. Choose above ψ with

$$\psi(y, z) = \gamma_k(y) w(z) \quad (y \in \mathbb{R}^N, z \in \mathbb{R}),$$

where $\gamma_k(y) = e^{2\pi i k \cdot y}$ and $w(z) = \sum_{l \in \mathbb{Z}} \varphi(z - l)$ (a locally finite sum). Since w coincides with φ in Z (see Remark 2.15), it follows by the arbitrariness of φ that

$$\sum_{j=1}^N \int_Y \frac{\partial u(\cdot, z)}{\partial y_j} \overline{\frac{\partial \gamma_k}{\partial y_j}} dy = 0$$

for a.e. $z \in \mathbb{R}$ (because of the Z-periodicity), or more explicitly

$$(-2\pi i)^2 |k|^2 \int_Y u(y, z) e^{-2\pi i k \cdot y} dy = 0$$

for a.e. $z \in \mathbb{R}$. Recalling that $u(\cdot, z)$ has mean zero, we deduce that all of the Fourier coefficients of $u(\cdot, z)$ are zero, and so $u = 0$. The lemma follows. ■

We are now in a position to establish the global homogenization theorem for the initial-boundary value problem (6.4).

Theorem 6.1 *For each $\varepsilon > 0$, let u_ε be the unique function in V^2 satisfying (6.4). Then, as $\varepsilon \to 0$,*

$$u_\varepsilon \to u_0 \text{ in } L^2(0, T; H_0^1(\Omega))\text{-weak}, \tag{6.15}$$

$$\frac{\partial u_\varepsilon}{\partial t} \to \frac{\partial u_0}{\partial t} \text{ in } L^2(0, T; H^{-1}(\Omega))\text{-weak}, \tag{6.16}$$

$$\frac{\partial u_\varepsilon}{\partial x_j} \to \frac{\partial u_0}{\partial x_j} + \frac{\partial u_1}{\partial y_j} \text{ in } L^2(Q)\text{-weak } \Sigma \quad (1 \leq j \leq N), \tag{6.17}$$

where $\mathbf{u} = (u_0, u_1)$ is the unique solution of (6.13).

Proof Let $\varepsilon > 0$. Clearly $u_\varepsilon \in V_0^2$ and

$$\int_0^T [u_\varepsilon'(t), v(t)]dt + \int_0^T a^\varepsilon(t; u_\varepsilon(t), v(t))dt = \int_0^T [f(t), v(t)]dt \tag{6.18}$$

for all $v \in L^2(0, T; H_0^1(\Omega))$. By choosing in particular $v = u_\varepsilon$ and then using (6.10) together with (6.14), we see by mere routine that the (generalized) sequence $(u_\varepsilon)_{\varepsilon>0}$ is norm bounded in $L^2(0, T; H_0^1(\Omega))$. By (6.18) and use of (6.9), this implies at once that the sequence $(u_\varepsilon')_{\varepsilon>0}$ is norm bounded in $L^2(0, T; H^{-1}(\Omega))$. Hence, given any fundamental sequence E, it follows by Lemma 6.2 that there are some subsequence E' from E and some $\mathbf{u} = (u_0, u_1) \in V^2 \times L^2(Q; \mathcal{H})$ such that (6.15)–(6.17) hold as $E' \ni \varepsilon \to 0$. Consequently, if we check that $\mathbf{u}$ satisfies (6.13), then thanks to Lemma 6.5, we will be justified in stating that the preceding convergence results stand when $E \ni \varepsilon \to 0$, hence when $0 < \varepsilon \to 0$, which therefore completes the proof.

To begin, observe that $u_0 \in V_0^2$, as is immediate by (6.15)–(6.16) (when $E' \ni \varepsilon \to 0$, of course) and use of Remark 6.2 in combination with the continuous embedding $V^2 \hookrightarrow \mathcal{C}([0, T]; L^2(\Omega))$. Thus, $\mathbf{u} = (u_0, u_1) \in \mathbb{F}_{\mathcal{H}}^{1,2}$. The next step is to show that $\mathbf{u}$ satisfies the variational equation in (6.13). To this end, take in (6.18) $v = \Phi_\varepsilon$, where Φ_ε is given by (6.7)–(6.8), and then seek to pass to the limit (as $E' \ni \varepsilon \to 0$) on both sides of the resultant equality. In view of Lemma 6.4 and part (iv) of Lemma 6.3 this is feasible if we can verify that

$$\int_0^T [u_\varepsilon'(t), \Phi_\varepsilon(t)]dt \to \int_0^T [u_0'(t), \psi_0(t)]dt \tag{6.19}$$

as $E' \ni \varepsilon \to 0$. For that purpose we begin by noting that

$$\int_0^T [u_\varepsilon'(t), \Phi_\varepsilon(t)]dt = -\int_Q u_\varepsilon \overline{\frac{\partial \Phi_\varepsilon}{\partial t}} dxdt.$$

Next, we observe that by part (iii) of Lemma 6.3 we have

$$\frac{\partial \Phi_\varepsilon}{\partial t} \to \frac{\partial \psi_0}{\partial t} \text{ in } L^2(Q)\text{-weak}$$

as $\varepsilon \to 0$. On the other hand, $u_\varepsilon \to u_0$ in $L^2(Q)$ as $E' \ni \varepsilon \to 0$, as is easily seen by (6.15)–(6.16) (when $E' \ni \varepsilon \to 0$) and use of the compactness of the embedding $V^2 \hookrightarrow L^2(Q)$. Hence (6.19) follows when $E' \ni \varepsilon \to 0$. Consequently, by passing to the limit as stated above we get

$$\int_0^T [u_0'(t), \psi_0(t)]dt + \int_0^T a_\Omega(t; \mathbf{u}(t), \mathbf{\Phi}(t))dt = \int_0^T [f(t), \psi_0(t)]dt$$

and that for any arbitrary $\mathbf{\Phi} = (\psi_0, \psi_1) \in \mathcal{F}_0^\infty$. Thanks to Lemma 6.6, this still holds with $L^2(0, T; \mathbb{V})$ in place of $\mathcal{F}_0^\infty$. Hence (6.13) follows, since $\mathbb{F}_{\mathcal{H}}^{1,2} \subset L^2(0, T; \mathbb{V})$. ■

Remark 6.4 Theorem 6.1 and its proof remain valid if f, instead of lying in $L^2(Q)$, is taken in the larger space $L^2(0, T; H^{-1}(\Omega))$.

6.1.4 Macroscopic Homogenized Problem

We begin by computing the homogenized coefficients. For fixed $(x, t) \in \overline{Q}$, let

$$a(x, t; u, v) = \sum_{i,j=1}^N \int_Y \int_Z a_{ij}(x, t, y, z) \frac{\partial u}{\partial y_j}(y, z) \overline{\frac{\partial v}{\partial y_i}}(y, z) dy dz$$

for $u, v \in \mathcal{H}$. This defines a sesquilinear form $a(x, t; \cdot, \cdot)$ on $\mathcal{H} \times \mathcal{H}$ which, according to (6.1)–(6.3), is continuous, Hermitian and coercive, i.e.,

$$\operatorname{Re} a(x, t; v, v) \geq \alpha \|v\|_{\mathcal{H}}^2 \quad (v \in \mathcal{H}),$$

where α is precisely that constant in (6.3). In view of these properties of $a(x, t; \cdot, \cdot)$, for any given index $1 \leq j \leq N$ the variational problem

$$\begin{cases} \chi^j(x, t) \in \mathcal{H} : \\ a(x, t; \chi^j(x, t), v) = \displaystyle\sum_{k=1}^N \int_Y \int_Z a_{kj}(x, t, y, z) \overline{\frac{\partial v}{\partial y_k}}(y, z) dy dz \\ \text{for all } v \in \mathcal{H} \end{cases}$$

has one and only one solution. This yields a mapping $(x, t) \mapsto \chi^j(x, t)$ of $\overline{Q}$ into $\mathcal{H}$, hence a family $(\chi^j)_{1 \leq j \leq N}$ by means of which the homogenized coefficients are usually defined.

Specifically, for $1 \leq i, j \leq N$ and $(x, t) \in \overline{Q}$, let

$$q_{ij}(x, t) = \int_Y \int_Z a_{ij}(x, t, y, z) dy dz - \sum_{l=1}^{N} \int_Y \int_Z a_{il}(x, t, y, z) \frac{\partial \chi^j(x, t)}{\partial y_l}(y, z) dy dz.$$

This gives a family of mappings $q_{ij} : \overline{Q} \to \mathbb{C}$ $(1 \leq i, j \leq N)$ called the *homogenized coefficients* for the homogenization problem under consideration. The main properties of the said coefficients are collected below.

Proposition 6.2 *The following statements hold:*

(i) $\chi^j \in \mathcal{C}(\overline{Q}; \mathcal{H})$.
(ii) $q_{ij} \in \mathcal{C}(\overline{Q})$.
(iii) $q_{ji} = \overline{q}_{ij}$.
(iv) There is some $\alpha_1 > 0$ *such that*

$$Re \sum_{i,j=1}^{N} q_{ij}(x, t) \xi_j \overline{\xi}_i \geq \alpha_1 |\xi|^2$$

for all $(x, t) \in \overline{Q}$ *and all* $\xi \in \mathbb{C}^N$.

Proof This is a mere adaptation of the proof of Proposition 5.3. ∎

By the same argument as followed before (see Chap. 5) one is easily led to the macroscopic homogenized problem (the details are left to the reader), viz.

Proposition 6.3 *The function* u_0 *in Theorem 6.1 is the weak solution of the initial-boundary value problem*

$$\frac{\partial u_0}{\partial t} - \sum_{i,j=1}^{N} \frac{\partial}{\partial x_i} \left(q_{ij} \frac{\partial u_0}{\partial x_j} \right) = f \text{ in } Q = \Omega \times (0, T),$$
$$u_0 = 0 \text{ on } \partial\Omega \times (0, T),$$
$$u_0(x, 0) = 0 \text{ in } \Omega.$$

The initial-boundary value problem in Proposition 6.3 is referred to as the *macroscopic homogenized problem* for (6.4). Thanks to the properties of the coefficients q_{ij} (see in particular parts (ii) and (iv) of Proposition 6.2), this problem has one and only one weak solution.

6.2 Quasilinear Parabolic Operators: The Two-Scale Setting

6.2.1 Setting of the Problem and Preliminaries

Let $2 \leq p < \infty$. Let $\mathbf{a} = (a_i)_{1\leq i\leq N}$ be a function of $\mathbb{R}_y^N \times \mathbb{R}_z \times \mathbb{R}_\mu^N$ into $\mathbb{R}^N$ with the following properties:

$$\begin{aligned}&\text{For fixed } \mu \in \mathbb{R}^N, \text{ the function } (y, z) \mapsto \mathbf{a}(y, z, \mu) \text{ is measurable}\\ &\text{from } \mathbb{R}^N \times \mathbb{R} \text{ into } \mathbb{R}^N.\end{aligned} \tag{6.20}$$

$$\mathbf{a}(y, z, \omega) = \omega \text{ for a.e. } (y, z) \in \mathbb{R}^N \times \mathbb{R} \;\; (\omega \text{ the origin in } \mathbb{R}^N). \tag{6.21}$$

$$\begin{cases}\text{There are some constants } \alpha_0, \alpha_1 > 0 \text{ such that}\\ \text{(1) } (\mathbf{a}(y, z, \mu) - \mathbf{a}(y, z, \nu)) \cdot (\mu - \nu) \geq \alpha_0 |\mu - \nu|^p\\ \text{(2) } |\mathbf{a}(y, z, \mu) - \mathbf{a}(y, z, \nu)| \leq \alpha_1 (|\mu| + |\nu|)^{p-2} |\mu - \nu|\\ \text{for all } \mu, \nu \in \mathbb{R}^N \text{ and almost all } (y, z) \in \mathbb{R}^N \times \mathbb{R}.\end{cases} \tag{6.22}$$

Here and in the sequel the basic notation is the same as used in Sects. 5.5 and 6.1. Let Ω and T be as in the preceding section, and let

$$f \in L^{p'}(0, T; W^{-1,p'}(\Omega; \mathbb{R}))$$

with $\frac{1}{p} + \frac{1}{p'} = 1$. For given $\varepsilon > 0$, we consider the initial-boundary value problem

$$\begin{cases}\dfrac{\partial u_\varepsilon}{\partial t} - \operatorname{div} \mathbf{a}\left(\dfrac{x}{\varepsilon}, \dfrac{t}{\varepsilon}, Du_\varepsilon\right) = f \text{ in } Q,\\ u_\varepsilon = 0 \text{ on } \partial\Omega \times (0, T),\\ u_\varepsilon(x, 0) = 0 \text{ in } \Omega.\end{cases} \tag{6.23}$$

Our goal is to investigate the limiting behaviour, as $\varepsilon \to 0$, of u_ε under the periodicity hypothesis on $\mathbf{a}$ in the variables y and z. Before we can accomplish this, however, we require a few preliminaries and in particular a rigorous definition of the diffusion term of the differential operator in (6.23).

Let $\mathbf{\Phi} \in \mathcal{C}_\mathbb{R}(\overline{Q})^N$. Let $1 \leq i \leq N$. A quick survey of (6.20)–(6.22) reveals that the function $(x, t, y, z) \mapsto a_i(y, z, \mathbf{\Phi}(x, t))$ of $\overline{Q} \times \mathbb{R}^N \times \mathbb{R}$ into $\mathbb{R}$ lies in $\mathcal{C}(\overline{Q}; L^\infty(\mathbb{R}^{N+1}))$. Therefore, the real function $(x, t) \mapsto a_i(\frac{x}{\varepsilon}, \frac{t}{\varepsilon}, \mathbf{\Phi}(x, t))$ on Q denoted farther by $a_i^\varepsilon(\cdot, \cdot, \mathbf{\Phi})$, is defined as in Proposition 1.8, and it belongs to $L^\infty_\mathbb{R}(Q)$. However, we need to define this function for $\mathbf{\Phi} \in L^p_\mathbb{R}(Q)^N$.

For that purpose, consider a function $G : \mathbb{R}^N \times \mathbb{R} \times \mathbb{R}^N \to \mathbb{R}$ with the following properties:

$$\text{For fixed } \mu \in \mathbb{R}^N, \text{ the function } (y, z) \mapsto G(y, z, \mu) \text{ of } \mathbb{R}^N \times \mathbb{R} \text{ into } \mathbb{R} \text{ is measurable.} \tag{6.24}$$

$$G(y, z, \omega) = 0 \text{ for a.e. } (y, z) \in \mathbb{R}^N \times \mathbb{R}. \tag{6.25}$$

$$\begin{cases} \text{There is a constant } \alpha_1 > 0 \text{ such that} \\ |G(y, z, \mu) - G(y, z, \nu)| \leq \alpha_1(|\mu| + |\nu|)^{p-2}|\mu - \nu| \quad (\mu, \nu \in \mathbb{R}^N) \\ \text{for a.e. } (y, z) \in \mathbb{R}^N \times \mathbb{R}. \end{cases} \tag{6.26}$$

If $\mathbf{\Phi} \in \mathcal{C}_\mathbb{R}(\overline{Q})^N$ then, as seen above, by (6.24), (6.25), (6.26) and use of Proposition 1.8 one can define the real-valued function $(x, t) \mapsto G(\frac{x}{\varepsilon}, \frac{t}{\varepsilon}, \mathbf{\Phi}(x, t))$ on Q denoted by $G^\varepsilon(\cdot, \cdot, \mathbf{\Phi})$, which lies in $L^\infty(Q)$. This can be extended to the whole of $L^p_\mathbb{R}(Q)^N$, as shown below.

Lemma 6.7 *The transformation $\mathbf{\Phi} \mapsto G^\varepsilon(\cdot, \cdot, \mathbf{\Phi})$ of $\mathcal{C}_\mathbb{R}(\overline{Q})^N$ into $L^\infty(Q)$ extends by continuity to a mapping , still denoted by $\mathbf{\Phi} \mapsto G^\varepsilon(\cdot, \cdot, \mathbf{\Phi})$, of $L^p_\mathbb{R}(Q)^N$ into $L^{p'}(Q)$ with the property*

$$\begin{aligned} &\|G^\varepsilon(\cdot, \cdot, \mathbf{\Phi}) - G^\varepsilon(\cdot, \cdot, \mathbf{\Psi})\|_{L^{p'}(Q)} \leq \\ &\alpha_1 N^{\frac{1}{2}-\frac{1}{p}} \||\mathbf{\Phi}| + |\mathbf{\Psi}|\|^{p-2}_{L^p(Q)} \|\mathbf{\Phi} - \mathbf{\Psi}\|_{L^p(Q)^N} \textit{ for } \mathbf{\Phi}, \mathbf{\Psi} \in L^p_\mathbb{R}(Q)^N. \end{aligned} \tag{6.27}$$

Proof Let us observe beforehand that Lemma 5.11 and its proof are rigorously valid with Q in place of Ω and $\mathbb{R}^{N+1}$ in that of $\mathbb{R}^N$. This being so, let $\mathbf{\Phi}, \mathbf{\Psi} \in \mathcal{C}_\mathbb{R}(\overline{Q})^N$. By applying the said lemma with

$$\begin{aligned} u(x, t, y, z) = \alpha_1(&|\mathbf{\Phi}(x, t)| + |\mathbf{\Psi}(x, t)|)^{p-2}|\mathbf{\Phi}(x, t) - \mathbf{\Psi}(x, t)| \\ &- |G(y, z, \mathbf{\Phi}(x, t)) - G(y, z, \mathbf{\Psi}(x, t))| \end{aligned}$$

we are readily led by (6.26) to

$$|G^\varepsilon(\cdot, \cdot, \mathbf{\Phi}) - G^\varepsilon(\cdot, \cdot, \mathbf{\Psi})|^{p'} \leq \alpha_1^{p'}(|\mathbf{\Phi}| + |\mathbf{\Psi}|)^{(p-2)p'}|\mathbf{\Phi} - \mathbf{\Psi}|^{p'}.$$

By considering the functions $|\mathbf{\Phi} - \mathbf{\Psi}|^{p'}$ and $(|\mathbf{\Phi}| + |\mathbf{\Psi}|)^{(p-2)p'}$ as elements of $L^q(Q)$ (with $q = \frac{p}{p'}$) and $L^{q'}(Q)$ (with $q' = \frac{q}{q-1}$), respectively, and using Hölder's inequality, one easily arrives at (6.27) with $\mathcal{C}_\mathbb{R}(\overline{Q})^N$ in place of $L^p_\mathbb{R}(Q)^N$. With this in mind, let

$$B_r = \left\{\mathbf{v} | \mathbf{v} \in L^p_\mathbb{R}(Q)^N, \|\mathbf{v}\|_{L^p(Q)^N} < \frac{r}{2}\right\} \quad \text{with } r > 0.$$

Let g_r be the restriction to $B_r \cap \mathcal{C}_{\mathbb{R}}(\overline{Q})^N$ of the mapping $\mathbf{v} \mapsto G^\varepsilon(\cdot, \cdot, \mathbf{v})$ (where $\varepsilon > 0$ is fixed, of course). It is clear that

$$\begin{aligned} &\|g_r(\boldsymbol{\Phi}) - g_r(\boldsymbol{\Psi})\|_{L^{p'}(Q)} \leq \alpha_1 N^{(\frac{1}{2}-\frac{1}{p})(p-1)} r^{p-2} \|\boldsymbol{\Phi} - \boldsymbol{\Psi}\|_{L^p(Q)^N} \\ &\text{for all } \boldsymbol{\Phi}, \boldsymbol{\Psi} \in B_r \cap \mathcal{C}_{\mathbb{R}}(\overline{Q})^N. \end{aligned} \tag{6.28}$$

Since $B_r \cap \mathcal{C}_{\mathbb{R}}(\overline{Q})^N$ is dense in B_r (the verification is an elementary exercise), it follows that g_r extends by continuity to a continuous mapping, still denoted for convenience by g_r, of B_r into $L^{p'}(Q)$ such that (6.28) holds with B_r in place of $B_r \cap \mathcal{C}_{\mathbb{R}}(\overline{Q})^N$. Whence we deduce a sequence $(g_n)_{n\geq 1}$ of continuous mappings $g_n : B_n \to L^{p'}(Q)$ with (6.28) and $g_n(\boldsymbol{\Phi}) = G^\varepsilon(\cdot, \cdot, \boldsymbol{\Phi})$ for $\boldsymbol{\Phi} \in B_n \cap \mathcal{C}_{\mathbb{R}}(\overline{Q})^N$. Noticing that $L^p_{\mathbb{R}}(Q)^N$ is the union of the balls B_n (integers $n \geq 1$) and, on the other hand, that $g_{n+1}(\boldsymbol{\Phi}) = g_n(\boldsymbol{\Phi})$ for $\boldsymbol{\Phi} \in B_n$, we are readily led to a uniquely defined continuous mapping $g : L^p_{\mathbb{R}}(Q)^N \to L^{p'}(Q)$ such that $g(\boldsymbol{\Phi}) = G^\varepsilon(\cdot, \cdot, \boldsymbol{\Phi})$ for any $\boldsymbol{\Phi} \in \mathcal{C}_{\mathbb{R}}(\overline{Q})^N$. Hence the whole lemma follows by the density of $\mathcal{C}_{\mathbb{R}}(\overline{Q})^N$ in $L^p_{\mathbb{R}}(Q)^N$. ■

It is a matter of course that we may take $G = a_i$ in the above lemma. Thus, for $u \in L^p(0, T; W^{1,p}(\Omega; \mathbb{R}))$, the function $a_i^\varepsilon(\cdot, \cdot, Du)$ is well defined as an element of $L^{p'}_{\mathbb{R}}(Q)$. By letting $\mathbf{a}^\varepsilon(\cdot, \cdot, Du) = (a_i^\varepsilon(\cdot, \cdot, Du))_{1\leq i\leq N}$, we have therefore $\mathbf{a}^\varepsilon(\cdot, \cdot, Du) \in L^{p'}(Q)^N = L^{p'}(0, T; L^{p'}(\Omega)^N)$. Hence, div $\mathbf{a}^\varepsilon(\cdot, \cdot, Du)$ actually stands for the function $t \mapsto$ div $\mathbf{a}^\varepsilon(\cdot, t, Du(\cdot, t))$ of $(0, T)$ into $W^{-1,p'}(\Omega)$ which lies in $L^{p'}(0, T; W^{-1,p'}(\Omega))$. Thus is understood the diffusion term in (6.23).

Now, it is obvious that the operator $u \mapsto \mathbf{a}^\varepsilon(\cdot, \cdot, Du)$ of $L^p(0, T; W^{1,p}(\Omega; \mathbb{R}))$ into $L^{p'}(Q)^N$ attains the value $\mathbf{0}$ (the origin in $L^{p'}(Q)^N$) on the function $u = 0$. Furthermore, the following hold for $u, v \in L^p(0, T; W^{1,p}(\Omega))$:

$$\begin{aligned} &\|\mathbf{a}^\varepsilon(\cdot, \cdot, Du) - \mathbf{a}^\varepsilon(\cdot, \cdot, Dv)\|_{L^{p'}(Q)^N} \leq \\ &\alpha_1 N^{\frac{3}{2}-\frac{2}{p}} \||Du| + |Dv|\|^{p-2}_{L^p(Q)} \|Du - Dv\|_{L^p(Q)^N} \end{aligned} \tag{6.29}$$

$$\begin{aligned} &\left[\mathbf{a}\left(\frac{x}{\varepsilon}, \frac{t}{\varepsilon}, Du(x,t)\right) - \mathbf{a}\left(\frac{x}{\varepsilon}, \frac{t}{\varepsilon}, Dv(x,t)\right)\right] \cdot (Du(x,t) - Dv(x,t)) \\ &\geq \alpha_0 |Du(x,t) - Dv(x,t)|^p \quad \text{for a.e. } (x,t) \in Q. \end{aligned} \tag{6.30}$$

Thanks to Lemma 6.7 it is now apparent that (6.23) is an initial-boundary value problem for a quasilinear parabolic monotone operator meeting the requirements of an existence and uniqueness theorem. One of the main references for such a topic is very likely Zeidler [147], where the reader can find among other things the proof of the following

Proposition 6.4 *Under the hypotheses (6.20)–(6.22), for fixed* $\varepsilon > 0$ *there is a unique function* $u_\varepsilon \in L^p(0, T; W_0^{1,p}(\Omega; \mathbb{R})) \cap \mathcal{C}([0, T]; L^2(\Omega))$ *satisfying (6.23).*

Let us make this a little more precise. First, let

$$V^p = \left\{ u | u \in L^p(0,T; W_0^{1,p}(\Omega;\mathbb{R})), u' = \frac{\partial u}{\partial t} \in L^{p'}(0,T; W^{-1,p'}(\Omega;\mathbb{R})) \right\}.$$

Equipped with the norm

$$\|u\|_{V^p} = \|u\|_{L^p(0,T;W_0^{1,p}(\Omega))} + \|u'\|_{L^{p'}(0,T;W^{-1,p'}(\Omega))} \quad (u \in V^p),$$

V^p is a reflexive Banach space. It is of great interest to note that, since $2 \le p < \infty$, the Banach space $W_0^{1,p}(\Omega;\mathbb{R})$ is continuously and densely embedded in $L^2_{\mathbb{R}}(\Omega)$. Hence, $L^2_{\mathbb{R}}(\Omega)$ being identified with its dual, it follows $W_0^{1,p}(\Omega;\mathbb{R}) \subset L^2_{\mathbb{R}}(\Omega) \subset W^{-1,p'}(\Omega;\mathbb{R})$ with continuous embeddings. Consequently:

(1) V^p is continuously embedded in $\mathcal{C}([0,T]; L^2_{\mathbb{R}}(\Omega))$ and compactly embedded in $L^p(0,T; L^2_{\mathbb{R}}(\Omega))$ (see Lions [72, pp. 57–58 and 156]).

(2) We will use the same symbol, $[\cdot,\cdot]$, to denote the inner product in $L^2_{\mathbb{R}}(\Omega)$ and the duality pairing between $W^{-1,p'}(\Omega;\mathbb{R})$ and $W_0^{1,p}(\Omega;\mathbb{R})$.

Having made this point, it is clear that a function $u_\varepsilon \in L^p(0,T; W^{1,p}(\Omega;\mathbb{R}))$ satisfying the equation in (6.23) verifies $\frac{\partial u_\varepsilon}{\partial t} \in L^{p'}(0,T; W^{-1,p'}(\Omega;\mathbb{R}))$. We deduce that the solution of (6.23) lies in V^p. Moreover, $u_\varepsilon \in V_0^p$, where

$$V_0^p = \{u | u \in V^p, u(0) = 0\}.$$

In the sequel V_0^p is considered equipped with the V^p norm, which makes it into a Banach space.

We assume from now that for fixed $1 \le i \le N$ and $\mu \in \mathbb{R}^N$, the function $(y,z) \mapsto a_i(y,z,\mu)$ is $Y \times Z$-periodic (see Sect. 6.1).

Attention is drawn to the fact that the forthcoming notion of Σ-convergence (in the periodic setting) is here relevant to the action exhibited in Sect. 6.1 and to the algebra $A = \mathcal{C}_{per}(Y \times Z)$. In the sequel we will put for convenience $A_{\mathbb{R}} = \mathcal{C}_{per}(Y \times Z;\mathbb{R})$ and $A_{\mathbb{R}}$ will be equipped with the supremum norm on $\mathbb{R}^N \times \mathbb{R}$, which makes it into a Banach algebra. The Banach spaces $L^r_{per}(Y \times Z)$ (resp. $L^r_{per}(Y \times Z;\mathbb{R})$), where $1 \le r \le \infty$, are defined in an obvious fashion (see Chaps. 2 and 3). See Sect. 6.1 for the cases $r = 2, \infty$.

Finally, it is worth recalling once for all the assumption $2 \le p < \infty$ which stands throughout the rest of this section. Also, $p' = \frac{p}{p-1}$.

We shall now pursue our series of preliminary results.

Proposition 6.5 *Let* $\boldsymbol{\Psi} = (\psi_j) \in \mathcal{C}(\overline{Q}; A^N_{\mathbb{R}})$, $A^N_{\mathbb{R}} = A_{\mathbb{R}} \times \cdots \times A_{\mathbb{R}}$ *(N-times). Then, as* $\varepsilon \to 0$,

$$a_i^\varepsilon(\cdot,\cdot,\boldsymbol{\Psi}^\varepsilon) \to a_i(\cdot,\cdot,\boldsymbol{\Psi}) \;\; \textit{in } L^{p'}(Q)\textit{-weak } \Sigma \;\; (1 \le i \le N), \tag{6.31}$$

where $\boldsymbol{\Psi}^\varepsilon = (\psi_j^\varepsilon)_{1\leq j\leq N}$ *with* $\psi_j^\varepsilon(x,t) = \psi_j(x,t,\frac{x}{\varepsilon},\frac{t}{\varepsilon})$ *for* $(x,t) \in Q$, *and where* $a_i(\cdot,\cdot,\boldsymbol{\Psi})$ *denotes the function* $(x,t,y,z) \mapsto a_i(y,z,\boldsymbol{\Psi}(x,t,y,z))$ *of* $\overline{Q} \times \mathbb{R}^N \times \mathbb{R}$ *into* $\mathbb{R}$, *which belongs to* $\mathcal{C}(\overline{Q}; L^\infty_{per}(Y \times Z;\mathbb{R}))$.

Proof For fixed $(x,t) \in \overline{Q}$, the function $(y,z) \mapsto a_i(y,z,\boldsymbol{\Psi}(x,t,y,z))$ is measurable from $\mathbb{R}^N \times \mathbb{R}$ into $\mathbb{R}$, as is readily seen by a classical property of Carathéodory functions (see Sect. 5.5). One deduces at once by (6.21) and (6.22)(2) that this function lies in $L^\infty(\mathbb{R}^{N+1};\mathbb{R})$, hence in $L^\infty_{per}(Y \times Z;\mathbb{R})$. Therefore, it follows by (6.22)(2), once again, that $a_i(\cdot,\cdot,\boldsymbol{\Psi}) \in \mathcal{C}(\overline{Q}; L^\infty_{per}(Y \times Z;\mathbb{R}))$. But

$$a_i^\varepsilon(\cdot,\cdot,\boldsymbol{\Psi}^\varepsilon) = (a_i(\cdot,\cdot,\boldsymbol{\Psi}))^\varepsilon, \tag{6.32}$$

as is immediate by an obvious adaptation of the proof of Lemma 5.12. Hence (6.31) follows by Corollary 3.1 (with Q in place of Ω and $Y \times Z$ in that of Y, of course). ■

As a consequence of Proposition 6.5 we have

Corollary 6.1 *Let* Φ_ε *be given by (6.7) and (6.8) with* ψ_0 *and* ψ_1 *assuming real values. As* $\varepsilon \to 0$,

$$a_i^\varepsilon(\cdot,\cdot,D\Phi_\varepsilon) \to a_i(\cdot,\cdot,D_x\psi_0 + D_y\psi_1) \quad \text{in } L^{p'}(Q)\text{-weak } \Sigma \tag{6.33}$$

for $1 \leq i \leq N$. *Furthermore, if* E *is a fundamental sequence and* $(v_\varepsilon)_{\varepsilon\in E}$ *is a sequence in* $L^p(Q)$ *such that* $v_\varepsilon \to v_0$ *in* $L^p(Q)$*-weak* Σ, *then*

$$\int_Q a_i^\varepsilon(\cdot,\cdot,D\Phi_\varepsilon)v_\varepsilon dxdt \to \int_Q\int_Y\int_Z a_i(\cdot,\cdot,D_x\psi_0 + D_y\psi_1)v_0 dxdtdydz$$

as $E \ni \varepsilon \to 0$ *for* $1 \leq i \leq N$.

Proof We have by (6.27) (with $G = a_i$)

$$\|a_i^\varepsilon(\cdot,\cdot,D\Phi_\varepsilon) - a_i^\varepsilon(\cdot,\cdot,D_x\psi_0 + (D_y\psi_1)^\varepsilon)\|_{L^{p'}(Q)} \leq c\varepsilon\|D_x\psi_1\|_{L^p(Q;A)}$$

for $0 < \varepsilon \leq 1$, where c is a positive constant depending only on ψ_0 and ψ_1. Hence (6.33) follows by Proposition 6.5. The rest of the corollary follows by mere routine and use of the preceding inequality (see also (6.32) with $\boldsymbol{\Psi} = D_x\psi_0 + D_y\psi_1$) together with Proposition 3.6. ■

The next preliminary result is the following

Lemma 6.8 *For* $\boldsymbol{\Phi} \in L^p(Q; L^p_{per}(Y\times Z;\mathbb{R})^N)$, *let* $a_i(\cdot,\cdot,\boldsymbol{\Phi})$ *stand for the function* $(x,t,y,z) \mapsto a_i(y,z,\boldsymbol{\Phi}(x,t,y,z))$ *of* $Q \times \mathbb{R}^N \times \mathbb{R}$ *into* $\mathbb{R}$ $(1 \leq i \leq N)$. *We have* $a_i(\cdot,\cdot,\boldsymbol{\Phi}) \in L^{p'}(Q; L^{p'}_{per}(Y \times Z))$ *and further the mapping* $\boldsymbol{\Phi} \mapsto \mathbf{a}(\cdot,\cdot,\boldsymbol{\Phi}) = (a_i(\cdot,\cdot,\boldsymbol{\Phi}))_{1\leq i\leq N}$ *of* $L^p(Q; L^p_{per}(Y \times Z;\mathbb{R})^N)$ *into* $L^{p'}(Q; L^{p'}_{per}(Y \times Z)^N)$ *has the following properties, where* $\boldsymbol{\Phi}$ *and* $\boldsymbol{\Psi}$ *are arbitrary in* $L^p(Q; L^p_{per}(Y \times Z;\mathbb{R})^N)$:

$$(i)\quad \begin{aligned}&\|\boldsymbol{a}(\cdot,\cdot,\boldsymbol{\Phi})-\boldsymbol{a}(\cdot,\cdot,\boldsymbol{\Psi})\|_{L^{p'}(Q;L^{p'}(Y\times Z)^N)}\leq\\ &\alpha_1 N^{\frac{3}{2}-\frac{2}{p}}\||\boldsymbol{\Phi}|+|\boldsymbol{\Psi}|\|^{p-2}_{L^p(Q\times Y\times Z)}\|\boldsymbol{\Phi}-\boldsymbol{\Psi}\|_{L^p(Q;L^p(Y\times Z)^N)}\end{aligned}$$

$$(ii)\ (\boldsymbol{a}(\cdot,\cdot,\boldsymbol{\Phi})-\boldsymbol{a}(\cdot,\cdot,\boldsymbol{\Psi}))\cdot(\boldsymbol{\Phi}-\boldsymbol{\Psi})\geq\alpha_0|\boldsymbol{\Phi}-\boldsymbol{\Psi}|^p\quad a.e.\ in\ Q\times\mathbb{R}^N\times\mathbb{R}.$$

Proof Given $\boldsymbol{\Phi}$ as above, the same argument (about Carathéodory functions) as used in proving Proposition 6.5 combines with (6.21) and (6.22)(2) to show that the function $(x,t)\mapsto a_i(\cdot,\cdot,\boldsymbol{\Phi})$ sends Q into $L^{p'}_{per}(Y\times Z)$. Furthermore, it is an easy exercise to verify that this function lies in $L^{p'}(Q;L^{p'}_{per}(Y\times Z))$, as claimed. The rest of the lemma follows readily by (6.22). ■

At the present time, let

$$\mathcal{H}_p=L^p_{per}(Z;W^{1,p}_{per}(Y;\mathbb{R})/\mathbb{R}),$$

$\mathcal{H}_p$ equipped with the norm

$$\|u\|_{\mathcal{H}_p}=\left(\int_Z\|u(z)\|^p_{W^{1,p}(Y)/\mathbb{C}}dz\right)^{\frac{1}{p}}\quad\text{(see Sect. 2.5)},$$

which makes it into a Banach space. The next lemma is the keystone of the homogenization approach followed here.

Lemma 6.9 *Given a norm bounded ordinary sequence $(u_\varepsilon)_{\varepsilon\in E}$ in V^p (thus, E is here a fundamental sequence), there are a subsequence E' from E and some $\boldsymbol{u}=(u_0,u_1)\in V^p\times L^p(Q;\mathcal{H}_p)$ such that as $E'\ni\varepsilon\to 0$,*

$$u_\varepsilon\to u_0\ in\ V^p\text{-}weak,$$

and

$$\frac{\partial u_\varepsilon}{\partial x_j}\to\frac{\partial u_0}{\partial x_j}+\frac{\partial u_1}{\partial y_j}\ in\ L^p(Q)\text{-}weak\ \Sigma\quad(1\leq j\leq N).$$

Proof The Banach space V^p is reflexive and compactly embedded in $L^p(0,T;L^2(\Omega))$, hence in $L^2(Q)$ also, since the former space is continuously embedded in the latter (recall that $p\geq 2$!). Therefore it is readily seen that the proof of Lemma 6.2 carries over mutatis mutandis to the present setting with, in particular, Lemma 2.10 substituted by Lemma 2.11. ■

Finally, let

$$\mathcal{C}^\infty_{per}(\widetilde{Y}\times Z;\mathbb{R})=\left\{\psi|\psi\in\mathcal{C}^\infty_{per}(Y\times Z;\mathbb{R}),\ \int_Y\psi(y,z)dy=0\text{ for all }z\in\mathbb{R}\right\}.$$

Lemma 6.10 *$\mathcal{F}_0^\infty = \mathcal{D}_\mathbb{R}(Q) \times [\mathcal{D}_\mathbb{R}(Q) \otimes \mathcal{C}_{per}^\infty(\widetilde{Y} \times Z; \mathbb{R})]$ is dense in the space $L^p(0, T; W_0^{1,p}(\Omega; \mathbb{R})) \times L^p(Q; \mathcal{H}_p)$.*

Proof Considering that $\mathcal{D}_\mathbb{R}(Q)$ is dense in $L^p(0, T; W_0^{1,p}(\Omega; \mathbb{R}))$, we see by the same line of proceeding as followed in proving Lemma 6.6 that the whole problem reduces to showing that $\mathcal{C}_{per}^\infty(\widetilde{Y} \times Z; \mathbb{R})$ is dense in $\mathcal{H}_p$. But this is straightforward. Indeed, given $u \in \mathcal{H}_p$, let $u_n = u * \theta_n$ with $\theta_n \in \mathcal{D}(\mathbb{R}^{N+1})$, where $(\theta_n)_{n\geq 1}$ is a mollifier on $\mathbb{R}^{N+1}$ (see the proof of Proposition 2.24). It is clear that $u_n \in \mathcal{C}_{per}^\infty(\widetilde{Y} \times Z; \mathbb{R})$ and

$$\frac{\partial u_n}{\partial y_j} \to \frac{\partial u}{\partial y_j} \quad \text{in } L_{per}^p(Y \times Z) \text{ as } n \to \infty,$$

and that for $1 \leq j \leq N$. Hence $u_n \to u$ in $\mathcal{H}_p$ as $n \to \infty$, which completes the proof. ■

6.2.2 Homogenization Results

Let $\mathbb{D}\mathbf{v} = D_x v_0 + D_y v_1$ for $\mathbf{v} = (v_0, v_1) \in L^p(0, T; W_0^{1,p}(Q; \mathbb{R})) \times L^p(Q; \mathcal{H}_p)$. We will need the following

Lemma 6.11 *The variational problem*

$$\begin{cases} \boldsymbol{u} = (u_0, u_1) \in \mathbb{F}_{\mathcal{H}_p}^{1,p} : \\ \displaystyle\int_0^T [u_0'(t), v_0(t)]dt + \int_Q \int_Y \int_Z \boldsymbol{a}(\cdot, \cdot, \mathbb{D}\boldsymbol{u}) \cdot \mathbb{D}\boldsymbol{v}\, dxdtdydz \\ \displaystyle = \int_0^T [f(t), v_0(t)]dt \quad \textit{for all } \boldsymbol{v} = (v_0, v_1) \in \mathbb{F}_{\mathcal{H}_p}^{1,p} \end{cases} \tag{6.34}$$

admits at most one solution.

Proof Assuming that $\mathbf{u} = (u_0, u_1)$ and $\mathbf{w} = (w_0, w_1)$ satisfy (6.34), a routine line of proceeding using (6.14) (with V^p in place of V^2) and inequality (ii) of Lemma 6.8 leads to

$$\frac{1}{2}\|u_0(T)-w_0(T)\|_{L^2(\Omega)}^2+\alpha_0 \int_Q\int_Y\int_Z |D_x(u_0-w_0)+D_y(u_1-w_1)|^p dxdtdydz = 0.$$

Letting $\varphi_0 = u_0 - w_0$ and $\varphi_1 = u_1 - w_1$, we deduce

$$\frac{\partial \varphi_0}{\partial x_j} + \frac{\partial \varphi_1}{\partial y_j} = 0 \qquad (1 \leq j \leq N),$$

or more precisely

$$\frac{\partial \varphi_0}{\partial x_j}(x,t) + \frac{\partial \varphi_1(x,t,\cdot,z)}{\partial y_j} = 0 \quad (1 \leq j \leq N)$$

for a.e. $(x,t) \in Q$ and a.e. $z \in Z$ (see the proof of Lemma 6.2). By integrating over Y it follows $\frac{\partial \varphi_0}{\partial x_j}(x,t) = 0$ for a.e. $(x,t) \in Q$, which amounts to $\frac{\partial \varphi_0(\cdot,t)}{\partial x_j} = 0$ for a.e. $0 < t < T$, and that for $1 \leq j \leq N$. One deduces at once that $\varphi_0 = 0$, hence $\varphi_1 = 0$. The lemma follows thereby. ∎

We are now in a position to achieve the global homogenization theorem for (6.23) in the periodic setting.

Theorem 6.2 *For each real $\varepsilon > 0$, let u_ε be the weak solution of the initial-boundary value problem (6.23). As $\varepsilon \to 0$, we have*

$$u_\varepsilon \to u_0 \textit{ in } L^p(0,T;W_0^{1,p}(\Omega))\textit{-weak}, \tag{6.35}$$

$$\frac{\partial u_\varepsilon}{\partial t} \to \frac{\partial u_0}{\partial t} \textit{ in } L^{p'}(0,T;W^{-1,p'}(\Omega))\textit{-weak}, \tag{6.36}$$

$$\frac{\partial u_\varepsilon}{\partial x_j} \to \frac{\partial u_0}{\partial x_j} + \frac{\partial u_1}{\partial y_j} \textit{ in } L^p(Q)\textit{-weak } \Sigma \ (1 \leq j \leq N), \tag{6.37}$$

where $\boldsymbol{u} = (u_0, u_1)$ is the solution of (6.34).

Proof To start with, let us check that the sequence $(u_\varepsilon)_{\varepsilon>0}$ is norm bounded in V^p. It is clear that for fixed $\varepsilon > 0$ one has $u_\varepsilon \in V_0^p$, as observed earlier, and further

$$\begin{aligned} &\int_0^T [u'_\varepsilon(t), v(t)]dt + \int_Q \mathbf{a}^\varepsilon(x,t,Du_\varepsilon(x,t)) \cdot Dv(x,t)dxdt \\ &= \int_0^T [f(t), v(t)]dt \end{aligned} \tag{6.38}$$

for all $v \in V_0^p$. By choosing in particular $v = u_\varepsilon$ and using (6.14) (with V^p instead of V^2) together with (6.30), one easily gets

$$\sup_{\varepsilon>0} \|u_\varepsilon\|_{L^p(0,T;W_0^{1,p}(\Omega))} < \infty.$$

It follows by (6.29) that

$$\sup_{\varepsilon>0} \|\mathbf{a}^\varepsilon(\cdot,\cdot,Du_\varepsilon)\|_{L^{p'}(Q)^N} < \infty,$$

hence $\sup_{\varepsilon>0} \|\mathrm{div}\, \mathbf{a}^{\varepsilon}(\cdot,\cdot,Du_{\varepsilon})\|_{L^{p'}(0,T;W^{-1,p'}(\Omega))} < \infty$, and further by (6.23)

$$\sup_{\varepsilon>0} \left\| \frac{\partial u_{\varepsilon}}{\partial t} \right\|_{L^{p'}(0,T;W^{-1,p'}(\Omega))} < \infty.$$

We deduce that the sequence $(u_{\varepsilon})_{\varepsilon>0}$ is bounded in V_0^p.

Thus, given an arbitrary fundamental sequence E, Lemma 6.9 assures us that there are some subsequence E' from E and some $\mathbf{u} = (u_0, u_1) \in \mathbb{F}^{1,p}_{\mathcal{H}_p} = V_0^p \times L^p(Q;\mathcal{H}_p)$ such that as $E' \ni \varepsilon \to 0$, (6.35)–(6.37) hold. Therefore, thanks to Lemma 6.11, the theorem is proved once we have shown that $\mathbf{u} = (u_0, u_1)$ satisfies the variational equation in (6.34). For that purpose, let $\mathbf{\Phi} = (\psi_0, \psi_1) \in \mathcal{F}_0^{\infty}$ (see Lemma 6.10), and let Φ_{ε} be given by (6.7). Take $v = \Phi_{\varepsilon}$ in (6.38) and then use (6.30) to obtain

$$0 \le \int_0^T [f(t) - u_{\varepsilon}'(t), u_{\varepsilon}(t) - \Phi_{\varepsilon}(t)]dt - \int_Q \mathbf{a}^{\varepsilon}(\cdot,\cdot,D\Phi_{\varepsilon}) \cdot (Du_{\varepsilon} - D\Phi_{\varepsilon})dxdt$$

or by (6.14), once again,

$$\begin{cases} \frac{1}{2}\|u_{\varepsilon}(T)\|^2_{L^2(\Omega)} \le \int_0^T [f(t), u_{\varepsilon}(t) - \Phi_{\varepsilon}(t)]dt + \int_0^T [u_{\varepsilon}'(t), \Phi_{\varepsilon}(t)]dt \\ -\int_Q \mathbf{a}^{\varepsilon}(\cdot,\cdot,D\Phi_{\varepsilon}) \cdot (Du_{\varepsilon} - D\Phi_{\varepsilon})dxdt \end{cases} \tag{6.39}$$

and that for any $\varepsilon > 0$. The aim now is to pass to the limit when $E' \ni \varepsilon \to 0$. First, we have by (6.35) and Lemma 6.3

$$\int_0^T [f(t), u_{\varepsilon}(t) - \Phi_{\varepsilon}(t)]dt \to \int_0^T [f(t), u_0(t) - \psi_0(t)]dt. \tag{6.40}$$

Next, we consider the obvious equality

$$\int_0^T [u_{\varepsilon}'(t), \Phi_{\varepsilon}(t)]dt = -\int_Q u_{\varepsilon}\frac{\partial \Phi_{\varepsilon}}{\partial t}dxdt, \tag{6.41}$$

where, on one hand, $u_{\varepsilon} \to u_0$ in $L^2(Q)$ as $E' \ni \varepsilon \to 0$, as is readily seen by (6.35)–(6.36) and use of the compactness of the embedding $V^p \hookrightarrow L^2(Q)$ (for convenience we still call E' some suitable extracted subsequence), on the other hand, $\frac{\partial \Phi_{\varepsilon}}{\partial t} \to \frac{\partial \psi_0}{\partial t}$ in $L^2(Q)$-weak, as a consequence of part (iii) of Lemma 6.3 (the details are left to the reader). Hence

$$\int_0^T [u_{\varepsilon}'(t), \Phi_{\varepsilon}(t)]dt \to \int_0^T [u_0'(t), \psi_0(t)]dt \text{ as } E' \ni \varepsilon \to 0.$$

Now, based on (6.37) as $E' \ni \varepsilon \to 0$, and on part (ii) of Lemma 6.3, appeal to Corollary 6.1 with $v_\varepsilon = \frac{\partial u_\varepsilon}{\partial x_i} - \frac{\partial \Phi_\varepsilon}{\partial x_i}$ (for arbitrary fixed $1 \leq i \leq N$) yields

$$\int_Q \mathbf{a}^\varepsilon(\cdot, \cdot, D\Phi_\varepsilon) \cdot (Du_\varepsilon - D\Phi_\varepsilon)dxdt \to \int_Q \int_Y \int_Z \mathbf{a}(\cdot, \cdot, \mathbb{D}\mathbf{\Phi}) \cdot \mathbb{D}(\mathbf{u} - \mathbf{\Phi})dxdtdydz$$

as $E' \ni \varepsilon \to 0$.

Finally, we have

$$\|u_0(T)\|^2_{L^2(\Omega)} \leq \liminf_{E' \ni \varepsilon \to 0} \|u_\varepsilon(T)\|^2_{L^2(\Omega)} \tag{6.42}$$

as is easily seen by the facts that $u_\varepsilon \to u_0$ in V_0^p-weak as $E' \ni \varepsilon \to 0$, and the mapping $v \mapsto \|v(T)\|^2_{L^2(\Omega)}$ is convex and continuous on V_0^p.

Hence, by taking the $\liminf_{E' \ni \varepsilon \to 0}$ of both sides of (6.39) and using

$$\frac{1}{2}\|u_0(T)\|^2_{L^2(\Omega)} = \int_0^T [u_0'(t), u_0(t)]dt, \tag{6.43}$$

it follows

$$0 \leq \int_0^T [f(t) - u_0'(t), u_0(t) - \psi_0(t)]dt - \int_Q \int_Y \int_Z \mathbf{a}(\cdot, \cdot, \mathbb{D}\mathbf{\Phi}) \cdot \mathbb{D}(\mathbf{u} - \mathbf{\Phi})dxdtdydz$$

and that for any $\mathbf{\Phi} \in \mathcal{F}_0^\infty$, hence for all $\mathbf{\Phi} \in \mathbb{F}^{1,p}_{\mathcal{H}_p}$, thanks to Lemma 6.10. Finally, by taking in particular $\mathbf{\Phi} = \mathbf{u} - r\mathbf{v}$ with $r > 0$ and $\mathbf{v} = (v_0, v_1) \in \mathbb{F}^{1,p}_{\mathcal{H}_p}$, then dividing both sides of the resultant inequality by r and passing to the limit as $r \to 0$ using Lemma 6.8(i) we see readily that $\mathbf{u}$ satisfies the variational equation in (6.34), as claimed. ■

The variational problem (6.34) is the so-called *global homogenized problem* for (6.23) in the periodic setting. By choosing in (6.34) the test function $\mathbf{v} = (v_0, v_1)$ such that $v_0 = 0$ and $v_1(x, t) = \varphi(x, t)w$ with $\varphi \in \mathcal{D}_\mathbb{R}(Q)$ and $w \in \mathcal{H}_p$, we obtain the variational equation satisfied by $u_1(x, t)$ for fixed $(x, t) \in Q$, commonly known as the *local equation* at (x, t), viz.

$$\int_Y \int_Z \mathbf{a}(\cdot, \cdot, Du_0(x, t) + D_y u_1(x, t)) \cdot D_y w \, dydz = 0 \tag{6.44}$$

for all $w \in \mathcal{H}_p$.

On the other hand, for any arbitrary $r \in \mathbb{R}^N$, consider the so-called *cell problem*

$$\begin{cases} \chi(r) \in \mathcal{H}_p : \\ \int_Y \int_Z \mathbf{a}(\cdot, \cdot, r + D_y\chi(r)) \cdot D_y w \, dydz = 0 \\ \text{for all } w \in \mathcal{H}_p \end{cases}$$

which uniquely determines $\chi(r)$ thanks to Lemma 6.8 (see, e.g., Kinderlehrer and Stampacchia [67]). Then, taking in particular $r = Du_0(x, t)$, where (x, t) is freely fixed in Q, and comparing with (6.44), we see that $u_1(x, t) = \chi(Du_0(x, t))$. Substituting this in (6.34) and then choosing there $\mathbf{v} = (v_0, v_1)$ with $v_1 = 0$, we obtain the *macroscopic homogenized problem* for (6.23), namely

$$\begin{cases} \frac{\partial u_0}{\partial t} - \operatorname{div} q(Du_0) = f \text{ in } Q, \\ u_0 = 0 \text{ on } \partial\Omega \times (0, T), \\ u_0(x, 0) = 0 \text{ in } \Omega, \end{cases} \tag{6.45}$$

where

$$q(r) = \int_Y \int_Z \mathbf{a}(\cdot, \cdot, r + D_y\chi(r)) dydz \quad (r \in \mathbb{R}^N),$$

and where $q(Du_0)$ stands for the function $(x, t) \mapsto q(Du_0(x, t))$ of Q into $\mathbb{R}^N$.

Conversely it can be shown that if u_0 is a function in V_0^p satisfying (6.45), and u_1 is a function in $L^p(Q; \mathcal{H}_p)$ such that $u_1(x, t)$ solves (6.44) for a.e. $(x, t) \in Q$, then $\mathbf{u} = (u_0, u_1)$ satisfies (6.34). Hence the unicity in (6.45) follows readily by Lemma 6.11.

6.3 Quasilinear Parabolic Operators: The Three-Scale Setting

6.3.1 Setting of the Problem and Preliminaries

Let $\mathbf{a} = (a_i)$ be a function of $\mathbb{R}_y^N \times \mathbb{R}_z \times \mathbb{R}_\mu^N$ into $\mathbb{R}^N$ satisfying (6.20)–(6.22). However, we shall restrict the present section to the case $p = 2$. We further assume that for fixed $\mu \in \mathbb{R}^N$, the function $(y, z) \mapsto \mathbf{a}(y, z, \mu)$ is $Y \times Z$-periodic.

Under the preceding hypotheses, let

$$f \in L^2(0, T; H^{-1}(\Omega; \mathbb{R})) \text{ with } H^{-1}(\Omega; \mathbb{R}) = W^{-1,2}(\Omega; \mathbb{R}),$$

where Ω and T are as in the preceding section. Our purpose is to study the limiting behaviour of u_ε as $\varepsilon \to 0$, where u_ε (for fixed real $\varepsilon > 0$) is the weak solution of the initial-boundary value problem

$$\begin{cases} \dfrac{\partial u_\varepsilon}{\partial t} - \operatorname{div} \mathbf{a}\left(\dfrac{x}{\varepsilon}, \dfrac{t}{\varepsilon^2}, Du_\varepsilon\right) = f \text{ in } Q = \Omega \times (0, T), \\ u_\varepsilon = 0 \text{ on } \partial\Omega \times (0, T), \\ u_\varepsilon(x, 0) = 0 \text{ in } \Omega. \end{cases} \tag{6.46}$$

Let us observe that the homogenization problem under consideration is stated in a three-scale setting, contrary to the preceding ones which were studied in the classical two-scale setting. However, we will still put

$$u^\varepsilon(x, t) = u\left(\frac{x}{\varepsilon}, \frac{t}{\varepsilon^2}\right) \quad (x \in \mathbb{R}^N, t \in \mathbb{R}) \tag{6.47}$$

for $u \in L^1_{loc}(\mathbb{R}^N_y \times \mathbb{R}_z)$, and more generally

$$u^\varepsilon(x, t) = u\left(x, t, \frac{x}{\varepsilon}, \frac{t}{\varepsilon^2}\right), \quad (x, t) \in Q, \tag{6.48}$$

for $u \in L^1_{loc}(Q \times \mathbb{R}^N_y \times \mathbb{R}_z)$ whenever the right-hand side in the latter case has meaning, e.g., when $u \in \mathcal{C}(\overline{Q}; L^\infty(\mathbb{R}^N \times \mathbb{R}))$ or $u \in L^2(Q; A)$ with $A = \mathcal{C}_{per}(Y \times Z)$. In this connection, attention must be drawn to the choice of the test functions Φ_ε (see (6.7)–(6.8)). As we shall see later, appropriate such functions in the present setting are given by

$$\Phi_\varepsilon(x, t) = \psi_0(x, t) + \varepsilon\psi_1\left(x, t, \frac{x}{\varepsilon}, \frac{t}{\varepsilon^2}\right), \quad (x, t) \in Q \tag{6.49}$$

with

$$\psi_0 \in \mathcal{D}_\mathbb{R}(Q), \psi_1 \in \mathcal{D}_\mathbb{R}(Q) \otimes \mathcal{C}^\infty_{per}(Y \times Z; \mathbb{R}). \tag{6.50}$$

Likewise, given $\boldsymbol{\Psi} \in \mathcal{C}(\overline{Q}; A^N_\mathbb{R})$ (see Proposition 6.5), if $a_i(\cdot, \cdot, \boldsymbol{\Psi})$ denotes the function $(x, t, y, z) \mapsto a_i(y, z, \boldsymbol{\Psi}(x, t, y, z))$ of $\overline{Q} \times \mathbb{R}^N \times \mathbb{R}$ into $\mathbb{R}$, which function lies in $\mathcal{C}(\overline{Q}; L^\infty_{per}(Y \times Z; \mathbb{R}))$ (see the proof of Proposition 6.5), then in the notation (6.48) we still have the equality (6.32), of course.

With this in mind, there is no difficulty in verifying that all of the results and observations pointed out in Sect. 6.2.1 carry over mutatis mutandis to the present three-scale setting.

The case is however different with part (iii) of Lemma 6.3 when Φ_ε is given by (6.49) and (6.50). Indeed, in the notation (6.48), we have here

$$\frac{\partial \Phi_\varepsilon}{\partial t} = \frac{\partial \psi_0}{\partial t} + \varepsilon \left(\frac{\partial \psi_1}{\partial t}\right)^\varepsilon + \frac{1}{\varepsilon}\left(\frac{\partial \psi_1}{\partial z}\right)^\varepsilon \tag{6.51}$$

and it is therefore clear that the sequence $(\frac{\partial \Phi_\varepsilon}{\partial t})_{\varepsilon>0}$ does not weakly converge in $L^2(Q)$ when $\varepsilon \to 0$. Hence it is neither strongly nor weakly Σ-convergent in $L^2(Q)$ as $\varepsilon \to 0$.

Thus, there is a great change with regard to those of the properties that are relevant to the time variable. This will come to light in the sequel.

Now, it is worth perhaps recalling that the space

$$L^2_{per}(Y;\mathbb{R})/\mathbb{R} = \left\{u|u \in L^2_{per}(Y;\mathbb{R}), \int_Y u(y)dy = 0\right\}$$

is equipped with the $L^2(Y)$ norm making it into a Hilbert space. Furthermore, since $H^1_{per}(Y;\mathbb{R})/\mathbb{R}$ is continuously and densely embedded in $L^2_{per}(Y;\mathbb{R})/\mathbb{R}$, by identifying the latter with its dual it follows

$$H^1_{per}(Y;\mathbb{R})/\mathbb{R} \subset L^2_{per}(Y;\mathbb{R})/\mathbb{R} \subset [H^1_{per}(Y;\mathbb{R})/\mathbb{R}]'$$

with continuous embeddings, where $[H^1_{per}(Y;\mathbb{R})/\mathbb{R}]'$ stands for the strong dual of $H^1_{per}(Y;\mathbb{R})/\mathbb{R}$. Hence, $[\cdot,\cdot]$ denoting the duality pairing between $[H^1_{per}(Y;\mathbb{R})/\mathbb{R}]'$ and $H^1_{per}(Y;\mathbb{R})/\mathbb{R}$, it follows

$$[f, g] = \int_Y f(y)g(y)dy \tag{6.52}$$

whenever $f \in L^2_{per}(Y;\mathbb{R})/\mathbb{R}$ and $g \in H^1_{per}(Y;\mathbb{R})/\mathbb{R}$.

On the other hand, let

$$\mathcal{H}_2 = L^2_{per}(Z; H^1_{per}(Y;\mathbb{R})/\mathbb{R}) \quad \text{(see Sect. 6.2.1).}$$

Equipped with the norm $\|\cdot\|_{\mathcal{H}_2}$, $\mathcal{H}_2$ is a Hilbert space. Its strong dual, $\mathcal{H}_2'$, is given by

$$\mathcal{H}_2' = L^2_{per}(Z; [H^1_{per}(Y;\mathbb{R})/\mathbb{R}]').$$

As a result of this, the duality pairing between $\mathcal{H}_2'$ and $\mathcal{H}_2$ is given by

$$\langle g, v\rangle_{\mathcal{H}_2',\mathcal{H}_2} = \int_Z [g(z), v(z)]dz \quad (g \in \mathcal{H}_2', v \in \mathcal{H}_2).$$

However, the space $\mathcal{H}_2$ is not sufficient to describe the homogenization process in the present setting. This leads us to introduce

$$\mathcal{U} = \left\{ v | v \in \mathcal{H}_2, \frac{dv}{dz} \in \mathcal{H}_2' \right\},$$

where the derivative is computed in the sense of $\mathcal{D}'(\mathbb{R}; H^1_{per}(Y; \mathbb{R})/\mathbb{R})$ (see Schwartz [125]). Equipped with the norm

$$\|v\|_{\mathcal{U}} = \left(\|v\|^2_{\mathcal{H}_2} + \left\| \frac{dv}{dz} \right\|^2_{\mathcal{H}_2'} \right)^{\frac{1}{2}} \quad (v \in \mathcal{U}),$$

$\mathcal{U}$ is a Hilbert space. Indeed, the completeness of this norm follows readily by some simple facts: $\mathcal{H}_2$ is continuously embedded in $\mathcal{D}'(\mathbb{R}; H^1_{per}(Y; \mathbb{R})/\mathbb{R})$, hence also in $\mathcal{D}'(\mathbb{R}; [H^1_{per}(Y; \mathbb{R})/\mathbb{R}]')$; likewise, $\mathcal{H}_2'$ is continuously embedded in the latter space, and finally, the mapping $u \mapsto \frac{du}{dz}$ sends continuously $\mathcal{D}'(\mathbb{R}; H^1_{per}(Y; \mathbb{R})/\mathbb{R})$ into itself.

In order to avoid some unfortunate confusion we wish to make a remark before embarking upon proving a series of preliminary results.

Remark 6.5 In the context of periodic homogenization one speaks of Σ-convergence if, and only if, two basic tools are fixed beforehand without the slightest ambiguity: on one hand, an action of the multiplicative group $\mathbb{R}^*_+$, on the other hand an algebra of periodic continuous complex-valued functions. It is clear that in this section as well as in the preceding ones of the present chapter the appropriate such algebra on $\mathbb{R}^N_y \times \mathbb{R}_z$ is $A = \mathcal{C}_{per}(Y \times Z)$. However, whereas the action attached to the two-scale setting in Sects. 6.1 and 6.2 was naturally the action $(H_\varepsilon)_{\varepsilon>0}$ of $\mathbb{R}^*_+$ on $\mathbb{R}^N \times \mathbb{R}$ defined by $H_\varepsilon(x, t) = (\frac{x}{\varepsilon}, \frac{t}{\varepsilon})$, it is a matter of course that the relevant action in this section is that given by $H_\varepsilon(x, t) = (\frac{x}{\varepsilon}, \frac{t}{\varepsilon^2})$. Therefore, throughout the present section, the notation u^ε has the exclusive meaning pointed out in (6.47) or (6.48).

The first preliminary result is a simple but basic one.

Lemma 6.12 *For $\theta \in \mathcal{D}^m_{\mathbb{R}}(\mathbb{R}) = \mathcal{K}(\mathbb{R}) \cap \mathcal{C}^m_{\mathbb{R}}(\mathbb{R})$ with $m \in \mathbb{N} \cup \{\infty\}$, let*

$$\overline{\omega}\theta = \sum_{k \in \mathbb{Z}} \tau_k \theta \quad \text{(a locally finite sum)},$$

where $\tau_k\theta(z) = \theta(z - k)$, $z \in \mathbb{R}$. The following statements hold:

(i) One has $\overline{\omega}\theta \in \mathcal{C}^m_{per}(Z; \mathbb{R}) = \mathcal{C}_{per}(Z) \cap \mathcal{C}^m_{\mathbb{R}}(\mathbb{R})$, and the mapping $\theta \mapsto \overline{\omega}\theta$ of $\mathcal{D}^m_{\mathbb{R}}(\mathbb{R})$ into $\mathcal{C}^m_{per}(Z; \mathbb{R})$, denoted by $\overline{\omega}$, is a surjective continuous linear mapping.

(*ii*) *We have* $\frac{d\overline{\omega}\theta}{dz}=\overline{\omega}\frac{d\theta}{dz}$ *for* $\theta\in\mathcal{D}^m_{\mathbb{R}}(\mathbb{R})$ $(m>1)$.

(*iii*) *For* $u\in L^1_{per}(Z;\mathbb{R})$ *and* $\theta\in\mathcal{D}^m_{\mathbb{R}}(\mathbb{R})$, *we have*

$$\int_{\mathbb{R}}u(z)\theta(z)dz=\int_Z u(z)\overline{\omega}\theta(z)dz.$$

Proof It is an easy matter to check that for each $\theta\in\mathcal{D}^m_{\mathbb{R}}(\mathbb{R})$, $\overline{\omega}\theta$ is a Z-periodic function of class C^m, i.e., $\overline{\omega}\theta\in\mathcal{C}^m_{per}(Z;\mathbb{R})$. Let us check the continuity of $\overline{\omega}$, the linearity being evident. Specifically, we have to check that for any given compact set $K\subset\mathbb{R}$, $\overline{\omega}$ maps continuously the Fréchet space $\mathcal{D}^m_K(\mathbb{R};\mathbb{R})$ (the space of those $\theta\in\mathcal{D}^m_{\mathbb{R}}(\mathbb{R})$ with supports in K) into $\mathcal{C}^m_{per}(Z;\mathbb{R})$. But this is straightforward because the restriction of $\overline{\omega}$ to $\mathcal{D}^m_K(\mathbb{R};\mathbb{R})$ is a finite sum of translation operators, and it is clear that each of these operators sends continuously $\mathcal{D}^m_K(\mathbb{R};\mathbb{R})$ into $\mathcal{C}^m_{per}(Z;\mathbb{R})$. We refer to Vo-Khac [139, p. 63] for the surjectivity. This completes the proof of part (i). We turn now to the proof of part (ii). Fix freely some $\theta\in\mathcal{D}^m_{\mathbb{R}}(\mathbb{R})$. We have

$$\frac{1}{r}(\tau_{-r}\theta-\theta)\to\frac{d\theta}{dz}\quad\text{in }\mathcal{D}^0(\mathbb{R})\equiv\mathcal{K}(\mathbb{R})$$

as $r\to 0$ (see, e.g., Vo-Khac [138, p. 143], Schwartz [126, p. 56]). By part (i) it follows

$$\frac{1}{r}(\tau_{-r}\overline{\omega}\theta-\overline{\omega}\theta)\to\overline{\omega}\frac{d\theta}{dz}\quad\text{in }\mathcal{C}_{per}(Z;\mathbb{R})\quad\text{as }r\to 0.$$

But the left-hand side converges uniformly on $\mathbb{R}$ to $\frac{d\overline{\omega}\theta}{dz}$ when $r\to 0$, as is easily seen by the Taylor formula. Hence, part (ii) follows thereby. Let us finally verify part (iii). Let u and θ be as stated. Since $\mathbb{R}$ is the union of all $k+\overline{Z}$ as k ranges over $\mathbb{Z}$, one has

$$\begin{aligned}\int_{\mathbb{R}}u(z)\theta(z)dz&=\sum_{k\in\mathbb{Z}}\int_{k+Z}u(z)\theta(z)dz\\&=\sum_{k\in\mathbb{Z}}\int_Z u(z)\theta(z+k)dz\\&=\int_Z u(z)\overline{\omega}\theta(z)dz,\end{aligned}$$

as stated. ■

For further details about the transformation $\overline{\omega}$, the interested reader is referred to Vo-Khac [139, pp. 61–65].

Next comes the following proposition in which $\mathcal{C}_{per}^{\infty}(Z; H_{per}^1(Y; \mathbb{R})/\mathbb{R})$ denotes the space of all Z-periodic indefinitely differentiable functions of $\mathbb{R}$ into $H_{per}^1(Y; \mathbb{R})/\mathbb{R}$.

Proposition 6.6 *$\mathcal{C}_{per}^{\infty}(Z; H_{per}^1(Y; \mathbb{R})/\mathbb{R})$ is dense in $\mathcal{U}$.*

Proof Let $(\theta_n)_{n\geq 1}$ be a mollifier on $\mathbb{R}$, i.e., $\theta_n \in \mathcal{D}(\mathbb{R})$, $\theta_n \geq 0$, $\int \theta_n(z)dz = 1$, θ_n having support in $[-\varepsilon_n, \varepsilon_n]$, where $0 < \varepsilon_n \to 0$ as $n \to \infty$. Let $u \in \mathcal{U}$. For each integer $n \geq 1$, put $u_n = u * \theta_n$, i.e.,

$$u_n(z) = \int_{\mathbb{R}} \theta_n(r)u(z-r)dr \quad (z \in \mathbb{R}).$$

Clearly $u_n \in \mathcal{C}_{per}^{\infty}(Z; H_{per}^1(Y; \mathbb{R})/\mathbb{R})$. Next, let $z \in \mathbb{R}$ be fixed and apply Hölder's inequality to the functions $r \mapsto \theta_n(r)^{\frac{1}{2}}$ and $r \mapsto \theta_n(r)^{\frac{1}{2}}[u(z-r) - u(z)]$, which lie in $L^2_{\mathbb{R}}(\mathbb{R})$ and $L^2(\mathbb{R}; H_{per}^1(Y; \mathbb{R})/\mathbb{R})$, respectively. It follows

$$\|u_n(z) - u(z)\|^2_{H^1(Y)/\mathbb{C}} \leq \int_{\mathbb{R}} \theta_n(r)\|u(z-r) - u(z)\|^2_{H^1(Y)/\mathbb{C}}dr.$$

Therefore

$$\|u_n - u\|^2_{\mathcal{H}_2} \leq \int_{\mathbb{R}} \theta_n(r)\|\tau_r u - u\|^2_{\mathcal{H}_2}dr,$$

where the translate $\tau_r u$ is defined as usual. Hence $u_n \to u$ in $\mathcal{H}_2$ as $n \to \infty$, since $\tau_r u \to u$ in $L^2_{loc}(\mathbb{R}; H_{per}^1(Y; \mathbb{R})/\mathbb{R})$ as $r \to 0$. Finally, thanks to the classical property

$$\frac{d}{dz}(u * \theta_n) = \theta_n * \frac{du}{dz},$$

the same argument mutatis mutandis as used above leads us readily to $u'_n \to u'$ in $\mathcal{H}'_2$ as $n \to \infty$, where $u' = \frac{du}{dz}$ (idem for u_n). In conclusion we have $u_n \to u$ in $\mathcal{U}$ as $n \to \infty$, which shows the proposition. ■

The next result is actually the main raison d'être of the preceding proposition.

Corollary 6.2 *One has*

$$\int_Z [u'(z), v(z)]dz + \int_Z [v'(z), u(z)]dz = 0 \tag{6.53}$$

for all $u, v \in \mathcal{U}$, where $u' = \frac{du}{dz}$ (idem for v).

Proof Thanks, on one hand to Proposition 6.6, on the other hand to an obvious continuity argument, it is quite enough to check (6.53) for $u, v \in$

$\mathcal{C}_{per}^{\infty}(Z; H_{per}^{1}(Y; \mathbb{R})/\mathbb{R})$. But this is straightforward. Indeed, according to a classical result, for any couple of such functions, the real function $z \mapsto [u(z), v(z)]$ (use (6.52) if necessary) denoted below by $[u, v]$, is continuously (or even indefinitely) differentiable with derivative given by

$$\frac{d}{dz}[u, v] \equiv [u, v]' = [u', v] + [v', u].$$

Hence (6.53) follows at once by integrating over $Z = (-\frac{1}{2}, \frac{1}{2})$ and then using the equality $[u, v](\frac{1}{2}) = [u, v](-\frac{1}{2})$ arising from the Z-periodicity. ■

We are now in a position to prove one of the major preliminary results.

Proposition 6.7 $\mathfrak{A} = [\mathcal{C}_{per}^{\infty}(Y; \mathbb{R})/\mathbb{R}] \otimes \mathcal{C}_{per}^{\infty}(Z; \mathbb{R})$ *is dense in* $\mathcal{U}$.

Proof The proposition is proved if we can show that an $l \in \mathcal{U}'$ (strong dual of $\mathcal{U}$) such that $l(v) = 0$ for all $v \in \mathfrak{A}$ is necessarily zero. We shall accomplish this in two steps.

Step 1. The aim here is to prove that each $l \in \mathcal{U}'$ is expressible in the form

$$l(v) = \int_Z [g(z), v(z)]dz + \int_Z [v'(z), f(z)]dz \quad (v \in \mathcal{U}) \tag{6.54}$$

where $g \in \mathcal{H}_2'$ and $f \in \mathcal{H}_2$. To do this, introduce the mapping

$$P : \mathcal{U} \to \mathcal{H}_2 \times \mathcal{H}_2', \quad Pv = (v, v') \quad \text{for } v \in \mathcal{U},$$

where the product space $\mathcal{H}_2 \times \mathcal{H}_2'$ is equipped with the norm

$$\|\mathbf{v}\| = (\|v_0\|_{\mathcal{H}_2}^2 + \|v_1\|_{\mathcal{H}_2'}^2)^{\frac{1}{2}}, \quad \mathbf{v} = (v_0, v_1) \in \mathcal{H}_2 \times \mathcal{H}_2'.$$

P carries $\mathcal{U}$ isometrically over $P(\mathcal{U})$ (the range of P).
Now, given $l \in \mathcal{U}'$, set $L = l \circ P^{-1}$ (usual composition). This is a continuous linear form on $P(\mathcal{U})$ (with the $\mathcal{H}_2 \times \mathcal{H}_2'$ norm). According to the Hahn-Banach theorem, there is some $L' \in (\mathcal{H}_2 \times \mathcal{H}_2')'$ such that $L'(Pv) = l(v)$ for all $v \in \mathcal{U}$. But $(\mathcal{H}_2 \times \mathcal{H}_2')' = \mathcal{H}_2' \times \mathcal{H}_2''$. Hence, $\mathcal{H}_2''$ being identified with $\mathcal{H}_2$, we see readily that (6.54) follows (see, e.g., Kato [65, p. 164] for some precise details).

Step 2. At the present time, let $l \in \mathcal{U}'$ be such that $l(\varphi) = 0$ for all $\varphi \in \mathfrak{A}$. Let a couple $(f, g) \in \mathcal{H}_2 \times \mathcal{H}_2'$ be attached to l such that (6.54) holds. Then

$$\int_Z [g(z), \varphi(z)]dz + \int_Z [\varphi'(z), f(z)]dz = 0 \quad (\varphi \in \mathfrak{A}).$$

This being so, fix freely some $\theta \in \mathcal{D}_{\mathbb{R}}(\mathbb{R})$ and some $\psi \in \mathcal{C}_{per}^{\infty}(Y; \mathbb{R})/\mathbb{R}$, and choose above $\varphi = \psi \otimes \overline{\omega}\theta$. An easy calculation using Lemma 6.12 (see also

Bourbaki [22, p. 143]) leads us to

$$\left[\int_{\mathbb{R}} \theta(z)g(z)dz - \left\langle \frac{df}{dz}, \theta \right\rangle, \psi \right] = 0,$$

where the derivative is taken in the sense of $\mathcal{D}'(\mathbb{R}; H^1_{per}(Y;\mathbb{R})/\mathbb{R})$ and the brackets $[\cdot,\cdot]$ stand for the duality pairing between $[H^1_{per}(Y;\mathbb{R})/\mathbb{R}]'$ and $H^1_{per}(Y;\mathbb{R})/\mathbb{R}$. In view of the arbitrariness of ψ and then of that of θ, it follows $g = \frac{df}{dz}$. Therefore $f \in \mathcal{U}$. By choosing in Corollary 6.2 $u = f$, one deduces that $l(v) = 0$ for all $v \in \mathcal{U}$, as claimed.

■

As a consequence of Proposition 6.7, there is the following corollary.

Corollary 6.3 $\mathcal{F}_0^\infty = \mathcal{D}_\mathbb{R}(Q) \times (\mathcal{D}_\mathbb{R}(Q) \otimes [\mathcal{C}^\infty_{per}(Y;\mathbb{R})/\mathbb{R}] \otimes \mathcal{C}^\infty_{per}(Z;\mathbb{R}))$ *is dense in* $L^2(0,T; H^1_0(\Omega;\mathbb{R})) \times L^2(Q;\mathcal{U})$.

Proof The same line of proceeding as followed in the proof of Lemma 6.6 shows that the whole problem reduces to showing that $[\mathcal{C}^\infty_{per}(Y;\mathbb{R})/\mathbb{R}] \otimes \mathcal{C}^\infty_{per}(Z;\mathbb{R})$ is dense in $\mathcal{U}$. The desired conclusion follows by Proposition 6.7. ■

Finally comes the following fundamental lemma.

Lemma 6.13 *Given a bounded ordinary sequence* $(u_\varepsilon)_{\varepsilon\in E}$ *in* V^2 *(thus, E is a fundamental sequence), there are a subsequence* E' *from E and some* $\boldsymbol{u} = (u_0, u_1) \in V^2 \times L^2(Q;\mathcal{U})$ *such that as* $E' \ni \varepsilon \to 0$,

$$u_\varepsilon \to u_0 \text{ in } V^2\text{-weak}, \tag{6.55}$$

$$\frac{\partial u_\varepsilon}{\partial x_j} \to \frac{\partial u_0}{\partial x_j} + \frac{\partial u_1}{\partial y_j} \text{ in } L^2(Q)\text{-weak } \Sigma, \tag{6.56}$$

$$\int_Q \frac{1}{\varepsilon} u_\varepsilon \Phi^\varepsilon \, dxdt \to \int_Q \int_Y \int_Z u_1 \Phi \, dxdtdydz \tag{6.57}$$

for all $\Phi \in \mathcal{D}_\mathbb{R}(Q) \otimes [\mathcal{C}^\infty_{per}(Y;\mathbb{R})/\mathbb{R}] \otimes \mathcal{C}^\infty_{per}(Z;\mathbb{R})$.

Proof To start with, according to Lemma 6.2, there are a subsequence E' from E and some couple $(u_0, u_1) \in V^2 \times L^2(Q;\mathcal{H}_2)$ such that as $E' \ni \varepsilon \to 0$, (6.55) and (6.56) hold. Hence (6.57) follows at once by an obvious adaptation of the proof of Corollary 3.3. Thus, the whole problem reduces to proving that u_1 lies in $L^2(Q;\mathcal{U})$.

To this end, let Φ be as in the statement of the lemma. Then, as noted above, one has, when $E' \ni \varepsilon \to 0$,

$$\int_Q u_\varepsilon \frac{1}{\varepsilon} \left(\frac{\partial \Phi}{\partial z}\right)^\varepsilon dxdt \to \int_Q \int_Y \int_Z u_1 \frac{\partial \Phi}{\partial z} dxdtdydz. \tag{6.58}$$

According to the equality

$$\frac{\partial \Phi^\varepsilon}{\partial t} = \left(\frac{\partial \Phi}{\partial t}\right)^\varepsilon + \frac{1}{\varepsilon^2}\left(\frac{\partial \Phi}{\partial z}\right)^\varepsilon,$$

we have on the other hand

$$\int_Q u_\varepsilon \frac{1}{\varepsilon}\left(\frac{\partial \Phi}{\partial z}\right)^\varepsilon dxdt = -\varepsilon\left\langle \frac{\partial u_\varepsilon}{\partial t}, \Phi^\varepsilon \right\rangle - \varepsilon \int_Q u_\varepsilon \left(\frac{\partial \Phi}{\partial t}\right)^\varepsilon dxdt \tag{6.59}$$

where the brackets stand for the duality pairing between $\mathcal{D}'(Q)$ and $\mathcal{D}(Q)$. But

$$\varepsilon\left\langle \frac{\partial u_\varepsilon}{\partial t}, \Phi^\varepsilon \right\rangle = \int_0^T [u'_\varepsilon(t), \varepsilon\Phi^\varepsilon(t)]dt, \tag{6.60}$$

where the brackets on the right denote the duality pairing between $H^{-1}(\Omega; \mathbb{R})$ and $H_0^1(\Omega; \mathbb{R})$, Φ^ε being viewed as a function of $(0, T)$ into $\mathcal{D}_\mathbb{R}(\Omega)$. This leads us to introduce, for fixed $0 < t < T$, the function $w_\varepsilon(t) \in H_0^1(\Omega; \mathbb{R})$ such that

$$\int_\Omega Dw_\varepsilon(t) \cdot Dv dx = [u'_\varepsilon(t), v] \tag{6.61}$$

for all $v \in H_0^1(\Omega; \mathbb{R})$. It is an elementary exercise to verify that the function $w_\varepsilon : (0, T) \to H_0^1(\Omega; \mathbb{R})$ thus defined belongs to the space $L^2(0, T; H_0^1(\Omega; \mathbb{R}))$ and satisfies $\|w_\varepsilon\|_{L^2(0,T;H_0^1(\Omega))} \leq \|u'_\varepsilon\|_{L^2(0,T;H^{-1}(\Omega))}$. It follows that the sequence $(Dw_\varepsilon)_{\varepsilon\in E'}$ is bounded in $L^2(Q)^N$. Therefore one can extract some subsequence from E' (still called E' for convenience) such that as $E' \ni \varepsilon \to 0$,

$$\frac{\partial w_\varepsilon}{\partial x_j} \to f_j \text{ in } L^2(Q)\text{-weak } \Sigma \quad (1 \leq j \leq N)$$

with $f_j \in L^2(Q; L^2_{per}(Y \times Z; \mathbb{R}))$. Thus,

$$\int_Q \frac{\partial w_\varepsilon}{\partial x_j}\left(\frac{\partial \Phi}{\partial y_j}\right)^\varepsilon dxdt \to \int_Q \int_Y \int_Z f_j \frac{\partial \Phi}{\partial y_j} dxdtdydz \quad (1 \leq j \leq N). \tag{6.62}$$

Now, recalling that

$$D\Phi^\varepsilon = (D_x\Phi)^\varepsilon + \frac{1}{\varepsilon}(D_y\Phi)^\varepsilon,$$

we deduce from (6.59),(6.60) and (6.61) that

$$\int_Q u_\varepsilon \frac{1}{\varepsilon}\left(\frac{\partial \Phi}{\partial z}\right)^\varepsilon dxdt = -\varepsilon \int_Q Dw_\varepsilon \cdot (D_x\Phi)^\varepsilon dxdt - \varepsilon \int_Q u_\varepsilon \left(\frac{\partial \Phi}{\partial t}\right)^\varepsilon dxdt$$
$$- \int_Q Dw_\varepsilon \cdot (D_y\Phi)^\varepsilon dxdt.$$

The next step is to pass to the limit when $E' \ni \varepsilon \to 0$. It is clear that the first two terms on the right go to zero. Hence, by passing to the limit on both sides we obtain by (6.58) and (6.62):

$$-\sum_{j=1}^N \int_Q \int_Y \int_Z f_j \frac{\partial \Phi}{\partial y_j} dxdtdydz = \int_Q \int_Y \int_Z u_1 \frac{\partial \Phi}{\partial z} dxdtdydz.$$

By choosing in particular $\Phi = \varphi \otimes \psi \otimes w$ with $\varphi \in \mathcal{D}_\mathbb{R}(Q)$, $\psi \in \mathcal{C}^\infty_{per}(Y;\mathbb{R})/\mathbb{R}$ and $w \in \mathcal{C}^\infty_{per}(Z;\mathbb{R})$ and turning the arbitrariness of φ to account it follows

$$\begin{aligned} &-\sum_{j=1}^N \int_Y \int_Z f_j(x,t,y,z)\frac{\partial \psi}{\partial y_j}(y)w(z)dydz = \\ &\int_Y \int_Z u_1(x,t,y,z)\psi(y)\frac{dw}{dz}(z)dydz \end{aligned} \tag{6.63}$$

for a.e. $(x,t) \in Q$. This being so, for fixed (x,t) in Q and z in $\mathbb{R}$, let $g(x,t,z) \in [H^1_{per}(Y;\mathbb{R})/\mathbb{R}]'$ be defined by

$$[g(x,t,z),\phi] = \sum_{j=1}^N \int_Y f_j(x,t,y,z)\frac{\partial \phi}{\partial y_j}(y)dy$$

for $\phi \in H^1_{per}(Y;\mathbb{R})/\mathbb{R}$. It is an easy matter to check that the function $z \mapsto g(x,t,z)$ of $\mathbb{R}$ into $[H^1_{per}(Y;\mathbb{R})/\mathbb{R}]'$ thus defined, and denoted by $g(x,t)$, lies in $\mathcal{H}'_2$ and further the function $(x,t) \mapsto g(x,t)$ of Q into $\mathcal{H}'_2$ lies in $L^2(Q;\mathcal{H}'_2)$.

With this in mind, let $\theta \in \mathcal{D}_\mathbb{R}(\mathbb{R})$ be freely given. Take $w = \overline{\omega}\theta$ in (6.63) and then use Lemma 6.12 to arrive at

$$\left[\int_\mathbb{R} g(x,t,z)\theta(z)dz, \psi\right] = \int_Y \left\langle \frac{d}{dz}u_1(x,t), \theta\right\rangle \psi dy$$

thanks to a classical result enabling the duality brackets and the integral to permute (see, e.g., Bourbaki [22, p. 143]). For the benefit of the reader it is worth noting that the integral on the left lies in $[H^1_{per}(Y;\mathbb{R})/\mathbb{R}]'$ whereas the derivative is understood in the sense of $\mathcal{D}'(\mathbb{R}; H^1_{per}(Y;\mathbb{R})/\mathbb{R})$.

Having made this point, by the arbitrariness of ψ and use of the fact that $C^\infty_{per}(Y;\mathbb{R})/\mathbb{R}$ is a dense subspace of $H^1_{per}(Y;\mathbb{R})/\mathbb{R}$ we deduce that

$$\left\langle \frac{d}{dz}u_1(x,t), \theta \right\rangle = \int_{\mathbb{R}} g(x,t,z)\theta(z)dz$$

for all $\theta \in \mathcal{D}_{\mathbb{R}}(\mathbb{R})$. Hence $\frac{d}{dz}u_1(x,t) = g(x,t) \in \mathcal{H}_2'$ and that for a.e. $(x,t) \in Q$. It follows readily that u_1 lies in $L^2(Q;\mathcal{U})$, since $g \in L^2(Q;\mathcal{H}_2')$. ■

Remark 6.6 For $u \in L^2(Q;\mathcal{U})$, $\frac{du}{dz}$ is naturally defined to be the function $(x,t) \mapsto \frac{d}{dz}u(x,t)$ of Q into $\mathcal{H}_2'$ which lies in $L^2(Q;\mathcal{H}_2')$. Thus, $\frac{du}{dz}(x,t) = \frac{d}{dz}u(x,t)$ for a.e. $(x,t) \in Q$. If $u(x,t) \in C^\infty_{per}(\widetilde{Y} \times Z;\mathbb{R})$ (see Sect. 6.2) or simply if $u(x,t) \in C^1_{per}(\widetilde{Y} \times Z;\mathbb{R})$ for a.e. $(x,t) \in Q$, then $\frac{du}{dz} = \frac{\partial u}{\partial z}$ (this follows at once by Lemma 6.12).

6.3.2 Homogenization Results

Before we can state and prove the homogenization theorem for (6.46) we need one more lemma. Let $\mathbb{D}\mathbf{v} = D_x v_0 + D_y v_1$ for $\mathbf{v} = (v_0, v_1) \in \mathbb{F}^{1,2}_{\mathcal{U}} = V_0^2 \times L^2(Q;\mathcal{U})$.

Lemma 6.14 *The variational problem*

$$\begin{cases} \boldsymbol{u} = (u_0, u_1) \in \mathbb{F}^{1,2}_{\mathcal{U}} = V_0^2 \times L^2(Q;\mathcal{U}) : \\ \int_0^T [u_0'(t), v_0(t)]dt + \int_Q \left\langle \frac{du_1}{dz}(x,t), v_1(x,t) \right\rangle_{\mathcal{H}_2',\mathcal{H}_2} dxdt \\ + \int_Q \int_Y \int_Z \boldsymbol{a}(\cdot,\cdot,\mathbb{D}\boldsymbol{u}) \cdot \mathbb{D}\boldsymbol{v}\, dxdtdydz = \int_0^T [f(t), v_0(t)]dt \\ \text{for all } \boldsymbol{v} = (v_0, v_1) \in \mathbb{F}^{1,2}_{\mathcal{U}} \end{cases} \tag{6.64}$$

admits at most one solution.

Proof Thanks to the property $\langle \frac{dv}{dz}, v \rangle_{\mathcal{H}_2',\mathcal{H}_2} = 0$ for $v \in \mathcal{U}$ (see Corollary 6.2), this follows by the same argument as used in proving Lemma 6.11. ■

Theorem 6.3 *For each real $\varepsilon > 0$, let u_ε be the weak solution of the initial-boundary value problem (6.46). As $\varepsilon \to 0$, we have*

$$u_\varepsilon \to u_0 \quad in\ L^2(0,T; H_0^1(\Omega;\mathbb{R}))\text{-}weak, \tag{6.65}$$

$$\frac{\partial u_\varepsilon}{\partial t} \to \frac{\partial u_0}{\partial t} \quad \text{in } L^2(0, T; H^{-1}(\Omega; \mathbb{R}))\text{-weak}, \tag{6.66}$$

$$\frac{\partial u_\varepsilon}{\partial x_j} \to \frac{\partial u_0}{\partial x_j} + \frac{\partial u_1}{\partial y_j} \quad \text{in } L^2(Q)\text{-weak } \Sigma \quad (1 \le j \le N), \tag{6.67}$$

where $\mathbf{u} = (u_0, u_1)$ *is the unique solution of (6.64).*

Proof A moment's reflection reveals that the proof of Theorem 6.2 carries over mutatis mutandis, up to (6.39), to the present setting. Thus, given an arbitrary fundamental sequence E, appeal to Lemma 6.13 yields a subsequence E' from E and some $\mathbf{u} = (u_0, u_1) \in \mathbb{F}^{1,2}_{\mathcal{U}} = V^2_0 \times L^2(Q; \mathcal{U})$ such that as $E' \ni \varepsilon \to 0$, (6.65)–(6.67) hold true. Therefore we are through if it can be proved that $\mathbf{u} = (u_0, u_1)$ satisfies the variational equation in (6.64). For that purpose, let $\boldsymbol{\Phi} = (\psi_0, \psi_1) \in \mathcal{F}^\infty_0$ (see Corollary 6.3), and let us attach to $\boldsymbol{\Phi}$ the sequence of functions $\Phi_\varepsilon \in \mathcal{D}_{\mathbb{R}}(Q)$ $(\varepsilon > 0)$ given by (6.49) with however $(\psi_0, \psi_1) \in \mathcal{F}^\infty_0$ instead of being as in the general setting of (6.50). Then, as claimed above, we have (6.39) with $\mathbf{a}^\varepsilon(\cdot, \cdot, D\Phi_\varepsilon) = (a^\varepsilon_i(\cdot, \cdot, D\Phi_\varepsilon))_{1 \le i \le N}$ where it is worth recalling that $a^\varepsilon_i(\cdot, \cdot, D\Phi_\varepsilon)$ stands this time for the function $(x, t) \mapsto a_i(\frac{x}{\varepsilon}, \frac{t}{\varepsilon^2}, D\Phi_\varepsilon(x, t))$ of Q into $\mathbb{R}$. The next step is to pass to the limit in (6.39) when $E' \ni \varepsilon \to 0$. To this end let us note that the argument we used to assert part (iv) of Lemma 6.3 is not valid here. We nevertheless have $\Phi_\varepsilon \to \psi_0$ in $L^2(0, T; H^1_0(\Omega; \mathbb{R}))$-weak as $\varepsilon \to 0$, that is,

$$\int_0^T [g(t), \Phi_\varepsilon(\cdot, t)]dt \to \int_0^T [g(t), \psi_0(\cdot, t)]dt \quad \text{as } \varepsilon \to 0$$

for all $g \in L^2(0, T; H^{-1}(\Omega; \mathbb{R}))$. Indeed, this is straightforward if g lies in $L^2(0, T; L^2(\Omega; \mathbb{R}))$, from which one deduces the general statement by the density of the latter space in $L^2(0, T; H^{-1}(\Omega; \mathbb{R}))$. Next, by substituting (6.51) in the equality

$$\int_0^T [u'_\varepsilon(t), \Phi_\varepsilon(\cdot, t)]dt = -\int_Q u_\varepsilon \frac{\partial \Phi_\varepsilon}{\partial t} dxdt$$

we obtain

$$\int_0^T [u'_\varepsilon(t), \Phi_\varepsilon(\cdot, t)]dt = -\int_Q u_\varepsilon \frac{\partial \psi_0}{\partial t} dxdt - \varepsilon \int_Q u_\varepsilon \left(\frac{\partial \psi_1}{\partial t}\right)^\varepsilon dxdt$$
$$- \int_Q \frac{1}{\varepsilon} u_\varepsilon \left(\frac{\partial \psi_1}{\partial z}\right)^\varepsilon dxdt.$$

Hence, by Lemma 6.13 and use of obvious facts it follows that as $E' \ni \varepsilon \to 0$, one has

$$\int_0^T [u_\varepsilon'(t), \Phi_\varepsilon(\cdot, t)]dt \to \int_0^T [u_0'(t), \psi_0(t)]dt - \int_Q \int_Y \int_Z u_1 \frac{\partial \psi_1}{\partial z} dxdtdydz.$$

But by a simple calculation based on an appropriate application of Lemma 6.12 (the details are left as an exercise to the reader) we see that

$$-\int_Q \int_Y \int_Z u_1 \frac{\partial \psi_1}{\partial z} dxdtdydz = \int_Q \left\langle \frac{du_1}{dz}(x,t), \psi_1(x,t) \right\rangle_{\mathcal{H}_2', \mathcal{H}_2} dxdt.$$

With this in mind, we are now in a position to pass to the limit in (6.39), where attention is drawn to the fact that the limit of the third integral on the right is furnished by (the analogue of) Corollary 6.1. Thus, by taking the $\liminf_{E' \ni \varepsilon \to 0}$ of both sides of (6.39) and then using Corollary 6.3, we obtain

$$0 \leq \int_0^T [f(t) - u_0'(t), u_0(t) - w_0(t)]dt + \int_Q \left\langle \frac{du_1}{dz}(x,t), w_1(x,t) \right\rangle_{\mathcal{H}_2', \mathcal{H}_2} dxdt$$
$$- \int_Q \int_Y \int_Z \mathbf{a}(\cdot, \cdot, \mathbb{D}\mathbf{w}) \cdot \mathbb{D}(\mathbf{u} - \mathbf{w}) dxdtdydz$$

for all $\mathbf{w} = (w_0, w_1) \in \mathbb{F}_{\mathcal{U}}^{1,2}$, where $\mathbb{D}\mathbf{w} = D_x w_0 + D_y w_1$. Finally, by taking in particular $\mathbf{w} = \mathbf{u} - r\mathbf{v}$ with $r > 0$ and $\mathbf{v} = (v_0, v_1) \in \mathbb{F}_{\mathcal{U}}^{1,2}$, next dividing both sides of the resultant inequality by r, and recalling at last that $\langle \frac{dv}{dz}, v \rangle_{\mathcal{H}_2', \mathcal{H}_2} = 0$ for $v \in \mathcal{U}$, we see that (6.64) follows at once by letting $r \to 0$. This completes the proof thanks to Lemma 6.14. ■

6.4 Comments and Problems

It is worth mentioning that the periodic homogenization of parabolic differential operators is a classical theory now. By the way a detailed bibliography on the subject can be found for example in Bensoussan et al. [14], Bakhvalov and Panesenko [13], Clark and Packer [37, 38], Jikov et al. [63], Svanstedt [130], Pankov [114].

Thus, our concern here is to concentrate on the sole use of Σ-convergence in the periodic homogenization of parabolic differential operators, and not to comment on this topic from a general point of view.

Whether it is named *two-scale* or *Σ-convergence* method, the approach followed in Sects. 6.1 and 6.2 is likely to fit various periodic homogenization problems relevant to parabolic operators.

In this connection we refer to Allaire and Piatnitski [9] for reaction-diffusion problems governed by parabolic operators with large potentials.

The study of Sects. 6.1 and 6.2 can also be adapted to the periodic homogenization of degenerate parabolic operators. Results along these lines are given in Clark and Packer [37, 38].

Attention is also drawn to the system of partial differential equations governing the exact micro-model for diffusion in a periodic partially fissured medium. It is proved that the solution of the micro-model is two-scale convergent to that of a macro-model as the parameter $\varepsilon > 0$ goes to zero. See Clark and Showalter [39] for results in this direction.

See also Holmbom [59] for an adaptation of the two-scale convergence method to the study of some specific problems relevant to the periodic homogenization of parabolic operators.

The results of Sects. 6.1 and 6.2 were announced in Nguetseng [98]. Those of Sect. 6.3 are given here for the first time.

One may reasonably wonder why the study of Sect. 6.3 is confined to the case $p = 2$. This needs some clarification.

Given $2 \leq p < \infty$ and $f \in L^{p'}(0, T; W^{-1,p'}(\Omega; \mathbb{R}))$, suppose our purpose is to investigate the limiting behaviour, as $\varepsilon \to 0$, of the solution u_ε of (6.46) (with $\mathbf{a} = (a_i)_{1 \leq i \leq N}$ as in Sect. 6.2.1) under the periodicity hypothesis: for fixed $1 \leq i \leq N$ and $\mu \in \mathbb{R}^N$, the function $(y, z) \mapsto a_i(y, z, \mu)$ is $Y \times Z$-periodic (see Sect. 6.1).

As a preliminary step, let us note that the space $W^{1,p}_{per}(Y; \mathbb{R})/\mathbb{R}$ is continuously and densely embedded in $L^2_{per}(Y; \mathbb{R})/\mathbb{R}$, so that, $L^2_{per}(Y; \mathbb{R})/\mathbb{R}$ being identified with its dual, one has

$$W^{1,p}_{per}(Y; \mathbb{R})/\mathbb{R} \subset L^2_{per}(Y; \mathbb{R})/\mathbb{R} \subset [W^{1,p}_{per}(Y; \mathbb{R})/\mathbb{R}]'$$

with continuous embedding, and further (6.52) holds for $f \in L^2_{per}(Y; \mathbb{R})/\mathbb{R}$ and $g \in W^{1,p}_{per}(Y; \mathbb{R})/\mathbb{R}$, the brackets denoting the duality pairing between $W^{1,p}_{per}(Y; \mathbb{R})/\mathbb{R}$ and its dual.

Finally, let

$$\mathcal{U} = \left\{ v | v \in \mathcal{H}_p, \frac{dv}{dz} \in \mathcal{H}'_p \right\} \quad (\mathcal{H}_p \text{ defined in Sect. 6.2.1}), \tag{6.68}$$

where $\mathcal{H}'_p = L^{p'}_{per}(Z; [W^{1,p}_{per}(Y; \mathbb{R})/\mathbb{R}]')$ (dual of $\mathcal{H}_p$), and where the derivative is computed in the sense of $\mathcal{D}'(\mathbb{R}; W^{1,p}_{per}(Y; \mathbb{R})/\mathbb{R})$.

This being so, one can easily verify the following statements:

(1) Proposition 6.6 still holds with $W^{1,p}_{per}(Y; \mathbb{R})/\mathbb{R}$ in place of $H^1_{per}(Y; \mathbb{R})/\mathbb{R}$ and $\mathcal{U}$ given by (6.68). Hence Corollary 6.2 follows readily.

(2) Proposition 6.7 holds true when $\mathcal{U}$ is given by (6.68).

(3) Corollary 6.3 stands with $L^p(0, T; W^{1,p}_0(\Omega; \mathbb{R}))$ in place of $L^2(0, T; H^1_0(\Omega; \mathbb{R}))$ and $L^p(Q; \mathcal{U})$ ($\mathcal{U}$ given by (6.68)) in that of $L^2(Q; \mathcal{U})$.

One is tempted to extend the list to Lemma 6.13, that is, to change therein V^2 into V^p, $L^2(Q;\mathcal{U})$ into $L^p(Q;\mathcal{U})$ ($\mathcal{U}$ given by (6.68)), and $L^2(Q)$ into $L^p(Q)$. Unfortunately one is not justified in doing so. Indeed, a quick survey of the proof of Lemma 6.13 reveals that the keystone of the said proof is precisely Eq. (6.61). Let this be clarified. For fixed $\varepsilon > 0$, let us note beforehand that $u'_\varepsilon = \frac{\partial u_\varepsilon}{\partial t} \in L^{p'}(0, T; W^{-1,p'}(\Omega; \mathbb{R}))$, and then consider the following subsidiary variational problem:

$$\begin{aligned}&\text{Find a measurable function } w_\varepsilon : (0, T) \to W_0^{1,p}(\Omega; \mathbb{R}) \text{ such that}\\&\text{(6.61) holds for all } v \in W_0^{1,p}(\Omega; \mathbb{R}) \text{ and for almost every } t \in (0, T).\end{aligned} \tag{6.69}$$

It is clear that such a w_ε, if it exists, is unique and further $w_\varepsilon \in L^p(0, T; W_0^{1,p}(\Omega; \mathbb{R}))$ with $\|w_\varepsilon\|_{L^p(0,T;W_0^{1,p}(\Omega))} \leq \|u'_\varepsilon\|_{L^{p'}(0,T;W^{-1,p'}(\Omega;\mathbb{R}))}$.

Thus, the resolution of (6.69) beyond $p = 2$ would make it possible to extend Sect. 6.3 to $2 \leq p < \infty$. But this is pure fiction. *The question that arises is whether (6.69) has a solution for* $2 < p < \infty$.

Chapter 7
Homogenization of Parabolic Operators II

This chapter generalizes the study of Chap. 6 in two main directions: on one hand, it enables the parameter k to run through the positive real numbers; on the other hand, it extends the results of Chap. 6 to a general setting covering a wide range of concrete behaviours with respect to $y = \frac{x}{\varepsilon}$ and $z = \frac{t}{\varepsilon^k}$, referred to as structure hypotheses, such as the almost periodicity hypothesis, and many others relevant to rapid oscillations.

7.1 Introduction

7.1.1 Setting of the Problem

The basic notation and hypotheses are those of Sect. 6.2.1. In particular, $\mathbf{a} = (a_i)_{1 \le i \le N}$ is a given function of $\mathbb{R}_y^N \times \mathbb{R}_z \times \mathbb{R}_\mu^N$ into $\mathbb{R}^N$ satisfying conditions (6.20)–(6.22), and $f \in L^{p'}(0, T; W^{-1,p'}(\Omega; \mathbb{R}))$, $2 \le p < \infty$.

Let k be a positive real number. For each real $\varepsilon > 0$, we consider the initial-boundary value problem

$$\begin{cases} \dfrac{\partial u_\varepsilon}{\partial t} - \operatorname{div} \mathbf{a}\left(\dfrac{x}{\varepsilon}, \dfrac{t}{\varepsilon^k}, Du_\varepsilon\right) = f \text{ in } Q = \Omega \times (0, T), \\ u_\varepsilon = 0 \text{ on } \partial\Omega \times (0, T), \\ u_\varepsilon(x, 0) = 0 \text{ in } \Omega, \end{cases} \tag{7.1}$$

which uniquely determines u_ε (see Proposition 6.4). The main purpose of this chapter is to discuss the limiting behaviour of u_ε as $\varepsilon \to 0$ under a suitable structure hypothesis on $a_i(y, z, \mu)$ $(1 \le i \le N)$ with respect to (y, z) for fixed μ.

G. Nguetseng, *Homogenization Algebras and Applications*, Springer Monographs in Mathematics, https://doi.org/10.1007/978-3-031-84705-9_7

7.1.2 *Orientation*

Contrary to Chap. 6 where the study of (7.1) is restricted to the periodic setting and the parameter k is considered in the sole cases $k = 1, 2$, the present chapter deals with the global situation in which k runs through the positive real numbers while the homogenization process takes place in a general setting characterized by an abstract assumption on $\mathbf{a}(y, z, \mu)$ (for fixed μ) covering a wide range of concrete behaviours with respect to $y = \frac{x}{\varepsilon}$ and $z = \frac{t}{\varepsilon^k}$. Different basic homogenization theorems are proved according as $0 < k \leq 1$ (Sect. 7.2.4), $1 < k < 2$ (Sect. 7.2.5), $k = 2$ (Sect. 7.2.6), and $k > 2$ (Sect. 7.2.7). The notion of a quasi V^p-proper H-algebra (Sect. 7.2.2) and that of a V^p-proper H-algebra (Sect. 7.2.6) turn out to be of major interest. As will presently become apparent, except for some small details, the difference in proceeding between the four cases is inherent in the determination of the limit (in a sense to be specified) of $u'_\varepsilon = \frac{\partial u_\varepsilon}{\partial t} \in L^{p'}(0, T; W^{-1,p'}(\Omega; \mathbb{R}))$ as $\varepsilon \to 0$. Finally, in order to illustrate the wide scope of the theory developed in Sect. 7.2, a number of examples of concrete homogenization problems for (7.1) are worked out in Sect. 7.3 on the lines of Chap. 5.

7.2 Abstract Homogenization Problem

7.2.1 *Setting of the Problem and Preliminaries*

Let $(\overline{H}_\varepsilon)_{\varepsilon>0}$ be the action of $\mathbb{R}^*_+$ on $\mathbb{R}^N$ given by $\overline{H}_\varepsilon(x) = \frac{x}{\varepsilon}$ $(x \in \mathbb{R}^N)$ and $({}^k H_\varepsilon)_{\varepsilon>0}$ the action of $\mathbb{R}^*_+$ on $\mathbb{R}$ given by ${}^k H_\varepsilon(t) = \frac{t}{\varepsilon^k}$ $(t \in \mathbb{R})$. Put $H_\varepsilon = \overline{H}_\varepsilon \times {}^k H_\varepsilon$ (direct product), i.e., $H_\varepsilon(x, t) = (\frac{x}{\varepsilon}, \frac{t}{\varepsilon^k})$ for $(x, t) \in \mathbb{R}^N \times \mathbb{R}$, which defines an action of $\mathbb{R}^*_+$ on $\mathbb{R}^N \times \mathbb{R}$ (see Example 1.6).

Now, let $\{A_y, A_z\}$ be a couple of H-algebras A_y on $\mathbb{R}^N$ under the action $(\overline{H}_\varepsilon)_{\varepsilon>0}$, and A_z on $\mathbb{R}$ under the action $({}^k H_\varepsilon)_{\varepsilon>0}$. As usual, we assume that A_y^∞ (resp. A_z^∞) is dense in A_y (resp. A_z). Points in $\Delta(A_y)$ (resp. $\Delta(A_z)$) are denoted by s (resp. s_0). The compact space $\Delta(A_y)$ (resp. $\Delta(A_z)$) is provided with the so-called M-measure β_y (resp. β_z) for A_y (resp. A_z). This being so, let A be the closed product of A_y and A_z, that is, the closure of $A_y \otimes A_z$ in $\mathcal{B}(\mathbb{R}^N \times \mathbb{R})$. It is worth perhaps recalling that A is an H-algebra on $\mathbb{R}^N \times \mathbb{R}$ under the action $(H_\varepsilon)_{\varepsilon>0}$, and A^∞ is dense in A. Furthermore $\Delta(A) = \Delta(A_y) \times \Delta(A_z)$ and $\beta = \beta_y \otimes \beta_z$, where β is the M-measure on $\Delta(A)$ (see Chap. 2 for further details). The same symbol, $\mathcal{G}$, will denote the Gelfand transformation on A_y, A_z and A, as well, and M will stand indifferently for the mean value on $\mathbb{R}^N$, $\mathbb{R}$ and $\mathbb{R}^N \times \mathbb{R}$ generated by the actions $(\overline{H}_\varepsilon)_{\varepsilon>0}$, $({}^k H_\varepsilon)_{\varepsilon>0}$ and $(H_\varepsilon)_{\varepsilon>0}$, respectively.

For the benefit of the reader it is well to recall the functional setting underlying the statement of the abstract structure hypothesis for the problem under consideration. To begin, for any $1 \leq p < \infty$, let Ξ^p denote the space of those

$u \in L^p_{loc}(\mathbb{R}^N_y \times \mathbb{R}_z)$ for which the (generalized) sequence $(u^\varepsilon)_{0<\varepsilon\leq 1}$ is bounded in the Fréchet space $L^p_{loc}(\mathbb{R}^N_x \times \mathbb{R}_t)$, where $u^\varepsilon(x,t) = u(\frac{x}{\varepsilon}, \frac{t}{\varepsilon^k})$ for $(x,t) \in \mathbb{R}^N \times \mathbb{R}$. Equipped with the norm

$$\|u\|_{\Xi^p} = \sup_{0<\varepsilon\leq 1} \left(\int_{B_{N+1}} \left| u\left(\frac{x}{\varepsilon}, \frac{t}{\varepsilon^k}\right)\right|^p dxdt \right)^{\frac{1}{p}} \quad (u \in \Xi^p),$$

where B_{N+1} is the open unit ball in $\mathbb{R}^{N+1}$, Ξ^p is a Banach space. Next, let $\mathfrak{X}^p_A$ be the closure of A in Ξ^p. Equipped with the Ξ^p-norm, $\mathfrak{X}^p_A$ is a Banach space and further, the Gelfand transformation $\mathcal{G} : A \to \mathcal{C}(\Delta(A))$ extends by continuity to a continuous linear mapping, still denoted by $\mathcal{G}$, of $\mathfrak{X}^p_A$ into $L^p(\Delta(A))$. This is referred to as the *canonical mapping* of $\mathfrak{X}^p_A$ into $L^p(\Delta(A))$.

We are now in a position to state the abstract homogenization problem for (7.1). Let $A_\mathbb{R} = A \cap \mathcal{C}_\mathbb{R}(\mathbb{R}^N \times \mathbb{R})$. The abstract homogenization problem for (7.1) deals with the mathematical analysis of the limiting behaviour, as $\varepsilon \to 0$, of u_ε (the solution of (7.1)) under the hypothesis

$$a_i(\cdot,\cdot,\boldsymbol{\Psi}) \in \mathfrak{X}^{p'}_A \ (1 \leq i \leq N) \text{ for all } \boldsymbol{\Psi} \in (A_\mathbb{R})^N, \tag{7.2}$$

where $p' = \frac{p}{p-1}$ for some suitable $2 \leq p < \infty$, and where $a_i(\cdot,\cdot,\boldsymbol{\Psi})$ denotes the function $(y,z) \mapsto a_i(y,z,\boldsymbol{\Psi}(y,z))$ from $\mathbb{R}^N \times \mathbb{R}$ to $\mathbb{R}$, which lies in $L^\infty_\mathbb{R}(\mathbb{R}^N \times \mathbb{R})$, as can be seen by a classical argument relevant to Carathéodory functions (see Sect. 5.5.2 together with the proof of Proposition 6.5). Before we can achieve homogenization results for this problem, however, we need some preliminaries. Let $\varepsilon > 0$. For $u \in L^1_{loc}(Q \times \mathbb{R}^N_y \times \mathbb{R}_z)$, we formally set (as in (6.48))

$$u^\varepsilon(x,t) = u\left(x, t, \frac{x}{\varepsilon}, \frac{t}{\varepsilon^k}\right) \quad ((x,t) \in Q) \tag{7.3}$$

provided the right-hand side has meaning (see Chap. 1). In this connection, if $\boldsymbol{\Phi} \in L^p_\mathbb{R}(Q)^N$, we define $a^\varepsilon_i(\cdot,\cdot,\boldsymbol{\Phi})$ to be the function $(x,t) \mapsto a_i(\frac{x}{\varepsilon}, \frac{t}{\varepsilon^k}, \boldsymbol{\Phi}(x,t))$ of Q into $\mathbb{R}$, which lies in $L^{p'}(Q)$ (for $2 \leq p < \infty$) with in particular $a^\varepsilon_i(\cdot,\cdot,\boldsymbol{\Phi}) \in L^\infty(Q)$ if $\boldsymbol{\Phi} \in \mathcal{C}_\mathbb{R}(\overline{Q})^N$ (this follows by Lemma 6.7).

In what follows it is assumed that (7.2) is satisfied. Let the index $1 \leq i \leq N$ be fixed. On the lines of Sect. 5.5, let

$$b_i(\boldsymbol{\varphi}) = \mathcal{G}(a_i(\cdot,\cdot,\mathcal{G}^{-1}\boldsymbol{\varphi}))$$

for $\boldsymbol{\varphi} = (\varphi_i)_{1\leq i\leq N}$ in $\mathcal{C}_\mathbb{R}(\Delta(A))^N$, where $\mathcal{G}^{-1}\boldsymbol{\varphi} = (\mathcal{G}^{-1}\varphi_i)_{1\leq i\leq N}$. This defines a mapping $\boldsymbol{\varphi} \mapsto b_i(\boldsymbol{\varphi})$ of $\mathcal{C}_\mathbb{R}(\Delta(A))^N$ into $L^\infty(\Delta(A))$ denoted by b_i.

Let

$$\mathbf{b}(\boldsymbol{\Phi}) = (b_i \circ \boldsymbol{\Phi})_{1\leq i\leq N} \text{ for } \boldsymbol{\Phi} \in \mathcal{C}(\overline{Q}; \mathcal{C}_\mathbb{R}(\Delta(A))^N).$$

The next proposition and its corollary can be established without difficulty by the same lines of argument as followed in proving Proposition 5.19 and Corollary 5.21 (see also the proof of Lemma 6.7 as far as part (ii) of the forthcoming Proposition 7.1 is concerned).

Proposition 7.1 *Let* $2 \leq p < \infty$. *Suppose (7.2) holds. For* $\boldsymbol{\Psi} = (\psi_j)_{1\leq j\leq N} \in \mathcal{C}(\overline{Q}; (A_{\mathbb{R}})^N)$, *let* $\widehat{\boldsymbol{\Psi}} = (\widehat{\psi}_j)_{1\leq j\leq N}$ *with* $\widehat{\psi}_j = \mathcal{G} \circ \psi_j$. *The following statements hold:*

(i) One has $b_i \circ \widehat{\boldsymbol{\Psi}} \in \mathcal{C}(\overline{Q}; L^\infty(\Delta(A)))$ *and, as* $\varepsilon \to 0$,

$$a_i^\varepsilon(\cdot,\cdot,\boldsymbol{\Psi}^\varepsilon) \to b_i \circ \widehat{\boldsymbol{\Psi}} \text{ in } L^{p'}(Q)\text{-weak } \Sigma \quad (1 \leq i \leq N),$$

where $\boldsymbol{\Psi}^\varepsilon = (\psi_j^\varepsilon)_{1\leq j\leq N}$, ψ_j^ε *defined as in (7.3).*

(ii) The mapping $\boldsymbol{\Phi} \mapsto \boldsymbol{b}(\boldsymbol{\Phi})$ *of* $\mathcal{C}(\overline{Q}; \mathcal{C}_{\mathbb{R}}(\Delta(A))^N)$ *into* $L^\infty(Q \times \Delta(A))^N$ *extends by continuity to a mapping, still denoted by* $\boldsymbol{b}$, *of* $L^p(Q; L^p_{\mathbb{R}}(\Delta(A))^N)$ *into* $L^{p'}(Q \times \Delta(A))^N$ *such that*

$$\|\boldsymbol{b}(\boldsymbol{u}) - \boldsymbol{b}(\boldsymbol{v})\|_{L^{p'}(Q\times\Delta(A))^N} \leq$$

$$\alpha_1 N^{\frac{3}{2}-\frac{2}{p}} \| |\boldsymbol{u}| + |\boldsymbol{v}| \|^{p-2}_{L^p(Q\times\Delta(A))} \|\boldsymbol{u} - \boldsymbol{v}\|_{L^p(Q;L^p(\Delta(A))^N)}$$

and

$$(\boldsymbol{b}(\boldsymbol{u}) - \boldsymbol{b}(\boldsymbol{v})) \cdot (\boldsymbol{u} - \boldsymbol{v}) \geq \alpha_0 |\boldsymbol{u} - \boldsymbol{v}|^p \text{ a.e. in } Q \times \Delta(A)$$

for all $\boldsymbol{u}, \boldsymbol{v} \in L^p(Q; L^p_{\mathbb{R}}(\Delta(A))^N)$.

Corollary 7.1 *For* $\varepsilon > 0$, *let*

$$\Phi_\varepsilon = \psi_0 + \varepsilon \psi_1^\varepsilon \quad (\psi_1^\varepsilon \text{ defined as in (7.3)}) \tag{7.4}$$

with $\psi_0 \in \mathcal{D}_{\mathbb{R}}(Q)$ *and* $\psi_1 \in \mathcal{D}_{\mathbb{R}}(Q) \otimes A^\infty_{\mathbb{R}}$. *Then, as* $\varepsilon \to 0$,

$$a_i^\varepsilon(\cdot,\cdot,D\Phi_\varepsilon) \to b_i \circ (D_x\psi_0 + \nabla\widehat{\psi}_1) \text{ in } L^{p'}(Q)\text{-weak } \Sigma \ (1 \leq i \leq N),$$

where $\nabla\widehat{\psi}_1 = (\partial_j\widehat{\psi}_1)_{1\leq j\leq N}$. *Furthermore, if* E *is a fundamental sequence, and* $(v_\varepsilon)_{\varepsilon\in E}$ *a sequence in* $L^p(Q)$ *such that* $v_\varepsilon \to v_0$ *in* $L^p(Q)$*-weak* Σ, *then*

$$\int_Q a_i^\varepsilon(\cdot,\cdot,D\Phi_\varepsilon)v_\varepsilon \, dxdt \to \int\int_{Q\times\Delta(A))} b_i \circ (D_x\psi_0 + \nabla\widehat{\psi}_1)v_0 \, dxdtd\beta$$

as $E \ni \varepsilon \to 0$ $(1 \leq i \leq N)$.

7.2.2 Quasi V^p-Proper Homogenization Algebras

The Σ-convergence of parabolic differential operators requires the use of appropriate H-algebras. For $1 \leq p < \infty$, we set

$$\mathcal{H}_p = L^p(\Delta(A_z); W_{\#}^{1,p}(\Delta(A_y); \mathbb{R})),$$

where the space $W_{\#}^{1,p}(\Delta(A_y); \mathbb{R})$ and basic related properties are introduced in Sect. 5.5.3. Equipped with the norm

$$\|u\|_{\mathcal{H}_p} = \left(\int_{\Delta(A_z)} \|u(s_0)\|^p_{W_{\#}^{1,p}(\Delta(A_y))} d\beta_z(s_0) \right)^{\frac{1}{p}} \quad (u \in \mathcal{H}_p),$$

$\mathcal{H}_p$ is a Banach space.

Definition 7.1 The H-algebra A is said to be quasi V^p-proper (V^p defined in Chap. 6) for some real $1 < p < \infty$ if the following two conditions are fulfilled:

(QP1) $\mathcal{D}(\Delta(A_y))$ is dense in $W^{1,p}(\Delta(A_y))$.
(QP2) Given a bounded sequence $(u_\varepsilon)_{\varepsilon \in E}$ in V^p, where E is any fundamental sequence, there are some subsequence E' from E and some $\mathbf{u} = (u_0, u_1) \in V^p \times L^p(Q; \mathcal{H}_p)$ such that as $E' \ni \varepsilon \to 0$,

$$u_\varepsilon \to u_0 \text{ in } V^p\text{-weak} \tag{7.5}$$

$$\frac{\partial u_\varepsilon}{\partial x_j} \to \frac{\partial u_0}{\partial x_j} + \partial_j u_1 \text{ in } L^p(Q)\text{-weak } \Sigma \ (1 \leq j \leq N). \tag{7.6}$$

Remark 7.1 The ∂_j's denote the $N+1$ basic partial differential operators on $\Delta(A) = \Delta(A_y) \times \Delta(A_z)$, namely ∂_j ($j = 1, \cdots, N$) with respect to $\Delta(A_y)$, and ∂_0 with respect to $\Delta(A_z)$. It is worth recalling that they are initially defined in $\mathcal{D}^1(\Delta(A))$, hence in $\mathcal{D}(\Delta(A))$, and then by duality in $\mathcal{D}'(\Delta(A))$. In particular for $1 \leq j \leq N$, the operator ∂_j (understood here in the distribution sense on $\Delta(A_y)$) is naturally extended from the locally convex space $W^{1,p}(\Delta(A_y))/\mathbb{C}$ to its separated completion, $W_{\#}^{1,p}(\Delta(A_y))$ (see Chap. 2, Proposition 2.12). So, for $u \in \mathcal{H}_p$, $\partial_j u$ actually denotes $\partial_j \circ u$, i.e., the function $s_0 \mapsto \partial_j u(s_0)$ from $\Delta(A_z)$ to $L^p(\Delta(A_y))$, which lies in $L^p(\Delta(A_z); L^p(\Delta(A_y))) = L^p(\Delta(A))$; whereas for $u_1 \in L^p(Q; \mathcal{H}_p)$, $\partial_j u_1$ stands for the function $(x,t) \mapsto \partial_j u_1(x,t)$ from Q to $L^p(\Delta(A))$, which lies in $L^p(Q \times \Delta(A))$. Finally, it is well to note that in the above definition the notion of Σ-convergence is relative to the H-algebra A.

Throughout this chapter, the notation u^ε refers exclusively to (7.3).

We will later need the following proposition which is interesting by itself.

Proposition 7.2 *Let $2 \leq p < \infty$. The following statements are equivalent:*

(i) The H-algebra A satisfies condition (QP2) of Definition 7.1.

(ii) For any given bounded sequence $(u_\varepsilon)_{\varepsilon\in E}$ in V^p, where E is a fundamental sequence, there are some subsequence E' from E and some $u_1 \in L^p(Q; \mathcal{H}_p)$ such that as $E' \ni \varepsilon \to 0$,

$$\int_Q \frac{u_\varepsilon}{\varepsilon}\left(\frac{\partial \psi}{\partial y_j}\right)^\varepsilon dxdt \to \int\int_{Q\times\Delta(A)} u_1 \partial_j \widehat{\psi}\, dxdtd\beta \tag{7.7}$$

for $\psi \in \mathcal{D}(Q) \otimes A^\infty$ and $1 \leq j \leq N$, where $\widehat{\psi} = \mathcal{G} \circ \psi$.

Proof Let $(u_\varepsilon)_{\varepsilon\in E}$ be a bounded sequence in V^p, where E is as above. Fix freely $1 \leq j \leq N$. Let $\psi \in \mathcal{D}(Q) \otimes A^\infty$. Clearly

$$\int_Q \frac{u_\varepsilon}{\varepsilon}\left(\frac{\partial \psi}{\partial y_j}\right)^\varepsilon dxdt = -\int_Q u_\varepsilon \left(\frac{\partial \psi}{\partial x_j}\right)^\varepsilon dxdt - \int_Q \frac{\partial u_\varepsilon}{\partial x_j}\psi^\varepsilon\, dxdt \tag{7.8}$$

and that for any arbitrary real $\varepsilon > 0$. Now, suppose (i) holds, so that we have at our disposal a subsequence E' from E and a couple (u_0, u_1) as in Definition 7.1, such that (7.5) and (7.6) hold as $E' \ni \varepsilon \to 0$. The aim now is to pass to the limit in (7.8). By recalling that one can extract from E' a subsequence (bearing the same name for convenience) such that $u_\varepsilon \to u_0$ in $L^2(Q)$ as $E' \ni \varepsilon \to 0$ (see Chap. 6) and taking account of (7.6), one has

$$\begin{aligned}\lim_{E'\ni\varepsilon\to 0} \int_Q \frac{u_\varepsilon}{\varepsilon}\left(\frac{\partial \psi}{\partial y_j}\right)^\varepsilon dxdt &= -\int_Q u_0 M\left(\frac{\partial \psi}{\partial x_j}\right) dxdt \\ &\quad - \int\int_{Q\times\Delta(A)} \left(\frac{\partial u_0}{\partial x_j} + \partial_j u_1\right) \widehat{\psi}\, dxdtd\beta.\end{aligned}$$

By taking in particular $\psi = \varphi \otimes w$ with $\varphi \in \mathcal{D}(Q)$ and $w \in A^\infty$, and noting that

$$M\left(\frac{\partial \psi}{\partial x_j}\right) = \frac{\partial \varphi}{\partial x_j} \int_{\Delta(A)} \widehat{w}\, d\beta,$$

it follows

$$\begin{aligned}\lim_{E'\ni\varepsilon\to 0} \int_Q \frac{u_\varepsilon}{\varepsilon}\left(\frac{\partial \psi}{\partial y_j}\right)^\varepsilon dxdt &= -\int_Q \left(\int_{\Delta(A)} \widehat{w}\partial_j u_1(x,t) d\beta\right) \varphi(x,t) dxdt \\ &= \int_Q \left(\int_{\Delta(A)} u_1(x,t)\partial_j \widehat{w} d\beta\right) \varphi(x,t) dxdt \\ &= \int\int_{Q\times\Delta(A)} u_1 \partial_j \widehat{\psi}\, dxdtd\beta,\end{aligned}$$

and so (7.7) is established for any elementary ψ of the preceding form. Hence the whole result follows readily by linearity.

Reciprocally let us assume that (ii) holds. Based on one hand, on the reflexivity of V^p together with the compactness of the embedding $V^p \hookrightarrow L^2(Q)$, on the other hand, on the Σ-reflexivity of $L^p(Q)$ (see Theorem 4.1), one can find a subsequence E' from E, a function $u_0 \in V^p$ and a family $f_j \in L^p(Q \times \Delta(A))$ $(1 \leq j \leq N)$ such that as $E' \ni \varepsilon \to 0$, we have (7.5), $u_\varepsilon \to u_0$ in $L^2(Q)$ and

$$\frac{\partial u_\varepsilon}{\partial x_j} \to f_j \text{ in } L^p(Q)\text{-weak } \Sigma \quad (1 \leq j \leq N).$$

By considering an arbitrary elementary function $\psi = \varphi \otimes w$ as above and passing to the limit (as $E' \ni \varepsilon \to 0$) in (7.8) using (7.7), we obtain

$$\int\int_{Q\times\Delta(A)} f_j \widehat{\psi}\, dxdtd\beta = \int\int_{Q\times\Delta(A)} \left(\frac{\partial u_0}{\partial x_j} + \partial_j u_1\right) \widehat{\psi}\, dxdtd\beta$$

for any elementary function ψ as above, hence for all $\psi \in \mathcal{D}(Q) \otimes A^\infty$ by linearity, and that for $1 \leq j \leq N$. Therefore the equality

$$f_j = \frac{\partial u_0}{\partial x_j} + \partial_j u_1 \quad (1 \leq j \leq N)$$

follows by the density of $\mathcal{D}(Q) \otimes \mathcal{D}(\Delta(A))$ in $L^p(Q \times \Delta(A))$. This shows (7.6), which completes the proof. ∎

Example 7.1 The periodic H-algebra $\mathcal{C}_{per}(Y \times Z)$ considered in Chap. 6 turns out to be the closure of $\mathcal{C}_{per}(Y) \otimes \mathcal{C}_{per}(Z)$ in $\mathcal{B}(\mathbb{R}^N \times \mathbb{R})$. Thus, we have here $A_y = \mathcal{C}_{per}(Y)$, $A_z = \mathcal{C}_{per}(Z)$ and $A = \mathcal{C}_{per}(Y \times Z)$. We will later see from Lemmas 6.2 and 6.9 (see also Example 4.2) that $\mathcal{C}_{per}(Y \times Z)$ is quasi V^p-proper for $2 \leq p < \infty$. This will be clarified in Sect. 7.3.2.

The preceding example is actually a particular case within a larger family. In the sequel we put $\mathcal{G}(\mathbf{u}) = (\mathcal{G}(u_j))_{1\leq j\leq N}$, $\mathbf{u} = (u_j) \in A_y^N$.

Definition 7.2 The H-algebra A_y is termed p-harmonic for some given $1 \leq p < \infty$ if the following conditions are fulfilled:

(H1) $\mathcal{D}(\Delta(A_y))$ is dense in $W^{1,p}(\Delta(A_y))$.
(H2) To any $\mathbf{f} = (f_j)_{1\leq j\leq N} \in L^p(\Delta(A_y))^N$ satisfying

$$\sum_{j=1}^{N} \int_{\Delta(A_y)} f_j \Phi_j d\beta_y = 0$$

for all $\mathbf{\Phi} = (\Phi_j) \in \mathcal{G}(\mathcal{U}_y)$ with $\mathcal{U}_y = \{\mathbf{u} | \mathbf{u} \in (A_y^\infty)^N, \operatorname{div}_y \mathbf{u} = 0\}$, there is attached a unique $q \in W_\#^{1,p}(\Delta(A_y))$ such that $f_j = \partial_j q$ $(1 \leq j \leq N)$.

This leads us to the following

Proposition 7.3 *Suppose the H-algebra A_y is p-harmonic for some $2 \leq p < \infty$. Then A is quasi V^p-proper.*

Proof This follows readily by an obvious adaptation of the proof of Lemma 6.9. ■

We shall illustrate this by two examples.

Example 7.2 The H-algebra $\mathcal{C}_{per}(Y)$ is p-harmonic for any $1 < p < \infty$, as is readily seen by Corollaries 2.13 and 2.15 and Lemma 2.11. A quick survey of the proofs of Lemmas 6.2 and 6.9 reveals that Example 7.1 is nothing else than a direct consequence of the p-harmonicity of $\mathcal{C}_{per}(Y)$.

Example 7.3 For any countable subgroup $\mathcal{R}$ of $\mathbb{R}^N_y$, the H-algebra $AP_{\mathcal{R}}(\mathbb{R}^N_y)$ is 2-harmonic (see Proposition 2.16 and Lemma 2.7). It follows by Proposition 7.3 that the H-algebra A with $A_y = AP_{\mathcal{R}}(\mathbb{R}^N)$ is quasi V^2-proper.

Remark 7.2 Before embarking upon the homogenization of the initial-boundary value problem (7.1) as such, let us observe that all of the notation and statements of Chap. 6 that are not relevant to the periodicity hypothesis carry over without change to the present chapter.

Remark 7.3 In the sequel, and unless otherwise specified, the Cartesian product, $F = F_1 \times F_2$, of two normed spaces F_1 and F_2 is tacitly equipped with the norm $\|\mathbf{u}\|_F = \|u_1\|_{F_1} + \|u_2\|_{F_2}$, $\mathbf{u} = (u_1, u_2) \in F$.

7.2.3 *On Some Basic Notation*

Let us first recall that the closed subspace $W^{1,p}(\Delta(A_y))/\mathbb{C}$ of $W^{1,p}(\Delta(A_y))$ is considered equipped not with the $W^{1,p}(\Delta(A_y))$ norm as one might expect, but with the (in general) non complete and non separated seminorm $\|\cdot\|_{W^{1,p}(\Delta(A_y))/\mathbb{C}}$. The separated completion of $W^{1,p}(\Delta(A_y))/\mathbb{C}$ is denoted by $W^{1,p}_{\#}(\Delta(A_y))$, and the canonical mapping of $W^{1,p}(\Delta(A_y))/\mathbb{C}$ into $W^{1,p}_{\#}(\Delta(A_y))$ is denoted by J.

Now, let $W^{1,p}(\Delta(A_y); \mathbb{R}) = \{u | u, \partial_j u \in L^p_{\mathbb{R}}(\Delta(A_y); \mathbb{R}), 1 \leq j \leq N\}$, a real Banach space with the $W^{1,p}(\Delta(A_y))$ norm. Its closed vector subspace $W^{1,p}(\Delta(A_y); \mathbb{R})/\mathbb{R}$ is considered equipped with the $W^{1,p}(\Delta(A_y))/\mathbb{C}$ seminorm. The separated completion, $W^{1,p}_{\#}(\Delta(A_y); \mathbb{R})$, of $W^{1,p}(\Delta(A_y); \mathbb{R})/\mathbb{R}$ is identified with the closed vector subspace of $W^{1,p}_{\#}(\Delta(A_y))$ made up of all $u \in W^{1,p}_{\#}(\Delta(A_y))$ such that $\partial_j u \in L^p_{\mathbb{R}}(\Delta(A_y)), 1 \leq j \leq N$. To conclude these preliminaries, let us note that the restriction of J to $W^{1,p}(\Delta(A_y); \mathbb{R})/\mathbb{R}$ is precisely the canonical mapping of the latter into its separated completion (the reader is referred to Sects. 5.2.2 and 5.5.3 for further details).

Having made this point, let us come to the heart of the matter. Let us fix freely some $u \in [\mathcal{D}_{\mathbb{R}}(\Delta(A_y))/\mathbb{R}] \otimes \mathcal{D}_{\mathbb{R}}(\Delta(A_z))$. Considered as a function of the variable

$s_0 \in \Delta(A_z)$, u writes as

$$u(s_0) = \sum_{finite} \theta_i(s_0)\psi_i \quad (s_0 \in \Delta(A_z))$$

where $\theta_i \in \mathcal{D}_{\mathbb{R}}(\Delta(A_z))$ and $\psi_i \in \mathcal{D}_{\mathbb{R}}(\Delta(A_y))/\mathbb{R}$. Thus, u is a mapping of $\Delta(A_z)$ into $\mathcal{D}_{\mathbb{R}}(\Delta(A_y))/\mathbb{R}$, therefore of $\Delta(A_z)$ into $W^{1,p}_{\mathbb{R}}(\Delta(A_y))/\mathbb{R}$, where $W^{1,p}_{\mathbb{R}}(\Delta(A_y) = W^{1,p}(\Delta(A_y);\mathbb{R})$. We then consider the mapping $s_0 \mapsto Ju(s_0)$ of $\Delta(A_z)$ into $W^{1,p}_{\#}(\Delta(A_y);\mathbb{R})$, commonly denoted by $J \circ u$, which sends continuously $\Delta(A_z)$ into $W^{1,p}_{\#}(\Delta(A_y);\mathbb{R})$, and so $J \circ u$ lies in $\mathcal{C}(\Delta(A_z); W^{1,p}_{\#}(\Delta(A_y);\mathbb{R}))$, thus in $\mathcal{H}_p$.

Finally, if $\Phi \in \mathcal{E} = \mathcal{D}_{\mathbb{R}}(Q) \otimes [\mathcal{D}_{\mathbb{R}}(\Delta(A_y))/\mathbb{R}] \otimes \mathcal{D}_{\mathbb{R}}(\Delta(A_z))$, by reasoning as above we see readily that

$$\Phi(x,t) \in [\mathcal{D}_{\mathbb{R}}(\Delta(A_y))/\mathbb{R}] \otimes \mathcal{D}_{\mathbb{R}}(\Delta(A_z)),$$

hence $J \circ \Phi(x,t) \in \mathcal{C}(\Delta(A_z); W^{1,p}_{\#}(\Delta(A_y);\mathbb{R}))$ for all $(x,t) \in Q$, and further the mapping $(x,t) \mapsto J \circ \Phi(x,t)$ of Q into the Banach space $\mathcal{C}(\Delta(A_z); W^{1,p}_{\#}(\Delta(A_y);\mathbb{R}))$ is continuous and of compact support.

Thus, throughout this chapter, J_Q denotes the linear mapping of $\mathcal{E}$ into
$\mathcal{K}(Q; \mathcal{C}(\Delta(A_z); W^{1,p}_{\#}(\Delta(A_y);\mathbb{R})))$ defined on each given $\Phi \in \mathcal{E}$ by
$J_Q(\Phi)(x,t) = J \circ \Phi(x,t)$ for all $(x,t) \in Q$.

Remark 7.4 For $\Phi \in \mathcal{D}_{\mathbb{R}}(Q) \otimes [\mathcal{D}_{\mathbb{R}}(\Delta(A_y))/\mathbb{R}] \otimes \mathcal{D}_{\mathbb{R}}(\Delta(A_z))$, we have:

(1) $J_Q(\Phi) \in \mathcal{K}(Q \times \Delta(A_z); W^{1,p}_{\#}(\Delta(A_y);\mathbb{R}))$.
(2) $J_Q(\Phi) \in L^p(Q;\mathcal{H}_p)$.

We shall conclude this J-topic with a simple but useful result. It is worth perhaps recalling beforehand that the tensor product of a scalar function, say $\xi : Q \times \Delta(A_z) \to \mathbb{R}$, and a «vector» $u \in W^{1,p}_{\#}(\Delta(A_y);\mathbb{R})$ is defined to be the function $\xi \otimes u$ of $Q \times \Delta(A_z)$ into $W^{1,p}_{\#}(\Delta(A_y);\mathbb{R})$ given by $\xi \otimes u(x,t,s_0) = \xi(x,t,s_0)u$ for $(x,t) \in Q$ and $s_0 \in \Delta(A_z)$.

Proposition 7.4 *The following identity holds true.*

$$J_Q(\mathcal{D}_{\mathbb{R}}(Q) \otimes [\mathcal{D}_{\mathbb{R}}(\Delta(A_y))/\mathbb{R}] \otimes \mathcal{D}_{\mathbb{R}}(\Delta(A_z))) =$$

$$\mathcal{D}_{\mathbb{R}}(Q) \otimes \mathcal{D}_{\mathbb{R}}(\Delta(A_z)) \otimes J([\mathcal{D}_{\mathbb{R}}(\Delta(A_y))/\mathbb{R}]).$$

Proof In view of the linearity of the mappings J and J_Q, it is enough to show the equality $J_Q(\varphi \otimes \psi \otimes \theta) = \varphi \otimes \theta \otimes J(\psi)$ for any triple $\varphi \in \mathcal{D}_{\mathbb{R}}(Q)$, $\psi \in [\mathcal{D}_{\mathbb{R}}(\Delta(A_y))/\mathbb{R}]$, $\theta \in \mathcal{D}_{\mathbb{R}}(\Delta(A_z))$. This is straightforward. Indeed, given φ, ψ and θ as stated, let $\Phi = \varphi \otimes \psi \otimes \theta$. According to point (1) of Remark 7.4, $J_Q(\Phi)$ may be viewed as a function of the three variables x, t and s_0, which justifies the first equality of the following series: $J_Q(\Phi)(x,t,s_0) = [J_Q(\Phi)(x,t)](s_0) = J(\Phi(x,t,\cdot,s_0)) = J(\varphi(x,t)\theta(s_0)\psi) = [\varphi \otimes \theta](x,t,s_0)J\psi = [\varphi \otimes \theta \otimes J(\psi)](x,t,s_0)$. Hence, the desired equality follows by the arbitrariness of x, t and s_0. ■

Finally, another clarification worth recalling is about the Gelfand transformation. Let $\mathcal{G}$ be the Gelfand transformation on A. If $\psi \in A$, for one reason or another, anyhow for convenience as far as this book is concerned, one usually puts $\widehat{\psi} = \mathcal{G}(\psi) \in \mathcal{C}(\Delta(A))$. Now, if ψ is a function of Q into A, e.g., $\psi \in \mathcal{C}(Q;A)$ or $\psi \in L^p(Q;A)$ for some $1 \le p < \infty$, $\widehat{\psi}$ stands this time for the function $(x,t) \mapsto \mathcal{G}(\psi(x,t))$ of Q into $\mathcal{C}(\Delta(A))$, that is, $\widehat{\psi} = \mathcal{G} \circ \psi$, with, e.g., $\widehat{\psi} \in \mathcal{C}(Q;\mathcal{C}(\Delta(A)))$ if $\psi \in \mathcal{C}(Q;A)$, or $\widehat{\psi} \in L^p(Q;\mathcal{C}(\Delta(A)))$ if $\psi \in L^p(Q;A)$.

Suppose now that ψ lies in, say $\mathcal{C}(Q) \otimes A_y \otimes A_z$, as often happens in the next sections. Thus ψ is introduced in terms of a function of the variables x, t, y and z, which seems to be beyond the preceding framework. One can nevertheless define $\widehat{\psi}$ as above provided ψ is regarded as a function of Q into A, which then belongs to $\mathcal{C}(Q;A)$. To conclude, attention must be especially drawn to this last case.

7.2.4 *Homogenization of (7.1) for* $0 < k \le 1$

In what follows, we set ${}_{\mathbb{R}}A_z^\infty = A_z^\infty \cap \mathcal{C}_{\mathbb{R}}(\mathbb{R})$, ${}_{\mathbb{R}}A_y^\infty = A_y^\infty \cap \mathcal{C}_{\mathbb{R}}(\mathbb{R}^N)$, ${}_{\mathbb{R}}A_y^\infty/\mathbb{R} = \{\psi | \psi \in {}_{\mathbb{R}}A_y^\infty, M(\psi) = 0\}$. Thus, $\mathcal{D}_{\mathbb{R}}(\Delta(A_y))/\mathbb{R} = \mathcal{G}({}_{\mathbb{R}}A_y^\infty/\mathbb{R})$. Attention must be drawn to the following:

(1) $[\mathcal{D}_{\mathbb{R}}(\Delta(A_y))/\mathbb{R}] \otimes \mathcal{D}_{\mathbb{R}}(\Delta(A_z)) = \mathcal{G}([{}_{\mathbb{R}}A_y^\infty/\mathbb{R}] \otimes {}_{\mathbb{R}}A_z^\infty)$.

(2) If $\mathcal{D}(\Delta(A_y))$ is dense in $W^{1,p}(\Delta(A_y))$, then $J(\mathcal{D}_{\mathbb{R}}(\Delta(A_y))/\mathbb{R})$ is dense in $W_\#^{1,p}(\Delta(A_y);\mathbb{R})$, where J denotes the canonical mapping of $W^{1,p}(\Delta(A_y))/\mathbb{C}$ into $W_\#^{1,p}(\Delta(A_y))$ (see Remark 4.6 together with Sect. 5.5.3).

Throughout this section it is assumed that A is quasi V^p-proper for some $1 < p < \infty$.

Lemma 7.1 *Let* $\mathcal{F}_0^\infty = \mathcal{D}_{\mathbb{R}}(Q) \times [\mathcal{D}_{\mathbb{R}}(Q) \otimes \mathcal{D}_{\mathbb{R}}(\Delta(A_z)) \otimes J(\mathcal{D}_{\mathbb{R}}(\Delta(A_y))/\mathbb{R})]$. *Then* $\mathcal{F}_0^\infty$ *is dense in* $L^p(0,T;W_0^{1,p}(\Omega;\mathbb{R})) \times L^p(Q;\mathcal{H}_p)$.

Proof The same argument as followed in proving Lemma 6.10 leads us to realize that the whole problem reduces to showing that $\mathcal{D}_{\mathbb{R}}(\Delta(A_z)) \otimes J(\mathcal{D}_{\mathbb{R}}(\Delta(A_y))/\mathbb{R})$ is dense in $\mathcal{H}_p$. But, since $\mathcal{D}_{\mathbb{R}}(\Delta(A_z))$ is dense in $\mathcal{C}_{\mathbb{R}}(\Delta(A_z))$, this follows by the density of $J(\mathcal{D}(\Delta(A_y))/\mathbb{R})$ in $W_\#^{1,p}(\Delta(A_y);\mathbb{R})$, as a consequence of (QP1) (Definition 7.1). ■

Before we can prove the main homogenization result in the present section, we need one last lemma. Let us recall beforehand a notation, viz.,

$$\mathbb{D}_j \mathbf{v} = \frac{\partial v_0}{\partial x_j} + \partial_j v_1 \text{ for } \mathbf{v} = (v_0, v_1) \in W^{1,p}(\Omega) \times W_{\#}^{1,p}(\Delta(A_y)),\ \mathbb{D} = (\mathbb{D}_j)_{1 \leq j \leq N}.$$

Lemma 7.2 *The variational problem:*

$$\begin{cases} \boldsymbol{u} = (u_0, u_1) \in \mathbb{F}^{1,p}_{\mathcal{H}_p} = V_0^p \times L^p(Q; \mathcal{H}_p) : \\ \int_0^T [u_0'(t), v_0(t)]dt + \int\int_{Q \times \Delta(A)} \boldsymbol{b}(\mathbb{D}\boldsymbol{u}) \cdot \mathbb{D}\boldsymbol{v}\, dxdtd\beta \\ = \int_0^T [f(t), v_0(t)]dt \textit{ for all } \boldsymbol{v} = (v_0, v_1) \in \mathbb{F}^{1,p}_{\mathcal{H}_p} \end{cases} \tag{7.9}$$

has at most one solution.

Proof This follows by the same argument as used in proving Lemma 5.15 (see also Lemma 6.11) and use of Proposition 7.1. ∎

We are now able to prove the following statement.

Theorem 7.1 *Let $2 \leq p < \infty$. Let $0 < k \leq 1$. Suppose (7.2) holds and further A is quasi V^p-proper. For each real $\varepsilon > 0$, let u_ε be the weak solution of the initial-boundary value problem (7.1). Then, as $\varepsilon \to 0$, we have*

$$u_\varepsilon \to u_0 \textit{ in } L^p(0, T; W_0^{1,p}(\Omega))\textit{-weak} \tag{7.10}$$

$$\frac{\partial u_\varepsilon}{\partial t} \to \frac{\partial u_0}{\partial t} \textit{ in } L^{p'}(0, T; W^{-1,p'}(\Omega))\textit{-weak} \tag{7.11}$$

$$\frac{\partial u_\varepsilon}{\partial x_j} \to \mathbb{D}_j \boldsymbol{u} = \frac{\partial u_0}{\partial x_j} + \partial_j u_1 \textit{ in } L^p(Q)\textit{-weak } \Sigma \quad (1 \leq j \leq N), \tag{7.12}$$

where $\boldsymbol{u} = (u_0, u_1)$ is the unique solution of (7.9).

Proof As noted earlier in proving Theorem 6.3, it is a matter of course that the proof of Theorem 6.2 carries over mutatis mutandis (in particular, by changing $\frac{t}{\varepsilon}$ into $\frac{t}{\varepsilon^k}$) up to (6.39), to the present case. Thus, in particular the sequence $(u_\varepsilon)_{\varepsilon>0}$ is bounded in V_0^p. Therefore, given an arbitrary fundamental sequence E, the quasi V^p-properness of the H-algebra A guarantees the existence of a subsequence E' from E and of some $\mathbf{u} = (u_0, u_1) \in \mathbb{F}^{1,p}_{\mathcal{H}_p} = V_0^p \times L^p(Q, \mathcal{H}_p)$ such that (7.10)–(7.12) hold as $E' \ni \varepsilon \to 0$. For further needs it is worth bearing in mind that in addition, $u_\varepsilon \to u_0$ in $L^2(Q)$, thanks to the compactness of the embedding $V^p \hookrightarrow L^2(Q)$. The aim now is to pass to the limit (as $E' \ni \varepsilon \to 0$) in what should be referred to as $(6.39)_k$, that is, inequality (6.39) with $\frac{t}{\varepsilon}$ changed into $\frac{t}{\varepsilon^k}$ and Φ_ε given by equality (7.4). For that purpose, fix freely some $\boldsymbol{\Phi} = (\psi_0, J_Q(\widehat{\psi}_1) \in \mathcal{F}_0^\infty$ (see Lemma 7.1 and Proposition 7.4), where it should be recalled that $\psi_0 \in \mathcal{D}_{\mathbb{R}}(Q)$ and $\psi_1 \in \mathcal{D}_{\mathbb{R}}(Q) \otimes ({}_{\mathbb{R}}A_y^\infty/\mathbb{R}) \otimes {}_{\mathbb{R}}A_z^\infty$. Next, attach to $\boldsymbol{\Phi}$ the sequence of functions

$\Phi_\varepsilon \in \mathcal{D}_{\mathbb{R}}(Q)$ given by (7.4). First, since $\left(\frac{\partial \psi_1}{\partial z}\right)^\varepsilon \to M\left(\frac{\partial \psi_1}{\partial z}\right) = 0$ in $L^p(Q)$-weak as $\varepsilon \to 0$ (see Chap. 4), we see by the equality

$$\frac{\partial \Phi_\varepsilon}{\partial t} = \frac{\partial \psi_0}{\partial t} + \varepsilon \left(\frac{\partial \psi_1}{\partial t}\right)^\varepsilon + \varepsilon^{1-k} \left(\frac{\partial \psi_1}{\partial z}\right)^\varepsilon$$

that

$$\frac{\partial \Phi_\varepsilon}{\partial t} \to \frac{\partial \psi_0}{\partial t} \text{ in } L^p(Q)\text{-weak as } \varepsilon \to 0. \tag{7.13}$$

By the same argument as followed in proving Lemma 6.3 (iv) we deduce that $\Phi_\varepsilon \to \psi_0$ in $L^p(0, T; W_0^{1,p}(\Omega))$-weak as $\varepsilon \to 0$. Now, let $E' \ni \varepsilon \to 0$. Then (6.40) follows by the latter convergence result and use of (7.10). Next, by the equality (6.41) and use of (7.13) we have

$$\int_0^T [u_\varepsilon'(t), \Phi_\varepsilon(t)]dt \to \int_0^T [u_0'(t), \psi_0(t)]dt, \tag{7.14}$$

as is readily seen, on one hand, by the strong convergence result in $L^2(Q)$ pointed out earlier, on the other hand, by the fact that (7.13) still holds with $L^2(Q)$ instead of $L^p(Q)$, since the latter is continuously embedded in the former. Finally, based on Corollary 7.1, the same line of reasoning as followed in proving Theorem 6.2 leads us by (7.12) to

$$\int_Q \mathbf{a}^\varepsilon(\cdot, \cdot, D\Phi_\varepsilon) \cdot (Du_\varepsilon - D\Phi_\varepsilon)dxdt \to \iint_{Q \times \Delta(A)} \mathbf{b}(\mathbb{D}\mathbf{\Phi}) \cdot \mathbb{D}(\mathbf{u} - \mathbf{\Phi})dxdtd\beta$$

as $E' \ni \varepsilon \to 0$, where the reader is reminded that $\mathbb{D}\mathbf{v} = D_x v_0 + \nabla v_1$ for $\mathbf{v} = (v_0, v_1) \in W^{1,p}(\Omega) \times W_\#^{1,p}(\Delta(A_y))$.

Hence, by taking the $\liminf_{E' \ni \varepsilon \to 0}$ of $(6.39)_k$ and recalling inequality (6.42) as well as equality (6.43), one arrives at

$$0 \leq \int_0^T [f(t) - u_0'(t), u_0(t) - \psi_0(t)]dt - \iint_{Q \times \Delta(A)} \mathbf{b}(\mathbb{D}\mathbf{\Phi}) \cdot \mathbb{D}(\mathbf{u} - \mathbf{\Phi})dxdtd\beta,$$

and that for any arbitrary $\mathbf{\Phi} \in \mathcal{F}_0^\infty$. Thanks to Lemma 7.1, this still holds for $\mathbf{\Phi} \in \mathbb{F}_{\mathcal{H}_p}^{1,p}$. Therefore, the theorem follows by the same classical argument as used to complete the proof of Theorem 6.2, of which argument Lemma 7.2 is here the chief point. ∎

The cell problem and the macroscopic homogenized problem are framed as usual (see, e.g., Sect. 5.5.3). Summarily, let $\mathcal{S}_A$ denote the set of all those $r \in \mathbb{R}^N$ for which the variational problem

$$\begin{cases} \pi(r) \in \mathcal{H}_p : \\ \int_{\Delta(A)} \mathbf{b}(r + \nabla\pi(r)) \cdot \nabla w \, d\beta = 0 \text{ for all } w \in \mathcal{H}_p \end{cases}$$

admits a solution. Let us note that $\pi(r)$ is then unique, thanks to the strong monotonicity property in Proposition 7.1. Whence a mapping $r \mapsto \pi(r)$ of $\mathcal{S}_A$ into $\mathcal{H}_p$, referred to as π. The important relation $u_1 = \pi \circ D_x u_0$ can be easily obtained by means of the so-called local problem derived from (7.9) (see, e.g., Sect. 5.5.3). It should be stressed that u_0 and u_1 in the above relation are precisely the components of the solution, $\mathbf{u}$, of (7.9). Substituting this relation in (7.9) and then choosing there the particular test functions $\mathbf{v} = (v_0, v_1)$ with $v_1 = 0$ leads to the initial-Boundary value problem (6.45) with this time

$$q(r) = \int_{\Delta(A)} \mathbf{b}(r + \nabla\pi(r)) d\beta \quad (r \in \mathcal{S}_A).$$

7.2.5 *Homogenization of (7.1) for* $1 < k < 2$

In this section it is required of the H-algebra A_y that it has the following property.

(P1) *There is a vector subspace $\mathcal{V}_{\infty y}$ of A_y^∞ that preserves complex conjugation and further, that satisfies:*

(1) $\mathcal{G}(\mathcal{V}_{\infty y})$ is dense in $W^{1,p}(\Delta(A_y))$ $(1 < p < \infty)$,
(2) for any $\psi \in \mathcal{V}_{\infty y}/\mathbb{C} = \{w | w \in \mathcal{V}_{\infty y}, M(w) = 0\}$, there is some function $\theta \in \mathcal{V}_{\infty y}/\mathbb{C}$ such that $\psi = \Delta_y \theta$ (Δ_y the Laplace operator in $\mathbb{R}_y^N$).

Remark 7.5 Since $\mathcal{V}_{\infty y}$ preserves complex conjugation, it is immediate that the statements (1) and (2) still hold true if $\mathcal{V}_{\infty y}$ is replaced by ${}_{\mathbb{R}}\mathcal{V}_{\infty y} = \mathcal{V}_{\infty y} \cap \mathcal{C}(\mathbb{R}^N; \mathbb{R})$ and $W^{1,p}(\Delta(A_y))$ is replaced by $W^{1,p}(\Delta(A_y); \mathbb{R})$.

The H-algebra $A_y = \mathcal{C}_{per}(Y)$ satisfies (P1) with $\mathcal{V}_{\infty y} = \mathcal{C}_{per}^\infty(Y)$ (see Sect. 2.5 and the proof of Corollary 3.3).

We need a preliminary result.

Lemma 7.3 *The space $\mathcal{E}_0^\infty = \mathcal{D}_{\mathbb{R}}(Q) \times [\mathcal{D}_{\mathbb{R}}(Q) \otimes \mathcal{D}_{\mathbb{R}}(\Delta(A_z)) \otimes J(\mathcal{G}({}_{\mathbb{R}}\mathcal{V}_{\infty y}/\mathbb{R}))]$ is dense in $L^p(0, T; W_0^{1,p}(\Omega; \mathbb{R})) \times L^p(Q; \mathcal{H}_p)$ $(1 \le p < \infty)$.*

Proof The lemma follows by an obvious line of argument (see the proof of Lemma 7.1) with taking account of Remark 7.5. ■

We finally turn to the proof of the next homogenization theorem.

Theorem 7.2 *Let $2 \leq p < \infty$. Let $1 < k < 2$. Suppose (7.2) holds and further A is quasi V^p-proper and satisfies (P1). For any given real $\varepsilon > 0$, let u_ε be the weak solution of (7.1). Then, as $\varepsilon \to 0$, we have (7.10)–(7.12), where $\mathbf{u} = (u_0, u_1)$ is the unique solution of the variational problem (7.9).*

Proof A quick survey of the proof of Theorem 7.1 reveals that the analysis of the case $0 < k \leq 1$ carries over to the present case save that Lemma 7.1 (resp. $\mathcal{F}_0^\infty$) shall be substituted by Lemma 7.3 (resp. $\mathcal{E}_0^\infty$) and the convergence result (7.14) established differently. We recall that at our disposal there are a subsequence E' from an arbitrary given fundamental sequence E , and some $\mathbf{u} = (u_0, u_1) \in \mathbb{F}^{1,p}_{\mathcal{H}_p}$ such that as $E' \ni \varepsilon \to 0$, one has (7.10)–(7.12), hence $u_\varepsilon \to u_0$ in $L^2(Q)$.

This being so, let us fix freely some $\mathbf{\Phi} = (\psi_0, J_Q(\widehat{\psi_1})) \in \mathcal{E}_0^\infty$ (see Sect. 7.2.3) to which we attach the sequence $(\Phi_\varepsilon)_{\varepsilon>0}$ given by (7.4) with $\psi_0 \in \mathcal{D}_\mathbb{R}(Q)$ and $\psi_1 \in \mathcal{D}_\mathbb{R}(Q) \otimes ({}_\mathbb{R}\mathcal{V}_{\infty y}/\mathbb{R}) \otimes {}_\mathbb{R}A_z^\infty$. We are through if we can establish (7.14) when $E' \ni \varepsilon \to 0$.

Clearly

$$\int_0^T [u_\varepsilon'(t), \Phi_\varepsilon(\cdot, t)]dt = -\int_Q u_\varepsilon \frac{\partial \psi_0}{\partial t} dxdt - \varepsilon \int_Q u_\varepsilon \left(\frac{\partial \psi_1}{\partial t}\right)^\varepsilon dxdt$$
$$- \varepsilon^{1-k} \int_Q u_\varepsilon \left(\frac{\partial \psi_1}{\partial z}\right)^\varepsilon dxdt.$$

As $E' \ni \varepsilon \to 0$, it is immediate that the first term on the right tends to $-\int_Q u_0 \frac{\partial \psi_0}{\partial t} dxdt = \int_0^T [u_0'(t), \psi_0(\cdot, t)]dt$ whereas the next one goes to zero. We claim that the last term on the right also goes to zero. Indeed, we have

$$\varepsilon^{1-k} \int_Q u_\varepsilon \left(\frac{\partial \psi_1}{\partial z}\right)^\varepsilon dxdt = \varepsilon^{2-k} \int_Q \frac{1}{\varepsilon} u_\varepsilon (\Delta_y \chi)^\varepsilon dxdt,$$

where, according to property (P1), $\chi \in \mathcal{D}_\mathbb{R}(Q) \otimes ({}_\mathbb{R}\mathcal{V}_{\infty y}/\mathbb{R}) \otimes {}_\mathbb{R}A_z^\infty$ with $\Delta_y \chi = \frac{\partial \psi_1}{\partial z}$. Hence the claimed result follows by summing (7.7) in $1 \leq j \leq N$ with $\psi = \frac{\partial \chi}{\partial y_j}$ and then using the fact that ε^{2-k} goes to zero. Therefore (7.14) follows, which completes the proof. ■

7.2.6 *Homogenization of (7.1) for k = 2*

As was pointed out in the periodic setting (see Chap. 6), the space $\mathcal{H}_p$ does not fit the case $k = 2$ and so we need some preliminaries. We begin with the notion of a vector distribution on $\Delta(A_z)$.

Let B be a Banach space, B' its strong dual. We assume that B' is separable. A distribution on $\Delta(A_z)$ with values in B is defined to be any continuous linear mapping on the Fréchet space $\mathcal{D}(\Delta(A_z))$ (see Sect. 2.1) into B. The space of all such distributions will be denoted by $\mathcal{D}'(\Delta(A_z); B)$, and the value of $f \in \mathcal{D}'(\Delta(A_z); B)$ on $\theta \in \mathcal{D}(\Delta(A_z))$ denoted by $\langle f, \theta \rangle$. We provide $\mathcal{D}'(\Delta(A_z); B)$ with the locally convex topology defined by the family of seminorms

$$N_\Phi(f) = \sup_{\theta \in \Phi} \|\langle f, \theta \rangle\| \quad (f \in \mathcal{D}'(\Delta(A_z); B)),$$

where $\|\cdot\|$ is the norm on B, and Φ ranges over the bounded subsets of $\mathcal{D}(\Delta(A_z))$ (see Sect. 2.1). $\mathcal{D}'(\Delta(A_z); B)$ so topologized is a complete separated locally convex space.

By the derivative (of the first order) of a distribution f on $\Delta(A_z)$ with values in B is meant the unique $f' = \frac{df}{ds_0} \in \mathcal{D}'(\Delta(A_z); B)$ such that $\langle f', \theta \rangle = -\langle f, \theta' \rangle$ for all $\theta \in \mathcal{D}(\Delta(A_z))$, where $\theta' = \frac{d\theta}{ds_0}$. Furthermore, the operator $f \mapsto f'$ maps continuously and linearly $\mathcal{D}'(\Delta(A_z); B)$ into itself. The derivative of order $m \geq 2$ is defined in an obvious way, and has the same properties of continuity and linearity.

To each function $u \in L^p(\Delta(A_z); B)$ $(1 \leq p \leq \infty)$ there is attached a distribution $f_u \in \mathcal{D}'(\Delta(A_z); B)$ given by

$$\langle f_u, \theta \rangle = \int_{\Delta(A_z)} u(s_0)\theta(s_0) d\beta_z(s_0) \quad (\theta \in \mathcal{D}(\Delta(A_z))).$$

This yields a mapping $u \mapsto f_u$ which is continuous linear of $L^p(\Delta(A_z); B)$ into $\mathcal{D}'(\Delta(A_z); B)$. Moreover this mapping is injective thanks to the fact that $\mathcal{D}(\Delta(A_z))$ is dense in $\mathcal{C}(\Delta(A_z))$. Consequently one may identify f_u with u, so that

$$L^p(\Delta(A_z); B) \subset \mathcal{D}'(\Delta(A_z); B) \text{ with continuous embedding.}$$

This enables us for further needs to define $u' = \frac{du}{ds_0} \in \mathcal{D}'(\Delta(A_z); B)$ for $u \in L^p(\Delta(A_z); B)$.

Throughout the rest of this section it is required of the H-algebra A_y that it satisfies the following condition.

(P2) There is a constant $c > 0$ such that $\|u\|_{L^p(\Delta(A_y))} \leq c\|u\|_{W^{1,p}(\Delta(A_y))/\mathbb{C}}$ for all $u \in W^{1,p}(\Delta(A_y))/\mathbb{C}$ $(1 < p < \infty)$.

Remark 7.6 Property (P2) has two major consequences:

(1) The seminorm $\|\cdot\|_{W^{1,p}(\Delta(A_y))/\mathbb{C}}$ is a norm on $W^{1,p}(\Delta(A_y))/\mathbb{C}$ equivalent to the $W^{1,p}(\Delta(A_y))$-norm, and $W^{1,p}(\Delta(A_y))/\mathbb{C}$ with the said norm is therefore a Banach space. Hence it follows that $W^{1,p}(\Delta(A_y))/\mathbb{C}$ may be identified with its separated completion, $W^{1,p}_{\#}(\Delta(A_y))$.

(2) By means of the mapping $u \mapsto \nabla u$, the space $W^{1,p}(\Delta(A_y))/\mathbb{C}$ is algebraically and topologically identifiable with a closed vector subspace of the separable

reflexive Banach space $[L^p(\Delta(A_y))/\mathbb{C}]^N$, and is therefore itself a separable reflexive Banach space provided $1 < p < \infty$, of course.

In the sequel it is assumed that $2 \leq p < \infty$. The duality pairing between $W^{1,p}_{\mathbb{R}}(\Delta(A_y))/\mathbb{R}$ and its dual, $[W^{1,p}_{\mathbb{R}}(\Delta(A_y))/\mathbb{R}]'$, will be denoted by $[\cdot, \cdot]$.

Now, it is clear that $W^{1,p}_{\mathbb{R}}(\Delta(A_y))/\mathbb{R} \subset L^2_{\mathbb{R}}(\Delta(A_y))/\mathbb{R}$, with continuous embedding. Furthermore, $W^{1,p}_{\mathbb{R}}(\Delta(A_y))/\mathbb{R}$ is dense in $L^2_{\mathbb{R}}(\Delta(A_y))/\mathbb{R}$, as is readily seen by the density of $\mathcal{D}_{\mathbb{R}}(\Delta(A_y))$ in $\mathcal{C}_{\mathbb{R}}(\Delta(A_y))$ (recall that by hypothesis A^∞_y is basically dense in A_y). Hence, by identifying $L^2_{\mathbb{R}}(\Delta(A_y))/\mathbb{R}$ with its dual, it follows

$$W^{1,p}_{\mathbb{R}}(\Delta(A_y))/\mathbb{R} \subset L^2_{\mathbb{R}}(\Delta(A_y))/\mathbb{R} \subset [W^{1,p}_{\mathbb{R}}(\Delta(A_y))/\mathbb{R}]'$$

with continuous embeddings. Consequently, we have

$$[u, w] = \int_{\Delta(A_y)} uw \, d\beta_y \tag{7.15}$$

whenever $u \in L^2_{\mathbb{R}}(\Delta(A_y))/\mathbb{R}$ and $w \in W^{1,p}_{\mathbb{R}}(\Delta(A_y))/\mathbb{R}$.

This being so, as a consequence of (P2) (see Remark 7.6), the space $\mathcal{H}_p$ introduced in Sect. 7.2.2 has a simpler expression, viz.

$$\mathcal{H}_p = L^p(\Delta(A_z); W^{1,p}_{\mathbb{R}}(\Delta(A_y))/\mathbb{R}),$$

and the same is true of its strong dual,

$$\mathcal{H}'_p = L^{p'}(\Delta(A_z); [W^{1,p}_{\mathbb{R}}(\Delta(A_y))/\mathbb{R}]').$$

Hence, the duality pairing between $\mathcal{H}'_p$ and $\mathcal{H}_p$ is given by

$$\langle u, v\rangle_{\mathcal{H}'_p,\mathcal{H}_p} = \int_{\Delta(A_z)} [u(s_0), v(s_0)] d\beta_z(s_0) \;\; (u \in \mathcal{H}'_p, v \in \mathcal{H}_p).$$

As we mentioned before, the space $\mathcal{H}_p$ is hardly likely to help us by itself achieve our purpose. Thus, on the lines of Chap. 6 we introduce the space

$$\mathcal{U} = \left\{ v | v \in \mathcal{H}_p, v' = \frac{dv}{ds_0} \in \mathcal{H}'_p \right\},$$

where the derivative is understood in the sense of vector distributions $\mathcal{D}'(\Delta(A_z); W^{1,p}_{\mathbb{R}}(\Delta(A_y))/\mathbb{R})$. Provided with the norm

$$\|v\|_{\mathcal{U}} = \|v\|_{\mathcal{H}_p} + \left\| \frac{dv}{ds_0} \right\|_{\mathcal{H}'_p} \quad (v \in \mathcal{U}),$$

$\mathcal{U}$ is a Banach space (one arrives at once at the result by the same argument as used to show the analogous statement in Chap. 6).

Definition 7.3 The H-algebra A is said to be V^p-proper for some real $p \geq 2$ if there exist a vector subspace $\mathcal{V}_{\infty y}$ of A_y^∞ that preserves complex conjugation, and a closed vector subspace $\mathcal{W}$ of $\mathcal{U}$ including $\mathcal{D}_{\mathbb{R}}(\Delta(A_z)) \otimes \mathcal{G}(\,_{\mathbb{R}}\mathcal{V}_{\infty y}/\mathbb{R})$, such that the following hold true.

(PR1) The following two conditions are satisfied:

$$\begin{aligned} &(1) \quad \mathcal{G}(\mathcal{V}_{\infty y}) \text{ is dense in } W^{1,p}(\Delta(A_y)).\\ &(2) \quad \mathcal{D}_{\mathbb{R}}(\Delta(A_z)) \otimes \mathcal{G}(\,_{\mathbb{R}}\mathcal{V}_{\infty y}/\mathbb{R}) \text{ is dense in } \mathcal{W}. \end{aligned}$$

(PR2) Given any norm-bounded sequence $(u_\varepsilon)_{\varepsilon\in E}$ in V^p, where E is an arbitrary fundamental sequence, there are some subsequence E' from E and some $\mathbf{u} = (u_0, u_1) \in V^p \times L^p(Q; \mathcal{W})$ such that as $E' \ni \varepsilon \to 0$,

$$\begin{aligned} &u_\varepsilon \to u_0 \text{ in } V^p\text{-weak},\\ &\frac{\partial u_\varepsilon}{\partial x_j} \to \frac{\partial u_0}{\partial x_j} + \partial_j u_1 \text{ in } L^p(Q)\text{-weak } \Sigma \quad (1 \leq j \leq N). \end{aligned}$$

Remark 7.7 Under condition (P2), it is clear that if A is V^p-proper, then A is quasi V^p-proper.

We need a few preliminary results.

Lemma 7.4 *Let $\Phi \in \mathcal{D}_{\mathbb{R}}(\Delta(A_z)) \otimes [\mathcal{D}_{\mathbb{R}}(\Delta(A_y))/\mathbb{R}]$. Let u be the function of $\Delta(A_z)$ into $\mathcal{D}_{\mathbb{R}}(\Delta(A_y))/\mathbb{R}$ given by $u(s_0) = \Phi(\cdot, s_0)$ for $s_0 \in \Delta(A_z)$, where, for fixed $s_0 \in \Delta(A_z)$, $\Phi(\cdot, s_0)$ stands for the function $s \mapsto \Phi(s, s_0)$ of $\Delta(A_y)$ into $\mathbb{R}$. One has $u \in \mathcal{C}(\Delta(A_z); W_{\mathbb{R}}^{1,p}(\Delta(A_y))/\mathbb{R}) \subset \mathcal{H}_p$ and*

$$\frac{du}{ds_0} = \partial_0 \Phi, \tag{7.16}$$

where the derivative on the left-hand side is to be understood in the sense of $\mathcal{D}'(\Delta(A_z); W_{\mathbb{R}}^{1,p}(\Delta(A_y))/\mathbb{R})$, whereas $\partial_0 \Phi$ is nothing else than the partial derivative of index $j = 0$ (corresponding to the factor $\Delta(A_z)$ of $\Delta(A) = \Delta(A_y) \times \Delta(A_z)$) of $\Phi \in \mathcal{D}(\Delta(A))$.

Proof That u lies in $\mathcal{C}(\Delta(A_z); W_{\mathbb{R}}^{1,p}(\Delta(A_y))/\mathbb{R})$ is evident. In the sequel u is considered as a distribution on $\Delta(A_z)$ with values in $W_{\mathbb{R}}^{1,p}(\Delta(A_y))/\mathbb{R}$. Fix freely some $\theta \in \mathcal{D}(\Delta(A_z))$. Recalling that $\langle \frac{du}{ds_0}, \theta \rangle \in W^{1,p}_{\mathbb{R}}(\Delta(A_y))/\mathbb{R} \subset L^2_{\mathbb{R}}(\Delta(A_y))/\mathbb{R}$, we have, for any arbitrary given $\psi \in \mathcal{D}_{\mathbb{R}}(\Delta(A_y))/\mathbb{R}$,

$$\begin{aligned}
&\int_{\Delta(A_y)} \left\langle \frac{du}{ds_0}, \theta \right\rangle (s)\psi(s)d\beta_y(s) \\
&= -\int_{\Delta(A_y)} \left[\int_{\Delta(A_z)} \Phi(s, s_0)\frac{d\theta}{ds_0}(s_0)d\beta_z(s_0) \right] \psi(s)d\beta_y(s) \\
&= -\iint_{\Delta(A)} \Phi \partial_0(\theta \otimes \psi)d\beta \\
&= \iint_{\Delta(A)} (\theta \otimes \psi)\partial_0\Phi \, d\beta \\
&= \int_{\Delta(A_y)} \left[\int_{\Delta(A_z)} \theta(s_0)\partial_0\Phi(s, s_0)d\beta_z(s_0) \right] \psi(s)d\beta_y(s).
\end{aligned}$$

By the arbitrariness of ψ and use of the density of $\mathcal{D}_{\mathbb{R}}(\Delta(A_y))/\mathbb{R}$ in $L^2_{\mathbb{R}}(\Delta(A_y))/\mathbb{R}$ we deduce

$$\left\langle \frac{du}{ds_0}, \theta \right\rangle = \int_{\Delta(A_z)} \theta(s_0)\partial_0\Phi(\cdot, s_0)d\beta_z(s_0),$$

and that for all $\theta \in \mathcal{D}(\Delta(A_z))$. Hence (7.16) follows. ■

Remark 7.8 In particular if the mapping $u : \Delta(A_z) \to \mathcal{D}_{\mathbb{R}}(\Delta(A_y))/\mathbb{R}$ is given by $u(s_0) = \theta(s_0)\psi$ $(s_0 \in \Delta(A_z))$, where $\psi \in \mathcal{D}_{\mathbb{R}}(\Delta(A_y))/\mathbb{R}$ and $\theta \in \mathcal{D}_{\mathbb{R}}(\Delta(A_z))$, then $\frac{du}{ds_0}(s_0) = \partial_0(\theta \otimes \psi)(\cdot, s_0) = (\frac{d\theta}{ds_0} \otimes \psi)(\cdot, s_0) = \frac{d\theta}{ds_0}(s_0)\psi$.

Next comes a fundamental lemma.

Lemma 7.5 *Suppose the H-algebra A satisfies (P1)–(P2) and further is V^p-proper for some real $p \geq 2$. Let $\mathcal{V}_{\infty y}$ and $\mathcal{W}$ be as in Definition 7.3. Then the following statements hold.*

- ***(i)*** *The space $\mathcal{E}_0^\infty = \mathcal{D}_{\mathbb{R}}(Q) \times (\mathcal{D}_{\mathbb{R}}(Q) \otimes \mathcal{G}([\,_{\mathbb{R}}\mathcal{V}_{\infty y}/\mathbb{R}] \otimes \,_{\mathbb{R}}A_z^\infty))$ is dense in $L^p(0, T; W_0^{1,p}(\Omega)) \times L^p(Q; \mathcal{W})$.*
- ***(ii)*** *$\int_{\Delta(A_z)}[u'(s_0), v(s_0)]d\beta_z(s_0) + \int_{\Delta(A_z)}[v'(s_0), u(s_0)]d\beta_z(s_0) = 0$ for all $u, v \in \mathcal{W}$, where $u' = \frac{du}{ds_0}$ (idem for v).*
- ***(iii)*** *Given any arbitrary bounded sequence $(u_\varepsilon)_{\varepsilon\in E}$ in V^p, where E is a fundamental sequence, let E' and $\mathbf{u} = (u_0, u_1)$ be as in (PR2). Then, as $E' \ni \varepsilon \to 0$,*

$$\int_Q \frac{1}{\varepsilon} u_\varepsilon \psi^\varepsilon \, dxdt \to \iint_{Q\times\Delta(A)} u_1\widehat{\psi} \, dxdtd\beta \tag{7.17}$$

for all $\psi \in \mathcal{D}(Q) \otimes (\mathcal{V}_{\infty y}/\mathbb{C}) \otimes A_z^\infty$.

Proof (i) follows by a routine argument (see the proof of Lemma 7.1) and use of (PR1)(2). Let us turn now to the proof of (ii). It is quite clear that by density, continuity and linearity, the problem reduces to showing (ii) for $u, v \in \mathcal{W}$ given by $u(s_0) = \widehat{\theta}(s_0)f$ and $v(s_0) = \widehat{\varphi}(s_0)g$ for $s_0 \in \Delta(A_z)$, where $f, g \in \mathcal{G}({}_{\mathbb{R}}\mathcal{V}_{\infty y}/\mathbb{R})$ and $\theta, \varphi \in {}_{\mathbb{R}}A_z^\infty, \widehat{\theta} = \mathcal{G}(\theta)$ (idem for φ), $\mathcal{G}$ the Gelfand transformation on A_z and on A_y, as well. Let us consider the mappings u_1 and v_1 of $\mathbb{R}$ into $W_{\mathbb{R}}^{1,p}(\Delta(A_y))/\mathbb{R}$ given by $u_1(z) = \theta(z)f$ and $v_1(z) = \varphi(z)g$ for $z \in \mathbb{R}$. These functions are continuously (or even indefinitely) differentiable, with derivatives $u_1'(z) = \theta'(z)f$ and $v_1'(z) = \varphi'(z)g$ for $z \in \mathbb{R}$. Since $W_{\mathbb{R}}^{1,p}(\Delta(A_y))/\mathbb{R} \subset L_{\mathbb{R}}^2(\Delta(A_y))/\mathbb{R}$ with continuous embedding, we deduce that each of the two functions is continuously differentiable on $\mathbb{R}$ into $L_{\mathbb{R}}^2(\Delta(A_y))/\mathbb{R}$ with the same derivative as above. Hence, by a classical argument, it follows that the function $z \mapsto [u_1(z), v_1(z)]$ (see (7.15)) of $\mathbb{R}$ into $\mathbb{R}$ denoted by $[u_1, v_1]$ is continuously differentiable with derivative given by

$$[u_1, v_1]'(z) = [u_1'(z), v_1(z)] + [v_1'(z), u_1(z)] \quad (z \in \mathbb{R}),$$

or by substitution

$$[u_1, v_1]'(z) = (\theta'(z)\varphi(z) + \theta(z)\varphi'(z))[f, g] \quad (z \in \mathbb{R}),$$

or again, for convenience,

$$[u_1, v_1]' = (\theta'\varphi + \theta\varphi')[f, g].$$

But note that this is an equality between two functions in A_z. Indeed, this is immediate as regards the right-hand side, whereas concerning the left-hand side it suffices to remember that $[u_1, v_1] = \theta\varphi[f, g] \in A_z^\infty$. Thus, $\mathcal{G}$ being viewed as the Gelfand transformation on A_z, we deduce that

$$\frac{d}{ds_0}\mathcal{G}([u_1, v_1]) = \left(\frac{d\widehat{\theta}}{ds_0}\widehat{\varphi} + \widehat{\theta}\frac{d\widehat{\varphi}}{ds_0}\right)[f, g],$$

or in pointwise writing (see Remark 7.8)

$$\begin{aligned}\frac{d}{ds_0}\mathcal{G}([u_1, v_1])(s_0) &= \left[\frac{d\widehat{\theta}}{ds_0}(s_0)f, \widehat{\varphi}(s_0)g\right] + \left[\frac{d\widehat{\varphi}}{ds_0}(s_0)g, \widehat{\theta}(s_0)f\right] \\ &= \left[\frac{du}{ds_0}(s_0), v(s_0)\right] + \left[\frac{dv}{ds_0}(s_0), u(s_0)\right]\end{aligned}$$

and that for all $s_0 \in \Delta(A_z)$. Hence, the desired result follows by integrating over $\Delta(A_z)$ and using Proposition 2.9. Finally, since A is quasi V^p-proper (see Remark 7.7), one sees readily that (iii) follows by Proposition 7.2 and use of (P1) (Sect. 7.2.5). ■

The last preliminary result is the following

Lemma 7.6 *Let the hypotheses be those of Lemma 7.5. Then, the variational problem*

$$\begin{cases} \boldsymbol{u} = (u_0, u_1) \in \mathbb{F}^{1,p}_{\mathcal{W}} = V^p_0 \times L^p(Q; \mathcal{W}) : \\ \int_0^T [u_0'(t), v_0(t)]dt + \int_Q \left\langle \frac{du_1(x,t)}{ds_0}, v_1(x,t) \right\rangle_{\mathcal{H}_p', \mathcal{H}_p} dxdt \\ + \iint_{Q\times\Delta(A)} \boldsymbol{b}(\mathbb{D}\boldsymbol{u}) \cdot \mathbb{D}\boldsymbol{v}\, dxdtd\beta = \int_0^T [f(t), v_0(t)]dt \\ \text{for all } \boldsymbol{v} = (v_0, v_1) \in \mathbb{F}^{1,p}_{\mathcal{W}} \end{cases} \tag{7.18}$$

admits at most one solution.

Proof Thanks to the property

$$\left\langle \frac{dv_1(x,t)}{ds_0}, v_1(x,t) \right\rangle_{\mathcal{H}_p', \mathcal{H}_p} = 0 \text{ a.e. in } (x,t) \in Q,\ v_1 \in L^p(Q; \mathcal{W}) \tag{7.19}$$

which is derived from Lemma 7.5 (ii), the lemma follows by a routine argument whose chief point is the strong monotony pointed out in Proposition 7.1 (see also the proof of Lemma 7.2). ■

We are now ready to prove the main result of this section.

Theorem 7.3 *Let $2 \le p < \infty$. Let $k = 2$. Suppose (7.2) holds and further A satisfies (P1)–(P2) and is V^p-proper. For each $\varepsilon > 0$, let u_ε be the weak solution of (7.1). Then, as $\varepsilon \to 0$, we have (7.10)–(7.12), where $\boldsymbol{u} = (u_0, u_1)$ is the unique solution of (7.18).*

Proof As we mentioned before, with the exception of the way the limit (as $\varepsilon \to 0$) of $\int_0^T [u_\varepsilon'(t), \Phi_\varepsilon(t)]dt$ is computed, the line of argument and the basic convergence results are the same mutatis mutandis as in the proof of Theorem 7.1. Therefore, we will not go too deeply into details.

To begin, let us name $(6.39)_2$ the inequality (6.39) after $\frac{t}{\varepsilon}$ therein has been changed into $\frac{t}{\varepsilon^2}$ (thus, in particular, Φ_ε in $(6.39)_2$ is given by equality (7.4) with ψ_1^ε defined as in (7.3) with $k = 2$). Next, the sequence $(u_\varepsilon)_{\varepsilon>0}$ being norm-bounded in V_0^p, let a fundamental sequence E be freely fixed. Then, thanks to the V^p-properness of A (Definition 7.3), there are some subsequence E' from E and some $\mathbf{u} = (u_0, u_1) \in V_0^p \times L^p(Q; \mathcal{W})$ such that as $E' \ni \varepsilon \to 0$, (7.10)–(7.12) hold true with further $u_\varepsilon \to u_0$ in $L^2(Q)$. This being so, let $\Phi = (\psi_0, \widehat{\psi}_1) \in \mathcal{E}_0^\infty$, where $\widehat{\psi}_1 = \mathcal{G} \circ \psi_1$ with $\psi_1 \in \mathcal{D}_{\mathbb{R}}(Q) \otimes [\,{}_{\mathbb{R}}\mathcal{V}_{\infty y}/\mathbb{R}] \otimes {}_{\mathbb{R}}A_z^\infty$, of course (see Sect. 7.2.2). Define Φ_ε as pointed out above. The aim is to pass to the limit in $(6.39)_2$ as $E' \ni \varepsilon \to 0$. Clearly the limit of the first integral on the right is available (see, e.g., Sect. 7.2.4), since it does not depend on the value of the parameter $k > 0$. The same is true mutatis mutandis of the limit of the third integral (see the proof of Theorem 7.1) on the same side, the chief point being here Corollary 7.1 (joined with (7.12), of course). Therefore, before we can pass to the limit in $(6.39)_2$, only the limit of the second integral on the right needs calculating. Clearly

$$\int_0^T [u'_\varepsilon(t), \Phi_\varepsilon(t)]dt = -\int_Q u_\varepsilon \frac{\partial \psi_0}{\partial t} dxdt - \varepsilon \int_Q u_\varepsilon \left(\frac{\partial \psi_1}{\partial t}\right)^\varepsilon dxdt$$
$$- \int_Q \frac{1}{\varepsilon} u_\varepsilon \left(\frac{\partial \psi_1}{\partial z}\right)^\varepsilon dxdt$$

with

$$\int_Q u_\varepsilon \frac{\partial \psi_0}{\partial t} dxdt \to \int_Q u_0 \frac{\partial \psi_0}{\partial t} dxdt$$

and

$$\varepsilon \int_Q u_\varepsilon \left(\frac{\partial \psi_1}{\partial t}\right)^\varepsilon dxdt \to 0$$

as $E' \ni \varepsilon \to 0$. On the other hand, by (7.17) we have

$$\int_Q \frac{1}{\varepsilon} u_\varepsilon \left(\frac{\partial \psi_1}{\partial z}\right)^\varepsilon dxdt \to \int_{Q\times\Delta(A)} u_1 \partial_0 \widehat{\psi}_1 \, dxdtd\beta$$

as $E' \ni \varepsilon \to 0$.

Let us concentrate on computing the latter limit. First, we shall identify any $\Phi \in \mathcal{D}_\mathbb{R}(\Delta(A_z)) \otimes [\mathcal{D}_\mathbb{R}(\Delta(A_y))/\mathbb{R}]$ with the function $s_0 \mapsto \Phi(\cdot, s_0)$ on $\Delta(A_z)$ which lies, say in $\mathcal{C}(\Delta(A_z); \mathcal{D}_\mathbb{R}(\Delta(A_y))/\mathbb{R})$. Consequently, according to Lemma 7.4, we have $\frac{d\Phi}{ds_0} = \partial_0 \Phi$. With this in mind, for fixed (x, t) we have, step by step,

$$\int_{\Delta(A)} u_1(x,t)\partial_0\widehat{\psi}_1(x,t)d\beta$$
$$= \int_{\Delta(A_z)} \left(\int_{\Delta(A_y)} u_1(x,t)\partial_0\widehat{\psi}_1(x,t)d\beta_y \right) d\beta_z$$
$$= \int_{\Delta(A_z)} \left[\frac{d\widehat{\psi}_1(x,t)}{ds_0}(s_0), u_1(x,t,s_0) \right] d\beta_z(s_0)$$
$$= -\int_{\Delta(A_z)} \left[\frac{du_1(x,t)}{ds_0}(s_0), \widehat{\psi}_1(x,t,s_0) \right] d\beta_z(s_0)$$
$$= -\left\langle \frac{du_1(x,t)}{ds_0}, \widehat{\psi}_1(x,t) \right\rangle_{\mathcal{H}'_p, \mathcal{H}_p},$$

where the second and third equalities have put (7.15) and Lemma 7.5, respectively, to account. Therefore, by taking the $\liminf_{E'\ni\varepsilon\to 0}$ of both sides of $(6.39)_2$ (see (6.42) and (6.43)) and using the density result of Lemma 7.5 we get

$$0 \leq \int_0^T [f(t) - u_0'(t), u_0(t) - w_0(t)]dt + \int_Q \left\langle \frac{du_1(x,t)}{ds_0}, w_1(x,t) \right\rangle_{\mathcal{H}_p', \mathcal{H}_p} dxdt$$

$$- \int\int_{Q\times\Delta(A)} \mathbf{b}(\mathbb{D}\mathbf{w}) \cdot \mathbb{D}(\mathbf{u} - \mathbf{w}) dxdtd\beta$$

and that for all $\mathbf{w} = (w_0, w_1) \in \mathbb{F}_{\mathcal{W}}^{1,p}$. Taking in particular $\mathbf{w} = \mathbf{u} - r\mathbf{v}$ with $r > 0$ and $\mathbf{v} = (v_0, v_1) \in \mathbb{F}_{\mathcal{W}}^{1,p}$, dividing both sides of the resultant inequality by r and using (7.19) with $v_1 = u_1$, one arrives at once at the variational equation in (7.18) by letting r tend to zero. Hence the theorem follows by Lemma 7.6. ■

7.2.7 *Homogenization of (7.1) for* $k > 2$

First of all, it is worth recalling that the preliminary Sects. 7.2.1–7.2.3 are valid here, since they do not depend on any particular value of k.

In what follows, the algebra $\mathcal{C}_{\mathbb{R}}(\Delta(A_y))$ is justifiably identified with the subalgebra of $\mathcal{C}_{\mathbb{R}}(\Delta(A))$ made up of all those functions that do not depend on the variable $s_0 \in \Delta(A_z)$. Having made this point, let us fix any arbitrary index $1 \leq i \leq N$, and let us consider the transformation $b_i : \mathcal{C}_{\mathbb{R}}(\Delta(A))^N \to L^\infty(\Delta(A))$ introduced in Sect. 7.2.1. For each $\boldsymbol{\Psi} \in \mathcal{C}_{\mathbb{R}}(\Delta(A_y))^N$, let

$$\widetilde{b}_i(\boldsymbol{\Psi}) = \int_{\Delta(A_z)} b_i(\boldsymbol{\Psi}) d\beta_z,$$

which yields a transformation $\widetilde{b}_i : \mathcal{C}_{\mathbb{R}}(\Delta(A_y))^N \to L^\infty(\Delta(A_y))$. Finally, put

$$\widetilde{\mathbf{b}}(\boldsymbol{\Phi}) = (\widetilde{b}_i \circ \boldsymbol{\Phi})_{1\leq i\leq N} \text{ for } \boldsymbol{\Phi} \in \mathcal{C}_{\mathbb{R}}(\overline{Q}; \mathcal{C}_{\mathbb{R}}(\Delta(A_y))^N).$$

The verification of the following remark is a simple exercise left to the reader.

Remark 7.9 Proposition 7.1 and Corollary 7.1 remain valid when A is changed into A_y, $A_{\mathbb{R}}$ is changed into ${}_{\mathbb{R}}A_y$ and $\mathbf{b}$ (resp. b_i) is changed into $\widetilde{\mathbf{b}}$ (resp. $\widetilde{b}_i$).

As a direct consequence of this remark, there is the following lemma.

Lemma 7.7 *The variational problem*

$$\begin{cases} \boldsymbol{u} = (u_0, u_1) \in \mathbb{F}_0^{1,p} = V_0^p \times L^p(Q; W_{\mathbb{R}}^{1,p}(\Delta(A_y))/\mathbb{R}) : \\ \int_0^T [u_0'(t), v_0(t)]dt + \iint_{Q\times\Delta(A_y)} \widetilde{\boldsymbol{b}}(\mathbb{D}\boldsymbol{u}) \cdot \mathbb{D}\boldsymbol{v}\, dxdtd\beta_y \\ = \int_0^T [f(t), v_0(t)]dt \text{ for all } \boldsymbol{v} = (v_0, v_1) \in \mathbb{F}_0^{1,p} \end{cases} \tag{7.20}$$

has at most one solution.

Before we can prove a homogenization theorem for (7.1) under the hypothesis $k > 2$, we require one further property of the H-algebra A_z.

(P3) Given a Banach space B as in Sect. 7.2.6, if $u \in L^p(\Delta(A_z); B)$ $(1 \leq p < \infty)$ satisfies $\frac{du}{ds_0} = 0$ in the sense of $\mathcal{D}'(\Delta(A_z); B)$, then u is constant, i.e., there is some $\xi \in B$ such that $u(s_0) = \xi$ for a.e. $s_0 \in \Delta(A_z)$.

We will later give a few examples of H-algebras satisfying (P3).

We are now in a position to prove the claimed result.

Theorem 7.4 *Let $2 \leq p < \infty$. Let $k > 2$. Suppose (7.2) holds, and further A is quasi V^p-proper and has properties (P1), (P2) and (P3). For any given real $\varepsilon > 0$, let u_ε be the solution of (7.1). Then, as $E' \ni \varepsilon \to 0$, we have (7.10)–(7.12), where $\boldsymbol{u} = (u_0, u_1)$ is uniquely defined by (7.20).*

Proof Given a fundamental sequence E, it is clear that there are a subsequence E' from E and some $\mathbf{u} = (u_0, u_1)$ in $V_0^p \times L^p(Q; \mathcal{H}_p)$ with $\mathcal{H}_p = L^p(\Delta(A_z); W_{\mathbb{R}}^{1,p}(\Delta(A_y))/\mathbb{R})$, such that (7.10)–(7.12) hold as $E' \ni \varepsilon \to 0$, with further $u_\varepsilon \to u_0$ in $L^2(Q)$. Therefore, the whole problem reduces to verifying that $\mathbf{u}$ satisfies (7.20) (as the reader can see, this proof is being framed on the lines of that of Theorem 7.1 and so certain details are not worth repeating). To start with, let us show that $u_1(x, t)$ lies in $W_\#^{1,p}(\Delta(A_y); \mathbb{R}) = W_{\mathbb{R}}^{1,p}(\Delta(A_y))/\mathbb{R}$ (i.e., $u_1(x, t)$ does not depend on $s_0 \in \Delta(A_z)$) for a.e. $(x, t) \in Q$. To do this, let us consider a real mapping ψ on $Q \times \mathbb{R}_y^N \times \mathbb{R}_z$ given by

$$\psi(x, t, y, z) = \varphi(x, t)\chi(y)\theta(z)$$

with $\varphi \in \mathcal{D}_{\mathbb{R}}(Q)$, $\chi \in {}_{\mathbb{R}}\mathcal{V}_{\infty y}/\mathbb{R}$ (see (P1)) and $\theta \in {}_{\mathbb{R}}A_z^\infty$. Put

$$\psi_\varepsilon(x, t) = \varepsilon^{k-1}\psi\left(x, t, \frac{x}{\varepsilon}, \frac{t}{\varepsilon^k}\right), \quad (x, t) \in Q,$$

which defines a function $\psi_\varepsilon \in \mathcal{D}_{\mathbb{R}}(Q)$. Then

$$\begin{aligned}&\int_0^T [u'_\varepsilon(t), \psi_\varepsilon(\cdot, t)]dt + \int_Q \mathbf{a}^\varepsilon(\cdot, \cdot, Du_\varepsilon) \cdot D\psi_\varepsilon \, dxdt \\ &= \int_0^T [f(t), \psi_\varepsilon(\cdot, t)]dt.\end{aligned} \tag{7.21}$$

The sequence $\{\mathbf{a}^\varepsilon(\cdot, \cdot, Du_\varepsilon)\}_{\varepsilon>0}$ being bounded in $L^{p'}(Q)^N$ (recall that the estimates in the beginning of the proof of Theorem 6.2 remain when $\frac{t}{\varepsilon}$ is changed into $\frac{t}{\varepsilon^k}$), one sees readily that the second integral on the left of (7.21) goes to zero as $\varepsilon \to 0$. Clearly the same is true of the right-hand side of the same equality. Let us compute the first term on the left of (7.21), which is evidently destined to tend to zero with ε. An elementary calculation gives

$$\int_0^T [u'_\varepsilon(t), \psi_\varepsilon(\cdot, t)]dt = -\varepsilon^k \int_Q \frac{1}{\varepsilon} u_\varepsilon \left(\frac{\partial \psi}{\partial t}\right)^\varepsilon dxdt - \int_Q \frac{1}{\varepsilon} u_\varepsilon \left(\frac{\partial \psi}{\partial z}\right)^\varepsilon dxdt.$$

With a view to passing to the limit on the right-hand side let us note that, thanks to Proposition 7.2, (7.17) remains valid even when A is not V^p-proper provided it satisfies (P1). Hence, on the right-hand side of the latter equality, as $E' \ni \varepsilon \to 0$, the first term goes to zero whereas the next one tends to

$$-\iint_{Q \times \Delta(A)} u_1 \partial_0 \widehat{\psi} \, dxdtd\beta,$$

which is zero as claimed. In view of the arbitrariness of φ and χ, it follows

$$\int_{\Delta(A_z)} u_1(x, t) \frac{d\widehat{\theta}}{ds_0} d\beta_z = 0 \text{ for a.e. } (x, t) \in Q$$

and that for any $\theta \in {}_{\mathbb{R}}A_z^\infty$, the density of ${}_{\mathbb{R}}\mathcal{V}_{\infty y}/\mathbb{R}$ in $W_{\mathbb{R}}^{1,p}(\Delta(A_y))/\mathbb{R}$ (this follows from property (1) of (P1)) having been put to account. Therefore, for a.e. (x, t), we have $\frac{d}{ds_0} u_1(x, t) = 0$ in the sense of distributions on $\Delta(A_z)$ with values in $W_{\mathbb{R}}^{1,p}(\Delta(A_y))/\mathbb{R}$. Hence $u_1(x, t) \in W_{\mathbb{R}}^{1,p}(\Delta(A_y))/\mathbb{R}$ for a.e. $(x, t) \in Q$, according to (P3).

Having done this, and noting that the space $\mathcal{D}_{\mathbb{R}}(Q) \times [\mathcal{D}_{\mathbb{R}}(Q) \otimes \mathcal{G}({}_{\mathbb{R}}\mathcal{V}_{\infty y}/\mathbb{R})]$ is dense in $\mathbb{F}_0^{1,p}$, we see that the same line of reasoning as followed in proving Theorem 7.1 shows at once that $\mathbf{u} = (u_0, u_1)$ satisfies the variational equation in (7.20). Hence the theorem follows by Lemma 7.7. ■

7.3 Some Examples of Concrete Homogenization Problems for (7.1)

Our main purpose in this section is to work out a few concrete homogenization problems for the initial-boundary value problem (7.1). Indeed, in practice the statement of a homogenization problem makes no mention of the notion of a homogenization algebra, still less of such abstract assumptions as (7.2). A concrete homogenization problem for, say (7.1) is set under a so-called structure hypothesis on $\mathbf{a}(y, z, \mu)$ (for fixed μ), e.g., the periodicity hypothesis, or the almost periodicity hypothesis, on both y and z. Anyhow, a structure hypothesis is stated in a natural fashion as opposed to the abstract assumption (7.2), and its form depends on the underlying physical phenomenon under examination.

As might be expected, our approach will consist in reducing the problem under consideration to the abstract setting based on (7.2) and then apply an appropriate result among those that we achieved earlier in Sect. 7.2. Before we can approach this, however, we need a few preliminaries.

7.3.1 Preliminaries

Let the basic notation and notions be as before. In addition to the H-algebras A_y, A_z and $A = \overline{A_y \otimes A_z}$ (the bar denotes here the closure in $\mathcal{B}(\mathbb{R}^{N+1})$), we consider here an H-algebra A_{2y} on $\mathbb{R}^N$ (under the same action as before) admitting A_{2y}^∞ as a dense subalgebra, and we put $A_2 = \overline{A_{2y} \otimes A_z}$. All we said in Sect. 7.2.1 about the triple $\{A_y, A_z, A\}$ carries over mutatis mutandis to the triple $\{A_{2y}, A_z, A_2\}$. In particular, we have $\Delta(A_2) = \Delta(A_{2y}) \times \Delta(A_z)$. The basic Radon measure on $\Delta(A_{2y})$ (resp. $\Delta(A_2)$) is the so-called M-measure for A_{2y} (resp. A_2) denoted by β_{2y} (resp. β_2), and we have $\beta_2 = \beta_{2y} \otimes \beta_z$. Finally, the Gelfand transformation on any H-algebra considered here is denoted indifferently by $\mathcal{G}$.

Now, let $1 < p < \infty$. We assume that the pair $\{A_y, A_{2y}\}$ satisfies the following hypothesis.

(H) *There exist an isometric isomorphism L_y of $L^p(\Delta(A_y))$ onto $L^p(\Delta(A_{2y}))$ mapping $L^p_{\mathbb{R}}(\Delta(A_y))$ onto $L^p_{\mathbb{R}}(\Delta(A_{2y}))$, a dense vector subspace $\mathcal{V}_y$ of A_y, a surjective linear mapping $l_y : \mathcal{V}_y \to A_{2y}$ sending ${}_{\mathbb{R}}\mathcal{V}_y$ onto ${}_{\mathbb{R}}A_{2y}$, and a vector subspace $\mathcal{V}_{\infty y}$ of $A_y^\infty \cap \mathcal{V}_y$ preserving complex conjugation, such that:*

(1) $L_y(\mathcal{G}(v)) = \mathcal{G}(l_y(v)) \ \ (v \in \mathcal{V}_y)$

(2) $L_y(\widehat{v}u) = L_y(\widehat{v})L_y(u)$ (with $\widehat{v} = \mathcal{G}(v)$) for $u \in L^p(\Delta(A_y))$, $v \in \mathcal{V}_y$

(3) $(v - l_y v)^\varepsilon \to 0$ in $L^p_{loc}(\mathbb{R}^N_x)$ as $\varepsilon \to 0 \ \ (v \in \mathcal{V}_y)$

(4) If $v \in \mathcal{V}_{\infty y}$, then $D_y^\alpha v \in \mathcal{V}_y \ \ (\alpha \in \mathbb{N}^N)$

(5) l_y maps $\mathcal{V}_{\infty y}$ onto A_{2y}^{∞}

(6) $D_y^{\alpha}(l_y v) = l_y(D_y^{\alpha} v)$ $(v \in \mathcal{V}_{\infty y}, \alpha \in \mathbb{N}^N)$.

The aim is to show that, under hypothesis (H), a number of interesting properties of A_{2y} (resp. A_2) carry over to A_y (resp. A), which enables us in the sequel to tackle outstanding concrete homogenization problems for (7.1). The symbol J_y throughout denotes the canonical mapping of $W^{1,p}(\Delta(A_y))/\mathbb{C}$ into its separated completion, $W_{\#}^{1,p}(\Delta(A_y))$, whereas that of $W^{1,p}(\Delta(A_{2y}))/\mathbb{C}$ into $W_{\#}^{1,p}(\Delta(A_{2y}))$ is denoted by J_{2y}.

The following lemma is a collection of basic results whose proofs can be either merely found among the preliminaries in Sect. 4.2.2 (see in particular Lemmas 4.1 and 4.2) or easily deduced from the latter.

Lemma 7.8 *Suppose $\mathcal{D}(\Delta(A_{2y}))$ is dense in $W^{1,p}(\Delta(A_{2y}))$. Then, the following statements are true.*

- ***(i)*** *The operator L_y carries $W^{1,p}(\Delta(A_y))$ (resp. $W_{\mathbb{R}}^{1,p}(\Delta(A_y))$) isometrically over $W^{1,p}(\Delta(A_{2y}))$ (resp. $W_{\mathbb{R}}^{1,p}(\Delta(A_{2y}))$), and $W^{1,p}(\Delta(A_y))/\mathbb{C}$ (resp. $W_{\mathbb{R}}^{1,p}(\Delta(A_y))/\mathbb{R}$) isometrically over $W^{1,p}(\Delta(A_{2y}))/\mathbb{C}$(resp. $W_{\mathbb{R}}^{1,p}(\Delta(A_{2y}))/\mathbb{R}$). Furthermore, $\partial_j(L_y(u)) = L_y(\partial_j u)$ $(1 \leq j \leq N)$ for all $u \in W^{1,p}(\Delta(A_y))$.*
- ***(ii)*** *There exists an isometric isomorphism $L_y^{\#}$ of $W_{\#}^{1,p}(\Delta(A_y))$ onto $W_{\#}^{1,p}(\Delta(A_{2y}))$ mapping $W_{\#}^{1,p}(\Delta(A_y);\mathbb{R})$ onto $W_{\#}^{1,p}(\Delta(A_{2y});\mathbb{R})$, and such that $L_y^{\#}(J_y u) = J_{2y}(L_y(u))$ for $u \in W^{1,p}(\Delta(A_y))/\mathbb{C}$, and $\partial_j L_y^{\#}(u) = L_y(\partial_j u)$ for $u \in W_{\#}^{1,p}(\Delta(A_y))$ $(1 \leq j \leq N)$.*
- ***(iii)*** *$\mathcal{G}(\mathcal{V}_{\infty y})$ [hence $\mathcal{D}(\Delta(A_y))$] is dense in $W^{1,p}(\Delta(A_y))$.*

Remark 7.10 Actually, only part of Lemma 7.8 is proved in Chap. 4, namely that part consisting of those of the results that are relevant to vector spaces over $\mathbb{C}$. But it is an elementary exercise (left to the reader) to deduce the rest; see Chap. 4, Remark 4.7.

Now, for $u \in L^p(\Delta(A_z); L^p(\Delta(A_y)))$, let

$$L(u) = L_y \circ u \text{ (usual composition).}$$

In other words, $L(u)$ denotes the function $s_0 \mapsto L_y(u(s_0))$ of $\Delta(A_z)$ into $L^p(\Delta(A_{2y}))$, which clearly lies in $L^p(\Delta(A_z); L^p(\Delta(A_{2y}))) = L^p(\Delta(A_2))$. This yields an isometric isomorphism L of $L^p(\Delta(A_z); L^p(\Delta(A_y)))$ onto $L^p(\Delta(A_z); L^p(\Delta(A_{2y})))$ that carries $L^p(\Delta(A_z); L_{\mathbb{R}}^p(\Delta(A_y)))$ isometrically over $L^p(\Delta(A_z); L_{\mathbb{R}}^p(\Delta(A_{2y})))$, $L^p(\Delta(A_z); L^p(\Delta(A_y))/\mathbb{C})$ isometrically over $L^p(\Delta(A_z); L^p(\Delta(A_{2y}))/\mathbb{C})$, and $L^p(\Delta(A_z); L_{\mathbb{R}}^p(\Delta(A_y))/\mathbb{R})$ isometrically over $L^p(\Delta(A_z); L_{\mathbb{R}}^p(\Delta(A_{2y}))/\mathbb{R})$.

It is worth recalling that the last two preceding statements follow by the equality

$$\int_{\Delta(A_{2y})} L_y(u)\, d\beta_{2y} = \int_{\Delta(A_y)} u\, d\beta_y \quad (u \in L^p(\Delta(A_y))) \tag{7.22}$$

which is shown in Chap. 4.

Lemma 7.9 *Suppose $\mathcal{D}(\Delta(A_{2y}))$ is dense in $W^{1,p}(\Delta(A_{2y}))$. Then, the following statements are true:*

(i) *The operator L carries $L^p(\Delta(A_z); W^{1,p}(\Delta(A_y)))$ isometrically over $L^p(\Delta(A_z); W^{1,p}(\Delta(A_{2y})))$, $L^p(\Delta(A_z); W^{1,p}_{\mathbb{R}}(\Delta(A_y)))$ isometrically over $L^p(\Delta(A_z); W^{1,p}_{\mathbb{R}}(\Delta(A_{2y})))$, $L^p(\Delta(A_z); W^{1,p}(\Delta(A_y))/\mathbb{C})$ isometrically over $L^p(\Delta(A_z); W^{1,p}(\Delta(A_{2y}))/\mathbb{C})$ and finally, $L^p(\Delta(A_z); W^{1,p}_{\mathbb{R}}(\Delta(A_y))/\mathbb{R})$ isometrically over $L^p(\Delta(A_z); W^{1,p}_{\mathbb{R}}(\Delta(A_{2y}))/\mathbb{R})$.*

(ii) *For $u \in L^p(\Delta(A_z); W^{1,p}(\Delta(A_y)))$, we have*

$$\partial_j L(u) = L(\partial_j u) \quad (1 \le j \le N), \tag{7.23}$$

where the partial derivative on the left (resp. right) is computed in the distribution sense on $\Delta(A_{2y})$ (resp. $\Delta(A_y)$).

Proof This follows readily by Lemma 7.8. ■

Remark 7.11 In (7.23), $\partial_j u$ stands for the function $s_0 \mapsto \partial_j u(s_0)$ [distribution derivative of $u(s_0) \in W^{1,p}(\Delta(A_y))$] of $\Delta(A_z)$ into $L^p(\Delta(A_y))$, which lies in $L^p(\Delta(A_z); L^p(\Delta(A_y)))$; whereas $\partial_j L(u)$ stands for the function $s_0 \mapsto [\partial_j L(u)](s_0) = \partial_j L_y(u(s_0))$ of $\Delta(A_z)$ into $L^p(\Delta(A_{2y}))$, which lies in $L^p(\Delta(A_z); L^p(\Delta(A_{2y})))$.

Another result worth pointing out is the following

Lemma 7.10 *Suppose $\mathcal{D}(\Delta(A_{2y}))$ is dense in $W^{1,p}(\Delta(A_{2y}))$. Then there exists an isometric isomorphism $L^\#$ of $L^p(\Delta(A_z); W^{1,p}_\#(\Delta(A_y)))$ onto $L^p(\Delta(A_z); W^{1,p}_\#(\Delta(A_{2y})))$ mapping $L^p(\Delta(A_z); W^{1,p}_\#(\Delta(A_y); \mathbb{R}))$ onto $L^p(\Delta(A_z); W^{1,p}_\#(\Delta(A_{2y}); \mathbb{R}))$, and such that*

$$L^\#(J(u)) = J_2(L(u)) \text{ for } u \in L^p(\Delta(A_z); W^{1,p}(\Delta(A_y))/\mathbb{C})$$

and

$$\partial_j L^\#(u) = L(\partial_j u) \quad (1 \le j \le N) \text{ for } u \in L^p(\Delta(A_z); W^{1,p}_\#(\Delta(A_y))),$$

where $J(u) = J_y \circ u$ for $u \in L^p(\Delta(A_z); W^{1,p}(\Delta(A_y))/\mathbb{C})$ and $J_2(v) = J_{2y} \circ v$ for $v \in L^p(\Delta(A_z); W^{1,p}(\Delta(A_{2y}))/\mathbb{C})$.

Proof Let $L^{\#}(u) = L_y^{\#} \circ u$ for $u \in L^p(\Delta(A_z); W_{\#}^{1,p}(\Delta(A_y)))$. Clearly $L^{\#}(u) \in L^p(\Delta(A_z); W_{\#}^{1,p}(\Delta(A_{2y})))$, and the mapping $L^{\#}$ thus defined meets the requirements of the lemma, as is straightforward by part (ii) of Lemma 7.8. ■

Before we can come to the heart of the matter we still need a few notation and one further lemma. Throughout the rest of this section it is assumed without further ado that $\mathcal{D}(\Delta(A_{2y}))$ is dense in $W^{1,p}(\Delta(A_{2y}))$.

To start with, let

$$L_Q(u) = L \circ u \text{ for } u \in L^p(Q; L^p(\Delta(A_z); L^p(\Delta(A_y)))).$$

It is immediate by Lemma 7.9 that the transformation L_Q thus defined is an isometric isomorphism of $L^p(Q; L^p(\Delta(A_z); L^p(\Delta(A_y)))) = L^p(Q \times \Delta(A))$ onto $L^p(Q; L^p(\Delta(A_z); L^p(\Delta(A_{2y})))) = L^p(Q \times \Delta(A_2))$.

Now, recall that

$$\mathcal{H}_p = L^p(\Delta(A_z); W_{\#}^{1,p}(\Delta(A_y)); \mathbb{R}) \text{ (see Sect. 7.2.2).}$$

Likewise

$$\mathcal{H}_{2p} = L^p(\Delta(A_z); W_{\#}^{1,p}(\Delta(A_{2y})); \mathbb{R}).$$

By letting

$$L_Q^{\#}(u) = L^{\#} \circ u \quad \text{for } u \in L^p(Q; L^p(\Delta(A_z); W_{\#}^{1,p}(\Delta(A_y))))$$

one defines a transformation $L_Q^{\#}$ which, according to Lemma 7.10, is an isometric isomorphism of $L^p(Q; L^p(\Delta(A_z); W_{\#}^{1,p}(\Delta(A_y))))$ onto $L^p(Q; L^p(\Delta(A_z); W_{\#}^{1,p}(\Delta(A_{2y}))))$ carrying $L^p(Q; \mathcal{H}_p)$ isometrically over $L^p(Q; \mathcal{H}_{2p})$ and further satisfying

$$\partial_j L_Q^{\#}(u) = L_Q(\partial_j u) \ (1 \le j \le N) \text{ for } u \in L^p(Q; L^p(\Delta(A_z); W_{\#}^{1,p}(\Delta(A_y))))$$

and

$$L_Q^{\#}(J_Q(u)) = J_{2Q}(L_Q(u)) \text{ for } u \in L^p(Q; L^p(\Delta(A_z); W^{1,p}(\Delta(A_y))/\mathbb{C})),$$

where $J_Q(u) = J \circ u$ for $u \in L^p(Q; L^p(\Delta(A_z); W^{1,p}(\Delta(A_y))/\mathbb{C}))$ and $J_{2Q}(v) = J_2 \circ v$ for $v \in L^p(Q; L^p(\Delta(A_z); W^{1,p}(\Delta(A_{2y}))/\mathbb{C}))$.

At the present time, let

$$l(w) = l_y \circ w \in A_z \otimes A_{2y} \quad \text{for } w \in A_z \otimes \mathcal{V}_y,$$

where w **is viewed as a function of** $\mathbb{R}$ **into** $\mathcal{V}_y$ (this is similar to the approach followed in Sect. 7.2.3). From parts (1) and (3) of hypothesis (H) it follows at once that $L(\mathcal{G}(w)) = \mathcal{G}(l(w))$ and

$$(w - lw)^{\varepsilon} \to 0 \text{ in } L^p_{loc}(\mathbb{R}^{N+1}_{x,t}) \text{ as } \varepsilon \to 0$$

for all $w \in A_z \otimes \mathcal{V}_y$. Whence, letting

$$l_Q(\psi) = l \circ \psi \in \mathcal{K}(Q) \otimes A_z \otimes A_{2y} \text{ for } \psi \in \mathcal{K}(Q) \otimes A_z \otimes \mathcal{V}_y,$$

where ψ **is viewed as a function of** Q **into** $A_z \otimes \mathcal{V}_y$, we deduce that

$$L_Q(\mathcal{G}_Q(\psi)) = \mathcal{G}_Q(l_Q\psi) \in \mathcal{K}(Q) \otimes \mathcal{C}(\Delta(A_z)) \otimes \mathcal{G}(\mathcal{V}_y) \tag{7.24}$$

and

$$(\psi - l_Q\psi)^{\varepsilon} \to 0 \text{ in } L^p(Q) \text{ as } \varepsilon \to 0 \tag{7.25}$$

for all $\psi \in \mathcal{K}(Q) \otimes A_z \otimes \mathcal{V}_y$, where $\mathcal{G}_Q(\psi) = \mathcal{G} \circ \psi$ (we usually put $\widehat{\psi} = \mathcal{G}_Q(\psi)$ for $\psi \in L^1_{loc}(Q; A)$).

Remark 7.12 Considering that in the sequel we shall be dealing with two leading H-algebras on $\mathbb{R}^{N+1}$, namely A and A_2, for the sake of clearness the term «Σ_2-convergence» (resp. «Σ_2-reflexivity») shall be relative to A_2, the same wording using Σ (instead of Σ_2) being reserved for the H-algebra A.

Finally comes the last preliminary lemma.

Lemma 7.11 *Let $(u_\varepsilon)_{\varepsilon\in E}$ be a sequence in $L^p(Q)$ ($1 < p < \infty$), E being a fundamental sequence. Suppose as $E \ni \varepsilon \to 0$, $u_\varepsilon \to v_0$ in $L^p(Q)$-weak Σ_2. Then, $u_\varepsilon \to u_0 = L_Q^{-1}(v_0)$ in $L^p(Q)$-weak Σ.*

Proof In view of the density of $\mathcal{K}(Q) \otimes A_z \otimes \mathcal{V}_y$ in $L^{p'}(Q; A)$ we see readily by (7.25) that the whole problem reduces to showing that as $E \ni \varepsilon \to 0$, one has

$$\int_Q u_\varepsilon (l_Q\psi)^{\varepsilon} dxdt \to \int\int_{Q\times\Delta(A)} u_0 \mathcal{G}_Q(\psi) dxdtd\beta$$

for all $\psi \in \mathcal{K}(Q) \otimes A_z \otimes \mathcal{V}_y$. But this follows by (7.24) and use of the equality $L_Q(\widehat{\psi} u_0) = L_Q(\widehat{\psi}) L_Q(u_0)$ [derived from (2) in (H)] together with (7.22) (see the proof of Lemma 4.3). ∎

We are now in a position to prove a series of results whose interest will come to light in the next sections.

Theorem 7.5 *Let hypothesis (H) be satisfied. Suppose the H-algebra A_2 is quasi V^p-proper for some $1 < p < \infty$. Then A is quasi V^p-proper too.*

Proof In view of part (iii) of Lemma 7.8, only property (QP2) remains to be verified. So, given a norm-bounded sequence $(u_\varepsilon)_{\varepsilon\in E}$ in V^p (with $1 < p < \infty$ and E fundamental), appeal to the quasi V^p-properness of A_2 yields some subsequence E' from E and some $\mathbf{v} = (u_0, v_1) \in V^p \times L^p(Q; \mathcal{H}_{2p})$ such that as $E' \ni \varepsilon \to 0$,

$$u_\varepsilon \to u_0 \text{ in } V^p\text{-weak} \tag{7.26}$$

$$\frac{\partial u_\varepsilon}{\partial x_j} \to \frac{\partial u_0}{\partial x_j} + \partial_j v_1 \text{ in } L^p(Q)\text{-weak } \Sigma_2 \;\; (1 \le j \le N). \tag{7.27}$$

It follows by Lemma 7.11 that

$$\frac{\partial u_\varepsilon}{\partial x_j} \to L_Q^{-1}\left(\frac{\partial u_0}{\partial x_j}\right) + L_Q^{-1}(\partial_j v_1) \text{ in } L^p(Q)\text{-weak } \Sigma \;\; (1 \le j \le N).$$

But $L_Q^{-1}\left(\frac{\partial u_0}{\partial x_j}\right) = \frac{\partial u_0}{\partial x_j}$, since $\frac{\partial u_0}{\partial x_j}$ does not depend on $(s, s_0) \in \Delta(A_2)$. On the other hand, let $u_1 \in L^p(Q; \mathcal{H}_p)$ with $L_Q^{\#}(u_1) = v_1$. As seen earlier, u_1 is uniquely determined and further $L_Q^{-1}(\partial_j v_1) = \partial_j u_1$. Hence the theorem follows. ■

Proposition 7.5 *Let hypothesis (H) be satisfied, and $\mathcal{D}(\Delta(A_{2y}))$ be dense in $W^{1,p}(\Delta(A_{2y}))$.*

(i) *Suppose the H-algebra A_{2y} verifies*

(P1)' *For each $f \in A_{2y}^\infty/\mathbb{C} = \{\varphi | \varphi \in A_{2y}^\infty, M(\varphi) = 0\}$, there is some $g \in A_{2y}^\infty/\mathbb{C}$ such that $f = \Delta_y g$.*

Then A_y verifies property (P1) (Sect. 7.2.5).

(ii) *Suppose A_{2y} verifies*

(P2)' *There is a constant $c > 0$ such that $\|u\|_{L^p(\Delta(A_{2y}))} \le c\|u\|_{W^{1,p}(\Delta(A_{2y}))/\mathbb{C}}$ for $u \in W^{1,p}(\Delta(A_{2y}))/\mathbb{C}$.*

Then A_y verifies the analogous property (P2) (Sect. 7.2.6).

(iii) *If A_{2y} is p-harmonic, then the following hold:*

(7) To any arbitrary given $\boldsymbol{f} = (f_j) \in L^p(\Delta(A_y))^N$ satisfying $\sum_{j=1}^N \int_{\Delta(A_y)} f_j \widehat{\Phi}_j \, d\beta_y = 0$ for all $\mathbf{\Phi} = (\Phi_j) \in \mathcal{U}_{\infty y} = \{\boldsymbol{u} | \boldsymbol{u} \in (\mathcal{V}_{\infty y})^N, \operatorname{div}_y \boldsymbol{u} = 0\}$, there is attached a unique $q_y \in W_\#^{1,p}(\Delta(A_y))$ such that $f_j = \partial_j q_y$ $(1 \le j \le N)$.

(8) A_y is p-harmonic.

Proof Part (ii) is immediate by Lemma 7.8 (part (i)) and use of the fact that L_y is an isometric isomorphism of $L^p(\Delta(A_y))$ onto $L^p(\Delta(A_{2y}))$. Concerning part (i), only part (2) of property (P1) needs proving, since we know from Lemma 7.8 that $\mathcal{G}(\mathcal{V}_{\infty y})$ is dense in $W^{1,p}(\Delta(A_y))$. So, fix freely some $\psi \in \mathcal{V}_{\infty y}/\mathbb{C}$. Put $f = l_y(\psi) \in A^\infty_{2y}$. Then $\widehat{f} = L_y(\widehat{\psi})$, where $\widehat{\psi} = \mathcal{G}(\psi)$ (idem for f). It follows by (7.22) that $M(f) = 0$, and so $f \in A^\infty_{2y}/\mathbb{C}$. We deduce by (P1)' that some g exists in $A^\infty_{2y}/\mathbb{C}$ such that $f = \Delta_y g$. Hence $\widehat{f} = \Delta_s \widehat{g}$, where $\Delta_s = \sum_{j=1}^N \partial_j^2$. But, combination of properties (1) and (5) of (H) together with use of (7.22) yields some $\theta \in \mathcal{V}_{\infty y}/\mathbb{C}$ such that $\widehat{g} = L_y(\widehat{\theta})$. It follows by property (6) of (H) that $\widehat{f} = L_y(\Delta_s\widehat{\theta})$. Hence $\widehat{\psi} = \Delta_s\widehat{\theta}$, or equivalently $\psi = \Delta_y\theta$. This shows (i).

To finish the proof, we turn now to the verification of (iii). For that purpose, let us fix freely some $\mathbf{f} = (f_j)$ as in (7). Let $\mathcal{U}_{2y} = \{\mathbf{v} | \mathbf{v} \in (A^\infty_{2y})^N,\ \mathrm{div}_y\mathbf{v} = 0\}$. We begin by verifying that $\mathcal{U}_{2y} = l_y(\mathcal{U}_{\infty y})$, where for convenience one sets $l_y(\mathbf{u}) = (l_y u_j)$ for $\mathbf{u} = (u_j)$, $u_j \in \mathcal{V}_y$. Let us fix freely some $\mathbf{v} = (v_j) \in \mathcal{U}_{2y}$. According to part (5) of hypothesis (H) there is some $\mathbf{u} \in (\mathcal{V}_{\infty y})^N$ such that $\mathbf{v} = l_y(\mathbf{u})$. Furthermore, by parts (6) and (1) of (H) one has $\mathcal{G}(\mathrm{div}_y\mathbf{v}) = L_y\mathcal{G}(\mathrm{div}_y\mathbf{u})$, from which we see that $\mathrm{div}_y\mathbf{u} = 0$, hence $\mathbf{u} \in \mathcal{U}_{\infty y}$. Therefore the claimed relation follows. As an immediate consequence of this, we have by part (2) of (H) and use of (7.22)

$$\sum_{j=1}^N \int_{\Delta(A_{2y})} L_y(f_j)\widehat{v}_j \, d\beta_{2y} = \sum_{j=1}^N \int_{\Delta(A_y)} f_j\widehat{u}_j \, d\beta_y = 0,$$

where $\mathbf{v} = (v_j)$ is arbitrary in $\mathcal{U}_{2y}$ whereas $\mathbf{u} = (u_j)$ is some element of $\mathcal{U}_{\infty y}$ such that $v_j = l_y u_j$ $(1 \le j \le N)$. By using the p-harmonicity of A_{2y} one deduces that there is a unique $q_{2y} \in W^{1,p}_\#(\Delta(A_{2y}))$ such that $L_y(f_j) = \partial_j q_{2y}$ $(1 \le j \le N)$. Now, by Lemma 7.8 there exists a unique $q_y \in W^{1,p}_\#(\Delta(A_y))$ such that $q_{2y} = L^\#_y(q_y)$ and further $\partial_j q_{2y} = L_y(\partial_j q_y)$ $(1 \le j \le N)$. Hence $f_j = \partial_j q_y$ $(1 \le j \le N)$ as claimed, since L_y is injective. This shows part (7) of (iii). Finally, the first portion of (8) is provided by Lemma 7.8(iii), whereas the next one follows readily by (7) and use of the relation $\mathcal{U}_{\infty y} \subset \mathcal{U}_y$ (see Definition 7.2). This completes the proof of the proposition. ∎

From now we assume that the H-algebra A_{2y} verifies property (P2)' of Proposition 7.5. On the other hand, we suppose in the sequel that $2 \le p < \infty$. Then, the basic notation and definition of Sect. 7.2.6 are still valid when y is changed into $2y$, A (resp. β) is changed into A_2 (resp. β_2), $\mathcal{H}_p$ (resp. $\mathcal{U}$) is changed into $\mathcal{H}_{2p}$ (resp. $\mathcal{U}_2$), and Σ is changed into Σ_2. For the benefit of the reader, it is worth perhaps specifying that we have here:

$$\mathcal{H}_{2p} = L^p(\Delta(A_z); W^{1,p}_{\mathbb{R}}(\Delta(A_{2y}))/\mathbb{R}) \text{ (a separable reflexive Banach space)}$$

$$\mathcal{H}'_{2p} = L^{p'}(\Delta(A_z); [W^{1,p}_{\mathbb{R}}(\Delta(A_{2y}))/\mathbb{R}]') \text{ (the strong dual of } \mathcal{H}_{2p})$$

$$\mathcal{U}_2 = \left\{ u | u \in \mathcal{H}_{2p}, \frac{du}{ds_0} \in \mathcal{H}'_{2p} \right\} \text{ (a Banach space),}$$

where the derivative is understood in the sense of distributions on $\Delta(A_z)$ with values in $W^{1,p}_{\mathbb{R}}(\Delta(A_{2y}))/\mathbb{R}$. Finally, for convenience we set

$$\mathcal{E} = \mathcal{D}_{\mathbb{R}}(\Delta(A_z)) \otimes \mathcal{G}({}_{\mathbb{R}}\mathcal{V}_{\infty y}/\mathbb{R})$$

$$\mathcal{E}_2 = \mathcal{D}_{\mathbb{R}}(\Delta(A_z)) \otimes [\mathcal{D}_{\mathbb{R}}(\Delta(A_{2y}))/\mathbb{R}].$$

For further needs let us bear the equality $\mathcal{E}_2 = L(\mathcal{E})$ in mind. Indeed, this follows from properties (1) and (5) of (H) and use of (7.22).

Having made this point, we turn now to the proof of the next theorem.

Theorem 7.6 *Let $2 \leq p < \infty$. Suppose hypothesis (H) is satisfied and further the H-algebra A_2 verifies property (P2)' of Proposition 7.5. If A_2 is V^p-proper, then so also is A.*

Proof Let us assume that A_2 is V^p-proper. This yields a vector subspace $\mathcal{V}_{\infty 2y}$ of A^{∞}_{2y} that preserves complex conjugation, and a closed vector subspace $\mathcal{W}_2$ of $\mathcal{U}_2$ including $\mathcal{D}_{\mathbb{R}}(\Delta(A_z)) \otimes \mathcal{G}({}_{\mathbb{R}}\mathcal{V}_{\infty 2y}/\mathbb{R})$, such that: **(a)** $\mathcal{G}(\mathcal{V}_{\infty 2y})$ [hence $\mathcal{D}(\Delta(A_{2y}))$] is dense in $W^{1,p}(\Delta(A_{2y}))$; **(b)** $\mathcal{D}_{\mathbb{R}}(\Delta(A_z)) \otimes \mathcal{G}({}_{\mathbb{R}}\mathcal{V}_{\infty 2y}/\mathbb{R})$ is dense in $\mathcal{W}_2$. Thus, it is clear that $\mathcal{E}_2$ is dense in $\mathcal{W}_2$.

Now, property (P2)' of A_{2y} implies by Proposition 7.5 the analogous property, (P2), for the H-algebra A_y. This leads us to Remark 7.6 and the subsequent developments up to Definition 7.3. With this in mind, our purpose is precisely to prove that the H-algebra A meets the requirements of the said definition. Thanks to Lemma 7.8, part (1) of condition (PR1) is fulfilled. It remains to check part (2) of the same condition together with the whole of (PR2). We accomplish this in two steps.

Step 1. The aim here is to show that the mapping L sends $\mathcal{U}$ isometrically into $\mathcal{U}_2$. We already know by Lemma 7.9 that L maps $\mathcal{U}$ into $\mathcal{H}_{2p}$. The next point is to show that

$$\frac{d}{ds_0}Lu \in \mathcal{H}'_{2p} \text{ and } \left\| \frac{d}{ds_0}Lu \right\|_{\mathcal{H}'_{2p}} = \left\| \frac{du}{ds_0} \right\|_{\mathcal{H}'_p} \quad (u \in \mathcal{U}). \tag{7.28}$$

To this end, fix freely some $u \in \mathcal{U}$. Let $\varphi \in \mathcal{D}_{\mathbb{R}}(\Delta(A_z))$. Recalling that

$$\left\langle \frac{d}{ds_0}Lu, \varphi \right\rangle \in W^{1,p}_{\mathbb{R}}(\Delta(A_{2y}))/\mathbb{R} \subset L^2_{\mathbb{R}}(\Delta(A_{2y}))/\mathbb{R} \subset [W^{1,p}_{\mathbb{R}}(\Delta(A_{2y}))/\mathbb{R}]'$$

we have, for any arbitrary given $\psi \in \mathcal{D}_{\mathbb{R}}(\Delta(A_{2y})/\mathbb{R}$,

$$\begin{aligned} \left[\left\langle \frac{d}{ds_0}Lu, \varphi \right\rangle, \psi \right] &= \int_{\Delta(A_{2y})} \left\langle \frac{d}{ds_0}Lu, \varphi \right\rangle \psi \, d\beta_{2y} \\ &= -\int_{\Delta(A_{2y})} \left(\int_{\Delta(A_z)} L_y(u(s_0)) \frac{d\varphi}{ds_0}(s_0) d\beta_z(s_0) \right) \psi \, d\beta_{2y} \end{aligned}$$

$$= -\int_{\Delta(A_z)} \left(\int_{\Delta(A_{2y})} L_y \left(\frac{d\varphi}{ds_0}(s_0)\widehat{v}u(s_0) \right) d\beta_{2y} \right) d\beta_z(s_0)$$

$$= -\int_{\Delta(A_z)} \left(\int_{\Delta(A_y)} u(s_0)\frac{d\varphi}{ds_0}(s_0)\widehat{v}d\beta_y \right) d\beta_z(s_0)$$

$$= \int_{\Delta(A_z)} \left[\frac{du}{ds_0}(s_0), \widehat{v} \right] \varphi(s_0)d\beta_z(s_0),$$

where $v \in {}_{\mathbb{R}}\mathcal{V}_{\infty y}/\mathbb{R}$, $L_y\widehat{v} = \psi$ (this follows by (1) and (5) of (H) and use of (7.22)), and where we have made use of (7.22), once again. The reader has no reason to bother about the use of the same brackets here and there. There is no risk of confusion.

At the present time, let L_y be viewed as a mapping of $W_{\mathbb{R}}^{1,p}(\Delta(A_y))/\mathbb{R}$ into $W_{\mathbb{R}}^{1,p}(\Delta(A_{2y}))/\mathbb{R}$, which mapping is then an isometric isomorphism. Let ${}^tL_y$ be the transposed of L_y, an isometric isomorphism of $[W_{\mathbb{R}}^{1,p}(\Delta(A_{2y}))/\mathbb{R}]'$ onto $[W_{\mathbb{R}}^{1,p}(\Delta(A_y))/\mathbb{R}]'$. Furthermore, let us for convenience introduce $f \in \mathcal{H}'_{2p}$ given by

$$f(s_0) = ({}^tL_y)^{-1}\frac{du}{ds_0}(s_0) \text{ for } s_0 \in \Delta(A_z). \tag{7.29}$$

Then, returning to the last equality thereinbefore, an elementary calculation taking account of the equality ${}^t(L_y^{-1}) = ({}^tL_y)^{-1}$ gives

$$\left[\left\langle \frac{d}{ds_0}Lu, \varphi \right\rangle, \psi \right] = \left[\int_{\Delta(A_z)} f\varphi \, d\beta_z, \psi \right]$$

and that for all $\psi \in \mathcal{D}_{\mathbb{R}}(\Delta(A_{2y}))/\mathbb{R}$ (a dense subspace of $W_{\mathbb{R}}^{1,p}(\Delta(A_{2y}))/\mathbb{R}$). By the arbitrariness of ψ and then of that of φ, it follows $\frac{d}{ds_0}Lu = f$ (given by (7.29)) and so $Lu \in \mathcal{U}_2$.

More precisely, considered as defined on $\mathcal{H}_p$, the operator L is an isometric isomorphism of $\mathcal{H}_p$ onto $\mathcal{H}_{2p}$. Its transposed, ${}^tL$, given by ${}^tLf = {}^tL_y \circ f$ ($f \in \mathcal{H}'_{2p}$) is an isometric isomorphism of $\mathcal{H}'_{2p}$ onto $\mathcal{H}'_p$ and one has $\frac{d}{ds_0}Lu = ({}^tL)^{-1}\frac{du}{ds_0}$ for all $u \in \mathcal{U}$. Hence (7.28) follows and further L maps isometrically $\mathcal{U}$ **into** $\mathcal{U}_2$. Consequently, by the density of $\mathcal{E}_2$ in $\mathcal{W}_2$ and use of the equality $\mathcal{E}_2 = L(\mathcal{E})$, we see that $\mathcal{W}_2 \subset L(\mathcal{U})$. By letting $\mathcal{W} = L^{-1}(\mathcal{W}_2)$ (a closed vector subspace of $\mathcal{U}$), it follows that L carries $\mathcal{W}$ isometrically over $\mathcal{W}_2$.

Step 2. The preceding result has two major consequences:

(1) $\mathcal{E}$ is a dense subspace of $\mathcal{W}$, and so the whole of (PR1) follows.
(2) The linear operator L_Q introduced thereinbefore carries $L^p(Q;\mathcal{W})$ isometrically over $L^p(Q;\mathcal{W}_2)$, and further $\partial_j L_Q(u) = L_Q(\partial_j u)$ ($1 \leq j \leq N$) for $u \in \mathcal{W}$ (see (7.23)).

Thus, given a bounded sequence $(u_\varepsilon)_{\varepsilon\in E}$ in V^p (E a fundamental sequence) and considering, by the V^p-properness of A_2, a subsequence E' from E and some $\mathbf{v} = (u_0, v_1) \in V^p \times L^p(Q; \mathcal{W}_2)$ such that (7.26) and (7.27) hold as $E' \ni \varepsilon \to 0$, we see at once by the same line of argument (based on Lemma 7.11) as followed in proving Theorem 7.5 that as $E' \ni \varepsilon \to 0$, one has

$$\frac{\partial u_\varepsilon}{\partial x_j} \to \frac{\partial u_0}{\partial x_j} + \partial_j u_1 \text{ in } L^p(Q)\text{-weak } \Sigma \ (1 \leq j \leq N),$$

where $u_1 = L_Q^{-1}(v_1)$. Hence (PR2) follows, which completes the proof. ∎

We end these preliminaries with the analog of Lemma 5.16. In what follows, G denotes a function from $\mathbb{R}_y^N \times \mathbb{R}_z \times \mathbb{R}_\mu^N$ to $\mathbb{R}$ satisfying (6.26).

Lemma 7.12 *Suppose we have*

$$G(\cdot,\cdot,\mu) \in A \textit{ for fixed } \mu \in \mathbb{R}^N. \tag{7.30}$$

Then

$$G(\cdot,\cdot,\boldsymbol{\Psi}) \in A \textit{ for all } \boldsymbol{\Psi} \in (A_{\mathbb{R}})^N. \tag{7.31}$$

Proof It is clear by (6.26) and (7.30) that the function $\mu \mapsto G(\cdot,\cdot,\mu)$ lies in $\mathcal{C}(\mathbb{R}^N; A)$. With this in mind, fix freely some $\boldsymbol{\Psi} \in (A_{\mathbb{R}})^N$. Following the same line of argument as in proving Lemma 5.16, consider a compact set $K \subset \mathbb{R}^N$ such that $\boldsymbol{\Psi}(y,z) \in K$ for all $(y,z) \in \mathbb{R}^N \times \mathbb{R}$. Let G_K denote the function of $\mathbb{R}^N \times \mathbb{R} \times K$ into $\mathbb{R}$ defined by $G_K(y,z,\mu) = G(y,z,\mu)$ $(y \in \mathbb{R}^N, z \in \mathbb{R}, \mu \in K)$. Then $G_K \in \mathcal{C}(K; A_{\mathbb{R}})$ ($A_{\mathbb{R}}$ considered as a real Banach algebra). Clearly the lemma is proved once we have shown that $G_K(\cdot,\cdot,\boldsymbol{\Psi}) \in A$. To this end, according to the density of $\mathcal{C}_{\mathbb{R}}(K) \otimes A_{\mathbb{R}}$ in $\mathcal{C}(K; A_{\mathbb{R}})$, let us consider a sequence of functions $q_n \in \mathcal{C}_{\mathbb{R}}(K) \otimes A_{\mathbb{R}}$ $(n \in \mathbb{N})$ such that

$$\sup_{\mu\in K} \|q_n(\cdot,\cdot,\mu) - G_K(\cdot,\cdot,\mu)\|_\infty \to 0 \quad \text{as } n \to \infty.$$

Hence, it follows that $q_n(\cdot,\cdot,\boldsymbol{\Psi}) \to G_K(\cdot,\cdot,\boldsymbol{\Psi})$ in $\mathcal{B}(\mathbb{R}^N \times \mathbb{R})$ as $n \to \infty$. Therefore (7.31) is proved if we can verify that $q_n(\cdot,\cdot,\boldsymbol{\Psi}) \in A$ for each integer $n \geq 0$. It clearly suffices to check that one has $q(\cdot,\cdot,\boldsymbol{\Psi}) \in A$ for any $q \in \mathcal{C}_{\mathbb{R}}(K) \otimes A_{\mathbb{R}}$ of the form $q = \chi \otimes \varphi$, i.e., $q(y,z,\mu) = \chi(\mu)\varphi(y,z)$ $(y \in \mathbb{R}^N, z \in \mathbb{R}, \mu \in K)$, where $\chi \in \mathcal{C}_{\mathbb{R}}(K)$ and $\varphi \in A_{\mathbb{R}}$. But then this follows readily by the Stone-Weierstrass theorem, as pointed out in proving Lemma 5.16. ∎

7.3.2 *Problem I: Periodic Setting*

The present section is not intended to complement what was accomplished in Sects. 6.2 and 6.3, still less to propose an alternative approach to the periodic homogenization of (7.1). Indeed, as the cases $k = 1, 2$ were worked out in Chap. 6, so could the remaining cases be achieved within the same chapter, as well. Actually, the aim here is to lay emphasis on the fact that periodic homogenization is only one of the possible sides of a great unity, namely deterministic homogenization theory. We shall start with some preliminaries.

7.3.2.1 Preliminaries

Let the homogenization problem for (7.1) be considered under the structure hypothesis

$$a_i(\cdot, \cdot, \mu) \text{ is } Y \times Z\text{-periodic } (1 \leq i \leq N) \text{ for fixed } \mu \in \mathbb{R}^N, \qquad (7.32)$$

as stated in Chap. 6. Suppose that, for a reason whose importance will come to light later, our purpose is to study this problem not in a self-contained manner as in Chap. 6 but within the abstract framework of Sect. 7.2 instead. We are thus led to set $A = \mathcal{C}_{per}(Y \times Z)$, $A_y = \mathcal{C}_{per}(Y)$ and $A_z = \mathcal{C}_{per}(Z)$ (see Example 7.1), and it then remains to be seen whether the H-algebra A has appropriate properties (see Sect. 7.2.1 for the basic notation).

It is not perhaps out of interest to remind the reader of the main function spaces we will be dealing with in the present section. For convenience we shall make a change of notation. For $2 \leq p < \infty$, let

$$\mathcal{H}_{per} = L^p_{per}(Z; W^{1,p}_{per}(Y; \mathbb{R})/\mathbb{R}) \text{ (Sect. 6.2)}$$

and

$$\mathcal{U}_{per} = \left\{ u | u \in \mathcal{H}_{per}, u' = \frac{du}{dz} \in \mathcal{H}'_{per} \right\},$$

where the derivative is understood in the sense of $\mathcal{D}'(\mathbb{R}; W^{1,p}_{per}(Y; \mathbb{R})/\mathbb{R})$ and the dual of $\mathcal{H}_{per}$ is given by

$$\mathcal{H}'_{per} = L^{p'}_{per}(Z; [W^{1,p}_{per}(Y; \mathbb{R})/\mathbb{R}]').$$

In terms of the Gelfand representation theory we have on the other hand

$$\mathcal{H}_A = L^p(\Delta(A_z); W^{1,p}_{\mathbb{R}}(\Delta(A_y)/\mathbb{R}))$$

$$\mathcal{H}'_A = L^{p'}(\Delta(A_z); [W^{1,p}_{\mathbb{R}}(\Delta(A_y))/\mathbb{R}]')$$

and

$$\mathcal{U}_A = \left\{ u | u \in \mathcal{H}_A, u' = \frac{du}{ds_0} \in \mathcal{H}'_A \right\}.$$

Before we can prove one basic lemma, we need the following remark.

Remark 7.13 Lemma 6.12 together with its proof carries over mutatis mutandis to functions of several variables: For $\theta \in \mathcal{D}^m(\mathbb{R}^{N+1})$ $(m \in \mathbb{N} \cup \{\infty\})$, the function $\overline{\omega}\theta = \sum_{k\in\mathbb{Z}^{N+1}} \tau_k\theta$ lies in $\mathcal{C}^m_{per}(Y \times Z)$. The mapping $\theta \mapsto \overline{\omega}\theta$ of $\mathcal{D}^m(\mathbb{R}^{N+1})$ into $\mathcal{C}^m_{per}(Y \times Z)$ is continuous linear and surjective. Furthermore, the following hold for $\theta \in \mathcal{D}^m(\mathbb{R}^{N+1})$.

(1) $\dfrac{\partial \overline{\omega}\theta}{\partial y_j} = \overline{\omega}\dfrac{\partial \theta}{\partial y_j}$ $(1 \le j \le N)$

(2) $\dfrac{\partial \overline{\omega}\theta}{\partial z} = \overline{\omega}\dfrac{\partial \theta}{\partial z}$

(3) $\displaystyle\int\int_{\mathbb{R}^N\times\mathbb{R}} u\theta \, dydz = \int\int_{Y\times Z} u\overline{\omega}\theta \, dydz, \; u \in L^1_{per}(Y \times Z).$

The classical Fourier analysis in $\mathcal{D}'(\mathbb{T}^N)$ carries over readily to $\mathcal{D}'(\Delta(A_y))$ by means of the topological isomorphism between the two spaces of distributions (see Proposition 2.8 together with Vo-Khac [139, p. 64]). So, the Fourier coefficients of $u \in \mathcal{D}'(\Delta(A_y))$ are given by $\mathcal{F}_k(u) = \langle u, \overline{\widehat{\gamma}_k}\rangle$ $(k \in \mathbb{Z}^N)$, and one has, in the $\mathcal{D}'(\Delta(A_y))$ topology, $u = \sum_{k\in\mathbb{Z}^N} \mathcal{F}_k(u)\widehat{\gamma}_k$. It follows that $u = 0$ if and only if $\mathcal{F}_k(u) = 0$ for all $k \in \mathbb{Z}^N$. Finally, it is worth bearing in mind that $\mathcal{F}_k(\partial_j u) = 2\pi i k_j \mathcal{F}_k(u)$ for $k = (k_1, \cdots, k_N) \in \mathbb{Z}^N$. For further details see Vo-Khac [139, pp. 64–70], Schwartz [126, p. 225].

In what follows, $\mathcal{G}$ stands for the canonical mapping of $L^p(Y \times Z)$ into $L^p(\Delta(A))$, which is an isometric isomorphism (see Sect. 2.5). On the other hand, for $w \in L^p(\Delta(A))$ and $s_0 \in \Delta(A_z)$, one writes $w(\cdot, s_0)$ for the function $s \mapsto w(s, s_0)$ of $\Delta(A_y)$ into $\mathbb{C}$, which lies in $L^p(\Delta(A_y))$.

Lemma 7.13 *Let $u \in \mathcal{H}_{per}$. The following hold true for $1 \le j \le N$.*

(i) $\partial_j \mathcal{G}(u) = \mathcal{G}\left(\dfrac{\partial u}{\partial y_j}\right)$

(ii) $\partial_j[\mathcal{G}(u)(\cdot, s_0)] = \mathcal{G}\left(\dfrac{\partial u}{\partial y_j}\right)(\cdot, s_0)$ *for a.e.* $s_0 \in \Delta(A_z)$

(iii) $\partial_j \mathcal{G}(u)(s, s_0) = \partial_j[\mathcal{G}(u)(\cdot, s_0)](s)$ *for a.e.* $(s, s_0) \in \Delta(A)$.

Proof We begin by proving part (i). Let $\psi \in \mathcal{C}^\infty_{per}(Y \times Z)$. Let $\theta \in \mathcal{D}(\mathbb{R}^{N+1})$, $\psi = \overline{\omega}\theta$ (Remark 7.13). Put $\mathcal{G}(\psi) = \widehat{\psi}$. One has

$$
\begin{aligned}
\langle \partial_j \mathcal{G}(u), \widehat{\psi} \rangle &= - \int_{\Delta(A)} \mathcal{G}(u) \partial_j \widehat{\psi} \, d\beta \\
&= - \int_{\Delta(A)} \mathcal{G}\left(u \frac{\partial \psi}{\partial y_j} \right) d\beta \\
&= - \int \int_{Y \times Z} u \, \overline{\omega} \frac{\partial \theta}{\partial y_j} \, dydz \\
&= - \int \int_{\mathbb{R}^N \times \mathbb{R}} u \frac{\partial \theta}{\partial y_j} \, dydz \\
&= \int \int_{\mathbb{R}^N \times \mathbb{R}} \frac{\partial u}{\partial y_j} \theta \, dydz \quad \left(\text{note that } \frac{\partial u}{\partial y_j} \in L^p_{per}(Y \times Z) \right) \\
&= \int \int_{Y \times Z} \frac{\partial u}{\partial y_j} \psi \, dydz \\
&= \int_{\Delta(A)} \mathcal{G}\left(\frac{\partial u}{\partial y_j} \right) \widehat{\psi} \, d\beta.
\end{aligned}
$$

Hence (i) follows by the arbitrariness of ψ. We next prove (ii). Fix freely $\varphi \in \mathcal{C}^\infty_{per}(Y)$ and $\theta \in \mathcal{C}^\infty_{per}(Z)$. By a simple calculation leant on Remark 7.13 one quickly arrives at

$$
\begin{aligned}
&\int_{\Delta(A_z)} \int_{\Delta(A_y)} \mathcal{G}(u)[\partial_j \mathcal{G}(\varphi) \otimes \mathcal{G}(\theta)] d\beta_y d\beta_z \\
&= - \int_{\Delta(A_z)} \int_{\Delta(A_y)} \mathcal{G}\left(\frac{\partial u}{\partial y_j} \right) [\mathcal{G}(\varphi) \otimes \mathcal{G}(\theta)] d\beta_y d\beta_z.
\end{aligned}
$$

By choosing in particular $\varphi = \overline{\gamma}_k$ ($k \in \mathbb{Z}^N$) one is readily led to

$$
\int_{\Delta(A_z)} \left[- \int_{\Delta(A_y)} 2\pi i k_j \overline{\gamma}_k \mathcal{G}(u) d\beta_y + \int_{\Delta(A_y)} \overline{\gamma}_k \mathcal{G}\left(\frac{\partial u}{\partial y_j} \right) d\beta_y \right] \mathcal{G}(\theta) d\beta_z = 0.
$$

In view of the arbitrariness of θ, we deduce that to each $k \in \mathbb{Z}^N$ there is assigned a β_z-negligible set $\mathcal{N}_k$ such that

$$
-2\pi i k_j \mathcal{F}_k(\mathcal{G}(u)(\cdot, s_0)) + \mathcal{F}_k\left(\mathcal{G}\left(\frac{\partial u}{\partial y_j} \right) (\cdot, s_0) \right) = 0 \;\; (1 \leq j \leq N)
$$

for every $s_0 \in \Delta(A_z) \setminus \mathcal{N}_k$. By letting $\mathcal{N} = \bigcup_{k \in \mathbb{Z}^N} \mathcal{N}_k$ (a β_z-negligible set), it follows

$$
\mathcal{F}_k \left[\mathcal{G}\left(\frac{\partial u}{\partial y_j} \right) (\cdot, s_0) - \partial_j (\mathcal{G}(u)(\cdot, s_0)) \right] = 0 \;\; (1 \leq j \leq N)
$$

for all $s_0 \in \Delta(A_z) \setminus \mathcal{N}$ and all $k \in \mathbb{Z}^N$, where the derivative ∂_j is understood in the sense of $\mathcal{D}'(\Delta(A_y))$, of course. Hence (ii) follows by the arbitrariness of k (see Remark 7.13). Finally, (iii) is an obvious consequence of (i) and (ii). This completes the proof. ■

We are now able to prove an interesting isomorphism result.

Proposition 7.6 *Let us still denote by $\mathcal{G}$ the restriction to $\mathcal{H}_{per}$ of the canonical mapping of $L^p_{per}(Y \times Z)$ into $L^p(\Delta(A))$. Then, the following statements hold.*

(i) $\mathcal{G}$ is an isometric isomorphism of $\mathcal{H}_{per}$ ***onto*** *$\mathcal{H}_A$*
(ii) $\mathcal{G}$ sends $\mathcal{U}_{per}$ isometrically ***into*** *$\mathcal{U}_A$.*

Proof To start with, let us establish (i). First of all, it is an easy exercise (left to the reader) to show that $\mathcal{G}$ maps $\mathcal{H}_{per}$ into $L^p(\Delta(A_z); L^p_{\mathbb{R}}(\Delta(A_y))/\mathbb{R})$. By joining this with part (ii) of Lemma 7.13 one sees readily that $\mathcal{G}$ sends isometrically $\mathcal{H}_{per}$ into $\mathcal{H}_A$. It remains to show that this mapping is surjective. To this end, let $\mathfrak{A} = [\mathcal{C}^\infty_{per}(Y; \mathbb{R})/\mathbb{R}] \otimes \mathcal{C}^\infty_{per}(Z; \mathbb{R})$. We have $\mathcal{G}(\mathfrak{A}) = [\mathcal{D}_{\mathbb{R}}(\Delta(A_y))/\mathbb{R}] \otimes \mathcal{D}_{\mathbb{R}}(\Delta(A_z))$. It follows that $\mathcal{G}(\mathfrak{A})$ is dense in $\mathcal{H}_A$ for three reasons:

(a) $\mathcal{C}_{\mathbb{R}}(\Delta(A_z)) \otimes [W^{1,p}_{\mathbb{R}}(\Delta(A_y))/\mathbb{R}]$ is dense in $\mathcal{H}_A$ (a familiar fact).
(b) $\mathcal{D}_{\mathbb{R}}(\Delta(A_z))$ is dense in $\mathcal{C}_{\mathbb{R}}(\Delta(A_z))$.
(c) $\mathcal{D}_{\mathbb{R}}(\Delta(A_y))/\mathbb{R}$ is dense in $W^{1,p}_{\mathbb{R}}(\Delta(A_y))/\mathbb{R}$ (Corollaries 2.13 and 2.15).

But $\mathcal{G}(\mathfrak{A}) \subset \mathcal{G}(\mathcal{H}_{per}) \subset \mathcal{H}_A$. We deduce that $\mathcal{G}(\mathcal{H}_{per})$ is dense in $\mathcal{H}_A$. Hence $\mathcal{G}(\mathcal{H}_{per}) = \mathcal{H}_A$ as desired, since $\mathcal{G}(\mathcal{H}_{per})$ is closed in $\mathcal{H}_A$. This shows part (i). We turn now to the proof of part (ii).

Let $u \in \mathcal{U}_{per}$. By part (i) we have $\mathcal{G}(u) \in \mathcal{H}_A$. Let us show that $\frac{d}{ds_0}\mathcal{G}(u)$ lies in $\mathcal{H}'_A$, where it should be recalled that the derivative is understood in the sense of $\mathcal{D}'(\Delta(A_z); W^{1,p}_{\mathbb{R}}(\Delta(A_y))/\mathbb{R})$. Let us fix freely some $\varphi \in \mathcal{C}^\infty_{per}(Z; \mathbb{R})$, $\psi \in \mathcal{C}^\infty_{per}(Y; \mathbb{R})/\mathbb{R}$. Let $\theta \in \mathcal{D}_{\mathbb{R}}(\mathbb{R})$ be so that $\overline{\omega}\theta = \varphi$ (Lemma 6.12). We have

$$
\begin{aligned}
\left[\left\langle \frac{d}{ds_0}\mathcal{G}(u), \widehat{\varphi}\right\rangle, \widehat{\psi}\right] &= -\int_{\Delta(A)} \mathcal{G}\left(u\frac{d\varphi}{dz}\psi\right) d\beta \\
&= -\iint_{Y\times Z} u\frac{d\varphi}{dz}\psi\, dydz \\
&= -\int_Y \left(\int_{\mathbb{R}} u(\cdot, z)\frac{d\theta}{dz}(z)dz\right)\psi\, dy \\
&= \left[\int_{\mathbb{R}} \frac{du}{dz}(z)\theta(z)dz, \psi\right] \\
&= \int_Z \left[\frac{du}{dz}(z), \varphi(z)\psi\right] dz \\
&= \int_Z \left[\frac{du}{dz}(z), (\mathcal{G}^{-1}(\widehat{\varphi}\otimes\widehat{\psi}))(z)\right] dz
\end{aligned}
$$

$$= \left\langle \frac{du}{dz}, \mathcal{G}^{-1}(\widehat{\varphi} \otimes \widehat{\psi}) \right\rangle_{\mathcal{H}'_{per}, \mathcal{H}_{per}}$$

$$= \left\langle {}^t\mathcal{G}^{-1}\frac{du}{dz}, \widehat{\varphi} \otimes \widehat{\psi} \right\rangle_{\mathcal{H}'_A, \mathcal{H}_A}$$

$$= \left\langle ({}^t\mathcal{G})^{-1}\frac{du}{dz}, \widehat{\varphi} \otimes \widehat{\psi} \right\rangle_{\mathcal{H}'_A, \mathcal{H}_A}$$

$$= \int_{\Delta(A_z)} \left[\left(({}^t\mathcal{G})^{-1}\frac{du}{dz} \right)(s_0), \widehat{\varphi}(s_0)\widehat{\psi} \right] d\beta_z(s_0)$$

$$= \left[\int_{\Delta(A_z)} \widehat{\varphi}({}^t\mathcal{G})^{-1}\frac{du}{dz} d\beta_z, \widehat{\psi} \right],$$

where, as a consequence of part (i), ${}^t\mathcal{G}$ (the transposed of $\mathcal{G}$) is an isometric isomorphism of $\mathcal{H}'_A$ onto $\mathcal{H}'_{per}$. We deduce by the arbitrariness of ψ, and then of that of φ, that

$$\frac{d}{ds_0}\mathcal{G}(u) = ({}^t\mathcal{G})^{-1}\frac{du}{dz}.$$

By joining this with part (i) one concludes that $\mathcal{G}$ maps $\mathcal{U}_{per}$ isometrically into $\mathcal{U}_A$. This completes the proof. ∎

7.3.2.2 Application to the Study of the Homogenization Problem for (7.1) Under Hypothesis (7.32)

To begin, for arbitrary given $\boldsymbol{\Psi} \in (A_{\mathbb{R}})^N$ we have $a_i(\cdot, \cdot, \boldsymbol{\Psi}) \in \mathfrak{X}_A^{p'}$, as is straightforward from (7.32) by use of the relations $a_i(\cdot, \cdot, \boldsymbol{\Psi}) \in L^{\infty}_{per}(Y \times Z) \subset L^{p'}_{per}(Y \times Z) = \mathfrak{X}_A^{p'}$ (Corollary 2.14). Next, there is the following

Proposition 7.7 *The following statements hold for the homogenization algebra* $A = \mathcal{C}_{per}(Y \times Z)$*:*

(i) *A is quasi* V^p*-proper for all* $2 \leq p < \infty$*.*
(ii) *A is* V^2*-proper.*
(iii) *A verifies properties (P1),(P2) and (P3).*

Proof It is worth recalling that the quasi V^p-properness of $A = \mathcal{C}_{per}(Y \times Z)$ was pointed out earlier in Sect. 7.2.2 (see Proposition 7.3 and Examples 7.1 and 7.2). Actually, a self-contained proof of it is available in the periodic setting (see Chap. 6), and the aim here is only to «embed» the latter in the Gelfand representation setting. First of all, condition (QP1) of Definition 7.1 follows readily by Corollaries 2.13 and 2.15. It remains to verify (QP2). Let $2 \leq p < \infty$. Let $(u_\varepsilon)_{\varepsilon \in E}$ be a norm-bounded sequence in V^p, where E is any arbitrary given fundamental sequence. According to Lemma 6.9 there are a subsequence E' from E and some $\mathbf{v} = (u_0, v_1) \in V^p \times L^p(Q; \mathcal{H}_{per})$ such that as $E' \ni \varepsilon \to 0$, we have $u_\varepsilon \to u_0$ in V^p-weak and, for $1 \leq j \leq N$,

$$\int_Q \frac{\partial u_\varepsilon}{\partial x_j} \psi^\varepsilon \, dxdt \to \int_Q \left[\iint_{Y \times Z} \left(\frac{\partial u_0}{\partial x_j}(x,t) + \frac{\partial}{\partial y_j} v_1(x,t) \right) \psi(x,t) dydz \right] dxdt$$

for all $\psi \in L^{p'}(Q; A)$, where $\psi^\varepsilon(x,t) = \psi(x, t, \frac{x}{\varepsilon}, \frac{t}{\varepsilon^k})$, of course. But, thanks to Proposition 7.6, the transformation $v \mapsto \mathcal{G} \circ v$ of $L^p(Q; \mathcal{H}_{per})$ into $L^p(Q; \mathcal{H}_A)$ is an isometric isomorphism. Thus, there is some $u_1 \in L^p(Q; \mathcal{H}_A)$ such that $v_1(x,t) = \mathcal{G}^{-1} u_1(x,t)$ for a.e. $(x,t) \in Q$, and further $\frac{\partial}{\partial y_j} v_1(x,t) = \mathcal{G}^{-1} \partial_j u_1(x,t)$ (Lemma 7.13). Hence, the desired result follows by the equality

$$\iint_{Y \times Z} \left(\frac{\partial u_0}{\partial x_j}(x,t) + \frac{\partial}{\partial y_j} v_1(x,t) \right) \psi(x,t) dydz =$$
$$\int_{\Delta(A)} \left(\frac{\partial u_0}{\partial x_j}(x,t) + \partial_j u_1(x,t) \right) \widehat{\psi}(x,t) d\beta.$$

We turn now to the verification of the V^2-properness of A. So, let $p = 2$. Let $\mathcal{V}_{\infty y} = A_y^\infty = \mathcal{C}_{per}^\infty(Y)$ and $\mathcal{W} = \mathcal{G}(\mathcal{U}_{per})$. As an H-algebra , $\mathcal{V}_{\infty y}$ preserves complex conjugation and further $\mathcal{G}(\mathcal{V}_{\infty y}) = \mathcal{D}(\Delta(A_y))$ is dense in $W^{1,2}(\Delta(A_y))$ (see the proof of part (i) above). On the other hand, part (ii) of Proposition 7.6 implies that $\mathcal{G}$ viewed as defined on $\mathcal{U}_{per}$ is an isometric isomorphism of $\mathcal{U}_{per}$ onto $\mathcal{W}$. Consequently, not only $\mathcal{W}$ is a closed vector subspace of $\mathcal{U}_A$, but also the space $\mathcal{D}_\mathbb{R}(\Delta(A_z)) \otimes \mathcal{G}(\,_\mathbb{R}\mathcal{V}_{\infty y}/\mathbb{R})$ is dense in $\mathcal{W}$, as is seen by Proposition 6.7. Thus, condition (PR1) of Definition 7.3 is fulfilled. It remains to check (PR2). Based on Lemma 6.13 and on the fact that the transformation $v \mapsto \mathcal{G} \circ v$ is an isometric isomorphism of $L^2(Q; \mathcal{U}_{per})$ onto $L^2(Q; \mathcal{W})$, we see readily that (PR2) is achieved by the same line of argument as followed in proving part (i). This proves part (ii). We finally prove part (iii). Concerning (P1), the choice $\mathcal{V}_{\infty y} = \mathcal{C}_{per}^\infty(Y)$ meets both of the requirements (1) and (2) (see the proof of Corollary 3.3). Property (P2) follows by the classical Poincarré type inequality for $W_{per}^{1,p}(Y)/\mathbb{C}$ and use of Corollaries 2.14 and 2.15. It remains to check (P3). Let $u \in L^p(\Delta(A_z); B)$ with $\int_{\Delta(A_z)} u(s_0)\psi'(s_0) d\beta_z(s_0) = 0$ for all $\psi \in \mathcal{D}(\Delta(A_z))$. Fixing freely some $\zeta \in B'$ (strong dual of B), it follows that

$$\int_{\Delta(A_z)} \langle \zeta, u(s_0) \rangle \psi'(s_0) d\beta_z(s_0) = 0$$

for all $\psi \in \mathcal{D}(\Delta(A_z))$. Let $f \in L_{per}^p(Z)$ and $\varphi \in \mathcal{C}_{per}^\infty(Z)$ be such that $\psi = \mathcal{G}(\varphi)$ and $\mathcal{G}(f)(s_0) = \langle \zeta, u(s_0) \rangle$ for a.e. $s_0 \in \Delta(A_z)$. Then $\int_Z f \varphi' dz = 0$, and that for all $\varphi \in \mathcal{C}_{per}^\infty(Z)$. By Lemma 6.12 it follows at once that $\int_\mathbb{R} f \theta' \, dz = 0$ for all $\theta \in \mathcal{D}(\mathbb{R})$. Hence f is constant, i.e., there is a constant $c(\zeta) \in \mathbb{C}$ such that $f(z) = c(\zeta)$ for a.e. $z \in \mathbb{R}$. But $c(\zeta) = M(f) = \left\langle \zeta, \int_{\Delta(A_z)} u \, d\beta_z \right\rangle = \langle \zeta, u(s_0) \rangle$ for a.e. $s_0 \in \Delta(A_z)$. Since this holds for any arbitrary $\zeta \in B'$, one deduces that $u(s_0) = \int_{\Delta(A_z)} u \, d\beta_z$ for a.e. s_0. This shows (P3). ■

This leads us immediately to the following statement.

Corollary 7.2 *Under the structure hypothesis (7.32), the homogenization problem for (7.1) is solved for $A = \mathcal{C}_{per}(Y \times Z)$ and $2 \leq p < \infty$ by Theorems 7.1, 7.2 and 7.4 in the cases $0 < k \leq 1$, $1 < k < 2$ and $k > 2$, respectively, and by Theorem 7.3 in the case $k = 2$ provided $p = 2$.*

Remark 7.14 Thanks to different basic results collected throughout Sect. 2.5, it is an easy exercise to have the statements of the preceding theorems (together with their proofs) rewritten in terms of periodic homogenization theory (see Remark 5.20).

7.3.3 *Problem II: Almost Periodic Setting*

7.3.3.1 Preliminaries

We study here the homogenization of the initial-boundary value problem (7.1) in the case where the functions $a_i(\cdot, \cdot, \mu)$ (for fixed $\mu \in \mathbb{R}^N$) are almost periodic. We will first consider the case where the latter are almost periodic continuous functions in the sense of Bohr (see Chaps. 1 and 2). More precisely, we assume that

$$a_i(\cdot, \cdot, \mu) \in AP(\mathbb{R}^{N+1}) \ \ (1 \leq i \leq N) \text{ for fixed } \mu \in \mathbb{R}^N. \tag{7.33}$$

Thus, each function $\mu \mapsto a_i(\cdot, \cdot, \mu)$ lies in $\mathcal{C}(\mathbb{R}^N; AP(\mathbb{R}^{N+1}))$. Hence, a trivial adaptation of the proof of Lemma 5.3 yields a countable subgroup G of $\mathbb{R}^{N+1}$ such that $a_i(\cdot, \cdot, \mu) \in AP_G(\mathbb{R}^{N+1})$ for all $\mu \in \mathbb{R}^N$ and all $1 \leq i \leq N$. Next, let $\mathcal{R}_y = pr_y(G)$ and $\mathcal{R}_z = pr_z(G)$, where pr_y (resp. pr_z) is the natural projection of $\mathbb{R}^{N+1} = \mathbb{R}^N_y \times \mathbb{R}_z$ onto $\mathbb{R}^N_y$ (resp. $\mathbb{R}_z$). Clearly $\mathcal{R}_y$ (resp. $\mathcal{R}_z$) is a countable subgroup of $\mathbb{R}^N$ (resp. $\mathbb{R}$). It follows that the product $\mathcal{R} = \mathcal{R}_y \times \mathcal{R}_z$ is a countable subgroup of $\mathbb{R}^{N+1}$ and further $G \subset \mathcal{R}$. Hence

$$a_i(\cdot, \cdot, \mu) \in A = AP_{\mathcal{R}}(\mathbb{R}^{N+1}) \ \ (1 \leq i \leq N) \text{ for fixed } \mu \in \mathbb{R}^N$$

and further

$$AP_{\mathcal{R}}(\mathbb{R}^{N+1}) = \overline{AP_{\mathcal{R}_y}(\mathbb{R}^N) \otimes AP_{\mathcal{R}_z}(\mathbb{R})} \ \ \text{(the closure in } \mathcal{B}(\mathbb{R}^{N+1})).$$

Therefore, the problem falls within the abstract setting of Sect. 7.2 with $A_y = AP_{\mathcal{R}_y}(\mathbb{R}^N)$, $A_z = AP_{\mathcal{R}_z}(\mathbb{R})$ and $A = AP_{\mathcal{R}}(\mathbb{R}^{N+1})$.

We need to know about the main properties of A.

Proposition 7.8 *The following hold true for* $A = AP_{\mathcal{R}}(\mathbb{R}^{N+1})$ *and* $p = 2$.

(i) A *is quasi* V^2*-proper.*
(ii) A *verifies (P1) and (P3).*
(iii) A *verifies (P2) if and only if* $\mathcal{R}_y$ *is a discrete subgroup of* $\mathbb{R}^N$.

Proof Part (i) follows by Example 7.3. We turn next to the proof of (ii). Let $\mathcal{V}_{\infty y}$ be the space of all functions of the form

$$\psi = \sum_{k \in F} c_k \gamma_k \quad (c_k \in \mathbb{C}),$$

where $\gamma_k(y) = e^{2\pi i k \cdot y}$ $(y \in \mathbb{R}^N)$, and F is a finite subset of $\mathcal{R}_y$ depending on ψ. $\mathcal{V}_{\infty y}$ is a subalgebra of A_y^∞ that preserves complex conjugation. Let $(,)$ stand for the inner product in $H^1(\Delta(A_y)) = W^{1,2}(\Delta(A_y))$, i.e.,

$$(u, v) = \int_{\Delta(A_y)} \left[u\overline{v} + \sum_{j=1}^{N} \partial_j u \overline{\partial_j v} \right] d\beta_y$$

for $u, v \in H^1(\Delta(A_y))$ (see Sect. 2.3 and in particular Sect. 2.3.3). Given $u \in H^1(\Delta(A_y))$, an elementary calculation gives

$$(u, \widehat{\gamma_k}) = (1 + 4\pi^2 |k|^2) \mathcal{F}_k(u) \quad (k \in \mathcal{R}_y)$$

where $\widehat{\gamma_k} = \mathcal{G}(\gamma_k)$, $\mathcal{F}_k(u)$ the Fourier coefficient of u at k. It follows at once that if $(u, \widehat{\psi}) = 0$ for all $\psi \in \mathcal{V}_{\infty y}$, then $u = 0$. This shows that $\mathcal{G}(\mathcal{V}_{\infty y})$ is dense in $H^1(\Delta(A_y))$. Finally, by noticing that

$$-\Delta_y \left(\frac{1}{4\pi^2 |k|^2} \gamma_k \right) = \gamma_k$$

for all $k \in \mathbb{R}^N \setminus \{\omega\}$ (ω the origin in $\mathbb{R}^N$), we see readily that any ψ as above with $F \subset \mathcal{R}_y \setminus \{\omega\}$ writes as $\psi = \Delta_y \theta$ with

$$\theta = -\sum_{k \in F} \frac{c_k}{4\pi^2 |k|^2} \gamma_k.$$

This completes the proof of (P1). The next point is to show (P3). Suppose $u \in L^2(\Delta(A_z); B)$ and $\int_{\Delta(A_z)} u(s_0) \varphi'(s_0) d\beta_z(s_0) = 0$ for all $\varphi \in \mathcal{D}(\Delta(A_z))$. Let $f \in L^2(\Delta(A_z))$, $f(s_0) = \langle \zeta, u(s_0) \rangle$ $(s_0 \in \Delta(A_z))$, where ζ is fixed in B' (the dual of B). By taking above $\varphi = \widehat{\overline{\gamma_k}}$ with $k \neq 0$, it follows $\mathcal{F}_k(f) = 0$ for all $k \in \mathbb{R}$, $k \neq 0$. In other words, one has $f(s_0) = \int_{\Delta(A_z)} f d\beta_z$, i.e., $\langle \zeta, u(s_0) \rangle = \langle \zeta, \int_{\Delta(A_z)} u d\beta_z \rangle$

for a.e. $s_0 \in \Delta(A_z)$. Hence (P3) follows by the arbitrariness of ζ. Finally, part (iii) follows readily by Lemma 2.6. ■

7.3.3.2 Homogenization of (7.1) in a General Almost Periodic Setting

Returning now to the homogenization of (7.1) under hypothesis (7.33) and noting that (7.2) holds (thanks to Lemma 7.12), we see in the light of Proposition 7.8 that for $p = 2$ and A as above, the hypotheses of Theorems 7.1 and 7.2 are satisfied without further ado, whereas the application of Theorem 7.4 requires that $\mathcal{R}_y$ should be a discrete subgroup of $\mathbb{R}^N$.

At the present time we consider the more general case where the function $a_i(\cdot, \cdot, \mu)$ (for fixed μ) is almost periodic in the sense of Stepanov (see Sect. 2.4). Specifically, we assume that

$$a_i(\cdot, \cdot, \mu) \in L^1_{AP}(\mathbb{R}^{N+1}) \ \ (1 \leq i \leq N) \text{ for fixed } \mu \in \mathbb{R}^N. \tag{7.34}$$

The same line of argument as followed before leads us to a countable product subgroup $\mathcal{R} = \mathcal{R}_y \times \mathcal{R}_z$ of $\mathbb{R}^{N+1}$ as above such that

$$a_i(\cdot, \cdot, \mu) \in L^1_{\mathcal{R}}(\mathbb{R}^{N+1}) \ \ (1 \leq i \leq N) \text{ for fixed } \mu \in \mathbb{R}^N, \tag{7.35}$$

where $L^1_{\mathcal{R}}(\mathbb{R}^{N+1}) = \{u | u \in L^1_{AP}(\mathbb{R}^{N+1}), Sp(u) \subset \mathcal{R}\}$ (Sect. 2.4). To $\mathcal{R}$ we attach the H-algebra $A = AP_{\mathcal{R}}(\mathbb{R}^{N+1})$ discussed earlier, and whose main properties are collected in Proposition 7.8. Thus, in order to reduce the problem to a mere application of the results achieved in Sect. 7.2, we only need to verify that the abstract hypothesis (7.2) is satisfied. To accomplish this, we require the following property which is nothing but the «evolution» version of Definition 5.1.

Definition 7.4 Given $1 \leq q < \infty$, the function $\mathbf{a} = (a_i)$ of $\mathbb{R}^N \times \mathbb{R} \times \mathbb{R}^N$ into $\mathbb{R}^N$ is said to amalgamate the spaces $\mathcal{B}(\mathbb{R}^{N+1})$ and $(L^q, l^\infty)(\mathbb{R}^{N+1})$ if for any $\boldsymbol{\Psi} \in \mathcal{B}(\mathbb{R}^{N+1}; \mathbb{R})^N$, the number

$$\sup_{k \in \mathbb{Z}^N \times \mathbb{Z}} \iint_{k+(Y \times Z)} |\mathbf{a}(y - r, z - \sigma, \boldsymbol{\Psi}(y, z)) - \mathbf{a}(y, z, \boldsymbol{\Psi}(y, z))|^q dydz$$

tends to zero as $|r| + |\sigma| \to 0$, where $Y = (-\frac{1}{2}, \frac{1}{2})^N$ and $Z = (-\frac{1}{2}, \frac{1}{2})$.

Remark 7.15 Sometimes this convergence property will shortly be written as

$$\lim_{|r|+|\sigma| \to 0} \|\tau_{(r,\sigma)} a_i(\cdot, \cdot, \boldsymbol{\Psi}) - a_i(\cdot, \cdot, \boldsymbol{\Psi})\|_{q,\infty} = 0 \ \ (1 \leq i \leq N),$$

where $\tau_{(r,\sigma)} a_i$ denotes the function $(y, z, \mu) \mapsto a_i(y - r, z - \sigma, \mu)$ of $\mathbb{R}^N \times \mathbb{R} \times \mathbb{R}^N$ into $\mathbb{R}^N$.

Remark 7.16 Examples 5.21–5.23 carry over mutatis mutandis to the present setting.

In the sequel it is assumed that the function $\mathbf{a} = (a_i)$ amalgamates $\mathcal{B}(\mathbb{R}^{N+1})$ and $(L^q, l^\infty)(\mathbb{R}^{N+1})$ for $q = 2$. This can be illustrated by an obvious adaptation of Example 5.24 (see also Example 5.25).

Proposition 7.9 *Under the structure hypothesis (7.34), the homogenization problem for (7.1) is solved for* $p = 2$ *and* $A = AP_{\mathcal{R}}(\mathbb{R}^{N+1})$ *($\mathcal{R}$ as above) by Theorems 7.1 and 7.2 in cases* $0 < k \leq 1$ *and* $1 < k < 2$*, respectively, and by Theorem 7.4 in the case* $k > 2$ *provided* $\mathcal{R}_y$ *is a discrete subgroup of* $\mathbb{R}^N$.

Proof In view of the preceding developments (see in particular Proposition 7.8), only (7.2) (for $p = 2$) remains to be established. The line of proceeding is an obvious adaptation of the argument followed in Sect. 5.5.4. To begin, we introduce a mollifier (θ_n) on $\mathbb{R}^{N+1}$, where θ_n has support in $\varepsilon_n \overline{B}_{N+1}$, (ε_n) being a fundamental sequence and $\overline{B}_{N+1}$ the closed unit ball in $\mathbb{R}^{N+1}$. For fixed $1 \leq i \leq N$ and $n \geq 0$, let

$$\xi_n^i(y, z, \mu) = \int\int \theta_n(r, \sigma) a_i(y - r, z - \sigma, \mu) dr d\sigma \tag{7.36}$$

with $y \in \mathbb{R}^N$, $z \in \mathbb{R}$ and $\mu \in \mathbb{R}^N$. One has

$$\xi_n^i(\cdot, \cdot, \mu) \in A_{\mathbb{R}} \text{ for fixed } \mu \in \mathbb{R}^N \tag{7.37}$$

and

$$\begin{aligned} &|\xi_n^i(y, z, \mu) - \xi_n^i(y, z, \nu)| \leq \alpha_1(|\mu| + |\nu|)^{p-2} |\mu - \nu| \\ &\text{for a.e. } (y, z) \in \mathbb{R}^N \times \mathbb{R} \text{ and for all } \mu, \nu \in \mathbb{R}^N. \end{aligned} \tag{7.38}$$

Indeed, (7.37) follows by (7.35) and use of Lemma 2.9, whereas (7.38) (with $p = 2$ here) results from the growth condition (6.22)(2). Hence, appeal to Lemma 7.12 yields

$$\xi_n^i(\cdot, \cdot, \mathbf{\Psi}) \in A \text{ for all } \mathbf{\Psi} \in (A_{\mathbb{R}})^N. \tag{7.39}$$

Bearing this in mind, one has on the other hand, by Hölder's inequality,

$$|\xi_n^i(y, z, \mu) - a_i(y, z, \mu)|^2 \leq \int\int \theta_n(r, \sigma) |a_i(y - r, z - \sigma, \mu) - a_i(y, z, \mu)|^2 dr d\sigma$$

and that for a.e. $(y, z) \in \mathbb{R}^N \times \mathbb{R}$ and for all $\mu \in \mathbb{R}^N$. Now, fix freely $\mathbf{\Psi} \in (A_{\mathbb{R}})^N$, and take $\mu = \mathbf{\Psi}(y, z)$ (for fixed $(y, z) \in \mathbb{R}^N \times \mathbb{R}$) in the preceding inequality to arrive at

$$\|\xi_n^i(\cdot, \cdot, \mathbf{\Psi}) - a_i(\cdot, \cdot, \mathbf{\Psi})\|_{2,\infty}^2 \leq \int\int \theta_n(r, \sigma) \|\tau_{(r,\sigma)} a_i(\cdot, \cdot, \mathbf{\Psi}) - a_i(\cdot, \cdot, \mathbf{\Psi})\|_{2,\infty}^2 dr d\sigma$$

and that for any integer $n \geq 0$. By the amalgamation hypothesis, we deduce

$$\xi_n^i(\cdot,\cdot,\Psi) \to a_i(\cdot,\cdot,\Psi) \text{ in } (L^2, l^\infty)(\mathbb{R}^{N+1}) \text{ as } n \to \infty.$$

It follows by (7.39) that $a_i(\cdot,\cdot,\Psi) \in L^2_{\mathcal{R}}(\mathbb{R}^{N+1}) \subset \mathfrak{X}^2_A$ (see Proposition 2.22 and Corollary 2.11), thereby proving the proposition. ■

7.3.4 Problem III: $L^1_{per}(Z; L^{1,\infty}_{per}(\mathbb{R}^N))$ Setting

First of all, we define $L^{1,\infty}_{per}(\mathbb{R}^N)$ as the closure of $\mathcal{B}_{\infty,\mathbb{Z}^N}(\mathbb{R}^N)$ in the amalgam space $(L^1, l^\infty)(\mathbb{R}^N)$ (see Examples 2.5 and 4.4 together with Sect. 2.4.1). Equipped with the $(L^1, l^\infty)(\mathbb{R}^N)$-norm, $L^{1,\infty}_{per}(\mathbb{R}^N)$ is a Banach space. Let us keep in mind that one may as well define $L^{1,\infty}_{per}(\mathbb{R}^N)$ as the closure of $\mathcal{K}(\mathbb{R}^N) \oplus \mathcal{C}_{per}(Y)$ in $(L^1, l^\infty)(\mathbb{R}^N)$. Also, note that $L^q(\mathbb{R}^N) + L^q_{per}(Y) \subset L^{1,\infty}_{per}(\mathbb{R}^N)$ $(1 \leq q < \infty)$.

One useful result is the following

Lemma 7.14 *Let $f \in L^{1,\infty}_{per}(\mathbb{R}^N)$ and $\theta \in \mathcal{K}(\mathbb{R}^N)$. Then $f * \theta \in \mathcal{B}_{\infty,\mathbb{Z}^N}(\mathbb{R}^N)$.*

Proof As a preliminary, for $u \in (L^1, l^\infty)(\mathbb{R}^N)$, on one hand we have $u * \theta \in \mathcal{C}(\mathbb{R}^N)$ (a classical fact), on the other hand, by the same argument as followed in proving part (ii) of Proposition 2.20 (see especially the inequality therein relevant to Lemma 2.8), one has $|(u * \theta)(y)| \leq c\|\theta\|_\infty \|u\|_{1,\infty}$ for all $y \in \mathbb{R}^N$, where the constant $c > 0$ depends only on θ. This being so, let $\eta > 0$. Fix some $\xi \in \mathcal{C}_{per}(Y)$ and $\varphi \in \mathcal{K}(\mathbb{R}^N)$ such that $\|f - \xi - \varphi\|_{1,\infty} \leq \frac{\eta}{c\|\theta\|_\infty}$. Then, by choosing above $u = f - \xi - \varphi$, one has $|(f - \xi) * \theta(y)| \leq \eta + |\varphi * \theta(y)|$ for $y \in \mathbb{R}^N$. Hence, by considering some compact set K in $\mathbb{R}^N$ containing the support of $\varphi * \theta$, it follows $|(f - \xi) * \theta(y)| \leq \eta$ for $y \in \mathbb{R}^N \setminus K$, from which one readily deduces that $f * \theta \in \mathcal{B}_0(\mathbb{R}^N) + \mathcal{C}_{per}(Y)$, since $\xi * \theta \in \mathcal{C}_{per}(Y)$. Therefore $f * \theta \in \mathcal{B}_{\infty,\mathbb{Z}^N}(\mathbb{R}^N)$. This completes the proof. ■

Now, the homogenization problem for (7.1) is here stated under the structure hypothesis

$$a_i(\cdot,\cdot,\mu) \in L^1_{per}(Z; L^{1,\infty}_{per}(\mathbb{R}^N)) \;\; (1 \leq i \leq N) \text{ for fixed } \mu \in \mathbb{R}^N. \tag{7.40}$$

We begin by analyzing the particular case where the a_i's are smooth functions of the variables y, z. More precisely, we assume that

$$a_i(\cdot,\cdot,\mu) \in \mathcal{C}_{per}(Z; \mathcal{B}_{\infty,\mathbb{Z}^N}(\mathbb{R}^N)) \;\; (1 \leq i \leq N) \;\; \text{for fixed } \mu \in \mathbb{R}^N, \tag{7.41}$$

where $\mathcal{C}_{per}(Z; \mathcal{B}_{\infty,\mathbb{Z}^N}(\mathbb{R}^N))$ denotes the space of Z-periodic continuous functions of $\mathbb{R}$ into $\mathcal{B}_{\infty,\mathbb{Z}^N}(\mathbb{R}^N)$. By identifying $\mathcal{C}_{per}(Z)$ with $\mathcal{C}(\mathbb{T})$ ($\mathbb{T}$ denotes the 1-torus) and $\mathcal{C}_{per}(Z; \mathcal{B}_{\infty,\mathbb{Z}^N}(\mathbb{R}^N))$ with $\mathcal{C}(\mathbb{T}; \mathcal{B}_{\infty,\mathbb{Z}^N}(\mathbb{R}^N))$ (see, e.g., Bourbaki [25, VII, p. 11]), we see that the space $\mathcal{C}_{per}(Z; \mathcal{B}_{\infty,\mathbb{Z}^N}(\mathbb{R}^N))$ is exactly the closure of $\mathcal{C}_{per}(Z) \otimes \mathcal{B}_{\infty,\mathbb{Z}^N}(\mathbb{R}^N)$ in $\mathcal{B}(\mathbb{R}^{N+1})$. Thus, the present case falls within the scope of Sect. 7.2.1 with

$$A = \mathcal{C}_{per}(Z; \mathcal{B}_{\infty,\mathbb{Z}^N}(\mathbb{R}^N)), \tag{7.42}$$

$A_y = \mathcal{B}_{\infty,\mathbb{Z}^N}(\mathbb{R}^N)$ and $A_z = \mathcal{C}_{per}(Z)$. By recalling that the pair $\{A_y = \mathcal{B}_{\infty,\mathbb{Z}^N}(\mathbb{R}^N), A_{2y} = \mathcal{C}_{per}(Y)\}$ satisfies hypothesis (H) (see Example 4.4 and Remark 4.7) so that Theorems 7.5 and 7.6 together with Proposition 7.5 apply with $A_2 = \mathcal{C}_{per}(Y \times Z)$ and A given by (7.42), we deduce at once the following from Proposition 7.7.

Proposition 7.10 *The H-algebra A (given by (7.42)) is V^2-proper and quasi V^p-proper for $2 \leq p < \infty$. Furthermore, A verifies properties (P1), (P2) and (P3).*

As a result, there is the following

Corollary 7.3 *Under the structure hypothesis (7.41), the homogenization problem for (7.1) is solved for $2 \leq p < \infty$ and A (in (7.42)) by Theorems 7.1, 7.2 and 7.4 if $0 < k \leq 1$, $1 < k < 2$ and $k > 2$, respectively, and by Theorem 7.3 if $k = 2$ provided $p = 2$.*

Proof One only needs to check (7.2). But this follows by Lemma 7.12 and use of the trivial relation $A \subset \mathfrak{X}_A^{p'}$. ■

At the present time, our purpose is to discuss the homogenization problem for (7.1) under the structure hypothesis (7.40). It is required this time of the function $\mathbf{a} = (a_i)$ that it amalgamates the spaces $\mathcal{B}(\mathbb{R}^{N+1})$ and $(L^q, l^\infty)(\mathbb{R}^{N+1})$ not only for $q = 2$ as in Problem II, but for $1 < q \leq 2$. We will need some preliminaries.

Lemma 7.15 *Let $\psi \in \mathcal{D}(\mathbb{R}_y^N)$ and $\varphi \in \mathcal{D}(\mathbb{R}_z)$. Put $\theta = \psi \otimes \varphi \in \mathcal{D}(\mathbb{R}^{N+1})$. Let $f \in L^1_{per}(Z; L^{1,\infty}_{per}(\mathbb{R}^N))$. Then $f * \theta \in \mathcal{C}^\infty(\mathbb{R}^{N+1}) \cap \mathcal{C}_{per}(Z; \mathcal{B}_{\infty,\mathbb{Z}^N}(\mathbb{R}^N))$.*

Proof It is a classical fact that $f * \theta$ is of class $\mathcal{C}^\infty$. We next check that $f * \theta \in \mathcal{C}_{per}(Z; \mathcal{B}_{\infty,\mathbb{Z}^N}(\mathbb{R}^N))$. For fixed $z \in \mathbb{R}$, let $f(\cdot, z)$ stand for the function $y \mapsto f(y, z)$ of $\mathbb{R}^N$ into $\mathbb{C}$, which lies in $L^{1,\infty}_{per}(\mathbb{R}^N)$. By Lemma 7.14 we have $f(\cdot, z) * \psi \in \mathcal{B}_{\infty,\mathbb{Z}^N}(\mathbb{R}^N)$ with $\|f(\cdot, z) * \psi\|_\infty \leq c\|\psi\|_\infty\|f(\cdot, z)\|_{1,\infty}$ for a.e. $z \in \mathbb{R}$, where the positive constant c does not depend on z. Now, let $F(z) = f(\cdot, z) * \psi$ $(z \in \mathbb{R})$. This defines a function $F : \mathbb{R} \to \mathcal{B}_{\infty,\mathbb{Z}^N}(\mathbb{R}^N)$ that lies in $L^1_{per}(Z; \mathcal{B}_{\infty,\mathbb{Z}^N}(\mathbb{R}^N))$ (the verification is an easy exercise left to the reader). It follows at once that $F * \varphi \in \mathcal{C}_{per}(Z; \mathcal{B}_{\infty,\mathbb{Z}^N}(\mathbb{R}^N))$. Thus, we are through if we can show the equality $f * \theta = F * \varphi$. This is straightforward. Indeed,

$$
\begin{aligned}
f * \theta(y, z) &= \int \varphi(\sigma) \left(\int \psi(r) f(y - r, z - \sigma) dr \right) d\sigma \\
&= \int \varphi(\sigma)[f(\cdot, z - \sigma) * \psi](y) d\sigma \\
&= \int \varphi(\sigma) F(y, z - \sigma) d\sigma
\end{aligned}
$$

and that for all $y \in \mathbb{R}^N$. It follows

$$
\begin{aligned}
f * \theta(\cdot, z) &= \int F(\cdot, z - \sigma)\varphi(\sigma) d\sigma \\
&= F * \varphi(z)
\end{aligned}
$$

for all $z \in \mathbb{R}$, hence $f * \theta = F * \varphi$, as claimed. ■

We are now ready to come to the heart of the matter.

Proposition 7.11 *Under the structure hypothesis (7.40), the conclusions of Corollary 7.3 carry over without the slightest change to the present general setting.*

Proof Thanks to Proposition 7.10, only (7.2) remains to be verified. The line of proceeding is the same as followed before. To begin with, we consider a mollifier $(\theta_n)_{n \in \mathbb{N}}$ of the form $\theta_n = \psi_n \otimes \varphi_n$, where (ψ_n) (resp. (φ_n)) is a mollifier on $\mathbb{R}^N$ (resp. $\mathbb{R}$) with Supp $\psi_n \subset \varepsilon_n \overline{B}_N$ (resp. Supp $\varphi_n \subset [-\varepsilon_n, \varepsilon_n]$), $0 < \varepsilon_n \leq 1$, $\varepsilon_n \to 0$ as $n \to \infty$. Next, for fixed $n \in \mathbb{N}$ and $1 \leq i \leq N$, we introduce the family $\{\xi_n^i(\cdot, \cdot, \mu)\}$ $(\mu \in \mathbb{R}^N)$ given by (7.36), with (7.37) and (7.38). Whereas there is no comment to make on (7.38), it is well to make it clear that (7.37) (with A given by (7.42)) follows by Lemma 7.15. From all that we deduce at once (7.39) by Lemma 7.12. On the other hand, by Hölder's inequality (as used earlier in proving Proposition 2.20) one easily gets

$$
|\xi_n^i(y, z, \mu) - a_i(y, z, \mu)|^{p'} \leq \int \int \theta_n(r, \sigma) |a_i(y - r, z - \sigma, \mu) - a_i(y, z, \mu)|^{p'} dr d\sigma,
$$

and that for a.e. $(y, z) \in \mathbb{R}^N \times \mathbb{R}$ and for all $\mu \in \mathbb{R}^N$. Hence

$$
\begin{aligned}
&\|\xi_n^i(\cdot, \cdot, \boldsymbol{\Psi}) - a_i(\cdot, \cdot, \boldsymbol{\Psi})\|_{p',\infty}^{p'} \leq \\
&\int \int \theta_n(r, \sigma) \|\tau_{(r,\sigma)} a_i(\cdot, \cdot, \boldsymbol{\Psi}) - a_i(\cdot, \cdot, \boldsymbol{\Psi})\|_{p',\infty}^{p'} dr d\sigma,
\end{aligned}
$$

where Ψ is arbitrary fixed in $(A_{\mathbb{R}})^N$. According to the amalgamation hypothesis, as $n \to \infty$ it follows that $\xi_n^i(\cdot,\cdot,\Psi) \to a_i(\cdot,\cdot,\Psi)$ in $(L^{p'}, l^\infty)(\mathbb{R}^{N+1})$, hence in $\Xi^{p'}(\mathbb{R}^{N+1})$ also (see Proposition 2.19). Therefore (7.2) follows by (7.39). ■

Let us illustrate this.

Example 7.4 In view of the relation $\mathcal{B}_{\infty,\mathbb{Z}^N}(\mathbb{R}^N) \subset L_{per}^{1,\infty}(\mathbb{R}^N)$ with continuous embedding, we see immediately that the homogenization of (7.1) under the structure hypothesis

$$a_i(\cdot,\cdot,\mu) \in L_{per}^q(Z; \mathcal{B}_{\infty,\mathbb{Z}^N}(\mathbb{R}^N)) \ (1 \le i \le N) \text{ for fixed } \mu \in \mathbb{R}^N,$$

where $1 \le q < \infty$, reduces to Problem III.

Example 7.5 Let $1 \le q < \infty$. Equipped with its natural norm, the Banach space $L_{per}^q(Y) \oplus L^q(\mathbb{R}^N)$ is continuously embedded in $L_{per}^{1,\infty}(\mathbb{R}^N)$. Hence, the homogenization problem for (7.1) under the structure hypothesis (for fixed $\mu \in \mathbb{R}^N$)

$$a_i(\cdot,\cdot,\mu) \in L_{per}^1(Z; L_{per}^q(Y) \oplus L^q(\mathbb{R}^N)) \ (1 \le i \le N)$$

reduces to Problem III.

7.3.5 *Problem IV: $L^{1,\infty}(\mathbb{R}; L_{per}^1(Y))$ Setting*

7.3.5.1 Preliminaries

Before we can state the problem on hand in this section, some preliminaries are necessary. In what follows, B denotes a Banach space. The notion of an amalgam space introduced in Sect. 2.4.1 (in the case of scalar functions) extends quite naturally to vector-valued functions. We are concerned here with the particular case of $(L^1, l^\infty)(\mathbb{R}; B)$ which is defined as the space of all $u \in L_{loc}^1(\mathbb{R}; B)$ such that

$$\|u\|_{1,\infty;B} = \sup_{k\in\mathbb{Z}} \int_{k+Z} \|u(z)\| dz < \infty, \ Z = \left(-\frac{1}{2}, \frac{1}{2}\right),$$

where $\|\cdot\|$ is the norm on B. It is an easy exercise (left to the reader) to show that this is a norm on $(L^1, l^\infty)(\mathbb{R}; B)$ making the latter into a Banach space. It is worth noting that $L^p(\mathbb{R}; B) \subset (L^1, l^\infty)(\mathbb{R}; B)$ $(1 \le p \le \infty)$ with continuous embedding. On the other hand, it is clear that $(L^1, l^\infty)(\mathbb{R}; B)$ is continuously embedded in $L_{loc}^1(\mathbb{R}; B)$.

Now, we define $L^{1,\infty}(\mathbb{R}; B)$ as the closure of $\mathcal{B}_\infty(\mathbb{R}; B)$ in $(L^1, l^\infty)(\mathbb{R}; B)$. We equip $L^{1,\infty}(\mathbb{R}; B)$ with the $(L^1, l^\infty)(\mathbb{R}; B)$-norm, which makes it into a Banach space. By the same line of proceeding mutatis mutandis as followed in proving

Lemma 7.14 one easily verifies that $f * \theta \in \mathcal{B}_\infty(\mathbb{R}; B)$ for $f \in L^{1,\infty}(\mathbb{R}; B)$ and $\theta \in \mathcal{K}(\mathbb{R})$.

7.3.5.2 Homogenization of (7.1) in the $L^{1,\infty}(\mathbb{R}; L^1_{per}(Y))$ Setting

We begin by discussing the homogenization of (7.1) under a «smooth» structure hypothesis connected with $L^{1,\infty}(\mathbb{R}; L^1_{per}(Y))$, namely

$$a_i(\cdot, \cdot, \mu) \in \mathcal{B}_\infty(\mathbb{R}; \mathcal{C}_{per}(Y)) \ \ (1 \le i \le N) \text{ for fixed } \mu \in \mathbb{R}^N. \tag{7.43}$$

Let us note at once that the H-algebra $\mathcal{B}_\infty(\mathbb{R}; \mathcal{C}_{per}(Y))$ is the closure of $\mathcal{C}_{per}(Y) \otimes \mathcal{B}_\infty(\mathbb{R})$ in $\mathcal{B}(\mathbb{R}^{N+1})$ (Example 2.4), so that the problem thus stated falls within the scope of Sect. 7.2 with

$$A = \mathcal{B}_\infty(\mathbb{R}; \mathcal{C}_{per}(Y)), \tag{7.44}$$

$A_y = \mathcal{C}_{per}(Y)$ and $A_z = \mathcal{B}_\infty(\mathbb{R})$; all these H-algebras having, besides, the required properties. Moreover, A_y is p-harmonic for all $1 < p < \infty$ (Example 7.2), from which it follows by Proposition 7.3 that A is quasi V^p-proper for any real $p \ge 2$. Also, the H-algebra A verifies properties (P1) and (P2). Finally, (7.2) follows in an evident fashion from (7.43)–(7.44) and use of Lemma 7.12. One deduces readily the following

Proposition 7.12 *Under the structure hypothesis (7.43), the homogenization problem for (7.1) is solved by Theorem 7.1 (case $0 < k \le 1$) and Theorem 7.2 (case $1 < k < 2$) for A given by (7.44) and for any real $p \ge 2$.*

We assume now that the function $\mathbf{a} = (a_i)$ amalgamates $\mathcal{B}(\mathbb{R}^{N+1})$ and $(L^q, l^\infty)(\mathbb{R}^{N+1})$ for any arbitrary $1 < q \le 2$ and, further, verifies

$$a_i(\cdot, \cdot, \mu) \in L^{1,\infty}(\mathbb{R}; L^1_{per}(Y)) \ \ (1 \le i \le N) \text{ for fixed } \mu \in \mathbb{R}^N. \tag{7.45}$$

We will need the analog of Lemma 7.15, viz.

Lemma 7.16 *Let ψ, φ and θ be as in Lemma 7.15. Let $f \in L^{1,\infty}(\mathbb{R}; L^1_{per}(Y))$. Then $f * \theta \in \mathcal{C}^\infty(\mathbb{R}^{N+1}) \cap \mathcal{B}_\infty(\mathbb{R}; \mathcal{C}_{per}(Y))$.*

Proof The property $f * \theta \in \mathcal{C}^\infty(\mathbb{R}^{N+1})$ being a classical fact, we shall concentrate our attention on verifying that $f * \theta$ lies in $\mathcal{B}_\infty(\mathbb{R}; \mathcal{C}_{per}(Y))$. The argument is the same as followed in proving Lemma 7.15 with, as chief point, the fact that there exists a constant $c = c(\psi) > 0$ such that

$$\|u * \psi\|_\infty \le c\|u\|_{1,\infty} \quad \text{for all } u \in (L^1, l^\infty)(\mathbb{R}^N). \tag{7.46}$$

Now, introduce the function $F : \mathbb{R} \to \mathcal{C}_{per}(Y)$, $F(z) = f(\cdot, z) * \psi$ $(z \in \mathbb{R})$. It is a rather simple exercise to see by (7.46) that the behaviour of f with respect to the variable z (e.g., in matters of measurability, growth, etc.) carries over to F, so that one has $F \in (L^1, l^\infty)(\mathbb{R}; \mathcal{C}_{per}(Y))$. Let us check that F lies in $L^{1,\infty}(\mathbb{R}; \mathcal{C}_{per}(Y))$. To this end, let $\eta > 0$. By definition there is some $v \in \mathcal{B}_\infty(\mathbb{R}; L^1_{per}(Y))$ such that $\|f - v\|_{1,\infty;L^1_{per}(Y)} \leq \frac{\eta}{c}$. Let $w(z) = v(\cdot, z) * \psi$ $(z \in \mathbb{R})$. This defines a function w of $\mathbb{R}$ into $\mathcal{C}_{per}(Y)$ which lies in $\mathcal{B}_\infty(\mathbb{R}; \mathcal{C}_{per}(Y))$; indeed, it is readily seen by (7.46) that w is continuous and further $w(z) \to \xi * \psi$ in $\mathcal{C}_{per}(Y)$ as $|z| \to \infty$, where ξ stands for the limit of $v(\cdot, z)$ in $L^1_{per}(Y)$ as $|z| \to \infty$.
Furthermore, by (7.46), once again, one readily arrives at $\|F - w\|_{1,\infty;\mathcal{C}_{per}(Y)} \leq \eta$, which shows that $F \in L^{1,\infty}(\mathbb{R}; \mathcal{C}_{per}(Y))$, as claimed. But then, as pointed out in the preliminaries, we have $F * \varphi \in \mathcal{B}_\infty(\mathbb{R}; \mathcal{C}_{per}(Y))$. Hence, the lemma follows by joining this with the equality $f * \theta = F * \varphi$. ■

Finally comes the proposition providing a solution to the problem before us.

Proposition 7.13 *Under the structure hypothesis (7.45), the conclusions of Proposition 7.12 carry over without the slightest change to the present general setting.*

Proof The different properties of A stated earlier show that only (7.2) needs proving. But, thanks to Lemma 7.16 it is easily seen that the proof of Proposition 7.11 carries over mutatis mutandis to the present situation. ■

This is worth illustrating.

Example 7.6 The solution to the homogenization problem (7.1) under the structure hypothesis

$$a_i(\cdot, \cdot, \mu) \in \mathcal{B}_\infty(\mathbb{R}; L^1_{per}(Y)) \ (1 \leq i \leq N) \text{ for fixed } \mu \in \mathbb{R}^N$$

is provided by Proposition 7.13. Indeed, this structure hypothesis reduces trivially to (7.45).

7.3.6 *Problem V: $L^{1,\infty}(\mathbb{R}; L^1_{AP}(\mathbb{R}^N))$ Setting*

This section deals with the homogenization of (7.1) under the structure hypothesis

$$a_i(\cdot, \cdot, \mu) \in L^{1,\infty}(\mathbb{R}; L^1_{AP}(\mathbb{R}^N)) \ (1 \leq i \leq N) \text{ for fixed } \mu \in \mathbb{R}^N. \tag{7.47}$$

It is clear that this problem generalizes the preceding one. Moreover, the line of proceeding, with which the reader must be familiar now, is the same as before. We will start by considering the homogenization problem not under (7.47), but under a suitable so-called smooth structure hypothesis, viz.

$$a_i(\cdot, \cdot, \mu) \in \mathcal{B}_\infty(\mathbb{R}; AP(\mathbb{R}^N)) \ (1 \leq i \leq N) \text{ for fixed } \mu \in \mathbb{R}^N. \tag{7.48}$$

Besides the fact that it is quite interesting by itself, the problem is actually a preliminary step to the solution of the original problem. Indeed, it helps have an idea of what the algebra that fits the whole study must look like.

Now, the same argument as used in proving Lemma 5.3 yields a countable subgroup $\mathcal{R}$ of $\mathbb{R}^N$ such that one has $a_i(\cdot,\cdot,\mu) \in \mathcal{B}_\infty(\mathbb{R}; AP_{\mathcal{R}}(\mathbb{R}^N))$ $(1 \leq i \leq N)$ for fixed $\mu \in \mathbb{R}^N$. Recalling that the latter space coincides with the closure in $\mathcal{B}(\mathbb{R}^{N+1})$ of $AP_{\mathcal{R}}(\mathbb{R}^N) \otimes \mathcal{B}_\infty(\mathbb{R})$ (Example 2.4), we see that the present problem falls within the abstract framework of Sect. 7.2.1 with

$$A = \mathcal{B}_\infty(\mathbb{R}; AP_{\mathcal{R}}(\mathbb{R}^N)), \tag{7.49}$$

$A_y = AP_{\mathcal{R}}(\mathbb{R}^N)$ and $A_z = \mathcal{B}_\infty(\mathbb{R})$. Furthermore, the H-algebra $AP_{\mathcal{R}}(\mathbb{R}^N)$ being 2-harmonic (Example 7.3), it follows by Proposition 7.3 that A is quasi V^2-proper. On the other hand, A verifies property (P1). Finally, by Lemma 7.12 we see that (7.2) holds. In conclusion, under the structure hypothesis (7.48), all of the hypotheses of Theorems 7.1 and 7.2 are satisfied for $p = 2$ and A given by (7.49), and so the solution of the problem before us is provided by the latter theorems.

We shall now discuss the homogenization of (7.1) under the structure hypothesis (7.47). We need two preliminary results.

Lemma 7.17 *Let $(f_\alpha)_{\alpha\in\mathfrak{A}}$ be a finite or a countable family of functions in $L^{1,\infty}(\mathbb{R}; L^1_{AP}(\mathbb{R}^N))$. There exists a countable subgroup $\mathcal{R}$ of $\mathbb{R}^N$ such that $f_\alpha \in L^{1,\infty}(\mathbb{R}; L^1_{\mathcal{R}}(\mathbb{R}^N))$ for all $\alpha \in \mathfrak{A}$.*

Proof In view of the density of $\mathcal{B}_\infty(\mathbb{R}; L^1_{AP}(\mathbb{R}^N))$ in $L^{1,\infty}(\mathbb{R}; L^1_{AP}(\mathbb{R}^N))$, to each index $\alpha \in \mathfrak{A}$ there is attached a sequence $(f_{(\alpha,n)})_{n\in\mathbb{N}}$ of functions in $\mathcal{B}_\infty(\mathbb{R}; L^1_{AP}(\mathbb{R}^N))$ such that as $n \to \infty$, we have $f_{(\alpha,n)} \to f_\alpha$ in $L^{1,\infty}(\mathbb{R}; L^1_{AP}(\mathbb{R}^N))$. On the whole, this yields a family $\{f_{(\alpha,n)}\}$ with (α, n) ranging over $\mathfrak{A} \times \mathbb{N}$. This is of course a countable family. By a result used several times in this work (see Nguetseng and Woukeng [103] for a more general statement), there is a countable subgroup $\mathcal{R}$ of $\mathbb{R}^N$ such that $f_{(\alpha,n)} \in \mathcal{B}_\infty(\mathbb{R}; L^1_{\mathcal{R}}(\mathbb{R}^N))$ for all $(\alpha, n) \in \mathfrak{A} \times \mathbb{N}$.

Now, fix freely some index $\alpha \in \mathfrak{A}$. Let $k \in \mathbb{R}^N$. For a.e. $z \in \mathbb{R}$, we have

$$|M(\overline{\gamma}_k f_{(\alpha,n)}(z)) - M(\overline{\gamma}_k f_\alpha(z))| \leq c\|f_{(\alpha,n)}(z) - f_\alpha(z)\|_{1,\infty} \tag{7.50}$$

for all integers $n \geq 0$, where c is a positive constant depending on non variable quantities (see Proposition 2.19 and Corollary 2.10). This being so, in view of the continuity of the embedding $(L^1, l^\infty)(\mathbb{R}; L^1_{AP}(\mathbb{R}^N)) \hookrightarrow L^1_{loc}(\mathbb{R}; L^1_{AP}(\mathbb{R}^N))$, one has $f_{(\alpha,n)} \to f_\alpha$ in $L^1_{loc}(\mathbb{R}; L^1_{AP}(\mathbb{R}^N))$ as $n \to \infty$. Therefore, given any arbitrary bounded open interval $I \subset \mathbb{R}$, by a well known result from integration theory, one can extract from the preceding sequence some subsequence still denoted for convenience by $(f_{(\alpha,n)})_{n\in\mathbb{N}}$, such that when $n \to \infty$, one has $f_{(\alpha,n)}(z) \to f_\alpha(z)$ in $(L^1, l^\infty)(\mathbb{R}^N)$ for a.e. $z \in I$. So, assuming that k lies outside $\mathcal{R}$, which implies $M(\overline{\gamma}_k f_{(\alpha,n)}(z)) = 0$ for all $n \in \mathbb{N}$ and all $z \in \mathbb{R}$, we see immediately by letting

$n \to \infty$ in (7.50) that $M(\overline{\gamma}_k f_\alpha(z)) = 0$ for a.e. $z \in I$. Hence the lemma follows by the arbitrariness of $I, k \in \mathbb{R}^N \setminus \mathcal{R}$ and $\alpha \in \mathfrak{A}$. ■

Lemma 7.18 *Let ψ, φ and θ be given as in Lemma 7.15. Let $\mathcal{R}$ be a countable subgroup of $\mathbb{R}^N$. For any given $f \in L^{1,\infty}(\mathbb{R}; L^1_{\mathcal{R}}(\mathbb{R}^N))$, one has $f * \theta \in \mathcal{C}^\infty(\mathbb{R}^{N+1}) \cap \mathcal{B}_\infty(\mathbb{R}; AP_{\mathcal{R}}(\mathbb{R}^N))$.*

Proof One only needs to show that $f * \theta$ lies in $\mathcal{B}_\infty(\mathbb{R}; AP_{\mathcal{R}}(\mathbb{R}^N))$. The line of proceeding is the same mutatis mutandis as before (see especially the proof of Lemma 7.16), and we will not go deeply into detail. To begin, for a.e. $z \in \mathbb{R}$ one has $f(\cdot, z) * \psi \in AP_{\mathcal{R}}(\mathbb{R}^N)$ (see Lemma 2.9). Therefore, letting $F(z) = f(\cdot, z) * \psi$ $(z \in \mathbb{R})$ yields a function $F : \mathbb{R} \to AP_{\mathcal{R}}(\mathbb{R}^N)$ that proves to lie in $L^{1,\infty}(\mathbb{R}; AP_{\mathcal{R}}(\mathbb{R}^N))$, thanks to (7.46). Consequently, $F * \varphi \in \mathcal{B}_\infty(\mathbb{R}; AP_{\mathcal{R}}(\mathbb{R}^N))$. Hence, the desired result follows by the equality $f * \theta = F * \psi$, that is, precisely, $f * \theta(\cdot, z) = F * \psi(z)$ for all $z \in \mathbb{R}$. ■

We are now in a position to be through with the question before us. We assume that the function $\mathbf{a} = (a_i)$ amalgamates $\mathcal{B}(\mathbb{R}^{N+1})$ and $(L^2, l^\infty)(\mathbb{R}^{N+1})$.

Proposition 7.14 *The homogenization problem for (7.1) under the structure hypothesis (7.47) is solved by Theorem 7.1 (case $0 < k \leq 1$) and Theorem 7.2 (case $1 < k < 2$) for $p = 2$ and A given by (7.49) for some suitable countable subgroup $\mathcal{R}$ of $\mathbb{R}^N$.*

Proof It is clear that only (7.2) remains to be verified. The line of proceeding is familiar now. From (7.47) we see by Lemma 7.17 that there exists a countable subgroup $\mathcal{R}$ of $\mathbb{R}^N$ such that $a_i(\cdot, \cdot, \mu) \in L^{1,\infty}(\mathbb{R}; L^1_{\mathcal{R}}(\mathbb{R}^N))$ $(1 \leq i \leq N)$ for fixed $\mu \in \mathbb{R}^N$. To $\mathcal{R}$ we attach the H-algebra A on $\mathbb{R}^{N+1}$ given by (7.49). This H-algebra is well known for enabling us to work out the homogenization of (7.1) under the structure hypothesis (7.48), and so we have only to concentrate on showing (7.2) for $p = 2$. For that purpose, we consider a mollifier as in the proof of Proposition 7.11. Next, for fixed $n \in \mathbb{N}$ and $1 \leq i \leq N$, we define the family $\{\xi^i_n(\cdot, \cdot, \mu)\}$ $(\mu \in \mathbb{R}^N)$ by (7.36), which family proves to satisfy (7.37) and (7.38) (for $p = 2$). It is worth noting that (7.37) results from the relation $a_i(\cdot, \cdot, \mu) \in L^{1,\infty}(\mathbb{R}; L^1_{\mathcal{R}}(\mathbb{R}^N))$ (established earlier) and use of Lemma 7.18. By exactly proceeding as in the proof of Proposition 7.11 (in which we put $p = 2$), one arrives at once at (7.2), as claimed. ■

The following generalizes Example 7.6.

Example 7.7 The structure hypothesis

$$a_i(\cdot, \cdot, \mu) \in \mathcal{B}_\infty(\mathbb{R}; L^1_{AP}(\mathbb{R}^N)) \ \ (1 \leq i \leq N) \text{ for fixed } \mu \in \mathbb{R}^N$$

reduces in a trivial fashion to (7.47).

7.4 Comments and Problems

The deterministic homogenization problem for (7.1) beyond the periodic setting was considered for the first time in Nguetseng and Woukeng [103, 104]. However, the results of this chapter beyond Sect. 7.2.5 are given here for the first time.

The study of the linear version of (7.1), i.e., with

$$a_i(y, z, \mu) = \sum_{j=1}^{N} a_{ij}(y, z)\mu_j \quad (1 \leq i \leq N), \tag{7.51}$$

goes back to Bensoussan et al. [14]. More precisely, under the periodicity hypothesis (in both y and z) on the a_{ij}'s (see Sect. 6.1.2), and the assumption that k ranges over the positive real numbers, the authors developed a thorough mathematical analysis of the corresponding homogenization problem following the well-known approach combining the use of multiple scales asymptotic expansions with so-called energy methods. Later on, Pankov [114] used G-convergence to study, under the periodicity hypothesis, the homogenization of the full quasilinear problem (7.1).

Though this chapter includes implicitly (through (7.51)) the linear setting, the deterministic homogenization, beyond the periodic setting, of linear parabolic operators deserves to be studied in a self-contained fashion on the lines of Sect. 6.1.

Let us consider the initial boundary value problem (6.4) with this time $a_{ij}^{\varepsilon}(x, t) = a_{ij}(x, t, \frac{x}{\varepsilon}, \frac{t}{\varepsilon^k})$ for $(x, t) \in Q$, where $k > 0$, and where for convenience the functions a_{ij} assume real values. Let $f \in L^2(0, T; H^{-1}(\Omega; \mathbb{R}))$. A quick survey of the proof of Lemma 5.2 reveals that Lemma 6.1 is valid here, so that to each $\varepsilon > 0$ there is assigned a unique function $u_\varepsilon \in L^2(0, T; H_0^1(\Omega; \mathbb{R})) \cap \mathcal{C}([0, T]; L^2(\Omega; \mathbb{R}))$ satisfying (6.4).

Let $A = \overline{A_y \otimes A_z}$ (closure in $\mathcal{B}(\mathbb{R}^N \times \mathbb{R})$) be as in Sect. 7.2.1. One assumes that A is a quasi V^2-proper homogenization algebra. Assuming further that A^∞ is dense in A, one is interested in investigating the limiting behaviour of u_ε as $\varepsilon \to 0$ under the abstract hypothesis

$$a_{ij}(x, t, \cdot, \cdot) \in \mathfrak{X}_A^2(\mathbb{R}^N \times \mathbb{R}) \quad (1 \leq i, j \leq N) \quad \text{for fixed } (x, t) \in \overline{Q}, \tag{7.52}$$

where $a_{ij}(x, t, \cdot, \cdot)$ stands for the function $(y, z) \mapsto a_{ij}(x, t, y, z)$ from $\mathbb{R}^N \times \mathbb{R}$ to $\mathbb{R}$.

To summarize, one can accomplish this in four cases under hypothesis (7.52).

(1) **The case** $0 < k \leq 1$. One can show that under the preceding hypotheses we have, as $\varepsilon \to 0$,

$$u_\varepsilon \to u_0 \quad \text{in } L^2(0, T; H_0^1(\Omega; \mathbb{R}))\text{-weak} \tag{7.53}$$

$$\frac{\partial u_\varepsilon}{\partial t} \to \frac{\partial u_0}{\partial t} \quad \text{in } L^2(0, T; H^{-1}(\Omega; \mathbb{R}))\text{-weak} \tag{7.54}$$

$$\frac{\partial u_\varepsilon}{\partial x_j} \to \mathbb{D}_j \mathbf{u} = \frac{\partial u_0}{\partial x_j} + \partial_j u_1 \quad \text{in } L^2(Q)\text{-weak } \Sigma \ \ (1 \le j \le N), \tag{7.55}$$

where $\mathbf{u} = (u_0, u_1)$ is uniquely determined by the variational problem

$$\begin{cases} \mathbf{u} = (u_0, u_1) \in \mathbb{F}^{1,2}_{\mathcal{H}_2} = V_0^2 \times L^2(Q; \mathcal{H}_2) : \\ \displaystyle\int_0^T [u_0'(t), v_0(t)]\, dt + \int_0^T a_\Omega(t; \mathbf{u}(t), \mathbf{v}(t))\, dt \\ \displaystyle = \int_0^T [f(t), v_0(t)]\, dt \text{ for all } \mathbf{v} = (v_0, v_1) \in \mathbb{F}^{1,2}_{\mathcal{H}_2} \end{cases} \tag{7.56}$$

with

$$a_\Omega(t; \mathbf{w}, \boldsymbol{\Phi}) = \sum_{i,j=1}^N \int\int_{\Omega \times \Delta(A)} \widehat{a}_{ij}(x, t, \cdot, \cdot) \mathbb{D}_j \mathbf{w}(x, \cdot, \cdot) \mathbb{D}_i \boldsymbol{\Phi}(x, \cdot, \cdot)\, dx d\beta$$

for $\mathbf{w}, \boldsymbol{\Phi} \in H_0^1(\Omega; \mathbb{R}) \times L^2(\Omega; \mathcal{H}_2)$ ($\mathcal{H}_2$ defined in Sect. 7.2.2).

(2) **The case** $1 < k < 2$. Assuming furthermore that the H-algebra A is subject to condition (P1) (see Sect. 7.2.5) one can see readily that (7.53)–(7.55) hold as $\varepsilon \to 0$, where $\mathbf{u} = (u_0, u_1)$ is uniquely determined by (7.56).

(3) **The case** $k = 2$. One assumes here that the H-algebra A satisfies conditions (P1) and (P2) (see Sect. 7.2.6) and further is V^2-proper. One can check that (7.53)–(7.55) hold true when $\varepsilon \to 0$, where $\mathbf{u} = (u_0, u_1)$ is the unique solution of the variational problem

$$\begin{cases} \mathbf{u} = (u_0, u_1) \in \mathbb{F}^{1,2}_{\mathcal{W}} = V_0^2 \times L^2(Q; \mathcal{W}) : \\ \displaystyle\int_0^T [u_0'(t), v_0(t)]\, dt + \int_0^T a_\Omega(t; \mathbf{u}(t), \mathbf{v}(t))\, dt \\ \displaystyle + \int_Q \left\langle \frac{d}{ds_0} u_1(x, t), v_1(x, t) \right\rangle_{\mathcal{H}_2', \mathcal{H}_2} dx dt = \int_0^T [f(t), v_0(t)]\, dt \\ \text{for all } \mathbf{v} = (v_0, v_1) \in \mathbb{F}^{1,2}_{\mathcal{W}}. \end{cases}$$

It is well to note that the Banach space $\mathcal{U} = \{v | v \in \mathcal{H}_2, \frac{dv}{ds_0} \in \mathcal{H}_2'\}$ (see Sect. 7.2.6) is a Hilbert space with the equivalent norm $\|v\| = (\|v\|^2_{\mathcal{H}_2} + \|\frac{dv}{ds_0}\|^2_{\mathcal{H}_2'})^{\frac{1}{2}}$ $(v \in \mathcal{U})$. Thus, equipped with this norm, $\mathcal{W}$ is a Hilbert space.

(4) **The case** $k > 2$. One assumes here that the H-algebra A satisfies the further conditions (P1), (P2) and (P3). By the same line of proceeding as followed in Sect. 7.2.7 one can show that (7.53)–(7.55) hold as $\varepsilon \to 0$, where $\mathbf{u} = (u_0, u_1)$ is uniquely determined by the variational problem

$$\begin{cases} \mathbf{u} = (u_0, u_1) \in \mathbb{F}_0^{1,2} = V_0^2 \times L^2(Q; H^1(\Delta(A_y); \mathbb{R})/\mathbb{R}) : \\ \int_0^T [u_0'(t), v_0(t)]\, dt + \int_0^T \widetilde{a}_\Omega(t; \mathbf{u}(t), \mathbf{v}(t))\, dt = \int_0^T [f(t), v_0(t)]\, dt \\ \text{for all } \mathbf{v} = (v_0, v_1) \in \mathbb{F}_0^{1,2} \end{cases}$$

with

$$\widetilde{a}_\Omega(t; \mathbf{w}, \mathbf{\Phi}) = \sum_{i,j=1}^{N} \int\int_{\Omega \times \Delta(A_y)} \widehat{b}_{ij}(x, t, s)\mathbb{D}_j \mathbf{w}(x, s)\mathbb{D}_i \mathbf{\Phi}(x, s)\, dx d\beta_y(s)$$

for $\mathbf{w}, \mathbf{\Phi} \in H_0^1(\Omega) \times L^2(\Omega; H^1(\Delta(A_y); \mathbb{R})/\mathbb{R})$, where

$$\widehat{b}_{ij}(x, t, s) = \int_{\Delta(A_z)} \widehat{a}_{ij}(x, t, s, s_0) d\beta_z(s_0) \quad (s \in \Delta(A_y)).$$

The various concrete homogenization problems that are worked out in Sect. 7.3 confirm the fact that homogenization beyond the periodic setting may involve difficulties that are beyond imagination.

Deterministic homogenization problems for evolution operators have been solved in various fields using the Σ-convergence method. For example, for evolution Navier-Stokes equations in fissured media see Douanla et al. [48]; for hyperbolic differential operators see Nguetseng et al. [105, 108]; for Vlasov equations see Kenne et al. [66].

Chapter 8
Reiterated Homogenization

Reiterated homogenization is beyond the common setting of Chaps. 5–7, and an appropriate adaptation of the Σ-convergence method is therefore required. Thus, the present topic needs to be discussed apart. Our approach is carried out on a Dirichlet boundary value problem for a quasilinear elliptic operator.

8.1 Introduction

8.1.1 Setting of the Problem

Let $1 < p < \infty$. Let $\mathbf{a} = (a_i)_{1 \le i \le N}$ be a function of $\mathbb{R}_y^N \times \mathbb{R}_z^N \times \mathbb{R}_\mu^N$ into $\mathbb{R}^N$ with the following properties:

$$\begin{aligned}&\text{For fixed } \mu \in \mathbb{R}^N, \text{ the function } (y, z) \mapsto \mathbf{a}(y, z, \mu) \text{ of } \mathbb{R}^N \times \mathbb{R}^N\\ &\text{into } \mathbb{R}^N \text{possesses the Carathéodory property in the following}\\ &\text{sense:}\\ &\textbf{(1)} \text{ for fixed } z \in \mathbb{R}^N, \text{ the function } y \mapsto \mathbf{a}(y, z, \mu) \text{ is}\\ &\text{measurable from } \mathbb{R}^N \text{ to } \mathbb{R}^N,\\ &\textbf{(2)} \text{ for a.e. } y \in \mathbb{R}^N, \text{ the function } z \mapsto \mathbf{a}(y, z, \mu) \text{ maps}\\ &\text{continuously } \mathbb{R}^N \text{ into } \mathbb{R}^N.\end{aligned} \tag{8.1}$$

$$\mathbf{a}(y, z, \omega) = \omega \text{ (the origin in } \mathbb{R}^N) \text{ for all } z \in \mathbb{R}^N \text{ and a.e. } y \in \mathbb{R}^N. \tag{8.2}$$

G. Nguetseng, *Homogenization Algebras and Applications*, Springer Monographs in Mathematics, https://doi.org/10.1007/978-3-031-84705-9_8

There are four constants $c_1, c_2 > 0, 0 < \alpha_1 \leq \min\{1, p-1\}$ and

$\alpha_2 \geq \max\{p, 2\}$ such that for all $z \in \mathbb{R}^N$ and a.e. $y \in \mathbb{R}^N$ one has

$$\begin{aligned}&\textbf{(3)}\ |\mathbf{a}(y,z,\mu) - \mathbf{a}(y,z,\nu)| \leq c_1(1+|\mu|+|\nu|)^{p-1-\alpha_1}|\mu-\nu|^{\alpha_1}\\ &\textbf{(4)}\ (\mathbf{a}(y,z,\mu) - \mathbf{a}(y,z,\nu))\cdot(\mu-\nu)\\ &\qquad \geq c_2(1+|\mu|+|\nu|)^{p-\alpha_2}|\mu-\nu|^{\alpha_2}\end{aligned} \tag{8.3}$$

for all $\mu, \nu \in \mathbb{R}^N$, where the dot denotes the inner product in $\mathbb{R}^N$.

For given $f \in W^{-1,p'}(\Omega;\mathbb{R})$, where Ω is a smooth bounded open set in $\mathbb{R}^N_x$, we consider the boundary value problem

$$-\operatorname{div}\mathbf{a}\left(\frac{x}{\varepsilon}, \frac{x}{\varepsilon^2}, Du_\varepsilon\right) = f \text{ in } \Omega,\ u_\varepsilon \in W_0^{1,p}(\Omega;\mathbb{R}), \tag{8.4}$$

where the real $\varepsilon > 0$ is arbitrary fixed. Provided the left-hand side of (8.4) has meaning (see Sect. 8.2), an existence and uniqueness result for this boundary value problem can be achieved by the same line of proceeding as sketched in Sect. 5.5 (see, e.g., Zeidler [147] for details). The aim is then to investigate the limiting behaviour of u_ε as $\varepsilon \to 0$ under a suitable structure hypothesis on $a_i(y,z,\mu)$ $(1 \leq i \leq N)$ with respect to (y,z) for fixed μ.

8.1.2 Orientation

Actually, the main purpose of this chapter is to attract attention to the capability of the Σ-convergence theory of carrying out the mathematical analysis of reiterated homogenization problems as well. Beyond the homogenization of (8.4) as a «model problem» we are thus led to introduce in Sect. 8.2 the new concept of a *reiteration* H-*algebra* underlying the other new one of (weak and strong) *reiterated* Σ-*convergence* together with that of $W^{1,p}$-*properness* in the specific context of reiterated homogenization. Section 8.3 is devoted to a thorough study of the algebra $\mathcal{C}_{per}(Y \times Z)$ with $Y = Z = (-\frac{1}{2}, \frac{1}{2})^N$, which algebra proves to be the perfect reiteration H-algebra in all respects. Particular attention is drawn to Lemma 8.4 justifiably considered as the keystone of reiterated homogenization in the periodic setting. In Sect. 8.4 is revisited the crucial notion of a $W^{1,p}$-proper reiteration H-algebra illustrated by a number of examples. Section 8.5 is devoted to the homogenization problem for (8.4) in two directions: on one hand, under an abstract assumption on $\mathbf{a}(y,z,\mu)$ (for fixed μ) underlying a general deterministic setting covering a variety of concrete cases (see Sect. 8.6); on the other hand, under a periodicity hypothesis on $\mathbf{a}(y,z,\mu)$ (for fixed μ) relevant to the self-contained periodic setting of Sect. 8.3. Finally, in order to illustrate the preceding abstract setting, Sect. 8.6 provides a number of concrete examples considered in various contexts such as the periodic setting, the almost periodic setting and others.

8.2 Preliminaries

8.2.1 *Traces*

Let Ω be an open set in $\mathbb{R}_x^N$ (Ω is not necessarily bounded). Let $\varepsilon > 0$. For $u \in L^1_{loc}(\Omega \times \mathbb{R}_y^N \times \mathbb{R}_z^N)$, we set

$$u^\varepsilon(x) = u\left(x, \frac{x}{\varepsilon}, \frac{x}{\varepsilon^2}\right) \quad (x \in \Omega) \tag{8.5}$$

whenever the right-hand side has meaning. That is obviously the case if u is continuous on $\Omega \times \mathbb{R}_y^N \times \mathbb{R}_z^N$. However, in most cases we shall be dealing with less smooth functions, and there will be need for giving meaning to the right of (8.5).

8.2.1.1 The Spaces $L^p(\Omega; F)$ $(1 \leq p \leq \infty)$

Let F be a closed vector subspace of $\mathcal{B}(\mathbb{R}_y^N \times \mathbb{R}_z^N)$ equipped with the supremum norm on $\mathbb{R}_y^N \times \mathbb{R}_z^N$.

Proposition 8.1 *Let $1 \leq p < \infty$. For $u \in \mathcal{K}(\Omega; F)$, let the function $u^\varepsilon \in \mathcal{K}(\Omega)$ be defined by (8.5). Then, the mapping $u \mapsto u^\varepsilon$ of $\mathcal{K}(\Omega; F)$ into $\mathcal{K}(\Omega)$ extends by continuity to a continuous linear mapping, still denoted by $u \mapsto u^\varepsilon$, of $L^p(\Omega; F)$ into $L^p(\Omega)$ such that $\|u^\varepsilon\|_{L^p(\Omega)} \leq \|u\|_{L^p(\Omega;F)}$ for all $u \in L^p(\Omega; F)$.*

Proof This follows in a classical way by the density of $\mathcal{K}(\Omega; F)$ in $L^p(\Omega; F)$ (see Proposition 1.6). ■

The next proposition follows by an elementary adaptation of the proof of Proposition 1.7 (the details are left to the reader).

Proposition 8.2 *For $u \in L^\infty(\Omega; F)$, the right-hand side of (8.5) is defined for almost every $x \in \Omega$, which yields a function $u^\varepsilon \in L^\infty(\Omega)$ and further a continuous linear mapping $u \mapsto u^\varepsilon$ of $L^\infty(\Omega; F)$ into $L^\infty(\Omega)$ such that $\|u^\varepsilon\|_{L^\infty(\Omega)} \leq \|u\|_{L^\infty(\Omega;F)}$.*

8.2.1.2 The Space $\mathcal{C}(\overline{\Omega}; L^\infty(\mathbb{R}_y^N; \mathcal{B}(\mathbb{R}_z^N)))$

For simplicity we suppose here that Ω is bounded. To begin, let $\psi \in L^\infty(\mathbb{R}_y^N; \mathcal{B}(\mathbb{R}_z^N))$. Put ${}^\varepsilon\psi(y) = \psi(y, \frac{y}{\varepsilon})$ for $y \in \mathbb{R}^N$, where the right-hand side is defined as in Proposition 1.7. This yields a function ${}^{\varepsilon}\psi \in L^{\infty}(\mathbb{R}^N_y)$ satisfying $\| {}^{\varepsilon}\psi \|_{L^{\infty}(\mathbb{R}^N)} \leq \| \psi \|_{L^{\infty}(\mathbb{R}^N;\mathcal{B}(\mathbb{R}^N))}$.

With this in mind, let us denote by $({}^1H_{\varepsilon})_{\varepsilon>0}$ the usual action of $\mathbb{R}^*_+$ on $\mathbb{R}^N$ given by ${}^1H_{\varepsilon}(x) = \frac{x}{\varepsilon}$. It is worth recalling we have $w \circ {}^1H_{\varepsilon} \in L^{\infty}(\mathbb{R}^N_x)$ for $w \in L^{\infty}(\mathbb{R}^N_y)$, and further $\|w \circ {}^1H_{\varepsilon}\|_{L^{\infty}(\mathbb{R}^N)} \leq \|w\|_{L^{\infty}(\mathbb{R}^N)}$. By choosing in particular $w = {}^{\varepsilon}\psi$ we are led to set

$$\psi^{\varepsilon} = {}^{\varepsilon}\psi \circ {}^1H_{\varepsilon} \tag{8.6}$$

or in a formal manner

$$\psi^{\varepsilon}(x) = \psi\left(\frac{x}{\varepsilon}, \frac{x}{\varepsilon^2}\right) \quad (x \in \mathbb{R}^N), \tag{8.7}$$

and it is clear that $\psi^{\varepsilon} \in L^{\infty}(\mathbb{R}^N_x)$ with

$$\|\psi^{\varepsilon}\|_{L^{\infty}(\mathbb{R}^N_x)} \leq \|\psi\|_{L^{\infty}(\mathbb{R}^N_y;\mathcal{B}(\mathbb{R}^N_z))}.$$

The following lemma will prove to be of major interest in the sequel.

Lemma 8.1 *Let $u \in \mathcal{C}(\overline{\Omega}) \otimes L^{\infty}(\mathbb{R}^N_y; \mathcal{B}(\mathbb{R}^N_z))$. There is some negligible set $\mathcal{N} \subset \mathbb{R}^N$ such that*

$$|u(x, y, z)| \leq \sup_{\zeta \in \overline{\Omega}} \|u(\zeta)\|_{L^{\infty}(\mathbb{R}^N;\mathcal{B}(\mathbb{R}^N))}$$

for any triple (x, y, z) with $x \in \overline{\Omega}$, $y \in \mathbb{R}^N/\mathcal{N}$ and $z \in \mathbb{R}^N$, where $u(\zeta) = u(\zeta, \cdot, \cdot)$.

Proof We proceed as in the proof of Lemma 1.6. To start with, we fix a dense countable subset D of $\overline{\Omega}$. Let $a \in D$. Since $u(a, \cdot, \cdot) \in L^{\infty}(\mathbb{R}^N_y; \mathcal{B}(\mathbb{R}^N_z))$, it is clear that there is a negligible set $T_a \subset \mathbb{R}^N$ such that $u(a, y, \cdot) \in \mathcal{B}(\mathbb{R}^N_z)$ for all $y \in \mathbb{R}^N \setminus T_a$. It follows that $|u(a, y, z)| \leq \|u(a, y, \cdot)\|_{\infty}$ for all $y \in \mathbb{R}^N \setminus T_a$ and all $z \in \mathbb{R}^N$. But there is a negligible set $N_a \subset \mathbb{R}^N$ such that $\|u(a, y, \cdot)\|_{\infty} \leq ess.sup_{t\in\mathbb{R}^N} \|u(a, t, \cdot)\|_{\infty} = \|u(a, \cdot, \cdot)\|_{L^{\infty}(\mathbb{R}^N;\mathcal{B}(\mathbb{R}^N))}$ for all $y \in \mathbb{R}^N \setminus N_a$. Hence, by letting $R = \bigcup_{a\in D}(T_a \cup N_a)$, it follows

$$|u(x, y, z)| \leq \sup_{\zeta \in \overline{\Omega}} \|u(\zeta)\|_{L^{\infty}(\mathbb{R}^N;\mathcal{B}(\mathbb{R}^N))}$$

for any triple (x, y, z) with $x \in D$, $y \in \mathbb{R}^N \setminus R$ and $z \in \mathbb{R}^N$. Having made this point, it becomes apparent that the rest of the proof proceeds by the same reasoning as followed in proving Lemma 1.6. ■

Remark 8.1 If $u \in \mathcal{C}(\overline{\Omega}) \otimes L^\infty(\mathbb{R}^N_y; \mathcal{B}(\mathbb{R}^N_z))$, i.e., if $u = \sum_{finite} \varphi_i \otimes u_i$ with $\varphi_i \in \mathcal{C}(\overline{\Omega})$ and $u_i \in L^\infty(\mathbb{R}^N_y; \mathcal{B}(\mathbb{R}^N_z))$, then $u^\varepsilon = \sum_{finite} \varphi_i u_i^\varepsilon$, where u_i^ε is defined as in (8.6) (or (8.7)). Thus $u^\varepsilon(x) = u(x, \frac{x}{\varepsilon}, \frac{x}{\varepsilon^2})$ for a.e. $x \in \Omega$.

This leads us to the following

Proposition 8.3 *Suppose Ω is bounded. The trace operator $u \mapsto u^\varepsilon$ of $\mathcal{C}(\overline{\Omega}) \otimes L^\infty(\mathbb{R}^N_y; \mathcal{B}(\mathbb{R}^N_z))$ into $L^\infty(\Omega)$ extends by continuity to a continuous linear mapping, still denoted by $u \mapsto u^\varepsilon$, of $\mathcal{C}(\overline{\Omega}; L^\infty(\mathbb{R}^N_y; \mathcal{B}(\mathbb{R}^N_z)))$ into $L^\infty(\Omega)$ such that*

$$\|u^\varepsilon\|_{L^\infty(\Omega)} \leq \sup_{x \in \overline{\Omega}} \|u(x)\|_{L^\infty(\mathbb{R}^N; \mathcal{B}(\mathbb{R}^N))}$$

for all $u \in \mathcal{C}(\overline{\Omega}; L^\infty(\mathbb{R}^N_y; \mathcal{B}(\mathbb{R}^N_z)))$.

Proof Indeed, thanks to Lemma 8.1 (see also Remark 8.1), this follows at once by extension by continuity, $\mathcal{C}(\overline{\Omega}) \otimes L^\infty(\mathbb{R}^N; \mathcal{B}(\mathbb{R}^N))$ being dense in $\mathcal{C}(\overline{\Omega}; L^\infty(\mathbb{R}^N; \mathcal{B}(\mathbb{R}^N)))$. ∎

Remark 8.2 In case Ω is unbounded, the above proposition is still valid provided $\mathcal{C}(\overline{\Omega})$ is substituted by $\mathcal{K}(\overline{\Omega})$ and $\mathcal{C}(\overline{\Omega}; L^\infty(\mathbb{R}^N; \mathcal{B}(\mathbb{R}^N)))$ is substituted by $\mathcal{K}(\overline{\Omega}; L^\infty(\mathbb{R}^N; \mathcal{B}(\mathbb{R}^N)))$.

Thus, for u in either $L^p(\Omega; F)$ $(1 \leq p \leq \infty)$ or $\mathcal{C}(\overline{\Omega}; L^\infty(\mathbb{R}^N; \mathcal{B}(\mathbb{R}^N)))$, the trace u^ε is well defined and the right-hand side of (8.5) has the meaning stated in Propositions 8.1, 8.2 or 8.3, as the case may be.

Our concern now is to give meaning to the formal writing $\mathbf{a}(\frac{x}{\varepsilon}, \frac{x}{\varepsilon^2}, Du(x))$ $(x \in \Omega)$ for $u \in W^{1,p}(\Omega; \mathbb{R})$. We will need the following lemma which is interesting by itself.

Lemma 8.2 *Let A_z be a closed subalgebra of $\mathcal{B}(\mathbb{R}^N_z)$ and f be a function of $\mathbb{R}^N_y \times \mathbb{R}^N_z$ into $\mathbb{R}$ with the following properties:*

(1) Equipped with the supremum norm, A_z is a separable Banach algebra.
(2) A_z contains the constants.
(3) A_z preserves complex conjugation.
(4) For each fixed $z \in \mathbb{R}^N$, the function $y \mapsto f(y, z)$ is measurable from $\mathbb{R}^N$ to $\mathbb{R}$.
(5) For almost every $y \in \mathbb{R}^N$, the function $z \mapsto f(y, z)$ of $\mathbb{R}^N$ into $\mathbb{R}$, denoted below by $f(y, \cdot)$, lies in A_z.

Then, the function $y \mapsto f(y, \cdot)$ is measurable from $\mathbb{R}^N$ to $\mathcal{B}(\mathbb{R}^N)$.

Proof It is clear that A_z is a commutative C^*-algebra with identity. Its spectrum, $\Delta(A_z)$, is a metrizable compact space admitting $\{\delta_z | z \in \mathbb{R}^N\}$ (δ_z the Dirac measure on $\mathbb{R}^N$ at z) as a dense subset (see Lemma 1.1). Furthermore, the Gelfand transformation on A_z, denoted below by $\mathcal{G}$, is an isometric C^*-isomorphism of A_z onto $\mathcal{C}(\Delta(A_z))$. Having recalled these preliminaries, let us fix freely some $s \in \Delta(A_z)$. Let $(z_n)_{n \in \mathbb{N}}$ be a sequence in $\mathbb{R}^N$ such that as $n \to \infty$, we have $\delta_{z_n} \to s$

in $\Delta(A_z)$. Then, as $n \to \infty$,

$$\langle \delta_{z_n}, f(y, \cdot) \rangle \to \langle s, f(y, \cdot) \rangle = \mathcal{G}(f(y, \cdot))(s) \quad \text{for a.e. } y \in \mathbb{R}^N,$$

where the brackets denote the duality pairing between A_z' (the strong dual of A_z) and A_z. It follows that the function $y \mapsto \mathcal{G}(f(y, \cdot))(s)$ is measurable from $\mathbb{R}^N$ to $\mathbb{R}$, since the same is true of each of the functions $y \mapsto \langle \delta_{z_n}, f(y, \cdot) \rangle = f(y, z_n)$ (n ranging over $\mathbb{N}$). In other words, if ε_s denotes the Dirac measure on $\Delta(A_z)$ at s, and g denotes the function $y \mapsto \mathcal{G}(f(y, \cdot))$ from $\mathbb{R}^N$ to $\mathcal{C}(\Delta(A_z))$, then the real function $y \mapsto \langle \varepsilon_s, g(y) \rangle$ is measurable on $\mathbb{R}^N$, and that for any arbitrary $s \in \Delta(A_z)$. Therefore, given any arbitrary Radon measure ρ on $\Delta(A_z)$ with finite support, i.e., ρ of the form

$$\rho = \sum_{finite} c_i \varepsilon_{s_i} \quad (c_i \in \mathbb{C}, s_i \in \Delta(A_z)),$$

the function $y \mapsto \langle \rho, g(y) \rangle$ is measurable from $\mathbb{R}^N$ to $\mathbb{C}$, the brackets standing this time for the duality pairing between $\mathcal{M}(\Delta(A_z)) = \mathcal{C}'(\Delta(A_z))$ and $\mathcal{C}(\Delta(A_z))$. With this in mind, fix freely some Radon measure η on $\Delta(A_z)$. Note that η is bounded, since $\Delta(A_z)$ is compact. Thus, we may assume without loss of generality that η lies in the closed unit ball $B' \subset \mathcal{M}(\Delta(A_z))$. But, as is classical, B' with the relative weak $*$ topology on $\mathcal{M}(\Delta(A_z))$ is a metrizable compact space. Hence, by a classical result (see, e.g., Bourbaki [22, III, p. 71, Corol. 1]) one may consider a sequence

$$(\rho_n)_{n \in \mathbb{N}}, \rho_n \in B', \rho_n \text{ with finite support,}$$

such that $\rho_n \to \eta$ in $\mathcal{M}(\Delta(A_z))$ weak $*$ as $n \to \infty$. Therefore, for a.e. $y \in \mathbb{R}^N$ we have $\langle \rho_n, g(y) \rangle \to \langle \eta, g(y) \rangle$ as $n \to \infty$. We deduce readily that the function $y \mapsto \langle \eta, g(y) \rangle$ is measurable from $\mathbb{R}^N$ into $\mathbb{C}$. Hence, the lemma follows by the arbitrariness of η and use of Bourbaki [22, IV, p. 182, Corol. 2 and p. 174, Thm. 1]. ∎

At the present time we assume that

$$a_i(y, \cdot, \mu) \in A_z \ (1 \le i \le N) \text{ for all } \mu \in \mathbb{R}^N \text{ and for a.e. } y \in \mathbb{R}^N, \tag{8.8}$$

where $a_i(y, \cdot, \mu)$ stands for the function $z \mapsto a_i(y, z, \mu)$ of $\mathbb{R}^N$ into $\mathbb{R}$, and A_z is as in Lemma 8.2. We will see that hypothesis (8.8) is fulfilled in practice.

Remark 8.3 For all $\mu \in \mathbb{R}^N$ and for a.e. $y \in \mathbb{R}^N$ we have $a_i(y, \cdot, \mu) \in \mathcal{B}(\mathbb{R}_z^N)$ $(1 \le i \le N)$, as a direct consequence of (8.1)–(8.3). «Unfortunately» Lemma 8.2 does not apply because $\mathcal{B}(\mathbb{R}^N)$ is not separable. However, this is far from being a «misfortune». We would further add that it is in the natural order of events that it be so. Indeed, as we shall see later, in practice an assumption such as (8.8) arises in a quite natural fashion from the concrete structure hypothesis on $a_i(\cdot, \cdot, \mu)$ $(1 \leq i \leq N)$ for fixed $\mu \in \mathbb{R}^N$.

Now, under assumption (8.8), it is clear that Lemma 8.2 applies with $f(y, z) = a_i(y, z, \mu)$ (for fixed $1 \leq i \leq N$ and $\mu \in \mathbb{R}^N$) and we get

$$a_i(\cdot, \cdot, \mu) \in L^\infty(\mathbb{R}^N_y; \mathcal{B}(\mathbb{R}^N_z)) \quad (1 \leq i \leq N) \text{ for fixed } \mu \in \mathbb{R}^N.$$

Consequently, for $\mathbf{v} \in \mathcal{C}(\overline{\Omega}; \mathbb{R})^N$, the function $(x, y, z) \mapsto a_i(y, z, \mathbf{v}(x))$ of $\overline{\Omega} \times \mathbb{R}^N \times \mathbb{R}^N$ into $\mathbb{R}$ lies in $\mathcal{C}(\overline{\Omega}; L^\infty(\mathbb{R}^N_y; \mathcal{B}(\mathbb{R}^N_z)))$. Hence, by Proposition 8.3 we can define the trace function

$$x \mapsto a_i\left(\frac{x}{\varepsilon}, \frac{x}{\varepsilon^2}, \mathbf{v}(x)\right) \quad \text{of } \Omega \text{ into } \mathbb{R}$$

denoted below by $a_i^\varepsilon(\cdot, \cdot, \mathbf{v})$, which lies in $L^\infty_{\mathbb{R}}(\Omega)$.

Proposition 8.4 *Let $1 < p < \infty$. Suppose (8.8) holds. Then, for $1 \leq i \leq N$, the transformation $\boldsymbol{v} \mapsto a_i^\varepsilon(\cdot, \cdot, \boldsymbol{v})$ of $\mathcal{C}(\overline{\Omega}; \mathbb{R})^N$ into $L^\infty_{\mathbb{R}}(\Omega)$ extends by continuity to a continuous mapping, still denoted for convenience by $\boldsymbol{v} \mapsto a_i^\varepsilon(\cdot, \cdot, \boldsymbol{v})$, of $L^p_{\mathbb{R}}(\Omega)^N$ into $L^{p'}_{\mathbb{R}}(\Omega)$ $(p' = \frac{p}{p-1})$. Furthermore, by letting*

$$\boldsymbol{a}^\varepsilon(\cdot, \cdot, \boldsymbol{v}) = (a_i^\varepsilon(\cdot, \cdot, \boldsymbol{v}))_{1 \leq i \leq N} \quad \text{for } \boldsymbol{v} \in L^p_{\mathbb{R}}(\Omega)^N,$$

we have

$$\|\boldsymbol{a}^\varepsilon(\cdot, \cdot, \boldsymbol{v}) - \boldsymbol{a}^\varepsilon(\cdot, \cdot, \boldsymbol{w})\|_{L^{p'}(\Omega)^N} \leq$$
$$c_1 N^{\alpha_1(\frac{1}{2} - \frac{1}{p}) + \frac{1}{p'}} \|1 + |\boldsymbol{v}| + |\boldsymbol{w}|\|^{p-1-\alpha_1}_{L^p(\Omega)} \|\boldsymbol{v} - \boldsymbol{w}\|^{\alpha_1}_{L^p(\Omega)^N}$$

and

$$(\boldsymbol{a}^\varepsilon(\cdot, \cdot, \boldsymbol{v}) - \boldsymbol{a}^\varepsilon(\cdot, \cdot, \boldsymbol{w})) \cdot (\boldsymbol{v} - \boldsymbol{w})$$
$$\geq c_2(1 + |\boldsymbol{v}| + |\boldsymbol{w}|)^{p-\alpha_2} |\boldsymbol{v} - \boldsymbol{w}|^{\alpha_2} \text{ a.e. in } \Omega$$

for all $\boldsymbol{v}, \boldsymbol{w} \in L^p_{\mathbb{R}}(\Omega)^N$.

Proof This follows by a simple adaptation of the proof of Lemma 6.7 together with the subsequent development (the use of Hölder's inequality as we did in proving the said lemma is possible here with $q = \frac{p}{\alpha_1 p'}$). ■

As a direct consequence of this, we have the following

Corollary 8.1 *Let the hypotheses of Proposition 8.4 be satisfied. For* $u \in W^{1,p}(\Omega;\mathbb{R})$, *let* $\boldsymbol{a}^{\varepsilon}(\cdot,\cdot,Du) = (a_i^{\varepsilon}(\cdot,\cdot,Du))_{1\leq i\leq N}$ *be defined as above, which gives a mapping* $u \mapsto \boldsymbol{a}^{\varepsilon}(\cdot,\cdot,Du)$ *of* $W^{1,p}(\Omega;\mathbb{R})$ *into* $L^{p'}(\Omega)^N$. *Then,*

$$\|\boldsymbol{a}^{\varepsilon}(\cdot,\cdot,Du)-\boldsymbol{a}^{\varepsilon}(\cdot,\cdot,Dv)\|_{L^{p'}(\Omega)^N} \leq$$

$$c_1 N^{\alpha_1(\frac{1}{2}-\frac{1}{p})+\frac{1}{p'}}\|1+|Du|+|Dv|\|_{L^p(\Omega)}^{p-1-\alpha_1}\|Du-Dv\|_{L^p(\Omega)^N}^{\alpha_1}$$

and

$$\begin{aligned}&(\boldsymbol{a}^{\varepsilon}(\cdot,\cdot,Du)-\boldsymbol{a}^{\varepsilon}(\cdot,\cdot,Dv))\cdot(Du-Dv)\\&\geq c_2(1+|Du|+|Dv|)^{p-\alpha_2}|Du-Dv|^{\alpha_2}\ \text{a.e. in}\ \Omega\end{aligned}$$

for all $u, v \in W^{1,p}(\Omega;\mathbb{R})$.

Thanks to Proposition 8.4 and its corollary the left of (8.4) is justified now. Furthermore, an existence and unicity result for the said boundary value problem can be achieved following the classical line of proceeding sketched in Sect. 5.5. Finally, let us observe that Proposition 8.4 implicitly assumes the open set Ω to be bounded, which assumption clearly appears in the statement of Proposition 8.3 by which the latter results.

8.2.2 Reiterated Σ-Convergence. $W^{1,p}$-proper RH-algebras

Let ${}^i\mathcal{H} = ({}^iH_{\varepsilon})_{\varepsilon>0}$ $(i = 1, 2)$ be the couple of actions of $\mathbb{R}_+^*$ on $\mathbb{R}^N$ given by ${}^iH_{\varepsilon}(x) = \frac{x}{\varepsilon^i}$ for $x \in \mathbb{R}^N$ (where $\varepsilon^1 = \varepsilon$, of course) and their product $\mathcal{H} = {}^1\mathcal{H} \times {}^2\mathcal{H}$, which is precisely the action $\mathcal{H} = (H_{\varepsilon})_{\varepsilon>0}$ of $\mathbb{R}_+^*$ on $\mathbb{R}^N \times \mathbb{R}^N$ given by $H_{\varepsilon} = {}^1H_{\varepsilon} \times {}^2H_{\varepsilon}$ (direct product), i.e.,

$$H_{\varepsilon}(x,x') = \left(\frac{x}{\varepsilon},\frac{x'}{\varepsilon^2}\right) \quad \text{for } (x,x') \in \mathbb{R}^N \times \mathbb{R}^N.$$

Now, let A_y, A_z be a couple of H-algebras on $\mathbb{R}^N$ under ${}^1\mathcal{H}$ and ${}^2\mathcal{H}$, respectively. We assume that A_y^{∞} (resp. A_z^{∞}) is dense in A_y (resp. A_z). Points in $\Delta(A_y)$ (resp. $\Delta(A_z)$) are denoted by s (resp. r). The compact space $\Delta(A_y)$ (resp. $\Delta(A_z)$) is provided with the M-measure β_y (resp. β_z). At the present time, let A be the closed product of A_y and A_z, that is, the closure of $A_y \otimes A_z$ in $\mathcal{B}(\mathbb{R}_y^N \times \mathbb{R}_z^N)$. In the sequel we will sometimes use the notation $A = \overline{A_y \otimes A_z}$, where the bar stands for the closure in $\mathcal{B}(\mathbb{R}^N \times \mathbb{R}^N)$. It is worth perhaps recalling that A is an H-algebra on $\mathbb{R}^N \times \mathbb{R}^N$ under the product action $\mathcal{H}$, and the subalgebra A^{∞} is dense in A. Furthermore, we have $\Delta(A) = \Delta(A_y) \times \Delta(A_z)$ and $\beta = \beta_y \otimes \beta_z$, where β denotes the M-measure on $\Delta(A)$ (see Chap. 2 for further details). Finally, the same symbol, $\mathcal{G}$, will denote the Gelfand transformation on A_y, A_z and A, as well; M will denote the mean value generated by the action ${}^i\mathcal{H}$ on $\mathbb{R}^N$ $(i = 1, 2)$ as well as that generated by the product action $\mathcal{H}$ on $\mathbb{R}^N \times \mathbb{R}^N$.

Before we can introduce a new kind of Σ-convergence, we need one further notion.

Definition 8.1 The H-algebra $A = \overline{A_y \otimes A_z}$ on $\mathbb{R}^N_y \times \mathbb{R}^N_z$ (under $\mathcal{H}$) is called a *reiteration* H-algebra or, for brevity, an RH-*algebra*, if for each $\psi \in A$ we have $\psi^\varepsilon \to M(\psi)$ in $L^\infty(\mathbb{R}^N_x)$-weak $*$ as $\varepsilon \to 0$, where ψ^ε is given by (8.7).

Remark 8.4 According to property (HA4) of Definition 2.1, we have $\psi \circ H_\varepsilon \to M(\psi)$ in $L^\infty(\mathbb{R}^N_x \times \mathbb{R}^N_{x'})$-weak $*$ as $\varepsilon \to 0$, which is quite another matter.

Let us give a few examples of RH-algebras.

Example 8.1 To begin, we have

$$\mathcal{C}_{per}(Y \times Z) = \overline{\mathcal{C}_{per}(Y) \otimes \mathcal{C}_{per}(Z)}, \tag{8.9}$$

where $Y = Z = (-\frac{1}{2}, \frac{1}{2})^N$, and where $\mathcal{C}_{per}(Y)$ (resp. $\mathcal{C}_{per}(Z)$) is defined as before. However, a little clarification is perhaps needed as regards the space on the left of (8.9). By a $Y \times Z$-periodic continuous complex function on $\mathbb{R}^N_y \times \mathbb{R}^N_z$ is meant a function $\psi \in \mathcal{C}(\mathbb{R}^N_y \times \mathbb{R}^N_z)$ satisfying $\psi(y+k, z+l) = \psi(y, z)$ for all $k, l \in \mathbb{Z}^N$ and all $y, z \in \mathbb{R}^N$. In other words, $\psi(y, \cdot)$ (resp. $\psi(\cdot, z)$) is Z (resp. Y)-periodic for fixed y (resp. z) $\in \mathbb{R}^N$. We claim that $\mathcal{C}_{per}(Y \times Z)$ is an RH-algebra. To see that, let $\gamma_{(k,l)} = \gamma_k \otimes \gamma_l$ for $k, l \in \mathbb{Z}^N$, where $\gamma_k(y) = e^{2\pi i k\cdot y}$ and $\gamma_l(z) = e^{2\pi i l\cdot z}$ for $y, z \in \mathbb{R}^N$. As is classical, the functions of the form

$$\psi = \sum_{(k,l)\in F} c_{(k,l)}\gamma_{(k,l)} \quad (c_{(k,l)} \in \mathbb{C}), \tag{8.10}$$

where F is a finite subset of $\mathbb{Z}^N \times \mathbb{Z}^N$ depending on ψ, form a dense subset of $\mathcal{C}_{per}(Y \times Z)$. Therefore, we only need to check that $\gamma^\varepsilon_{(k,l)} \to M(\gamma_{(k,l)})$ in $L^\infty(\mathbb{R}^N_x)$-weak $*$ for any couple of $k, l \in \mathbb{Z}^N$. Clearly the case $k = l = \omega = (0, \cdots, 0)$ being trivial, we assume in the sequel that $(k, l) \neq (\omega, \omega)$. So, fix freely one such couple and let $\varphi \in L^1(\mathbb{R}^N_x)$. We see readily that

$$\int \gamma^\varepsilon_{(k,l)}\varphi\, dx = \mathcal{F}\varphi\left(-\frac{k}{\varepsilon} - \frac{l}{\varepsilon^2}\right),$$

where $\mathcal{F}\varphi$ denotes the Fourier transform of φ. But, according to the Riemann-Lebesgue lemma, the right-hand side goes to zero as $\varepsilon \to 0$, since $|\frac{k}{\varepsilon} + \frac{l}{\varepsilon^2}| \to \infty$ as $\varepsilon \to 0$. Hence, the claimed property follows by the fact that $M(\gamma_{(k,l)}) = M(\gamma_k)M(\gamma_l) = 0$.

Example 8.2 Let $\mathcal{R}_y$ (resp. $\mathcal{R}_z$) be a countable subgroup of $\mathbb{R}^N$. Clearly the product $\mathcal{R} = \mathcal{R}_y \times \mathcal{R}_z$ is a countable subgroup of $\mathbb{R}_y^N \times \mathbb{R}_z^N$ and

$$AP_{\mathcal{R}}(\mathbb{R}_y^N \times \mathbb{R}_z^N) = \overline{AP_{\mathcal{R}_y}(\mathbb{R}_y^N) \otimes AP_{\mathcal{R}_z}(\mathbb{R}_z^N)}.$$

Moreover, the functions of the form (8.10) with $F \subset \mathcal{R}$, F finite, form a dense subset of $AP_{\mathcal{R}}(\mathbb{R}_y^N \times \mathbb{R}_z^N)$ (see Chap. 2). Finally, by the same line of proceeding as followed in Example 8.1 one sees readily that $AP_{\mathcal{R}}(\mathbb{R}_y^N \times \mathbb{R}_z^N)$ is an RH-algebra on $\mathbb{R}_y^N \times \mathbb{R}_z^N$ (under $\mathcal{H}$).

Example 8.3 Let A_z be as stated above. To start with, observe that

$$\mathcal{B}_\infty(\mathbb{R}_y^N; A_z) = \overline{\mathcal{B}_\infty(\mathbb{R}_y^N) \otimes A_z},$$

as it results by an elementary adaptation of the line of argument followed in Example 2.4. This is an H-algebra on $\mathbb{R}_y^N \times \mathbb{R}_z^N$ under the action $\mathcal{H}$ (Example 2.1 and Proposition 2.3), admitting as a dense subalgebra the space of all $\psi \in \mathcal{B}_\infty(\mathbb{R}_y^N; A_z) \cap \mathcal{C}^\infty(\mathbb{R}^N \times \mathbb{R}^N)$ such that $D^\alpha \psi \in \mathcal{B}_\infty(\mathbb{R}_y^N; A_z)$ for all $\alpha \in \mathbb{N}^N \times \mathbb{N}^N$. The next point is to check that $\mathcal{B}_\infty(\mathbb{R}_y^N; A_z)$ is an RH-algebra. But, as observed above, it is a simple routine exercise to check that this holds if and only if the convergence property in Definition 8.1 is satisfied by any elementary function $\psi = \theta \otimes w$ ($\theta \in \mathcal{B}_\infty(\mathbb{R}_y^N)$, $w \in A_z$). Fix freely one such ψ. Let $\varphi \in L^1(\mathbb{R}_x^N)$. Recalling that $M(\psi) = M(\theta)M(w)$, we have

$$\int \psi^\varepsilon \varphi dx - M(\psi) \int \varphi\, dx = \int \left[\theta\left(\frac{x}{\varepsilon}\right) - M(\theta)\right] w\left(\frac{x}{\varepsilon^2}\right) \varphi(x) dx$$
$$+ M(\theta) \int \left[w\left(\frac{x}{\varepsilon^2}\right) - M(w)\right] \varphi(x) dx.$$

Either integral on the right goes to zero as $\varepsilon \to 0$: the first one by the fact that for $x \neq \omega, \theta(\frac{x}{\varepsilon}) \to M(\theta)$ as $\varepsilon \to 0$, and use of Lebesgue dominated convergence; the next one for the very reason that w lies in A_z. The desired conclusion follows thereby.

The next result is worth pointing out.

Proposition 8.5 *Suppose A is an RH-algebra. Then, the following statements are true when* $\varepsilon \to 0$*:*

(i) *If* $u \in \mathcal{C}(\overline{\Omega}; A)$ *and* Ω *bounded, then* $u^\varepsilon \to \widetilde{u}$ *in* $L^\infty(\Omega)$*-weak* $*$*, where* $\widetilde{u}(x) = M(u(x))$ *for* $x \in \Omega$.

(ii) *If* $u \in L^p(\Omega; A)$ $(1 \leq p < \infty)$*, then* $u^\varepsilon \to \widetilde{u}$ *in* $L^p(\Omega)$*-weak.*

Proof First of all, let us note that in (i) u^ε is merely defined by (8.5), whereas in (ii) u^ε is given by Proposition 8.1. This being so, (i) is easily deduced from Definition 8.1 if $u \in \mathcal{C}(\overline{\Omega}) \otimes A$, whereas the general case follows by the density of

$\mathcal{C}(\overline{\Omega})\otimes A$ in $\mathcal{C}(\overline{\Omega}; A)$. As regards (ii), the preceding argument stands with $\mathcal{K}(\Omega)\otimes A$ in place of $\mathcal{C}(\overline{\Omega}) \otimes A$ and $\mathcal{K}(\Omega; A)$ in that of $\mathcal{C}(\overline{\Omega}; A)$. Hence the general case follows by the density of $\mathcal{K}(\Omega; A)$ in $L^p(\Omega; A)$. ∎

We are now in a position to introduce the concepts of weak and strong reiterated Σ-convergence in $L^p(\Omega)$. Unless otherwise specified, Ω stands for an open set in $\mathbb{R}^N_x$. The letter E throughout is as stated in the beginning of Chap. 3. Finally, it is assumed from now that A **is an RH-algebra**.

Definition 8.2 Let $1 \leq p < \infty$. A sequence $(u_\varepsilon)_{\varepsilon\in E}$ in $L^p(\Omega)$ is said to:

(1) weakly RΣ-converge in $L^p(\Omega)$ to some $u_0 \in L^p(\Omega \times \Delta(A))$ if as $E \ni \varepsilon \to 0$, we have

$$\int_\Omega u_\varepsilon \psi^\varepsilon \, dx \to \iint_{\Omega\times\Delta(A)} u_0 \widehat{\psi} \, dx d\beta \tag{8.11}$$

for all $\psi \in L^{p'}(\Omega; A)$ $(\frac{1}{p'} = 1 - \frac{1}{p})$, where ψ^ε is defined in either Proposition 8.1 (if $p > 1$) or Proposition 8.2 (if $p = 1$), and $\widehat{\psi} = \mathcal{G} \circ \psi$.

(2) strongly RΣ-converge in $L^p(\Omega)$ to some $u_0 \in L^p(\Omega \times \Delta(A))$ if the following condition is fulfilled:

(SC) Given $\eta > 0$ and $\psi \in L^p(\Omega; A)$ with $\|u_0 - \widehat{\psi}\|_{L^p(\Omega\times\Delta(A))} \leq \frac{\eta}{2}$, there is some $\alpha > 0$ such that $\|u_\varepsilon - \psi^\varepsilon\|_{L^p(\Omega)} \leq \eta$ for $E \ni \varepsilon \leq \alpha$.

We express this by writing $u_\varepsilon \to u_0$ in $L^p(\Omega)$-weak RΣ in case (1), and $u_\varepsilon \to u_0$ in $L^p(\Omega)$-strong RΣ in case (2). The unicity of u_0 is easily shown in case (1), whereas case (2) in this matter follows from case (1) by the forthcoming Proposition 8.8. However, a direct proof of unicity in case (2) can also be given (see the proof of Proposition 3.1).

Each of the following results admits a proof on the lines of its analogue in Chaps. 3 and 4 (the verification is left to the reader).

(3) *Let $u_0 \in L^p(\Omega; A)$. Given a sequence $(u_\varepsilon)_{\varepsilon\in E}$ in $L^p(\Omega)$, we shall have $u_\varepsilon \to \widehat{u}_0$ in $L^p(\Omega)$-strong RΣ if and only if $\|u_\varepsilon - u_0^\varepsilon\|_{L^p(\Omega)} \to 0$ as $E \ni \varepsilon \to 0$.*

(4) *For $u \in L^p(\Omega; A)$, we have $u^\varepsilon \to \widehat{u}$ in $L^p(\Omega)$-strong RΣ.*

(5) *If $u_\varepsilon \to u$ in the $L^p(\Omega)$ norm, then $u_\varepsilon \to u$ in $L^p(\Omega)$-strong RΣ (note that a function in $L^p(\Omega)$ may be viewed as a function in $L^p(\Omega \times \Delta(A))$ that does not depend on the variables $(s, r) \in \Delta(A)$).*

Also, the proof of the next proposition is a simple exercise left to the reader.

Proposition 8.6 *Let $1 \leq p < \infty$. Suppose a sequence $(u_\varepsilon)_{\varepsilon\in E}$ in $L^p(\Omega)$ is weakly RΣ-convergent in $L^p(\Omega)$ to some $u_0 \in L^p(\Omega \times \Delta(A))$. Define $u_0^* \in L^p(\Omega \times \Delta(A_y))$ as $u_0^*(x, s) = \int_{\Delta(A_z)} u_0(x, s, r) d\beta_z(r)$ $(x \in \Omega, s \in \Delta(A_y))$ and $\widetilde{u}_0 \in$*

$L^p(\Omega)$ as $\widetilde{u}_0(x) = \int_{\Delta(A_z)} \int_{\Delta(A_y)} u_0(x,s,r) d\beta_y(s) d\beta_z(r)$ $(x \in \Omega)$. Then, as $E \ni \varepsilon \to 0$,

(i) $u_\varepsilon \to u_0^*$ in $L^p(\Omega)$-weak Σ_y (relevant to the H-algebra A_y).
(ii) $u_\varepsilon \to \widetilde{u}_0$ in $L^p(\Omega)$-weak.

The results of the Σ-convergence setting (Chaps. 3 and 4) carry over mutatis mutandis, together with their proofs, to the present setting. We state below the most important of them as far as this chapter is concerned. The details are left to the reader.

Proposition 8.7 *Let $1 < p < \infty$. Given a fundamental sequence E and a sequence $(u_\varepsilon)_{\varepsilon \in E}$ which is norm bounded in $L^p(\Omega)$, a subsequence E' can be extracted from E such that the sequence $(u_\varepsilon)_{\varepsilon \in E'}$ weakly $R\Sigma$-converges in $L^p(\Omega)$.*

Proposition 8.8 *Suppose a sequence $(u_\varepsilon)_{\varepsilon \in E}$ strongly $R\Sigma$-converges in $L^p(\Omega)$ to some $u_0 \in L^p(\Omega \times \Delta(A))$. Then, as $E \ni \varepsilon \to 0$,*

(i) $u_\varepsilon \to u_0$ in $L^p(\Omega)$-weak $R\Sigma$.
(ii) $\|u_\varepsilon\|_{L^p(\Omega)} \to \|u_0\|_{L^p(\Omega \times \Delta(A))}$.

Reciprocally, if $p = 2$ and if (i) and (ii) hold, then $u_\varepsilon \to u_0$ in $L^p(\Omega)$-strong $R\Sigma$.

Proposition 8.9 *Let $1 < p, q < \infty$ with $\frac{1}{\nu} = \frac{1}{p} + \frac{1}{q} \leq 1$. Let $u_0 \in L^p(\Omega \times \Delta(A))$ and $v_0 \in L^q(\Omega \times \Delta(A))$; $u_\varepsilon \in L^p(\Omega)$ and $v_\varepsilon \in L^q(\Omega)$, ε ranging over E. Finally suppose $u_\varepsilon \to u_0$ in $L^p(\Omega)$-strong $R\Sigma$ and $v_\varepsilon \to v_0$ in $L^q(\Omega)$-weak $R\Sigma$. Then $u_\varepsilon v_\varepsilon \to u_0 v_0$ in $L^\nu(\Omega)$-weak $R\Sigma$.*

The notion of a $W^{1,p}(\Omega)$-proper RH-algebra will play a fundamental role in this chapter.

Definition 8.3 The RH-algebra A is said to be $W^{1,p}(\Omega)$-proper for some $1 < p < \infty$ if the following conditions are satisfied.

(RP1) $\mathcal{D}(\Delta(A_y))$ is dense in $W^{1,p}(\Delta(A_y))$
(RP2) $\mathcal{D}(\Delta(A_z))$ is dense in $W^{1,p}(\Delta(A_z))$
(RP3) $W^{1,p}(\Omega)$ is RΣ-reflexive in the following sense: given a fundamental sequence E and a norm bounded sequence $(u_\varepsilon)_{\varepsilon \in E}$ in $W^{1,p}(\Omega)$, there are some subsequence E' from E and some three functions $u_0 \in W^{1,p}(\Omega)$, $u_1 \in L^p(\Omega; W^{1,p}_\#(\Delta(A_y)))$ and $u_2 \in L^p(\Omega; L^p(\Delta(A_y); W^{1,p}_\#(\Delta(A_z))))$, such that, as $E' \ni \varepsilon \to 0$,

$$u_\varepsilon \to u_0 \quad \text{in } W^{1,p}(\Omega)\text{-weak}$$

and

$$\frac{\partial u_\varepsilon}{\partial x_j} \to \frac{\partial u_0}{\partial x_j} + \partial_j u_1 + \partial_j u_2 \quad \text{in } L^p(\Omega)\text{-weak R}\Sigma \quad (1 \leq j \leq N).$$

Remark 8.5 In the preceding definition the derivatives $\partial_j u_1$ and $\partial_j u_2$ are understood in the distribution sense on $\Delta(A_y)$ and $\Delta(A_z)$, respectively (see Remark 4.4 for details).

8.3 Reiterated Σ-Convergence: The Periodic Setting

In this section is summarized a self-contained reiterated Σ-convergence theory in the periodic setting. The basic notation is as in Sect. 8.2 (see in particular Example 8.1).

8.3.1 Definitions and Basic Results

Definition 8.4 Let $1 \leq p < \infty$. A sequence $(u_\varepsilon)_{\varepsilon\in E}$ in $L^p(\Omega)$ is said to:

(1) weakly RΣ-converge in $L^p(\Omega)$ to some $u_0 \in L^p(\Omega; L^p_{per}(Y \times Z))$ if as $E \ni \varepsilon \to 0$, one has

$$\textbf{(W)} \quad \int_\Omega u_\varepsilon(x)\psi^\varepsilon(x)dx \to \int_\Omega \int_Y \int_Z u_0(x, y, z)\psi(x, y, z)dxdydz$$

for all $\psi \in L^{p'}(\Omega; \mathcal{C}_{per}(Y \times Z))$, where ψ^ε is as in Definition 8.2

(2) strongly RΣ-converge in $L^p(\Omega)$ to $u_0 \in L^p(\Omega; L^p_{per}(Y \times Z))$ if the following holds true:

(S) Given $\eta > 0$ and $\psi \in L^p(\Omega; \mathcal{C}_{per}(Y \times Z))$ with $\|u_0 - \psi\|_{L^p(\Omega\times Y\times Z)} \leq \frac{\eta}{2}$, there is some $\alpha > 0$ such that $\|u_\varepsilon - \psi^\varepsilon\|_{L^p(\Omega)} \leq \eta$ for $E \ni \varepsilon \leq \alpha$.

We shall symbolically express this by writing $u_\varepsilon \to u_0$ in $L^p(\Omega)$-weak RΣ in case **(1)**, and $u_\varepsilon \to u_0$ in $L^p(\Omega)$-strong RΣ in case **(2)**. The unicity of u_0 in either case can be shown as in Chap. 3.

As the reader can see, Definition 8.4 is self-contained in so far as it totally ignores the Gelfand representation theory. However, it is equivalent to Definition 8.2 in which the same $A = \mathcal{C}_{per}(Y \times Z)$ is viewed this time as an RH-algebra on $\mathbb{R}^N \times \mathbb{R}^N$ (under the action $\mathcal{H}$). By the way, it is worth noting that the equivalence between the two definitions can be readily established by means of the canonical mapping $\mathcal{G} : L^p_{per}(Y \times Z) \to L^p(\Delta(A))$ and use of various facts to be found in Chap. 2 (see Sect. 2.5). In the same way it is a rather simple exercise to verify that the results lying above between Definitions 8.2 and 8.3 carry over mutatis mutandis to the present periodic setting. This can also be accomplished directly by a simple adaptation of Sect. 3.1.

Let us note in particular that for $\psi \in L^p(\Omega; \mathcal{C}_{per}(Y \times Z))$ $(1 \leq p < \infty)$, as $\varepsilon \to 0$ one has $\psi^\varepsilon \to \psi$ in $L^p(\Omega)$-weak RΣ as well as in $L^p(\Omega)$-strong RΣ.

The following proposition and its corollaries will play a crucial role later on.

Proposition 8.10 *Let $1 < p < \infty$. Let the open set $\Omega \subset \mathbb{R}^N_x$ be bounded. Suppose a sequence $(u_\varepsilon)_{\varepsilon \in E}$ weakly RΣ-converges in $L^p(\Omega)$ to some $u_0 \in L^p(\Omega; L^p_{per}(Y \times Z))$. Then, the statement (W) in Definition 8.4 still holds for $\psi \in \mathcal{C}(\overline{\Omega}; L^\infty_{per}(Y; \mathcal{C}_{per}(Z)))$.*

Proof Thanks to Proposition 8.3 and further, by considering the fact that there is a constant $c_0 > 0$ such that $\|w^\varepsilon\|_{L^{p'}(\Omega)} \leq c_0 \|w\|_{L^{p'}(Y;\mathcal{B}(\mathbb{R}^N_z))}$ $(0 < \varepsilon \leq 1)$ for all $w \in L^{p'}_{per}(Y; \mathcal{C}_{per}(Z))$ (this is immediate by Lemma 1.9 and use of the density of $\mathcal{C}_{per}(Y \times Z)$ in $L^{p'}_{per}(Y; \mathcal{C}_{per}(Z))$), the desired result follows by an obvious adaptation of the proof of Proposition 3.6. ■

The same lines of argument as followed in proving Corollaries 3.1 and 3.2 lead us to the following

Corollary 8.2 *Let p and Ω be as in Proposition 8.10. Then, the following statements are equivalent for a function $u_0 \in L^p(\Omega; L^p_{per}(Y \times Z))$ and a norm bounded sequence $(u_\varepsilon)_{\varepsilon \in E}$ in $L^p(\Omega)$.*

(i) $u_\varepsilon \to u_0$ *in $L^p(\Omega)$-weak RΣ*
(ii) Property (W) in Definition 8.4 stands for $\psi \in \mathcal{C}(\overline{\Omega}; L^\infty_{per}(Y; \mathcal{C}_{per}(Z)))$.

Corollary 8.3 *Let p and Ω be as in Proposition 8.10. Then, for any function $\psi \in \mathcal{C}(\overline{\Omega}; L^\infty_{per}(Y; \mathcal{C}_{per}(Z)))$ one has $\psi^\varepsilon \to \psi$ in $L^p(\Omega)$-weak RΣ as $\varepsilon \to 0$.*

8.3.2 $\mathcal{C}_{per}(Y \times Z)$ as the Prototype of a $W^{1,p}$-proper RH-algebra

The aim here is to establish the counterpart, in the periodic setting, of the triptyque (RP1)-(RP2)-(RP3) of Definition 8.3. To this end we need some preliminaries.

8.3.2.1 Preliminaries

Let

$$\mathcal{V}_{re}(Y \times \widetilde{Z}) = \mathbb{C} \otimes \mathcal{V}_{per}(Y) + \mathcal{C}^\infty_{per}(Y) \otimes \mathcal{V}_{per}(\widetilde{Z})$$

with

$$\mathcal{V}_{per}(\widetilde{Z}) = \left\{ \mathbf{v} | \mathbf{v} \in \mathcal{V}_{per}(Z), \int_Z \mathbf{v}(z)dz = 0 \right\},$$

where $\mathcal{V}_{per}(Y)$ (resp. $\mathcal{V}_{per}(Z)$) is defined in Sect. 2.5, and $\mathbb{C}$ is identified with the constant functions in $\mathcal{C}_{per}^{\infty}(Z)$. We observe in passing that the sum above is actually a direct sum. In the sequel div_y (resp. div_z) denotes the divergence operator on $\mathbb{R}_y^N$ (resp. $\mathbb{R}_z^N$) whereas D_y (resp. D_z) is the gradient operator on $\mathbb{R}_y^N$ (resp. $\mathbb{R}_z^N$). Finally, for $\mathbf{u} = (u_j) \in L_{per}^p(Y \times Z)^N$ and $\mathbf{v} = (v_j) \in L_{per}^{p'}(Y \times Z)^N$, where $1 \leq p < \infty$ and $\frac{1}{p} + \frac{1}{p'} = 1$, one sets

$$\int\int_{Y\times Z} \mathbf{u} \cdot \mathbf{v}\, dydz = \sum_{j=1}^{N} \int\int_{Y\times Z} u_j v_j\, dydz.$$

Lemma 8.3 *The following statements are equivalent for a vector function $\boldsymbol{F} = (F_j) \in L_{per}^p(Y \times Z)^N$ with $1 \leq p < \infty$.*

(i) One has

$$\int\int_{Y\times Z} \boldsymbol{F} \cdot \boldsymbol{v}\, dydz = 0 \tag{8.12}$$

for all $\boldsymbol{v} = (v_j) \in \mathcal{V}_{re}(Y \times \widetilde{Z})$.

(ii) There is a unique couple of functions $u_1 \in W_{per}^{1,p}(Y)/\mathbb{C}$ and $u_2 \in L_{per}^p(Y; W_{per}^{1,p}(Z)/\mathbb{C})$ such that

$$\boldsymbol{F} = D_y u_1 + D_z u_2. \tag{8.13}$$

Proof It is evident that (ii) implies (i). Let us concentrate on the proof of the converse statement. To start with, take in (8.12) the particular $\mathbf{v} = 1 \otimes \mathbf{w}$, i.e., $\mathbf{v}(y, z) = \mathbf{w}(y)$ for $y, z \in \mathbb{R}^N$, where $\mathbf{w}$ ranges over $\mathcal{V}_{per}(Y)$. It follows readily by Lemma 2.11 that there exists a unique $u_1 \in W_{per}^{1,p}(Y)/\mathbb{C}$ such that $\mathbf{F}^1 = D_y u_1$, where $\mathbf{F}^1 \in L_{per}^p(Y)^N$, $\mathbf{F}^1(y) = \int_Z \mathbf{F}(y, z)dz$. This being so, put now $\mathbf{F}^2 = \mathbf{F} - \mathbf{F}^1$, and choose in (8.12) the particular $\mathbf{v} = \psi \otimes \mathbf{w}$ with $\psi \in \mathcal{C}_{per}^{\infty}(Y)$, $\mathbf{w} \in \mathcal{V}_{per}(\widetilde{Z})$. It follows

$$\int_Y \left(\int_Z \mathbf{F}^2(y, z) \cdot \mathbf{w}(z)dz \right) \psi(y)dy = 0,$$

hence, by the arbitrariness of ψ,

$$\int_Z \mathbf{F}^2(y, \cdot) \cdot \mathbf{w}\, dz = 0 \quad \text{for a.e. } y \in \mathbb{R}^N, \tag{8.14}$$

and that for all $\mathbf{w} \in \mathcal{V}_{per}(\widetilde{Z})$. But, since $\int_Z \mathbf{F}^2(y, z)dz = 0$, one sees readily that (8.14) holds for all $\mathbf{w} \in \mathcal{V}_{per}(Z)$. Hence, appealing once again to the preceding reference yields a function $y \mapsto u_2(y,\cdot)$ from $\mathbb{R}^N$ to $W^{1,p}_{per}(Z)/\mathbb{C}$ such that $\mathbf{F}^2(y,\cdot) = D_z u_2(y,\cdot)$ for a.e. $y \in \mathbb{R}^N$. It is an elementary exercise to check that this function lies in $L^p_{loc}(\mathbb{R}^N_y; W^{1,p}_{per}(Z)/\mathbb{C})$ and further, satisfies $u_2(y+k,\cdot) = u_2(y,\cdot)$ for all $k \in \mathbb{Z}^N$ and for a.e. $y \in \mathbb{R}^N$. Moreover, (8.13) is satisfied and so (ii) is established. This completes the proof. ■

Before we can establish the properness of $\mathcal{C}_{per}(Y \times Z)$ as claimed, we need one further lemma.

Lemma 8.4 *Let Ω be a smooth bounded open set in $\mathbb{R}^N_x$. Let $(u_\varepsilon)_{\varepsilon>0}$ be a norm bounded sequence in $W^{1,p}(\Omega)$ for some real $p > 1$. As $\varepsilon \to 0$, we have*

$$\int_\Omega \frac{1}{\varepsilon} u_\varepsilon(x)\psi\left(\frac{x}{\varepsilon}, \frac{x}{\varepsilon^2}\right)\varphi(x)dx \to 0 \tag{8.15}$$

for all $\varphi \in \mathcal{D}(\Omega)$ and all $\psi \in \mathcal{C}^\infty_{per}(Y) \otimes [\mathcal{C}^\infty_{per}(Z)/\mathbb{C}]$.

Proof We proceed in two steps. Let us fix freely a couple of functions φ, ψ as stated above.

Step 1. Let us show that there is some $\theta \in \mathcal{C}^\infty_{per}(Y) \otimes [\mathcal{C}^\infty_{per}(Z)/\mathbb{C}]$ such that $\psi = \Delta_z \theta$, where $\Delta_z = \sum_{j=1}^N \frac{\partial^2}{\partial z_j^2}$ (Laplace operator in $\mathbb{R}^N_z$). Let $\psi = \sum_{i=1}^m \psi_i \otimes \chi_i$ with $\psi_i \in \mathcal{C}^\infty_{per}(Y)$ and $\chi_i \in \mathcal{C}^\infty_{per}(Z)/\mathbb{C}$ $(1 \le i \le m$, integer $m \ge 1)$. It is a classical fact that there is a (unique) $\theta_i \in W^{1,2}_{per}(Z)/\mathbb{C}$ such that $\chi_i = \Delta_z \theta_i$. But then, by the hypoellipticity of Δ_z, one has $\theta_i \in \mathcal{C}^\infty_{per}(Z)/\mathbb{C}$. It follows readily that the function $\theta = \sum_{i=1}^m \psi_i \otimes \theta_i$ meets the requirements stated above.

Step 2. This being so, the aim now is to show (8.15) when $\varepsilon \to 0$. Clearly it is enough to establish (8.15) for an arbitrary given fundamental sequence E. With this in mind, by a simple calculation one gets at once

$$\frac{1}{\varepsilon}\psi^\varepsilon = \varepsilon \sum_{i=1}^N \frac{\partial}{\partial x_i}\left(\frac{\partial \theta}{\partial z_i}\right)^\varepsilon - \sum_{i=1}^N \left(\frac{\partial^2 \theta}{\partial y_i \partial z_i}\right)^\varepsilon,$$

where the notation in (8.7) is used in a systematic manner. Therefore

$$\int_\Omega \frac{1}{\varepsilon} u_\varepsilon \psi^\varepsilon \varphi \, dx = -\varepsilon \sum_{i=1}^N \int_\Omega \left(\frac{\partial \theta}{\partial z_i}\right)^\varepsilon \frac{\partial}{\partial x_i}(\varphi u_\varepsilon)dx - \sum_{i=1}^N \int_\Omega \left(\frac{\partial^2 \theta}{\partial y_i \partial z_i}\right)^\varepsilon \varphi u_\varepsilon \, dx,$$

and so the problem reduces to computing the limit of each of the two terms on the right. It is easily seen that the first one goes to zero as $\varepsilon \to 0$, since

$$\left|\sum_{i=1}^N \int_\Omega \left(\frac{\partial \theta}{\partial z_i}\right)^\varepsilon \frac{\partial}{\partial x_i}(\varphi u_\varepsilon)dx\right| \le c\|\varphi\|_{W^{1,p'}(\Omega)}\|u_\varepsilon\|_{W^{1,p}(\Omega)}$$

with $c = 2N \max_{1\leq i\leq N} \|\frac{\partial\theta}{\partial z_i}\|_\infty$ and $\|u_\varepsilon\|_{W^{1,p}(\Omega)} = \|u_\varepsilon\|_{L^p(\Omega)} + \|Du_\varepsilon\|_{L^p(\Omega)^N}$ bounded for $\varepsilon > 0$. As regards the next term, according to Definition 8.1 (see also Example 8.1) one has, as $\varepsilon \to 0$,

$$\left(\frac{\partial^2\theta}{\partial y_i \partial z_i}\right)^\varepsilon \to M\left(\frac{\partial^2\theta}{\partial y_i \partial z_i}\right) = 0 \ \text{ in } L^\infty(\mathbb{R}^N_x)\text{-weak } *,$$

hence

$$\left(\frac{\partial^2\theta}{\partial y_i \partial z_i}\right)^\varepsilon \to 0 \ \text{ in } L^{p'}(\Omega)\text{-weak.}$$

On the other hand, by weak compactness ($W^{1,p}(\Omega)$ being reflexive) and then the Rellich-Kondrachov theorem, one has $u_\varepsilon \to u$ in $L^p(\Omega)$ as $E \ni \varepsilon \to 0$, where $u \in W^{1,p}(\Omega)$, and where for convenience the extracted subsequence is not relabelled. Therefore, it follows that the next term on the right of the preceding equality also goes to zero when $E \ni \varepsilon \to 0$. Hence the lemma follows.

■

8.3.2.2 A Properness Result

Proposition 8.11 *Let $1 < p < \infty$. Let Ω be a smooth bounded open set in $\mathbb{R}^N_x$. The following statements hold.*

(i) *$\mathcal{C}^\infty_{per}(Y)$ (resp. $\mathcal{C}^\infty_{per}(Z)$) is dense in $W^{1,p}_{per}(Y)$ (resp. $W^{1,p}_{per}(Z)$).*
(ii) *Given a fundamental sequence E and a norm bounded sequence $(u_\varepsilon)_{\varepsilon\in E}$ in $W^{1,p}(\Omega)$, there are some (non-relabelled) subsequence from E and some triple $u_0 \in W^{1,p}(\Omega)$, $u_1 \in L^p(\Omega; W^{1,p}_{per}(Y)/\mathbb{C})$, $u_2 \in L^p(\Omega; L^p_{per}(Y; W^{1,p}_{per}(Z)/\mathbb{C}))$ such that, as $E \ni \varepsilon \to 0$,*

$$u_\varepsilon \to u_0 \ \textit{ in } W^{1,p}(\Omega)\textit{-weak,} \tag{8.16}$$

$$\frac{\partial u_\varepsilon}{\partial x_j} \to \frac{\partial u_0}{\partial x_j} + \frac{\partial u_1}{\partial y_j} + \frac{\partial u_2}{\partial z_j} \ \textit{ in } L^p(\Omega)\textit{-weak } R\Sigma \tag{8.17}$$

for $1 \leq j \leq N$.

Proof In view of Corollary 2.13, only statement (ii) needs verifying. To this end, given any arbitrary fundamental sequence E, let $(u_\varepsilon)_{\varepsilon\in E}$ be a bounded sequence in $W^{1,p}(\Omega)$. By appealing on one hand, to a classical argument relevant to the reflexivity of $W^{1,p}(\Omega)$, on the other hand, to Proposition 8.7 (in its periodic version) we are readily led to a subsequence from E (still denoted by E for convenience) and some $u_0 \in W^{1,p}(\Omega)$, $U_j \in L^p(\Omega; L^p_{per}(Y \times Z))$ $(1 \leq j \leq N)$ such that as $E \ni \varepsilon \to 0$, one has (8.16) and

$$\frac{\partial u_\varepsilon}{\partial x_j} \to U_j \quad \text{in } L^p(\Omega)\text{-weak R}\Sigma. \tag{8.18}$$

At the present time, let $\mathbf{\Phi} = (\Phi_j) \in \mathcal{V}_{re}(Y \times \widetilde{Z})$ and $\varphi \in \mathcal{D}(\Omega)$. An elementary calculation gives

$$\sum_{j=1}^{N} \int_\Omega \left(u_\varepsilon \Phi_j^\varepsilon \frac{\partial \varphi}{\partial x_j} + \frac{\partial u_\varepsilon}{\partial x_j} \Phi_j^\varepsilon \varphi \right) dx = - \int_\Omega \frac{1}{\varepsilon} u_\varepsilon (div_y \mathbf{\Phi})^\varepsilon \varphi \, dx. \tag{8.19}$$

By recalling that $\Phi_j^\varepsilon \to M(\Phi_j)$ in $L^{p'}(\Omega)$-weak as $\varepsilon \to 0$, and that $u_\varepsilon \to u_0$ in $L^p(\Omega)$ as $E \ni \varepsilon \to 0$, one can pass to the limit (as $E \ni \varepsilon \to 0$) using (8.18) and Lemma 8.4; and the result is that

$$\int_\Omega \left(\int\int_{Y \times Z} \mathbf{F}(x, y, z) \cdot \mathbf{\Phi}(y, z) dy dz \right) \varphi(x) dx = 0,$$

where $\mathbf{F} = \mathbf{U} - D_x u_0$ (D_x the gradient on Ω) and $\mathbf{U} = (U_j)$. Since φ is arbitrary, it follows

$$\int\int_{Y \times Z} \mathbf{F}(x, y, z) \cdot \mathbf{\Phi}(y, z) dy dz = 0 \quad \text{for a.e. } x \in \Omega.$$

By Lemma 8.3 and use of a routine line of proceeding we deduce the existence of $u_1 \in L^p(\Omega; W^{1,p}_{per}(Y)/\mathbb{C})$ and $u_2 \in L^p(\Omega; L^p_{per}(Y; W^{1,p}_{per}(Z)/\mathbb{C}))$ such that $\mathbf{F} = D_y u_1 + D_z u_2$. Hence (8.17) (as $E \ni \varepsilon \to 0$) follows, thereby completing the proof of the proposition. ■

Remark 8.6 Concerning Proposition 8.11, the statement **(i)** is equivalent to conditions (RP1) and (RP2) of Definition 8.3 whereas the statement **(ii)** is equivalent to (RP3), where $A = \mathcal{C}_{per}(Y \times Z)$. Indeed, this follows by Corollaries 2.13 and 2.15 together with Lemma 7.13 and part (i) of Proposition 7.6 (the latter two results were established earlier with $Z = (-\frac{1}{2}, \frac{1}{2})$, and they are still manisfestly valid with $Z = (-\frac{1}{2}, \frac{1}{2})^N$). In conclusion, considered as an RH-algebra on $\mathbb{R}^N \times \mathbb{R}^N$, $\mathcal{C}_{per}(Y \times Z)$ is $W^{1,p}$-proper in the sense of Definition 8.3.

8.4 Further Examples of $W^{1,p}$-proper RH-algebras

8.4.1 A General Properness Result

Let the basic notation be as before (see in particular Sect. 8.2.2). The character Ω throughout denotes a smooth bounded open set in $\mathbb{R}^N_x$. We will later have to do with the following

Definition 8.5 One says that the algebra A_z solves Poisson equation in A_z^∞ if for each $f \in A_z^\infty/\mathbb{C} = \{w | w \in A_z^\infty, M(w) = 0\}$, there exists $u \in A_z^\infty$ such that $\Delta_z u = f$.

With this in mind, let us state one result of major interest.

Proposition 8.12 *Let $1 < p < \infty$. Suppose the following conditions are fulfilled by the H-algebra $A = \overline{A_y \otimes A_z}$ on $\mathbb{R}^N \times \mathbb{R}^N$ (under $\mathcal{H}$).*

(C1) A is a reiteration H-algebra.
(C2) The H-algebra A_y is p-harmonic (Chap. 7, Definition 7.2).
(C3) There exists a vector subspace $\mathcal{V}_{\infty z}$ of A_z^∞ that preserves complex conjugation and further, that satisfies:

(1) $\mathcal{G}(\mathcal{V}_{\infty z})$ is dense in $W^{1,p}(\Delta(A_z))$.
(2) For any $f \in \mathcal{V}_{\infty z}/\mathbb{C} = \{w | w \in \mathcal{V}_{\infty z}, M(w) = 0\}$, there is some $u \in \mathcal{V}_{\infty z}$ such that $f = \Delta_z u$.
(3) For any vector function $\boldsymbol{f} = (f_j) \in L^p(\Delta(A_z))^N$ satisfying

$$\int_{\Delta(A_z)} \boldsymbol{f} \cdot \widehat{\boldsymbol{\Phi}} d\beta_z \equiv \sum_{j=1}^{N} \int_{\Delta(A_z)} f_j \widehat{\Phi}_j \, d\beta_z = 0 \;\; (\widehat{\Phi}_j = \mathcal{G}(\Phi_j), \widehat{\boldsymbol{\Phi}} = (\widehat{\Phi}_j))$$

for all $\boldsymbol{\Phi} = (\Phi_j) \in \mathcal{U}_{\infty z} = \{\boldsymbol{w} | \boldsymbol{w} \in \mathcal{V}_{\infty z}^N, \mathrm{div}_z \boldsymbol{w} = 0\}$, there is a unique $q_z \in W_\#^{1,p}(\Delta(A_z))$ such that $f_j = \partial_j q_z$ $(1 \le j \le N)$.

Then A is $W^{1,p}$-proper.

Remark 8.7 If A_z is p-harmonic and if in addition A_z solves Poisson equation in A_z^∞, then (C3) is satisfied with $\mathcal{V}_{\infty z} = A_z^\infty$.

Remark 8.8 The H-algebra $A = \mathcal{C}_{per}(Y \times Z)$ satisfies all the conditions required by Proposition 8.12.

The proof of this proposition requires some preliminaries. In all that follows it is assumed that the hypotheses of Proposition 8.12 are satisfied.

Let now

$$\mathcal{V}_A = \mathbb{C} \otimes \mathcal{U}_y + A_y^\infty \otimes (\mathcal{U}_{\infty z}/\mathbb{C}),$$

where

$$\mathcal{U}_y = \{\boldsymbol{\Phi} | \boldsymbol{\Phi} \in (A_y^\infty)^N, \mathrm{div}_y \boldsymbol{\Phi} = 0\},$$

$$\mathcal{U}_{\infty z}/\mathbb{C} = \{\boldsymbol{\Phi} | \boldsymbol{\Phi} = (\Phi_j) \in \mathcal{U}_{\infty z}, M(\Phi_j) = 0, \; 1 \le j \le N\}$$

$\mathcal{U}_{\infty z}$ defined above (see Proposition 8.12).

Lemma 8.5 *The following are equivalent for a vector function* $\boldsymbol{F} = (F_j) \in L^p(\Delta(A))^N$.

(i) For any $\boldsymbol{v} = (v_j) \in \mathcal{V}_A$, *one has*

$$\int_{\Delta(A)} \boldsymbol{F} \cdot \widehat{\boldsymbol{v}}\, d\beta = 0$$

(ii) There is a unique couple of functions $u_1 \in W^{1,p}_{\#}(\Delta(A_y))$, $u_2 \in L^p(\Delta(A_y); W^{1,p}_{\#}(\Delta(A_z)))$ *such that* $\boldsymbol{F} = \nabla_s u_1 + \nabla_r u_2$, *where* ∇_s *(resp.* ∇_r*) denotes the gradient on* $\Delta(A_y)$ *(resp.* $\Delta(A_z)$*).*

Proof This is a mere reproduction mutatis mutandis of the proof of Lemma 8.3. ■

Lemma 8.6 *Let* $1 < p < \infty$. *Let* $\psi \in A^\infty_y \otimes (\mathcal{V}_{\infty z}/\mathbb{C})$. *For any bounded sequence* $(u_\varepsilon)_{\varepsilon>0}$ *in* $W^{1,p}(\Omega)$, *and for any* $\varphi \in \mathcal{D}(\Omega)$, *we have (8.15) as* $\varepsilon \to 0$.

Proof In view of property (2) of (C3) one has $\psi = \Delta_z \theta, \theta \in A^\infty_y \otimes (\mathcal{V}_{\infty z}/\mathbb{C})$. Hence the lemma follows by reproducing Step 2 of the proof of Lemma 8.4 without the slightest modification. ■

We are now able to prove our main result.

Proof of Proposition 8.12 In view of (C1), (C2) and property (1) of (C3) (implying that $\mathcal{D}(\Delta(A_z))$ is dense in $W^{1,p}(\Delta(A_z))$), we only need to check that condition (RP3) of Definition 8.3 is fulfilled. Given a fundamental sequence E and a norm bounded sequence $(u_\varepsilon)_{\varepsilon\in E}$ in $W^{1,p}(\Omega)$, the same line of argument as followed in proving Proposition 8.11 yields a subsequence E' and some functions $u_0 \in W^{1,p}(\Omega)$ and $U_j \in L^p(\Delta(A))$ $(1 \leq j \leq N)$ such that as $E' \ni \varepsilon \to 0$ one has (8.16) and $\frac{\partial u_\varepsilon}{\partial x_j} \to U_j$ in $L^p(\Omega)$-weak RΣ $(1 \leq j \leq N)$. Next, by letting $\boldsymbol{\Phi} = (\Phi_j) \in \mathcal{V}_A$ and $\varphi \in \mathcal{D}(\Omega)$, and passing to the limit in (8.19) using in particular Lemma 8.6, one readily arrives at

$$\int\int_{\Delta(A_y)\times\Delta(A_z)} \mathbf{F}(x, s, r) \cdot \widehat{\boldsymbol{\Phi}}(s, r) d\beta_y(s) d\beta_z(r) = 0$$

for a.e. $x \in \Omega$, where $\mathbf{F} = \mathbf{U} - D_x u_0, \mathbf{U} = (U_j)$. Hence the proposition follows by Lemma 8.5 and use of the same argument mutatis mutandis as followed in completing the proof of Proposition 8.11. ■

8.4.2 Some Concrete Examples

The preceding proposition enables us to give some examples of $W^{1,p}$-proper RH-algebras.

8.4.2.1 Almost Periodic RH-algebras

Let us consider the RH-algebra of Example 8.2. We assume moreover that the subgroup $\mathcal{R}_z$ of $\mathbb{R}^N$ is discrete, i.e., there is some $c > 0$ such that $|k| \geq c$ for all $k \in \mathcal{R}_z \setminus \{\omega\}$, where $|\cdot|$ stands for the Euclidean norm in $\mathbb{R}^N$. Our purpose is to show that under this hypothesis, the RH-algebra $AP_{\mathcal{R}}(\mathbb{R}_y^N \times \mathbb{R}_z^N)$ is $W^{1,2}$-proper. As is already known (see Lemma 2.7 and Proposition 2.16), either of the factors $AP_{\mathcal{R}_y}(\mathbb{R}^N)$ and $AP_{\mathcal{R}_z}(\mathbb{R}^N)$ is 2-harmonic. Thus, according to Proposition 8.12, we are through if we can show that $A_z = AP_{\mathcal{R}_z}(\mathbb{R}^N)$ solves Poisson equation in A_z^∞ (see Definition 8.5 and Remark 8.7).

Placing ourselves within a general setting, let G be a discrete subgroup of $\mathbb{R}^N$. As is classical (see Bourbaki [25, VII, p. 2]), there is a unique integer $1 \leq d < N$ (d is the rank of G) and a bijective linear mapping $\phi : \mathbb{R}^N \to \mathbb{R}^N$ (thus, ϕ is an automorphism of the topological additive group $\mathbb{R}^N$) such that $\phi(G) = \mathbb{Z}^d$ (for convenience $\mathbb{Z}^d$ is here identified with the discrete subgroup of $\mathbb{R}^d$ generated by the canonical basis of the vector space $\mathbb{R}^d$). It follows that the topological group G is isomorphic to $\mathbb{Z}^d$. Consequently, it goes without saying that the topological group G is isomorphic to the product $\mathbb{Z}^d \times \{\mathbf{0}\}$, where $\mathbf{0}$ denotes the origin in $\mathbb{R}^{N-d}$. Therefore we are justified in systematically identifying the latter two groups, i.e., in merely writing $G = \mathbb{Z}^d \times \{\mathbf{0}\}$ in the sequel.

We are thus led to consider the factorization $\mathbb{R}^N = \mathbb{R}^d \times \mathbb{R}^{N-d}$, and to introduce the natural projection σ of $\mathbb{R}^N$ onto $\mathbb{R}^d$. So, one has $z = (\sigma(z), z')$ for $z = (z_1, \cdots, z_d, z_{d+1}, \cdots, z_N) \in \mathbb{R}^N$ with $\sigma(z) = (z_1, \cdots, z_d)$ and $z' = (z_{d+1}, \cdots, z_N)$.

In the sequel we put $Z = (-\frac{1}{2}, \frac{1}{2})^d \subset \mathbb{R}_z^d$, and we write $\mathcal{C}_{per}(Z)$ for the algebra of Z-periodic continuous complex functions on $\mathbb{R}_z^d$. Finally the symbol $\langle G\rangle$ (resp. $\langle \mathbb{Z}^d\rangle$) denotes in a generical fashion a finite subset of G (resp. $\mathbb{Z}^d$).

We turn now to the statement and proof of one basic result.

Lemma 8.7 *There is an isometric isomorphism Λ of $\mathcal{C}_{per}(Z)$ (as a Banach space) onto $AP_G(\mathbb{R}^N)$ such that*

$$\Lambda u(z) = u(\sigma(z)) \quad (z \in \mathbb{R}^N) \tag{8.20}$$

for all $u \in \mathcal{C}_{per}(Z)$.

Proof Let $T_G(\mathbb{R}^N)$ (resp. $T_{\mathbb{Z}^d}(\mathbb{R}^d)$) denote the algebra of all finite linear combinations (with complex coefficients) of characters γ_k with $k \in G$ (resp. $k \in \mathbb{Z}^d$), where $\gamma_k(z) = e^{2\pi i k\cdot z}$ for $z \in \mathbb{R}^N$ (resp. $z \in \mathbb{R}^d$). Let the mapping $\Lambda : T_{\mathbb{Z}^d}(\mathbb{R}^d) \to T_G(\mathbb{R}^N)$ be defined as follows:

$$\text{for } u = \sum_{k\in\langle\mathbb{Z}^d\rangle} a_k\gamma_k \quad (a_k \in \mathbb{C}),$$

put

$$\Lambda u = \sum_{k \in \langle \mathbb{Z}^d \rangle} a_k \gamma_{(k,\mathbf{0})}$$

with the same $\langle \mathbb{Z}^d \rangle$ as above, of course. The mapping so defined is linear and injective. Furthermore, it is surjective. Indeed, given $f \in T_G(\mathbb{R}^N)$,

$$f = \sum_{k \in \langle G \rangle} b_k \gamma_k,$$

it is easily checked that the function

$$u = \sum_{l \in \langle \mathbb{Z}^d \rangle} a_l \gamma_l$$

with $\langle \mathbb{Z}^d \rangle = \sigma(\langle G \rangle)$, $a_l = b_{(l,\mathbf{0})}$, is such that $\Lambda u = f$. Finally, one has $\Lambda u(z) = u(\sigma(z))$ $(z \in \mathbb{R}^N)$. It follows that $\|\Lambda u\|_\infty = \|u\|_\infty$ for all $u \in T_{\mathbb{Z}^d}(\mathbb{R}^d)$. Since $T_{\mathbb{Z}^d}(\mathbb{R}^d)$ is dense in $\mathcal{C}_{per}(Z)$, we deduce that Λ extends by continuity to an isometric mapping of $\mathcal{C}_{per}(Z)$ into $AP_G(\mathbb{R}^N)$, still denoted by Λ. But then Λ is surjective, as is readily seen by the density of $T_G(\mathbb{R}^N)$ in $AP_G(\mathbb{R}^N)$ and use of the fact that Λ maps $T_{\mathbb{Z}^d}(\mathbb{R}^d)$ onto $T_G(\mathbb{R}^N)$. Thus, Λ is an isometric isomorphism of $\mathcal{C}_{per}(Z)$ onto $AP_G(\mathbb{R}^N)$. It remains to verify (8.20). But this follows at once by the density of $T_{\mathbb{Z}^d}(\mathbb{R}^d)$ in $\mathcal{C}_{per}(Z)$, the continuity of Λ, and use of pointwise convergence. ■

Now, returning to the case that concerns us, it is clear that we may take $G = \mathcal{R}_z = \mathbb{Z}^d \times \{\mathbf{0}\}$ in the preceding development and keep the rest of the notation unchanged. The main properties of the operator $\Lambda : \mathcal{C}_{per}(Z) \to AP_{\mathcal{R}_z}(\mathbb{R}_z^N)$ are collected below. Let $A_z = AP_{\mathcal{R}_z}(\mathbb{R}_z^N)$.

Lemma 8.8 *The following statements hold.*

(i) $M(\Lambda u) = M(u)$ *for all* $u \in \mathcal{C}_{per}(Z)$.
(ii) Λ *carries* $\mathcal{C}^1_{per}(Z)$ *over* A^1_z *and further, for* $u \in \mathcal{C}^1_{per}(Z)$,

$$\frac{\partial}{\partial z_j}\Lambda u = \Lambda \frac{\partial u}{\partial z_j} \quad (1 \le j \le d) \tag{8.21}$$

$$\frac{\partial}{\partial z_j}\Lambda u = 0 \quad (d < j \le N). \tag{8.22}$$

Proof Part (i) is evident with $T_{\mathbb{Z}^N}(\mathbb{R}^d)$ instead of $\mathcal{C}_{per}(Z)$. Hence the whole result follows by density and use of the continuity of Λ together with that of the mean value operator. As regards part (ii), let us fix freely $u \in \mathcal{C}^1_{per}(Z)$. Put $f = \Lambda u$. Fix freely some $a \in \mathbb{R}^N$. Let $1 \leq j \leq N$. Put $h = te_j$, where e_j is the j-th element of the canonical basis of $\mathbb{R}^N$, and $t \in \mathbb{R}$, $t \neq 0$. One has

$$\frac{f(a+h) - f(a)}{t} = \frac{u(\sigma(a) + \sigma(h)) - u(\sigma(a))}{t}.$$

Suppose $d < j \leq N$. Then the right-hand side cancels, since $\sigma(h) = \sigma(\omega)$ (the origin in $\mathbb{R}^d$), and that for any $t \neq 0$. In view of the arbitrariness of a, one deduces (8.22). Suppose now that $1 \leq j \leq d$. Then $\sigma(h) = t\sigma(e_j)$. Since $\sigma(e_j)$ is the j-th vector of the canonical basis of $\mathbb{R}^d$, it follows that

$$\lim_{t \to 0} \frac{f(a+h) - f(a)}{t} = \frac{\partial u}{\partial z_j}(\sigma(a)).$$

Consequently, the derivative $\frac{\partial f}{\partial z_j}$ is defined on the whole $\mathbb{R}^N$ and $\frac{\partial f}{\partial z_j}(z) = \frac{\partial u}{\partial z_j}(\sigma(z))$, which shows (8.21); from which it follows that $\Lambda u \in A_z^1$. It remains to verify that Λ^{-1} map A_z^1 into $\mathcal{C}_{per}^1(Z)$. Let $f \in A_z^1$. Let $u \in \mathcal{C}_{per}(Z)$ be such that $\Lambda u = f$. The aim is to check that $u \in \mathcal{C}_{per}^1(Z)$. Fix freely some $a \in \mathbb{R}^d$. Let $1 \leq j \leq d$. Let h be as above. Let b be any given point in $\mathbb{R}^N$ such that $\sigma(b) = a$. Then

$$\frac{u(a + \sigma(h)) - u(a)}{t} = \frac{f(b+h) - f(b)}{t}.$$

It follows that

$$\lim_{t \to 0} \frac{u(a + \sigma(h)) - u(a)}{t} = \frac{\partial f}{\partial z_j}(b).$$

Thus, u has a partial derivative with respect to z_j at $a \in \mathbb{R}^d$ and

$$\frac{\partial u}{\partial z_j}(a) = \frac{\partial f}{\partial z_j}(b),$$

where a is an arbitrary given point in $\mathbb{R}^d$, and b is any point in $\mathbb{R}^N$ such that $\sigma(b) = a$. Hence $\frac{\partial u}{\partial z_j}(\sigma(z)) = \frac{\partial f}{\partial z_j}(z)$ for all $z \in \mathbb{R}^N$. Let us finally verify that the function $\frac{\partial u}{\partial z_j}$, which is Z-periodic, is continuous on $\mathbb{R}^d$. Fix freely some point $a \in \mathbb{R}^d$. For any $h \in \mathbb{R}^d$, one has

$$\frac{\partial u}{\partial z_j}(a+h) - \frac{\partial u}{\partial z_j}(a) = \frac{\partial f}{\partial z_j}(\widetilde{a} + \widetilde{h}) - \frac{\partial f}{\partial z_j}(\widetilde{a}),$$

where $\widetilde{a} = (a, \mathbf{0})$ and $\widetilde{h} = (h, \mathbf{0})$. Hence, the continuity of $\frac{\partial u}{\partial z_j}$ at a follows from that of $\frac{\partial f}{\partial z_j}$ at $\widetilde{a}$, since the Euclidean norm of $\widetilde{h}$ in $\mathbb{R}^N$ is equal to that of h in $\mathbb{R}^d$. This completes the proof. ■

We are now able to prove the desired result, viz.,

Proposition 8.13 *The RH-algebra $AP_{\mathcal{R}}(\mathbb{R}^N_y \times \mathbb{R}^N_z)$ of Example 8.2 is $W^{1,2}$-proper provided $\mathcal{R}_z$ is a discrete subgroup of $\mathbb{R}^N_z$.*

Proof As pointed out thereinbefore, the present proposition is proved if we can verify that the algebra $A_z = AP_{\mathcal{R}_z}(\mathbb{R}^N)$ solves Poisson equation in A_z^∞. First of all, it is an easy exercise (left to the reader) to verify that Λ maps $\mathcal{C}^\infty_{per}(Z)$ onto A_z^∞ and further

$$\frac{\partial^2}{\partial z_j^2}\Lambda w = \Lambda \frac{\partial^2 w}{\partial z_j^2} \quad (1 \leq j \leq d)$$

and

$$\frac{\partial^2}{\partial z_j^2}\Lambda w = 0 \quad (d < j \leq N)$$

for all $w \in \mathcal{C}^\infty_{per}(Z)$ (see Lemma 8.8). Hence

$$\Delta_z \Lambda w = \Lambda \Delta_{\sigma(z)} w \tag{8.23}$$

for all $w \in \mathcal{C}^\infty_{per}(Z)$, where Δ_z (resp. $\Delta_{\sigma(z)}$) is the Laplace operator on $\mathbb{R}^N_z$ (resp. $\mathbb{R}^d$).

Having made this point, let $f \in A_z^\infty/\mathbb{C}$. Put $g = \Lambda^{-1} f$. One has $g \in \mathcal{C}^\infty_{per}(Z)/\mathbb{C}$ (see part (i) of Lemma 8.8). By recalling that $\mathcal{C}_{per}(Z)$ solves Poisson equation in $\mathcal{C}^\infty_{per}(Z)$ (see Step 1 of the proof of Lemma 8.4), we see that there exists $w \in \mathcal{C}^\infty_{per}(Z)/\mathbb{C}$ such that $g = \Delta_{\sigma(z)} w$. By letting $u = \Lambda w \in A_z^\infty/\mathbb{C}$, we see readily by (8.23) that $f = \Delta_z u$, as claimed. ■

8.4.2.2 The RH-algebra $\mathcal{C}_{per}(Y; \mathcal{B}_{\infty,\mathbb{Z}^N}(\mathbb{R}^N_z))$

Let Y and Z be as in Example 8.1, where we recall that Y (resp. Z) is considered as the open unit cube in $\mathbb{R}^N_y$ (resp. $\mathbb{R}^N_z$). We also recall that if F is, say a Banach space, we denote by $\mathcal{C}_{per}(Y; F)$ the space of Y-periodic continuous functions of $\mathbb{R}^N_y$ into F. Finally, $\mathcal{B}_{\infty,\mathbb{Z}^N}(\mathbb{R}^N_z)$ denotes the closure of $\mathcal{B}_\infty(\mathbb{R}^N_z) + \mathcal{C}_{per}(Z)$ in $\mathcal{B}(\mathbb{R}^N_z)$. This is an H-algebra (for the action $({}^1H_\varepsilon)_{\varepsilon>0}$ defined in Sect. 8.2.2). For more detail see Examples 2.5 and 4.4.

Now, by Bourbaki [25, VII, p. 11] and Bourbaki [22, p. 46], we see that

$$\mathcal{C}_{per}(Y; \mathcal{B}_{\infty,\mathbb{Z}^N}(\mathbb{R}^N_z)) = \overline{\mathcal{B}_{\infty,\mathbb{Z}^N}(\mathbb{R}^N_z) \otimes \mathcal{C}_{per}(Y)},$$

where the bar stands for the closure in $\mathcal{B}(\mathbb{R}^N_y \times \mathbb{R}^N_z)$. Thus, as the closed product of the H-algebras $A_z = \mathcal{B}_{\infty,\mathbb{Z}^N}(\mathbb{R}^N_z)$ and $A_y = \mathcal{C}_{per}(Y)$ on $\mathbb{R}^N$, the space $A = \mathcal{C}_{per}(Y; \mathcal{B}_{\infty,\mathbb{Z}^N}(\mathbb{R}^N_z))$ is an H-algebra on $\mathbb{R}^N_y \times \mathbb{R}^N_z$ under the product action $\mathcal{H} = (H_\varepsilon)_{\varepsilon>0}$ defined in Sect. 8.2.2 (see Proposition 2.3). Therefore, the present case falls under the setting of Sect. 8.2.2 with further A^∞ dense in A, since A_y and A_z verify each an analogous property.

Proposition 8.14 *$\mathcal{C}_{per}(Y; \mathcal{B}_{\infty,\mathbb{Z}^N}(\mathbb{R}^N_z))$ is a $W^{1,p}$-proper RH-algebra for any $1 < p < \infty$.*

Proof To begin with, let us check that $\mathcal{C}_{per}(Y; \mathcal{B}_{\infty,\mathbb{Z}^N}(\mathbb{R}^N_z))$ is an RH-algebra. As we proceeded before (see Examples 8.1–8.3), it is enough to verify that as $\varepsilon \to 0$, one has $\psi^\varepsilon \to M(\psi)$ in $L^\infty(\mathbb{R}^N_x)$-weak $*$ for $\psi = \theta \otimes w$ with $\theta \in \mathcal{B}_{\infty,\mathbb{Z}^N}(\mathbb{R}^N_z)$ and $w \in \mathcal{C}_{per}(Y)$, or better still with $\theta \in \mathcal{B}_\infty(\mathbb{R}^N_z) + \mathcal{C}_{per}(Z)$ and $w \in \mathcal{C}_{per}(Y)$. So, fixing one such ψ, let $\theta = \theta_1 + \theta_2$ with $\theta_1 \in \mathcal{B}_\infty(\mathbb{R}^N_z)$ and $\theta_2 \in \mathcal{C}_{per}(Z)$. Then $\psi^\varepsilon = (\theta_1 \otimes w)^\varepsilon + (\theta_2 \otimes w)^\varepsilon$. But $\theta_1 \otimes w \in \mathcal{B}_\infty(\mathbb{R}^N_z; \mathcal{C}_{per}(Y))$ (see Example 8.3). Hence $(\theta_1 \otimes w)^\varepsilon \to M(\theta_1 \otimes w)$ in $L^\infty(\mathbb{R}^N)$-weak $*$ as $\varepsilon \to 0$. On the other hand, $\theta_2 \otimes w \in \mathcal{C}_{per}(Y \times Z)$ (Example 8.1), so that $(\theta_2 \otimes w)^\varepsilon \to M(\theta_2 \otimes w)$ in $L^\infty(\mathbb{R}^N)$-weak $*$ as $\varepsilon \to 0$. The desired property follows thereby and so $\mathcal{C}_{per}(Y; \mathcal{B}_{\infty,\mathbb{Z}^N}(\mathbb{R}^N_z))$ is an RH-algebra. With a view to applying Proposition 8.12 let us observe that $\mathcal{C}_{per}(Y)$ is p-harmonic (this is a well known fact now), so that conditions (C1) and (C2) of the said proposition are fulfilled. It remains to check that (C3) is also satisfied. As was established earlier in Sect. 4.2.3 (see also Sect. 7.3.1), the couple $\{A_z = \mathcal{B}_{\infty,\mathbb{Z}^N}(\mathbb{R}^N_z), A_{2z} = \mathcal{C}_{per}(Z)\}$ satisfies hypothesis (H) with in particular $\mathcal{V}_{\infty z} = \mathcal{D}(\mathbb{R}^N_z) + \mathcal{C}^\infty_{per}(Z)$, and further $\mathcal{D}(\Delta(A_{2z}))$ is dense in $W^{1,p}(\Delta(A_{2z}))$. Therefore (C3) follows by Lemma 7.8 (as regards part (1)) and Proposition 7.5 (as far as (2) and (3) are concerned). In conclusion, $\mathcal{C}_{per}(Y; \mathcal{B}_{\infty,\mathbb{Z}^N}(\mathbb{R}^N_z))$ is $W^{1,p}$-proper for $1 < p < \infty$ by virtue of Proposition 8.12. ■

8.4.2.3 The RH-algebra $\mathcal{C}_{per}(Y; \mathcal{B}_{\infty,\mathcal{R}}(\mathbb{R}^N_z))$

Let $\mathcal{R}$ be a discrete subgroup of $\mathbb{R}^N_z$. We define $\mathcal{B}_{\infty,\mathcal{R}}(\mathbb{R}^N_z)$ as the closure of $\mathcal{B}_\infty(\mathbb{R}^N_z) + AP_{\mathcal{R}}(\mathbb{R}^N_z)$ in $\mathcal{B}(\mathbb{R}^N_z)$. This is an H-algebra on $\mathbb{R}^N$ (see Sect. 4.2.3). Our purpose is to verify that $\mathcal{C}_{per}(Y; \mathcal{B}_{\infty,\mathcal{R}}(\mathbb{R}^N_z))$ is a $W^{1,2}$-proper RH-algebra.

First, one has

$$\mathcal{C}_{per}(Y; \mathcal{B}_{\infty,\mathcal{R}}(\mathbb{R}^N_z)) = \overline{\mathcal{C}_{per}(Y) \otimes \mathcal{B}_{\infty,\mathcal{R}}(\mathbb{R}^N_z)},$$

so that the present case is embedded in the framework of Sect. 8.2.2 with

$$A_y = \mathcal{C}_{per}(Y),\ A_z = \mathcal{B}_{\infty,\mathcal{R}}(\mathbb{R}^N_z) \text{ and } A = \mathcal{C}_{per}(Y; \mathcal{B}_{\infty,\mathcal{R}}(\mathbb{R}^N_z)).$$

Proposition 8.15 *$\mathcal{C}_{per}(Y; \mathcal{B}_{\infty,\mathcal{R}}(\mathbb{R}_z^N))$ is a $W^{1,2}$-proper RH-algebra.*

Proof We shall first verify that $\mathcal{C}_{per}(Y; \mathcal{B}_{\infty,\mathcal{R}}(\mathbb{R}_z^N))$ is an RH-algebra. This amounts to showing that as $\varepsilon \to 0$, we have $\psi^\varepsilon \to M(\psi)$ in $L^\infty(\mathbb{R}_x^N)$-weak $*$ for any $\psi = \theta \otimes w$, where $w \in \mathcal{C}_{per}(Y)$ and $\theta = \theta_1 + \theta_2$ with $\theta_1 \in \mathcal{B}_0(\mathbb{R}_z^N)$ and $\theta_2 \in AP_{\mathcal{R}}(\mathbb{R}_z^N)$. By the dominated convergence theorem we have $(\theta_1 \otimes w)^\varepsilon \to 0$ in $L^\infty(\mathbb{R}^N)$-weak $*$ as $\varepsilon \to 0$. On the other hand, by recalling that $\mathcal{C}_{per}(Y) = AP_{\mathbb{Z}^N}(\mathbb{R}_y^N)$, we see that $\theta_2 \otimes w \in AP_{\mathcal{R}}(\mathbb{R}_z^N) \otimes AP_{\mathbb{Z}^N}(\mathbb{R}_y^N) \subset AP_G(\mathbb{R}_y^N \times \mathbb{R}_z^N)$ with $G = \mathcal{R} \times \mathbb{Z}^N$. Therefore $(\theta_2 \otimes w)^\varepsilon \to M(\theta_2 \otimes w)$ in $L^\infty(\mathbb{R}^N)$-weak $*$ as $\varepsilon \to 0$ (Example 8.2). The desired result follows, since $M(\theta_1 \otimes w) = 0$. It remains to check that $\mathcal{C}_{per}(Y; \mathcal{B}_{\infty,\mathcal{R}}(\mathbb{R}_z^N))$ is $W^{1,2}$-proper. The line of argument is the same mutatis mutandis as followed in treating of the case of $\mathcal{C}_{per}(Y; \mathcal{B}_{\infty,\mathbb{Z}^N}(\mathbb{R}_z^N))$. Therefore, we shall not carry out the details but instead refer the reader to the said case and to the appropriate properties of $AP_{\mathcal{R}}(\mathbb{R}_z^N)$ (see the proof of Proposition 8.13, see also Proposition 2.16 and Lemma 2.7). ■

8.4.2.4 The RH-algebra $\mathcal{C}_{per}(Y; \mathcal{B}_\infty(\mathbb{R}; \mathcal{C}_{per}(Z')))$

Clearly

$$\mathcal{C}_{per}(Y; \mathcal{B}_\infty(\mathbb{R}; \mathcal{C}_{per}(Z'))) = \overline{\mathcal{C}_{per}(Y) \otimes \mathcal{B}_\infty(\mathbb{R}; \mathcal{C}_{per}(Z'))}$$

with $Z' = (-\frac{1}{2}, \frac{1}{2})^{N-1}$ $(N \geq 2)$, where the H-algebra $\mathcal{B}_\infty(\mathbb{R}; \mathcal{C}_{per}(Z'))$ is introduced and thoroughly studied in Chap. 4 (see Examples 4.6 and 4.7). Thus, this is an H-algebra on $\mathbb{R}_y^N \times \mathbb{R}_z^N$ (under the action $\mathcal{H}$), and we are then led to put $A_y = \mathcal{C}_{per}(Y)$, $A_z = \mathcal{B}_\infty(\mathbb{R}; \mathcal{C}_{per}(Z'))$ and $A = \mathcal{C}_{per}(Y; \mathcal{B}_\infty(\mathbb{R}; \mathcal{C}_{per}(Z')))$ in order to have this case lying within the framework of Sect. 8.2.2.

Proposition 8.16 *$\mathcal{C}_{per}(Y; \mathcal{B}_\infty(\mathbb{R}; \mathcal{C}_{per}(Z')))$ is a $W^{1,2}$-proper RH-algebra.*

Proof We shall check beforehand that this is indeed an RH-algebra. It clearly amounts to verifying that $\psi^\varepsilon \to M(\psi)$ in $L^\infty(\mathbb{R}_x^N)$-weak $*$ as $\varepsilon \to 0$ for all $\psi = \theta_0 \otimes w + \theta_1 \otimes w$ with $w \in \mathcal{C}_{per}(Y)$, $\theta_0 \in \mathcal{B}_0(\mathbb{R}; \mathcal{C}_{per}(Z'))$ and $\theta_1 \in \mathbb{C} \otimes \mathcal{C}_{per}(Z')$. By the density of $\mathcal{K}(\mathbb{R}) \otimes \mathcal{C}_{per}(Z')$ in $\mathcal{B}_0(\mathbb{R}; \mathcal{C}_{per}(Z'))$ one readily sees that for $x \in \mathbb{R}^N \setminus \{x_N = 0\}$, we have that $\theta_0(\frac{x}{\varepsilon^2}) \to 0$ as $\varepsilon \to 0$. This implies by dominated convergence that $(\theta_0 \otimes w)^\varepsilon \to 0$ in $L^\infty(\mathbb{R}^N)$-weak $*$ as $\varepsilon \to 0$. On the other hand, it is clear that $\theta_1 \otimes w \in \mathcal{C}_{per}(Y \times Z)$. It follows that $(\theta_1 \otimes w)^\varepsilon \to M(\theta_1 \otimes w)$ in $L^\infty(\mathbb{R}^N)$-weak $*$ as $\varepsilon \to 0$ (Example 8.1). Hence $\psi^\varepsilon \to M(\psi)$ in $L^\infty(\mathbb{R}^N)$-weak $*$ as $\varepsilon \to 0$, since $M(\theta_0 \otimes w) = 0$.

Now, let $A_{2z} = AP_{\mathcal{R}}(\mathbb{R}^N)$ with $\mathcal{R} = \mathbb{Z}^{N-1} \times \{0\}$. As pointed out in Chap. 4 (see in particular Example 4.6), the pair $\{A_z, A_{2z}\}$ satisfies hypothesis (H) (see Sect. 7.3.1). Moreover, A_{2z} is 2-harmonic and it solves Poisson equation in A_{2z}^∞ (see the proof of Proposition 8.15). Therefore, by joining Proposition 7.5 with Lemma 7.8 we see that A_z satisfies condition (C3) of Proposition 8.12 with $\mathcal{V}_{\infty z} = A_{2z}^\infty + \mathcal{C}_{per}^\infty(Z') \otimes \mathcal{D}(\mathbb{R})$. Hence the $W^{1,2}$-properness of the RH-algebra $\mathcal{C}_{per}(Y; \mathcal{B}_\infty(\mathbb{R}; \mathcal{C}_{per}(Z')))$ follows by the said proposition. ■

8.5 The Deterministic Homogenization Problem for (8.4)

In this section we study the deterministic homogenization problem for (8.4) in two settings: on one hand in the abstract setting relevant to the Gelfand representation theory, on the other hand in the classical periodic setting.

8.5.1 The Abstract Setting

8.5.1.1 Setting of the Problem and Preliminaries

The way we defined ψ^ε (in (8.6) or formally in (8.7)) for $\psi \in L^\infty(\mathbb{R}_y^N; \mathcal{B}(\mathbb{R}_z^N))$ can be extended to $L^p_{loc}(\mathbb{R}_y^N; \mathcal{B}(\mathbb{R}_z^N))$ in a quite natural manner.

Lemma 8.9 *Let $1 \leq p < \infty$. For fixed $\varepsilon > 0$, the mapping $\psi \mapsto \psi^\varepsilon$ of $\mathcal{K}(\mathbb{R}_y^N; \mathcal{B}(\mathbb{R}_z^N))$ into $\mathcal{K}(\mathbb{R}_x^N)$ extends by continuity to a continuous linear mapping, still denoted by $\psi \mapsto \psi^\varepsilon$, of $L^p_{loc}(\mathbb{R}_y^N; \mathcal{B}(\mathbb{R}_z^N))$ into $L^p_{loc}(\mathbb{R}_x^N)$.*

Proof For any bounded open set $U \subset \mathbb{R}_x^N$, we have

$$\|\psi^\varepsilon\|_{L^p(U)} \leq \varepsilon^{\frac{N}{p}} \|\psi\|_{L^p(\frac{1}{\varepsilon}U;\mathcal{B}(\mathbb{R}_z^N))}$$

for all $\psi \in \mathcal{K}(\mathbb{R}_y^N; \mathcal{B}(\mathbb{R}_z^N))$, where the same letter is utilized to denote a function defined on $\mathbb{R}^N$ and its restriction to some open subset of $\mathbb{R}^N$. This shows that the linear operator $\psi \mapsto \psi^\varepsilon$ from the Fréchet space $L^p_{loc}(\mathbb{R}_y^N; \mathcal{B}(\mathbb{R}_z^N))$ to the Fréchet space $L^p_{loc}(\mathbb{R}_x^N)$ with domain $\mathcal{K}(\mathbb{R}_y^N; \mathcal{B}(\mathbb{R}_z^N))$ is continuous (on the said domain). Since the latter is dense in $L^p_{loc}(\mathbb{R}_y^N; \mathcal{B}(\mathbb{R}_z^N))$ (this fact is classical), this operator extends by continuity to a unique continuous linear operator of $L^p_{loc}(\mathbb{R}_y^N; \mathcal{B}(\mathbb{R}_z^N))$ into $L^p_{loc}(\mathbb{R}_x^N)$. ■

Meaning is thus given to the right-hand side of (8.7) for ψ in the larger space $L^p_{loc}(\mathbb{R}_y^N; \mathcal{B}(\mathbb{R}_z^N))$, and this will enable us to enlarge the concept of RΣ-convergence.

Now, let $1 \leq p < \infty$. We define $\Xi^p(\mathbb{R}_y^N; \mathcal{B}(\mathbb{R}_z^N))$, or simply Ξ^p when there is no danger of confusion, as the space of all functions $u \in L^p_{loc}(\mathbb{R}_y^N; \mathcal{B}(\mathbb{R}_z^N))$ such that

$$\sup_{0<\varepsilon\leq 1} \int_U \left\| u\left(\frac{x}{\varepsilon}, \cdot\right)\right\|_\infty^p dx < \infty$$

for any bounded open set U in $\mathbb{R}_x^N$. It is a rather simple exercise (proceed as in the proof of Lemma 2.3) to verify that Ξ^p is a Banach space with norm

$$\|u\|_{\Xi^p} = \sup_{0<\varepsilon\leq 1} \left(\int_{B_N} \left\| u\left(\frac{x}{\varepsilon}, \cdot\right) \right\|_\infty^p dx \right)^{\frac{1}{p}} \quad (u \in \Xi^p),$$

where B_N is the open unit ball in $\mathbb{R}_x^N$.

Remark 8.9 To any bounded open set U in $\mathbb{R}_x^N$ there is attached some integer $m \geq 1$ such that

$$\sup_{0<\varepsilon\leq 1} \left(\int_U \left\| u\left(\frac{x}{\varepsilon}, \cdot\right) \right\|_\infty^p dx \right)^{\frac{1}{p}} \leq m^{\frac{N}{p}} \|u\|_{\Xi^p} \quad (u \in \Xi^p).$$

More precisely, m is any positive integer such that $U \subset mB_N$.

Remark 8.10 $\Xi^p(\mathbb{R}_y^N; \mathcal{B}(\mathbb{R}_z^N))$ is continuously imbedded in $L^p_{loc}(\mathbb{R}_y^N; \mathcal{B}(\mathbb{R}_z^N))$. Indeed, this follows at once by Remark 8.9.

The basic notation and hypotheses being those of Sect. 8.2.2, let $A = \overline{A_y \otimes A_z}$ be an RH-algebra on $\mathbb{R}_y^N \times \mathbb{R}_z^N$ (under the action $\mathcal{H}$). Let $1 \leq p < \infty$. We define $\mathfrak{X}_A^p(\mathbb{R}_y^N; \mathcal{B}(\mathbb{R}_z^N))$, or $\mathfrak{X}_A^p$ for short when there is no risk of confusion, as the closure of A in $\Xi^p(\mathbb{R}_y^N; \mathcal{B}(\mathbb{R}_z^N))$. We provide $\mathfrak{X}_A^p$ with the Ξ^p-norm, which makes it into a Banach space. All the results of Sect. 2.2.1 carry over with more or less modification to the present setting. We begin with the basic ones, which follow readily by an obvious adaptation of the proofs of their analogs of Chap. 2 (the details are left to the reader).

Proposition 8.17 *The mean value M on $\mathbb{R}_y^N \times \mathbb{R}_z^N$ (under the action $\mathcal{H}$) viewed as defined on A, extends by continuity to a unique positive continuous linear form on $\mathfrak{X}_A^p$ still denoted by M. Furthermore, for every $u \in \mathfrak{X}_A^p$ and any fixed bounded open set $\Omega \subset \mathbb{R}_x^N$, we have $u^\varepsilon \to M(u)$ in $L^p(\Omega)$-weak as $\varepsilon \to 0$, where u^ε (defined in Lemma 8.9) is formally identified with its restriction to Ω.*

Proposition 8.18 *The Gelfand transformation $\mathcal{G} : A \to \mathcal{C}(\Delta(A))$ extends by continuity to a unique continuous linear mapping of $\mathfrak{X}_A^p$ into $L^p(\Delta(A))$ still denoted by $\mathcal{G}$.*

These two propositions enable us to get a series of useful properties.

Corollary 8.4 *Let $u \in \mathfrak{X}_A^p$ $(1 \leq p < \infty)$. The following hold.*

(i) $M(u) = \int_{\Delta(A)} \mathcal{G}(u) d\beta$.

(ii) $\overline{u} \in \mathfrak{X}_A^p$ *and* $\mathcal{G}(\overline{u}) = \overline{\mathcal{G}(u)}$.

(iii) $|u|^p \in \mathfrak{X}_A^1$ *and* $\mathcal{G}(|u|^p) = |\mathcal{G}(u)|^p$.

(iv) $M(|u|^p) = \int_{\Delta(A)} |\mathcal{G}(u)|^p d\beta$.

(v) *If* $\psi \in A$, *then* $\psi u \in \mathfrak{X}_A^p$ *and* $\mathcal{G}(\psi u) = \mathcal{G}(\psi)\mathcal{G}(u)$.

Proof Each of these properties is obtained by the same line of proceeding as followed in establishing its analog in Sect. 2.2. ■

Corollary 8.5 *Let* $1 < p, q < \infty$ *with* $\frac{1}{p} + \frac{1}{q} = \frac{1}{\nu} \leq 1$. *Let* $u \in \mathfrak{X}_A^p$ *and* $v \in \mathfrak{X}_A^q$. *Suppose* v *does not depend on the variable* z. *Then,* $uv \in \mathfrak{X}_A^\nu$ *and* $\mathcal{G}(uv) = \mathcal{G}(u)\mathcal{G}(v)$.

Proof This follows by mere adaptation of the proof of Corollary 2.2. ■

Corollary 8.6 *Let* $u \in \mathfrak{X}_A^1$. *The following statements hold true.*

(i) *If* u *assumes real values, then so also does* $\mathcal{G}(u)$. *If moreover* $u \geq 0$ *a.e., then* $\mathcal{G}(u) \geq 0$ *a.e.*
(ii) *Suppose further that* $u \in L^\infty(\mathbb{R}_y^N; \mathcal{B}(\mathbb{R}_z^N))$. *Then, one has* $\mathcal{G}(u) \in L^\infty(\Delta(A))$ *and* $\|\mathcal{G}(u)\|_{L^\infty(\Delta(A))} \leq \|u\|_{L^\infty(\mathbb{R}_y^N;\mathcal{B}(\mathbb{R}_z^N))}$.

Proof Part (i) is immediate by the same argument as followed earlier in proving Corollary 2.6. It remains to check (ii). Since there is a negligible set $\mathcal{N} \subset \mathbb{R}_y^N$ such that $|u(y, z)| \leq \|u\|_{L^\infty(\mathbb{R}_y^N;\mathcal{B}(\mathbb{R}_z^N))}$ for all $(y, z) \in (\mathbb{R}^N \setminus \mathcal{N}) \times \mathbb{R}^N$ (the complement in $\mathbb{R}_y^N \times \mathbb{R}_z^N$ of the negligible set $\mathcal{N} \times \mathbb{R}_z^N$), one sees readily that the proof of Corollary 2.7 carries over mutatis mutandis to the present case. ■

Now, for $1 \leq q < \infty$, let

$$\mathfrak{X}_A^{q,\infty}(\mathbb{R}_y^N; \mathcal{B}(\mathbb{R}_z^N)) = \mathfrak{X}_A^q(\mathbb{R}_y^N; \mathcal{B}(\mathbb{R}_z^N)) \cap L^\infty(\mathbb{R}_y^N; \mathcal{B}(\mathbb{R}_z^N)).$$

In the sequel $\mathfrak{X}_A^{q,\infty}(\mathbb{R}_y^N; \mathcal{B}(\mathbb{R}_z^N))$, denoted for short by $\mathfrak{X}_A^{q,\infty}$ when there is no risk of ambiguity, is equipped with the $L^\infty(\mathbb{R}_y^N; \mathcal{B}(\mathbb{R}_z^N))$ norm, which makes it into a Banach space.

Proposition 8.19 *Let* $1 < p < \infty$. *Suppose a sequence* $(u_\varepsilon)_{\varepsilon\in E}$ *weakly* $R\Sigma$-*converges in* $L^p(\Omega)$ *to some* $u_0 \in L^p(\Omega \times \Delta(A))$, *where* Ω *is a bounded open set in* $\mathbb{R}_x^N$. *Then, as* $E \ni \varepsilon \to 0$, *(8.11) (in Definition 8.2) holds true for* $\psi \in \mathcal{C}(\overline{\Omega}; \mathfrak{X}_A^{p',\infty})$ *as well.*

Proof The proof of Proposition 4.4 carries over mutatis mutandis to the present case. ■

In each of the following corollaries it is assumed that $1 < p < \infty$.

Corollary 8.7 *Suppose the open set* $\Omega \subset \mathbb{R}^N_x$ *is bounded, and a sequence* $(u_\varepsilon)_{\varepsilon\in E}$ *is norm bounded in* $L^p(\Omega)$. *Then, one has* $u_\varepsilon \to u_0$ *in* $L^p(\Omega)$*-weak* $R\Sigma$ *if and only if as* $E \ni \varepsilon \to 0$, *(8.11) holds for all* $\psi \in \mathcal{C}(\overline{\Omega}; \mathfrak{X}^{p',\infty}_A)$.

Proof Indeed, if (8.11) holds for any $\psi \in \mathcal{C}(\overline{\Omega}; \mathfrak{X}^{p',\infty}_A)$, then it clearly holds for all $\psi \in \mathcal{K}(\Omega; A)$, hence by density for $\psi \in L^{p'}(\Omega; A)$. Therefore, the corollary follows by Proposition 8.19. ■

Corollary 8.8 *Under the hypotheses of Proposition 8.19, as* $\varepsilon \to 0$, *one has* $\psi^\varepsilon \to \widehat{\psi}$ *in* $L^p(\Omega)$*-weak* $R\Sigma$ *for any* $\psi \in \mathcal{C}(\overline{\Omega}; \mathfrak{X}^{p,\infty}_A)$.

Proof Since for any $u \in L^{p'}(\Omega; A)$ one has $u^\varepsilon \to \widehat{u}$ in $L^{p'}(\Omega)$-weak $R\Sigma$ as $\varepsilon \to 0$ (see Sect. 8.2), one sees readily that the corollary follows directly by Proposition 8.19. ■

We are now in a position to state the abstract homogenization problem for (8.4). Throughout the remainder of this section it is assumed that $1 < p < \infty$. Our main purpose is to investigate the limiting behaviour, as $\varepsilon \to 0$, of u_ε (the solution of (8.4) for fixed $\varepsilon > 0$) under the abstract hypothesis

$$a_i(\cdot, \cdot, \boldsymbol{\Psi}) \in \mathfrak{X}^{p'}_A(\mathbb{R}^N_y; \mathcal{B}(\mathbb{R}^N_z)) \quad (1 \le i \le N) \quad \text{for all } \boldsymbol{\Psi} \in (A_{\mathbb{R}})^N, \tag{8.24}$$

where $A_{\mathbb{R}} = A \cap \mathcal{C}(\mathbb{R}^N_y \times \mathbb{R}^N_z; \mathbb{R})$, and where $a_i(\cdot, \cdot, \boldsymbol{\Psi})$ stands for the function $(y, z) \mapsto a_i(y, z, \boldsymbol{\Psi}(y, z))$ from $\mathbb{R}^N \times \mathbb{R}^N$ to $\mathbb{R}$.

The problem thus stated is referred to as the abstract deterministic homogenization problem for (8.4) in connection with the reiteration H-algebra A. However, the resolution of the problem before us requires further preliminaries. We shall start with the following lemma.

Lemma 8.10 *Let (8.24) be satisfied. Then, the following hold for any arbitrary* $\boldsymbol{\Psi} \in (A_{\mathbb{R}})^N$.

(i) One has $a_i(\cdot, \cdot, \boldsymbol{\Psi}) \in L^\infty(\mathbb{R}^N_y; \mathcal{B}(\mathbb{R}^N_z)) \quad (1 \le i \le N)$.
(ii) For a.e. $y \in \mathbb{R}^N$, *we have* $a_i(y, \cdot, \boldsymbol{\Psi}(y, \cdot)) \in A_z \quad (1 \le i \le N)$.

Proof Let $\boldsymbol{\Psi} \in (A_{\mathbb{R}})^N$ and $1 \le i \le N$. It is clear from (8.24) that the function $a_i(\cdot, \cdot, \boldsymbol{\Psi})$ is measurable from $\mathbb{R}^N_y$ to $\mathcal{B}(\mathbb{R}^N_z)$. Hence (i) follows at once by (8.2) and (8.3)(3). Next, it is clear that we will be through with (ii) if we have shown that

$$\mathfrak{X}^q_A(\mathbb{R}^N_y; \mathcal{B}(\mathbb{R}^N_z)) \subset L^q_{loc}(\mathbb{R}^N_y; A_z) \qquad (1 \le q < \infty). \tag{8.25}$$

But then, this follows by three facts: **(1)** $A_y \otimes A_z \subset L^q_{loc}(\mathbb{R}^N_y; A_z)$. **(2)** $A_y \otimes A_z$ is dense in the space $\mathfrak{X}^q_A(\mathbb{R}^N_y; \mathcal{B}(\mathbb{R}^N_z))$, and the latter is continuously embedded in $L^q_{loc}(\mathbb{R}^N_y; \mathcal{B}(\mathbb{R}^N_z))$. **(3)** $L^q_{loc}(\mathbb{R}^N_y; A_z)$ is closed in $L^q_{loc}(\mathbb{R}^N_y; \mathcal{B}(\mathbb{R}^N_z))$. ■

Remark 8.11 Equation (8.24) implies (8.8), as is easily seen by part (ii) of Lemma 8.10. This corroborates Remark 8.3.

Throughout the rest of the present section it is assumed that (8.24) is satisfied. Also, Ω denotes a bounded open set in $\mathbb{R}_x^N$.

More generally, for $\boldsymbol{\Psi} \in \mathcal{C}(\overline{\Omega}; (A_{\mathbb{R}})^N)$ we put $a_i(\cdot,\cdot,\boldsymbol{\Psi})$ for the function $(x, y, z) \mapsto a_i(y, z, \boldsymbol{\Psi}(x, y, z))$ of $\overline{\Omega} \times \mathbb{R}^N \times \mathbb{R}^N$ into $\mathbb{R}$. By Lemma 8.10 one has $a_i(\cdot,\cdot,\boldsymbol{\Psi}(x,\cdot,\cdot)) \in L^\infty(\mathbb{R}_y^N; \mathcal{B}(\mathbb{R}_z^N))$ for all $x \in \overline{\Omega}$. Furthermore, appeal to (8.3)(3) yields

$$\|a_i(\cdot,\cdot,\boldsymbol{\Psi}(x,\cdot,\cdot)) - a_i(\cdot,\cdot,\boldsymbol{\Psi}(x',\cdot,\cdot))\|_{L^\infty(\mathbb{R}^N;\mathcal{B}(\mathbb{R}^N))} \leq c\|\boldsymbol{\Psi}(x,\cdot,\cdot) - \boldsymbol{\Psi}(x',\cdot,\cdot)\|_\infty^{\alpha_1}$$

for all $x, x' \in \overline{\Omega}$, where c is a positive constant that depends only on $\boldsymbol{\Psi}$. Hence

$$a_i(\cdot,\cdot,\boldsymbol{\Psi}) \in \mathcal{C}(\overline{\Omega}; L^\infty(\mathbb{R}_y^N; \mathcal{B}(\mathbb{R}_z^N))) \quad (1 \leq i \leq N).$$

This being so, let us bear in mind that

$$a_i^\varepsilon(\cdot,\cdot,\boldsymbol{\Psi}^\varepsilon) = (a_i(\cdot,\cdot,\boldsymbol{\Psi}))^\varepsilon \quad (1 \leq i \leq N) \quad \text{for } \boldsymbol{\Psi} \in \mathcal{C}(\overline{\Omega}; (A_{\mathbb{R}})^N),$$

where the left-hand side denotes the value of the mapping $\mathbf{v} \mapsto a_i^\varepsilon(\cdot,\cdot,\mathbf{v})$ on $\mathbf{v} = \boldsymbol{\Psi}^\varepsilon \in \mathcal{C}(\overline{\Omega}; \mathbb{R})^N$ (see Proposition 8.4), whereas the right-hand side arises from a straightforward application of Proposition 8.3 (see Sect. 8.7 for details).

Now, let the index $1 \leq i \leq N$ be freely fixed. For $\boldsymbol{\varphi} = (\varphi_j) \in \mathcal{C}_{\mathbb{R}}(\Delta(A))^N$, let

$$b_i(\boldsymbol{\varphi}) = \mathcal{G}(a_i(\cdot,\cdot,\mathcal{G}^{-1}\boldsymbol{\varphi})), \tag{8.26}$$

where $\mathcal{G}^{-1}\boldsymbol{\varphi} = (\mathcal{G}^{-1}\varphi_j)_{1\leq j\leq N}$. From (8.24) we see by Lemma 8.10 that $a_i(\cdot,\cdot,\mathcal{G}^{-1}\boldsymbol{\varphi}) \in \mathfrak{X}_A^{p',\infty}$. Hence, according to Corollary 8.6, equality (8.26) determines a mapping $\boldsymbol{\varphi} \mapsto b_i(\boldsymbol{\varphi})$ of $\mathcal{C}_{\mathbb{R}}(\Delta(A))^N$ into $L^\infty(\Delta(A))$ denoted by b_i. In the sequel we set

$$\mathbf{b}(\boldsymbol{\Phi}) = (b_i \circ \boldsymbol{\Phi})_{1\leq i\leq N} \quad \text{for } \boldsymbol{\Phi} \in \mathcal{C}(\overline{\Omega}; \mathcal{C}_{\mathbb{R}}(\Delta(A))^N),$$

defining so a mapping $\mathbf{b} : \mathcal{C}(\overline{\Omega}; \mathcal{C}_{\mathbb{R}}(\Delta(A))^N) \to \mathcal{C}(\overline{\Omega}; L^\infty(\Delta(A))^N)$.

Proposition 8.20 *For $\boldsymbol{\Psi} = (\Psi_j) \in \mathcal{C}(\overline{\Omega}; (A_{\mathbb{R}})^N)$, we have as $\varepsilon \to 0$,*

$$a_i^\varepsilon(\cdot,\cdot,\boldsymbol{\Psi}^\varepsilon) \to b_i \circ \widehat{\boldsymbol{\Psi}} \quad \textit{in } L^{p'}(\Omega)\textit{-weak } R\Sigma \quad (1 \leq i \leq N).$$

Proof One has, on one hand $b_i \circ \widehat{\boldsymbol{\Psi}} = \mathcal{G}(a_i(\cdot,\cdot,\boldsymbol{\Psi}))$, on the other hand $a_i(\cdot,\cdot,\boldsymbol{\Psi}) \in \mathcal{C}(\overline{\Omega}; \mathfrak{X}_A^{p',\infty})$. Hence, the proposition follows by Corollary 8.8. ■

The preceding proposition has two corollaries of considerable interest.

Corollary 8.9 *For fixed $\varepsilon > 0$, let $\Phi_\varepsilon \in \mathcal{D}_{\mathbb{R}}(\Omega)$ be given by*

$$\Phi_\varepsilon(x) = \psi_0(x) + \varepsilon\psi_1\left(x, \frac{x}{\varepsilon}\right) + \varepsilon^2\psi_2\left(x, \frac{x}{\varepsilon}, \frac{x}{\varepsilon^2}\right) \quad (x \in \Omega) \tag{8.27}$$

with $\psi_0 \in \mathcal{D}_{\mathbb{R}}(\Omega)$, $\psi_1 \in \mathcal{D}_{\mathbb{R}}(\Omega) \otimes {}_{\mathbb{R}}A_y^\infty$ and $\psi_2 \in \mathcal{D}_{\mathbb{R}}(\Omega) \otimes {}_{\mathbb{R}}A_y^\infty \otimes {}_{\mathbb{R}}A_z^\infty$, where ${}_{\mathbb{R}}A_y^\infty = A_y^\infty \cap \mathcal{C}(\mathbb{R}^N; \mathbb{R})$ (idem for A_z^∞). As $\varepsilon \to 0$, we have

$$a_i^\varepsilon(\cdot, \cdot, D\Phi_\varepsilon) \to b_i \circ (D_x\psi_0 + \nabla_s\widehat{\psi}_1 + \nabla_r\widehat{\psi}_2) \text{ in } L^{p'}(\Omega)\text{-weak } R\Sigma$$

for $1 \leq i \leq N$.

Furthermore, if $(v_\varepsilon)_{\varepsilon\in E}$ is any sequence in $L^p(\Omega)$ such that $v_\varepsilon \to v_0$ in $L^p(\Omega)$-weak $R\Sigma$, then as $E \ni \varepsilon \to 0$,

$$\int_\Omega a_i^\varepsilon(\cdot, \cdot, D\Phi_\varepsilon)v_\varepsilon\, dx \to \int\int_{\Omega\times\Delta(A)} b_i \circ (D_x\psi_0 + \nabla_s\widehat{\psi}_1 + \nabla_r\widehat{\psi}_2)v_0\, dxd\beta$$

for $1 \leq i \leq N$.

Proof By noting that $a_i(\cdot, \cdot, D_x\psi_0 + D_y\psi_1 + D_z\psi_2) \in \mathcal{C}(\overline{\Omega}; \mathfrak{X}_A^{p',\infty})$, one sees readily that the corollary follows by Proposition 8.20 together with Corollary 8.7 and use of the estimate

$$\|a_i^\varepsilon(\cdot, \cdot, D\Phi_\varepsilon) - a_i^\varepsilon(\cdot, \cdot, D_x\psi_0 + (D_y\psi_1)^\varepsilon + (D_z\psi_2)^\varepsilon)\|_{L^{p'}(\Omega)} \leq c\varepsilon^{\alpha_1}$$

for, say $0 < \varepsilon \leq 1$, where c is a positive constant that does not depend on ε. ■

Corollary 8.10 *The mapping $\boldsymbol{u} \mapsto \boldsymbol{b}(\boldsymbol{u}) = (b_i \circ \boldsymbol{u})_{1\leq i\leq N}$ of $\mathcal{C}(\overline{\Omega}; \mathcal{C}_{\mathbb{R}}(\Delta(A))^N)$ into $\mathcal{C}(\overline{\Omega}; L^\infty(\Delta(A))^N)$ extends by continuity to a mapping still denoted by $\boldsymbol{b}$ of $L^p(\Omega; L^p_{\mathbb{R}}(\Delta(A))^N)$ into $L^{p'}(\Omega \times \Delta(A))^N$ such that*

$$\begin{aligned} &\|\boldsymbol{b}(\boldsymbol{u}) - \boldsymbol{b}(\boldsymbol{v})\|_{L^{p'}(\Omega\times\Delta(A))^N} \leq \\ &c_1 N^{\alpha_1(\frac{1}{2}-\frac{1}{p})+\frac{1}{p'}} \|1 + |\boldsymbol{u}| + |\boldsymbol{v}|\|^{p-1-\alpha_1}_{L^p(\Omega\times\Delta(A))} \|\boldsymbol{u} - \boldsymbol{v}\|^{\alpha_1}_{L^p(\Omega;L^p(\Delta(A))^N)} \end{aligned} \tag{8.28}$$

and

$$(\boldsymbol{b}(\boldsymbol{u}) - \boldsymbol{b}(\boldsymbol{v}))\cdot(\boldsymbol{u} - \boldsymbol{v}) \geq c_2(1 + |\boldsymbol{u}| + |\boldsymbol{v}|)^{p-\alpha_2}|\boldsymbol{u} - \boldsymbol{v}|^{\alpha_2} \text{ a.e. in } \Omega \times \Delta(A). \tag{8.29}$$

Proof Based on Proposition 8.20, this is accomplished by the same line of reasoning as followed in proving Proposition 5.19 (it is a rather simple exercise to verify that Lemmas 5.11 and 5.14 have their analogs in the present setting). ■

Remark 8.12 In Corollary 8.9, ∇_s (resp. ∇_r) denotes the gradient on $\Delta(A_y)$ (resp. $\Delta(A_z)$).

8.5.1.2 The Abstract Homogenization Theorem

Let the basic notation be as before. We assume here that the RH-algebra A is $W^{1,p}$-proper.

Let

$$\mathbb{F}_0^{1,p}(\Omega \times \Delta(A); \mathbb{R}) = W_0^{1,p}(\Omega; \mathbb{R}) \times L^p(\Omega; W_\#^{1,p}(\Delta(A_y); \mathbb{R})) \\ \times L^p(\Omega; L^p(\Delta(A_y); W_\#^{1,p}(\Delta(A_z); \mathbb{R}))).$$

We equip $\mathbb{F}_0^{1,p}(\Omega \times \Delta(A); \mathbb{R})$ with the norm

$$\|\mathbf{v}\|_{\mathbb{F}_0^{1,p}} = \sum_{i=1}^{N} \left[\|D_{x_i} v_0\|_{L^p(\Omega)} + \|\partial_i v_1\|_{L^p(\Omega\times\Delta(A_y))} + \|\partial_i v_2\|_{L^p(\Omega\times\Delta(A))} \right]$$

(with $D_{x_i} = \frac{\partial}{\partial x_i}$) for $\mathbf{v} = (v_0, v_1, v_2) \in \mathbb{F}_0^{1,p}(\Omega \times \Delta(A); \mathbb{R})$, which makes it into a Banach space [the reader is referred to Chap. 5 for details about real Sobolev spaces such as $W_\#^{1,p}(\Delta(A_y); \mathbb{R})$]. We also need the space

$$\mathcal{F}_0^\infty(\Omega \times \Delta(A); \mathbb{R}) = \mathcal{D}_\mathbb{R}(\Omega) \times [\mathcal{D}_\mathbb{R}(\Omega) \otimes J_y(\mathcal{D}_\mathbb{R}(\Delta(A_y))/\mathbb{R})] \\ \times [\mathcal{D}_\mathbb{R}(\Omega) \otimes \mathcal{D}_\mathbb{R}(\Delta(A_y)) \otimes J_z(\mathcal{D}_\mathbb{R}(\Delta(A_z))/\mathbb{R})],$$

where J_y (resp. J_z) denotes the canonical mapping of $W^{1,p}(\Delta(A_y))/\mathbb{C}$ (resp. $W^{1,p}(\Delta(A_z))/\mathbb{C}$) into its separated completion $W_\#^{1,p}(\Delta(A_y))$ (resp. $W_\#^{1,p}(\Delta(A_z))$). The following lemma is a direct consequence of the $W^{1,p}$-properness of A.

Lemma 8.11 *$\mathcal{F}_0^\infty(\Omega \times \Delta(A); \mathbb{R})$ is dense in $\mathbb{F}_0^{1,p}(\Omega \times \Delta(A); \mathbb{R})$.*

Remark 8.13 $\mathcal{F}_0^\infty(\Omega \times \Delta(A); \mathbb{R})$ may as well be defined as the space of all

$\mathbf{\Phi} = (\psi_0, J_y \circ \widehat{\psi_1}, J_z \circ \widehat{\psi_2})$ with $\psi_0 \in \mathcal{D}_\mathbb{R}(\Omega)$, $\psi_1 \in \mathcal{D}_\mathbb{R}(\Omega) \otimes ({}_\mathbb{R}A_y^\infty/\mathbb{R})$ and $\psi_2 \in \mathcal{D}_\mathbb{R}(\Omega) \otimes {}_\mathbb{R}A_y^\infty \otimes ({}_\mathbb{R}A_z^\infty/\mathbb{R})$.

We will also need the variational problem

$$\begin{cases} \mathbf{u} = (u_0, u_1, u_2) \in \mathbb{F}_0^{1,p}(\Omega \times \Delta(A); \mathbb{R}) : \\ \displaystyle\int_{\Omega\times\Delta(A)} \mathbf{b}(\mathbb{D}\mathbf{u}) \cdot \mathbb{D}\mathbf{v}\, dx d\beta = \langle f, v_0 \rangle \\ \text{for all } \mathbf{v} = (v_0, v_1, v_2) \in \mathbb{F}_0^{1,p}(\Omega \times \Delta(A); \mathbb{R}), \end{cases} \tag{8.30}$$

where f denotes the right-hand side of (8.4), $\langle\ ,\ \rangle$ the duality pairing between $W^{-1,p'}(\Omega; \mathbb{R})$ and $W_0^{1,p}(\Omega; \mathbb{R})$, and $\mathbb{D}\mathbf{v} = D_x v_0 + \nabla_s v_1 + \nabla_r v_2$. Thanks to (8.29), this problem admits at most one solution (the argument leading to this conclusion is the same as followed in Sect. 5.5).

We are now able to prove the basic deterministic homogenization theorem for (8.4).

Theorem 8.1 *Let* $1 < p < \infty$. *Let* Ω *be a smooth bounded open set in* $\mathbb{R}^N_x$. *Suppose (8.24) holds and further the RH-algebra* A *is* $W^{1,p}$*-proper and it admits* A^∞ *as a dense subalgebra. For fixed* $\varepsilon > 0$, *let* u_ε *be the solution of (8.4). Then, as* $\varepsilon \to 0$,

$$u_\varepsilon \to u_0 \quad in\ W_0^{1,p}(\Omega)\text{-}weak, \tag{8.31}$$

$$\frac{\partial u_\varepsilon}{\partial x_i} \to \frac{\partial u_0}{\partial x_i} + \partial_i u_1 + \partial_i u_2 \quad in\ L^p(\Omega)\text{-}weak\ R\Sigma\ (1 \le i \le N), \tag{8.32}$$

where $\mathbf{u} = (u_0, u_1, u_2)$ *is uniquely determined by (8.30).*

Proof It is a classical exercise to check using Corollary 8.1 that the (generalized) sequence $(u_\varepsilon)_{\varepsilon>0}$ is norm bounded in $W_0^{1,p}(\Omega)$. Thus, given any arbitrary fundamental sequence E, appeal to the $W^{1,p}$-properness of A yields some subsequence E' from E and some triple $\mathbf{u} = (u_0, u_1, u_2) \in \mathbb{F}_0^{1,p}(\Omega \times \Delta(A); \mathbb{R})$ such that (8.31) and (8.32) hold as $E' \ni \varepsilon \to 0$. Therefore, the whole problem reduces to showing that this triple satisfies the variational equation in (8.30) (which admits at most one solution). To do this, let us fix freely some $\mathbf{\Phi} = (\psi_0, J_y \circ \widehat{\psi_1}, J_z \circ \widehat{\psi_2})$ as in Remark 8.13, and let us attach to $\mathbf{\Phi}$ the sequence $(\Phi_\varepsilon)_{\varepsilon>0}$ given by (8.27). By proceeding as in Chap. 5 (Sect. 5.5.3) one readily arrives at

$$\langle f, u_\varepsilon - \Phi_\varepsilon \rangle - \int_\Omega \mathbf{a}^\varepsilon(\cdot, \cdot, D\Phi_\varepsilon) \cdot (Du_\varepsilon - D\Phi_\varepsilon) dx \ge 0. \tag{8.33}$$

The next step is to pass to the limit when $E' \ni \varepsilon \to 0$. Before we can do this, however, we need to know that as $\varepsilon \to 0$,

$$D_{x_i}\Phi_\varepsilon \to D_{x_i}\psi_0 + \partial_i \widehat{\psi_1} + \partial_i \widehat{\psi_2} \text{ in } L^p(\Omega)\text{-strong } \mathrm{R}\Sigma\ (1 \le i \le N) \tag{8.34}$$

and

$$\Phi_\varepsilon \to \psi_0 \text{ in } W_0^{1,p}(\Omega)\text{-weak.} \tag{8.35}$$

We only need to verify (8.34) of which (8.35) is a simple corollary (the verification is left to the reader). By the obvious equality

$$D_{x_i}\Phi_\varepsilon = D_{x_i}\psi_0 + (D_{y_i}\psi_1)^\varepsilon + (D_{z_i}\psi_2)^\varepsilon + \varepsilon[(D_{x_i}\psi_1)^\varepsilon + (D_{y_i}\psi_2)^\varepsilon] + \varepsilon^2(D_{x_i}\psi_2)^\varepsilon$$

and use of (8.3)(3), one is readily led to

$$\|D_{x_i}\Phi_\varepsilon - D_{x_i}\psi_0 - (D_{y_i}\psi_1)^\varepsilon - (D_{z_i}\psi_2)^\varepsilon\|_\infty \to 0$$

as $\varepsilon \to 0$, from which (8.34) follows by the statement (3) under Definition 8.2. Finally, recalling (8.31)–(8.32) as $E' \ni \varepsilon \to 0$, and that strong $R\Sigma$-convergence implies weak $R\Sigma$-convergence, we can pass to the limit in (8.33) using (8.34), (8.35) and Corollary 8.9 to get

$$\langle f, u_0 - \psi_0\rangle - \int\int_{\Omega\times\Delta(A)} \mathbf{b}(\mathbb{D}\mathbf{\Phi})\cdot(\mathbb{D}\mathbf{u} - \mathbb{D}\mathbf{\Phi})dxd\beta \geq 0,$$

where $\mathbf{\Phi}$ ranges over $\mathcal{F}_0^\infty(\Omega \times \Delta(A); \mathbb{R})$. We deduce by density (see Lemma 8.11) and use of (8.28) that this inequality is satisfied by all $\mathbf{\Phi} \in \mathbb{F}_0^{1,p}(\Omega \times \Delta(A); \mathbb{R})$. Hence the theorem follows by mere routine (see Sect. 5.5). ■

8.5.2 *The Periodic Setting*

8.5.2.1 Setting of the Problem and Preliminaries

Throughout this section it is assumed that for fixed $\mu \in \mathbb{R}^N$ and $1 \leq i \leq N$, the function $(y, z) \mapsto a_i(y, z, \mu)$ satisfies the following hypothesis:

$$\begin{cases} \text{For a.e. } y \in \mathbb{R}^N, \text{ the mapping } z \mapsto a_i(y, z, \mu) \text{ of } \mathbb{R}^N \text{ into } \mathbb{R} \\ \text{denoted by } a_i(y, \cdot, \mu) \text{ is } Z\text{-periodic (thus, lies in } \mathcal{C}_{per}(Z)), \\ \text{and further the function } y \mapsto a_i(y, \cdot, \mu) \text{ from } \mathbb{R}^N \text{ to } \mathcal{C}_{per}(Z) \\ \text{is } Y\text{-periodic.} \end{cases} \tag{8.36}$$

Our purpose is to study the homogenization of the boundary value problem (8.4) under the structure hypothesis (8.36) commonly known as the *periodicity hypothesis*. Let us begin by noting that under this hypothesis, condition (8.8) is satisfied for $A_z = \mathcal{C}_{per}(Z)$, so that Proposition 8.4 and Corollary 8.1 are valid. The result is that the left-hand side of (8.4) has meaning.

We now record a few results underlying the approach we shall follow in the sequel.

Proposition 8.21 *Let $\mathbf{\Psi} \in \mathcal{C}_{per}(Y \times Z; \mathbb{R})^N$. For fixed $1 \leq i \leq N$, let $a_i(\cdot, \cdot, \mathbf{\Psi})$ stand for the function $(y, z) \mapsto a_i(y, z, \mathbf{\Psi}(y, z))$ from $\mathbb{R}^N \times \mathbb{R}^N$ into $\mathbb{R}$. Then $a_i(\cdot, \cdot, \mathbf{\Psi}) \in L^\infty_{per}(Y; \mathcal{C}_{per}(Z))$.*

Proof Let $z \in \mathbb{R}^N$ be fixed. In view of (8.1) and (8.3)(3), the function $(y, \mu) \mapsto h(y, \mu) = a_i(y, z, \mu)$ from $\mathbb{R}^N \times \mathbb{R}^N$ to $\mathbb{R}$ has the Carathéodory property. Hence, a classical result used earlier in Chap. 5 reveals that if u is any measurable function from $\mathbb{R}^N$ to $\mathbb{R}^N$, then the function $y \mapsto h(y, u(y))$ is measurable from $\mathbb{R}^N$ to $\mathbb{R}$. By choosing in particular $u(y) = \boldsymbol{\Psi}(y, z)$ it follows that the function $y \mapsto a_i(y, z, \boldsymbol{\Psi}(y, z))$ is measurable from $\mathbb{R}^N$ to $\mathbb{R}$, and that for any arbitrary given $z \in \mathbb{R}^N$. Furthermore, let y be freely fixed (outside some negligible set in $\mathbb{R}^N$). Let $z_0 \in \mathbb{R}^N$. By writing

$$\begin{aligned} a_i(y, z, \boldsymbol{\Psi}(y, z)) - a_i(y, z_0, \boldsymbol{\Psi}(y, z_0)) &= a_i(y, z, \boldsymbol{\Psi}(y, z)) - a_i(y, z, \boldsymbol{\Psi}(y, z_0)) \\ &\quad + a_i(y, z, \boldsymbol{\Psi}(y, z_0)) - a_i(y, z_0, \boldsymbol{\Psi}(y, z_0)) \end{aligned}$$

and using (8.1) and (8.3)(3), once again, one sees readily that for a.e. $y \in \mathbb{R}^N$, the function $z \mapsto a_i(y, z, \boldsymbol{\Psi}(y, z))$ maps continuously $\mathbb{R}^N$ into $\mathbb{R}$. We deduce by (8.36) that the said function denoted by $a_i(y, \cdot, \boldsymbol{\Psi}(y, \cdot))$ (for fixed y, of course), lies in $\mathcal{C}_{per}(Z)$. Hence, it follows by Lemma 8.2 that the function $y \mapsto a_i(y, \cdot, \boldsymbol{\Psi}(y, \cdot))$ again denoted by $a_i(\cdot, \cdot, \boldsymbol{\Psi})$, is measurable from $\mathbb{R}^N$ to $\mathcal{C}_{per}(Z)$. From this we deduce using (8.2), (8.3)(3) and (8.36) that $a_i(\cdot, \cdot, \boldsymbol{\Psi})$ lies in $L^\infty_{per}(Y; \mathcal{C}_{per}(Z))$, which completes the proof. ■

This leads us to the following corollary.

Corollary 8.11 *Let Ω be a bounded open set in $\mathbb{R}^N_x$. Let $\boldsymbol{\Psi}$ be given in $\mathcal{C}(\overline{\Omega}; \mathcal{C}_{per}(Y \times Z; \mathbb{R})^N)$. For fixed $1 \leq i \leq N$, let $a_i(\cdot, \cdot, \boldsymbol{\Psi})$ more generally denote the function $(x, y, z) \mapsto a_i(y, z, \boldsymbol{\Psi}(x, y, z))$ from $\overline{\Omega} \times \mathbb{R}^N \times \mathbb{R}^N$ to $\mathbb{R}$. Then $a_i(\cdot, \cdot, \boldsymbol{\Psi}) \in \mathcal{C}(\overline{\Omega}; L^\infty_{per}(Y; \mathcal{C}_{per}(Z)))$.*

Proof This follows readily by Proposition 8.21 and use of (8.3)(3). ■

In the next proposition, for $\mathbf{v} \in L^p(\Omega; L^p_{per}(Y \times Z; \mathbb{R})^N)$, $\mathbf{a}(\cdot, \cdot, \mathbf{v})$ denotes the function $(x, y, z) \mapsto \mathbf{a}(y, z, \mathbf{v}(x, y, z))$ from $\Omega \times \mathbb{R}^N \times \mathbb{R}^N$ to $\mathbb{R}^N$.

Proposition 8.22 *Let Ω be as before, and let $1 < p < \infty$. The following statements hold true for $\boldsymbol{u}, \boldsymbol{v} \in L^p(\Omega; L^p_{per}(Y \times Z; \mathbb{R})^N)$.*

$$(i) \quad \boldsymbol{a}(\cdot, \cdot, \boldsymbol{u}), \boldsymbol{a}(\cdot, \cdot, \boldsymbol{v}) \in L^{p'}(\Omega; L^{p'}_{per}(Y \times Z)^N), \quad \frac{1}{p} + \frac{1}{p'} = 1,$$

$$(ii) \quad \begin{aligned} &\|\boldsymbol{a}(\cdot, \cdot, \boldsymbol{u}) - \boldsymbol{a}(\cdot, \cdot, \boldsymbol{v})\|_{L^{p'}(\Omega \times Y \times Z)^N} \leq \\ &c_1 N^{\alpha_1(\frac{1}{2} - \frac{1}{p}) + \frac{1}{p'}} \|1 + |\boldsymbol{u}| + |\boldsymbol{v}|\|^{p-1-\alpha_1}_{L^p(\Omega \times Y \times Z)} \|\boldsymbol{u} - \boldsymbol{v}\|^{\alpha_1}_{L^p(\Omega \times Y \times Z)^N} \end{aligned}$$

$$(iii) \quad \begin{aligned} &(\boldsymbol{a}(\cdot, \cdot, \boldsymbol{u}) - \boldsymbol{a}(\cdot, \cdot, \boldsymbol{v})) \cdot (\boldsymbol{u} - \boldsymbol{v}) \\ &\geq c_2(1 + |\boldsymbol{u}| + |\boldsymbol{v}|)^{p-\alpha_2} |\boldsymbol{u} - \boldsymbol{v}|^{\alpha_2} \ \textit{a.e. in } \Omega \times Y \times Z. \end{aligned}$$

Proof (i) follows by (8.1), (8.2), (8.3)(3) and use of the property of Carathéodory functions pointed out before, whereas (ii) and (iii) follow from (8.3) by routine calculations (concerning (ii), put $q = \frac{p}{\alpha_1 p'}$ and proceed by Hölder's inequality as in proving Lemma 6.7). ■

The following proposition and its corollary will also be needed.

Proposition 8.23 *Let the open set* $\Omega \subset \mathbb{R}^N_x$ *be bounded, and let* $1 < p < \infty$. *Let* $\mathbf{\Psi} = (\Psi_j) \in \mathcal{C}(\overline{\Omega}; \mathcal{C}_{per}(Y \times Z; \mathbb{R})^N)$. *As* $\varepsilon \to 0$, *one has*

$$a_i^\varepsilon(\cdot, \cdot, \mathbf{\Psi}^\varepsilon) \to a_i(\cdot, \cdot, \mathbf{\Psi}) \ (1 \le i \le N) \text{ in } L^{p'}(\Omega)\text{-weak } R\Sigma,$$

where $\mathbf{\Psi}^\varepsilon = (\Psi_j^\varepsilon)$.

Proof One has $a_i(\cdot, \cdot, \mathbf{\Psi}) \in \mathcal{C}(\overline{\Omega}; L^\infty_{per}(Y; \mathcal{C}_{per}(Z)))$ (Corollary 8.11) and further $a_i^\varepsilon(\cdot, \cdot, \mathbf{\Psi}^\varepsilon) = a_i(\cdot, \cdot, \mathbf{\Psi})^\varepsilon$ (see Sect. 8.7). Hence the proposition follows by Corollary 8.3. ■

Corollary 8.12 *For fixed* $\varepsilon > 0$, *let* Φ_ε *be given by (8.27) with* $\psi_0 \in \mathcal{D}_\mathbb{R}(\Omega)$, $\psi_1 \in \mathcal{D}_\mathbb{R}(\Omega) \otimes \mathcal{C}^\infty_{per}(Y; \mathbb{R})$ *and* $\psi_2 \in \mathcal{D}_\mathbb{R}(\Omega) \otimes \mathcal{C}^\infty_{per}(Y \times Z; \mathbb{R})$. *Then, as* $\varepsilon \to 0$, *one has*

$$a_i^\varepsilon(\cdot, \cdot, D\Phi_\varepsilon) \to a_i(\cdot, \cdot, D_x\psi_0 + D_y\psi_1 + D_z\psi_2) \text{ in } L^{p'}(\Omega)\text{-weak } R\Sigma$$

for $1 \le i \le N$. *Furthermore, if* $(v_\varepsilon)_{\varepsilon \in E}$ *is an arbitrary given sequence in* $L^p(\Omega)$ *such that* $v_\varepsilon \to v_0$ *in* $L^p(\Omega)$*-weak* $R\Sigma$, *then as* $E \ni \varepsilon \to 0$,

$$\int_\Omega a_i^\varepsilon(\cdot, \cdot, D\Phi_\varepsilon) v_\varepsilon \, dx \to \int_\Omega \int_Y \int_Z a_i(\cdot, \cdot, D_x\psi_0 + D_y\psi_1 + D_z\psi_2) v_0 \, dxdydz$$

for $1 \le i \le N$.

Proof Since $a_i(\cdot, \cdot, D_x\psi_0 + D_y\psi_1 + D_z\psi_2) \in \mathcal{C}(\overline{\Omega}; L^\infty_{per}(Y; \mathcal{C}_{per}(Z)))$, we see by Propositions 8.10 and 8.23, and use of (8.52), that the corollary is immediate by the same argument as followed in proving Corollary 8.9. ■

8.5.2.2 The Periodic Homogenization Theorem

Starting with some preliminaries, let $1 \le p < \infty$. Put

$$\mathbb{F}_0^{1,p} = W_0^{1,p}(\Omega; \mathbb{R}) \times L^p(\Omega; W_\#^{1,p}(Y; \mathbb{R})) \times L^p(\Omega; L^p_{per}(Y; W_\#^{1,p}(Z; \mathbb{R}))),$$

where

$$W_\#^{1,p}(Y; \mathbb{R}) = W_{per}^{1,p}(Y; \mathbb{R})/\mathbb{R} \quad \text{and} \quad W_\#^{1,p}(Z; \mathbb{R}) = W_{per}^{1,p}(Z; \mathbb{R})/\mathbb{R}.$$

This is a Banach space under the norm

$$\|\mathbf{v}\|_{\mathbb{F}_0^{1,p}} = \sum_{i=1}^N \left[\|D_{x_i} v_0\|_{L^p(\Omega)} + \|D_{y_i} v_1\|_{L^p(\Omega \times Y)} + \|D_{z_i} v_2\|_{L^p(\Omega \times Y \times Z)} \right]$$

for $\mathbf{v} = (v_0, v_1, v_2) \in \mathbb{F}_0^{1,p}$.

Now, let

$$\mathcal{F}_0^\infty = \mathcal{D}_\mathbb{R}(\Omega) \times [\mathcal{D}_\mathbb{R}(\Omega) \otimes (\mathcal{C}_{per}^\infty(Y;\mathbb{R})/\mathbb{R})] \times [\mathcal{D}_\mathbb{R}(\Omega) \otimes \mathcal{C}_{per}^\infty(Y \times \widetilde{Z};\mathbb{R})],$$

where

$$\mathcal{C}_{per}^\infty(Y \times \widetilde{Z};\mathbb{R}) = \left\{\psi | \psi \in \mathcal{C}_{per}^\infty(Y \times Z;\mathbb{R}), \int_Z \psi(\cdot, z)dz = 0\right\}.$$

Lemma 8.12 *$\mathcal{F}_0^\infty$ is dense in $\mathbb{F}_0^{1,p}$.*

Proof Since $\mathcal{D}_\mathbb{R}(\Omega)$ is dense in $W_0^{1,p}(\Omega;\mathbb{R})$ and on the other hand, $\mathcal{D}_\mathbb{R}(\Omega) \otimes (\mathcal{C}_{per}^\infty(Y;\mathbb{R})/\mathbb{R})$ is dense in $L^p(\Omega; W_\#^{1,p}(Y;\mathbb{R}))$ (use Corollary 2.13), one sees readily that the lemma is proved once we have shown that $\mathcal{C}_{per}^\infty(Y \times \widetilde{Z};\mathbb{R})$ is dense in $L_{per}^p(Y; W_\#^{1,p}(Z;\mathbb{R}))$. But this follows by proceeding as in the proof of Lemma 6.10. ∎

We shall conclude this preliminary step by considering the variational problem

$$\begin{cases} \mathbf{u} = (u_0, u_1, u_2) \in \mathbb{F}_0^{1,p} : \\ \displaystyle\int_\Omega \int_Y \int_Z \mathbf{a}(y, z, \mathbb{D}\mathbf{u}(x, y, z)) \cdot \mathbb{D}\mathbf{v}(x, y, z)dxdydz = \langle f, v_0 \rangle \\ \text{for all } \mathbf{v} = (v_0, v_1, v_2) \in \mathbb{F}_0^{1,p}, \end{cases} \tag{8.37}$$

where $\mathbb{D}\mathbf{v} = D_x v_0 + D_y v_1 + D_z v_2$ for $\mathbf{v} = (v_0, v_1, v_2) \in \mathbb{F}_0^{1,p}$. This problem admits at most one solution, as is readily seen by part (iii) of Proposition 8.22.

We are now ready to prove the global periodic homogenization theorem for (8.4).

Theorem 8.2 *Let $1 < p < \infty$. Let Ω be a smooth bounded open set in $\mathbb{R}_x^N$. Suppose (8.36) holds. For arbitrary fixed $\varepsilon > 0$, let u_ε be the solution of the boundary value problem (8.4). Then, as $\varepsilon \to 0$,*

$$u_\varepsilon \to u_0 \textit{ in } W_0^{1,p}(\Omega)\textit{-weak} \tag{8.38}$$

$$\frac{\partial u_\varepsilon}{\partial x_i} \to \frac{\partial u_0}{\partial x_i} + \frac{\partial u_1}{\partial y_i} + \frac{\partial u_2}{\partial z_i} \textit{ in } L^p(\Omega)\textit{-weak } R\Sigma \ (1 \le i \le N), \tag{8.39}$$

where $\boldsymbol{u} = (u_0, u_1, u_2)$ is uniquely determined by (8.37).

Proof The line of proceeding is the same as followed in proving Theorem 8.1. As seen before, the sequence $(u_\varepsilon)_{\varepsilon>0}$ is norm bounded in $W_0^{1,p}(\Omega)$. Then, by fixing freely some fundamental sequence E, Proposition 8.11 ensures that there are some subsequence E' extracted from E and some triple $\mathbf{u} = (u_0, u_1, u_2) \in \mathbb{F}_0^{1,p}$ such that (8.38) and (8.39) hold true when $E' \ni \varepsilon \to 0$. Therefore we are through if we can establish that $\mathbf{u}$ satisfies the variational equation in (8.37). For that purpose, let $\mathbf{\Phi} = (\psi_0, \psi_1, \psi_2) \in \mathcal{F}_0^\infty$. Attach to $\mathbf{\Phi}$ the sequence $(\Phi_\varepsilon)_{\varepsilon>0}$ given by (8.27). Then, as shown before, one has (8.33), where ε ranges over E', and the aim is to pass to the limit in this inequality when $E' \ni \varepsilon \to 0$. To start with, recall that

$$\|D_{x_i}\Phi_\varepsilon - D_{x_i}\psi_0 - (D_{y_i}\psi_1)^\varepsilon - (D_{z_i}\psi_2)^\varepsilon\|_\infty \to 0 \;\; (1 \leq i \leq N)$$

as $\varepsilon \to 0$, where $D_{x_i} = \frac{\partial}{\partial x_i}$ and an analogous definition for D_{y_i} and D_{z_i}. One deduces readily that, as $\varepsilon \to 0$,

$$D_{x_i}\Phi_\varepsilon \to D_{x_i}\psi_0 + D_{y_i}\psi_1 + D_{z_i}\psi_2 \text{ in } L^p(\Omega)\text{-weak R}\Sigma \;\; (1 \leq i \leq N)$$

and

$$\Phi_\varepsilon \to \psi_0 \text{ in } W_0^{1,p}(\Omega)\text{-weak.}$$

With this in mind, as $E' \ni \varepsilon \to 0$ we can pass to the limit in (8.33) using (8.38)–(8.39) and Corollary 8.12, and we obtain

$$\langle f, u_0 - \psi_0\rangle - \int_\Omega \int_Y \int_Z \mathbf{a}(\cdot, \cdot, \mathbb{D}\mathbf{\Phi}) \cdot (\mathbb{D}\mathbf{u} - \mathbb{D}\mathbf{\Phi}) dx dy dz \geq 0,$$

and that for any arbitrary $\mathbf{\Phi} \in \mathcal{F}_0^\infty$. Thanks to Lemma 8.12 together with Proposition 8.22 (ii) this extends to all $\mathbf{\Phi} \in \mathbb{F}_0^{1,p}$. Hence the theorem follows by the same argument as used in proving Theorem 5.9. ■

8.6 Some Concrete Homogenization Problems for (8.4)

In this section we consider a few examples of homogenization problems for (8.4) in a concrete setting (as opposed to the abstract framework of Sect. 8.5.1) and we show how their study leads naturally to the abstract setting of Sect. 8.5.1 and so one may conclude by merely applying Theorem 8.1. The basic notation and hypotheses are those of Sects. 8.1 and 8.2.

8.6.1 Problem I: The Periodic Setting Revisited

Let the problem before us be to investigate the limiting behaviour (as $\varepsilon \to 0$) of the solution, u_ε, of (8.4) under the structure hypothesis (8.36). In fact, this problem should be considered in two different ways: **(1)** As relevant to periodic homogenization theory, in which case one is readily led to Sect. 8.5.2 where a thorough study culminating in Theorem 8.2 has been carried out. **(2)** As a concrete case falling within the scope of the abstract theory developed in Sect. 8.5.1. Specifically, the aim here is to show that the RH-algebra $A = \mathcal{C}_{per}(Y \times Z)$ (which is naturally attached to hypothesis (8.36)) meets the requirements of Theorem 8.1. It is a fact that this is a $W^{1,p}$-proper RH-algebra on $\mathbb{R}^N \times \mathbb{R}^N$ under the action $\mathcal{H}$ (see Proposition 8.11 together with Remark 8.8). It remains to verify that (8.24) holds with $A = \mathcal{C}_{per}(Y \times Z)$. In view of Proposition 8.21, one has $a_i(\cdot, \cdot, \boldsymbol{\Psi}) \in L^{p'}_{per}(Y; \mathcal{C}_{per}(Z))$ $(1 < p < \infty,\ 1 \leq i \leq N)$ for $\boldsymbol{\Psi} \in \mathcal{C}_{per}(Y \times Z; \mathbb{R})^N$. Thus, we are through if one can check that $L^{p'}_{per}(Y; \mathcal{C}_{per}(Z)) \subset \mathfrak{X}^{p'}_A(\mathbb{R}^N_y; \mathcal{B}(\mathbb{R}^N_z))$. But this is straightforward by the density of $\mathcal{C}_{per}(Y \times Z)$ in $L^{p'}_{per}(Y; \mathcal{C}_{per}(Z))$ and use of the fact that the latter space is continuously embedded in $\Xi^{p'}(\mathbb{R}^N_y; \mathcal{B}(\mathbb{R}^N_z))$.

Remark 8.14 It is readily seen by (8.25) that for $A = \mathcal{C}_{per}(Y \times Z)$ we have $\mathfrak{X}^p_A(\mathbb{R}^N_y; \mathcal{B}(\mathbb{R}^N_z)) = L^p_{per}(Y; \mathcal{C}_{per}(Z))$ $(1 \leq p < \infty)$.

In conclusion, both of the ways are consistent in so far as they lead to the same result. This confirms the fact that periodic homogenization is not an isolated theory but instead one specific aspect of a great unity, namely deterministic homogenization theory.

8.6.2 *Problem II: A General Setting*

In what follows, G denotes a function of $\mathbb{R}^N \times \mathbb{R}^N \times \mathbb{R}^N$ into $\mathbb{R}$ satisfying condition (8.3)(3) (where $\mathbf{a}$ is substituted by G), and A is a closed subalgebra of $\mathcal{B}(\mathbb{R}^N \times \mathbb{R}^N)$. We equip A with the supremum norm on $\mathbb{R}^N \times \mathbb{R}^N$. Let $A_{\mathbb{R}} = A \cap \mathcal{C}(\mathbb{R}^N \times \mathbb{R}^N; \mathbb{R})$.

Lemma 8.13 *Suppose we have*

$$G(\cdot, \cdot, \mu) \in A \quad \textit{for fixed } \mu \in \mathbb{R}^N. \tag{8.40}$$

Then

$$G(\cdot, \cdot, \boldsymbol{\Psi}) \in A \quad \textit{for all } \boldsymbol{\Psi} \in (A_{\mathbb{R}})^N. \tag{8.41}$$

Proof This is a copy mutatis mutandis of the proof of Lemma 7.12. ■

As an immediate consequence of this, there is the following

Proposition 8.24 *Let $1 < p < \infty$. Let A be a $W^{1,p}$-proper RH-algebra on $\mathbb{R}^N \times \mathbb{R}^N$ (under the action $\mathcal{H}$) admitting A^∞ as a dense subalgebra. Suppose we have*

$$a_i(\cdot, \cdot, \mu) \in A \ (1 \leq i \leq N) \quad \textit{for fixed } \mu \in \mathbb{R}^N. \tag{8.42}$$

Then, the homogenization problem for (8.4) under the structure hypothesis (8.42) is solved by Theorem 8.1.

Proof This follows readily by Lemma 8.13 and use of the trivial relation $A \subset \mathfrak{X}_A^{p'}(\mathbb{R}_y^N; \mathcal{B}(\mathbb{R}_z^N))$. ■

This proposition has an interesting corollary worth pointing out.

Corollary 8.13 *Let A_1 be an RH-algebra on $\mathbb{R}^N \times \mathbb{R}^N$ which is not necessarily $W^{1,p}$-proper. Suppose there is some $W^{1,p}$-proper RH-algebra A on $\mathbb{R}^N \times \mathbb{R}^N$ such that $A_1 \subset A$. Then, the homogenization problem for (8.4) under the structure hypothesis*

$$a_i(\cdot, \cdot, \mu) \in A_1 \quad (1 \leq i \leq N) \quad \text{for fixed } \mu \in \mathbb{R}^N \tag{8.43}$$

is solved by Theorem 8.1

Proof Indeed (8.43) implies (8.42), whence the corollary follows by Proposition 8.24. ■

Thus, thanks to Lemma 8.13, the homogenization problem for (8.4) under the structure hypothesis (8.42) is solvable when A ***denotes any one of the RH-algebras of Sect. 8.4.2 (provided the corresponding real*** $p > 1$ ***in Proposition 8.24 is the appropriate one). Better still, the problem remains solvable even when*** A ***is not*** $W^{1,p}$***-proper provided it is contained in a*** $W^{1,p}$***-proper RH-algebra-proper. It is worth noting that each RH-algebra*** A ***in Sect. 8.4.2 admits*** A^∞ ***as a dense subalgebra.***

This is worth illustrating.

Example 8.4 Let the problem on hand be the study of the homogenization problem for (8.4) under the structure hypothesis

$$a_i(\cdot, \cdot, \mu) \in \mathcal{C}_{per}(Y; \mathcal{B}_\infty(\mathbb{R}_z^N)) \quad (1 \leq i \leq N) \quad \text{for fixed } \mu \in \mathbb{R}^N.$$

This problem is solvable. Indeed, to begin with, $\mathcal{C}_{per}(Y; \mathcal{B}_\infty(\mathbb{R}_z^N))$ is the closure of $\mathcal{C}_{per}(Y) \otimes \mathcal{B}_\infty(\mathbb{R}_z^N)$ in $\mathcal{B}(\mathbb{R}^N \times \mathbb{R}^N)$. Moreover, it is an RH-algebra (see, e.g., Example 8.3). Finally, one has $\mathcal{C}_{per}(Y; \mathcal{B}_\infty(\mathbb{R}_z^N)) \subset A = \mathcal{C}_{per}(Y; \mathcal{B}_{\infty,\mathbb{Z}^N}(\mathbb{R}_z^N))$, and A is a $W^{1,p}$-proper RH-algebra (see Proposition 8.14). Hence the claimed result follows by Corollary 8.13.

8.6.3 Problem III: Almost Periodic Setting

We consider here the homogenization problem for (8.4) under the hypothesis that for fixed $\mu \in \mathbb{R}^N$, each function $(y, z) \mapsto a_i(y, z, \mu)$ $(1 \leq i \leq N)$ is an almost periodic continuous real function on $\mathbb{R}^N \times \mathbb{R}^N$. More precisely, we have here

$$a_i(\cdot, \cdot, \mu) \in AP(\mathbb{R}_y^N \times \mathbb{R}_z^N) \quad (1 \leq i \leq N) \quad \text{for fixed } \mu \in \mathbb{R}^N. \tag{8.44}$$

As was pointed out earlier (see Sect. 7.3.3), there is a countable subgroup R of $\mathbb{R}^N \times \mathbb{R}^N$ such that $a_i(\cdot, \cdot, \mu) \in AP_R(\mathbb{R}^N \times \mathbb{R}^N)$ $(\mu \in \mathbb{R}^N, 1 \leq i \leq N)$. It follows that

$$a_i(\cdot, \cdot, \mu) \in AP_{\mathcal{R}}(\mathbb{R}^N \times \mathbb{R}^N) \quad (1 \leq i \leq N) \quad \text{for fixed } \mu \in \mathbb{R}^N \tag{8.45}$$

with $\mathcal{R} = \mathcal{R}_y \times \mathcal{R}_z$, where $\mathcal{R}_y = pr_y(R)$ and $\mathcal{R}_z = pr_z(R)$, pr_y (resp. pr_z) being the natural projection of $\mathbb{R}_y^N \times \mathbb{R}_z^N$ onto $\mathbb{R}_y^N$ (resp. $\mathbb{R}_z^N$). It is clear that $\mathcal{R}_y$ and $\mathcal{R}_z$ are countable subgroups of $\mathbb{R}^N$. Furthermore, $AP_{\mathcal{R}}(\mathbb{R}_y^N \times \mathbb{R}_z^N)$ is an RH-algebra on $\mathbb{R}^N \times \mathbb{R}^N$ (see Example 8.2). According to Proposition 8.13, if $\mathcal{R}_z$ is a discrete subgroup of $\mathbb{R}^N$, then this RH-algebra is $W^{1,2}$-proper. Hence, it follows by Proposition 8.24 (note that (8.44) and (8.45) are equivalent) that the homogenization problem for (8.4) under the structure hypothesis (8.44) is solved by Theorem 8.1 provided $p = 2$.

8.6.4 Problem IV: A General Setting

We assume that the following condition is satisfied for $1 \leq i \leq N$:

$$\begin{cases} \text{For any given } \boldsymbol{\Psi} \in \mathcal{B}(\mathbb{R}_y^N \times \mathbb{R}_z^N; \mathbb{R})^N, \text{ the function} \\ y \mapsto a_i(y, \cdot, \boldsymbol{\Psi}(y, \cdot)) \text{ is measurable from } \mathbb{R}^N \text{ to } \mathcal{B}(\mathbb{R}^N) \end{cases} \tag{8.46}$$

Remark 8.15 Actually, this hypothesis has nothing surprising. Indeed, as seen in proving Proposition 8.21, for $\boldsymbol{\Psi}$ as above one has $a_i(y, \cdot, \boldsymbol{\Psi}(y, \cdot)) \in \mathcal{B}(\mathbb{R}_z^N)$ for a.e. $y \in \mathbb{R}^N$ (the boundedness is a direct consequence of (8.2) and (8.3)(3)). Subsequently, the measurability of the function $a_i(\cdot, \cdot, \boldsymbol{\Psi})$ (defined above) from $\mathbb{R}^N$ to $\mathcal{B}(\mathbb{R}^N)$ seems to be intrinsically connected to the behaviour of the differential operator under consideration (see Lemma 8.10)

This being so, let A_z be an H-algebra as in Sect. 8.2.2. In this section we discuss the homogenization of (8.4) under the structure hypothesis:

$$\begin{cases} \text{For a.e. } y \in \mathbb{R}^N, \text{ the mapping } z \mapsto a_i(y, z, \mu) \text{ denoted by} \\ a_i(y, \cdot, \mu) \text{ lies in } A_z \text{ and further, the function } y \mapsto a_i(y, \cdot, \mu) \\ \text{is } Y\text{-periodic,} \end{cases} \tag{8.47}$$

where $\mu \in \mathbb{R}^N$ and $1 \leq i \leq N$ are arbitrary fixed. It is clear by (8.1)(1), (8.2), (8.3)(3) and use of Lemma 8.2 that (8.47) is equivalent to

$$a_i(\cdot, \cdot, \mu) \in L^\infty_{per}(Y; A_z) \quad (1 \leq i \leq N) \quad \text{for fixed } \mu \in \mathbb{R}^N. \tag{8.48}$$

Proposition 8.25 *Let (8.46) be satisfied. Suppose $\mathcal{C}_{per}(Y; A_z)$ is a $W^{1,p}$-proper RH-algebra on $\mathbb{R}^N \times \mathbb{R}^N$ (under the action $\mathcal{H}$) for some $1 < p < \infty$. Then, the homogenization problem for (8.4) under the structure hypothesis (8.47) is solved by Theorem 8.1.*

Proof First of all, note that $\mathcal{C}_{per}(Y; A_z) = \overline{\mathcal{C}_{per}(Y) \otimes A_z}$, and so the problem on hand falls under the setting of Sect. 8.2.2 with $A_y = \mathcal{C}_{per}(Y)$ and $A = \mathcal{C}_{per}(Y; A_z)$. Next, in view of the hypothesis on A it is clear that one only needs to show that (8.24) holds with $A = \mathcal{C}_{per}(Y; A_z)$.

To this end, following a familiar line of proceeding (see, e.g., Sect. 7.3), we introduce a mollifier $(\theta_n)_{n\in\mathbb{N}}$ on $\mathbb{R}^N_y$ with $\mathrm{Supp}\,\theta_n \subset \varepsilon_n \overline{B}_N$, where (ε_n) is a fundamental sequence and $\overline{B}_N$ the closed unit ball in $\mathbb{R}^N$. Let $1 \le i \le N$ and $n \in \mathbb{N}$ be freely fixed. For $y, z, \mu \in \mathbb{R}^N$, let

$$\xi_n^i(y, z, \mu) = \int \theta_n(x) a_i(y - x, z, \mu) dx.$$

Clearly

$$\xi_n^i(\cdot, \cdot, \mu) \in A \quad \text{for fixed } \mu \in \mathbb{R}^N \tag{8.49}$$

and

$$|\xi_n^i(y, z, \mu) - \xi_n^i(y, z, \nu)| \le c_1(1 + |\mu| + |\nu|)^{p-1-\alpha_1}|\mu - \nu|^{\alpha_1} \tag{8.50}$$

for all $z, \mu, \nu \in \mathbb{R}^N$, but also for all $y \in \mathbb{R}^N$. Indeed, (8.49) follows by (8.48) whereas (8.50) follows readily by (8.3)(3). By Lemma 8.13 we deduce from (8.49) and (8.50) that

$$\xi_n^i(\cdot, \cdot, \mathbf{\Psi}) \in A \quad \text{for } \mathbf{\Psi} \in (A_{\mathbb{R}})^N. \tag{8.51}$$

Now, thanks to (8.48) and to the relation $L^\infty_{per}(Y; A_z) \subset L^{p'}_{per}(Y; A_z)$, we have by Hölder's inequality

$$|\xi_n^i(y, z, \mu) - a_i(y, z, \mu)|^{p'} \le \int \theta_n(x)|a_i(y - x, z, \mu) - a_i(y, z, \mu)|^{p'} dx$$

and that for arbitrary fixed $z, \mu \in \mathbb{R}^N$ and for a.e. $y \in \mathbb{R}^N$. Fix freely some $\mathbf{\Psi} \in (A_{\mathbb{R}})^N$, and note that by (8.2), (8.3)(3), (8.46) and (8.47), one has $a_i(\cdot, \cdot, \mathbf{\Psi}) \in L^\infty_{per}(Y; \mathcal{B}(\mathbb{R}^N)) \subset L^{p'}_{per}(Y; \mathcal{B}(\mathbb{R}^N))$. Taking $\mu = \mathbf{\Psi}(y, z)$ above yields

$$\|\xi_n^i(\cdot, \cdot, \mathbf{\Psi}) - a_i(\cdot, \cdot, \mathbf{\Psi})\|_{L^{p'}(Y;\mathcal{B})} \le$$

$$\left[\int \theta_n(x)\|a_i[x](\cdot, \cdot, \mathbf{\Psi}) - a_i(\cdot, \cdot, \mathbf{\Psi})\|^{p'}_{L^{p'}(Y;\mathcal{B})} dx\right]^{\frac{1}{p'}}$$

where $a_i[x](\cdot,\cdot,\Psi)$ denotes the function $(y,z)\mapsto a_i(y-x,z,\Psi(y,z))$, and where the norm in $L^{p'}_{per}(Y;\mathcal{B}(\mathbb{R}^N_z))$ is given by

$$\|u\|_{L^{p'}(Y;\mathcal{B})}=\left(\int_Y\|u(y,\cdot)\|_\infty^{p'}dy\right)^{\frac{1}{p'}}.$$

By the trivial equality

$$\begin{aligned}&a_i(y-x,z,\Psi(y,z))-a_i(y,z,\Psi(y,z))\\&=a_i(y-x,z,\Psi(y,z))-a_i(y-x,z,\Psi(y-x,z))\\&+a_i(y-x,z,\Psi(y-x,z))-a_i(y,z,\Psi(y,z)),\end{aligned}$$

and use of (8.3)(3), it follows that

$$\begin{aligned}&\|\xi^i_n(\cdot,\cdot,\Psi)-a_i(\cdot,\cdot,\Psi)\|_{L^{p'}(Y;\mathcal{B})}\\&\le\left(\int\theta_n(x)\|\tau_x a_i(\cdot,\cdot,\Psi)-a_i(\cdot,\cdot,\Psi)\|^{p'}_{L^{p'}(Y;\mathcal{B})}dx\right)^{\frac{1}{p'}}\\&+c\left(\int\theta_n(x)\|\,|\tau_x\Psi-\Psi|^{\alpha_1}\|^{p'}_{L^{p'}(Y;\mathcal{B})}dx\right)^{\frac{1}{p'}},\end{aligned}$$

where c is a positive constant that depends only on Ψ, and where $\tau_x a_i(\cdot,\cdot,\Psi)$ stands for the usual translate, that is, the function $y\mapsto a_i(y-x,\cdot,\Psi(y-x,\cdot))$ of $\mathbb{R}^N$ into $\mathcal{B}(\mathbb{R}^N_z)$, which lies in $L^{p'}_{per}(Y;\mathcal{B}(\mathbb{R}^N_z))$, of course.

Now, as $|x|\to 0$, one has $\tau_x a_i(\cdot,\cdot,\Psi)\to a_i(\cdot,\cdot,\Psi)$ and $|\tau_x\Psi-\Psi|^{\alpha_1}\to 0$ in the $L^{p'}_{per}(Y;\mathcal{B}(\mathbb{R}^N_z))$-norm. Indeed, whereas the former property is a classical fact, the latter one is easily deduced from the uniform continuity of Ψ on $\mathbb{R}^N$ into $\mathcal{B}(\mathbb{R}^N_z)^N$. Furthermore, since the preceding integrals are computed over $\varepsilon_n B_N$ (with $\varepsilon_n\to 0$ as $n\to\infty$), it follows at once that $\xi^i_n(\cdot,\cdot,\Psi)\to a_i(\cdot,\cdot,\Psi)$ in $L^{p'}_{per}(Y;\mathcal{B}(\mathbb{R}^N_z))$ as $n\to\infty$. This leads us to (8.24) by (8.51) and use of the fact that the space $L^{p'}_{per}(Y;\mathcal{B}(\mathbb{R}^N_z))$ is continuously embedded in $\Xi^{p'}(\mathbb{R}^N_y;\mathcal{B}(\mathbb{R}^N_z))$, as a direct consequence of Lemma 1.9. The proof is complete. ■

We shall conclude with a few concrete examples. It is assumed that condition (8.46) is fulfilled.

Example 8.5 The homogenization problem for (8.4) under the structure hypothesis (8.47) is solved by Theorem 8.1 with $A=\mathcal{C}_{per}(Y;A_z)$, where A_z is any of the H-algebra (on $\mathbb{R}^N_z$) $\mathcal{C}_{per}(Z)$, $AP_{\mathcal{R}_z}(\mathbb{R}^N_z)$ ($\mathcal{R}_z$ discrete), $\mathcal{B}_{\infty,\mathbb{Z}^N}(\mathbb{R}^N_z)$, $\mathcal{B}_{\infty,\mathcal{R}}(\mathbb{R}^N_z)$ ($\mathcal{R}$ discrete), $\mathcal{B}_\infty(\mathbb{R};\mathcal{C}_{per}(Z'))$. See Sect. 8.4.2 for details.

Example 8.6 The homogenization problem for (8.4) under the structure hypothesis (for fixed $\mu \in \mathbb{R}^N$ and $1 \leq i \leq N$)

$$\begin{cases} \text{For a.e. } y \in \mathbb{R}^N, \text{ the mapping } z \mapsto a_i(y, z, \mu) \text{ denoted by } a_i(y, \cdot, \mu) \\ \text{lies in } \mathcal{B}_\infty(\mathbb{R}_z^N) \text{ and further, the function } y \mapsto a_i(y, \cdot, \mu) \text{ is } Y\text{-periodic} \end{cases}$$

is solvable by Theorem 8.1 with $A = \mathcal{C}_{per}(Y; \mathcal{B}_{\infty,\mathbb{Z}^N}(\mathbb{R}^N))$. Indeed, by hypothesis, for fixed $\mu \in \mathbb{R}^N$ and for almost every $y \in \mathbb{R}^N$, one has $a_i(y, \cdot, \mu) \in A_z = \mathcal{B}_{\infty,\mathbb{Z}^N}(\mathbb{R}_z^N)$. Hence the claimed result follows by Example 8.5 and use of Proposition 8.14.

8.7 Comments and Problems

Reiterated homogenization goes back to Bensoussan et al. [14].

The homogenization problem for (8.4) was first studied by Lions et al. [75, 76] in the periodic setting, and later by Lukkassen et al. [78] in the deterministic setting using the Σ-convergence theory.

The results of Sect. 8.2 were first given in the latter reference, whereas those of the remaining sections are given here for the first time.

Reiterated homogenization considered in the general Σ-convergence setting requires a thorough adaptation of the Σ-convergence method in its primary version of Chap. 5. This is illustrated by the need of further tools such as the notion of a reiteration H-algebra, or that of reiterated Σ-convergence. In this connection it is worth noting that although the present study is focalized on the homogenization of the boundary value problem (8.4), the machinery framed above is the one underlying the study, via the Σ-convergence theory, of the reiterated homogenization of linear as well as nonlinear differential operators, at least in the three-scale setting.

The theory developed in the preceding sections readily extends to the boundary value problem

$$-\operatorname{div} \mathbf{a}\left(\frac{x}{\varepsilon}, \frac{x}{\varepsilon^k}, Du_\varepsilon\right) = f \text{ in } \Omega, \quad u_\varepsilon \in W_0^{1,p}(\Omega; \mathbb{R}),$$

where k is any arbitrary given real number with $k \geq 2$. Indeed, a quick survey of the preceding sections reveals that the whole study remains rigorously valid mutatis mutandis provided the trace in (8.6) is suitably redefined. This is straightforward. Given $\psi \in L^\infty(\mathbb{R}_y^N; \mathcal{B}(\mathbb{R}_z^N))$, by letting ${}^\varepsilon\psi(y) = \psi(y, \frac{y}{\varepsilon^{k-1}})$ $(y \in \mathbb{R}^N)$ in the sense of Proposition 1.7 one gets a function ${}^\varepsilon\psi \in L^\infty(\mathbb{R}_y^N)$ with $\| {}^\varepsilon\psi \|_{L^\infty(\mathbb{R}^N)} \leq \|\psi\|_{L^\infty(\mathbb{R}^N; \mathcal{B}(\mathbb{R}^N))}$. One may then define ψ^ε by (8.6), that is, formally written, $\psi^\varepsilon(x) = \psi(\frac{x}{\varepsilon}, \frac{x}{\varepsilon^k})$ $(x \in \mathbb{R}^N)$. One has $\psi^\varepsilon \in L^\infty(\mathbb{R}_x^N)$ with $\|\psi^\varepsilon\|_{L^\infty(\mathbb{R}_x^N)} \leq \|\psi\|_{L^\infty(\mathbb{R}_y^N; \mathcal{B}(\mathbb{R}_z^N))}$.

Therefore, Proposition 8.3 follows by Lemma 8.1 and use of the analogue of Remark 8.1. Furthermore, it is worth noting that (8.15) becomes

$$\int_\Omega \frac{1}{\varepsilon^{k-1}} u_\varepsilon(x)\psi\left(\frac{x}{\varepsilon}, \frac{x}{\varepsilon^k}\right) dx \to 0.$$

We shall conclude by revisiting the equality

$$a_i^\varepsilon(\cdot, \cdot, \boldsymbol{\Phi}^\varepsilon) = a_i(\cdot, \cdot, \boldsymbol{\Phi})^\varepsilon \quad (1 \le i \le N), \quad \boldsymbol{\Phi} \in \mathcal{C}(\overline{\Omega}; \mathcal{B}(\mathbb{R}^N \times \mathbb{R}^N; \mathbb{R})^N) \tag{8.52}$$

which is of considerable interest, as seen in proving Theorem 8.1 (see also Proposition 8.20 and Corollary 8.9). For the benefit of the reader this equality needs clarifying. First, in order to simplify the writing, in place of a_i we consider a function G of $\mathbb{R}_y^N \times \mathbb{R}_z^N \times \mathbb{R}_\mu^N$ into $\mathbb{R}$ satisfying

$$\begin{cases} \text{(1)} \quad G(y, z, \omega) = 0 \quad \text{for all } z \in \mathbb{R}^N \quad \text{and for a.e. } y \in \mathbb{R}^N, \\ \text{(2)} \quad |G(y, z, \mu) - G(y, z, \nu)| \le c_1(1 + |\mu| + |\nu|)^{p-1-\alpha_1}|\mu - \nu|^{\alpha_1} \\ \quad \text{for all } \mu, \nu, z \in \mathbb{R}^N \quad \text{and for a.e. } y \in \mathbb{R}^N \quad \text{(see Sect. 8.1.1).} \end{cases} \tag{8.53}$$

Finally, we make the hypothesis:

$$\begin{cases} \text{For any given } \boldsymbol{\Psi} \in \mathcal{B}(\mathbb{R}_y^N \times \mathbb{R}_z^N; \mathbb{R})^N, \text{ the function } y \mapsto G(y, \cdot, \boldsymbol{\Psi}(y, \cdot)) \\ \text{is measurable from } \mathbb{R}^N \text{ to } \mathcal{B}(\mathbb{R}^N). \end{cases}$$

It then follows by (8.53) that

$$G(\cdot, \cdot, \boldsymbol{\Psi}) \in L^\infty(\mathbb{R}_y^N; \mathcal{B}(\mathbb{R}_z^N)), \tag{8.54}$$

where $G(\cdot, \cdot, \boldsymbol{\Psi})$ stands for the function $(y, z) \mapsto G(y, z, \boldsymbol{\Psi}(y, z))$ of $\mathbb{R}^N \times \mathbb{R}^N$ into $\mathbb{R}$. In particular $G(\cdot, \cdot, \mu) \in L^\infty(\mathbb{R}^N; \mathcal{B}(\mathbb{R}^N))$ for $\mu \in \mathbb{R}^N$.

This being so, before we can show (8.52), that is,

$$G^\varepsilon(\cdot, \cdot, \boldsymbol{\Phi}^\varepsilon) = G(\cdot, \cdot, \boldsymbol{\Phi})^\varepsilon, \quad \boldsymbol{\Phi} \in \mathcal{C}(\overline{\Omega}; \mathcal{B}(\mathbb{R}^N \times \mathbb{R}^N; \mathbb{R})^N) \tag{8.55}$$

with $G = a_i$, let us give meaning to each side of the claimed equality. Fix freely some $\boldsymbol{\Phi}$ as in (8.55). By (8.54) with $\boldsymbol{\Psi} = \boldsymbol{\Phi}(x, \cdot, \cdot)$ (for fixed $x \in \overline{\Omega}$) and use of (8.53)(2) one sees readily that $G(\cdot, \cdot, \boldsymbol{\Phi})$ lies in $\mathcal{C}(\overline{\Omega}; L^\infty(\mathbb{R}^N; \mathcal{B}(\mathbb{R}^N)))$. It follows by Proposition 8.3 that $G(\cdot, \cdot, \boldsymbol{\Phi})^\varepsilon$ is well defined as a function in $L^\infty(\Omega)$. Likewise, for $\mathbf{v} \in \mathcal{C}(\overline{\Omega}; \mathbb{R})^N$, the function $(x, y, z) \mapsto G(y, z, \mathbf{v}(x))$ of $\overline{\Omega} \times \mathbb{R}^N \times \mathbb{R}^N$ into $\mathbb{R}$ denoted by $G(\cdot, \cdot, \mathbf{v})$ lies in $\mathcal{C}(\overline{\Omega}; L^\infty(\mathbb{R}^N; \mathcal{B}(\mathbb{R}^N)))$ and one has $G(\cdot, \cdot, \mathbf{v})^\varepsilon \in L^\infty(\Omega)$. For the sake of convenience one puts (by definition) $G^\varepsilon(\cdot, \cdot, \mathbf{v}) = G(\cdot, \cdot, \mathbf{v})^\varepsilon$ for $\mathbf{v} \in \mathcal{C}(\Omega; \mathbb{R})^N$. Thus (8.55) holds true for $\mathbf{\Phi} = \mathbf{v} \in \mathcal{C}(\overline{\Omega}; \mathbb{R})^N$, and the aim is then to extend this to the whole of $\mathcal{C}(\overline{\Omega}; \mathcal{B}(\mathbb{R}^N \times \mathbb{R}^N; \mathbb{R})^N)$. To do this we need the following result:

$$\left\{\begin{array}{l}\text{If } u \in \mathcal{C}(\overline{\Omega}; L^\infty(\mathbb{R}^N_y; \mathcal{B}(\mathbb{R}^N_z; \mathbb{R}))) \text{ is such that for fixed } x \in \overline{\Omega} \\ \text{one has } u(x, y, z) \geq 0 \text{ for a.e. } y \in \mathbb{R}^N \text{ and for all } z \in \mathbb{R}^N, \text{ then} \\ u^\varepsilon(x) \geq 0 \text{ for a.e. } x \in \Omega, \text{ where } u^\varepsilon \text{ is given by Proposition 8.3.}\end{array}\right. \tag{8.56}$$

This can be obtained by the same line of proceeding (based on the result analogous to Lemma 5.1) as followed in proving Lemma 5.11.

To conclude, let $\varepsilon > 0$ be fixed. Let $\mathbf{\Phi} \in \mathcal{C}(\overline{\Omega}; \mathcal{B}(\mathbb{R}^N \times \mathbb{R}^N; \mathbb{R})^N)$. Fix freely some triple (x, y, z) with $x \in \overline{\Omega}, z \in \mathbb{R}^N, y \in \mathbb{R}^N \setminus \mathcal{N}$, where $\mathcal{N}$ is a negligible set in $\mathbb{R}^N$ that depends on the sole G. In (8.53)(2), choose $\mu = \mathbf{\Phi}^\varepsilon(x)$ and $\nu = \mathbf{\Phi}(x, y, z)$ to see that

$$|G(y, z, \mathbf{\Phi}^\varepsilon(x)) - G(y, z, \mathbf{\Phi}(x, y, z))| \leq c|\mathbf{\Phi}^\varepsilon(x) - \mathbf{\Phi}(x, y, z)|^{\alpha_1},$$

where c is a positive constant depending only on $\mathbf{\Phi}$. By an argument followed before in proving Lemma 5.12 and whose chief point is (8.56) in the present context, and noting on the other hand that the function $\psi = |\mathbf{\Phi}^\varepsilon - \mathbf{\Phi}|^{\alpha_1}$ lies in $\mathcal{C}(\overline{\Omega}; \mathcal{B}(\mathbb{R}^N \times \mathbb{R}^N))$ and so ψ^ε is defined in a pointwise fashion, which gives $\psi^\varepsilon = 0$, one is readily led to (8.55), thereby proving (8.52).

For alternative approaches to reiterated homogenization see Bensoussan et al. [14], Allaire and Briane [8], Nguetseng et al. [107].

References

1. Abdulle, A.: A priori and a posteriori error analysis for numerical homogenization: a unified framework. Ser. Contemp. Appl. Math. CAM **16**, 280–305 (2011)
2. Abdulle, A.: Discontinuous Galerkin finite element heterogeneous multiscale method for elliptic problems with multiple scales. Math. Comput. **81**(278), 687–713 (2012)
3. Adams, R.: Sobolev Spaces. Academic Press, New York (1975)
4. Agmon, S., Douglis, A., Nirenberg, L.: Estimates near the boundary for solutions of elliptic partial differential equations satisfying general boundary conditions I. Commun. Pure Appl. Math. **12**, 623–727 (1959)
5. Aiyappan, S., Cardone, G., Perugia, C., Prakash, R.: Homogenization of a nonlinear monotone problem in a locally periodic domain via unfolding method. Nonlinear Anal.: RWA **66**, 103537 (2022)
6. Allaire, G.: Homogenization and two-scale convergence. SIAM J. Math. Anal. **23**, 1482–1518 (1992)
7. Allaire, G.: Shape Optimization by the Homogenization Method. Springer, New York (2002)
8. Allaire, G., Briane, M.: Multi-scale convergence and reiterated homogenization. Proc. R. Soc. Edinb. Sect. A **126**, 297–342 (1996)
9. Allaire, G., Piatnitski, A.: Homogenization of nonlinear reaction-diffusion equation with a large reaction term. Ann. Univ. Ferrara **56**, 141–161 (2010). Hal-00784058
10. Amar, M.: Two-scale convergence and homogenization on $BV(\Omega)$. Asymptot. Anal. **16**(1), 65–84 (1998)
11. Arbogast, T., Douglas, J., Hornung, U.: Derivation of the double porosity model of single phase flow via homogenization theory. SIAM J. Math. Anal. **21**, 823–836 (1990)
12. Bahlali, K., Elouaflin, A., Pardoux, E.: Averaging for BSDEs with null recurrent fast component. Application to Homogenization in a nonperiodic media. Stoch. Proces. Appl. **127**(4), 1321–1353 (2017)
13. Bakhvalov, N., Panashenko, G.: Homogenization: Averaging Processes in Periodic Media. Mathematics and Its Applications: BVil 36. Kluwer Acad. Publ., Dordrecht (1989)
14. Bensoussan, A., Lions, J.L., Papanicolaou, G.: Asymptotic Analysis for Periodic Structures. North-Holland, Amsterdam (1978)
15. Berger, M., Gostiaux, B.: Géométrie différentielle: variétés, courbes et surfaces. PUF, Paris (1987)
16. Bertrandias, J.P., Datry, C., Dupuis, C.: Unions et intersections d'espaces L^p invariantes par translation ou convolution. Ann. Inst. Fourier (Grenoble) **28**, 53–84 (1978)
17. Blanc, X., Le Bris, C.: Homogenization Theory for Multiscale Problems. Springer, Berlin (2023)

G. Nguetseng, *Homogenization Algebras and Applications*, Springer Monographs in Mathematics, https://doi.org/10.1007/978-3-031-84705-9

18. Blanc, X., Wolf, S.: Homogenization of the Poisson equation in a non-periodically perforated domain. Asymptot. Anal. **126**(1–2), 129–155 (2022)
19. Blanc, X., Le Bris, C., Lions, P.L.: Une variante de la théorie de l'homogénéisation stochastique des opérateurs elliptiques. C.R. Acad. Paris, Ser. I **343**, 717–724 (2006)
20. Blanc, X., Le Bris, C., Lions, P.L.: Stochastic homogenization and random lattices. J. Math. Pures Appl. **88**, 34–63 (2007)
21. Bourbaki, N.: Espaces vectoriels topologiques, Chap. I-II. Hermann, Paris (1965)
22. Bourbaki, N.: Intégration, Chap. 1–4. Hermann, Paris (1966)
23. Bourbaki, N.: Intégration, Chap. 5. Hermann, Paris (1967)
24. Bourbaki, N.: Topologie générale, Chap. I-IV. Hermann, Paris (1971)
25. Bourbaki, N.: Topologie générale, Chap. V-X. Hermann, Paris (1974)
26. Bourgeat, A., Mikelic, A., Wright, S.: Stochastic two-scale convergence in the mean and applications. J. Reine Angew. Math. **456**, 19–51 (1994)
27. Brézis, H.: Analyse Fonctionnelle. Masson, Paris (1983)
28. Cardone, G., Woukeng, J.L.: Corrector problem in the deterministic homogenization of nonlinear elliptic equations. Appl. Anal. **98**, 118–135 (2019)
29. Cardone, G., Jäger, W., Woukeng, J.L.: Derivation and analysis of a nonlocal Hele-Shaw-Cahn-Hilliard system for flow in thin heterogeneous layers. Math. Models Methods Appl. Sci. **34**(7), 1343–1400 (2024)
30. Casado-Diaz, J.: Two-scale convergence for nonlinear Dirichlet problems in perforated domains. Proc. R. Soc. Edin. Sect. A **130**(2), 249–276 (2000)
31. Casado-Diaz, J., Gayte, I.: A general compactness result and its application to the two-scale convergence of almost periodic functions. C.R. Acad. Sci. Paris Série I Math. **323**(4), 329–334 (1996)
32. Chechkin, G.A., Piatnitski, A.L.: Homogenization of boundary value problems in a locally periodic perforated domain. Appl. Anal. **71**, 215–235 (1999)
33. Chiado Piat, V., De Franceschi, A.: Asymptotic behaviour of quasi-linear problems with Neumann boundary conditions on perforated domains. Appl. Anal. **36**, 67–87 (1990)
34. Cioranescu, D., Donato, P.: An Introduction to Homogenization. Oxford Lecture Series in Mathematics and Its applications, vol. 17. Oxford University Press, New York (1999)
35. Cioranescu, D., Saint Jean Paulin, J.: Homogenization in open sets with holes. J. Math. Anal. Appl. **71**, 590–607 (1979)
36. Cioranescu, D., Damlamian, A., Griso, G.: The Periodic Unfolding Method. Theory and Applications to Partial Differential Problems. Springer, Singapore (2018)
37. Clark, G.W., Packer, L.A.: Two-scale homogenization of implicit degenerate evolution equations. J. Math. Anal. Appl. **214**(2), 420–438 (1997)
38. Clark, G.W., Packer, L.A.: Two-scale homogenization of nonlinear degenerate evolution equations. J. Math. Anal. Appl. **238**(1), 316–328 (1999)
39. Clark, G.W., Showalter, R.E.: Two-scale convergence of a model for flow in a partially fissured medium. Electron. J. Differ. Equ. **2**, 1–20 (1999)
40. Dal Maso, G.: An Introduction to Gamma-Convergence. Birkhauser, Boston (1993)
41. Damlamian, A.: The periodic unfolding method in homogenization. In: Multiscale Problems: Theory, Numerical Approximation and Applications, vol. 16. CAM. World Scientific, Singapore (2011)
42. De Giorgi, E., Spagnolo, S.: Sulla convergenza degli integrali dell'energia per operatori ellittici del secundo ordine. Bull. U.M.I. **8**, 391–411 (1973)
43. Deugoue, G., Woukeng, J.L.: Sigma-convergence of semilinear stochastic wave equations. Nonlinear Differ. Equ. Appl. **25**, 1–29 (2018)
44. Dieudonné, J.: Eléments d'analyse, Tome II, Chap. XII–XV. Gauthier-Villars, Paris (1968)
45. Dieudonné, J.: Eléments d'analyse, Tome VI, Chap. XXII. Gauthier-Villars, Paris (1975)
46. Dixmier, J.: Les C^*-algèbres et leurs représentations. Gauthier-Villars, Paris (1964)
47. Douanla, H., Svanstedt, N.: Reiterated homogenization of linear eigenvalue problems in multiscale perforated domains beyond the periodic setting. Commun. Math. Anal. **11**(1), 61–93 (2011)

48. Douanla, H., Nguetseng, G., Woukeng, J.L.: Incompressible viscous Newtonian flow in a fissured medium of general deterministic type. Problems in mathematical analysis, No70. J. Math. Sci. (N.Y.) **191**, 214–242 (2013)
49. Edwards, R.D.: Functional Analysis. Holt-Rinehart-Winston, New York (1965)
50. Fouegap, P., Kenne, R.B., Nguetseng, G., Dongo, D., Woukeng, J.L.: Homogenization of linear Boltzmann equations in the context of algebras with mean value. Zeitschrift Ang. Math. Phys. **71**(173), 1–22 (2020)
51. Fouetio, A., Woukeng, J.L.: Homogenization of hyperbolic damped stochastic wave equations. Acta Math. Sin. (English Series) **34**, 233–254 (2018)
52. Fouetio, A., Nguetseng, G., Woukeng, J.L.: Multiscale analysis of semilinear damped stochastic wave equations. Acta Math. Sin. (English Series) **39**(7), 1305–1331 (2023)
53. Fournier, J.F.F., Stewart, J.: Amalgams of L^p and l^q. Bull. Am. Math. Soc. **13**, 1–21 (1985)
54. Frénod, E.: Two-Scale Approach to Oscillatory Singularly Perturbed Transport Equations. Springer, Berlin (2017)
55. Fučik, S., Kufner, A.: Nonlinear Differential Equations. Elsevier, Amsterdam (1980)
56. Geers, M.G.D., Koutznetsova, V.G., Brekelmans, W.A.M.: Computational Homogenization. Springer, Berlin (2010)
57. Guichardet, A.: Analyse Harmonique Commutative. Dunod, Paris (1968)
58. Hewitt, E., Ross, K.A.: Abstract Harmonic Analysis, vol. II. Springer, Berlin (1970)
59. Holmbom, A.: Homogenization of parabolic equations: an alternative approach and some corrector type results. Appl. Math. **12**, 321–343 (1997)
60. Jacobs, K.: Measure and Integral. Academic Press, London (1978)
61. Jäger, W., Woukeng, J.L.: Homogenization of Richards equation in multiscale porous media with soft inclusions. J. Differ. Equ. **281**, 503–549 (2021)
62. Jäger, W., Tambue, A., Woukeng, J.L.: Approximation of homogenized coefficients in deterministic homogenization and convergence rates in the asymptotic almost periodic setting. Anal. Appl. **21**(05), 1311–1363 (2023)
63. Jikov, V.V., Koslov, S.M., Oleinik, O.A.: Homogenization of Differential Operators and Integral Functionals. Springer, Berlin (1994)
64. Junca, S.: A two-scale convergence result for a nonlinear scalar conservation law in one space variable. Asymptot. Anal. **17**(3), 221–238 (1998)
65. Kato, T.: Pertubation Theory for Linear Operators, 2nd edn. Springer, Berlin (1976)
66. Kenne, R.B., Nguetseng, G., Woukeng, J.L.: Deterministic homogenization of Vlasov equations. Math. Methods Appl. Sci. **43**(3), 1359–1379 (2019)
67. Kinderlehrer, D., Stampacchia, G.: An Introduction to Variational Inequalities and Their Applications. Academic Press, New York (1980)
68. Krasnosel'skii, M.A.: Topological Methods in the Theory of Nonlinear Integral Equations. Macmillan, New York (1964)
69. Larsen, R.: Banach Algebras. Marcel Dekker, New York (1973)
70. Larsen, R.: Functional Analysis: An Introduction. Dekker, New York (1973)
71. Lions, J.L.: Contrôle optimal de systèmes gouvernés par des équations aux dérivées partielles. Dunod, Paris (1968)
72. Lions, J.L.: Quelques méthodes de résolution des problèmes aux limites non linéaires. Dunod, Paris (1969)
73. Lions, J.L.: Some Methods in the Mathematical Analysis of Systems and Their Control. Gordon and Breach, New York (1981)
74. Lions, J.L., Magenes, E.: Problèmes aux limites non homogènes et applications, vol. 1. Dunod, Paris (1968)
75. Lions, J.L., Lukkassen, D., Persson, L.-E., Wall, P.: Reiterated homogenization of monotone operators. C.R. Acad. Sci. Paris, Sér.I Math. **330**, 675–680 (2000)
76. Lions, J.L., Lukkassen, D., Persson, L.-E., Wall, P.: Reiterated homogenization of monotone operators. Chin. Ann. Math. **22**B(1), 1–12 (2001)
77. Lukkassen, D., Nguetseng, G., Wall, P.: Two-scale convergence. Int. J. Pure Math. **2**, 35–86 (2002)

78. Lukkassen, D., Nguetseng, G., Nnang, H., Wall, P.: Reiterated homogenization of nonlinear elliptic operators in a general deterministic setting. J. Funct. Spaces Appl. **7**, 121–152 (2009)
79. Marcellini, P., Sbordone, C.: An approach to the asymptotic behaviour of elliptic-parabolic operators. J. Math. Pures Appl. **56**, 157–182 (1977)
80. Marcellini, P., Sbordone, C.: Sur quelques questions de G-convergence et d'homogénéisation non linéaire. C.R.A.S. Paris, t.**284**, 535–537 (1977)
81. Marusic-Paloka, E., Marusic, S.: Two-scale convergence for thin domains and its applications to some lower-dimensional models in fluid mechanics. Asymptot. Anal. **23**(1), 23–57 (2000)
82. Mawassy, N., Alavi, S.E., Reda, H., Ganghoffer, J.F.: Higher gradient homogenization of quasi-periodic media and applications to inclusion-based composites. Compos. Struct. **333**, 117912 (2024)
83. Meirmanov, A.: A description of seismic acoustic wave propagation in porous media via homogenization. Siam J. Math. Anal. **40**(3), 1272–1289 (2008)
84. Meirmanov, A.: Non-isothermal filtration and seismic acoustic in porous soils: thermo-viscoelastic and Lamé equations. J. Math. Sci. **160**, 33–44 (2009)
85. Miller, R.E.: Homogenization of time-dependant systems with Kelvin-Voigt damping by two-scale convergence. Discrete Contin. Dynam. Syst. **1**(4), 485–502 (1995)
86. Mohammed, M., Sango, M.: Homogenization of linear hyperbolic stochastic partial differential equation with rapidly oscillating coefficients: the two-scale convergence method. Asymptot. Anal. **91**(3–4), 341–371 (2015)
87. Mohammed, M., Sango, M.: Homogenization of nonlinear hyperbolic stochastic partial differential equations with nonlinear damping and forcing. Netw. Heterog. Media **14**(2), 341–369 (2019)
88. Murat, F.: H-convergence. Séminaire d'Analyse fonctionnelle et numérique, 1977/1978, Université d'Alger
89. Murat, F., Tartar, L.: H-convergence. In: Cherkaev, A.V., Kohn, R. (eds.) Topics in the Mathematical Modeling of Composite Materials. Modern Birkhäuser Classics. Birkhäuser, Cham (2018)
90. Neuss-Radu, M.: Some extensions of two-scale convergence. C.R. Acad. Sci. Paris Sér. I Math. **322**(9), 899–904 (1996)
91. Nguetseng, G.: Problèmes d'écrans perforés pour l'équation de Laplace. RAIRO Anal. Numér. **19**, 1–31 (1985)
92. Nguetseng, G.: A general convergence result for a functional related to the theory of homogenization. SIAM J. Math. Anal. **20**, 608–623 (1989)
93. Nguetseng, G.: Homogenization structures and applications I. Zeit. Ana. Anwend. **22**, 73–107 (2003)
94. Nguetseng, G.: Mean value on locally compact Abelian groups. Acta Sci. Math. (Szeged) **69**, 203–221 (2003)
95. Nguetseng, G.: Homogenization structures and applications II. Zeit. Ana. Anwend. **23**, 483–508 (2004)
96. Nguetseng, G.: Homogenization in perforated domains beyond the periodic setting. J. Math. Anal. Appl. **289**, 608–628 (2004)
97. Nguetseng, G.: Deterministic homogenization of a semilinear elliptic partial differential equation of order $2m$. Math. Rep. **8(58)**(2), 167–195 (2006)
98. Nguetseng, G.: Σ-Convergence of Parabolic Differential Operators. Multiple Scales in Biomathematics, Mechanics, Physics and Numerics. Gakuto Internat. Ser. Math. Sci. Appl., vol. 31, pp. 93–132. Gakkōtosho, Tokyo (2009)
99. Nguetseng, G.: Absorptive continuous $\mathbb{R}$-group actions on locally compact spaces. Acta Sci. Math. (Szeged) **77**, 633–667 (2011)
100. Nguetseng, G., Nnang, H.: Homogenization of nonlinear monotone operators beyond the periodic setting. Electron. J. Differ. Equ. **2003**, 1–24 (2003)
101. Nguetseng, G., Signing, L.: Sigma-convergence of stationary Navier-Stokes type equations. Electron. J. Differ. Equ. **2009**(74), 1–18 (2009)

102. Nguetseng, G., Svanstedt, N.: Sigma-convergence. Banach J. Math. Anal. **5**(1), 101–135 (2011)
103. Nguetseng, G., Woukeng, J.L.: Deterministic homogenization of parabolic monotone operators with time dependent coefficients. Electron. J. Differ. Equ. **2004**, 1–23 (2004)
104. Nguetseng, G., Woukeng, J.L.: Σ-convergence of nonlinear parabolic operators. Nonlinear Anal. **66**, 968–1004 (2007)
105. Nguetseng, G., Nnang, H., Svanstedt, N.: Asymptotic analysis for a weakly damped wave equation with application to a problem arising in elasticity. J. Funct. Spaces Appl. **8**(1), 17–54 (2010)
106. Nguetseng, G., Nnang, H., Woukeng, J.L.: Deterministic homogenization of integral functionals with convex integrands. Nonlinear Differ. Equ. Appl. **17**, 757–781 (2010)
107. Nguetseng, G., Sango, M., Woukeng, J.L.: Reiterated ergodic algebras and applications. Commun. Math. Phys. **300**, 835–876 (2010)
108. Nguetseng, G., Nnang, H., Svanstedt, N.: Deterministic homogenization of quasilinear damped hyperbolic equations. Acta Math. Sci. **31B**(5), 823–850 (2011)
109. Nguetseng, G., Showalter, R.E., Woukeng, J.L.: Diffusion of a single phase fluid through a deterministic partially fissured medium. Electron. J. Differ. Equ. **2014**(164), 26 pp. (2014)
110. Nnang, H.: Deterministic homogenization on problems of minimal hypersurface type. Nonlinear Anal. **75**, 5311–5347 (2012)
111. Nnang, H.: Deterministic homogenization of nonlinear degenerate elliptic operators with nonstandard growth. Acta Math. Sin. English Ser. **30**(9), 1621–1654 (2014)
112. Oleinik, O.A., Shaposhnikova, T.A.: On the homogenization of the biharmonic equation in a domain perforated along manifolds of small dimensions. Differentsial'nye Uravneniya **32**, 830–842 (1996)
113. Pak, H.C.: Geometric two-scale convergence on forms and its applications to Maxwell's equations. Proc. R. Soc. Edin. **135A**, 133–147 (2005)
114. Pankov, A.: G-convergence and Homogenization of Nonlinear Partial Differential Operators. Kluwer, Dordrecht (1997)
115. Pardoux, E., Piatnitski, A.: Homogenization of a nonlinear random parabolic partial differential equation. Stoch. Process. Appl. **104**, 1–27 (2000)
116. Pavliotis, G.A., Stuart, A.M.: Multiscale Methods, Averaging and Homogenization. Texts in Applied Mathematics. Springer, Berlin (2007)
117. Pokam Kakeu, A.L., Woukeng, J.L.: Homogenization of nonlinear parabolic equations with hysteresis. Zeit. Ang. Mat. Mec. **100**, 1–22 (2020)
118. Rosenblatt, J.: Uniqueness of invariant means for measure-preserving transformations. Trans. Am. Math. Soc. **265**, 623–636 (1981)
119. Rosenblatt, J.: Automatic continuity is equivalent to uniqueness of invariant means. Illinois J. Math. **35**, 339–348 (1991)
120. Rosenblatt, J., Yang, Z.: Functions with a unique mean value. Illinois J. Math. **34**, 744–764 (1990)
121. Sanchez-Palencia, E.: Nonhomogeneous Media and Vibration Theory. Springer, Berlin (1980)
122. Sango, M.: Homogenization of the Neumann problem for a quasi-linear elliptic equation in a perforated domain. Netw. Heterog. Media **5**(2), 361–384 (2010)
123. Sango, M.: Homogenization of stochastic semilinear parabolic equations with non-lipschitz forcings in domains with fine grained boundaries. Commun. Math. Sci. **12**(2), 345–382 (2014)
124. Sbordone, C., Su alcune applicazioni di un tipo di convergenza variazionale. Annali Scuola Norm. Sup. Piza **IV**, 617–638 (1975)
125. Schwartz, L.: Distributions à valeurs vectorielles I, II. Ann. Inst. Fourier **7**, 1–141 (1957); **8**, 1–209 (1958)
126. Schwartz, L.: Théorie des distributions. Hermann, Paris (1969)
127. Schwartz, L.: Topologie générale et analyse fonctionnelle. Hermann, Paris (1970)
128. Spagnolo, S.: Sulla convergenza di soluzioni di equazioni paraboliche ed ellittiche. Ann. Sc. Norm. Sup. Pisa, III **22**(4), 571–597 (1968)

129. Su, D.: Stochastic homogenization for wave equation with random potential and non-periodic coefficients. J. Math. Phys. **65**, 112706 (2024)
130. Svanstedt, N.: Correctors for the homogenization of monotone parabolic operators. J. Nonlinear Math. Phys. **7**, 268–283 (2000)
131. Tartar, L.: Cours Peccot. Collège de France, Paris (1977)
132. Tartar, L.: The General Theory of Homogenization: A Personalized Introduction. Springer, Berlin (2009)
133. Temam, R.: Navier-Stokes Equations. North Holland, Amsterdam (1977)
134. Vajnberg, M.N.: Variational Methods for the Study of Nonlinear Operators. Holden. Day, San Francisco (1964)
135. Visintin, A.: Some properties of two-scale convergence. Rendic. Accad. Lincei **XV**, 93–107 (2004)
136. Visintin, A.: Towards a two-scale calculus. ESAIM COCV **12**(3), 371–397 (2006)
137. Visintin, A.: Two-scale convergence of some integral functionals. Calc. Var. **29**, 239–265 (2007)
138. Vo-Khac, K.: Distributions, Analyse de Fourier, Opérateurs aux dérivées partielles, t.1. Vuibert, Paris (1972)
139. Vo-Khac, K.: Distributions, Analyse de Fourier, Opérateurs aux dérivées partielles, t.2. Vuibert, Paris (1972)
140. Voke, P.R., Collins, M.V.: Forms of the generalized Navier-Stokes equations. J. Eng. Math. **18**, 219–233 (1984)
141. Walters, P.: An Introduction to Ergodic Theory. Springer, Berlin (1980)
142. Woukeng, J.L.: Deterministic homogenization of nonlinear non-monotone degenerate elliptic operators. Adv. Math. **219**, 1608–1631 (2008)
143. Woukeng, J.L.: Homogenization of nonlinear degenerate non-monotone elliptic operators in domains perforated with tiny holes. Acta Appl. Math. **112**, 35–68 (2010)
144. Woukeng, J.L.: Corrector problem and homogenization of nonlinear elliptic monotone PDEs. In: Shulz V., Seck, D. (eds.) Shape Optimization, Homogenization and Optimal Control. International Series of Numerical Mathematics, vol. 169, pp. 57–73. Birkhäuser, Cham (2018)
145. Wright, S.: Homogenization of the Stokes equations with general random coefficients. Z. Anal. Anwendungen **13**(3), 405–416 (1994)
146. Wright, S.: On the steady-state flow of an incompressible fluid through a randomly perforated porous medium. J. Differ. Equ. **146**(2), 261–286 (1998)
147. Zeidler, E.: Nonlinear Functional Analysis and Its Applications. Springer, Berlin (1989). Part 2B: Nonlinear monotone operators
148. Zhang, X.: Two-scale convergence and homogenization for a class of quasilinear elliptic equations. J. Partial Differ. Equ. **9**(3), 263–276 (1996)
149. Zhikov, V.V.: On an extension and application of the two-scale convergence method. Sb. Math. **191**, 973–1014 (2000)
150. Zhikov, V.V.: On two-scale convergence. J. Math. Sci. **120**, 1328–1352 (2004)
151. Zhikov, V.V., Krivenko, E.V.: Homogenization of singularly perturbed elliptic operators. Matem. Zametki **33**(4), 571–582 (1983)
152. Zhuge, J.: Uniform boundary regularity in almost-periodic homogenization. J. Differ. Equ. **262**(1), 418–453 (2017)

Index

G. Nguetseng, *Homogenization Algebras and Applications*, Springer Monographs in Mathematics, https://doi.org/10.1007/978-3-031-84705-9

The manufacturer's authorised representative in the EU is Springer Nature Customer Service Centre GmbH, Europaplatz 3, 69115 Heidelberg, Germany. If you have any concerns regarding our products, please contact ProductSafety@springernature.com

Printed and bound by CPI Group (UK) Ltd, Croydon, CR0 4YY
30/09/2026
02225595-0017